Oral Microbiome

This book comprehensively reviews the intricate and multifaceted role of the oral microbiome in overall health and well-being. The introductory chapters present the diverse bacterial consortium inhabiting the oral cavity and the influence of dietary habits on the nature of the oral microbiome. The book investigates the link between the gut microbiome and susceptibility to disease, expanding the understanding of the interconnectedness of the human microbiota. Further, the book discusses the involvement of the oral microbiome in acute and chronic infections and addresses the concerning issue of drug resistance within this complex environment. It unveils the connection between the oral microbiome and mouth cancer and malignancies, presenting critical insights into disease prevention and management. The book also expands the scope, examining the role of the oral microbiome in systemic infections and revealing connections between oral bacteria and cardiac diseases. Towards the end, the book explores the pivotal role of oral microbiota in children's well-being and reviews the various diagnostic techniques associated with the oral microbiome. This book is intended for researchers in the fields of microbiology and immunology, dentists, oral health specialists, physicians, and health care practitioners.

Oral Microbiome

Hygiene and Health

Edited by

Rajesh Kumar and Shamsher S. Kanwar

CRC Press
Taylor & Francis Group
Boca Raton London New York

CRC Press is an imprint of the
Taylor & Francis Group, an **informa** business

First edition published 2025
by CRC Press
2385 NW Executive Center Drive, Suite 320, Boca Raton FL 33431

and by CRC Press
4 Park Square, Milton Park, Abingdon, Oxon, OX14 4RN

CRC Press is an imprint of Taylor & Francis Group, LLC

ISBN: 9781032764795 (hbk)
ISBN: 9781032764801 (pbk)
ISBN: 9781003478638 (ebk)

DOI: 10.1201/9781003478638

Typeset in Times LT Std
by KnowledgeWorks Global Ltd.

Contents

Preface

Microorganisms are omnipresent in both the biotic and abiotic worlds. They comprise bacteria, fungi, algae, protozoans, and viruses. The bacteria include both eubacteria and the archaea. The bacteria when characterized based on gram staining show the existence of gram-positive and gram-negative organisms. Bacteria that do not have a cell wall, often referred to as mycoplasma, are another variety of bacterial species. Since the discovery and coining of the term "cell" as a functional and structural unit of all living beings, the composition, variety, and diversity of cells have always been intriguing facets of research. Humans predominate the mammalian species on Earth. They live in villages and semi-urban and urban areas as well as unapproachable regions (tribes or indigenous peoples in voluntary isolation). Their food habits are also varied, vegetarian or non-vegetarian diets or both. The human body itself harbors diverse microbial species as normal or native microflora or, in diseased condition(s), the pathogenic microorganisms. The skin is abundantly and consistently exposed to the environment and, depending on the environmental factor, possesses a diverse microflora. Similarly, the mouth handles food and mixes it with saliva to make it amenable for digestion in the gut. The oral cavity and the denture offer a congenial surface to come into contact with the microbes present in the air, water, beverages, and diets. Any stale and infected food taken unintentionally may allow the pathogenic food-borne pathogens to enter, colonize, and invade tissue in situ or in the gastrointestinal tract of human beings. The oral cavity is also exposed to organic and inorganic contaminants and toxic metabolites as an ingredient of food and beverages and, hence, these contaminants and metabolites may influence the nature as well as the diversity of the oral microbiome. Similarly, the intake of probiotics including fermented dairy products does influence the oral microbiome. Thus, there is a great need to extend our knowledge about the oral microbiome holistically and comprehensively.

The book includes 19 chapters covering a broader view of the oral microbiome; the role of health and hygiene; oral microflora and its diversity; the influence of human diets; implications of oral microflora in health and disease; setting in of acute and chronic infections; the possibility of drug resistance in bacterial strains; development of malignancies and mouth cancer; the microbiome's role in cardiac diseases, systemic diseases, and also in disease prognosis; pediatric health; and the roles of oral microflora as probiotics and in longevity. The editors have assembled diverse information on oral microbiomes and their functional aspects by compiling a compendium of knowledge to serve the audience. The target primary audience shall be postgraduate science students, research scholars, clinicians, dietitians, and scientists. The book will provide comprehensive knowledge on the role of diet, both vegetarian and meat-rich; on the maintenance of gut microbes; and alterations, if any, because of eating/dietary habits, medication, and drug abuse including alcoholism. The content shall be both attractive as well as supplementing to alter our diets to stay healthy by manifesting a beneficial gut microbe that shall support disease resistance, healthy human life, and longevity!

Rajesh Kumar
Shamsher S. Kanwar
Editors

About the Editors

Dr. Rajesh Kumar presently serves as an Assistant Professor in the Department of Biosciences at Himachal Pradesh University, Shimla, Himachal Pradesh, India. With over 10 years of teaching and research experience, his primary areas of research interest are Animal Physiology and Applied Zoology. Dr. Kumar has a remarkable track record, having published more than 100 research articles in highly reputable journals. He has received 12 awards for his outstanding research contributions. Additionally, Dr. Kumar has successfully completed two Major Research Projects and two Startup Projects sponsored by government agencies.

Dr. Shamsher S. Kanwar is a Professor of Biotechnology and serves as the Coordinator of the Bioinformatics Centre at Himachal Pradesh University in Shimla, India. In the past, he has received several awards and fellowships, including a Visiting Fellowship from the Indian National Science Academy in 1998–1999, which allowed him to conduct research in the molecular epidemiology of tuberculosis at CDRI, Lucknow. He was also granted a Visiting Fellowship for the year 1999–2000 by the Science and Engineering Research Council (SERC), Department of Science & Technology in New Delhi, to research multidrug-resistant (MDR) strains of *Mycobacterium tuberculosis* at the National Institute of Mycobacterial Diseases in Agra. In 2005–2006, he received a Biotechnology Overseas Associateship from the Ministry of Science & Technology, Government of India, to work on the interaction of HIV-Tat and drug abuse on human neural cells at Kansas University Medical College in Kansas, United States. Furthermore, he served as a Visiting Scholar in a Program on Mouse Adult Pluripotent Cells at the Stem Cell Institute, University of Minnesota, Minneapolis, United States, in May–June 2008. Dr. Kanwar's research focuses on microbiology, immunology, cell culture, and fermentation, with an emphasis on developing bioproducts and bioprocesses, particularly for a range of extracellular microbial enzymes as an alternative to traditional antibiotics and antilipolytic drugs. Currently, Dr. Kanwar is dedicated to the development of anticancer therapeutics through the exploration of microbial ribonuclease, cholesterol oxidase, and L-methionase.

Contributors

Sandipan Acharjee
Department of Biotechnology
Himachal Pradesh University
Shimla, Himachal Pradesh, India

Arti
Research Lab-III, Department of Biotechnology, Multi-faculty Building Phase-I
Himachal Pradesh University
Shimla, Himachal Pradesh, India

Ekta Arya
Department of Biotechnology
Himachal Pradesh University
Shimla, Himachal Pradesh, India

Vedpriya Arya
Patanjali Herbal Research Department
Patanjali Research Foundation
Haridwar, Uttar Pradesh, India

Wamik Azmi
Department of Biotechnology
Himachal Pradesh University
Shimla, Himachal Pradesh, India

Megha Baghel
Department of Biotechnology
Motilal Nehru National Institute of Technology Allahabad
Prayagraj, Uttar Pradesh, India

Indu Bala
Sukhjinder Group of Institutes
VPO
Gurdaspur, Punjab, India

Ajay Balda
Department of Zoology
Baba Mastnath University
Rohtak, Haryana, India

Acharya Balkrishna
Patanjali Herbal Research Department
Patanjali Research Foundation
Haridwar, Uttar Pradesh, India

Hema Bhardwaj
Department of Biotechnology
Himachal Pradesh University
Shimla, Himachal Pradesh, India

Duni Chand
Department of Biotechnology
Himachal Pradesh University
Shimla, Himachal Pradesh, India

Dimple
Department of Zoology
Baba Mastnath University
Rohtak, Haryana, India

Divya
Department of Biosciences
Himachal Pradesh University
Shimla, Himachal Pradesh, India

Kirti Garg
Department of Biotechnology
Himachal Pradesh University
Shimla, Himachal Pradesh, India

Neha Gautam
Department of Food Science and Technology
Dr. Y. S. Parmar University of Horticulture and Forestry
Solan, Himachal Pradesh, India

Arup Giri
Department of Zoology
Baba Mastnath University
Rohtak, Haryana, India

Reena Gupta
Department of Biotechnology
Faculty of Life Sciences, Himachal Pradesh University
Shimla, Himachal Pradesh, India

Sanjana Gupta
Department of Biotechnology
Himachal Pradesh University
Shimla, Himachal Pradesh, India

Younis Ahmad Hajam
Department of Life Sciences and Allied Health
Sant Baba Bhag Singh University
Jalandhar, Punjab, India

Aishwarya Jain
Department of Life Sciences
Shiv Nadar Institution of Eminence
Greater Noida, Uttar Pradesh, India

Kiranjeet Kaur
Department of Allied Health Sciences
Chitkara School of Health Sciences
Chitkara University
Rajpura, Punjab, India

Sukhraj Kaur
Department of Microbiology
Guru Nanak Dev University
Amritsar, Punjab, India

Lalita Kaushal
Research Lab-III, Department of Biotechnology, Multi-faculty Building Phase-I
Himachal Pradesh University
Shimla, Himachal Pradesh, India

Shahzeb Khan
Department of Biotechnology
Himachal Pradesh University
Shimla, Himachal Pradesh, India

Bhanu Krishan
Department of Biotechnology
Himachal Pradesh University
Shimla, Himachal Pradesh, India

Ankita Kukreti
Patanjali Herbal Research Department
Patanjali Research Foundation
Haridwar, Uttar Pradesh, India

Anil Kumar
Department of Zoology
Baba Mastnath University
Rohtak, Haryana, India

Ashwani Kumar
Patanjali Herbal Research Department
Patanjali Research Foundation
Haridwar, Uttar Pradesh, India

Dinesh Kumar
School of Bioengineering and Food Technology
Shoolini University of Biotechnology and Management Sciences
Solan, Himachal Pradesh, India

Pradeep Kumar
Department of Forensic Science
Himachal Pradesh University
Shimla, Himachal Pradesh, India

Sourav Kumar
Department of Biotechnology and Bioinformatics
Jaypee University of Information Technology
Waknaghat, Himachal Pradesh, India

Alka Kumari
Research Lab-III, Department of Biotechnology, Multi-faculty Building Phase-I
Himachal Pradesh University
Shimla, Himachal Pradesh, India

Amita Kumari
Patanjali Herbal Research Department
Patanjali Research Foundation
Haridwar, Uttar Pradesh, India

Vandna Kumari
Department of Biotechnology
Himachal Pradesh University
Shimla, Himachal Pradesh, India

Gaytri Mahajan
Department of Biotechnology
Himachal Pradesh University
Shimla, Himachal Pradesh, India

Shivani Mahra
Department of Biotechnology
Motilal Nehru National Institute of Technology Allahabad
Prayagraj, Uttar Pradesh, India

Maneesh Kumar Maurya
Department of Biotechnology
Motilal Nehru National Institute of Technology Allahabad
Prayagraj, Uttar Pradesh, India

Arpit Mehrotra
Department of Allied Health Sciences
Chitkara School of Health Sciences
Chitkara University
Rajpura, Punjab, India

Vani Mishra
Department of Biotechnology
Motilal Nehru National Institute of Technology Allahabad
Prayagraj, Uttar Pradesh, India

Anmol Narang
Department of Microbiology
Guru Nanak Dev University
Amritsar, Punjab, India

Neelam
Department of Life Sciences and Allied Health
Sant Baba Bhag Singh University
Jalandhar, Punjab, India

Priyanka
Department of Biotechnology
Himachal Pradesh University
Shimla, Himachal Pradesh, India

Maneesha Rana
Patanjali Herbal Research Department
Patanjali Research Foundation
Haridwar, Uttar Pradesh, India

Ruby
Department of Zoology
Baba Mastnath University
Rohtak, Haryana, India

Savitri
Department of Biotechnology
Himachal Pradesh University
Shimla, Himachal Pradesh, India

Ishan Sen
Department of Biotechnology
Himachal Pradesh University
Shimla, Himachal Pradesh, India

Anupam Sharma
Department of Biotechnology
Himachal Pradesh University
Shimla, Himachal Pradesh, India

Divya Sharma
Department of Biotechnology
Himachal Pradesh University
Shimla, Himachal Pradesh, India

Jatin Sharma
Department of Biotechnology
Himachal Pradesh University
Shimla, Himachal Pradesh, India

Nivedita Sharma
Microbiology Research Laboratory
Department of Basic Sciences
Dr. Y. S. Parmar University of Horticulture and Forestry
Solan, Himachal Pradesh, India

Rahul Sharma
Department of Microbiology
Guru Nanak Dev University
Amritsar, Punjab, India

Shivesh Sharma
Department of Biotechnology
Motilal Nehru National Institute of Technology Allahabad
Prayagraj, Uttar Pradesh, India

Sohini Sharma
Department of Biotechnology
Himachal Pradesh University, Summer Hill
Shimla, Himachal Pradesh, India

Stuti Sharma
Department of Chemistry
Sardar Patel University
Mandi, Himachal Pradesh, India

Rahul Shrivastava
Department of Biotechnology and Bioinformatics
Jaypee University of Information Technology
Waknaghat, Himachal Pradesh, India

Shweta Singh
Department of Life Sciences and Allied Health
Sant Baba Bhag Singh University
Jalandhar, Punjab, India

Abhilasha Sood
Department of Allied Health Sciences
Chitkara School of Health Sciences
Chitkara University
Rajpura, Punjab, India

Deepika Srivastava
Patanjali Herbal Research Department
Patanjali Research Foundation
Haridwar, Uttar Pradesh, India

Diksha Suman
Department of Biotechnology and Bioinformatics
Jaypee University of Information Technology
Waknaghat, Himachal Pradesh, India

Ankita Thakur
Department of Biosciences
Himachal Pradesh University
Shimla, Himachal Pradesh, India

Pooja Thakur
Department of Biotechnology
Himachal Pradesh University
Shimla, Himachal Pradesh, India

Sneha Tripathi
Department of Biotechnology
Motilal Nehru National Institute of Technology Allahabad
Prayagraj, Uttar Pradesh, India

Pranjal Verma
Department of Biotechnology
Himachal Pradesh University
Shimla, Himachal Pradesh, India

Vicky
Department of Biotechnology
Himachal Pradesh University
Shimla, Himachal Pradesh, India

Vikrant
Department of Biotechnology
Himachal Pradesh University
Shimla, Himachal Pradesh, India

1 The Oral Microbiome

A Critical Biomechanics for the Protection against Bacterial Invasion and Colonization

Megha Baghel, Maneesh Kumar Maurya, Sneha Tripathi, Shivani Mahra, Vani Mishra, and Shivesh Sharma
Department of Biotechnology, Motilal Nehru National Institute of Technology Allahabad, Prayagraj, Uttar Pradesh, India

1.1 INTRODUCTION

Microbes, which are made up of thousands upon thousands of different species, are an essential component of the human microbiota and help to keep us healthy. Among them, the oral microbiota is one of the most crucial since it greatly influences our oral and overall health [1, 2]. Joshua Lederberg introduced the term "microbiome," which describes the microorganisms that reside within the human body. In a synergistic relationship with humans, they can be harmful, neutral, or helpful; this is referred to as a "holobiont" [2, 3]. With about more than 700 different types of microbial organisms, the human oral microbiota stands as the largest and most complex ecosystem [4]. The mouth is a habitat for a wide variety of microorganisms, the majority being bacteria [5]. Of these, the core members include prominent genera such as *Actinomyces*, *Neisseria*, *Veillonella*, and *Streptococcus* [6]. The oral microbial ecology exhibits unique microbe-host connections triggered by selection pressures due to its frequent exposure to exogenous chemicals. Certain distinctive characteristics allow pioneer colonists, such as *Streptococcus* species, to flourish in this environment. Gaining an understanding of these processes can help resolve the complexities of the oral microbiome, which is essential for sustaining oral and general health [7].

Dental plaque is the outcome of a complex polymicrobial biofilm community composed of several microbial colonizations that are initiated by a protective glycoprotein coating that develops on teeth [8]. These complex dental plaque matrices allow them to select different organisms than those growing directly on the tooth surface because they convey a variety of microenvironments, including both acidic and anaerobic. Throughout evolution there have been substantial changes in the oral microbiota in addition to notable nutritional modifications in the past. In addition to providing nourishment for the oral microbiota, diet acts as a selection pressure, enabling the organisms to better utilize certain nutrients derived from the host [9].

The Western diet, which emerged from lifestyle shifts brought about by the Neolithic Revolution and later the Industrial Revolution, significantly differs from pre-agricultural diets. It was characterized by staple foods like refined vegetable oils, dairy products, processed cereal grains, and animal meats raised on farms. Nowadays, these food components are commonplace in both industrialized and developing nations. Pathological alterations to the oral microbiota, such as a higher prevalence of periodontal pathogens and acid-producing, acid-tolerant organisms, correspond with such dietary modifications [7]. The oral microbiomes are affected by diet [10]. This association highlights the crucial role of dietary habits in shaping the oral microbial environment.

DOI: 10.1201/9781003478638-1

Current research indicates that the oral microbiome influences the food patterns of the host. Taste sensations like sweetness, sourness, saltiness, and bitterness have been linked to certain bacteria, including several species of *Prevotella* and *Clostridium*. According to [11], the oral microbiota affects what we consume to sustain its existence and longevity in the oral cavity. Many different organs, including the brain, heart, placenta, small intestines, and lungs, have been found to harbor oral cavity–associated microbes [12]. Numerous connections between oral bacteria, especially those connected to periodontal disease and other prevalent chronic illnesses like hypertension and cardiovascular disorders, have also been demonstrated [13].

According to [14], sequencing methods like 16S rRNA, pyrosequencing, metagenome, next-generation sequencing, third-generation sequencing, or metatranscriptomics can be used to analyze the diversity of oral microbes. Numerous techniques have been established in the field of oral health therapy and treatment to combat infections and improve oral health. Treatment options for oral microbial infections are constantly expanding, encompassing both conventional antibiotics and novel approaches such as quorum quenching therapy, photodynamic therapy, and phage therapy. Given the rising prevalence of antibiotic resistance, these alternative medications provide viable solutions to the problems of antimicrobial resistance and oral infection treatment.

In this chapter, we explore the intricate role of the oral microbiota in maintaining both oral and general health. The oral microbiome has as a key role in overall health, from its makeup and dynamics to its influence on diet and systemic health. Altogether, this chapter highlights its potential as a target for diagnosis and treatment, as well as its connections with different chronic conditions. Furthermore, a greater comprehension of the diversity of oral microbiota has been made possible by developments in sequencing techniques. Given the therapeutic implications for the treatment of infections associated with oral microbiome, the conventional methods utilizing antibiotics have been revamped by the novel cutting-edge techniques. Overall, the current work highlights the diverse characteristics of the oral microbiome and its significance for maintaining health and managing illnesses.

1.2 COMPOSITION OF THE ORAL MICROBIOME

The oral microbiome encompasses a diverse composition that includes fungi, viruses, bacteria, and archaea. It is categorized as a core and variable microbiome. The main species that flourish in a healthy environment at different body sites make up the core microorganism. The variable microbiome has developed in response to distinct lifestyle and genetic factors, making it exclusive to each person [15]. The genera *Actinomyces*, *Veillonella*, *Neisseria*, and *Streptococcus* are linked to the core microbiome [6]. There are about 700 prokaryote species, spread across 185 genera and 12 phyla, that have been found in the human oral cavity. According to [16], 54% of these species have official names, 14% are nameless but cultivated, and 32% are phylotypes that are not cultivated. Fusobacteria, Proteobacteria, Bacteroidetes, Actinobacteria, Chloroflexi, Spirochaetes, SR1, Synergistetes, Chlamydiae, Gracilibacteria (GN02), and Saccharibacteria (TM7) are among the 12 phyla. The most common bacterial species that are present in the oral cavity of healthy individuals include [15]:

Gram-positive: *Actinomyces*, *Peptostreptococcus*, *Abiotrophia*, *Streptococcus*, *Stomatococcus*, *Pseudoramibacter*, *Lactobacillus*, *Eubacterium*, *Bifidobacterium*, *Propionibacterium*, *Corynebacterium*, and *Rothia*

Gram-negative: *Veillonella*, *Neisseria*, *Moraxella*, *Fusobacterium*, *Campylobacter*, *Eikenella*, *Capnocytophaga*, *Desulfobacter*, *Hemophilus*, *Treponema*, *Desulfovibrio*, *Wolinella*, *Leptotrichia*, *Simonsiella*, *Prevotella*, and *Selemonas*

In the oral microbiome, a variety of non-bacterial microorganisms, including fungi, protozoa, and viruses also exist. Commonly reported protozoans that display saprophytic traits are

Entamoeba gingivalis and *Trichomonas tenax*. Twenty healthy hosts were used in culture-independent research to identify 85 fungal species. *Candida*, *Aureobasidium*, *Saccharomycetales*, *Cladosporium*, *Aspergillus*, *Fusarium*, and *Cryptococcus* were the prominent genera found [17].

In healthy mouth, there is a consistent presence of a core microbial community at the genus level. However, the diversity within the oral microbiome is unique to each individual and varies depending on the specific oral site. Despite similarities, the composition of the microbiome is individual-specific and site-specific [15]. The composition of the oral microbiome changes depending on the oral compartment, which includes the teeth, tongue, gingival sulcus, and palate. Each of these ecosystems provides specific niches for microbes.

1.3 ORAL NICHES

The oropharyngeal microbial ecosystem is extremely intricate and consists of various habitats, such as saliva, the hard tissue area of teeth, both above and below the gums (subgingival plaque), non-keratinized epithelia (such as the throat and buccal mucosa), keratinized epithelia (such as the attached gingiva and hard palate), and specialized epithelia (such as the tonsils and dorsal tongue) [18, 19]. Distinct surfaces attract different microbiomes because all niches supply a distinctive environment with the ideal stages and nourishment for its inhabiting microorganisms. Oral microbiomes sampled from similar anatomical positions in different individuals exhibited greater similarity compared to microbiomes sampled from distinct anatomical areas within the same individual, because of different environmental effects on the anatomical site of microbial niche.

Saliva continuously envelops the bacterial biofilms and tissue interfaces within the oral cavity. The primary source of bacteria in saliva comes from the biofilms on oral tissues. Following shedding, the epithelial surfaces release the associated bacteria which subsequently get immersed into the surrounding saliva [20]. Saliva and spongy tissues have similar microbial side views, but they differ significantly from dental biofilm in terms of microbial organization [21]. On the surfaces of teeth and mucosal membranes, certain microbial proteins can promote microbial adherence. On the other hand, as saliva is swallowed, other salivary proteins help in the separation, clumping, and removal of microbes [20]. Additionally, it is feasible to use the makeup of the salivary microbiota as a diagnostic and prognostic sign for several specific illnesses, such as mouth cancer and dental caries disease [21, 22].

The cheek and palate surface typically consist of single layers of bacteria that regularly originate and shed. In variance, the interface of the tongue features multiple layers of plaque-like bacteria. This is believed to be because the tongue harbors a more densely populated and diverse microbial community compared to other mucosal surfaces. Among healthy individuals, the most prevalent microbiota found on the dorsal surface of the tongue include *Rothia mucilaginosa*, *Streptococcus salivarius,* and an unidentified *Eubacterium* strain [23]. The microorganisms residing on the interface of the tongue are closely linked to the occurrence of halitosis, commonly known as bad breath. This is because the crypts or crevices on the tongue provide a favorable environment for anaerobic bacteria to thrive, and these bacteria are a well-known source of halitosis [24].

The outermost layers of teeth have sufficient oxygen as compared to the tooth margin in the periodontal gap because they are submerged in a substance called periodontal crevicular fluids, which is a protein that contains exudates from the gingival connective tissues [25]. According to Costalonga and Herzberg [20], the non-shedding surface of the tooth provides a stable and favorable environment for the development and growth of biofilms (Figure 1.1).

The complex oral microbiome, which inhabits a variety of habitats including saliva, tooth surfaces, and mucosal tissues, may be used as a diagnostic and prognostic marker for diseases including oral cancer and dental cavities. Gaining insight into the dynamics of these microbial communities

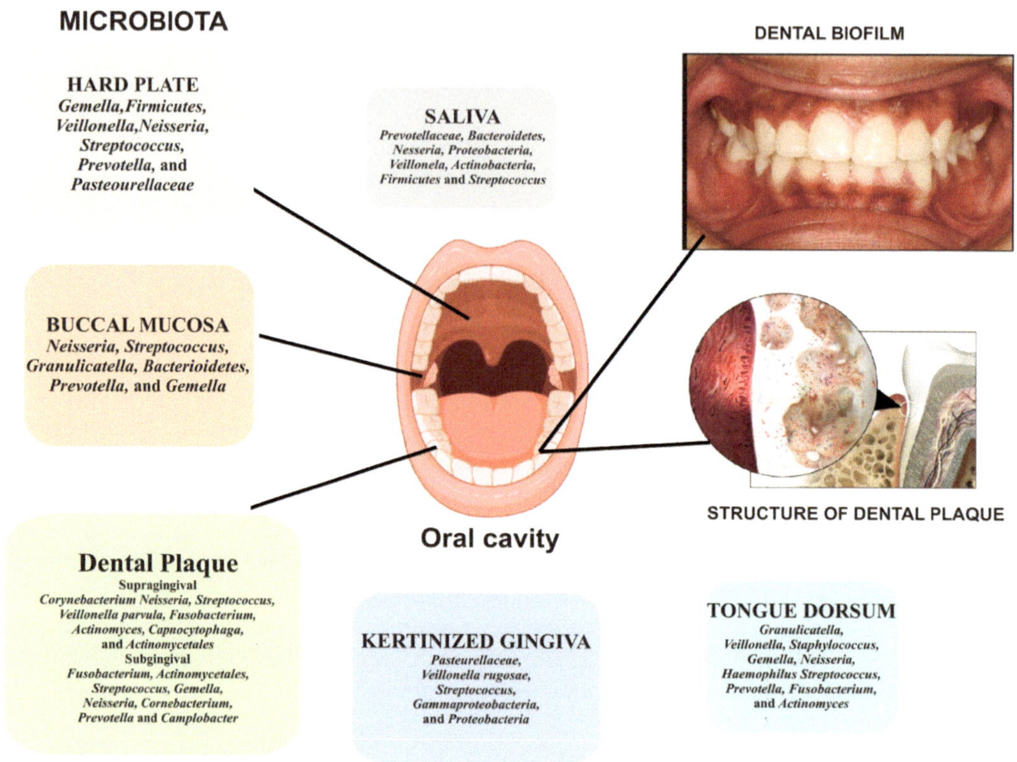

FIGURE 1.1 Niches of oral microbiota.

may help develop tailored interventions and therapies for problems related to oral health, ultimately leading to better oral health results.

1.4 ORAL BIOFILMS

Oral microbial biofilm, as defined by Velsko et al.[26], refers to the three-dimensional structure of oral bacterial species. It is fixed to solid surfaces such as tooth enamel, dental embeds, or the surface of teeth roots [27], and it is implanted in an exopolysaccharide matrix [28]. The intricate microbial community responsible for dental plaque formation forms a biofilm on enamel surfaces.

1.4.1 Understanding Oral Biofilms: Structural Features

Biofilms consist of small microbial colonies that are orderly distributed within a formed matrix [27]. The oral bacteria in plaques aggregate to create sessile, mushroom-like colonies. All individual microbial colonies are a free bunch, each with its specialized living habitat. The huge microbial growth followed by secretion of a large amount of extracellular polymer holds critical importance for the fast development of microbial biofilms.

Along with other organic and inorganic materials, a dense layer of microbes adheres to the polysaccharide matrix in the lower levels of the majority of biofilms. The subsequent layer extending into the surrounding medium is rather loose and often irregular. The liquid surface surrounding the bacterial biofilm may have a bit stationary subsurface and a liquid surface in flow. Gaps are found among the microbial colonies that were present in this film. The water channel acts as a crude "circulatory" structure allowing nutrients and more agents to move throughout the microbial biofilm [27]. Rather than coming from the matrix, nutrients diffuse from the waterway to the sessile (attached) small colonies. Each tiny colony is composed of a free community including thousands of synergistic bacteria. Non-identical tiny colonies may include distinct combinations of bacterial strains. Microbes in the middle of a tiny colony may survive in a severe anaerobic environmental condition, while other microbes at the side of liquid channels may survive in an oxygen-rich environment. Consequently, based on different physiological and nutritional conditions, an array of conditioned living entities is present in the biofilm structure. Due to fluid passing across the microbial biofilm, shear pressures dictate the numerous forms that tiny colonies take in biofilms. The microbial colonies have such umbrella or tower shapes at minimum shear pressure, yet they are stretched out and enable fast oscillation at huge shear forces [27].

1.4.1.1 Five-Step Model and Mechanistic Details of Biofilm Formation

The process of biofilm formation is difficult and multistage. To make sure that the entire process proceeds as intended, several important variables and circumstances must exist in the oral cavity. It is, therefore, distinguished into five distinct phases, known as the five-step model (Figure 1.2).

- *Step I*: The first step involves microbial attachment to a solid surface. Two mechanisms are used by *S. mutans* to adhere to the dental implants and tooth enamel: The sucrose-independent mechanism [27], which relies on the interaction between saliva agglutinins and microorganism adhesion particles, and the other sucrose-dependent mechanism, which depends on the activity of glycosyl transferase and glucan-binding proteins.
- *Step II*: The second is the irreversible bacterial attachment to a surface, resulting in the beginning of the exopolysaccharide matrix emergence. This matrix creates a barrier against desiccation and host defense mechanisms because of its open constructions of nutritional waterways, vacant spaces, and other characteristics like environmental variability (including oxygen and pH gradients, and co-adhesion) [29].
- *Step III*: Biofilm maturation occurs when a new bacterial species joins the biofilm while the matrix is still forming during its third stage [30]. Bacteria produce extracellular polymers that make up the plaque matrix, such as fructans, glucans, and soluble and insoluble heteropolymers. All biofilms have a matrix; however, the matrix is much greater than just the chemical stage that keeps the microbial plaque shaped. The bioactive forge keeps nutrients, water, and biological catalysts within the biofilm composition.

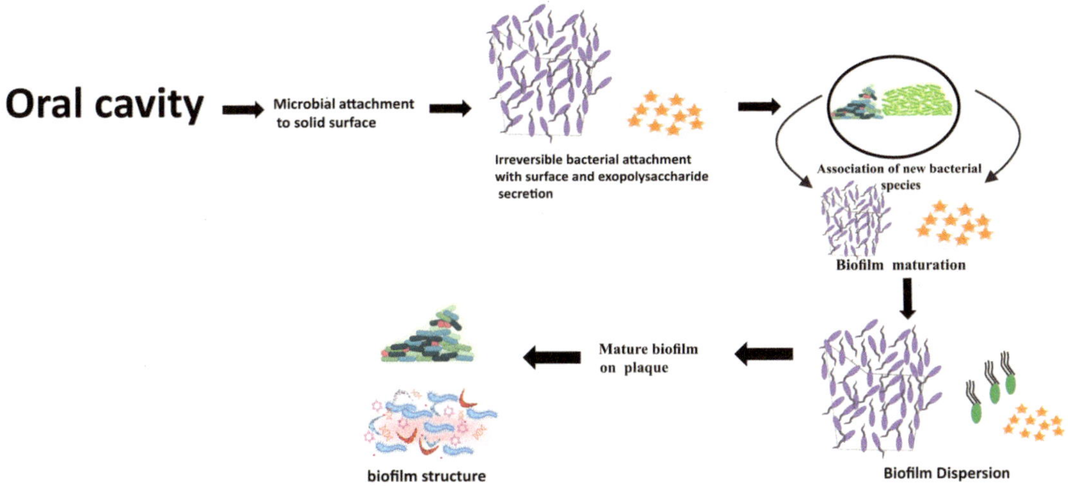

FIGURE 1.2 Five-step model of oral biofilm development.

- *Step IV*: The change in the early dominance of *Streptococcus* species toward *Actinomyces* and gram-positive bacterial cocci strains is linked to bacterial succession during stage four. Bacteria from newly discovered species adhere to those from earlier pioneering species. Other bacteria can survive in the newly favorable settings because one microorganism builds ecological niches for them. This phenomenon was first described as an assemblage by [31]. The removal of lactate by bacterial biofilm–developing *Streptococci*, which provides a source of organic compound carbon for the development of *Candida albicans*, is an illustration of this process. Consequently, this lowers oxygen availability for *Streptococci*, which promotes the development of non-obligatory anaerobic microbes [31].
- *Step V*: As a mature biofilm forms, the growth rate of a particular type of bacteria is known to decrease. The number of bacteria doubles every 1–2 hours in the early phases of mature biofilm, whereas it takes approximately 12–15 hours in dental plaque that has been there for 1–3 days. A three-dimensional structure is formed by the biofilm. By now, microbial interactions with the host immune system become essential for the development of mature biofilm structures and the separation of individual bacterial strains from these structures, which permits them to occupy a subsequent ecological niche within the oral microbiome [32]. Bacteria can sense negative changes in their environment and trigger the expression of genes linked to active separation.

For example, the significance and application of 338 microbial biofilms is one such resource. *Prevotella loescheii* is responsible for co-aggregating with *S. mitis*; it generates proteases that hydrolyze its adhesions associated with fimbriae. Thus, the equilibrium between microbe adherence, growth, and elimination produces a mature biofilm structure. When the biomass reaches a threshold size, or when shear forces prevent further expansion, the plague develops; nonetheless, structural development and rearrangement may continue. Dental calculus, or mineralized dental plaque, is a well-known illustration of an established biofilm structure. Each bacterium is surrounded by intracellular and external mineral deposits, which eventually combine to create a hard deposit that is covered in a layer of fresh bacteria. Both supragingival and subgingival calculus are seen, serving as an extra retention site for plaque buildup that results in gingivitis (which is linked, among other things, to the presence of the previously described *Prevotella loescheii*).

It has been reported that prevention of the biofilm formation is difficult in stages IV and V. This is caused by interactions between certain species that have collected in its anatomy and gestures from the outside world (originated from mouth bacteria). At this point, hyperadditive synergism could be discussed, where the synergistic effects of multiple species contributing to biofilm activity are observed.

It is crucial to take into account several critical factors that influence the creation of biofilms, such as surface type (smooth, rough, and their combinations), nutritional conditions [33], hydrodynamic conditions, and a host of other factors that are still unclear.

1.4.1.2 Factors Influencing the Development of Biofilms

1.4.1.2.1 Bacterial Surface Adhesions and Saliva

The acquired enamel pellicle layer of the tooth surface initiates the biofilm development process [34]. Proline-rich polymers of amino acids, amylase, lysozyme enzymes, histatin, oxidative enzymes like peroxides, and mucin are examples of salivary-origin components. Bacterial components, such as fructosyltransferase, and glucosyltransferase with a lipoteichoic organic acid, are specifically adsorbed on the hard surface of the tooth to generate the membrane [35]. The oral cavity is colonized by bacteria that produce biofilms based on acquired enamel pellicle. Two independent mechanisms have been elaborated for the binding of the bacterial cell either alone or in cluster to the pellicle; these are sucrose dependent and sucrose independent [29, 36, 37]. The establishment of a relationship between *S. mutans*, adherence particles, and acquired enamel pellicle, which functions as a spreading barrier sheltering the tooth from the straight action of carbonic acids, is referred to as the method of adhesion [35]. Agglutinins interact with adhesin P1, which is present in the bacterial cell wall, to facilitate adherence and aggregation [38].

In the biofilm formation process, the oral agglutinins hold equal importance to that of surface bacterial adhesins. The impact of salivary antigen on the potential of *S. mutans* to form the plaque was studied in the laboratory [36, 39]. The scientist used bacterial strains of *S. mutans* that had mutations in the *spaP* gene, which codes for protein P1, in addition to the strain which is wild type. This allowed the researchers to examine the variability and interconnection that were noticed at the time of biofilm development [40].

Moreover, salivary proteins that bind inorganic ions, primarily calcium, are also highly significant in the formation of biofilms. Approximately 33% of the calcium in dental biofilm is in loose form, 17% is mixed with phosphate ions with organic anions, and the remaining 50% is linked to particular proteins. Modifications in the protein orientation may lead to a decrease in Ca^{2+} levels in dental plaque, potentially leading to the development of caries. The presence of calcium ion–binding sites may be essential for amino acid proline-rich proteins (PRPs) in bacterial biofilms, which are important for the synthesis of useful proteins like statherins and amino acid histatins (which are involved in the formation of acquired enamel pellicle) and cysteine-rich phosphoproteins found in dental plaque. Furthermore, the smallest molecular weight peptides seen in oral saliva are linked to these locations. Moreover, statherins and proline-rich proteins prevent calcium phosphate from precipitating. Calcium-binding proteins are the reservoir for these ions and are therefore important for the remineralization of dental enamel [41]. Caries may happen as a result of less salivary proteins mixed up with inorganic ions, which might favor demineralization processes over remineralization activities.

1.4.1.2.2 Availability of Sucrose

Intensive studies on oral cavity microorganisms started when it was discovered that the pH of tooth plaque is prone to reduction after sucrose utilization and subsequently recovers to its previous level. The study holds significance for healthcare as well as the number of interspecies interactions that have been identified and the curiosity about the behavior of microbes sharing an ecological niche [30]. The glucosyltransferases (Gtfs B, C, D) are produced by *Streptococcus mutans* in conjunction with glucan-binding proteins and represent the sucrose-dependent mechanism of tooth plaque development [34, 37, 42]. Gtfs occupy the charge of glucan synthesis from sucrose. The synthetic glucans allow germs to stick together and allow bacteria to adhere to dental enamel. Microcolonies can be formed by this mechanism, which promotes the production of biofilms [34, 42].

During *in vivo* conditions, Gtfs rapidly adsorb to the hydroxypartite surface (sHA) and forms a strong bond with salivary film. Moreover, GtfC is a hydrophilic molecule. Its attachment to plaque, however, comes from the hydrophobic domain. It allows interactions with saliva proteins like α-amylase and lysozyme that are present in the salivary film to overlap. Glucosyltransferase B (GtfB) is primarily responsible for the ability of *S. mutans* to interact with other bacteria. The remarkably unusual tiny colonies that comprise the biofilm structure are under their control. When glucose is present, they become noticeably more active [42].

Five of the 65 amino acid sequences that are recurrent at the C-C terminal of the polymer chain provide GtfD its capacity to bind to the glucans. Even when bacteria are not producing their glycosyltransferases, they can attach to other bacteria through glycosyltransferases [34]. Among the several polymers that make up the plaque [alpha-(1→6)-, alpha-(1→4)-, and alpha-(1→3)-glucans and beta-(2→6)-fructans], alpha-(1→3)-glucan is crucial for understanding the etiopathology of dental caries. This polymer, which makes up about 1.3%–1.4% of the dry matter on plates, has several special properties that enable it to form a basic structure or criterion. These properties include its ability to readily adsorb to saliva or enamel, encourage the reciprocal adhering of bacteria (colonization), and significantly increase the steadiness of plaque. Because alpha-(1→3)-glucans have a fiber structure and are insoluble in water, oral cavity fluids cannot break them down and wash them away. Furthermore, the stability and longevity of the plaque are guaranteed by the fact that alpha-(1→3)-glucans are neither influenced by enzymes nor created by microbes living in the mouth cavity [43].

Glucan-binding proteins (Gbps) are another component facilitating binding of bacteria to glucans using a sucrose-dependent mechanism. GbpA, GbpB, GbpC, and GbpD are four known protein types [37, 42]. The cell wall of bacteria is associated with the GbpC polymer of an amino acid (and may be GbpB), and as a result, is a glucan-specific receptor. Although each of these four protein types contributes to the production of plaque and bacterial adherence, the amino acid GbpD polymer seems to be the most important. According to research, *S. mutans* strains GS5 (GbpB gene alteration by deletion) and UACA2 (bacteria producing a siRNA for the GbpB gene) exhibit altered cell architectures and sluggish development when the GbpB coding gene is absent or spontaneously altered [42]. As a result, the biofilm is prevented from growing normally, and instead of developing into a diverse, compact formation, it becomes an unevenly structured matrix with a peculiar structure around an irregular cell cluster.

To disrupt oral biofilm development, prevent dental caries, and provide individualized oral health care, it is essential to comprehend bacterial adhesion, saliva composition, and sucrose availability. Despite advancements, there are still gaps in our understanding of these intricate relationships and the identification of new therapeutic targets, which emphasizes the necessity of more study and clinical application.

1.5 FACTORS AFFECTING THE ORAL MICROBIOTA

1.5.1 Diet

The oral microbiome formation begins even before birth when the infant is still in the uterus, and the method of delivery also holds significance in determining the initial type of microorganism colonized in the infant. At 3 months of age, infants delivered veginally had more taxonomic diversity than those delivered by cesarean section. Research indicates that newborns fed with formula milk had higher proportions of *Actinomyces* and *Prevotella*, while breastfed infants showed higher proportions of *Streptococcus*. Infants fed with breast milk or formula exhibit lower levels of oral candidiasis than those weaned onto solid food [44]. In adults, modulation in dietary habits can influence the oral microbiota and serve as a preventive measure for oral disease. Consumption of refined carbohydrates and sugars like carbonated beverages are linked with an abundance of microbial community in oral microbiota. For example, the number of *Fusobacterium,* Bacteroidetes, *Veillonella, and Gammaproteobacteria* is favorably correlated with the consumption of sugary, carbonated beverages [45]. It is commonly known that eating foods high in sugar fosters the growth of bacteria that cause dental caries. It has been discovered that sugar alcohols, like xylitol, erythritol, and sorbitol, can be used as sugar substitutes to prevent dental caries [46].

Around 10,000 years ago, during the shift from the Mesolithic to the Neolithic era, the introduction of agriculture brought a drastic change in the dietary composition. Through the use of shotgun sequencing of salivary DNA, researchers discovered that the diets of hunter-gatherers, traditional farmers, those who follow a Western diet, and vegetarians have a substantial impact on the composition of oral bacteria. Traditional farmers fall between hunter-gatherers and Western diet followers in terms of the prevalence of *Haemophilus* and *Neisseria*. Hunter-gatherers have some oral infections, proving that consuming excessive amounts of meat increases the chance of developing oral diseases. Vegetarians have significant changes in all taxonomic levels of the oral microbiota, including respiratory tract bacteria (*Campylobacter* and *Porphyromonas*) and oral pathogens (*Neisseria* and *Haemophilus*) [47].

1.5.2 Smoking

Smoking has an impact on the oral microbiota composition [48] studied the effect of several substances, including cigarettes, shisha, and dokha, on the oral microbiome composition of 330 participants in the United Arab Emirates Healthy Future Study. The findings demonstrated that cigarettes and dokha had an impact on the structure of the oral microbiota, but shisha had no such effect. While *Porphyromonas*, *Lautropia*, *Actinobacillus*, and *Bifidobacterium* were much more prevalent in dokha users, many other types of bacteria were altered in cigarette smokers. Wu et al. [49] reported that the oral microbiomes of smokers had an increase in Actinobacteria and Firmicutes and a decrease in Proteobacteria. There are several ways that smoking affects the oral microbiota. These include the following: The effects of oxygen deprivation [50], antibiotics [51], bacterial attachment to the surface of mucosa [52], host immunological damage [53], and increasing saliva pH [54, 55]. Studies indicate that smoking-induced changes to the oral microbiome are temporary. After the individual quits smoking, the oral microbiota usually reverts to its pre-smoking state [49].

1.5.3 Drinking

Alcohol consumption may also impact the composition of oral microbiota, influence the functional microbial pathway, and cause a decrease in bacterial richness in the oral biofilm [56]. The enrichment of *Prevotella* and *Moryella* and the depletion of *Lautropia*, *Haemophilus*, and *Porphyromonas*, is significantly found in people consuming alcohol [57]. Red liquor and wine have different impacts on oral microbiota. A study conducted on 75 Swiss adults investigated the effects of red wine on the oral microbiota. The findings revealed that red wine could affect both the supragingival and the subgingival areas of the mouth [58]. The consumption of liquor appeared to correlate with the increased population of gram-positive bacteria, including *S. mutans* [59]. The bacteria of the oral community are capable of converting ethanol to acetaldehyde, which is considered harmful and has been identified as a human carcinogen [60]. According to [61], the general diversity and stability of significant bacteria detected in human saliva are unaffected by the regular and moderate consumption of red wine.

In summary, the oral microbiota is influenced by a variety of factors such as diet, smoking, and drinking, which can disrupt the balance of the oral microbiome and cause dysbiosis, which subsequently leads to an overgrowth of disease-associated bacteria, increased inflammation, and a greater risk of oral disorders such as tooth decay, gum disease, and other oral infections.

1.6 ORAL DISEASES

1.6.1 Diseases Associated with the Oral Microbiota

There is a large landscape of microbial diversity in the oral cavity. The microbes have been observed to significantly interact with each other and their microenvironments. This alters the balance between several immune components and human health subsequently leading to several disease etiologies (Table 1.1).

1.6.1.1 Dental Caries

Dental caries, generally known as tooth spoils or cavities, is the primary cause of buccal cavity pain and tooth loss. It is the result of a complex interaction between bacteria, sugars, and tooth enamel. It results from the interaction of bacteria and simple carbohydrates [62]. The consumption of high amounts of sugar regularly helps to increase acid formation, dysregulate salivary buffering activity, and decrease pH conditions. The main factor influencing the morphological and genome-wide changes in the microflora during caries development is environmental acidity [63]. IgA secreted from saliva and serum IgG from the gingivae may impact the expansions of enhanced dental caries microbiota during different stages of infection. In many patients with tooth caries, *Streptococcus mutans* and *Lactobacilli* were either completely absent or present in less amounts [64]. All these findings reveal that the beginning and development of cavities cannot be allocated to these microbes solely. Therefore, tooth cavities, a bacterial disorder, results from a consortium of germs, not just one [20]. The species of the genera *Actinomyces*, *Bifidobacterium*, *Propionibacterium*, *Veillonella*, *Streptococcus mutans*, *Streptococci*, *Lactobacillus*, *and Atopobium* also play a major

TABLE 1.1
Microbial Species Related to Oral Illnesses

Disease	Pathogen	References
Dental caries	*Actinomyces* spp., *Streptococcus mutans*, *Lactobacillus* spp., *Bifidobacterium*	[89]
Periodontitis	*Porphyromonas gingivalis*, *Aggregatibacter actinomycetemcomitans*, *Treponema denticola*, and *Tannerella forsythia*	[90]
Oral cancer	*Peptostreptococcus* spp., *Fusobacterium* spp., *Porphyromonas gingivalis*, *Prevotella* spp., *Capnocytophaga gingivalis*, and *Streptococcus* spp.	[91]
Peri-implantitis	*Tannerela forsythia*, *Porphiromonas gingivalis*, *Aggregatibacter actinomycetemcomitans*, *Prevotella intermedia*, *S. salivarius*, and *Treponema denticola*	[92]
Pulp peripheral	*Actinomyces*, *Fusobacterium*, and *Prevotella*	[93]
Recurrent aphthous stomatitis	*Acinetobacter johnsonii*, *Alloprevotella*, *E. coli*, and *Streptococcus* spp.	[94]
Halitosis	*Tannerella forsythia*, *Selenomonas* spp., *Bacteroides* spp., *Fusobacterium* spp., *Leptotrichia* spp., *Peptostreptococcus* spp., *Dialister* spp., *Porphyromonas* spp., *Eubacterium* spp., *Prevotella* spp., *Solobacterium* spp., *Actinomyces* spp., and *Veillonella* spp.	[95]

role in cavity development [64]. *Propionibacterium* FMA5 in a cavity of juvenile, enduring teeth were remarkably connected with caries development, although these were not developed at huge levels. Certain species, including *Eubacterium* IR009, *Corynebacterium matruchotii*, *Campylobacter rectus*, *Eubacterium gordonii*, *Lachnospiraceae* sp. C1, *Capnocytophaga gingivalis*, *Streptococcus cristatus*, *Streptococcus mitis*, *Streptococcus infantis*, and *Streptococcus pneumonia* group, showed a significant decline as caries progressed. Apart from bacteria that produce acid, certain bacteria can produce ammonia through the combination of arginine and urea. This enhances pH levels, aiding in maintaining the pH balance of oral biofilms and perhaps reducing the onset and advancement of dental caries [65]. Teng and colleagues [66] conducted a screening process to isolate 20 taxa, consisting of *Streptococcus mutans* and *Veillonella dispar/Veillonella parvula*, from plaque and saliva. This enabled them to cure early childhood caries (ECC) in healthy persons with 70% accuracy and forecast the onset of future ECC in samples that were clinically considered healthy with 81% accuracy.

1.6.1.2 Periodontal Diseases

Periodontal diseases, also known as gum disease, are a group of inflammatory conditions that influence the tissue surrounding and supporting the teeth. This disease can span from simple gum inflammation (gingivitis) to critical conditions that can lead to tooth degradation. A resident microbial plaque that forms at the gingival edge is the source of the reversible inflammatory illness gingivitis. Eight major taxa identified in the biofilm, TM7, *Streptococcus*, *Veillonella*, *Haemophilus*, *Selenomonas*, *Leptotrichia*, *Lautropia*, *and Prevotella*, have been linked to gingivitis and may act as biomarkers for the condition. Gum disease is the earliest phase of periodontal illnesses. If it remains untreated, it grows into periodontitis, which is a more prolonged chronic and irreversible inflammatory disease. Herein, the long-term immune cell infiltration causes alveolar bone degradation, vascular development, and connective tissue damage [67]. A shift in paradigm of the microbial population results in modifications to the host-microbe relationship that are potent enough to lead to inflammation and bone loss. The three bacteria known historically as the "red complex", *Tannerella forsythia, Treponema denticola,* and *Porphyromonas gingivalis*, have been connected to gum disease [68, 69].

A small bacterial population is associated with healthy gums but is believed to play a significant role in the occurrence and development of periodontitis. Following successful periodontal therapy, the red complex disappears or becomes less detectable [20]. Applying whole-genomic DNA probes, the red complex was later verified to be linked to periodontal diseases that were first identified by culture techniques. The variety of species associated with the occurrence of the disease has increased as a result of new sequencing methods. These organisms now include *Prevotella*, *Filifactor alocis*, *Synergistes*, *Selenomonas*, *Desulfobulbus*, *Peptostreptococcus stomatis*, and *Megasphaera* [70, 71]. It has been demonstrated that the presence of *Candida albicans* is directly related to the severity of chronic periodontitis, in addition to bacteria that were also present during subgingival colonization [72].

1.6.1.3 Recurrent Aphthous Stomatitis

Recurrent aphthous stomatitis (RAS), also referred to as canker sores, is a disorder in which tiny, shallow oral ulcers or sores regularly occur inside the mouth, affecting approximately 20% of the general population [73]. RAS is linked to mucosal and salivary microbiome dysbiosis [74, 75]. Numerous reasons have been proposed for the cause of RAS. These include genetic predisposition, immunologic disorders, bacterial and viral infections, dietary allergies, deficiencies in vitamins and microelements, systemic illnesses, hormonal imbalance, mechanical traumas, and stress [76]. Marchini et al. [74] reported that *Prevotella* is specific to the mucosal microbiota of RAS patients and were found to be absent from that of a healthy person. When comparing the mucosal microbiota of RAS patients to that of healthy controls, it was observed that the taxa Actinobacteria, particularly the *Rothia* species, is commonly associated with the former group. The authors in [77] and [78] conducted a comparative analysis of the taxonomic richness of each species found in the mucosal and salivary microbiota of patients with RAS. The results indicated dybiosis of the microbial community in the salivary and mucosal region. The irregularity during RAS risk was associated with a decrease in healthy microbiota like *Streptococcus salivarius* and an increase in rare species like *Acinetobacter johnsonii*. RAS is not a serious medical issue in itself, but it is significant because it affects everyday functioning, mental stability, and the dental health of an individual. Individuals suffering from RAS may find it easier to manage their condition with the assistance, diagnosis, and treatment of medical specialists.

1.6.1.4 Oral Tumor

There is mounting evidence providing a potential role of oral microbiota in the growth of mouth cancer; oral squamous cell carcinoma (OSCC) stands as the most prominent of all the known prevalent cancers [79, 80]. Employing culture-dependent techniques, scientists studied the oral malignancies and the associated microbial community. Biofilm samples were obtained from areas surrounded by lesions and from adjacent healthy mucosa, prior to any antibiotic or tumor therapy [81, 82]. Additionally, culture-independent 16S rRNA techniques have revealed clear variations in the microbiota composition between tumorous and non-tumorous mucosa. *Gemella morbillorum*, *Johnsonella ignava*, *Gemella haemolysans*, *P. stomatis*, *S. salivarius*, *S. gordonii*, *Streptococcus* sp. oral taxon058, *Streptococcus parasanguinis*, and *Granulicatell aadiacens* were more prevalent in non-tumor locations, which had a strong association with cancer sites [83, 84]. According to different research, individuals who had OSCC had larger salivary counts in the number of *P. melaninogenica, C. gingivalis, and S. mitis* [21]. According to Lee et al. [85], there are notable distinctions between cancer patients and those with epithelial precursor lesions in terms of the oral microbiome. Significant classification in the experimental clusters were marked by the difference of five genera, namely, *Paramonas, Bacillus, Peptostreptococcus, Enterococcus,* and *Slackia*. The discernible difference in the composition of the oral microbiome could potentially function as indicators for the growth, advancement, and reappearance of oral cancer.

It is yet not known how the oral microbiome relates to the etiology of oral malignancies. A few kinds of mechanisms have been proposed. First, microbes may cause long-term inflammatory reactions. The process results in the formation of chronic inflammatory mediators that either induce or enhance angiogenesis, mutagenesis, oncogene activation, and the growth of cells. Second, bacteria may directly affect the development and progression of cancers by secreting microbial effector proteins via type-3/type-4 secretion systems (T3SS/T4SS). These secretions may affect cytoskeletal changes, the growth of cells, NF-κB activation, and the prevention of tissue death [86]. Third, bacteria have the capacity to produce some compounds that are known to cause cancer. For example, they can transform ethanol into acetaldehyde, which has been identified as a carcinogenic agent [87, 88].

Investigations over the implications of the oral microbiome in disease pathophysiology has significantly contributed toward the knowledge related to illness prevention, diagnosis, and therapy; however, gaps still exist. Further studies are needed to build an understanding about the biological mechanisms that underlie microbial dysbiosis in diverse oral illnesses, including dental caries, periodontal illnesses, and oral cancer. Furthermore, it is critical to understand how host-microbe interactions influence the course of illness. Implementing the cutting-edge technology such as metagenomics and multi-omics techniques will further assist us to comprehend the intricate nature of the oral microbiome. Moreover, analyzing the impact of genetics, lifestyle choices, and environmental factors on oral microbial communities will be essential to creating specific treatment strategies and enhancing oral health wellness more broadly.

1.6.2 Oral Microbiota and Systemic Diseases

The oral cavity with its wide range of microbial diversity serves as a vital interface between the external environments and human health through complex signaling pathways. These processes not only affect the oral health but are also associated with substantial systemic diseases (Figure 1.3).

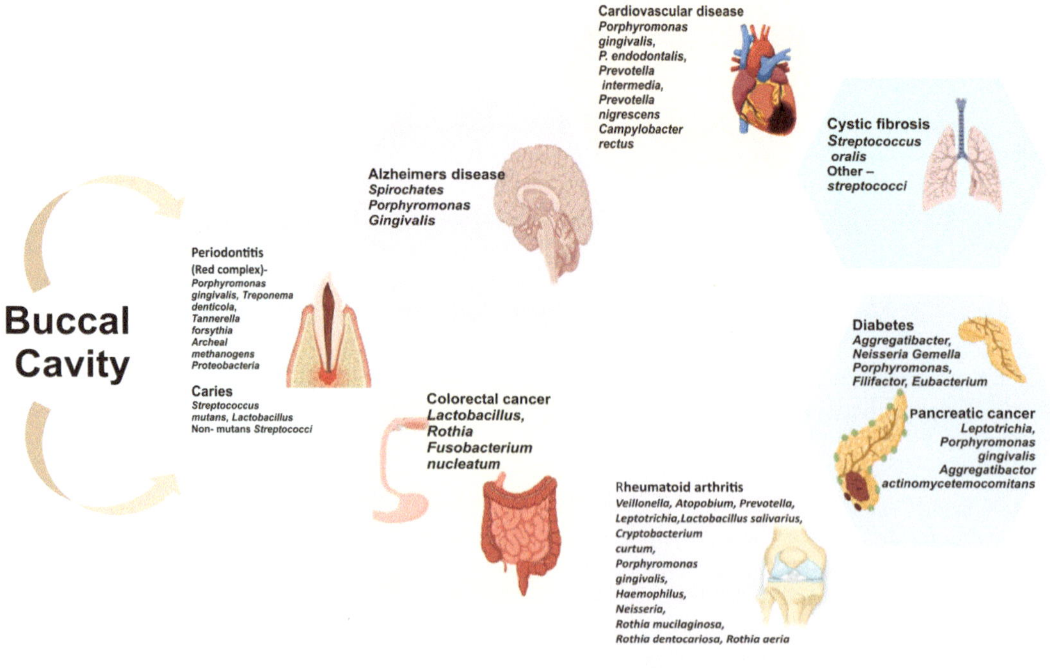

FIGURE 1.3 Oral microbiota associated with oral and systemic diseases.

1.6.2.1 Diabetes

It has been widely discussed recently how diabetes and periodontitis are "two-way" linked. Periodontitis is considered to be one of the consequences of uncontrolled diabetes [96]. Several studies have investigated the impact of diabetes on the composition of the oral microbiota [97, 98]. Both diabetic and non-diabetic participants were recovered with similar amounts of various periodontal pathogenic infections, including *Fusobacterium nucleatum*, *Campylobacter rectus*, *Capnocytophaga* spp., *Eikenella corrodens*, *Aggregatibacter actinomycetemcomitans*, and *Prevotella intermedia*; however, a notably higher number of diabetic participants harbored *P. gingivalis* [99]. Hintao et al. [97] found that while there were no notable changes in the microbial community between diabetic and non-diabetic individuals, oral rinse, subgingival plaques, saliva, and supragingival plaques of diabetic subjects showed higher concentrations of *Prevotella nigrescens*, *T. denticola*, *Streptococci oralis*, *Streptococcus sanguinis*, and *Streptococci intermedius*. Furthermore, the presence of periodontitis-related microbes complicates the management of glucose levels. By inducing the production of specific inflammatory cytokines, *P. gingivalis*, a primary pathogen responsible for periodontitis, produces lipopolysaccharide (LPS), which may lead to insulin resistance and subsequently impairing insulin function [100]. In addition, bacterial infections may trigger whole-body insulin resistance by lowering insulin-mediated metabolism of glucose in skeletal muscles. Moreover, glycemic control in diabetes is enhanced by antimicrobial therapy in periodontal compartments, probably as an outcome of reduced serum TNF-α and reduced insulin resistance [101].

1.6.2.2 Cardiovascular Diseases

Microbial infection has been linked to periodontal diseases, which stands as a significant risk factor for cardiovascular illnesses. Studies suggesting an association between tooth loss from periodontal disease and cardiovascular disease lent validity to the theory that periodontitis may be a contributing factor in the formation of atherosclerosis [102]. According to [103], there is a connection between the oral cavity and atherosclerotic plaques containing *Veillonella* and *Streptococcus*. Atherosclerotic plaques have been shown to include several oral bacteria, such as *Gingivalis*, *Fusobacterium nucleatum*, *Aggregatibacter actinomycetemcomitans*, *Tannerella forsythia*, *Eikenella corrodens*, and *Campylobacter rectus* [104]. Unstable atherosclerosis plaques have been reported to comprise *S. sanguinis* and *P. gingivalis*. In a comparable vein, it was observed by polymerase chain reaction (PCR) that *P. gingivalis*, *A. actinomycetemcomitans*, and *T. denticola* were detected in the thrombi of individuals suffering from acute coronary artery disease, indicating that oral microbiota could have an important role in plaque inflammation and instability. It is interesting to point out that some oral bacteria and cardiovascular disease biomarkers are associated. *Neisseria* indicated a high negative correlation with these indicators, while *Streptococcus* showed a substantial positive correlation with both high-density lipoprotein (HDL) cholesterol and ApoAI, an important component of HDL. The quantity of *Fusobacterium* has been associated with both total and low-density lipoprotein (LDL) cholesterol [103]. Accumulated research has revealed a significant relationship between periodontitis and cardiovascular disease with several potential factors associated with it. The host reaction to prolonged exposure to bacteria from periodontal disease is typically a significant determinant [100]. Through oral tissue, the microorganism passes through the circulatory system and flows to the arteries where it secretes inflammatory chemicals and lipopolysaccharides. This could result in several cardiovascular complications, which include platelet aggregation, lipid deposition, smooth muscle proliferation, and vascular injury to the endothelial cells [1].

1.6.2.3 Tumors of Distant Organs

Other organ cancers are also triggered by oral bacteria found in the mouth [105]. The immune-related mechanisms that trigger the emergence of cancer may be affected by the increased production of cytokines and other inflammatory chemicals brought on by oral microbes [106]. According to [107], squamous cell carcinoma of the head and neck may arise and spread as a result of *Streptococcus anginosus* infection. A study by Narikiyo et al. [108], utilizing a culture-independent molecular technique on saliva, revealed a correlation between esophageal cancer and oral periodontopathic spirochetes *S. mitis*, *T. denticola*, and *S. anginosus*. Interestingly, there was a significant difference in the ratios of *S. mitis* and *Neisseria elongata* between pancreatic cancer patients and the healthy controls [109]. Furthermore, evidence indicates a possible connection between oral *Streptococcus* and colon cancer.

The oral microbiota plays a pivotal role in systemic diseases like diabetes, cardiovascular diseases, and distant organ tumors. Understanding their influence on disease progression is vital for preventive and therapeutic strategies. However, gaps remain in elucidating specific mechanisms and establishing targeted interventions, warranting further research for improved clinical outcomes and disease management.

1.7 METHODS TO STUDY THE ORAL MICROBIOME

The study of microbiota can be categorized into culture-independent and culture-based methods. However, the emergence of molecular biology introduced new approaches like DNA microarrays and PCR-based methods. The most groundbreaking advancement came with massively parallel sequencing or next-generation sequencing (NGS), particularly in conjunction with the 16S rRNA. Furthermore, whole-genome shotgun metagenome sequencing, which involves sequencing all DNA present in a sample, became a valuable tool for studying the microbiome.

1.7.1 Culture and Microscopy

Previous methods for identifying bacterial taxa included culture-dependent methods like microscopy, biochemical testing, growth preferences, sugar metabolism, and antibiotic reactions [110]. Approximately half of the 700 known bacterial species are commonly found in the oral microbiome; research has been done on their isolation, cultivation, identification, characterization, and categorization [111]. The use of traditional culture and culture-based analytical techniques is not feasible for a large number of bacteria present in the biological sample. Its drawback renders the method unsuitable for research [112].

1.7.2 DNA Hybridization Techniques

DNA hybridization techniques involve the use of single-stranded DNA probes labeled with fluorescent dye, chemiluminescent tag, or radioactive isotope, which allow them to specifically bind to the complementary sequences in a target DNA strand resulting in duplex formation that can be identified and quantified using specialized equipment or chemical agents [113]. This approach makes use of either whole genome DNA probes or oligonucleotides. It is more common for whole genome probes to show cross-reactivity with non-target bacteria since different species have similar or homologous DNA sequences. In ideal conditions, more specific oligonucleotide probes show minimal to no cross-reactivity with non-target microorganisms [114].

Fluorescence in situ hybridization, reverse-capture oligonucleotide hybridization, and checkerboard DNA-DNA hybridization are only a few of the various DNA hybridization methods utilized in oral microbiological research [113].

1.7.2.1 Fluorescence *In Situ* Hybridization

Fluorescence in situ hybridization (FISH) was introduced in 1980. It involves the use of whole bacterial cells for analysis [115]. The technique utilizes oligonucleotide probes that are fluorescently tagged and selectively aimed at the ribosomal RNA (rRNA) gene. Further these probes are hybridized with partially fixed bacterial cells [113]. To investigate a variety of microbiological samples, the labeled probes are then detected using fluorescence microscopy or in conjunction with flow cytometry. FISH aids in the detection of cultivable as well as uncultivated bacteria and also provides information about their morphology, quantity, biofilm architecture, taxonomic identification, spatial distribution, and organization [113, 114]. The continuous advancement of FISH has brought new possibilities in the field and study of microbial communities. For example, combinational labeling and spectral imaging FISH can be used to detect the complex taxonomic spatial structures of biofilms like dental plaque, and it has expanded the range of microbial taxa that can be distinguished within a single sample [30, 114, 116]. FISH was used to identify bacteria found in pulp infections and confirmed the presence of *Fretibacterium fastidiosum* and *Pyramidobacter piscolens.* Detecting these specific bacteria in endodontic infections is difficult due to challenges in isolating and cultivating them [117].

1.7.2.2 Checkerboard DNA-DNA Hybridization

[118] introduced the checkerboard DNA-DNA hybridization technique. It involves the immobilization of DNA standards representing target bacterial species in an array configuration on a nylon membrane. Subsequently, these immobilized standards undergo simultaneous cross-hybridization with genomic DNA probes, frequently marked with radioactivity. Cultivable bacteria are required for constructing DNA probes used in the hybridization process. [119] introduced a modified PCR-based checkerboard method, known as the reverse-capture checkerboard hybridization. This methodology eliminates the requirement for bacterial culture, which is needed for preparing whole genomic probes. It involves the use of 30 reverse-capture oligonucleotide probes that are intended to target specific 16S rRNA gene regions. These probes are then placed onto a nylon membrane in individual horizontal lanes, with the help of Minislot apparatus. This technology was further modified by utilizing a microarray format instead of nylon membrane. When utilized in conjunction with sophisticated computer-based analysis and visualization scanners, this approach permits the simultaneous detection of numerous species. The modified checkerboard hybridization technique has been applied to studies on oral cancer, periodontitis, endodontics, dental caries, and HIV [113]. The DNA checkerboard hybridization approach can be effectively used to detect and quantify *Candida* species in the oral microbiome [120].

1.7.3 DNA Microarray

DNA microarrays consist of sequences of thousands of different genes fixed at specific locations on small supports like silicon chips, glass slides, or nylon membranes [121]. It is based on the principle of hybridization where complementary sequences will bind to each other. The DNA used for analysis is fragmented by restriction endonuclease and then labeled with fluorescent markers. The labeled DNA fragments are then exposed on DNA probes present on DNA chips. The complementary sequences of labeled DNA bind with a DNA probe and are detected by fluorescence emission [122]. According to [123], by utilizing DNA microarray, data genetic variations in the chromosomes of different categories, including species, biovars, serotypes, and strains, in *Yersinia pestis* and *Yersinia pseudotuberculosis* can be discerned.

1.7.4 Polymerase Chain Reaction Techniques

Millions of copies of the chosen gene segment can be produced in 2–3 hours using the potent, cutting-edge PCR technique. Its wide use in diagnostic microbiology has been linked to its excellent specificity and sensitivity. For the identification of microorganisms, PCR techniques include nested PCR, asymmetric PCR, reverse transcription PCR, and qualitative PCR [124]. For instance, to increase the positive rate of skeletal tuberculosis cases that were detected by histology, researchers have employed formalin-fixed paraffin-embedded tissues to study the application of qPCR [125]. In the COVID-19 pandemic, reverse transcriptase PCR was utilized to identify the coronavirus [126].

1.7.5 16S rRNA Sequencing

The 16S rRNA sequencing technique makes it possible to identify different taxa of bacteria and archaea. It is widely utilized to identify microbial diversity and perform phylogenetic analyses [127]. According to [128], 16S rRNA sequencing showed that newborns born via cesarean section and those delivered vaginally harbored different types of oral bacteria. In the cesarean section group, the dominant bacteria were *Staphylococcus*, *Petrimonas*, *Bacteroides*, *Pseudomonas,* and *Desulfovibrio*, whereas in the vaginal delivery group, *Lactobacillus*, *Prevotella*, and *Gardnerella* spp. were the most prominent genera. Viruses and fungi cannot be identified using 16s rRNA sequencing, while the identification of bacteria and archaea can be facilitated through this technique.

1.7.6 Next-Generation Sequencing

Next-generation sequencing (NGS) techniques have revolutionized the field of microbial diversity research by enabling the rapid completion of large-scale sequencing projects. The NGS method is also known as massively parallel sequencing and is characterized as sequencing by synthesis. It can sequence millions of nucleic acid fragments simultaneously. Numerous NGS methods utilize distinct sequencing technologies; these include 454 pyrosequencing, Illumina, etc. (Figure 1.4).

1.7.6.1 454 Pyrosequencing

In 2005, the first commercially available next-generation sequencing, known as 454 pyrosequencing by Roche, was launched. In this method, DNA fragments are clonally amplified on beads within the emulsion. With this method, reads longer than 250 base pairs (bp)can be produced, and each sequencing run can generate about 400,000 reads [110]. This method detects chemiluminescent signals generated when pyrophosphate is released during the incorporation of nucleotide [129].

1.7.6.2 Illumina

Illumina relies on the technique called bridge amplification. In this process, DNA molecules are ligated with special adapters at each end that are then amplified on a solid surface such as a glass slide. The surface contains complementary oligonucleotides on the adapters of the DNA. The repetitive amplification of bound DNA on the surface results in the formation of clonal clusters. Each glass slide supports millions of cluster reactions simultaneously. The four nucleotides used in Illumina are tagged with different fluorescent so that when the nucleotide is added to the growing DNA strand their fluorescence signal is used for DNA sequencing [130].

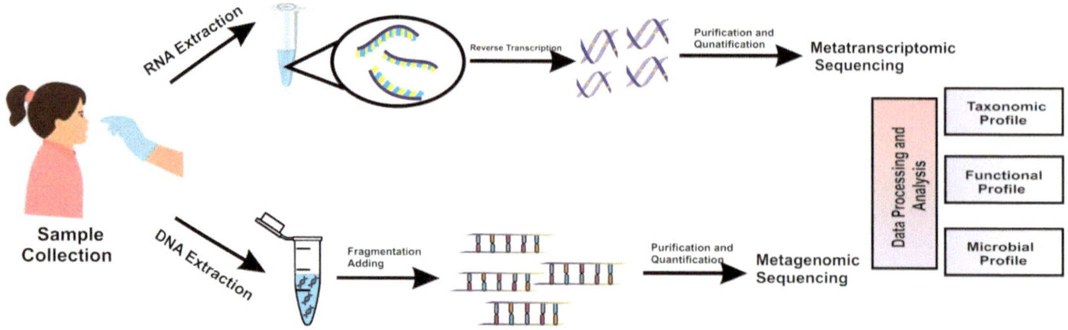

FIGURE 1.4 Schematic representation of next-generation sequencing.

1.7.7 Third-Generation Sequencing

Oxford Nanopore Technologies (ONT) and Pacific Biosciences (PacBio) debuted third-generation sequencing (TGS) in 2011. Single-molecule and real-time sequencing are accomplished using these sequencing technologies. [131] described the utilization of Single Molecule Real-Time (SMRT) sequencing by Pacific Biosciences. Nucleotide inclusion during the elongation of replicated or transcribed DNA can be detected in real time. This could be attributed to PacBio's sequencing technology. An electrically resistant membrane containing nanopores is used by ONT. Long sequences of read lengths for these sequencing platforms are approximately 10–15kb and 2–5kb for PacBio and Oxford Nanopore, respectively [131–133]. Previous studies indicated that TGS of the complete 16S rRNA gene (V1–V9 regions) with the PacBio technique achieved a level of accuracy that is sufficient for resolving subtle nucleotide variations. This accuracy is especially beneficial for taxonomic studies conducted at the species and even strain levels [134, 135].

These methods help to provide a deeper understanding of the composition and diversity of the oral microbiota. Identifying and characterizing the bacteria responsible for oral disorders is critical for early diagnosis and precise therapy.

1.8 HUMAN ORAL MICROBIOME DATABASE

The National Institute of Dental and Craniofacial Research launched the Human Oral Microbiome Database (HOMD) for oral microbiota in 2010 [136]. It preserves data on oral environment-derived microorganisms that are both cultivable and non-cultivable. HOMD is a clearinghouse with well-selected data about oral microorganisms. It uses a 16s rRNA identification approach and has genome sequences from oral bacteria [137].

The expanded Human Oral Microbiome Database (eHOMD) aimed to provide curated data regarding microorganisms found in the pharynx, sinuses, esophagus, aerodigestive tract, and mouth of humans. A total of 774 bacteria species have been identified, where 58% have been named, 16% cultivated but unnamed, and 26% are uncultivated phylotypes. It provides data on the classification, genetic makeup, morphology, bibliography, and evolutionary relationships of the human microbiome [138].

1.9 TREATMENT AND THERAPY

1.9.1 Antibiotics

Antibiotics can be highly effective in treating various odontogenic and periodontal diseases. Commonly used antibiotics for such dental issues include macrolides, lincosamides, nitroimidazoles, β-lactams, and tetracyclines [139]. Antibiotics can interfere with the signaling pathways that bacteria use to cause tooth caries. As an example, sulfated vizantin suppresses *S. mutans* biofilm growth, raises levels of cell-associated glucosyltransferase, and stops cell-free glucosyltransferase from being released extracellularly [140]. Research further suggests that the combination of metronidazole and amoxicillin with scale and root planing (SRP) produces better results than SRP alone. This combination improves saliva and subgingival parameters while minimizing clinical attachment loss and probing depth [141]. This is particularly important when treating patients with type 2 diabetes who have advanced periodontitis [142]. Antibiotics are effective, but their overuse has accelerated the rise of antibiotic resistance. Therefore, the development of novel, pathogen-specific medications is imperative to address this problem [143].

1.9.2 Photodynamic Therapy

The process of producing reactive oxygen species (ROS) in an oxygen-rich environment with light energy and a photosensitizer (PS) is known as photodynamic therapy (PDT) or photo antimicrobial chemotherapy. According to [144], this strategy successfully targets and eliminates dangerous pathogens without creating bacterial resistance. Trials conducted on patients with periapical periodontitis [145], peri-implantitis [146], pulpitis [147], oral leukoplakia [148], and harmful biofilm alterations caused by orthodontic brackets [149] have shown that PDT is superior to conventional treatments in these contexts. The potential of PDT as a substitute therapy for avoiding oral microbial infections must be properly assessed through more clinical studies.

1.9.3 Quorum Quenching Therapy

Quorum quenching (QQ) is a method that inhibits the quorum-sensing mechanism of microbes. The virulence factor generation process and the formation of biofilms are significantly influenced by quorum sensing. This tactic interferes with the synthesis of virulence components and can stop the growth of biofilms [150]. By preventing the creation of biofilms and stopping the growth of *Candida albicans* and *Streptococcus epidermidis* without upsetting other oral bacteria, quorum quenching is essential for preserving the balance of the oral microbiota [151, 152]. For instance, it was revealed that using AII20J changes the synthesis of polymicrobial biofilms without changing the structure of the microbiome. Quorum-sensing inhibitors, such as D-ribose and furanone compounds, have been demonstrated to lessen the harm that *Fusobacterium nucleatum* and *Porphyromonas gingivalis* cause to periodontal tissue and to lower bacterial infection [153]. This indicates that QQ therapy is a potentially effective treatment for oral infections.

1.9.4 Phage Therapy

Phages are viruses that have the ability to efficiently eliminate biofilms and even kill antibiotic-resistant bacteria [154]. Phage therapy is considered a unique approach to the prevention, management, and treatment of oral infections, as it utilizes only lytic phages [155]. Oral infectious illnesses have been effectively treated with phages linked to Firmicutes [156], Actinobacteria [157], and Fusobacteria [158].

1.10 CONCLUSION

This chapter provides an extensive overview of the oral microbiome, highlighting its critical function in preserving health as well as its connections to systemic and oral illnesses. Within the oral cavity, oral microbial populations predominantly live as biofilms, fostering an environment that promotes health. Diverse techniques have been devised to enhance our comprehension of these microbial populations. Many systemic and oral disorders are influenced by dysbiosis, which can be brought on by a variety of internal and environmental sources. Moreover, this chapter offers an overview of the oral microbiota makeup, variables that affect it, methods and instruments for studying it, and a discussion of oral and systemic disorders, along with alternatives for diagnosis and treatment. Recent developments in metagenomics, metatranscriptomics, and microbiomics have made it simpler to connect diverse oral and non-oral disorders with intricate oral microbial ecosystems. Moreover, keeping a healthier oral community has the potential to improve general health through the management of the oral microbiome by focusing on particular species. Despite progress, the field of oral microbiota research is still developing, requiring more investigation to fully understand the intricate relationships between microbial communities.

ACKNOWLEDGMENT

Authors are thankful to the Director of MNNIT Allahabad, Prayagraj, for providing the necessary facilities.

REFERENCES

1. Zarco, M. F.; Vess, T. J.; Ginsburg, G. S. The oral microbiome in health and disease and the potential impact on personalized dental medicine. *Oral Dis.* **2012**, *18*(2), 109–120.
2. Kilian, M.; Chapple, I. L. C.; Hannig, M. et al. The oral microbiome–an update for oral healthcare professionals. *Br. Dent. J.* **2016**, *221*(10), 657–666.
3. Chowdhry, A.; Kapoor, P.; Bhargava, D.; Bagga, D. K. Exploring the oral microbiome: an updated multidisciplinary oral healthcare perspective. *Discoveries* **2023**, *11*(2).
4. Lamont, R. J.; Koo, H.; Hajishengallis, G. The oral microbiota: dynamic communities and host interactions. *Nat. Rev. Microbiol.* **2018**, *16*(12), 745–759.
5. Kilian, M. The oral microbiome–friend or foe? *Eur. J. Oral Sci.* **2018**, *126*, 5–12.
6. Burcham, Z. M.; Garneau, N. L.; Comstock, S. S.; Tucker, R. M.; Knight, R.; Metcalf, J. L. Patterns of oral microbiota diversity in adults and children: a crowdsourced population study. *Sci. Rep.* **2020**, *10*(1), 2133.
7. Baker, J. L.; Edlund, A. Exploiting the oral microbiome to prevent tooth decay: has evolution already provided the best tools? *Front. Microbiol.* **2019**, *9*, 437902.
8. Rosan, B.; Lamont, R. J. Dental plaque formation. *Microbes Infect.* **2000**, *2*(13), 1599–1607.
9. Moye, Z. D.; Zeng, L.; Burne, R. A. Fueling the caries process: carbohydrate metabolism and gene regulation by *Streptococcus mutans*. *J. Oral Microbiol.* **2014**, *6*(1), 24878.
10. Woelber, J.P.; Tennert, C. Diet and periodontal diseases. In The impact of nutrition and diet on oral health, ed. **2020**, F. Zohoori and R. M. Duckworth,125–133. Karger Medical and Scientific Publishers.
11. Cattaneo, C.; Gargari, G.; Koirala, R. et al. New insights into the relationship between taste perception and oral microbiota composition. *Sci. Rep.* **2019**, *9*(1), 3549.
12. Atarashi, K.; Suda, W.; Luo, C. et al. Ectopic colonization of oral bacteria in the intestine drives TH1 cell induction and inflammation. *Science* **2017**, *358*(6361), 359–365.
13. Olsen, I. From the acta prize lecture 2014: The periodontal-systemic connection seen from a microbiological standpoint: Summary of the Acta Odontologica Scandinavia Price lecture 2014 presented at the meeting of the IADR/Pan European region in Dubrovnik, September 10–13. 2014. *Acta Odontologica Scandinavica* **2015**, *73*(8), 563–568.
14. Willis, J. R.; Gabaldón, T. The human oral microbiome in health and disease: from sequences to ecosystems. *Microorganisms* **2020**, *8*(2), 308.
15. Deo, P. N.; Deshmukh, R. Oral microbiome: unveiling the fundamentals. *J. Oral. Maxillofac. Surg. Med. Patho.* **2019**, *23*(1), 122–128.
16. Zhao, H., Chu, M., Huang, Z., et al. Variations in oral microbiota associated with oral cancer. *Sci. Rep.* **2017**, *7*(1), 11773.
17. Sharma, N.; Bhatia, S.; Sodhi, A. S.; Batra, N. Oral microbiome and health. *AIMS Microbiol.* **2018**, *4*(1), 42.
18. Schwiertz, A. Microbiota of the human body. *Adv. Exp. Med. Biol.* **2016**, *902*, 83–93.
19. Welch, J. L. M.; Ramírez-Puebla, S. T.; Borisy, G. G. Oral microbiome geography: micron-scale habitat and niche. *Cell Host Microbe* **2020**, *28*(2), 160–168.
20. Costalonga, M.; Herzberg, M. C. The oral microbiome and the immunobiology of periodontal disease and caries. *Immunol. Lett.* **2014**, *162*(2), 22–38.
21. Mager, D. L.; Haffajee, A. D.; Devlin, P. M.; Norris, C. M.; Posner, M. R.; Goodson, J. M. The salivary microbiota as a diagnostic indicator of oral cancer: a descriptive, non-randomized study of cancer-free and oral squamous cell carcinoma subjects. *J. Transl. Med.* **2005**, *3*, 1–8.
22. Crielaard, W.; Zaura, E.; Schuller, A. A.; Huse, S. M.; Montijn, R. C.; Keijser, B. J. Exploring the oral microbiota of children at various developmental stages of their dentition in the relation to their oral health. *BMC Medical Genomics* **2011**, *4*, 1–13.
23. Kazor, C. E.; Mitchell, P. M.; Lee, A. M.; Stokes, L. N. et al. Diversity of bacterial populations on the tongue dorsa of patients with halitosis and healthy patients. *J. Clin. Microbiol.* **2003**, *41*(2), 558–563.
24. Scully, C.; Greenman, J. Halitosis (breath odor). *Periodontol. 2000* **2008**, *48*(1), 66–75.
25. Jakubovics, N. S. Saliva as the sole nutritional source in the development of multispecies communities in dental plaque. *Microbiol. Spectr.* **2015**, *3*(3), 10–1128.
26. Velsko, I. M.; Fellows Yates, J. A.; Aron, F. et al. Microbial differences between dental plaque and historic dental calculus are related to oral biofilm maturation stage. *Microbiome* **2019**, *7*, 1–20.
27. Socransky, S. S.; Haffajee, A. D. Dental biofilms: difficult therapeutic targets. *Periodontol.2000* **2002**, *28*(1), 12–55.

28. Reese, S.; Guggenheim, B. A novel TEM contrasting technique for extracellular polysaccharides in in vitro biofilms. *Microsc. Res. Tech.* **2007**, *70*(9), 816–822.
29. Zijnge, V.; Van Leeuwen, M. B. M.; Degener, J. E. et al. Oral biofilm architecture on natural teeth. *PLoS One* **2010**, *5*(2), e9321.
30. Kolenbrander, P. E.; PalmerR. J.; Rickard, A. H. et al. Bacterial interactions and successions during plaque development. *Periodontol.2000* **2006**, *42*(1), 47–79.
31. Rickard, A. H.; Gilbert, P.; High, N. J.; Kolenbrander, P. E.; Handley, P. S. Bacterial coaggregation: an integral process in the development of multi-species biofilms. *Trends Microbiol.* **2003**, *11*(2), 94–100.
32. Klein, M.I.; Xiao, J.; Lu, B.; Delahunty, C.M.; Yates III, J.R.; Koo, H. *Streptococcus mutans* protein synthesis during mixed-species biofilm development by high-throughput quantitative proteomics. *PLoS One* **2012**, *7*(9), e45795.
33. Signoretto, C.; Marchi, A.; Bertoncelli, A. et al. Effects of mushroom and chicory extracts on the shape, physiology and proteome of the cariogenic bacterium *Streptococcus mutans. BMC Complement. Altern. Med.* **2013**, *13*, 1–14.
34. Bowen, W. H.; Koo, H. J. C. R. Biology of Streptococcus mutans-derived glucosyltransferases: role in extracellular matrix formation of cariogenic biofilms. *Caries Res.* **2011**, *45*(1), 69–86.
35. Siqueira, W. L.; Custodio, W.; McDonald, E. E. New insights into the composition and functions of the acquired enamel pellicle. *J. Dent. Res.* **2012**, *91*(12), 1110–1118.
36. Ahn, S.J.; Ahn, S.J.; Wen, Z.T.; Brady, L.J.; Burne, R.A. Characteristics of biofilm formation by Streptococcus mutans in the presence of saliva. *Infection and Immunity* **2008**, *76*(9), 4259–4268.
37. Wen, Z. T.; Yates, D.; Ahn, S. J.; Burne, R. A. Biofilm formation and virulence expression by *Streptococcus mutans* are altered when grown in dual-species model. *BMC Microbiol.* **2010**, *10*, 1–9.
38. Brady, L. J.; Maddocks, S. E.; Larson, M. R. et al. The changing faces of Streptococcus antigen I/II polypeptide family adhesins. *Mol. Microbiol.* **2010**, *77*(2), 276–286.
39. Khan, A.U.; Islam, B.; Khan, S.N.; Akram, M. A proteomic approach for exploring biofilm in Streptococcus mutans. *Bioinformation* **2011**, *5*(10), p. 440.
40. Gibbons, R.J. Bacterial adhesion to oral tissues: a model for infectious diseases. *Journal of Dental Res.* **1989**, *68*(5), 750–760.
41. Shashikiran, N.D.; Gugawad, S.; Gaonkar, N.; Hadakar, S.; Waghmode, S.; Pokharkar, P.; Patil, S. *The Emergence of Peptides and Dendrimers in Dentistry.* **2023**, Cambridge Scholars Publishing.
42. Duque, C.; Stipp, R. N.; Wang, B.; Smith, D. J. et al. Downregulation of GbpB, a component of the VicRK regulon, affects biofilm formation and cell surface characteristics of Streptococcus mutans. *Infect. Immun.* **2011**, *79*(2),786–796.
43. Hotz, P.; Guggenheim, B.; Schmid, R. Carbohydrates in pooled dental plaque. *Caries Res.* **1972**, *6*(2), 103–121.
44. Oba, P. M.; Holscher, H. D.; Mathai, R. A.; Kim, J.; Swanson, K. S. Diet influences the oral microbiota of infants during the first six months of life. *Nutrients* **2020**, *12*(11), 3400.
45. Chumponsuk, T.; Gruneck, L.; Gentekaki, E. et al. The salivary microbiota of Thai adults with metabolic disorders and association with diet. *Arch. Oral Biol.* **2021**, *122*, 105036.
46. Štšepetova, J.; Truu, J.; Runnel, R. et al. Impact of polyols on Oral microbiome of Estonian school children. *BMC Oral Health* **2019**, *19*, 1–10.
47. Lassalle, F.; Spagnoletti, M.; Fumagalli, M. et al. Oral microbiomes from hunter-gatherers and traditional farmers reveal shifts in commensal balance and pathogen load linked to diet. *Mol. Ecol.* **2018**, *27*(1), 182–195.
48. Vallès, Y.; Inman, C. K.; Peters, B. A. et al. Types of tobacco consumption and the oral microbiome in the United Arab Emirates Healthy Future (UAEHFS) Pilot Study. *Sci. Rep.* **2018**, *8*(1), 11327.
49. Wu, J.; Peters, B. A.; Dominianni, C. et al. Cigarette smoking and the oral microbiome in a large study of American adults. *The ISME Journal* **2016**, *10*(10), 2435–2446.
50. Kenney, E. B.; Saxe, S. R.; Bowles, R. D. The effect of cigarette smoking on anaerobiosis in the oral cavity. *J. Periodontol.* **1975**, *46*(2), 82–85.
51. Macgregor, I. D. Effects of smoking on oral ecology. A review of the literature. *Clin. Prev. Dent.* **1989**, *11*(1), 3–7.
52. Brook, I. The impact of smoking on oral and nasopharyngeal bacterial flora. *J. Dent. Res.* **2011**, *90*(6), 704–710.
53. Sopori, M. Effects of cigarette smoke on the immune system. *Nat. Rev. Immunol.* **2002**, *2*(5), 372–377.
54. Parvinen, T. Stimulated salivary flow rate, pH and lactobacillus and yeast concentrations in persons with different types of dentition. *Eur. J. Oral Sci.* **1984**, *92*(5), 412–418.

55. Kanwar, A.; Sah, K.; Grover, N. et al. Long-term effect of tobacco on resting whole mouth salivary flow rate and pH: an institutional based comparative study. *Eur. J. Dent.* **2013**, *2*(03), 296–299.
56. Thomas, A. M.; Gleber-Netto, F. O.; Fernandes, G. R. et al. Alcohol and tobacco consumption affects bacterial richness in oral cavity mucosa biofilms. *BMC Microbiol.* **2014**, *14*, 1–12.
57. Liao, Y.; Tong, X. T.; Jia, Y. J. et al. The effects of alcohol drinking on oral microbiota in the Chinese population. *Int. J. Environ. Res. Public Health* **2022**, *19*(9), 5729.
58. Signoretto, C.; Bianchi, F.; Burlacchini, G.; Sivieri, F.; Spratt, D.; Canepari, P. Drinking habits are associated with changes in the dental plaque microbial community. *J. Clin. Microbiol.* **2010**, *48*(2), 347–356.
59. Jabbour, Z.; do Nascimento, C.; dos Santos Kotake, B. G. et al. Assessing the oral microbiota of healthy and alcohol-treated rats using whole-genome DNA probes from human bacteria. *Arch. Oral Biol.* **2013**, *58*(3), 317–323.
60. Amer, A.; Whelan, A.; Al-Hebshi, N.N.; Healy, C.M.; Moran, G.P. Acetaldehyde production by *Rothia mucilaginosa* isolates from patients with oral leukoplakia. *J. Oral Microbiol.* **2020**, *12*(1), 1743066.
61. Barroso, E.; Martín, V.; Martínez-Cuesta, M. C.; Peláez, C.; Requena, T. Stability of saliva microbiota during moderate consumption of red wine. *Arch. Oral Biol.* **2015**, *60*(12), 1763–1768.
62. Selwitz, R. H.; Ismail, A. I.; Pitts, N. B. Dental caries. *The Lancet* **2007**, *369*(9555), 51–59.
63. Takahashi, N.; Nyvad, B. The role of bacteria in the caries process: ecological perspectives. *J. Dent. Res.* **2011**, *90*(3), 294–303.
64. Aas, J. A.; Griffen, A. L.; Dardis, S. R.; Lee, A. M.; Olsen, I.; Dewhirst, F. E. et al. Bacteria of dental caries in primary and permanent teeth in children and young adults. *J. Clin. Microbiol.* **2008**, *46*(4), 1407–1417.
65. Burne, R. A.; Marquis, R. E. Alkali production by oral bacteria and protection against dental caries. *FEMS Microbiol. Lett.* **2000**, *193*(1), 1–6.
66. Teng, F.; Yang, F.; Huang, S. et al. Prediction of early childhood caries via spatial-temporal variations of oral microbiota. *Cell Host Microbe* **2015**, *18*(3), 296–306.
67. Dubey, P.; Mittal, N. Periodontal diseases-a brief review. *Int J Oral Heal Dent.* **2020**, *6*(3), 177–87.
68. Socransky, S.S.; Haffajee, A.D.; Cugini, M.A.; Smith, C.K.J.R.; Kent Jr, R.L. Microbial complexes in subgingival plaque. *Journal of Clinical Periodontal* **1998**, *25*(2), 134–144.
69. Wei, Y.; Shi, M.; Zhen, M. et al. Comparison of subgingival and buccal mucosa microbiome in chronic and aggressive periodontitis: a pilot study. *Front. Cell. Infect. Microbiol.* **2019**, *9*, 53.
70. Kumar, P. S.; Griffen, A. L.; Barton, J. A. et al. New bacterial species associated with chronic periodontitis. *J. Dent. Res.* **2003**, *82*(5), 338–344.
71. Kumar, S. R.; Narayanan, M. A.; Jayanthi, D. Comparative assessment of the prevalence of periodontal disease in subjects with and without systemic autoimmune diseases: a case–control study. *Contemp. Clin. Dent.* **2016**, *7*(2), 170–175.
72. Canabarro, A.; Valle, C.; Farias, M. R.; Santos, F. B.; Lazera, M.; Wanke, B. Association of subgingival colonization of *Candida albicans* and other yeasts with severity of chronic periodontitis. *J. Periodontal Res.* **2013**, *48*(4), 428–432.
73. Akintoye, S. O; Greenberg, M. S. Recurrent aphthous stomatitis. *Dental Clinics* **2014**, *58*(2), 281–297.
74. Marchini, L.; Campos, M. S.; Silva, A. M.; Paulino, L. C.; Nobrega, F. G. Bacterial diversity in aphthous ulcers. *Oral Microbiol. Immun. Journal* **2007**, *22*(4), 225–231.
75. Hijazi, K.; Lowe, T.; Meharg, C. et al. Mucosal microbiome in patients with recurrent aphthous stomatitis. *J. Dent. Res.* **2015**, *94*(3_suppl), 87S–94S.
76. Slebioda, Z.; Szponar, E.; Kowalska, A. Etiopathogenesis of recurrent aphthous stomatitis and the role of immunologic aspects: literature review. *Arch. Immunol. Ther. Exp.* **2014**, *62*, 205–215.
77. Seoudi, N.; Bergmeier, L. A.; Drobniewski, F.; Paster, B.; Fortune, F. The oral mucosal and salivary microbial community of Behcet's syndrome and recurrent aphthous stomatitis. *J. Oral Microbiol.* **2015**, *7*(1), 27150.
78. Kim, Y. J.; Choi, Y. S.; Baek, K. J. et al. Mucosal and salivary microbiota associated with recurrent aphthous stomatitis. *BMC Microbiol.* **2016**, *16*, 1–10.
79. Markopoulos, A. K. Role of human papillomavirus in the pathogenesis of oral squamous cell carcinoma. *World J. Exp. Med.* **2012**, *2*(4), 65.
80. Safari, Z.; Firouzi, A.; Rezaeikalantari, N. et al. The salivary exosomal microRNA as a potential biomarker in patients with periodontitis and oral cancers. *Chem. Biol. Drug. Des.* **2023**, *101*(5), 1204–1215.
81. Nagy, K. N.; Sonkodi, I.; Szöke, I.; Nagy, E.; Newman, H. N. The microflora associated with human oral carcinomas. *Oral Oncol.* **1998**, *34*(4), 304–308.
82. Zhang, L.; Liu, Y.; Zheng, H. J.; Zhang, C. P. The oral microbiota may have influence on oral cancer. *Front. Cell. Infect. Microbiol.* **2020**, *9*, 476.

83. Hooper, S. J.; Crean, S. J.; Fardy, M. J. et al. A molecular analysis of the bacteria present within oral squamous cell carcinoma. *J. Med. Microbiol.* **2007**, *56*(12), 1651–1659.
84. Pushalkar, S.; Ji, X.; Li, Y. et al.; Comparison of oral microbiota in tumor and non-tumor tissues of patients with oral squamous cell carcinoma. *BMC Microbiol.* **2012**, *12*, 1–15.
85. Lee, W. H.; Chen, H. M.; Yang, S. F. et al. Bacterial alterations in salivary microbiota and their association in oral cancer. *Sci. Rep.* **2017**, *7*(1), 1–11.
86. Azman, S. N. Detection of Host-specific Immunogenic Proteins in the Sera of Oral Squamous Cell Carcinoma (OSCC) Patients. PhD diss., **2015**, Univ Sains Malaysia.
87. Meurman, J. H.; Uittamo, J. Oral micro-organisms in the etiology of cancer. *Acta Odontol. Scand.* **2008**, *66*(6), 321–326.
88. Makinen, A. I.; PappalardoV. Y.; Buijs, M. J. et al. Salivary microbiome profiles of oral cancer patients analyzed before and after treatment. *Microbiome* **2023**, *11*(1), 171.
89. Spatafora, G.; Li, Y.; He, X.; Cowan, A.; Tanner, A. C. The Evolving Microbiome of Dental Caries. *Microorganisms* **2024**, *12*(1), 121.
90. Gasner, N. S.; Schure, R. S. Periodontal Disease 2023 Periodontal disease. In: StatPearls [Internet]. https://www.ncbi.nlm.nih.gov/books/NBK554590/.
91. Karpiński, T. M. Role of oral microbiota in cancer development. *Microorganisms* **2019**, *7*(1), 20.
92. Alves, C.H.; Russi, K.L.; Rocha, N.C. Host-microbiome interactions regarding peri-implantitis and dental implant loss. *J. Transl. Med.* **2022**, *20*, 425.
93. Bronzato, J. D.; Bomfim, R. A.; Hayasida, G. Z. et al. Analysis of microorganisms in periapical lesions: a systematic review and meta-analysis. *Arch. Oral Biol.* **2021**, *124*, 105055.
94. Yang, Z.; Cui, Q.; An, R. et al. Comparison of microbiomes in ulcerative and normal mucosa of recurrent aphthous stomatitis (RAS)-affected patients. *BMC Oral Health* **2020**, *20*, 1–8.
95. Hampelska, K.; Jaworska, M. M.; Babalska, Z. Ł. et al. The role of oral microbiota in intra-oral halitosis. *J. Clin. Med.* **2020**, *9*(8), 2484.
96. Preshaw, P. M.; Alba, A. L.; Herrera, D. et al. Periodontitis and diabetes: a two-way relationship. *Diabetologia* **2012**, *55*, 21–31.
97. Hintao, J.; Teanpaisan, R.; Chongsuvivatwong, V.; Ratarasan, C.; Dahlen, G. The microbiological profiles of saliva, supragingival and subgingival plaque and dental caries in adults with and without type 2 diabetes mellitus. *Oral Microbiol Immunol.* **2007**, *22*(3), 175–181.
98. Casarin, R. C. V.; Barbagallo, A.; Meulman, T. et al. Subgingival biodiversity in subjects with uncontrolled type-2 diabetes and chronic periodontitis. *J. Periodontal Res.* **2013**, *48*(1), 30–36.
99. Thorstensson, H.; Dahlen, G.; Hugoson, A. Some suspected periodontopathogens and serum antibody response in adult long-duration insulin-dependent diabetics. *J. Clin. Periodontol.* **1995**, *22*(6), 449–458.
100. Kuo, L.C.; Polson, A.M.; Kang, T. Associations between periodontal diseases and systemic diseases: a review of the inter-relationships and interactions with diabetes, respiratory diseases, cardiovascular diseases and osteoporosis. *Public Health* **2008**, *122*(4), 417–433.
101. Teeuw, W. J.; Gerdes, V. E.; Loos, B.G. Effect of periodontal treatment on glycemic control of diabetic patients: a systematic review and meta-analysis. *Diabetes Care* **2010**, *33*(2), 421–427.
102. Mattila, K.J.; Valtonen, V.V.; Nieminen, M.S.; Asikainen, S. Role of infection as a risk factor for atherosclerosis, myocardial infarction, and stroke. *Clin. Infect. Dis.* **1998**, *26*(3), 719–734.
103. Koren, O.; Spor, A.; Felin, J.; Fåk, F.; Stombaugh, J.; Tremaroli, V.; Behre, C.J.; Knight, R.; Fagerberg, B.; Ley, R.E.; Bäckhed, F. Human oral, gut, and plaque microbiota in patients with atherosclerosis. *Proc. Natl. Acad. Sci. USA* **2011**, *108*(supplement_1), 4592–4598.
104. Figuero, E.; Sánchez-Beltrán, M.; Cuesta-Frechoso, S.; Tejerina, J.M.; del Castro, J.A.; Gutiérrez, J.M.; Herrera, D.; Sanz, M. Detection of periodontal bacteria in atheromatous plaque by nested polymerase chain reaction. *J. Periodontol.* **2011**, *82*(10), 1469–1477.
105. Ahn, J.; Chen, C. Y.; Hayes, R. B. Oral microbiome and oral and gastrointestinal cancer risk. *Cancer Causes & Control* **2012**, *23*, 399–404.
106. Sheu, B.C.; Chang, W.C.; Cheng, C.Y.; Lin, H.H.; Chang, D.Y.; Huang, S.C. Cytokine regulation networks in the cancer microenvironment. *Front Biosci.* **2008**, *13*(13), 6255–6268.
107. Shiga, K.; Tateda, M.; Saijo, S.; Hori, T.; Sato, I.; Tateno, H.; Matsuura, K.; Takasaka, T.; Miyagi, T. Presence of *Streptococcus* infection in extra-oropharyngeal head and neck squamous cell carcinoma and its implication in carcinogenesis. *Oncol. Rep.* **2001**, *8*(2), 245–248.
108. Narikiyo, M.; Tanabe, C.; Yamada, Y.; Igaki, H.; Tachimori, Y.; Kato, H.; Muto, M.; Montesano, R.; Sakamoto, H.; Nakajima, Y.; Sasaki, H. Frequent and preferential infection of *Treponema denticola*, *Streptococcus mitis*, and *Streptococcus anginosus* in esophageal cancers. *Cancer Science* **2004**, *95*(7), 569–574.

109. Farrell, J.J.; Zhang, L.; Zhou, H.; Chia, D.; Elashoff, D.; Akin, D.; Paster, B.J.; Joshipura, K.; Wong, D.T. Variations of oral microbiota are associated with pancreatic diseases including pancreatic cancer. *Gut* **2012**, *61*(4), 582–588.
110. Krishnan, K.; Chen, T.; Paster, B. J. A practical guide to the oral microbiome and its relation to health and disease. *Oral Dis.* **2017**, *23*(3), 276–286.
111. Aas, J. A.; Paster, B. J.; Stokes, L. N.; Olsen, I.; Dewhirst, F. E. Defining the normal bacterial flora of the oral cavity. *J. Clin. Microbiol.* **2005**, *43*(11), 5721–5732.
112. Pozhitkov, A. E.; Beikler, T.; Flemmig, T.; Noble, P. A. High-throughput methods for analysis of the human oral microbiome. *Periodontol. 2000* **2011**, *55*(1), 70–86.
113. Benn, A. M. L.; Heng, N. C. K.; Broadbent, J. M.; Thomson, W. M. Studying the human oral microbiome: challenges and the evolution of solutions. *Aust. Dent. J.* **2018**, *63*(1), 14–24.
114. Siqueira, J. F.; Rôças, I. N. The oral microbiota: general overview, taxonomy, and nucleic acid techniques. *Oral Biology: Molecular Techniques and Applications* **2010**, 55–69.
115. Levsky, J. M.; Singer, R. H. Fluorescence in situ hybridization: past, present and future. *J. Cell Biol.* **2003**, *116*(14), 2833–2838.
116. Valm, A. M.; Welch, J. L. M.; Rieken, C. W. et al. Systems-level analysis of microbial community organization through combinatorial labeling and spectral imaging. *Proc. Natl. Acad. Sci. USA* **2011**, *108*(10), 4152–4157.
117. do Cabo Fernandes, C., Rechenberg, D. K., Zehnder, M., and Belibasakis, G. N. et al. Identification of Synergistetes in endodontic infections. *Microb. Path.* **2014**, *73*, 1–6.
118. Socransky, S.S.; Smith, C.; Martin, L.; Paster, B.J.; Dewhirst, F.E.; Levin, A.E.; "Checkerboard" DNA-DNA hybridization. *Biotechniques* **1994**, *17*(4), 788–792.
119. Paster, B. J.; Olsen, I.; Aas, J. A.; Dewhirst, F. E. The breadth of bacterial diversity in the human periodontal pocket and other oral sites. *Periodontol.2000* **2006**, 42(1).
120. do Nascimento, C.; Ferreira de Albuquerque Junior, R.; Issa, J. P. M. et al. Use of the DNA Checkerboard hybridization method for detection and quantitation of Candida species in oral microbiota. *Can. J. Microbiol.* **2009**, *55*(5), 622–626.
121. Fesseha, H.; Tilahun, H. Principles and applications of deoxyribonucleic acid microarray: a review. *Pathol. Lab. Med. Open J.* **2020**, *1*, 54–62.
122. Govindarajan, R.; Duraiyan, J.; Kaliyappan, K.; Palanisamy, M. Microarray and its applications. *J. Pharm. Bioallied Sci.* **2012**, *4*(Suppl 2), S310–S312.
123. Garaizar, J.; Rementeria, A.; Porwollik, S. DNA microarray technology: a new tool for the epidemiological typing of bacterial pathogens? *FEMS Microbiol. Immunol.* **2006**, *47*(2), 178–189.
124. Sakamoto, M.; Umeda, M.; Benno, Y. Molecular analysis of human oral microbiota. *Journal of Periodontal Research* **2005**, *40*(3), 277–285.
125. He, G.; Chen, C. Y.; Zhang, X. et al. Clinical performance of quantitative PCR for the molecular identification of skeletal tuberculosis from formalin-fixed paraffin-embedded tissues. *BMC Infect. Dis.* **2022**, *22*(1), 651.
126. El-Kafrawy, S. A.; El-Daly, M. M.; Hassan, A. M. et al. A direct method for RT-PCR detection of SARS-CoV-2 in clinical samples. *In Healthcare* **2021**, *9*, 1–37.
127. Xiao, X.; Liu, S.; Deng, H.; Song, Y. et al. Advances in the oral microbiota and rapid detection of oral infectious diseases. *Front. Microbiol.* **2023**, *14*, 1121737.
128. Li, H.; Wang, J.; Wu, L. et al. The impacts of delivery mode on infant's oral microflora. *Sci. Rep.* **2018**, *8*(1), 11938.
129. Jo, J. H.; Kennedy, E. A.; Kong, H. H. Research techniques made simple: bacterial 16S ribosomal RNA gene sequencing in cutaneous research. *J. Invest. Dermatol.* **2016**, *136*(3), e23–e27.
130. Slatko, B. E.; Gardner, A. F.; Ausubel, F. M. Overview of next-generation sequencing technologies. *Curr. Protoc. Mol Biol.* **2018**, *122*(1), e59.
131. Van Dijk, E. L.; Jaszczyszyn, Y.; Naquin, D.; Thermes, C. The third revolution in sequencing technology. *Trends Genet.* **2018**, *34*(9), 666–681.
132. Mosher, J. J.; Bowman, B.; Bernberg, E. L.; Shevchenko, O.; Kan, J.; Korlach, J.; Kaplan, L. A. Improved performance of the PacBio SMRT technology for 16S rDNA sequencing. *J. Microbiol. Methods* **2014**, *104*, 59–60.
133. Laver, T.; HarrisonJ.; O'neill, P. A. et al. Assessing the performance of the oxford nanopore technologies minion. *Biomol detect quantif* **2015**, *3*, 1–8.
134. Earl, J. P.; Adappa, N. D.; Krol, J. et al. Species-level bacterial community profiling of the healthy sinonasal microbiome using Pacific Biosciences sequencing of full-length 16S rRNA genes. *Microbiome* **2018**, *6*, 1–26.

135. Johnson, J. S.; Spakowicz, D. J.; Hong, B. Y. et al. Evaluation of 16S rRNA gene sequencing for species and strain-level microbiome analysis. *Nat. Commun.* **2019**, *10*(1), 5029.
136. Chen, T.; Yu, W. H.; Izard, J.; Baranova, O. V.; Lakshmanan, A.; Dewhirst, F. E. The Human Oral Microbiome Database: a web accessible resource for investigating oral microbe taxonomic and genomic information. *Database* **2010**, baq013,
137. Wade, W. G. **2013**. The oral microbiome in health and disease. *Pharmacol. Res.*69(1), 137–143.
138. http://www.homd.org (Assessed on May 10, 2024)
139. Amato, A. Oral Microbiota, Bacterial Infections, Antibiotic Prescriptions, and Antimicrobial Resistance in Children. *Microorganisms* **2023**, *11*(8), 1927.
140. Oda, M.; Kurosawa, M.; Yamamoto, H. et al. Sulfated vizantin inhibits biofilm maturation by Streptococcus mutans. *Microbiol Immunol.* **2020**, *64*(7), 493–501.
141. Lu, H.; He, L.; Jin, D.; Zhu, Y.; Meng, H. Effect of adjunctive systemic antibiotics on microbial populations compared with scaling and root planing alone for the treatment of periodontitis: a pilot randomized clinical trial. *J. Periodontol.* **2022**, *93*(4), 570–583.
142. Cruz, D. F. D.; Duarte, P. M.; Figueiredo, L. C.; da Silva, H. D. P.; Retamal-Valdes, B.; Feres, M.; Miranda, T. S.; Metronidazole and amoxicillin for patients with periodontitis and diabetes mellitus: 5-year secondary analysis of a randomized controlled trial. *J. Periodontol.* **2021**, *92*(4), 479–487.
143. Li, X.; Liu, Y.; Yang, X.; Li, C.; Song, Z. The oral microbiota: community composition, influencing factors, pathogenesis, and interventions. *Front. Microbiol.* **2022**, *13*, 895537.
144. Theodoro, L. H.; Ferro-Alves, M. L.; Longo, M. et al. Curcumin photodynamic effect in the treatment of the induced periodontitis in rats. *Lasers Med. Sci.* **2017**, *32*(8), 1783–1791.
145. Moreira, M. S. N. A.; de Freitas Archilla, J. R.; Lascala, C. A.; Ramalho, K. M.; Gutknecht, N.; and Marques, M. M. Post-treatment apical periodontitis successfully treated with antimicrobial photodynamic therapy via sinus tract and laser phototherapy: report of two cases. *Photomed. Laser Surg.* **2015**, *33*(10), 524–528.
146. Habashneh, R. A.; Asaad, F. A.; Khader, Y. Photodynamic therapy in periodontal and peri-implant diseases. *Quintessence Int.* **2015**, *46*(8), 677.
147. da Mota, A. C. C.; Gonçalves, M. L. L.; Bortoletto, C. et al. Evaluation of the effectiveness of photodynamic therapy for the endodontic treatment of primary teeth: study protocol for a randomized controlled clinical trial. *Trials* **2015**, *16*, 1–6.
148. Li, Y.; Wang, B.; Zheng, S.; He, Y. Photodynamic therapy in the treatment of oral leukoplakia: A systematic review. *Photodiagnosis Photodyn. Ther.* **2019**, *25*, 17–22.
149. Rosa, E. P.; Murakami-Malaquias-Silva, Schalch, T. O. et al. Efficacy of photodynamic therapy and periodontal treatment in patients with gingivitis and fixed orthodontic appliances: Protocol of randomized, controlled, double-blind study. *Medicine* **2020**, *99*(14), e19429.
150. Paluch, E.; Rewak-Soroczyńska, J.; Jędrusik, I.; Mazurkiewicz, E.; Jermakow, K. J. A. M. Prevention of biofilm formation by quorum quenching. *Appl. Microbiol. Biotechnol.* **2020**, *104*, 1871–1881.
151. Weiland-Bräuer, N.; Malek, I.; Schmitz, R. A. Metagenomic quorum quenching enzymes affect biofilm formation of *Candida albicans* and *Staphylococcus epidermidis*. *PLoS One* **2019**, *14*(1), e0211366.
152. Parga, A.; Muras, A.; Otero-Casal, P. The quorum quenching enzyme Aii20J modifies in vitro periodontal biofilm formation. *Front. Cell. Infect. Microbiol.* **2023**, *13*, 1118630.
153. Ben Amara, H.; Song, H. Y.; Ryu, E. et al. Effects of quorum-sensing inhibition on experimental periodontitis induced by mixed infection in mice. *Eur. J. Oral Sci.* **2018**, *126*(6), 449–457.
154. Shlezinger, M.; Khalifa, L.; Houri-Haddad, Y. et al. Phage therapy: a new horizon in the antibacterial treatment of oral pathogens. *Curr. Top. Med. Chem.* **2017**, *17*(10), 1199–1211.
155. Harada, L. K.; Silva, E. C.; Campos, W. F. et al. Biotechnological applications of bacteriophages: State of the art. *Microbiol. Res.* **2018**, *212*, 38–58.
156. Wolfoviz-Zilberman, A.; Kraitman, R.; Hazan, R.; Friedman, M.; Houri-Haddad, Y.; Beyth, N. Phage targeting Streptococcus mutans in vitro and in vivo as a caries-preventive modality. *Antibiotics* **2021**, *10*(8), 1015.
157. Hatfull, G. F. Actinobacteriophages: genomics, dynamics, and applications. *Annu. Rev. Virol.* **2020**, *7*, 37–61.
158. Machuca, P.; Daille, L.; Vinés, E.; Berrocal, L.; Bittner, M. Isolation of a novel bacteriophage specific for the periodontal pathogen *Fusobacterium nucleatum*. *Applied and Environmental Microbiology* **2010**, *76*(21), 7243–7250.

2 The Oral Microbiome, Hygiene, and Health

Pooja Thakur, Jatin Sharma, and Savitri
Department of Biotechnology, Himachal Pradesh University, Shimla, Himachal Pradesh, India

2.1 INTRODUCTION

The oral cavity contains the body's second-largest microbiota. It has both adamantine and pliable anatomical structures such as teeth, attached gingiva, gingival sulcus, inner cheek, tongue, and palates. This unique and sharp anatomical structure provides a heterogeneous habitat for diverse microbial communities, including bacteria, viruses, fungi, and protozoa. These microorganisms comprise a complex ecosystem that survives in the dynamic oral environment in a symbiotic relationship with the human host [1]. However, due to the prevalence of bacteria in the oral microbiome, the concept of a "core microbiome" has been limited to the bacterial microbiome. The oral microbiome, with all its interactions throughout species, develops a core ecology that flourishes in the oral cavity, a very versatile habitat [2].

Interspecies and host-microbial interactions substantially impact microbial composition, threatening the health and disease status of the host [3]. Bacteria, fungi, and protozoa colonize themselves on various surfaces in the buccal cavity. Microbial biofilms are formed on the buccal mucosa, dorsum of the tongue, tooth surfaces, and gingiva due to the growth of microorganisms. Various studies demonstrate the relationship between unbalanced microflora and the development of diseases like tooth caries, periodontal diseases, type 2 diabetes, and circulatory system–related diseases [1, 4]. Transcriptome-based remodeling of microbial metabolism in health and disease-associated states has been well reported [5]. Human diets and habitat can trigger overall microbial interactions of the oral microbiome.

The overall systemic health of the host is significantly influenced by the oral microbiota. As it is said, "The mouth is the gateway to total body wellness; thus, the oral microbiome influences the overall health of an individual." For instance, some oral microorganisms also participate in the enterosalivary nitrate-nitrite-nitric oxide cycle, which converts dietary nitrate into nitrite and nitric oxide in the saliva. This cycle may help with immune modulatory effects, vasodilation, blood pressure modulation, cytoprotection, and antithrombotic effects and may improve the outcomes of myocardial infarction, pulmonary hypertension, and heart failure [6]. It is important to note that abundances of *Prevotella* and *Veillonella* genera were found to be harmful to homeostasis whereas abundances of *Rothia* and *Neisseria* genera were proven to be beneficial for the preservation of nitric oxide homeostasis and cardiovascular health in the host.

Although the oral microbiome is resilient, injuries or modifications, like smoking, can cause the eubiotic balance to change from one of mutualism/commensalism to an imbalanced parasitic/pathogenic state, which encourages illness in the host. Additionally, oral hygiene routines, including the periodic transit and mastication of ingested food as well as motions of the lips, cheeks, and tongue, cause more or less frequent disruptions in the mouth's environment. Along with external activities like eating, smoking, and brushing, salivary flow and electrolyte content also alter according to the circadian cycle. The risk of periodontitis and dental caries may be reduced by improving oral hygiene.

DOI: 10.1201/9781003478638-2

2.2 THE ORAL MICROBIOME

The oral microbiome was primarily introduced by the Dutchman Antonie van Leeuwenhoek, using a microscope of his invention. In the late 1670s, he reported to the British Royal Society that different microbes were present on tooth surface plaque. The microbial community found in plaque is mostly composed of viruses, bacteria, archaea, fungi, and protozoa (Figure 2.1). The candida genera are the most prevalent commensal fungi in the oral mycobiome, found in 70% of healthy individuals. The majority of viruses in the mouth cavity, often known as the virobiome, are bacteriophages. In healthy individuals, the main bacteriophage families found are *Siphoviridae, Myoviridae*, and *Podoviridae*, all belonging to the Caudovirales order. Specific periodontal pathogens are *Streptococcus mutans*, *Porphyromonas gingivalis*, *Staphylococcus* spp., *Treponema denticola*, *Aggregatibacter actinomycetemcomitans*, *Tannerella forsythia*, and *Lactobacillus* spp. *Streptococcus mutans* is a major component of dental plaque and the main component of the oral microbiota. Like *Lactobacillus*, this pathogen also causes dental caries, a bacterial infectious illness that affects the hard tissues of the teeth and has the highest prevalence among oral disorders [7]. In addition, *P. gingivalis*, also the most promising pathogen as a cause for periodontitis, affected periodontal tissue and may trigger the gums to slip off the teeth [8, 9]. This pathogen spreads from the buccal cavity to the stomach in different conditions including inflammatory bowel disease (IBD), diabetes, and colon cancer. This promotes dysbiosis by reducing innate host defenses and inducing inflammatory reactions in phagocytic cells. This microorganism can affect the connection between the host microbiota and the mucosa by influencing the innate immune system and signaling mechanisms.

2.2.1 Healthy Oral Microbiota (Eubiosis)

The human oral microbiome is an uncommon ecological niche from the highly studied gut microbiome and is formed early in development. According to the studies, more than half of microbial phenotypes are heritable [10]. The oral cavity contains several different habitats, including the teeth, gingival sulcus, tongue, cheeks, hard and soft palates, and tonsils, which are colonized by microbes and form a healthy oral microbial community. This condition is known as eubiosis (Figure 2.1). This healthy microbial community is disturbed by different lifestyle habits including diet, toothpaste, toothbrushes, oral health, mouthwashes, and overall health and disease.

A healthy oral microbiota or eubiotic condition contributes to shaping and driving a healthy gut microbiome, regulates the saliva buffering capacity (for high pH) to neutralize acids in the mouth, reduces numbers of pathogenic acid-producing bacteria via competitive exclusion (decreases the bacterial acid load responsible for tooth decay and cavities), helps to prevent gum diseases, initiates digestion, metabolizes nitrates into nitrites (a key molecule to reduce blood pressure), and prevents plaque and bad breath [11].

An inordinate use of antimicrobial mouthwashes and eating junk food are not the best ways to sustain a healthy oral microbiome and eubiotic condition of the oral cavity. On the contrary, these activities either starve or wipe them out, acidify the saliva, and allow sugar-loving pathogenic bacteria to take over, wreaking havoc on our teeth, gums, brain, and body [12].

2.2.2 Unhealthy Oral Microbiota (Dysbiosis)

Oral microbiome dysbiosis refers to an instability in the composition and role of the bacteria in the buccal cavity, which interrupts the natural oral ecology and leads to oral health problems. Although, the oral microbiome can deal with changes in the oral cavity without shifting its state of symbiosis. These changes are due to tobacco use and smoking, which shift the eubiotic equilibrium from mutualism to an unstable pathogenic state, thus nurturing disease in the host and, hence, creating an unbalanced, unhealthy microbiome or dysbiosis [13].

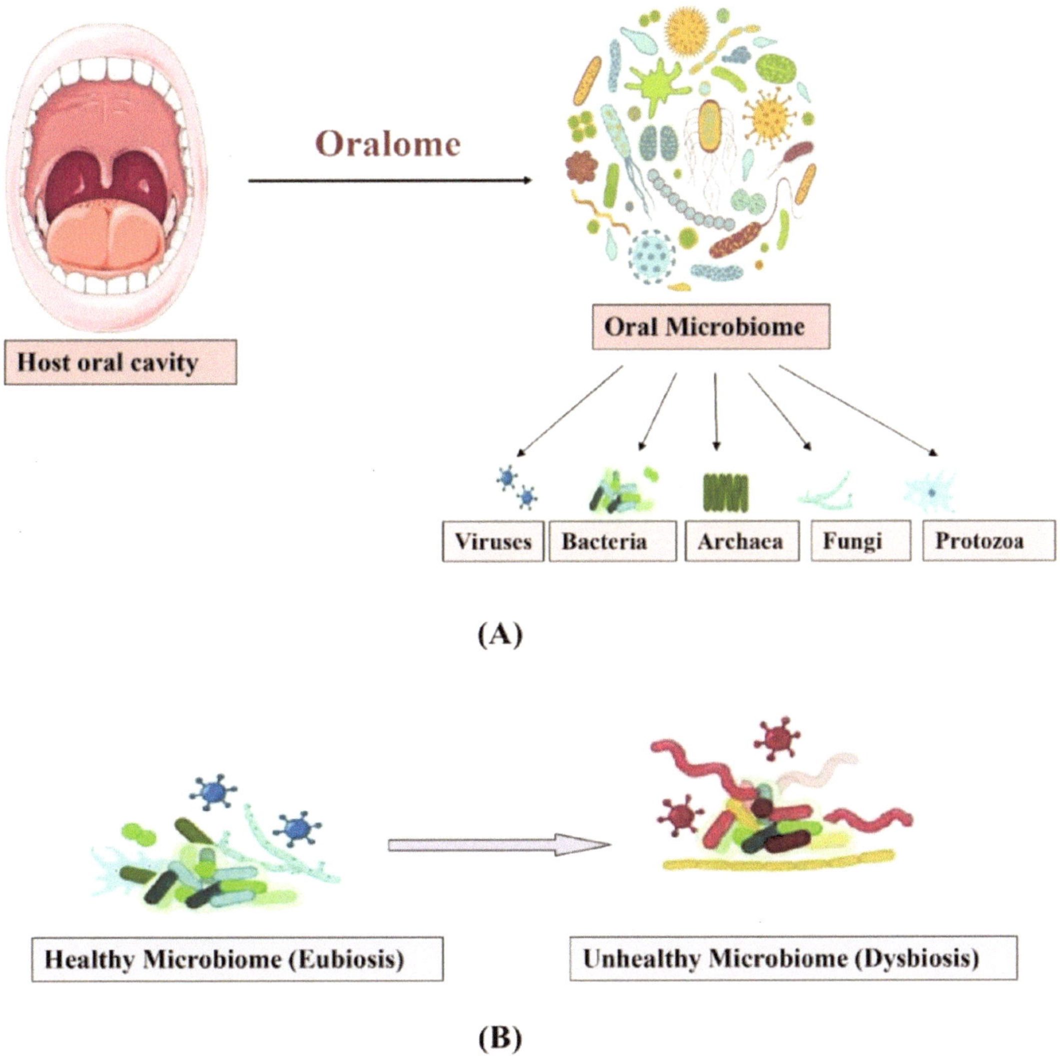

FIGURE 2.1 (A) The oral microbiome. (B) Healthy oral microbiome (eubiosis) and unhealthy oral microbiome (dysbiosis).

The overall loss of microbial diversity, the beneficial microbes, and the expansion of the pathogenic microbes leads to an unhealthy microbiome. Dysbiosis outcomes in some species cause harsh changes in the oral microbiota, ultimately resulting in deprivation of oral health [7]. Systemic problems are mostly caused due to transformations in the epithelial barriers in the oral cavity, which develop as a result of infections like *P. gingivalis* producing proinflammatory cytokines, such as tumor necrosis factor and interleukins [14].

2.3 ORAL MICROBIOME AND SYSTEMIC DISEASE

The buccal cavity is the initial point of entry for foreign entities to the gastrointestinal and pulmonary systems. The mouth cavity of an individual may contain hundreds of bacterial species [15]. Oral microbial imbalance is linked with oral inflammation and may contribute to systemic diseases via bacteremia (Figure 2.2). Oral microbial dysbiosis leads to various systemic illnesses

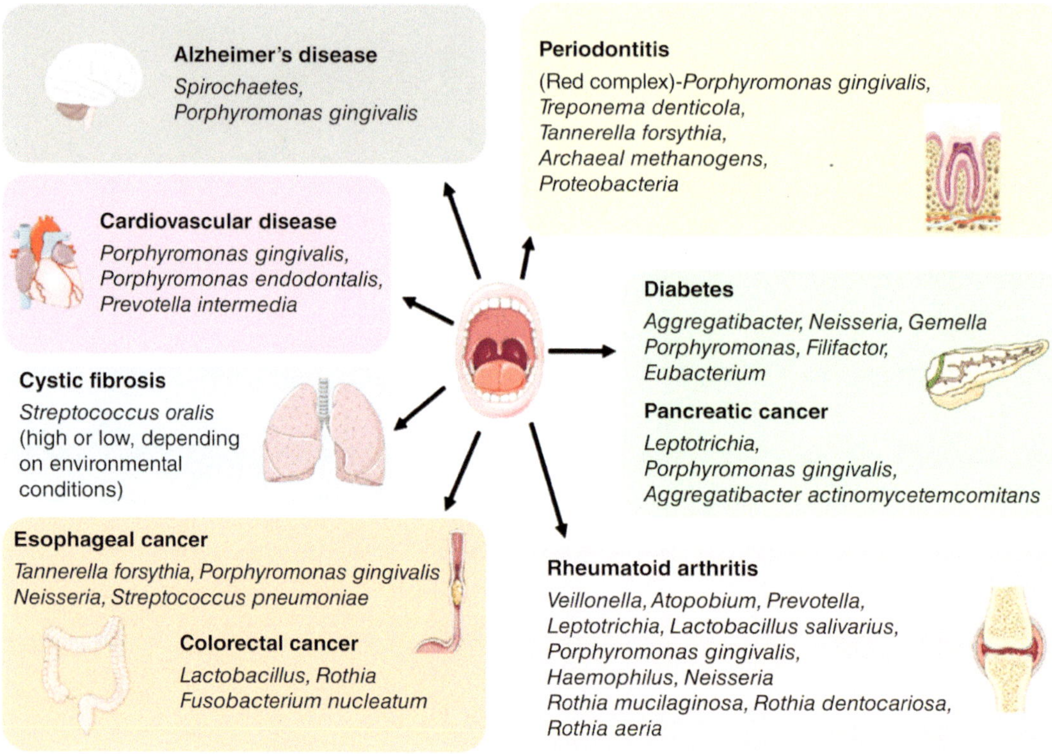

FIGURE 2.2 The microorganisms involved in dysbiosis that contribute to various systemic disorders.

including gastrointestinal system diseases like IBD, pancreatic cancer, liver cirrhosis, endocrine system diseases like diabetes, nervous system diseases like Alzheimer's disease, adverse pregnancy outcomes, obesity and polycystic ovary syndrome, cardiovascular system diseases like atherosclerosis, and immune system diseases like human immunodeficiency virus (HIV) infection and rheumatoid arthritis. The various microorganisms involved in causing these disorders are listed in Figure 2.2.

2.3.1 Oral Microbes and Inflammatory Bowel Disease

IBD is a complex disorder characterized by persistent inflammation of the gastrointestinal tract (GIT). It covers a wide range of disorders from ulcerative colitis (UC) to Crohn's disease (CD). UC is distinguished by persistent, diffuse, and acute colon inflammation [16]. CD is a chronic inflammatory illness that may impact any area of the digestive system, from the oral cavity to the rectum, but it generally affects the colon and terminal ileum [17]. Recurrent diarrhea, abdominal discomfort, gastrointestinal bleeding, and weight loss are the predominant symptoms of IBD. Many patients also experience extraintestinal problems, often in the form of oral, skin, joint, eye, and bone lesions.

The oral microorganisms can be translocated from the buccal cavity into the intestines causing microbial dysbiosis and inflammation, but on the other hand, it also suggested that mouth microbial dysbiosis will occur as a result of gut inflammation so that it can be utilized as a biomarker of IBD. Oral bacteria can enter the digestive system through swallowing, aspiration, or translocation, thus potentially influencing the microbial ecosystem in the gut.

The etiology of IBD is presently assumed to include an inappropriate and prolonged inflammatory reaction to commensal intestinal microbiomes in genetically vulnerable people. IBD has been

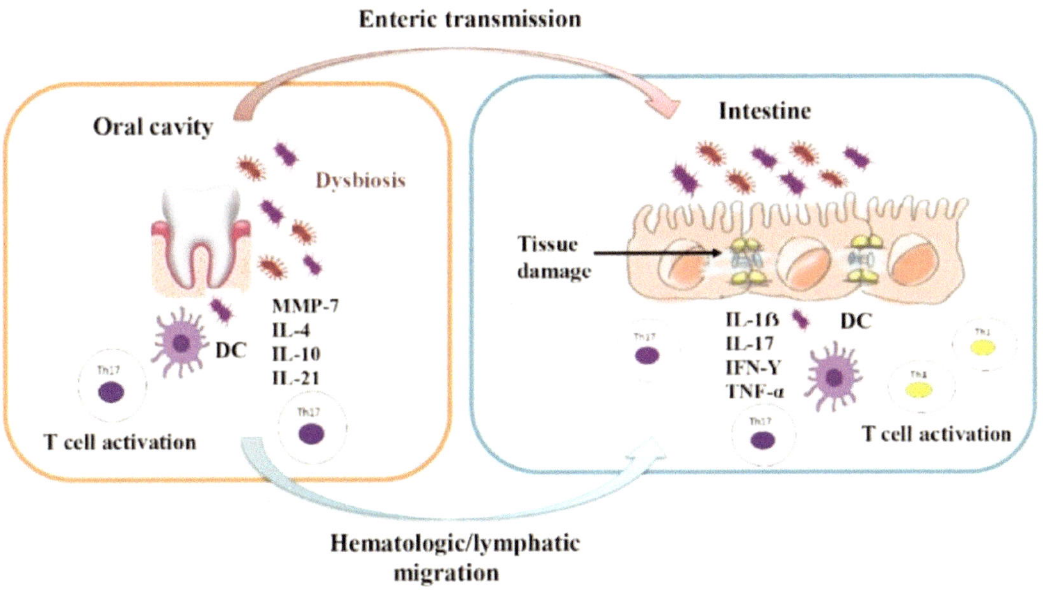

FIGURE 2.3 Pathogenic association between oral and gastrointestinal disease.

regarded as the consequence of the complicated interplay between genetic, immunological, microbiological, and environmental variables [18]. Loss of antigen tolerance and oral dysbiosis results in the formation of T-helper 17 (Th17) cells, the release of interleukins (IL-4, IL-10, IL-21), and the release of MMP-7 protease, which regulates cell trafficking and tissue repair (Figure 2.3). These all reach up to gut epithelial tissue by enteric and hematologic/lymphatic migration. In epithelial tissue of the gastrointestinal tract, activated Th17 cells result in the release of tumor necrosis factors (TNF-α) and interferons-γ, which in turn promotes dendritic cells. Interleukins and Th17 chemokines attract inflammatory cells, which subsequently produce non-specific inflammatory substances such as polyunsaturated omega-6 lipids, proteases, arachidonic acid, platelet activation factor, and free radicals, causing intestinal damage [19] and leading to IBD-like gastrointestinal diseases.

2.3.2 Oral Microbes and Neurological Illnesses

Nervous system disorders (NSDs) are on the peak and are very prominent these days yet the causal mechanisms behind this epidemic remain poorly understood. NSDs were shown to be connected to gut microbiota, which can communicate with and influence the brain through the gut-brain axis (GBA). Parallel studies examining oral microbiota and their connections to the brain also suggest that microbes in the mouth can similarly influence NSD outcomes.

A very common nervous system illness, Alzheimer's disease (AD), is a prevalent kind of dementia, accounting for 60%–80% of all cases, with cognitive deficits and a complicated, multiple etiology [20]. Some bacterial phyla (*Treponema*) and spirochetes have a stronger association with Alzheimer's disease. This is in line with the discovery of lipopolysaccharide and virulence factors from the oral anaerobe *Porphyromonas gingivalis* through the bloodstream to the brains of AD patients [21]. Poor oral health (dysbiosis), xerostomia (absence of saliva or dry mouth), free heme (cause of cellular injury), and low salivary lactoferrin (Figure 2.4) lead to oral biofilm dysbiosis, which disseminate virulence factors from the buccal cavity to the nervous system through the oral-brain axis. Alzheimer's patients have elevated levels of TNF-α and oral bacteria antibodies; all these factors reach the brain through hematologic and lymphatic migration.

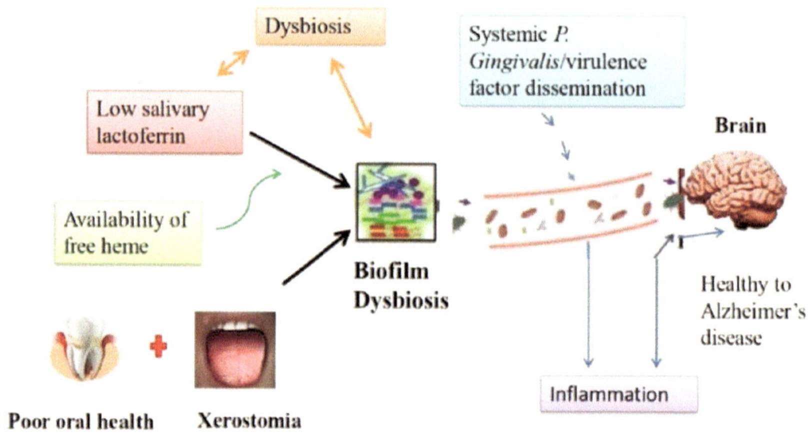

FIGURE 2.4 Oral dysbiosis and Alzheimer's disease.

2.3.3 Oral Microbes and Diabetes

Diabetes is a common metabolic and endocrine disease caused by impaired pancreatic islet function and insulin deficiency. Periodontitis is the sixth major complication of diabetes. Diabetes with poor blood sugar control is a risk factor for periodontal diseases. Compared with non-diabetic patients, the risk of chronic periodontitis in diabetic patients is increased by 2–3 times. The oral bacteria in the periodontal pockets with periodontitis have complex interactions with the human immune system, resulting in continuous chronic inflammation. The oral microbiome may also trigger insulin resistance by influencing the oxidative stress and immune inflammation, thereby affecting diabetes [22].

Oxidative stress is important for microbial killing and host immune defense. However, the large amount of reactive oxygen species (ROS) produced by oxidative stress, and the lack of a compensatory endogenous antioxidant response, can affect the prevalence and development of diabetes and cause body damage [23]. ROS interferes with the insulin signal transduction pathway and causes insulin resistance. Insulin resistance will enhance the level of ROS, which will in turn damage the vascular and metabolic factors in the insulin signal transduction pathway. Oxidative stress interferes with the insulin signaling pathway by activating a variety of serine/threonine kinase cascades through nuclear factor-kappa B (NF-κB) activating enzymes, p38 mitogen-activated protein kinases (p38MAPK), Jun amino-terminal kinases/stress-activated protein kinases (JNK/SAPK), and protein kinase C (PKC) isoenzymes and inhibiting tyrosine phosphorylation [4].

2.3.4 Oralome and Undesirable Pregnancy Outcomes

During pregnancy, a series of hormonal, metabolic, and immunological modifications are important for normal fetal development and delivery. These changes alter the mother's microbiota in different locations of the body including the oral cavity, gut, and vagina. Adverse pregnancy outcomes (APOs) have been linked to oral microbiota alterations. The prevalence of *Porphyromonas gingivalis*, *Bacteroides forsythus*, *Campylobacter rectus, Aggregatibacter actinomycetemcomitans*, *Streptococci*, *Staphylococci*, and *Candida* species is considerably higher during the first and second trimesters of pregnant women [24]. *Fusobacterium nucleatum*, which is related to periodontal (gum) disease, was discovered to be connected with APOs. *F. nucleatum* can be transferred hematogenously to the placenta, resulting in negative pregnancy outcomes [25]. Periodontitis and gingivitis seen during pregnancy are chronic mouth infections with systemic and local inflammatory responses that have been linked to negative pregnancy outcomes. Alterations in the oral microbial community during pregnancy elevate the risk of periodontal diseases and result in low birth weight, preterm birth, preeclampsia, and miscarriages.

2.3.5 Oral Microbiome and Obesity

Obesity is described as a chronic condition characterized by low-grade inflammation, which has been identified as a key connection between obesity and metabolic disease. Thus, numerous inflammatory pathways have been involved in the control of metabolic homeostasis. This relationship has been characterized as metabolic inflammation and has been observed in various metabolic tissues, with adipose tissue in particular. Supporting the association between metabolic dysregulation and inflammation, anti-inflammatory treatment is linked with metabolic improvement and weight reduction reduces systemic inflammation in individuals with obesity [26]. Characteristic changes in the gut microbiome underscore a causal link between inflammation and obesity. The oral microbial dysbiosis may contribute to inflammatory changes in obesity and metabolic inflammation in adipose tissue, both of which can lead to metabolic dysregulation [27].

2.3.6 Oralome and Polycystic Ovary Syndrome

Polycystic ovary syndrome (PCOS) is one of the most common metabolic and endocrine disorders and is characterized by hyperandrogenism, insulin resistance (IR), and oligo- or anovulatory (self-reported menstrual cycle length of 35 days or <10 menstrual periods per year), affecting 7%–8% of reproductive-age women [28]. Oral microbiota dysbiosis contributes to the pathogenesis of PCOS. Intestinal flora and oral flora are the two main components of the microecosystem in the human body affecting PCOS. The main etiological factor for periodontal diseases is microorganisms in the tooth plaque, such as *Porphyromonas gingivalis*, *Treponema denticola*, *Tannerella forsythia*, *Aggregatibacter actinomycetemcomitans*, and *Prevotella intermedia* species, which can also be found in saliva. *Bacteroides vulgatus* causes PCOS-like manifestations, such as polycystic ovaries and disorders of the estrous cycle [29]. The presence of PCOS increased *P. gingivalis*, *Prevotella intermedia*, and *S. oralis* blood antibody levels when gingivitis was also present. Mainly, the influence of female steroid hormones on the diversity of oral microbiota has been observed in puberty, menstruation, pregnancy, and oral contraceptive usage.

2.3.7 Oral Microbiome and Immune System

Oral microbiomes have a close relationship with human immune system activities and, hence, are correlated with human immune system disorders such as rheumatoid arthritis. They also have an impact on multisystem disease performance in the immune system, such as HIV. The etiopathogenesis of systemic autoimmune diseases is unknown, but a complex interplay of genetic and environmental factors linked with stochastic events leads to autoimmunity and loss of immunological tolerance.

2.3.7.1 Rheumatoid Arthritis

Rheumatoid arthritis (RA) is a persistent chronic autoimmune inflammatory illness distinguished by joint bone destruction and auto-antibody production such as rheumatoid factor and, especially, anticitrullinated protein antibodies (ACPAs), autoantibodies against citrullinated peptides and proteins. ACPAs may be reliable markers for the early diagnosis of RA and periodontal disease (PD), and tobacco use seems to be related to ACPA production and netosis (neutrophiles have the ability to mix and extrude their DNA and bactericidal molecules creating NET-like structures in a unique type of cell death) as reported by [30].

Oral dysbiosis increases the burden of citrullination (a posttranslational modification of protein in which arginine is converted into citrulline) through both direct and indirect pathways involving

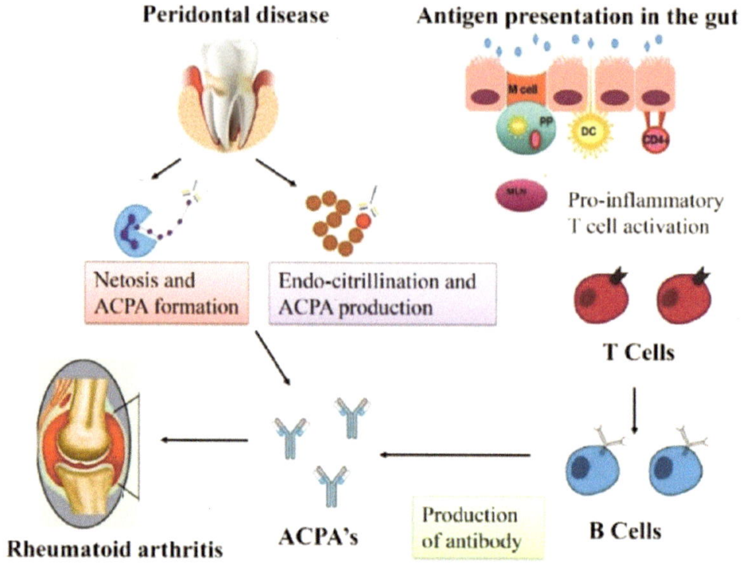

FIGURE 2.5 Illustration showing the influence of the oral microbiome on rheumatoid arthritis.

the important pathogens *Porphyromonas gingivalis* and *Aggregatibacter actinomycetemcomitans* [31]. The gut epithelia deliver antigens to the underlying immune cells via microfold cells (M cells) and then to mesenteric lymph nodes (MLN) via dendritic cells (DC) located in the payer's patches (PP) of the lamina propria (Figure 2.5). Follicular-associated enterocytes may directly activate alternative CD^{4+} T cells. This immunological response produces cytokines and stimulates pro-inflammatory T cells, which activate B cells and produce anticyclic citrullinated protein antibodies [32]. ACPAs generated in the mouth cavity and the stomach can reach the bloodstream, move to the joints, and lead to the progression of arthritis (Figure 2.5).

2.3.7.2 Oral Microbiota and HIV

Even with sustained use of antiretroviral therapy (ART), HIV-infected individuals are more likely to develop oral pathologies and systemic comorbid conditions including opportunistic infections, periodontal and gingival diseases, and oral mucosal inflammation. Untreated patients with elevated viremia had considerably greater proportions of potentially harmful *Veillonella*, *Prevotella, Megasphaera*, and *Campylobacter* species compared to healthy persons [33]. The greater prevalence of periodontal inflammation and dental caries found in HIV-infected individuals on ART is linked with oral microbial communities, which may increase susceptibility to inflammation. Microbial diversity in the buccal cavity of HIV-infected people was lower than in healthy people, and it was much lower after ART treatment [34].

2.3.8 Oralome and Cardiovascular Diseases

Cardiovascular diseases (CVDs) include different conditions affecting the blood vessels and heart and are often linked to oral microbes. Within CVDs, the following diseases are mainly included: atherosclerotic cardiovascular disease (coronary, cerebrovascular, and peripheral vascular disease), heart failure and cardiomyopathies, valvular heart disease, infective and autoimmune conditions (including infective endocarditis), arrhythmias (problem of heartbeat rate), and hypertension. Atherosclerosis is defined by the buildup of cholesterol and the migration of macrophages to the artery wall. It can thus be classified as both a metabolic and inflammatory condition. *Chryseomonas*, *Veillonella*, and *Streptococcus* are detected in atherosclerotic patients [35].

Oral bacteria can migrate into the gut via enteral and hematogenous routes. The enteral route is the direct entry of bacteria into the gut via saliva, and the hematogenous route is via the blood vessels (systemic circulation) around the periodontal and gingival tissues [31]. Numerous oral microorganisms such as *Prevotella intermedia*, *P. gingivalis*, *Fusobacterium nucleatum*, and *Helicobacter pylori* have been found in the arterial plaques, gut mucosal tissues, and pancreatic tissues. The migration of oral microorganisms from the buccal cavity into systemic circulation activates different immuno-inflammatory responses that cause a massive release of acute phase proteins, pro-inflammatory mediators (IL-1, IL-6, TNF-α, and MCP-1), fibrinogen, C-reactive protein (CRP), and free radicals into the systemic circulation. These mediators of inflammation affect the cardiovascular system and modify lipid metabolism, cause endothelial injury, and increase the hypercoagulability of the blood and platelets. This, in turn, increases the risk of arterial calcification, atherosclerosis, cardiovascular tissue injury, and fibrogenesis.

2.4 FACTORS INFLUENCING ORAL MICROBIOTA

Sugars, vitamins, and fats are among the most impacting nutrients in the growth of oral microbiota. Sugar intake has been linked with the presence of *Streptococcus mutans* and *Fusobacterium nucleatum*, while it was seen that fats and vitamin C act as promoters of *Fusobacterium* spp. Fiber and dairy products also help in maintaining oral bacterial homeostasis. In addition, various other factors influence the oral microbial composition such as age, dietary habits, tabagism (tobacco addiction), alcohol, medications (antibiotics), toxic substances, pregnancy, and genetic predisposition (Figure 2.6).

2.4.1 Age (Time)

The oral microbial community varies as the individual matures (for example, due to tooth eruption and age-related hormonal and immune system changes). In addition, horizontal transmission of microorganisms and oral microbiome microevolution occur throughout time [36].

The global population is rapidly aging as a result of declining fertility rates and continued growth of average lifespan. It is predicted that the proportion of the population aged 65 years and over might increase from 20% in the year 2000 to 40% by 2050. In the elderly, the incidence of oral diseases, such as dental caries, soft tissue diseases, and periodontal diseases, is increased. The oral microbiome composition, species identity and combinations, density, and distribution of bacteria may have a profound effect on how well a person ages. Age-related changes in oral microbiota, although they vary, underlie the complexity of the oral microbiome. Microbial composition of oral sites (gingival crevicular fluid, tongue back, and saliva) decreases in alpha diversity (diversity within a particular area) with age, while beta diversity (diversity within two adjacent ecosystems) increases with age.

2.4.2 Host and Environment

The host's environment has an impact on the oral microbiome composition by altering behaviors and eating. The oral microbiome cooperates with the host to reflect the status of metabolism and immunity through two-way communication along the systemic organs and the oral cavity. The oral cavity is a key interaction window between the human body and the environment. The different microenvironment in the buccal cavity has different microbial compositions, which are regulated by complex signaling, hosts, and external environmental factors.

2.4.3 Habitat

The environmental parameters of the mouth cavity vary according to oxygen levels, pH, oral hygiene, salivary flow, gingival crevicular fluid (GCF), and nutrition. GCF is an inflammatory

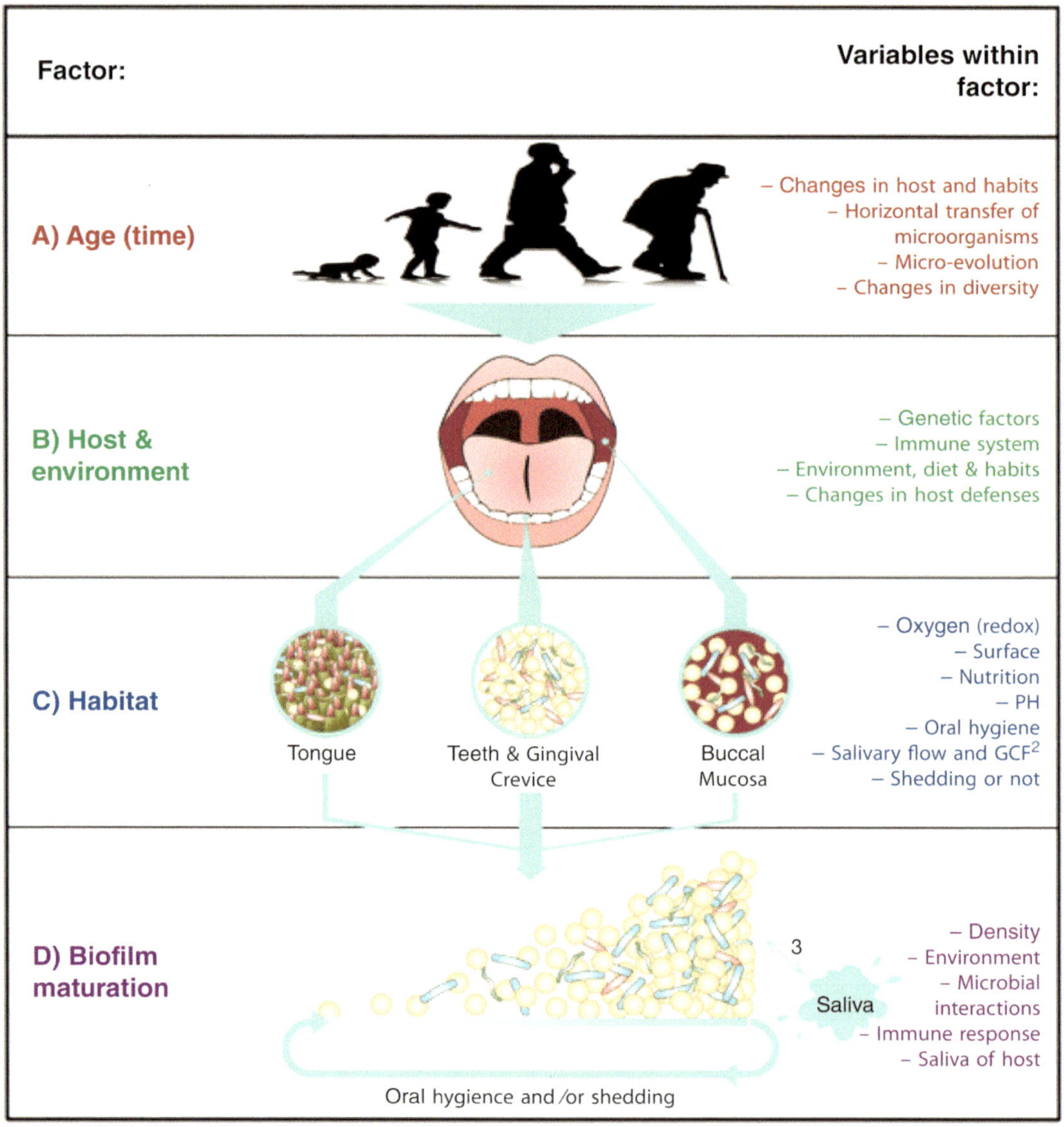

FIGURE 2.6 Factors influencing the makeup of the buccal microbiome.

exudate derived from the periodontal tissues, composed of serum, tissue breakdown products, inflammatory mediators, and antibodies directed against dental plaque bacteria. Importantly, oral mucosal surfaces undergo desquamation (the buccal mucosa sheds often to minimize biofilm formation), but dental surfaces do not. Changes caused by biofilm maturation are especially important to the teeth and gingival crevice in the context of the most frequent oral disorders (caries and periodontal diseases), as shown by the thicker arrow in the middle moving to row D in Figure 2.6 [37].

2.4.4 Biofilm Maturation

Chemical and physical alterations from oral hygiene, food, and in the case of oral mucosa, density changes impact the internal environment which in turn alter the microbial interactions (e.g., quorum sensing), and the host-induced immune response. Oral biofilm is an important etiologic factor in different oral diseases such as periodontal diseases, dental caries, oropharyngeal candidiasis, and implant-related infections [38].

2.5 LIFESTYLE AND ORAL HEALTH

A lifestyle is a way of life that reflects the attitudes and values of an individual or community. It is a broad concept that is frequently used to describe how people live, reflecting a wide range of social values, attitudes, and activities [39]. It is made up of cultural and behavioral patterns as well as lifelong personal habits such as physical activity, diet, smoking, and alcoholism, which develop through the process of socialization and are learned through social interactions with parents, peer groups, friends, and siblings, or school and mass media [40].

It also has an impact on health, as some of today's health problems, such as coronary heart disease, cancer, and drug addiction, have been linked to social disparities in adolescent oral health behavior in developing countries and elsewhere, with oral health detrimental behaviors being most common in subjects with lower socio-demographic status. Lifestyle habits like bad eating, tabagism (tobacco and smoking addiction), toothpaste, toothbrushes, mouthwashes, and antibiotics abuse lead to oral microbial community disturbances.

2.5.1 Impact of Tobacco on the Oral Microbiome

Tobacco is the most disease-causing substance, and its widespread addictive habit has influenced human behavior for centuries. Tobacco is absorbed orally in several forms, including smoking and chewable forms. Smoking is quickly expanding in the developing world, and it is one of the most serious dangers to the present and future global health. Tobacco products are commercially accessible in India with additional smells and flavoring chemicals, which appeal to both the rural and urban populations.

Tobacco smoking has been associated with numerous major ailments, including cardiovascular disorders, cancer, low birth weight, and a variety of other health issues, resulting in thousands of early deaths each year. When exposed to smoke, salivary behavior changes, and saliva loses its antioxidant capability, transforming into a powerful pro-oxidant milieu [41]. The majority of tobacco-related oral consequences reduce the quality of life, whether they are as basic as halitosis (bad-smelling breath), as complicated as oral birth abnormalities, as frequent as periodontal disease, or as problematic as wound healing difficulties.

Tobacco smoking and chewing cause not just tooth discoloration, periodontitis, dental cavities, changed taste, and nicotinic stomatitis, but also leukoplakia and cancer, both of which have a significant morbidity and death rate [42]. It is critical to remember that the detrimental effects of tobacco products are dose dependent, relying more on abuse than simple consumption.

2.5.2 Impact of Smoking on Oral Health

Smoking may modify microbial niches in various oral habitats, alter the makeup of oral bacteria, and prepare the mouth for harmful bacterial colonization. Compounds found in cigarette smoke come into direct contact with the oral microbiome and may disrupt microbial ecology through several mechanisms, such as influencing bacterial adherence to mucosal surfaces [37], forming unstable bacterial colonization in biofilms, increasing the acidity of saliva, depleting oxygen [43], featuring antibiotic resistance effects [36], and resistance to immune cell killing. Smoking also reduces microbial aerobic metabolic pathways, increasing subgingival anaerobic bacteria [42, 44, 45].

Importantly, the total oral microbiome makeup of former smokers and non-smokers did not differ significantly, showing that quitting smoking can restore normal dental microbiota [46]. A complete metagenomic sequencing of the oral microbiome in smokers as well as non-smokers revealed

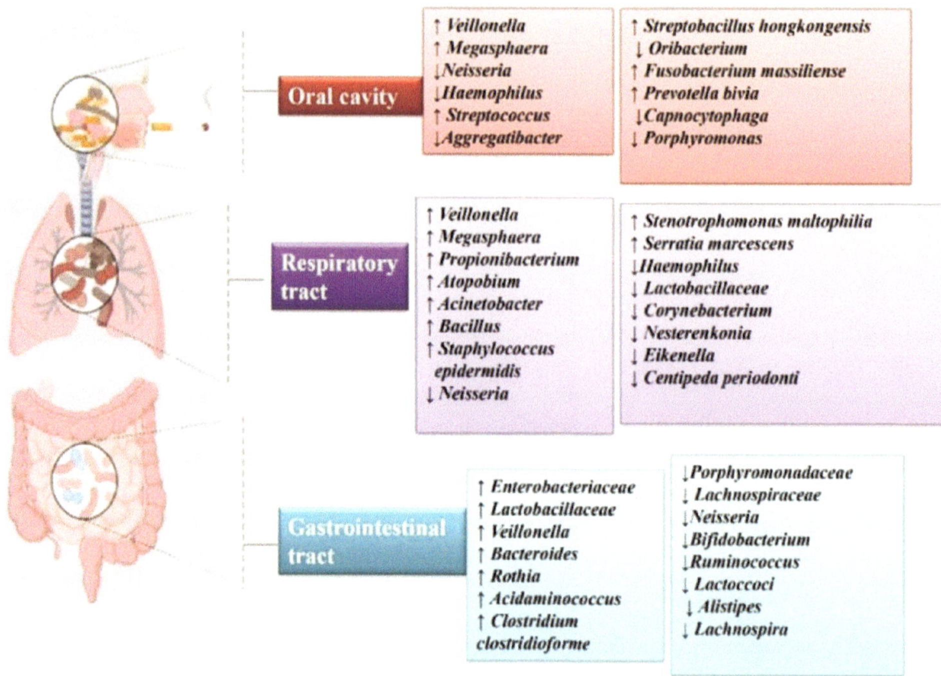

FIGURE 2.7 Microbial dysbiosis.

that smokers had a larger abundance of *Prevotella* and *Megasphaera*, while the taxa *Oribacterium*, *Capnocytophaga*, *Porphyromonas*, and *Neisseria* were dramatically decreased [41]. Alteration in the makeup of the oral microbiome can have a range of implications. For example, smoking-induced salivary microbial imbalance has been associated with lower task performance of network connections in smokers' brains [47]. Beyond these correlations between smoking and oral microbial dysbiosis (Figure 2.7), more research is needed to determine the potential causal involvement of oral microbial dysbiosis in smoking-related health concerns.

2.5.3 Effect of Toothbrushes on the Oral Microbiota

Throughout the nineteenth century, toothbrushes were commonly used to inhibit the expansion of dangerous germs, remove food residue, keep breath fresh, preserve enamel, and prevent and treat gingival inflammation [48]. However, the remaining moisture and food particles in toothbrushes create an ideal habitat for disease growth, making them a possible danger for different oral disorders [49]. The microbes *Acinetobacter baumannii*, *Staphylococcus aureus*, and *Candida albicans* are found on toothbrushes. The existence of these microorganisms raises the host's risk of contracting infectious illnesses, neurological disorders, cardiovascular diseases, and cancer (Figure 2.8). Medicinal and antibacterial toothpastes can not only prevent pathogen growth but also the growth of probiotics *Lactobacillus salivarius* and *Streptococcus salivarius*. As a result, the inhibitory effect of toothpaste on probiotics, along with the presence of germs in toothbrushes, suggests that teeth brushing may pose a risk to immunocompromised individuals [48].

2.5.4 Effect of Different Toothpastes on the Oral Microbiome

Dental caries is a chronic infectious disease. Its causes are diverse and complex, but its formation is due to the acid production in the mouth cavity, which leads to the dissolution and disturbance of

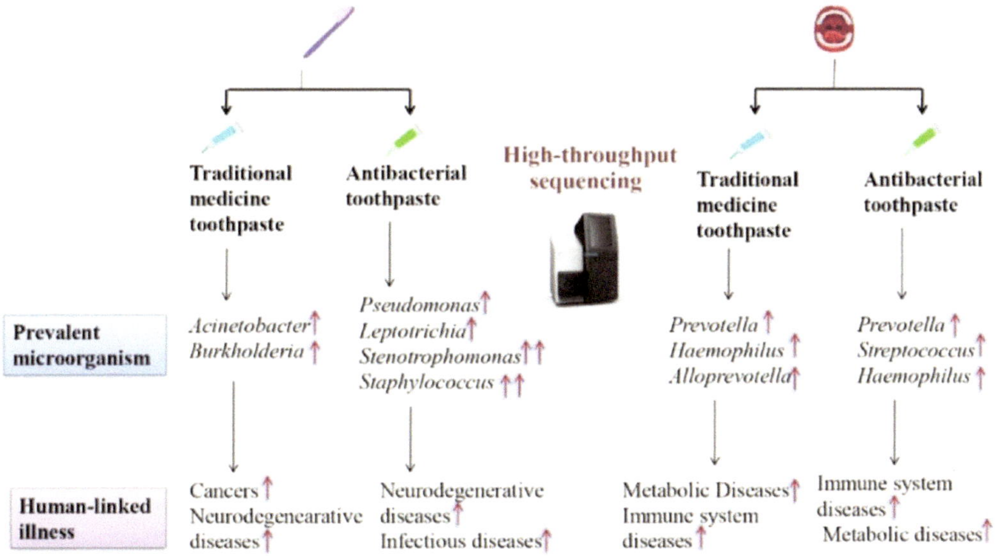

FIGURE 2.8 Effect of toothbrushes on the oral microbiota.

enamel (dental hard tissue). Due to modifications in the oral cavity, such as changes in diet, sugar intake, and pH, the healthy and balanced microbial community in biofilms can alter significantly. The microbiome balance of plaque is disturbed and can lead to dental caries. Long-term use of fluoride-containing toothpaste may also cause alterations in the oral microbial flora. Fluoride can prevent the growth of different oral microorganisms, such as *Streptococcus sialis*, *Lactobacillus*, *Porphyromonas gingivalis*, *Streptococcus mutans*, and *Candida albicans*, while the number of *Streptococcus sanguis* increases.

Modifications in the DNA and RNA functional profiles of the oral microorganisms have been observed following long-term usage of various fluoride and fluoride + arginine–containing toothpastes [50]. Long-term brushing increased lipopolysaccharide production, fatty acid biosynthesis, histidine metabolism, and macrolide transport [51]. Aside from stimulating the arginolytic pathway, the use of the flouride + arginine–containing toothpaste favored the transport of several amino acids and glutamate, the phosphotransferase systems (PTSs) of several sugars, ethanolamine utilization, and modified oxidative phosphorylation complexes [39]. The influence of fluoride and fluoride + arginine on arginine metabolism is shown in Figure 2.9. Also shown are the enzymatic pathways and genes involved in arginine production, arginine breakdown via citrulline (arginolytic pathway), and arginine conversion into agmatine and subsequent degradation [agmatine deiminase system (AgDS)]. Overexpressed genes after fluoride therapy are marked orange, whereas others that are overexpressed after fluoride + arginine treatment are colored blue [50].

The abundance of common commensal species, including health-related bacteria like *Neisseria*, and the brushing with a fluoride-containing toothpaste improves dental health. Lactate-utilizing *Veillonella* also decreased after brushing with the fluoride toothpaste [52], implying that acid generation was reduced, which is consistent with perhaps enhanced oral hygiene and sugar fermentation inhibition caused by fluoride.

2.5.5 Effect of Mouthwashes on the Oral Microbiome

For generations, oral hygiene and health have been maintained by brushing, nonsurgical therapy, and interdental cleaning. However, removing harmful microbes from the oral cavity with antiseptic

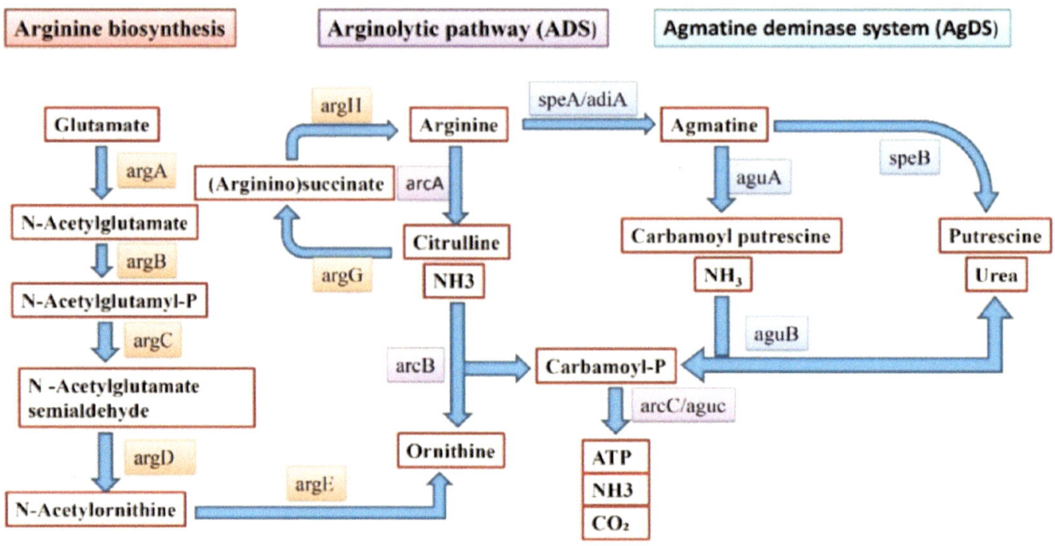

FIGURE 2.9 Influence of fluoride and fluoride + arginine on arginine metabolism.

mouthwashes also constitutes good oral hygiene. The oral microbiome exists as a complex and diverse community of bacteria, fungi, archaea, protozoa, and viruses in the buccal cavity. The antibacterial effect of mouthwashes on these microbes can be both advantageous and deleterious, depending on whether there is a good shift toward oral health or a bad shift toward disease (Figure 2.10) [53]. Different mouthwashes containing chlorhexidine gluconate, alcohol, and probiotic mouthwashes are discussed in this section.

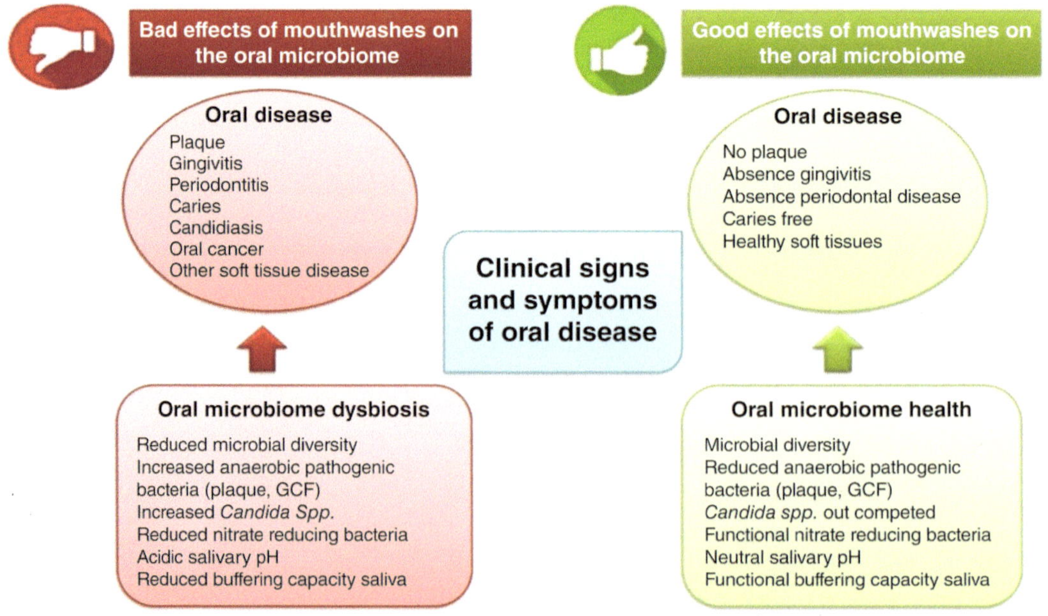

FIGURE 2.10 Effect of mouthwashes on the oral microbiota, as well as clinical signs of oral disease.

2.5.5.1 Chlorhexidine Mouthwashes

Chlorhexidine (CHX) reduces bacterial diversity and viability in saliva and on the tongue, and it is highly effective against *Streptococcus mitis*, *Fusobacterium nucleatum*, *Porphyromonas gingivalis*, *Veillonella*, *Actinomyces*, *Haemophilus*, *Rothia*, *Neisseria*, and *Aggregatibacter actinomycetemcomitans*. Chlorhexidine mouthwashes decrease plaque and gingivitis and can be used as an adjuvant to periodontal disease treatment [54].

These health-associated oral bacteria in saliva play a significant role in converting dietary nitrates to nitrite, which contributes to cardiovascular health by releasing nitric oxide. Because of its non-specific antibacterial action, chlorhexidine mouthwashes may counteract the positive benefits of nitrate-rich diets mediated by the oral microbiota [55].

2.5.5.2 Alcohol-Containing Mouthwashes

Alcohol is present in different mouthwashes. There is widespread agreement that alcohol kills both good and bad microbial communities in the oral microbiome indiscriminately. Indeed, alcohol appears to reduce the number of commensal bacteria in heavy drinkers while increasing the prevalence of *Actinomyces*, *Leptotrichia*, and *Neisseria* (some of which include oral pathogens) [56]. Alcohol-free mouthwashes are becoming accessible over the counter (OTC), for several reasons, including the possibility that alcohol may increase salivary acetaldehyde, which has been related to oral cancer in the context of mouthwashes.

2.5.5.3 Probiotic Mouthwashes

Probiotics are living microorganisms that are provided to the patient to help maintain a healthy microbiome. *Lactobacillus* and *Bifidobacterium* are two of the most frequent bacteria found in probiotic mouthwashes. *Lactobacillus salivarius* NK02 in mouthwash decreased plaque and bleeding scores in periodontitis patients, as well as pathogenic *Aggregatibacter actinomycetemcomitans* counts and commensal bacteria levels in saliva and gingival crevicular fluid [57]. A mouthwash comprising freeze-dried powder from *Lactobacillus rhamnosus* and *Bifidobacterium* species decreased *Streptococcus mutans* colonies; nevertheless, there is little data to support the efficacy of probiotics on the oral microbiome as a whole [40].

2.5.6 Antibiotic Abuse and the Oral Microbiome

Antibiotics are mostly used to treat systematic infections that do not affect the mouth cavity. Antibiotic use may have a severe effect on the healthy oral microbial community by disturbing the bacterial oral populations, thereby leading to dysbacteriosis [58]. Antibiotic abuse disturbs oral microbiota which in turn alters salivary antibody content [12]. Spiramycin (SP) therapy resulted in lesser changes in microbial diversity compared to other antibiotics, suggesting that macrolide antibiotics (azithromycin, clarithromycin, and erythromycin) have a greater impact on microbiota than β-lactam antibiotics (penicillins, cephalosporins, monobactams, and carbapenems).

Antibiotic administration increases bacterial diversity, independent of the kind of antibiotic administered, which is most likely due to the establishment of antibiotic-resistant bacterium communities in organisms. Increasing microbial variety in the oral cavity may result in more germs entering the dental treatment room via the aerosol produced by dental therapy, affecting medical professionals and patients [11]. The treatment of several antibiotics also altered the metabolic pathways of the oral microbiome.

2.6 CONCLUSION

The oral microbiome is important for health because an unhealthy oral microbiome can lead to systemic and oral diseases, so it is crucial to maintain a consistent oral hygiene routine to support a healthy oral microbiome. Dysbiosis of the oral cavity disturbs overall systemic health and causes various diseases like cardiovascular, nervous system, diabetes, periodontitis, rheumatoid arthritis, and many more. This oral cavity dysbiosis is caused by improper and undisciplined habits of lifestyle like eating, tabagism, brushing, and antibiotic abuse. Frequent and unnecessary use of antibiotics can disrupt the natural balance of the oral microbiome. Some of the most vital functions for human life tightly connect the oral cavity to the rest of the body and give it a powerful influence on many of the body's processes. Dysbiosis in the oral microbiome is linked to some local and systemic diseases, though particular associations and mechanisms of action require further study.

REFERENCES

1. Janak, L.; Pathak, I.; Yongyong, Y.I.; Zhang, Q.; Wang, L.; and Ge, L. The Role of Oral Microbiome in Respiratory Health and Diseases. *Respiratory Medicine* **2021**, *185*, 106475.
2. Radaic, A.; and Kapila, Yvonne L. The Oralome and Its Dysbiosis: New Insights into Oral Microbiome-host Interactions. *Computational and Structural Biotechnology Journal* **2021**, *19*, 1335–1360.
3. Choudhary, A.; Kapoor, P.; Bhargava, D.; and Bagga, D. K. Exploring the Oral Microbiome: and Updated Multidisciplinary Oral Healthcare Perspective. *Discoveries* **2023**, *11*, e165.
4. Emrich, L.J.; Shlossman, M.; and Genco, R.J. Periodontal Disease in Non-Insulin-Dependent Diabetes Mellitus. *Journal of Periodontology* **1991**, *62*, 123–132.
5. Jonathon, L.B.; Jessica, L.M.W.; Kathryn, M.K.; Jeffrey, S.M.; and He, X. The Oral Microbiome: Diversity, Biogeography and Human Health. *Nature Reviews Microbiology* **2023**, *22*, 89–104. https://doi.org/10.1038/s41579-023-00963-6.
6. Rosier, B.T.; Moya-Gonzalvez, E.M.; Corell-Escuin, P.; and Mira, A. Isolation and Characterization of Nitrate-Reducing Bacteria as Potential Probiotics for Oral and Systemic Health. *Frontiers in Microbiology* **2020**, *11*, 555465.
7. Cugini, C.; Ramasubbu, N.; Tsiagbe, V.K.; and Fine, D.H. Dysbiosis from a Microbial and Host Perspective Relative to Oral Health and Disease. *Frontiers in Microbiology* **2021**, *12*, 617485.
8. Lu, M.; Xuan, S.; and Wang, Z. Oral Microbiota: A New View to Body Health. *Food Science and Human Wellness*. **2019**, *8*, 8–15.
9. Yamashita, Y.; and Takeshita, T. The Oral Microbiome and Oral Health. *The Journal of Oral Science* **2017**, *59*, 201–206.
10. Kosciolek, T.; Victor, A.T.; Kuplicki, R.; Rossi, M.; Estaki, M.; Ackermann G, et al. Individuals with Substance use Disorders have a Distinct Oral Microbiome Pattern. *Brain, Behavior and Immunity – Health* **2021**, *15*, 100271.
11. Cheng, X.; Huang, F.; Zhang, K.; Yuan, X.; Song, C. Effects of None-Steroidal Anti-Inflammatory and Antibiotic Drugs on the Oral Immune System and Oral Microbial Composition in Rats. *Biochemical and Biophysical Research Communications* **2018**, *507* (1-4), 420–425.
12. Behr, C.; Kamp, H.; Fabian, E.; Krennrich, G.; Mellert, W.; Peter, E.; et al. Gut Microbiome-Related Metabolic Changes in Plasma of Antibiotic-Treated Rats. *Archieves of Toxicology* **2017**, *91*, 3439–3454.
13. Rebelo, M.B.; Oliveira, C.S.; and Tavaria, F.K. Novel Strategies for Preventing Dysbiosis in the Oral Cavity. *Frontiers in Biosciences* **2023**, *15*, 23.
14. Thomas, C.; Minty, M.; Vinel, A.; Canceill, T.; Loubieres, P.; Burcelin, R.; et al. Oral Microbiota: A Major Player in the Diagnosis of Systemic Diseases. *Diagnostics* **2021**, *11*, 1376.
15. Peng, X.; Cheng, L.; You, Y.; Tang, C.; Ren, B.; Li, Y.; *et al.* Oral Microbiota in Human Systematic Diseases. *International Journal of Oral Science* **2022**, *14*, 14.
16. Ford, A.C.; Moayyedi, P.; Hanauer, S.B. Ulcerative colitis. *British Medical Jouranal* **2013**, *346*, 432.
17. Baumgart, C.D.; and Carding, R.S. Inflammatory Bowel Disease: Cause and Immunology. *The Lancet* **2007**, *369*, 1627–1640.
18. Kaistha, A.; and Levine, J. Inflammatory Bowel Disease: the Classic Gastrointestinal Autoimmune Diseases. *Current Problems in Pediatric and Adolescent Health Care* **2014**, 44, 328334.
19. Radford-Smith, G. What is the Importance of Appendectomy in the Natural History of IBD? *Inflammatory Bowel Diseases* **2008**, *14*, 72–74.
20. Gaugler, J.E.; Reese, M.; and Mittelman, M.S. Effects of the Minnesota Adaptation of the NYU Caregiver Intervention on Primary Subjective Stress of Adult Child Caregivers of Persons with Dementia. *The Gerontologist* **2016**, *56*, 461–474.
21. Poole, S.; Singhrao, K.S.; Kesavalu, L.; Curtis, A.M.; Cream, S.J. Determining the Presence of Periodontopathic Virulence Factors in Short-term Postmortem Alzheimer's Disease Brain Tissue. *Journal of Alzheimer's Disease* **2013**, *36*, 665–672.

22. Ussar, S.; Fujisaka, S.; and Kahn, R.C. Interactions between Host Genetics and Gut Microbiome in Diabetes and Metabolic Syndrome. *Molecular Metabolism* **2016**, *5*, 795–801.
23. Cullinan, M.P.; Westerman, B.; Hamlet, S.M.; Palmer, E.; Faddy, J.M.; Lang, N.P. et al. A Longitudinal Study of Interleukin-1 Gene Polymorphisms and Periodontal Disease in a General Adult Population. *Journal of Clinical Periodontology* **2001**, *28*, 1137–1144.
24. Saadaoui, M.; Singh, P.; and Khodor, A.S. Oral Microbiome and Pregnancy: A Bidirectional Relationship. *Journal of Reproductive Immunology* **2021**, 145, 103293.
25. Madianos, P.N.; Bobetsis, Y.A.; and Offenbacher, S. Adverse Pregnancy Outcomes (APOs) and Periodontal Disease: Pathogenic Mechanisms. *Journal of Clinical Periodontology.* **2013**, *40*, 170–180.
26. Schamarek, I.; Andres, L.; Chakaroun, M.R.; Kovacs, P.; and Rohde-Zimmermann. The Role of the Oral Microbiome in Obesity and Metabolic Disease: Potential Systemic Implications and Effects on Taste Perception. *Nutrition Journal* **2023**, *22*, 28.
27. Goodson, J.M.; Groppo, D.; Halem, S.; and Carpino, E. Is Obesity an Oral Bacterial Disease. *Journal of Dental Research.* **2009**, 88, 519–523.
28. Li, N.; Li, Y.; Quin, C.; Liu, Q.; et al. Dysbiosis of the Saliva Microbiome in Patients with Polycystic Ovary Syndrome. *Frontier in cellular and infection microbiology* **2020**, *10*, 624504.
29. Lindheim, L.; Bashir, M.; Münzker, J.; Trummer, C.; Zachhuber, V.; et al. The Salivary Microbiome in Polycystic Ovary Syndrome (PCOS) and Its Association with Disease-Related Parameters: A Pilot Study. *Frontiers in Microbilogy* **2016**,*7*,1270.
30. McInnes, B.I.; and Schett, G. The Pathogenesis of Rheumatoid Arthritis. *The New England Journal of Medicine* **2011**, *365*, 2205–2219.
31. Calandrini, C.A.; Ribeiro, A.C.; Gonnelli, A.C.; and Ota-Tsuzuki, C.; et al. Microbial Composition of Atherosclerotic Plaques. *Oral Diseases* **2014**, *20*, 128–134.
32. Bellando-Randone; S. Russo, E.; Venerito, V.; Matacci-Cerinic, M.; et al. Exploring the Oral Microbiome in Rheumatic Diseases, State of Art and Future Prospective in Personalized Medicine with an AI approach. *Journal of Personalized Medicine* **2021**, *11*, 625.
33. Dang, T.A.; Cotton, S.; and Sankaran-Walters, S. et al. Evidence of an Increased Pathogenic Footprint in the Lingual Microbiome of Untreated HIV Infected Patients. *BMC Microbiology* **2012**, *12*, 153.
34. Torres, S.R.; Garzino-Demo, A.; Meiller, T.F.; Meeks, V.; and Jabra-Rizk, M.A. Salivary Histatin-5 and Oral Fungal Colonisation in HIV + Individuals. *Mycoses* **2009**, *52*, 11–15.
35. Koren, O.; Spor, A.; Felin, J.; and Backhed, F. Human Oral, Gut, and Plaque Microbiota in Patients with Atherosclerosis. *Proceedings of the National Academy of Sciences of the United States of America* **2011**, *15*, 108.
36. Chien, J.; Hwang H. J.; Nilaad, S.; Masso-Silva, A. J.; Ahn, J. S.; McEachern, K.E; *et al.* Cigarette Smoke Exposure Promotes Virulence of *Pseudomonas aeruginosa* and Induces Resistance to Neutrophil Killing. *Infection and Immunology* **2020**, *88*, 1–13.
37. Brook, I. The Impact of Smoking on Oral and Nasopharyngeal Bacterial Flora. *Journal of Dental Research* **2011**, *90*, 704–710.
38. Bertolini, M.; Costa, C.R.; Barao, R.A.V.; Villar, C.C.; et al. Oral Microorganisms and Biofilms: New Insights to Defeat the Main Etiologic Factor of Oral Diseases. *Microorganisms* **2022**, *10*, 2413.
39. Aas, J.A.; Griffen, A.L.; Dardis, S.R.; Lee, A.M.; Olsen, I.; Dewhirst, F.E.; et al. Bacteria of Dental Caries in Primary and Permanent Teeth in Children and Young Adults. *Journal Clinical Microbiology* **2008**, *46*, 1407–1417.
40. Gedam, K.Y.; and Katre, A.N. Efficacy of Probiotic, Chlorhexidine and Sodium Fluoride Mouthrinses on Mutans *Streptococci* in 8-to-12 Years Old Children: A Crossover Randomized Trial. *Lifestyle Genomics* **2022**, *15*, 35–44.
41. Aruni, A.W.; Roy, F. and Fletcher, H.M. *Filifactor alocis* has Virulence Attributes that can Enhance its Persistence Under Oxidative Stress Conditions and Mediate Invasion of Epithelial Cells by *Porphyromonas gingivalis. Infection Immunity* **2011**, *79*, 3872–3886.
42. Giovannucci, E. An Updated Review of the Epidemiological Evidence that Cigarette Smoking Increases Risk of Colorectal Cancer. *Cancer Epidemiology Biomarkers and Prevention* **2001**, *10*, 725–731.
43. Wu, J.; Peters, A.B.; Dominianni, C.; Zhang, Y.; Pei, Z.; Yang, L.; et al. Cigarette Smoking and the Oral Microbiome in a Large Study of American Adults. *ISME Journal* **2016**, *10*, 2435–2446.
44. Macgregor, I. D. Effects of Smoking on Oral Ecology: A Review of the Literature. *Clinical Preventive Dentistry* **1989**, *11*, 3–7.
45. McEachern, E. K.; Hwang, H. J.; Sladewski, M. K.; Nicatia, S.; Dewitz, C.; Mathew, P. D.; et al. Analysis of the Effects of Cigarette Smoke on *Staphylococcal* Virulence Phenotypes. *Infectious Immunology* **2015**, *83*, 2443–2452.

46. Kumar, P.S.; Matthews, C.R.; Joshi, V.; de, Jager, M.; and Aspiras, M. Tobacco Smoking Affects Bacterial Acquisition and Colonization in Oral Biofilms. *Infectious Immunulogy* **2011**, 79, 4730–4738.
47. Lin, D.; Kent, E. H.; Salvador, P.; Victor, V.; Jarrod, M.; Ellingson, J. M.; et al. Association Between the Oral Microbiome and Brain Resting State Connectivity in Smokers. *Neuroimage* **2019**, *200*, 121–131.
48. Shang, Q.; Gao, Y.; Qin, T.; Way, S.; Shi, Y.; and Chen, T. Interaction of Oral and Toothbrush Microbiota Affects Oral Cavity Health. *Frontier in Cellular and Infection Microbiology* **2020**, 10, 17.
49. Bik, E.M.; Long, C.D.; Armitage, G.C.; Loomer, P.; Emerson, J.; Mongodin, E.F.; et al. Bacterial Diversity in the Oral Cavity of 10 Healthy Individuals. *International Society for Microbial Ecology* **2010**, *4*, 962–974.
50. Carda-Dieguez, M.; Moazzez, R.; and Mira, A. Functional Changes in the Oral Microbiome After Use of Fluoride and Arginine Containing Dentifrices: A Metagenomic and Metatranscriptomic Study. *Microbiome* **2022** *10*, 159.
51. Nascimento, M.M.; Alvarez, A.J.; Huang, X.; Browngardt, C.; Jenkins, R.; Sinhoreti, M.C.; et al. Metabolic Profile of Supragingival Plaque Exposed to Arginine and Fluoride. *Journal of Dental Research* **2019**, *98*, 1245–52.
52. Koopman, J.E.; Hoogenkamp, M.A.; Buijs, M.J.; Brandt, B.W.; Keijser, B.J.F.; Crielaard, W.; et al. Changes in the Oral Ecosystem Induced by the Use of 8% Arginine Toothpaste. *Archieves of Oral Biology* **2017**, *73*, 79–87.
53. Collins, L. M.; and Dawes, C. The Surface Area of the Adult Human Mouth and Thickness of the Salivary Film Covering the Teeth and Oral Mucosa. *Journal Dental Research* **1987**, *66*, 1300–1302.
54. Teng, F.; He, T.; Huang, S.; Bo, C.; Li, Z.; Chang, J.; Liu, J.; et al. Cetylpyridinium Chloride Mouth Rinses Alleviate Experimental Gingivitis by Inhibiting Dental Plaque Maturation. *International Journal Oral Science* **2016**, *8*, 182–190.
55. Brookes, Z.; Teoh, L.; Cieplik, F.; and Kumar, P. Mouthwashes Effects on the Oral Microbiome: Are they Good, Bad or Balanced. *Dental Journal* **2023**, *73*, 74–81.
56. Sauerberi, A. Bacterial and Virucidal Activity of Ethanol and Povidone-Iodine. *MicrobiologyOpen* **2020**, *9*, e1097.
57. Sajedinejad, N.; Paknejad, M.; Houshmand, B.; Sharafi, H.; Jelodar, R.; Zahiri, S.H.; et al . *Lactibacillus salivarius* NK02: A Potent Probiotic for Clinical Application in Mouthwash. *Probiotics and Antimicrobial Proteins* **2018**, *10*, 485–495.
58. Cheng, X.; He, F.; Si, M.; Sun, P.; and Chen, Q. Effects of Antibiotic Use on Saliva Antibody Content and Oral Microbiota in Sprague Dawley Rats. *Microbiome in Health and Disease* **2022**, *12*, doi.org/10.3389.

3 Bacterial Consortium of the Oral Cavity

Sohini Sharma, Pranjal Verma, Vicky, and Shamsher S. Kanwar
Department of Biotechnology, Himachal Pradesh University,
Summer Hill, Shimla, Himachal Pradesh, India

3.1 INTRODUCTION

The microbial community that resides inside our bodies is referred to as the microbiome. The name "microbiome" was used by Nobel Prize winner Joshua Lederberg to describe the biological community of symbiotic, commensal, and pathogenic microorganisms. Our bodies directly share space with these microbes [1]. Our bodies are literally inhabited by these microbes. The bacterial count in our body is about equivalent or even higher than the count of cells [2]. In the previous 20 years, interest in the human microbiome has grown significantly, and hundreds of articles on the subject are being published annually. Countless resources have been devoted to study the human microbiome as a potential cause or treatment for a group of illnesses, such as inflammatory bowel disease and cardiometabolic disorders [3].

The terms oral microbiome, oral microbiota, and oral microflora refer to the microorganisms that exist in the human mouth cavity [4]. Oral disease conditions are closely related to oral microorganisms such as viruses, bacteria, archaea, protozoa, and fungi. Research indicates that children's oral microbiomes become more diverse as soon as they encounter their first colonizing bacterium [5]. The oral microflora was first detected by the Dutchman Antonie van Leeuwenhoek, who used a microscope to discover it. He was known as the "Father of Microbiology" and was credited for the discovery of both bacteria and protists and sighted (1683) his own tooth plaque and described it as "little living animalcules prettily moving" [1]. After the gut microbiome, the human oral microbiome (HOM) is the second-largest microbial community in the human body. The HOM is involved in the start of a variety of systemic and localized illnesses, including those caused by bacteria, viruses, and fungi [6]. Throughout a person's life, several bidirectional communication and regulatory systems keep the oral bacteria population and the host in a homeostatic balance. Oral microbiota dysbiosis, however, can result in oral infectious diseases such as dental cavities, periodontal disease, and oral candidiasis [7].

Bacteria find the oral cavity and adjacent nasopharyngeal regions to be the ideal environments for growth. The average temperature of the oral cavity is 37°C when there are no significant variations, providing bacteria with a stable environment in which to survive. Saliva also consistently maintains a pH of 6.5–7.0, which is optimal for most bacterial species. It maintains bacterial moisture levels and acts as a conduit for the delivery of nutrition to microorganisms [8]. This chapter will give a general summary of composition of the oral microbiota, bacterial interactions between different species, molecules responsible for interactions, and how microbial dysbiosis relates to both systemic and oral disorders.

3.2 MICROBIAL COMPOSITION OF THE ORAL CAVITY

Today, the composition of human oral microflora is still being explored, and it is now known that the mouth cavity is highly inhabited with a variety of microorganisms including bacteria, viruses, protozoa, archaea, and fungi [9]. The oral microbiota has the second most complex species diversity

DOI: 10.1201/9781003478638-3

after the colon. One of the most varied and distinctive populations of microorganisms in the human body is found in the mouth cavity [10]. The tongue, soft and hard palates, teeth, and buccal mucosa are just a few of the many microscopic microbial homes that make up the oral cavity and these habitats, which together make up a species-rich heterogeneous ecological system [11]. Due to oral microbes having unique adhesins on their surfaces that help them attach to corresponding receptors on oral surfaces, some parts of the oral cavity are preferentially colonized by oral microbes [12].

There were 772 bacteria species found in the human aerodigestive tract when the Human Oral Microbiome Database (eHOMD), www.homd.org (accessed on 9 December 2022), was last updated [13]. 16S rRNA amplicon sequencing by using targeted gene technique is the most used method for describing the oral bacteriome, as mentioned before [14]. In the normal oral cavity, the most common bacteria are *Lactobacillus*, *Veillonella*, *Lactococcus*, *Streptococcus*, *Enterococcus*, *Staphylococcus*, *Corynebacterium*, and *Bacteroids* [15, 16]. The mouth cavity contains different types of bacteria, mostly from the phyla Firmicutes, Proteobacteria, Fusobacteria, Spirochaetes, Chlamydia, Tenericutes, Bacteroidetes, and Euryarchaeota [17]. A few lesser-known phyla and potential divisions, including Synergistetes, Chloroflexi, Chlorobi, GN02, SR1, TM7, and WPS-2, are also present in the oral cavity [18]. Among these, TM7, GN02, and SR1 belong to the candidate phyla radiation (CPR). Oral CPR members are anticipated to affect the mouth microbial ecology [19], which has been linked to oral illnesses including periodontitis, by changing the structural hierarchy and functions of the oral microbiome [20] and to halitosis [21].

The following are the main bacterial genera that are present in a healthy mouth cavity [22]:

Gram negative

1. *Cocci*: *Moraxella*, *Neisseria*, *Veillonella*
2. *Rods*: *Campylobacter*, *Capnocytophaga*, *Desulfobacter*, *Desulfovibrio*, *Eikenella*, *Fusobacterium*, *Haemophilus*, *Leptotrichia*, *Prevotella*, *Selenomonas*, *Simonsiella*, *Treponema*, *Wolinella*

Gram positive

1. *Cocci*: *Abiotrophia*, *Peptostreptococcus*, *Streptococcus*, *Stomatococcus*
2. *Rods*: *Actinomyces*, *Bifidobacterium*, *Corynebacterium*, *Eubacterium*, *Lactobacillus*, *Propionibacterium*, *Pseudoramibacter*, *Rothia*

3.2.1 Bacterial Profiles for Each Site

The varied microbial flora live in equilibrium in the mouth cavity under normal circumstances. An imbalance in the flora of the oral cavity or by the colonization of additional microorganisms of a viral, fungal, or bacterial origin can lead to an infection of the mucosa [23]. The oral microflora contains greater than 700 distinct bacterial species. These species live inside the mouth and colonize the periodontal pockets, hard palate, and carious lesions (Figure 3.1). Because of the unique local environments these sites provide, for example, the anaerobic environment supplied by the periodontal pocket, most species prefer certain places over others in the oral cavity. Microbes can be found in both people with and without caries, as well as those who have both carries and periodontal pocket. Numerous clinical reports have revealed that patients having periodontitis or caries have more proportions of specific bacterial species like *Streptococcus mutans* or *Porphyromonas gingivalis*, respectively [24].

Overall, most oral cavity locations frequently contain both cultivable and non-cultivable species of *Veillonella*, *Gemella*, *Streptococcus*, and *Granulicatella*. *S. mitis* is the most frequently occurring species almost everywhere and in every individual [12].

The assemblage of several cell types to form biofilm enclosed in extracellular matrix is one of the specialized survival strategies that bacteria have developed as a result of variations in temperature, pH, oxygen content, shear pressures, and the availability of nutrients and energy in different conditions. Despite planktonic bacteria being present in the mouth cavity, the majority of the

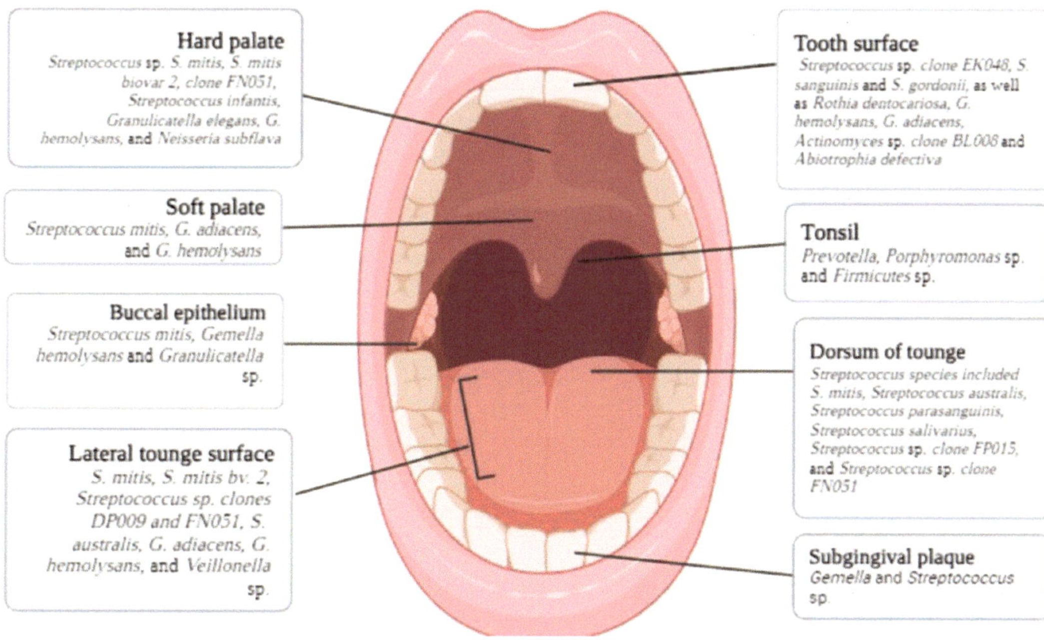

FIGURE 3.1 Bacterial distribution at different sites of the oral cavity.

oral microbes occur in an oral biofilm state [25]. These biofilms are intricate microecosystems of microbes that are adhered to surfaces and embedded in an exopolymeric matrix. The microbial inhabitants of oral biofilms establish distinctive chemical, metabolic, and physical interactions. These are possible inside a species of microorganisms (intraspecies interactions), between species (interspecies), or across kingdoms (cross-kingdom) [26]. Oral biofilms play a well-known function in promoting both a healthy oral cavity and illness by acting as a protective habitat for their microbial occupants [27]. Additionally, there is mounting evidence linking oral microbiome dysbiosis with a number of systemic disorders [28].

3.3 INTERACTION OF BACTERIA FOR BIOFILM FORMATION IN THE ORAL CAVITY

In the oral cavity, one of the predominant mechanisms of bacterial interaction is coaggregation, a process wherein genetically distinct bacteria attach to each other through specific molecular interactions (Table 3.1). These interactions can be categorized as intergeneric (between different genera of bacteria), intrageneric (within the same genus) and multigeneric (involving multiple genera) [29, 30]. Microbial coaggregation is prevalent among a range of bacterial species, including *Fusobacterium*, *Veillonella*, *Haemophilus*, *Campylobacter*, *Neisseria*, *Gemella*, *Granulicatella*, *Capnocytophaga*, and *Bacteroides*. These secondary colonizers, along with their interactions with tertiary colonizers comprising gram-negative anaerobic microorganisms, play a significant role in the development of oral and dental infections [31].

A pioneer bacterial community is dominated by *Streptococcus* spp. and *Actinomyces* spp. Other genera of this community include *Neisseria* spp., *Veillonella* spp., and *Granulicatella* spp. [32]. These communities of microorganisms create an inherent network of mutual coexistence and interspecies coaggregation [33]. Being the most abundant bacterial group in the oral cavity, streptococci play a crucial role in oral health. Commensal streptococci in the mouth assume diverse functions, serving as early colonizers and contributing to the oral ecosystem as secondary pathogens.

TABLE 3.1
Ligand-Receptor Interactions Inhabiting Oral Microbes

	Ligand Protein		Receptor Protein		
S. No.	**Protein**	**Occurrence**	**Protein**	**Occurrence**	**References**
1.	MUC7	Salivary pellicle	SRRP adhesins Type-2 fimbriae	Oral streptococci *Actinomyces* spp.	[36, 37]
2.	Statherin	Salivary pellicle	Type-1 fimbriae	*Actinomyces* spp.	[36, 37]
3.	Galβ1-3GalNAc	Cell surface proteins of oral streptococci	Type-2 fimbriae	*Actinomyces* spp.	[36, 37]
4.	VE-cadherin	Endothelial junction	FadA	*F. nucleatum*	[83]
5.	Has	*S. gordonii*	Hag-1	*Veillonella* spp.	[42]
6.	CRC-	CRC tumors	Fap2	*F. nucleatum*	[84]
7.	SspB	*S. gordonii*	α1,4-linked glucan	Cell surface proteins of *Actinomyces* spp.	[38]

The interaction among bacteria within the oral microbiota is frequently characterized by dynamic and cooperative relationships. Among streptococcus bacteria *S. mitis*, *S. parasanguinis*, and *S. salivarius* are considered pathobionts, which play a key role in increasing the likelihood of oral biofilm formation through their synergistic interactions with highly virulent pathogens [34].

3.3.1 Interaction of Salivary Proteins with *Streptococcus* Species

Human saliva contains glycosylated proteins, with salivary mucins, primarily MUC5B and MUC7, being significant components for interactions. Although MUC19 is present in small quantities, it is notable. Several streptococcal species, including *S. gordonii*, *S. sanguinis*, and *S. oralis*, have demonstrated an ability to bind to MUC7. This adhesion to MUC7 is reliant on the presence of sialic acid. These streptococcal species carry genetic information for serine-rich repeat protein (SRRP) adhesins, which contain domains similar to sialic acid–binding immunoglobulin-type lectins (Siglec). These domains facilitate the binding of the bacteria to sialic acid–containing glycan epitopes found on oral epithelial cells, platelets, and saliva. This interaction supports the infestation of these bacteria in the oral cavity [35]. Oral streptococci possess a wide array of adhesins that allow them to interact with the salivary pellicle. Meanwhile, *Actinomyces* employ type 1 fimbriae to engage with salivary proline-rich proteins and statherin. In contrast, *Actinomyces* employ type 2 fimbriae, which specifically interact with Galβ1-3GalNAc molecules found within the cell surface receptor polysaccharides (RPS) of mitis-group streptococci, salivary glycoprotein MUC7, and secretory immunoglobulin IgA. These type 2 fimbriae facilitate coaggregation with streptococci that express RPS and also seem to play a role in the formation of biofilms [36, 37]. SspB, found in *S. gordonii*, recognizes an α1,4-linked glucan in *Actinomyces oris* T14V, enabling coaggregation [38].

3.3.2 Interaction of *Fusobacterium nucleatum*

Fusobacterium nucleatum (Fn) serves as a structural bridge, connecting primary colonizers like *Streptococcus* species to predominantly anaerobic secondary colonizers, including *Aggregatibacter actinomycetemcomitans* and *Porphyromonas gingivalis*. It has the ability to coaggregate with nearly all bacterial species involved in the formation of oral plaque. This coaggregation is primarily facilitated by various fusobacterial adhesins such as Aid1, CmpA, Fap2, FadA, FomA, and RadD.

These adhesins play a pivotal role in promoting microbial interactions and invasions and the dissemination of bacteria within the oral environment [39]. Fn, a normally harmless anaerobic oral bacterium, enhances its pathogenicity by secreting an amyloid-like protein called FadA. The intact pre-FadA is a crucial element of this amyloid-like structure, and it is secreted through a mechanism involving the autotransporter protein Fap2. This secretion of amyloid-like FadA by Fn has significant implications, as it has been linked to promoting periodontal bone loss and is also associated with colorectal cancer. Interestingly, compounds that can bind to amyloid structures have shown promise in disrupting Fn's pathogenicity, and this disruption occurs in a FadA-dependent manner. Amyloid-like FadA serves multiple functions in supporting Fn's virulence, including facilitating biofilm formation, enhancing tolerance to acidic conditions, and aiding in the binding of Fn to host cells [40]. In addition to single-protein interactions, coaggregation processes also encompass complex interactions involving multiple proteins and carbohydrates. Recent findings suggest that the coaggregation between Fn and *Streptococcus gordonii* engages a minimum of three distinct outer membrane proteins within Fn. These proteins include two adhesins that can be inhibited by arginine, namely RadD and CmpA, along with an accessory protein known as Aid1. Aid1 plays a crucial role in facilitating the adhesive function of RadD. This multiprotein interaction mechanism underscores the intricate and cooperative nature of coaggregation processes between these oral bacteria [41].

3.3.3 Interaction of *Veillonella* Species

Veillonella spp. engage in physical coaggregation with streptococci. Within the *V. atypica* OK5 strain, researchers have discovered eight potential genes responsible for encoding adhesive proteins. One of these genes, *hag1*, encodes a Hag1 autotransporter protein that bears resemblance to *Yersinia* adhesin A (YadA). This protein plays a crucial role in coaggregation with *Streptococcus gordonii*, *S. oralis*, *S. cristatus*, *P. gingivalis*, and adhesion to epithelial cells of oral cavity. Due to its notably large size (7187 amino acids) and intricate structure, Hag1 likely utilizes distinct domains and mechanisms for coaggregation with various partners. The study of the interaction between *V. atypica* OK5 and *S. gordonii* has also helped in identifying Hsa adhesin as the target binding protein. This protein in *S. gordonii* is known for showing binding affinity toward sialic acid [42].

3.3.4 Antagonist Interaction and Colonization Resistance

The dynamics of these interactions is not just limited to synergism but also antagonism through colonization resistance. The mechanisms responsible for colonization resistance can be categorized into four main strategies: (a) vying for nutrients, (b) competing for surface attachment sites, (c) generating antagonistic substances, and (d) establishing unfavorable environmental conditions for incoming microorganisms [43]. One example of such interaction is between *S. mutans* and *S. sanguinis*. *S. mutans* can inhibit *S. sanguinis* by producing a significant amount of organic acid. Both bacteria can metabolize glucose and generate lactate when exposed to excess glucose. However, *S. mutans* significantly lowers the pH as it exhibits higher activity of ATP-glucose phosphotransferase as compared to *S. sanguinis*. This excessive lactic acid from *S. mutans* emanates in the intercellular surroundings, inhibiting *S. sanguinis* growth and leading to an increased ratio of *S. mutans* to *S. sanguinis*. Under acidic conditions (pH 5.5), *S. mutans* outperforms *S. sanguinis*, as acidic conditions suppress or damage the ATP-glucose phosphotransferase activity of both strains, with *S. mutans* being less affected than *S. sanguinis* [24]. Apart from this, antimicrobial agents are secreted by bacteria inhabiting the oral cavity to impede other bacteria, which includes bacteriocins, mutacins, hydrogen peroxide, and salivaracin [43]. Bacteriocins are long-chain peptides synthesized by bacterial ribosomes. They serve to inhibit or kill other bacteria, allowing for competition in the ecosystem. Bacteriocins are classified into two main groups: Class I lantibiotics and Class II non-lantibiotics. Both gram-positive and gram-negative bacteria can produce bacteriocins. Gram-positive bacteria typically produce shorter bacteriocins, around 60 amino acids in length, but they exhibit a broad spectrum of activity. Bacteriocin-producing gram-positive bacteria in the oral cavity include *Enterococcus*, *Lactobacillus*, *Lactococcus*, and *Streptococcus* [44]. Bacteriocins secreted by *S. mutans* are termed as mutacins. Mutacins have exhibited inhibitory properties against various *Streptococcus* species, encompassing those within the anginosus, bovis, mitis, pyogenic, and salivarius subgroups. These mutacins are further classified into lantibiotic and non-lantibiotic types. Notably, *S. mutans* has been documented to hinder the growth of *S. sanguinis* by producing both lantibiotic mutacin I and non-lantibiotic mutacin IV [34]. *S. sanguinis*, an early colonizer in biofilms, plays a valuable role in the oral biofilm ecosystem. It releases hydrogen peroxide (H_2O_2) to manage excess oxygen and acts as a general antimicrobial agent, effectively limiting the advancement of *S. mutans* and other anaerobic periodontal pathogens [24]. Salivaricin is a type of bacteriocin produced by *Streptococcus salivarius* [45]. Many salivaricins have been characterized till date, and amongst them salivaricin D has broad-spectrum antibacterial activity. It has displayed its efficacy by potentially inhibiting pathogenic bacteria like *S. pneumonia* and *S. pyogenes* [25, 46].

3.3.5 Polymicrobial Synergy and Dysbiosis Model

The interaction of microbes in oral microbiota can be best explained by the Polymicrobial Synergy and Dysbiosis Model [47]. Dysbiosis is a condition where the balance in symbiotic relationships has been disrupted. According to this concept, certain diseases may arise due to a decrease in the population of beneficial symbiotic organisms and an increase in the number of harmful pathogens [47, 48]. The oral microbiome differs between individuals with good oral health and those with periodontitis and gingivitis. In individuals with healthy gums, the gingival sulcus has a small, balanced population of gram-positive bacteria and facultative anaerobes, with less than 5% of spirochaetes and motile rods. As the disease worsens, there's a shift with more strict anaerobes, gram-negative bacteria, and motile microorganisms [49].

3.4 ORAL MICROBIOME IN HEALTH AND DISEASES

The human oral microbiome can influence the development and course of a variety of localized and systemic disorders, such as those that are viral in nature and specifically viruses that have oropharynx-mediated entry into the body. Because oral bacterial infections are opportunistic, infections which develop when the right factors come together, including diet, the host's immunological response, escalating systemic or genetic issues, pH, inadequate tooth care, and way of life [50]. The mouth cavity and adjacent nasopharyngeal regions offer the ideal environment for the growth of microorganisms. Because the oral cavity typically has a temperature of 37°C, germs may dwell there in a stable environment. Saliva also keeps its pH between 6.5 and 7.0 constant, which is optimal for the majority of microbes. It serves as a medium for the transportation of nutrients to microorganisms and keeps bacteria hydrated [8]. There is growing evidence that this microbiota contributes to periodontal diseases and dental caries, the two most prevalent oral diseases in humans. These conditions pose serious risks to human health, including cardiovascular disease, tumors, diabetes mellitus, preterm birth, and low birth weight in infants. The interactions between different species in the oral community are vital in deciding whether or not oral microbiota cause illnesses, as it is commonly believed that oral microbes primarily cause diseases in a cooperative or synergistic fashion [51].

3.4.1 Dental Caries

This disorder is a huge health care issue worldwide since it is very common [52] and people are susceptible to it for the entirety of their life [53]. It is one of the most typical bacterial illnesses in people, and it can result in tooth deterioration and even tooth loss. *Streptococcus mutans* has been extensively examined for its cariogenic capabilities and is even considered a particular causative microbe of dental caries. Other pathogens such as *Actinomyces*, *Thiomonas*, *Prevotella*, *Veillonella*,

Leptotrichia, *Bifidobacterium*, and *Granulicatella* were also found in increased concentrations in the plaque of children with caries [54]. Distinct bacterial components may have distinct roles in early enamel lesions versus caries progression into dentin. For demineralization, enamel with a high hydroxyapatite content could need a more acidic microbiota than dentin. Acidogenic non-mutans streptococci, *S. mutans*, *Actinomyces* species, and *Bifidobacterium/Scardovia* species are among the most acidogenic species [55]. Caries development into dentin may require proteolytic degradation of proteins, and this proteolysis is carried out by *Prevotella* species [56].

3.4.2 Periodontal Disease

Periodontal disease is one of the most prevalent illnesses worldwide. A number of inflammatory disorders that affect the tissues supporting the teeth are caused by periodontal disease. It extends from the gingiva into the periodontium's deeper supporting elements including the connective tissue, gum, and alveolar bone that encircles and sustains tooth [57], and in the worst case it can cause the loss of a tooth [58]. The oral microbiome is a dynamic, well-balanced ecosystem that typically retains its typical values in terms of health [59], but imbalances then progress to gingivitis and lastly periodontitis. In individuals with periodontitis, the subgingival microbial makeup experiences incredibly complicated alterations. Periodontitis is closely related to bacterial species such as *Eubacterium saphenum*, *Desulfobulbus* spp., *P. endodontalis*, *Prevotella denticola*, *Bacteroidetes* spp., *Parvimonas micra*, *Peptostreptococcus* spp., *Filifactoralocis*, *Dialister* spp., and *Synergistetes* spp [60]. Bacteria that are considered beneficial include *Neisseria*, *Streptococcus*, *Veillonella*, *Abiotrophia*, *Campylobacter*, *Gemella*, and *Capnocytophaga* [61]. Smoking, a significant periodontitis risk factor, may alter the subgingival composition of the microbiota. Smokers with periodontitis have been found to have higher abundances of species belonging to *Fusobacterium*, *Bacteroides*, *Porphyromonas*, *Campylobacter*, and *Parvimonas* and reduced abundances of *Veillonella*, *Streptococci*, and *Neisseriain* subgingival environment [62]. Additionally, there are some reports suggesting a link between periodontitis and the archaea domain [63].

3.4.3 Oral Cancer

There is mounting proof that various elements of the microbiome of humans are closely related to many different forms of cancer. Oral squamous cell carcinoma (SCC) is the most prevalent head and neck cancer in the world and is the most widespread and frequently investigated of all oral malignancies [64]. Over 300,000 individuals worldwide are affected with oral cancer each year, making it the sixth most common malignancy [65]. Numerous studies conducted across the world show that the majority of individuals with oral cancer have inadequate oral hygiene, which increases the risk of the cancer. Inflammatory reactions are frequently the first indication of poor oral health, and it worsens as health deteriorates. Pathogenic substances originating from inflammatory infections are found in around 15%–20% of human malignancies. Proper dental hygiene will keep such inflammatory agents, which contribute to oral cancer, under control [66]. There is increasing confirmation of a link between periodontal disorder and multiple cancers. Various studies have demonstrated that periodontal disease increases the likelihood of developing oral SCC [67, 68]. Periodontitis is a common infectious condition that results in persistent inflammation in the mouth cavity. In research that compared the microbiota of gingival SCC to the microbial composition of periodontitis, representatives of the genera *Prevotella*, *Fusobacterium*, and *Peptostreptococcus* were shown to be extra numerous in malignant and periodontal tissues [69]. Numerous investigations have also demonstrated that periodontal pathogens cause cancer. One of the primary periodontal pathogens is *P. gingivalis*, which elevates cancer cells resistance to chemotherapy drugs and invasiveness of oral cancer cells [70]. Moreover, the production of cytokines that promote inflammation, such as IL-1 and TNF, in periodontal disease has been associated with microbial-induced carcinogenesis [71].

Viruses have long been linked to an increased chance of getting oral SCC. According to many meta-analyses, human papilloma virus (HPV) infection increases the danger of oral SCC by up to 3 times. Around 25% of oral SCC patients have HPV; however, there is a regional component, with Africa and Asia having the greatest HPV prevalence [72, 73]. Additional carcinogenic virus species, such as herpes simplex virus (Type 1) and Epstein-Barr virus, have been associated with oral carcinogenesis [74]. Environmental variables, including the involvement of the microbiome, which has just recently been acknowledged as a potential factor, are crucial in the growth and advancement of oral SCC. Oral microbiota imbalance, colonization, and translocation are expected to have a substantial influence on the progression of localized or distant malignancies [75]. Oral anaerobes, notably *F. nucleatum* and *P. gingivalis*, are potentially harmful and have been linked to a variety of cancers [76]. Certain aerobes like *Parvimonas* have been related to cancer [77]. Tumor formation involves dysbiosis in the entire bacterial population, which is mostly based on persistent inflammation and immunosuppression. Chronic inflammation can also be brought on by variations in the relative plenty of specific species of bacteria, including *Prevotella*, *Peptostreptococcus*, and *Parvimonas*. Many cytokines (TNF-α, IL-1β, IL-6, IL-17, IL-23, etc.) and inflammatory mediators are up-regulated by these microorganisms (matrix metalloproteinases MMP-8 and MMP-9) [78]. Additionally, interactions between microbial metabolites and the immune system of the host (for example, lactate and tryptophan) may promote tumor growth. The antitumor immune responses can be suppressed by lactate by attracting and inducing myeloid-derived suppressor cells, immune-suppressive cell types like regulatory T cells, and tumor-associated macrophages [79]. A pancreatic ductal adenocarcinoma can develop because dietary tryptophan metabolites stimulate the myeloid cell's aryl hydrocarbon receptor and promote an immunosuppressive tumor environment [80].

Additionally, throughout the formation of a tumor, cancer cells can escape the immune system in various ways along with immunological evasion. Through interference with immune cell identification and T-cell depletion, carcinogens generated by microbiota aid in the immunological escape of malignancies [81]. While the tumor microenvironment aids in the colonization of the microbiota, certain oral bacteria and their metabolites have the ability to promote tumors. In conclusion, the oral microbiota's role in the development of tumors has garnered some interest. More research is required to give proof that oral microbial treatment may have a significant therapeutic advantage for cancer patients.

3.4.4 Colorectal Cancer

Fusobacterium nucleatum strains that are similar in almost 40% of individuals with colorectal cancer have been found in both saliva and colorectal cancer (CRC) samples, suggesting that the oral cavity is most likely the source of *F. nucleatum's* presence in CRC [82]. Adhesins FadA and Fap2 of *F. nucleatum* have a significant part in colorectal cancer. Fap2 hinders the antitumor activity of NK cells and T cells by directly engaging inhibitory receptors. It binds to the CRC-associated carbohydrate moiety Gal-GalNAc, facilitating *F. nucleatum* binding to CRC cells through the bloodstream. Additionally, Fap2 prompts the production of CXCL1 and proinflammatory cytokines IL-8 fostering CRC cell migration. In contrast, FadA plays a role in promoting CRC cell proliferation by activating the E-cadherin/β-catenin signaling pathway, which supports the development of colorectal carcinogenesis. Moreover, FadA contributes to CRC progression by cleaving VE-cadherin, leading to the disruption of endothelial junctions. This disruption facilitates the systemic spread of bacteria, causing infections. Furthermore, FadA is also involved in CRC development through its activation of the Wnt/β-catenin regulator, Annexin A1 [83, 84].

3.5 CONCLUSION

The microflora of the oral cavity is a large ecosystem composed of several types of microorganisms including bacteria, fungi, archaea, and viruses. We now have knowledge that was previously unattainable thanks to the invention of 16S rRNA gene sequencing, which allowed us to identify the oral bacteriome. But this has raised a lot of concerns for us and sparked our curiosity about studying not just the individual microbes but also their interactions with one another. Symbiotic connections abound in the oral microbiome, but when pathogens interfere with these partnerships and induce dysbiosis, a previously good condition can turn unhealthy and raise the risk of comorbidities.

Diverse locations in the oral cavity have diverse microenvironments that are controlled by intricate signals, hosts, and external factors. Since some health conditions appear to be linked to the makeup of oral bacteria, these processes may have an impact on human health. Identification of the microbiome in health is the first step in studying the human microbiome. This is followed by a knowledge of the microbiome's involvement in altering the metabolic and functional pathways linked to various disorders. In this chapter, we summarized the latest literature highlighting the bacterial composition of the oral cavity, the interaction within microbial communities, and some diseases associated with the imbalance of normal microbiota. More focus must be placed on the processes underpinning interactions between oral microorganisms within communities, as well as their interactions with their hosts, to assess the many aspects responsible for dysbiosis and, ultimately, disease.

REFERENCES

1. Deo, P. N.; Deshmukh, R. Oral microbiome: Unveiling the fundamentals. *Journal of Oral and Maxillofacial Pathology: JOMFP* **2019**, *23*(1), 122.
2. Scotti, E.; Boué, S.; Sasso, G. L.; Zanetti, F.; Belcastro, V.; Poussin, C.; Sierro, N.; Battey, J.; Gimalac, A.; Ivanov, N. V.; Hoeng, J. Exploring the microbiome in health and disease: Implications for toxicology. *Toxicology Research and Application* **2017**, *1*, 2397847317741884.
3. Walker, A. W.; Hoyles, L. Human microbiome myths and misconceptions. *Nature Microbiology* **2023**, *8*(8), 1392–1396.
4. Gao, L.; Xu, T.; Huang, G.; Jiang, S.; Gu, Y.; Chen, F. Oral microbiomes: more and more importance in oral cavity and whole body. *Protein & Cell* **2018**, *9*(5), 488–500.
5. Gomez, A.; Nelson, K. E. The oral microbiome of children: development, disease, and implications beyond oral health. *Microbial Ecology* **2017**, *73*, 492–503.
6. Soffritti, I.; D'Accolti, M.; Fabbri, C.; Passaro, A.; Manfredini, R.; Zuliani, G.; Libanore, M.; Franchi, M.; Contini, C.; Caselli, E. Oral microbiome dysbiosis is associated with symptoms severity and local immune/inflammatory response in COVID-19 patients: a cross-sectional study. *Frontiers in Microbiology* **2021**, *12*, 687513.
7. Lamont, R. J.; Koo, H.; Hajishengallis, G. The oral microbiota: dynamic communities and host interactions. *Nature Reviews Microbiology* **2018**, *16*(12), 745–759.
8. Lim, Y.; Totsika, M.; Morrison, M.; Punyadeera, C. Oral microbiome: a new biomarker reservoir for oral and oropharyngeal cancers. *Theranostics* **2017**, *7*(17), 4313.
9. Wade, W. G. The oral microbiome in health and disease. *Pharmacological Research* **2013**, *69*(1), 137–143.
10. Li, K.; Bihan, M.; Yooseph, S.; Methe, B. A. Analyses of the microbial diversity across the human microbiome. *PloS one* **2012**, *7*(6), e32118.
11. Kilian, M. The oral microbiome–friend or foe?. *European Journal of Oral Sciences* **2018**, *126*, 5–12.
12. Aas, J. A.; Paster, B. J.; Stokes, L. N.; Olsen, I.; Dewhirst, F. E. Defining the normal bacterial flora of the cavity. *Journal of Clinical Microbiology* **2005**, *43*(11), 5721–5732.
13. Verma, D.; Garg, P. K.; Dubey, A. K. Insights into the human oral microbiome. *Archives of Microbiology* **2018**, *200*, 525–540.
14. Wang, S.; Song, F.; Wang, Y.; Huang, Y.; Xie, B.; Luo, H. High resolution melting analysis (HRM) based on 16SrRNA as a tool for personal identification with the human oral microbiome. *Forensic Science International: Genetics Supplement Series* **2019**, *7*(1), 161–163.
15. Rogers, A. *Molecular oral microbiology. Caister Academic Press* **2008**.
16. Wang, Q. Q.; Zhang, C. F.; Chu, C. H.; Zhu, X. F. Prevalence of *Enterococcus faecalis* in saliva and filled root canals of teeth associated with apical periodontitis. *International Journal of Oral Science* **2012**, *4*(1), 19–23.

17. Dewhirst, F. E.; Chen, T.; Izard, J.; Paster, B. J.; Tanner, A. C.; Yu, W. H.; Lakshmanan, A.; Wade, W. G. The human oral microbiome. *Journal of Bacteriology* **2010**, *192*(19), 5002–5017.
18. Camanocha, A.; Dewhirst, F. E. Host-associated bacterial taxa from *Chlorobi, Chloroflexi,* GN02, *Synergistetes,* SR1, TM7, and WPS-2 Phyla/candidate divisions. *Journal of Oral Microbiology* **2014**, *6*(1), 25468.
19. Bor, B.; Bedree, J. K.; Shi, W.; McLean, J. S.; He, X. *Saccharibacteria* (TM7) in the human oral microbiome. *Journal of Dental Research* **2019**, *98*(5), 500–509.
20. Griffen, A. L.; Beall, C. J.; Campbell, J. H.; Firestone, N. D.; Kumar, P. S.; Yang, Z. K.; Podar M.; Leys, E. J. Distinct and complex bacterial profiles in human periodontitis and health revealed by 16S pyrosequencing. *The ISME Journal* **2012**, *6*(6), 1176–1185.
21. Takeshita, T.; Suzuki, N.; Nakano, Y.; Yasui, M.; Yoneda, M.; Shimazaki, Y.; Hirofiuji, T.; Yamashita, Y. Discrimination of the oral microbiota associated with high hydrogen sulfide and methyl mercaptan production. *Scientific Reports* **2012**, *2*(1), 215.
22. Marsh, P. D. Role of the oral microflora in health. *Microbial Ecology in Health and Disease* **2000**, *12*(3), 130–137.
23. Minhas, S.; Sajjad, A.; Kashif, M.; Taj, F.; Al Waddani, H.; Khurshid, Z. Oral ulcers presentation in systemic diseases: An update. *Open access Macedonian Journal of Medical Sciences* **2019**, *7*(19), 3341.
24. Huang, R.; Li, M.; Gregory, R. L. Bacterial interactions in dental biofilm. *Virulence* **2011**, *2*(5), 435–444.
25. Berger, D.; Rakhamimova, A.; Pollack, A.; Loewy, Z. Oral biofilms: development, control, and analysis. *High-throughput* **2018**, *7*(3), 24.
26. Bandara, H. M. H. N.; Lam, O. L. T.; Jin, L. J.; Samaranayake, L. Microbial chemical signaling: a current perspective. *Critical Reviews in Microbiology* **2012**, *38*(3), 217–249.
27. Marsh, P. D.; Zaura, E. Dental biofilm: ecological interactions in health and disease. *Journal of Clinical Periodontology* **2017**, *44*, S12–S22.
28. Georges, F. M.; Do, N. T.; Seleem, D. Oral dysbiosis and systemic diseases. *Frontiers in Dental Medicine* **2022**, *3*, 995423.
29. Rickard, A. H.; Gilbert, P.; High, N. J.; Kolenbrander, P. E.; Handley, P. S. Bacterial coaggregation: an integral process in the development of multi-species biofilms. *Trends in Microbiology* **2003**, *11*(2), 94–100.
30. Soria, S.; Angulo-Bejarano, P. I.; Sharma, A. Biofilms: Development and molecular interaction of microbiome in the human oral cavity. *New and Future Developments in Microbial Biotechnology and Bioengineering: Microbial Biofilms* **2020**, (pp. 61–75). Elsevier.
31. Chenicheri, S.; Usha, R.; Ramachandran, R.; Thomas, V.; Wood, A. Insight into oral biofilm: primary, secondary and residual caries and phyto-challenged solutions. *The Open Dentistry Journal* **2017**, *11*, 312.
32. Boisen, G.; Davies, J. R.; Neilands, J. Acid tolerance in early colonizers of oral biofilms. *BMC Microbiology* **2021**, *21*(1), 1–9.
33. Sterzenbach, T.; Helbig, R.; Hannig, C.; Hannig, M. Bioadhesion in the oral cavity and approaches for biofilm management by surface modifications. *Clinical Oral Investigations* **2020**, *24*, 4237–4260.
34. Baty, J. J.; Stoner, S. N.; Scoffield, J. A. Oral commensal streptococci: gatekeepers of the oral cavity. *Journal of Bacteriology* **2022**, *204*(11), e00257–22.
35. Chahal, G.; Quintana-Hayashi, M. P.; Gaytán, M. O.; Benktander, J.; Padra, M.; King, S. J.; Linden, S. K. *Streptococcus oralis* employs multiple mechanisms of salivary mucin binding that differ between strains. *Frontiers in Cellular and Infection Microbiology* **2022**, *12*, 889711.
36. Cavalcanti, I. M.; Del Bel Cury, A. A.; Jenkinson, H. F.; Nobbs, A. H. Interactions between *Streptococcus oralis*, *Actinomyces oris*, and *Candida albicans* in the development of multispecies oral microbial biofilms on salivary pellicle. *Molecular Oral Microbiology* **2017**, *32*(1), 60–73.
37. Cross, B. W.; Ruhl, S. Glycan recognition at the saliva - oral microbiome interface. *Cellular Immunology* **2018**, *333*, 19–33.
38. Back, C. R.; Douglas, S. K.; Emerson, J. E.; Nobbs, A. H.; Jenkinson, H. F. *Streptococcus gordonii* DL 1 adhesin SspB V-region mediates coaggregation via receptors polysaccharide of *Actinomyces oris T14V. Molecular Oral Microbiology* **2015**, *30*(5), 411–424.
39. Groeger, S.; Zhou, Y.; Ruf, S.; Meyle, J. Pathogenic Mechanisms of *Fusobacterium nucleatum* on Oral Epithelial Cells. *Frontiers in Oral Health* **2022**, *3*, 831607.
40. Meng, Q.; Gao, Q.; Mehrazarin, S.; Tangwanichgapong, K.; Wang, Y.; Huang, Y.; Pan, Y.; Robinson, S.; Liu, Z.; Zangiabadi, A.; Lux R.; Papapanou, P.; Guo, X.; Wang, H.; Berchowitz, L.; Han, Y. W. *Fusobacterium nucleatum* secretes amyloid-like FadA to enhance pathogenicity. *EMBO Reports* **2021**, *22*(7), e52891.

41. Lima, B. P.; Shi, W.; Lux, R. Identification and characterization of a novel *Fusobacterium nucleatum* adhesin involved in physical interaction and biofilm formation with *Streptococcus gordonii*. *Microbiologyopen* **2017**, *6*(3), e00444.
42. Zhou, P.; Manoil, D.; Belibasakis, G. N.; Kotsakis, G. A. *Veillonellae*: beyond bridging species in oral biofilm ecology. *Frontiers in Oral Health* **2021**, *2*, 774115.
43. Akimbekov, N. S.; Digel, I.; Yerezhepov, A. Y.; Shardarbek, R. S.; Wu, X.; Zha, J. Nutritional factors influencing microbiota-mediated colonization resistance of the oral cavity: A literature review. *Frontiers in Nutrition* **2022**, *9*, 1029324.
44. Alghamdi, S. Isolation and identification of the oral bacteria and their characterization for bacteriocin production in the oral cavity. *Saudi Journal of Biological Sciences* **2022**, *29*(1), 318–323.
45. Barbour, A.; Smith, L.; Oveisi, M.; Williams, M.; Huang, R. C.; Marks, C.; Fine, N.; Sun, C.; Younsei, F.; Zargaran, S.; Orugunty, R.; Horvath, T. D.; Haidacher, S. J.; Haag, A. M.; Sabharwal, A.; Hinz, B.; Glogauer, M. Discovery of phosphorylated lantibiotics with proimmune activity that regulate the oral microbiome. *Proceedings of the National Academy of Sciences* **2023**, *120*(22), e2219392120.
46. Birri, D. J.; Brede, D. A.; Nes, I. F. Salivaricin D, a novel intrinsically trypsin-resistant lantibiotic from *Streptococcus salivarius* 5M6c isolated from a healthy infant. *Applied and Environmental Microbiology* **2012**, *78*(2), 402–410.
47. Hajishengallis, G.; Lamont, R. J. Beyond the red complex and into more complexity: the polymicrobial synergy and dysbiosis (PSD) model of periodontal disease etiology. *Molecular Oral Microbiology* **2012** *27*(6), 409–419.
48. Shaikh, H. F. M.; Patil, S. H.; Pangam, T. S.; Rathod, K. V. Polymicrobial synergy and dysbiosis: An overview. *Journal of Indian Society of Periodontology* **2018**, *22*(2), 101.
49. Samaranayake, L. *Essential microbiology for dentistry-E-Book*. Elsevier Health Sciences **2018**.
50. Krishnan, K.; Chen, T.; Paster, B. J. A practical guide to the oral microbiome and its relation to health and disease. *Oral Diseases* **2017**, *23*(3), 276–286.
51. He, J.; Li, Y.; Cao, Y.; Xue, J.; Zhou, X. The oral microbiome diversity and its relation to human diseases. *Folia Microbiologica* **2015**, *60*, 69–80.
52. Schwendicke, F.; Dörfer, C. E.; Schlattmann, P.; Page, L. F. Thomson, W. M.; Paris, S. Socioeconomic inequality and caries: a systematic review and meta-analysis. *Journal of Dental Research* **2015**, *94*(1), 10–18.
53. Selwitz, R. H.; Ismail, A. I.; Pitts, N. B. Dental caries. *The Lancet* **2007**, *369*(9555), 51–59.
54. Tanner, A. C. R.; Kent Jr, R. L.; Holgerson, P. L.; Hughes, C. V.; Loo, C. Y.; Kanasi, E.; Chalmers, N. I.; Johansson, I. Microbiota of severe early childhood caries before and after therapy. *Journal of Dental Research* **2011**, *90*(11), 1298–1305.
55. Chalmers, N. I.; Oh, K.; Hughes, C. V.; Pradhan, N.; Kanasi, E.; Ehrlich, Y.; Dewhirst, F. E.; Tanner, A. C. Pulp and plaque microbiotas of children with severe early childhood caries. *Journal of Oral Microbiology* **2015**, *7*(1), 25951.
56. Hashimoto, K.; Sato, T.; Shimauchi, H.; Takahashi, N. Profiling of dental plaque microflora on root caries lesions and the protein-denaturing activity of these bacteria. *American Journal of Dentistry* **2011**, *24*(5), 295.
57. Hernandez, M.; Dutzan, N.; García-Sesnich, J.; Abusleme, L.; Dezerega, A.; Silva, N.; González, F. E.; Vernal, R.; Sorsa, T.; Gamonal, J. Host-pathogen interactions in progressive chronic periodontitis. *Journal of Dental Research* **2011**, *90*(10), 1164–1170.
58. Kirst, M. E.; Li, E. C.; Alfant, B.; Chi, Y. Y.; Walker, C.; Magnusson, I.; Wang, G. P. Dysbiosis and alterations in predicted functions of the subgingival microbiome in chronic periodontitis. *Applied and Environmental Microbiology* **2015**, *81*(2), 783–793.
59. Najmanova, L.; Sabova, L.; Lenartova, M.; Janatova, T.; Mysak, J.; Vetrovsky, T.; Tesinska, B.; Novotna, G. B.; Kobersk, M.; Roukal, Z.; Duskova, J.; Podzimek, S.; Janata, J. R/G Value—A Numeric Index of Individual Periodontal Health and Oral Microbiome Dynamics. *Frontiers in Cellular and Infection Microbiology* **2021**, *11*, 602643.
60. Paster, B. J.; Olsen, I.; Aas, J. A.; Dewhirst, F. E. The breadth of bacterial diversity in the human periodontal pocket and other oral sites. *Periodontology* **2006**, *42*(1), 80–87.
61. Kumar, P. S.; Griffen, A. L.; Moeschberger, M. L.; Leys, E. J. Identification of candidate periodontal pathogens and beneficial species by quantitative 16S clonal analysis. *Journal of Clinical Microbiology* **2005**, *43*(8), 3944–3955.
62. Shchipkova, A. Y.; Nagaraja, H. N.; Kumar, P. S. Subgingival microbial profiles of smokers with periodontitis. *Journal of Dental Research* **2010**, *89*(11), 1247–1253.

63. Pérez-Chaparro, P. J.; Gonçalves, C.; Figueiredo, L. C.; Faveri, M.; Lobão, E.; Tamashiro, N.; Duarte, P.; Feres, M. Newly identified pathogens associated with periodontitis: a systematic review. *Journal of Dental Research* **2014**, *93*(9), 846–858.
64. Irfan, M.; Delgado, R. Z. R.; Frias-Lopez, J. The oral microbiome and cancer. *Frontiers in Immunology* **2020**, *11*, 591088.
65. Gill, S. R.; Gill lab: research overview. *Department of Microbiology & Immunology, University of Rochester Medical Center: Rochester, NY* **2011**.
66. Meurman, J. H. Oral microbiota and cancer. *Journal of Oral Microbiology* **2010**, *2*(1), 5195.
67. Karmakar, S.; Kar, A.; Thakur, S.; Rao, V. U. Periodontitis and oral Cancer-A striking link. *Oral Oncology* **2020**, *106*, 104630.
68. Corbella, S.; Veronesi, P.; Galimberti, V.; Weinstein, R.; Del Fabbro, M.; Francetti, L. Is periodontitis a risk indicator for cancer? A meta-analysis. *PloS One* **2018**, *13*(4), e0195683.
69. Li, Y.; Tan, X.; Zhao, X.; Xu, Z.; Dai, W.; Duan, W.; Huang, S.; Zhang, E.; Liu, J.; Zhang, S.; Yiu, R.; Shi, X.; Lu, Z.; Pan, Y. Composition and function of oral microbiota between gingival squamous cell carcinoma and periodontitis. *Oral Oncology* **2020**, *107*, 104710.
70. Geng, F.; Liu, J.; Guo, Y.; Li, C.; Wang, H.; Wang, H.; Wang, H.; Zhao, H.; Pan, Y. Persistent exposure to *Porphyromonas gingivalis* promotes proliferative and invasion capabilities, and tumorigenic properties of human immortalized oral epithelial cells. *Frontiers in Cellular and Infection Microbiology* **2017**, *7*, 57.
71. Kipanyula, M. J.; Etet, P. F. S.; Vecchio, L.; Farahna, M.; Nukenine, E. N.; Kamdje, A. H. N. Signaling pathways bridging microbial-triggered inflammation and cancer. *Cellular Signaling* **2013**, *25*(2), 403–416.
72. Hardefeldt, H. A.; Cox, M. R.; Eslick, G. D. Association between human papillomavirus (HPV) and esophageal squamous cell carcinoma: a meta-analysis. *Epidemiology & Infection* **2014**, *142*(6), 1119–1137.
73. Petrick, J. L.; Wyss, A. B.; Butler, A. M.; Cummings, C.; Sun, X.; Poole, C.; Smith, J. S.; Olshan, A. F. Prevalence of human papillomavirus among oesophageal squamous cell carcinoma cases: systematic review and meta-analysis. *British Journal of Cancer* **2014**, *110*(9), 2369–2377.
74. Sand, L.; Jalouli, J. Viruses and oral cancer. Is there a link. *Microbes and Infection* **2014**, *16*(5), 371–378.
75. Mascitti, M.; Togni, L.; Troiano, G.; Caponio, V. C. A.; Gissi, D. B.; Montebugnoli, L.; Procaccini, M.; Muzio, L. L.; Santarelli, A. Beyond head and neck cancer: the relationship between oral microbiota and tumour development in distant organs. *Frontiers in Cellular and Infection Microbiology* **2019**, *9*, 232.
76. Wang, X.; Jia, Y.; Wen, L.; Mu, W.; Wu, X.; Liu, T.; Liu, X.; Fang, J.; Luan, Y.; Chen, P.; Gao, J.; guyen, K.; Cui, J.; Zeng, G.; Lan, P.; Chen, Q.; Cheng, B.; Wang, Z. *Porphyromonas gingivalis* promotes colorectal carcinoma by activating the hematopoietic *NLRP3* inflammasome. *Cancer Research* **2021**, *81*(10), 2745–2759.
77. Sun, J.; Tang, Q.; Yu, S.; Xie, M.; Xie, Y.; Chen, G.; Chen, L. Role of the oral microbiota in cancer evolution and progression. *Cancer Medicine* **2020**, *9*(17), 6306–6321.
78. Zhao, X.; Liu, J.; Zhang, C.; Yu, N.; Lu, Z.; Zhang, S.; Li, Y.; Li, Q.; Liu, J.; Liu, D.; Pan, Y. *Porphyromonas gingivalis* exacerbates ulcerative colitis via *Porphyromonas gingivalis* peptidylarginine deiminase. *International Journal of Oral Science* **2021**, *13*(1), 31.
79. Hayes, C.; Donohoe, C. L.; Davern, M.; Donlon, N. E. The oncogenic and clinical implications of lactate induced immunosuppression in the tumour microenvironment. *Cancer Letters* **2021**, *500*, 75–86.
80. Frankel, T. L.; di Magliano, M. P. Immune sensing of microbial metabolites: Action at the tumor. *Immunity* **2022**, *55*(2), 192–194.
81. Vinasco, K.; Mitchell, H. M.; Kaakoush, N. O.; Castano-Rodriguez, N. Microbial carcinogenesis: Lactic acid bacteria in gastric cancer. *Biochimica et Biophysica Acta (BBA)- Reviews on Cancer* **2019**, *1872*(2), 188309.
82. Komiya, Y.; Shimomura, Y.; Higurashi, T.; Sugi, Y.; Arimoto, J.; Umezawa, S.; Uchiyama, S.; Matsumoto, M.; Nakajima, A. Patients with colorectal cancer have identical strains of *Fusobacterium nucleatum* in their colorectal cancer and oral cavity. *Gut* **2019**, *68*(7), 1335–1337.
83. Dash, S.; Duraivelan, K.; Samanta, D. Cadherin-mediated host–pathogen interactions. *Cellular Microbiology* **2021**, *23*(5), e13316.
84. Ou, S.; Wang, H.; Tao, Y.; Luo, K.; Ye, J.; Ran, S.; Guan, Z.; Wang, Y.; Hu, H.; Huang, R. *Fusobacterium nucleatum* and colorectal cancer: From phenomenon to mechanism. *Frontiers in Cellular and Infection Microbiology* **2022**, *12*, 1778.

4 Dietary Habits and Nature of the Oral Microbiome

Ashwani Kumar
Patanjali Herbal Research Department,
Patanjali Research Foundation, Haridwar, Uttar Pradesh, India

Acharya Balkrishna, Deepika Srivastava, Maneesha Rana, Ankita Kukreti, Amita Kumari, and Vedpriya Arya
Patanjali Herbal Research Department, Patanjali Research Foundation, Haridwar, Uttar Pradesh, India

Dinesh Kumar
School of Bioengineering and Food Technology, Shoolini University of Biotechnology and Management Sciences, Solan, Himachal Pradesh, India

4.1 INTRODUCTION

The oral microbiome (OM) is the body's second-largest microbial diversity after the gut [1]. The OM contains a diverse array of microorganisms such as viruses, bacteria, fungi, and protozoa. From bacterial acquisition at birth to bacterial colonization in old age, an individual's oral microbiome changes. The earliest bacteria that colonize the oral cavity are typical species of *Streptococcus*, with *S. salivarius* being mostly found on the tongue dorsum and in saliva, *S. sanguinis* on the teeth, and *S. mitis* on the buccal mucosa [2]. By fending off infections, preserving homeostasis, and adjusting the immune system, these beneficial microbes contribute to host health. A rich habitat that is conducive to the growth of microorganisms is provided by the teeth, tongue, cheeks, gingival sulcus, tonsils, and hard and soft palates [3]. The human microbiome comprises a core microbiome that is shared by all people and a variable microbiome that is unique to each person due to physiological variances and lifestyle [1]. The Human Oral Microbiome Database (HOMD) included 774 oral bacterial species in 14 phyla, Firmicutes, Bacteroidetes, Actinobacteria, Proteobacteria, Fusobacteria, Spirochaetes, Synergistetes, Saccharibacteria, Chlamydia, SR1, Gracilibacteria, Tenericutes, Chloroflexi, and Chlorobi [4].

Oral microbial communities maintain a dynamic equilibrium over time, but significant changes in an individual's oral environment or lifestyle can disrupt this stability, leading to the overgrowth of insignificant components and potential illness [5]. For example, *Streptococcus mutans*, *S. sobrinus*, *Lactobacillus* spp., and *Actinomycetes* cause dental caries [6]. Oral cavity mucosal membrane candidiasis is frequently an indication of either local or systemic immunological dysfunction. In healthy people, *Candida albicans* is frequently detected in the oral microbiota and doesn't seem to have any negative clinical effects [7]. However, elevated amounts of *C. albicans* and local inflammation might be attributed to weakened immunity or widespread bacterial suppression (due to antibiotic therapy) [8]. OM imbalance and dysfunction, known as dysbiosis, is associated with caries, periodontal diseases, mucosal diseases, oral cancer, inflammatory bowel disease, nervous system diseases, and others [3].

Dietary or nutritive plans play an important role in maintaining the OM. Various studies have shown that a diet high in fiber, vitamins C and D, low in carbohydrates, and rich in omega-3 fatty

DOI: 10.1201/9781003478638-4

acids can dramatically decrease periodontal and gingival inflammation [9, 10]. Compared to non-vegetarians, vegetarians had considerably lower pocket depth, less bleeding on probing, and better oral cleanliness [11]. Sugar-containing food habits contribute significantly to dental caries and tooth loss [12]. An increased chance of developing periodontal lesions has been linked to low blood levels of vitamin B12 [13]. OM dysbiosis is associated with the pathogenesis and development of many oral and systemic diseases. In this chapter, we provide a brief overview of the OM and the effect of dietary habits on its modulation.

4.2 ORAL MICROBIOME AND HEALTH: A BRIEF OVERVIEW

The microbiome is the community of microorganisms that reside in the human body. Joshua Lederberg first used the term "microbiome" to describe the biological community of commensal, symbiotic, and pathogenic microorganisms. The OM refers to the microorganisms residing in the human oral cavity which comprises the teeth, buccal mucosa, soft and hard palates, and tongue [3]. The OM was first recognized by Antonie van Leeuwenhoek [14]; all microorganisms residing in the oral cavity work collectively to build a varied and dynamic environment [15]. Approximately 774 known bacterial species (classified into 14 phyla) inhabit the oral cavity [4, 16, 17]. An overview of the OM is presented in Figure 4.1.

In addition to these, 100 species of fungi (*Candida*, *Aureobasidium*, *Aspergillus*, *Cryptococcus*, *Cladosporium*, *Fusarium*, *Gibberella*, *Penicillium*, *Saccharomyces*, *Rhodotorula*, and *Schizophyllum*) [18, 19], some eukaryotic viruses (Herpesviridae, Anelloviridae, and Papillomaviridae), archaea (*Thermoplasmatales*, *Methanobrevibacter*, *Methanobacterium*, and *Methanosphaera*), and protozoans (*Entamoeba gingivalis* and *Trichomonas tenax*) [20–24] have also been identified as a part of the OM. Actinobacteria (*Rothia*, *Corynebacterium*, and *Actinomyces*),

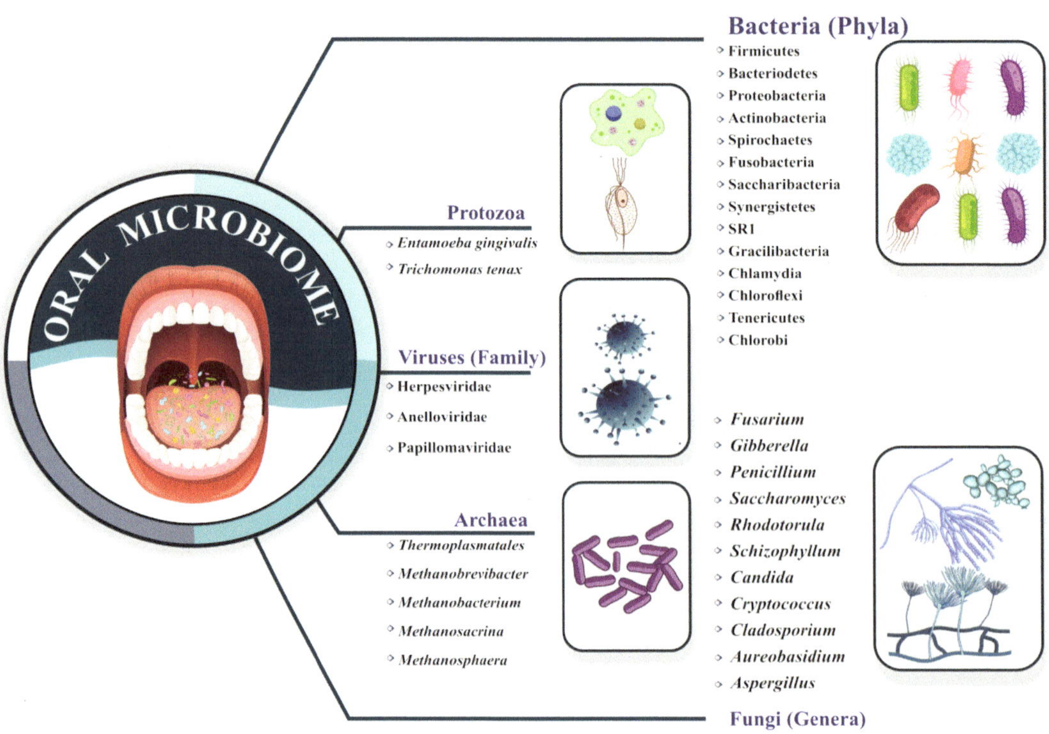

FIGURE 4.1 Diversity of the oral microbiome.

Bacteroidetes (*Capnocytophaga*, *Prevotella*, and *Porphyromonas*), Firmicutes (*Granulicatella* and *Streptococcus*), Fusobacteria (*Fusobacterium*), and Proteobacteria (*Haemophilus* and *Neisseria*) are predominant in healthy people [25]. Similarly, methanogens make up just a small number of the OM and may be found in healthy people, but their prevalence and numbers are higher in those who have periodontitis [20, 26]. Among individuals, the OM differs according to the oral niches and changes dynamically across life-history stages.

The OM plays a crucial role in sustaining oral health, oral tolerance, allergy prevention, and pH balance [27, 28]. Oral dysbiosis results in several disease indications, *Streptococcus* is the most prevalent genus that contributes to the production of dental plaque. *Veillonella* spp. are involved in metabolic activities such as lactate breakdown [29]. *Prevotella* and *Fusobacterium* species flourish in certain oral cavity habitats and are linked to illness as well as health [30, 31]. Although *Actinomyces* spp. are typically beneficial, they can cause tooth infections. Factors such as a weakened immune system and poor oral hygiene impact the shift of these bacteria to potential pathogenic forms contributing to the development of oral illnesses. For example, *S. mutans* is known for its participation in dental caries, where it uses carbohydrates to generate acids that damage tooth enamel [32]. *Porphyromonas gingivalis* is an anaerobic bacterium that infiltrates and damages gingival tissues, causing inflammation and destroying supporting structures around teeth [33]. *Aggregatibacter actinomycetemcomitans* and *Tannerella forsythia* have been linked to aggressive periodontitis, which rapidly damages periodontal tissues [34]. *C. albicans,* a fungus, can become pathogenic, causing oral thrush, especially in those with weakened immune systems [35]. Surprisingly, research is revealing links between an unbalanced oral microbiota and systemic illnesses such as cardiovascular disease and diabetes [36], rheumatoid arthritis [37], and adverse pregnancy outcomes [38], which emphasizes the larger significance of the OM.

Maintaining a healthy OM is critical for preventing the expansion of several diseases [39]. Several factors, including nutrition, dental hygiene habits, and medicines, affect the composition of the OM [40]. The balance between the host and the OM is essential for oral health throughout life stages. However, several factors such as eating habits, sugar consumption, and others affect this balance. In this regard, the next section provides an overview of current understanding regarding the nature of the OM and its connection to dietary habits.

4.3 EFFECT OF DIETARY HABITS ON THE ORAL MICROBIOME

The OM, a diverse community of microorganisms residing within the mouth, plays a pivotal role in maintaining oral health. Recent research has underscored the profound impact of dietary habits on the composition and function of the OM, shedding light on the intricate mechanisms governing oral health and disease. The dynamic interaction between the microbiome and the host is influenced by numerous facets of contemporary lifestyle, such as host diet, inflammatory reaction, systemic disturbances, and habits such as alcohol, tobacco, and stress, which can alter the OM and its properties and induce a state imbalance, known as dysbiosis [7, 39]. With advancements in technology, incredible strides have been made in understanding the intricate world of the OM. This has given us a fresh perspective on how it impacts our health, both in its equilibrium and when disrupted. Dietary habits serve as a mainstay in shaping the oral microbiome. The nutrient-rich diet influences the balance of the OM [1]. For instance, feeding practices at birth (within 48 hours after delivery) and at 2, 4, and 6 months of age have different effects on the OM. With growing age irrespective of feeding mode, like human milk, infant formula, mixed feeding (human milk + formula), and introduction to solid foods (mixture of solid foods + human milk/formula or both), the bacterial population increase (*Fusobacterium*, *Gemella*, *Granulicatella*, *Haemophilus*, *Leptotrichia*, *Neisseria*, *Propionibacterium*, *Porphyromonas*, *Prevotella*, and *Veillonella*) as well as decrease (*Cloacibacterium* and *Dechloromonas*) are what ultimately affect the OM. Babies fed with formula showed a greater prevalence of *Actinomyces* and *Prevotella*, whereas breastfed babies had a higher prevalence of *Streptococcus* [41].

Likewise, in a study, the mixing of breast milk and saliva resulted in a predominant inhibition of microbial growth, which persisted for up to 24 hours. Nevertheless, certain bacteria were able to recuperate and proliferate after being exposed to these micromolar concentrations of hydrogen peroxide (formed by the natural mixing of breast milk and saliva) [42]. The microbial community of breastfed versus formula-fed newborns differs, with the phylum Firmicutes having the greatest dominance. Also, the prevalence of the phylum Bacteroidetes in the mouths of formula-fed infants was greater than in breast-fed infants ($p = 0.01$), whereas breast-fed babies had a higher frequency of *Actinobacteria* and *Proteobacteria* in their saliva ($p = 0.04$) [43].

Similarly, increased intake of dairy, dietary fiber, and sugar throughout certain dietary phases led to changes in the oral bacterial microbiome that persisted for 3 months after going back to the initial baseline diet. Further, high dietary fiber consumption was associated with a high abundance of largely normal oral microbiota representatives; however, frequent consumption of milk and yogurt decreased the abundances of these taxa, whereas higher intake of sugar supported caries-promoting non-mutans *Streptococci*. Thus, dietary modifications that alter the oral biofilm bacteria, even in adults, may be a factor in preventing the formation of dental cavities [44]. Consumption of yogurt (>4 times/week) strongly correlated with a decreased incidence of dental caries compared to intake <1 time/week in young children. Intake of milk, cheese, bread, and butter was not significantly associated with the incidence of dental cavities [45]. Subjects with caries show significantly more frequent genus *Haemophilus*, whereas the group without caries experience had *Capnocytophaga*, *Corynebacterium*, *Fusobacterium*, *Leptotrichia*, *Porphyromonas*, and *Prevotella* as substantially more prevalent [46]. These studies indicate the role of dietary components in managing oral health.

On the other hand, the salivary microbiomes of omnivores and vegans when investigated using 16S rRNA gene amplicon sequencing indicate that certain nutrients and long-term dietary habits influence the salivary microbiome. The oral microbiome of vegans was shown to be significantly different from that of omnivores, concerning the community structure, taxonomic composition, and genomic potential [47]. There's a never-ending struggle going on inside the oral cavity. The symbiotic link between microflora, dental biofilms, pH balance, saliva, temperature, humidity, and nutrition helps to prevent tooth decay, oral diseases, and other illnesses by fostering a healthy environment in the mouth cavity. This oral world is ever-changing, experiencing rapid shifts in composition and activity, influenced by factors like diet, host habits, and the interactions among bacteria. Within this dynamic, there is a constant population fluctuation [48]. The microorganisms in the oral cavity coexist in a symbiotic relationship. While commensal populations shield the mucosa from damage and prevent pathogenic species from sticking to it, bacteria that penetrate these barriers and cause illness and disease are considered pathogenic [49].

The microorganisms residing in the oral region can either promote oral health or contribute to dental issues like cavities, gum disease, and bad breath. Due to their impact on the oral microbiome, diets high in glucose or high glycemic load may be linked to poor oral health outcomes (and maybe systemic health consequences) in older women [50]. Some key food components that contribute to a healthy oral microbiome are foods rich in high fiber, antioxidants, probiotics, calcium and phosphorus, polyphenols, xylitol, omega-3 fatty acids, vitamin C, and others, while food having high amounts of carbohydrates, proteins, fats, and lipids and processed food lead to oral dysbiosis. Figure 4.2 shows the impact of dietary habits on the composition of the oral microbiome and oral health.

4.3.1 Carbohydrates

Carbohydrates, essential nutrients found in whole foods like fruits, vegetables, legumes, and whole grains, are often regarded as supportive of overall health due to their complex nature, rich fiber content, and low glycemic impact. For instance, a prebiotic xylooligosaccharides enriched diet decreased *Candida* virulence and bacterial dysbiosis, increasing lactobacilli and decreasing enterococci on the oral mucosa [51]. On the other hand, carbohydrates that are highly processed,

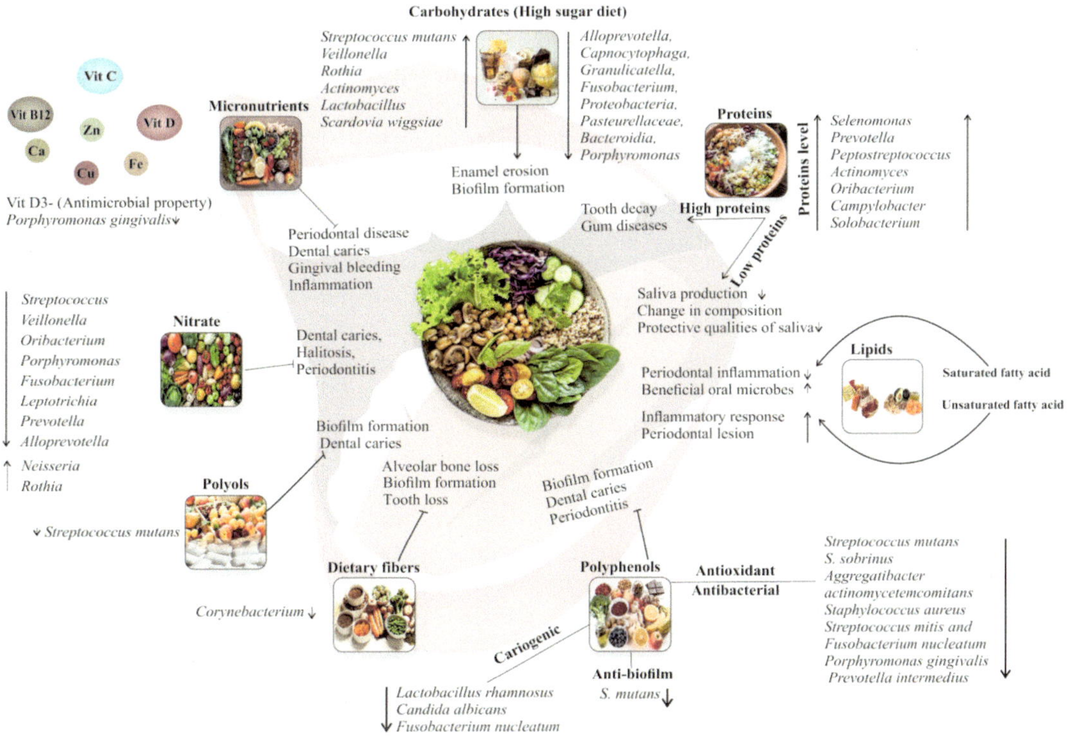

FIGURE 4.2 Effect of dietary habits on the oral microbiome and health.

fermentable, and often low in fiber, like refined sugars, white wheat flour, and sugary drinks, might have the potential to trigger chronic inflammation within the body [52]. Diets high in simple sugars can lower mouth pH more frequently and to more acidic levels compared to healthier diets. This shift promotes the growth of acid-tolerant, disease-causing *S. mutans*. With increased carbohydrate intake, *S. mutans* metabolizes sugars into enamel-eroding acids [53–55]. Increased consumption of fermentable carbohydrates, such as sucrose, fosters bacterial strains that produce acids, damaging tooth enamel and causing cavities. These bacterial strains possess virulence traits like acid tolerance and adhesion capabilities, hindering the oral immune response and saliva's buffering function, worsening the incidence of cavities [56–59]. The World Health Organization defines a high-sugar diet as consuming free sugars more than 5% of total daily energy intake, leading to caries [60].

High sugar intake not only affects acid production but also alters the oral microbiome's diversity. Moreover, alterations in bacterial populations in the oral cavity due to high sugar consumption are observed. *Streptococcus*, *Veillonella*, *Scardovia*, *Actinomyces*, *Rothia*, and *Lactobacillus* become abundant, while *Fusobacterium*, *Alloprevotella*, *Granulicatella*, *Capnocytophaga*, and *Porphyromonas* diminish in abundance [61–67]. *Streptococcus*, specifically *S. mutans*, prevails, negatively impacting the oral microbiota's diversity and reducing other beneficial Proteobacteria, Bacteroidia, *Porphyromonas*, and members of the Pasteurellaceae family [63, 64]. Additionally, bacteria like *Veillonella*, *Rothia*, *Lactobacillus*, *Scardovia wiggsiae*, and *Actinomyces* demonstrate acid resistance and fluoride tolerance, influencing their role in dental biofilms and saliva [68, 69].

4.3.2 Proteins

Insufficient protein intake significantly affects oral health by decreasing saliva production, altering its composition, and reducing its protective qualities. This deficiency might even lead to salivary gland atrophy and enamel issues, setting the stage for dental problems like cavities [70]. Higher protein intake seemed to correlate positively with various bacterial taxa like *Selenomonas*, *Johnsonella*, *Prevotella*, *Peptostreptococcus*, and *Actinomyces* [71]. The composition of certain bacterial groups such as *Actinomyces*, *Oribacterium*, *Campylobacter*, and *Solobacterium* on the tongue dorsum was directly associated with protein intake levels and helped in taste perception. There is a correlation between the presence of the *Clostridia* population and a diet high in protein. This suggests that altering protein intake could potentially modify these bacterial taxa, influencing the OM, and overall balance and potentially impacting oral health [39, 70]. In contrast, diets high in certain proteins could lead to an overgrowth of acid-producing bacteria, potentially causing dental issues like decay or gum disease. Proteins influence microbial interactions in the oral cavity. Some proteins act as adhesion sites for bacteria, facilitating their attachment to oral surfaces and potentially impacting biofilm formation, which is crucial in oral microbial communities [52]. Collectively, more research is required to determine the relationship between the nature of the OM and a balanced protein diet.

4.3.3 Lipids

OM composition can be influenced by dietary factors, and emerging evidence suggests that lipids play a significant role in shaping the microbial community. Diets rich in unsaturated fats, such as omega-3 fatty acids, have been associated with a more diverse and beneficial microbiome [72]. Studies have revealed that a diet high in omega-3 and low in processed carbohydrates reduces periodontal inflammation. Clinical trials have also reported that omega-3 supplementation, from various sources, decreases periodontal inflammation and pocket depth, benefiting periodontal therapy [52]. Subsequently, diets rich in saturated fatty acids (SFAs) produce an inflammatory response. Hyper-inflammation is a basic etiologic factor for periodontal disease [52, 73]. Moreover, there has been evidence of an association between periodontal disease and saturated fatty acids [74].

4.3.4 Polyphenols

Polyphenols are the bioactive compounds that are abundant in a healthy diet. Their health-beneficial effects are attributed to antibacterial, antioxidant, and anti-inflammatory activities. These properties play predominant roles in preventing periodontal diseases, halitosis, oral cancer, and dental caries [75]. *S. mutans* is a primary causative agent of dental caries, and polyphenols are reported to be effective against *S. mutans* and *S. sobrinus*. Likewise, other cariogenic microorganisms such as *Fusobacterium nucleatum*, *C. albicans*, and *Lactobacillus rhamnosus* are inhibited by polyphenols. In addition to their ability to prevent bacterial adherence to the tooth, polyphenols can also disrupt the development of biofilms [75]. Furthermore, polyphenols may prevent *S. mutans* from producing glucan and biofilm by inhibiting glycosyltransferase enzymes (GTF) [76–78]. Polyphenols from roasted green tea exhibited potent activity against *S. mutans* biofilm, which is mediated by GTF inhibitory action [78]. Furthermore, they possess broad-spectrum antimicrobial potential against *Streptococcus mitis*, *P. gingivalis*, *A. actinomycetemcomitans*, *F. nucleatum*, and *Staphylococcus aureus* [79–83]. Halitosis-related bacterial strains, *P. gingivalis*, *Prevotella intermedius*, and *F. nucleatum*, were also inhibited by polyphenols [84]. Polyphenols are essential for maintaining the proper balance between antioxidant activity and oxidative stress in the oral cavity, thereby halting the degradation of periodontal tissue.

4.3.5 Fibers

Fiber plays an essential role in sustaining a well-balanced OM. Diets abundant in fibrous fruits and vegetables promote a more varied OM, nurturing the proliferation of beneficial bacteria crucial for oral health. Functioning as a prebiotic, fiber fuels the expansion of advantageous microbes, contributing to the maintenance of a harmonious oral ecosystem. The act of chewing fibrous foods not only induces saliva production, aiding digestion, but also helps sustain a wholesome oral environment by rinsing away food particles and neutralizing acids [52]. Foods rich in dietary fiber can directly contribute to preventing oral health diseases by assisting in plaque removal from teeth, coupled with the stimulation of increased salivary flow through prolonged chewing. Studies have suggested that a higher intake of fiber, particularly from fruits, is linked to reduced alveolar bone loss and tooth loss [85]. Additionally, dietary fiber intake results in a decrease in *Corynebacterium* species, recognized as pivotal contributors to oral biofilm formation [39].

4.3.6 Nitrates

Consuming a diet rich in nitrates, commonly found in vegetables and fruits, can lead to major changes in the OM. This dietary nitrate is transformed into nitric oxide through the action of nitrate-reducing bacteria in the oral cavity. The low prevalence of dental caries is attributed to high salivary nitrite and nitrate levels (>1 mM) and the potent nitrate-reducing ability of oral commensal bacterial strains [86]. Nitric oxide can inhibit specific species and also alters the composition of oral biofilms [87]. Nitrate supplementation increases nitrate-reducing genera and results in oral eubiosis (dominance of beneficial species) [88]. After a nitrate-rich vegetable meal, salivary nitrate rises to 5–8 mM as salivary glands concentrate plasma nitrate. A nitrate-rich diet promotes the growth of nitrate-reducing *Neisseria* and *Rothia*, while inhibiting caries-associated *Veillonella, Streptococcus*, and *Oribacterium.* Further, this diet suppresses halitosis and periodontitis-associated *Leptotrichia*, *Porphyromonas*, *Alloprevotella, Fusobacterium*, and *Prevotella.* Substantial changes in the nature and functioning of the microbiome were observed when nitrate was added to oral communities; these changes may have positive effects on the host by promoting eubiosis or reducing dysbiosis [89].

4.3.7 Polyols

Polyols also impact the oral microbiome; in Estonia, the composition of school children's salivary microbiomes is specifically impacted by the daily intake of erythritol, xylitol, or control (sorbitol) candies. Erythritol has been linked to the lowest levels of clinical caries experience and the lowest frequency of caries-related mutans streptococci (MS) [90]. Subsequently, 173 mothers were divided into three groups (A–C) and instructed to chew gum containing sodium fluoride (C), chlorhexidine/xylitol (B), and xylitol (A). MS prevalence (MS colonization in mothers' 18-month-old infants) was 10%, 16%, and 28% in groups A, B, and C, respectively. Chewing gum containing xylitol and chlorhexidine/xylitol greatly decreased the salivary MS transmission from mother to child [91]. Furthermore, a meta-analysis revealed that regular xylitol consumption by mothers with high levels of MS significantly decreased mother-child transmission of salivary MS [92]. Conversely, a 3-year community intervention experiment conducted in Lithuania discovered that chewing gum without sugar had a caries-prevention effect that is related to the chewing process itself, not to gum sweeteners or additives like carbamide and polyols [93].

4.3.8 Micronutrients

Micronutrients also play a vital role in balancing the oral microbiome. The deficiency of micronutrients such as vitamins C, D, B, and A and minerals like calcium, iron, magnesium, iron, potassium, copper, manganese, selenium, and zinc are associated with oral diseases [94, 95]. Oral mucosa undergo a variety of degenerative alterations that have been linked to particular micronutrient deficits that lead to several types of problems such as periodontal disease, cheilitis, aphthous ulcers, stomatitis, glossitis, gingivitis, bleeding gums, and scurvy [96]. Increased levels of oxidative stress are typically a consequence of an imbalance between the virulence of the bacteria and the host's defense ability, which causes periodontal disease. Its progression may be linked to deficiencies in vitamins C, D, or B12 [97]. Subsequently, consuming fruits high in vitamin C, including grapefruits, peppers, kiwis, etc., can help reduce periodontal and gingival inflammation. Therefore, irrespective of maintaining oral hygiene, a reduction in vitamin C intake could trigger gingival bleeding [52]. Parthasarathy et al. [98] revealed the connection between dental caries and vitamin D3, pointing out that immune system cells have vitamin D3 receptors. These receptors bind to vitamin D3 and promote the production of antimicrobial proteins, which in turn prevent dental caries. Vitamin D3 receptors are present in the ameloblast and odontoblast cells, which produce dentin and enamel, respectively, and help reduce the incidence of dental decay [99]. Additionally, studies have also supported that deficiencies in vitamin D might result in the dysfunction of salivary glands [100]. Vitamin D could potentially inhibit *P. gingivalis* by promoting active autophagy and selectively suppressing the expression of critical virulence factors in these bacteria, such as adhesins (fimA, hagA, and hagB) and proteinases (rgpA, rgpB, and kgp) [10, 101]. Furthermore, it potentially decreases the inflammatory load associated with periodontitis by reducing inflammatory markers like RANKL, TNF-α, IL-1, and MMP-9 and inhibiting excessive IL-6 expression [101]. Compared to healthy individuals, patients with periodontal disease frequently have greater levels of vitamin D insufficiency (< 50 nmol/L) [102]. Similarly, 1,25(OH)2D3 could interfere with the pathogenesis of periodontitis by reducing *P. gingivalis*' virulence [103].

Minerals such as zinc, selenium, calcium, copper, and iron play key roles in maintaining periodontal health. Low dietary intake of calcium might contribute to the progression of periodontal diseases [104]. Combining calcium and vitamin D supplements can help with periodontal disease progression and management [105, 106]. A lack of zinc in the diet can potentially heighten the risk of gingival inflammation due to increased aggregation of IL-1 [107]. Iron deficiency commonly leads to reduced salivary secretion [108]. Compared to healthy persons, patients with periapical inflammation had lower serum copper levels [109].

The impact of dietary habits on the oral microbiome is profound and multifaceted. A balanced diet not only sustains overall health but also nurtures a diverse and harmonious oral microbial community, contributing significantly to oral health and potentially influencing systemic well-being. Exploring these mechanistic insights paves the way for targeted dietary interventions and personalized approaches to maintaining oral health and preventing associated diseases. Eating a balanced diet can help to create the ideal conditions for overall and oral health.

4.4 CONCLUSION AND FUTURE PERSPECTIVES

Diet is an important component in maintaining the balance of a healthy OM, which is essential for good health. However, studies on the influence of diet on the OM are meager. The aforementioned research makes it clear that nutrition has a significant role in the eubiosis and dysbiosis of OM because it fosters the growth and survival of beneficial microorganisms as well as specific periodontal pathogens. The main components influencing the OM and its dysbiosis-related indications include micronutrients, proteins, lipids, fibers, polyphenols, carbohydrates, nitrates, and polyols. Due to their anti-inflammatory and antioxidant capabilities, micronutrients are essential for the maintenance of the OM. Oral inflammatory illnesses can be triggered by deficiencies in specific micronutrients. Subsequently, the lowest frequency of caries-related mutans streptococci was observed with xylitol intake. Nitrate-rich diet supplementation inhibited caries-associated *Streptococcus*, *Veillonella*, and *Oribacterium*. It is becoming more evident how nutrition affects the composition and functions of oral microbes. It is now essential to conduct a thorough investigation of the stability and reproducibility of such dietary patterns. Furthermore, research on the oral microbiome is still in its early stages, in contrast to that on the gut microbiome. A complete understanding of the role that the OM plays in health and how different types of diet could influence this microbial diversity may provide a different approach to the prevention and treatment of diverse diseases. To provide useful suggestions for nutritional planning, substantial dietary studies are necessary.

CONFLICT OF INTEREST

None.

ACKNOWLEDGMENTS

The authors would like to express their gratitude to the revered Swami Ramdev for his inspiration. Further, the authors are thankful to the Ministry of AYUSH under Grant-in-Aid for the Establishment of the Centre of Excellence of Renovation and Upgradation of Patanjali Ayurveda Hospital, Haridwar. Sunil, a designer at Patanjali Research Institute, assisted the authors with the infographics.

REFERENCES

1. Deo, P.N.; Deshmukh, R. Oral microbiome: Unveiling the fundamentals. *J. Oral Maxillofac. Pathol.* **2019**, *23*(1), 122–128.
2. Krishnan, K.; Chen, T.; Paster, B.J. A practical guide to the oral microbiome and its relation to health and disease. *Oral Dis.* **2017**, *23*(3), 276–286.
3. Gao, L.; Xu, T.; Huang, G.; Jiang, S.; Gu, Y.; Chen, F. Oral microbiomes: more and more importance in oral cavity and whole body. *Protein Cell* **2018**, *9*(5), 488–500.
4. Human Oral Biome Database V3.1, **2017**–2022. Available online: http://www.homd.org.
5. Marsh, P.D. Percival, R.S.; The oral microflora-friend or foe? Can we decide?. *Int. Dent. J.* **2006**, *56*, 233–239.
6. Hardie, J.M. Oral microbiology: Current concepts in the microbiology of dental caries and periodontal disease. *Br. Dent. J.* **1992**, *172*(7), 271–278.
7. Kilian, M.; Chapple, I.L.C.; Hannig, M.; Marsh, P.D.; Meuric, V.; Pedersen, A.M.L.; Zaura, E. The oral microbiome–an update for oral healthcare professionals. *Br. Dent. J.* **2016**, *221*(10), 657–666.
8. Poulain, D. Candida, albicans: plasticity and pathogenesis. *Crit. Rev. Microbiol.* **2015**, *41*, 208–217.
9. Woelber, J.P.; Bremer, K.; Vach, K.; König, D.; Hellwig, E.; Ratka-Krüger, P.; Tennert, C.J.B.O.H. An oral health optimized diet can reduce gingival and periodontal inflammation in humans-a randomized controlled pilot study. *BMC oral health* **2017**, *17*(1), 1–8.
10. Lu, E.M.C. The role of vitamin D in periodontal health and disease. *J. Periodontal Res.* **2023**, *58*(2), 213–224.
11. Staufenbiel, I.; Weinspach, K.; Förster, G.; Geurtsen, W.; Günay, H. Periodontal conditions in vegetarians: A clinical study. *Eur. J. Clin. Nutr.* **2013**, *67*(8), 836–840.
12. Stein, C.; Cunha-Cruz, J.; Hugo, F.N. Is dietary pattern a mediator of the relationship between socioeconomic status and dental caries? *Clin. Oral Investig.* **2021**, *25*, 5441–5447.
13. Zong, G.; Holtfreter, B.; Scott, A.E.; Völzke, H.; Petersmann, A.; Dietrich, T.; Kocher, T. Serum vitamin B12 is inversely associated with periodontal progression and risk of tooth loss: A prospective cohort study. *J. Clin. Periodontol.* **2016**, *43*(1), 2–9.
14. Yamashita, Y.; Takeshita, T. The oral microbiome and human health. *J Oral Sci.* **2017**, *59*, 201–6.
15. Berg, G.; Rybakova, D.; Fischer, D.; Cernava, T.; Vergès, M.C.C.; Charles, T.; Schloter, M. Microbiome definition re-visited: old concepts and new challenges. *Microbiome* **2020**, *8*(1), 1–22.
16. Zhao, H.; Chu, M.; Huang, Z.; Yang, X.; Ran, S.; Hu, B.; Liang, J. Variations in oral microbiota associated with oral cancer. *Sci. Rep.* **2017**, *7*(1), 11773.
17. Willis, J.R.; Gabaldón, T. The human oral microbiome in health and disease: from sequences to ecosystems. *Microorganisms* **2020**, *8*(2), 308.
18. Ghannoum, M.A.; Jurevic, R.J.; Mukherjee, P.K.; Cui, F.; Sikaroodi, M.; Naqvi, A.; Gillevet, P. M. Characterization of the oral fungal microbiome (mycobiome) in healthy individuals. *PLoS Pathog.* **2010**, *6*, e1000713.
19. Peters, B.A.; Wu, J.; Hayes, R.B.; Ahn, J. The oral fungal mycobiome: characteristics and relation to periodontitis in a pilot study. *BMC Microbiol.* **2017**, *17*, 157. Doi: 10.1186/s12866-017-1064-9.
20. Lepp, P.W.; Brinig, M.M.; Ouverney, C.C.; Palm, K.; Armitage, G.C.; Relman, D.A. Methanogenic Archaea and human periodontal disease. *PNAS* **2004**, *101*(16), 6176–6181.
21. Bottalico, D.; Chen, Z.; Dunne, A.; Ostoloza, J.; McKinney, S.; Sun, C.; Schlecht, N.F.; Fatahzadeh, M.; Herrero, R.; Schiffman, M.; and Burk, R.D. The oral cavity contains abundant known and novel human papillomaviruses from the Betapapillomavirus and Gammapapillomavirus genera. *J. Infect. Dis.* **2011**, *204*(5), 787–792.

22. Wade, WG. The oral microbiome in health and disease. *Pharmacol. Res.* **2013**, *69*, 137–43.
23. Wylie, K.M.; Mihindukulasuriya, K.A.; Zhou, Y.; Sodergren, E.; Storch, G.A.; Weinstock, G.M. Metagenomic analysis of double-stranded DNA viruses in healthy adults. *BMC Biol.* **2014**, *12*, 71. Doi: 10.1186/s12915-014-0071-7.
24. Caselli, E.; Fabbri, C.; D'Accolti, M.; Soffritti, I.; Bassi, C.; Mazzacane, S.; Franchi, M. Defining the oral microbiome by whole-genome sequencing and resistome analysis: the complexity of the healthy picture. *BMC Microbiol.* **2020**, *20*(1), 1–19.
25. Zaura, E.; Keijser, B.; Huse, S.; Crielaard, W. Defining the healthy. *BMC Microbiol.* **2009**, *9*, 259. Doi: 10.1186/1471-2180-9-259.
26. Matarazzo, F., Ribeiro, A.C.; Feres, M.; Faveri, M.; Mayer, M.P.A. Diversity and quantitative analysis of Archaea in aggressive periodontitis and periodontally healthy subjects. *J. Clin. Periodontol.* **2011**, *38*(7), 621–627.
27. Lazar, V.; Ditu, L.M.; Curutiu, C.; Gheorghe, I.; Holban, A.; Popa, M.; Chifiriuc, C. Impact of dental plaque biofilms in periodontal disease: Management and future therapy. In: Arjunan, P., Ed, Periodontitis: A Useful Reference. InTech Open: London, UK, **2017**, 11–42.
28. Sharma, N.; Bhatia, S.; Sodhi, A.S.; Batra, N. Oral microbiome and health. *AIMS microbiology* **2018**, *4*(1), 42.
29. Scheiman, J.; Luber, J.M.; Chavkin, T.A.; MacDonald, T.; Tung, A.; Pham, L.D.; Kostic, A.D. Meta-omics analysis of elite athletes identifies a performance-enhancing microbe that functions via lactate metabolism. *Nat. Med.* **2019**, *25*(7), 1104–1109.
30. Zarco, M.F.; Vess, T.J.; Ginsburg, G.S. The oral microbiome in health and disease and the potential impact on personalized dental medicine. *Oral Dis.* **2012**, *18*(2), 109–120.
31. Daboor, S.M.; Masood, F.S.S.; Al-Azab, M.S.; Nori, E.E. A review on *Streptococcus mutans* with its diseases dental caries, dental plaque and endocarditis. *IJMR* **2015**, 2(2), 76–82.
32. Lemos, J.A.; Palmer, S.R.; Zeng, L.; Wen, Z.T.; Kajfasz, J.K.; Freires, I.A.; Brady, L.J. The biology of *Streptococcus mutans*. *Microbiol. Spectr.* **2019**, *7*(1), 10–1128.
33. Xu, W.; Zhou, W.; Wang, H.; and Liang, S. Roles of *Porphyromonas gingivalis* and its virulence factors in periodontitis. *Adv. Protein Chem. Struct. Biol.* **2020**, *120*, 45–84.
34. Kononen, E.; Kumar, P.S. Bacteriology of periodontal diseases. *Mol. Microbiol.* **2015**, 957–968. Academic Press.
35. Patel, M. Oral cavity and *Candida albicans*: Colonisation to the development of infection. *Pathogens* **2022**, *11*(3), 335.
36. Sudhakara, P.; Gupta, A.; Bhardwaj, A.; Wilson, A. Oral dysbiotic communities and their implications in systemic diseases. *J. Dent.* **2018**, *6*(2), 10.
37. Chen, B.; Zhao, Y.; Li, S.; Yang, L.; Wang, H.; Wang, T.; Bin, S.; Gai, Z.; Heng, X.; Zhang, C.; Yang, J.; Zhang, L. Variations in oral microbiome profiles in rheumatoid arthritis and osteoarthritis with potential biomarkers for arthritis screening. *Sci. Rep.* **2018**, *8*(1). Doi: 10.1038/s41598-018-35473-6.
38. Cobb, C.M.; Kelly, P.J.; Williams, K.B.; Babbar, S.; Angolkar, M.; Derman, R.J. The oral microbiome and adverse pregnancy outcomes. *Int. J. Women's Health* **2017**, *9*, 551–559
39. Sedghi, L.; DiMassa, V.; Harrington, A.; Lynch, S.V.; Kapila, Y.L. The oral microbiome: Role of key organisms and complex networks in oral health and disease. *Periodontology* **2021**, *87*(1), 107–131.
40. Butera, A.; Maiorani, C.; Morandini, A.; Simonini, M.; Morittu, S.; Trombini, J.; Scribante, A. Evaluation of children caries risk factors: A narrative review of nutritional aspects, oral hygiene habits, and bacterial alterations. *Children* **2022**, *9*(2), 262.
41. Oba, P.M.; Holscher, H.D.; Mathai, R.A.; Kim, J.; Swanson, K.S. Diet influences the oral microbiota of infants during the first six months of life. *Nutrients* **2020**, *12*(11), 3400
42. Sweeney, E.L.; Al-Shehri, S.S.; Cowley, D.M.; Liley, H.G.; Bansal, N.; Charles, B.G.; Knox, C.L. The effect of breastmilk and saliva combinations on the in vitro growth of oral pathogenic and commensal microorganisms. *Sci. Rep.* **2018**, *8*(1), 15112. Doi: 10.1038/s41598-018-33519-3.
43. Al-Shehri, S.S.; Sweeney, E.L.; Cowley, D.M.; Liley, H.G.; Ranasinghe, P.D.; Charles, B.G.; Knox, C.L. Deep sequencing of the 16S ribosomal RNA of the neonatal oral microbiome: a comparison of breast-fed and formula-fed infants. *Sci. Rep.* **2016**, *6*(1), 38309. Doi: 10.1038/srep38309
44. Anderson, A.C.; Rothballer, M.; Altenburger, M.J.; Woelber, J.P.; Karygianni, L.; Vach, K.; Al-Ahmad, A. Long-term fluctuation of oral biofilm microbiota following different dietary phases. *AEM* **2020**, *86*(20), e01421–20.
45. Tanaka, K.; Miyake, Y.; Sasaki, S. Intake of dairy products and the prevalence of dental caries in young children. *J. Dent.* **2010**, *38*(7), 579–583.

46. Schoilew, K.; Ueffing, H.; Dalpke, A.; Wolff, B.; Frese, C.; Wolff, D.; Boutin, S. Bacterial biofilm composition in healthy subjects with and without caries experience. *J. Oral Microbiol.* **2019**, *11*(1), 1633194. Doi: 10.1080/20002297.2019.1633194
47. Hansen, T.H.; Kern, T.; Bak, E.G.; Kashani, A.; Allin, K.H.; Nielsen, T.; Pedersen, O. Impact of a vegan diet on the human salivary microbiota. *Sci. Rep.* **2018**, *8*(1), 5847. 41598-018-24207-3.
48. McLean, J.S. Advancements toward a systems level understanding of the human oral microbiome. *Front Cell Infect Microbiol.* **2014**, *4*, 98.
49. Avila, M.; Ojcius, D.M.; Yilmaz, O. The oral microbiota: Living with a permanent guest. *DNA Cell Biol.* **2009**, *28*, 405 11.
50. Millen, A.E.; Dahhan, R.; Freudenheim, J.L.; Hovey, KM.; Li, L.; McSkimming, D.I.; Andrews, C.A.; Buck, M.J.; LaMonte, M.J.; Kirkwood, K.L.; Sun, Y.; Murugaiyan, V.; Tsompana, M.; Wactawski-Wende, J. Dietary carbohydrate intake is associated with the subgingival plaque oral microbiome abundance and diversity in a cohort of postmenopausal women. *Sci. Rep.* **2022**, *12*(1), 2643. Doi: 10.1038/s41598-022-06421-2.
51. Vazquez-Munoz, R.; Thompson, A.; Sobue, T.; Dongari-Bagtzoglou, A. A prebiotic diet modulates the oral microbiome composition and results in the attenuation of oropharyngeal candidiasis in mice. *Microbiol. Spectr.* **2023**, *11*(5), e0173423. Doi: 10.1128/spectrum.01734-23.
52. Santonocito, S.; Polizzi, A.; Palazzo, G.; Indelicato, F.; Isola, G. Dietary Factors Affecting the Prevalence and Impact of Periodontal Disease. *Clin. Cosmet. Investig. Dent.* **2021**, *13*, 283–292. Doi: 10.2147/CCIDE.S288137.
53. Gross, E.L.; Beall, C. J.; Kutsch, S.R.; Firestone, N.D.; Leys, E.J.; Griffen, A.L. Beyond *Streptococcus mutans*: Dental caries onset linked to multiple species by 16S rRNA community analysis. *PLoS One.* **2012**, *7*(10), e47722.
54. Bradshaw, D.J.; Lynch, R.J. Diet and the microbial aetiology of dental caries: new paradigms. *Int. Dent. J.* **2013**, *63*, 64–72.
55. Moye, Z.D.; Zeng, L.; Burne, R.A. Fueling the caries process: Carbohydrate metabolism and gene regulation by *Streptococcus mutans*. *J. Oral Microbiol.* **2014**, *6*, 24878
56. Marsh, P.D. Dental plaque as a biofilm and a microbial community—Implications for health and disease. *BMC Oral Health* **2006**, *6*(1), S14. Doi: 10.1186/1472-6831-6-S1-S14.
57. Dang, M.H.; Jung, J.E.; Choi, H.M.; Jeon, J.G. Difference in virulence and composition of a cariogenic biofilm according to substratum direction. *Sci. Rep.* **2018**, *8*(1), 6244. Doi: 10.1038/s41598-018-24626-2.
58. Rosier, BT.; Buetas, E.; Moya-Gonzalvez, EM.; Artacho, A.; Mira, A. Nitrate as a potential prebiotic for the oral microbiome. *Sci. Rep.* **2020**, *10*(1), 12895. Doi: 10.1038/s41598-020-69931-x.
59. Nyvad, B.; Takahashi, N. Integrated hypothesis of dental caries and periodontal diseases. *J. Oral Microbiol.* **2020**, *12*(1), 1710953. Doi: 10.1080/20002297.2019.1710953.
60. WHO (World Health Organization) (**2015**). Guideline: Sugars intake for adults and children. Retracted from https://www.who.int/publications/i/item/9789241549028.
61. Tian, J.; Qin, M.; Ma, W.; Xia, B.; Xu, H.; Zhang, Q.; Chen, F. Microbiome interaction with sugar plays an important role in relapse of childhood caries. *BBR* **2015**, *468*(1-2), 294–299.
62. Keller, M.K.; Kressirer, C.A.; Belstrøm, D.; Twetman, S.; Tanner, A.C.R. Oral microbial profiles of individuals with different levels of sugar intake. *J. Oral Microbiol.* **2017**, *9*(1), 1355207. Doi: 10.1080/20002297.2017.1355207
63. Minty, M.; Canceill, T.; Lê, S.; Dubois, P.; Amestoy, O.; Loubieres, P.; Christensen, J.E.; Champion, C.; Azalbert, V.; Grasset, E.; Hardy, S.; Loubes, J.M.; Mallet, J.P.; Tercé, F.; Vergnes, J.N.; Burcelin, R.; Serino, M.; Diemer, F.; Blasco-Baque, V. Oral health and microbiota status in professional rugby players: A case-control study. *J. Dent.* **2018**, *79*, 53–60.
64. Anderson, A.C.; Rothballer, M.; Altenburger, M.J.; Woelber, J.P.; Karygianni, L.; Lagkouvardos, I.; Hellwig, E.; Al-Ahmad, A. In-vivo shift of the microbiota in oral biofilm in response to frequent sucrose consumption. *Sci. Rep.* **2018**, *8*(1), 14202. Doi: 10.1038/s41598-018-32544-6.
65. Esberg, A.; Haworth, S.; Hasslöf, P.; Lif Holgerson, P.; Johansson, I.; Oral microbiota profile associates with sugar intake and taste preference genes. *Nutrients* **2020**, *12*(3), 681. Doi: 10.3390/nu12030681
66. Esberg, A.; Eriksson, L.; Hasslöf, P.; Haworth, S.; Holgerson, P.L.; Johansson, I. Using oral microbiota data to design a short sucrose intake index. *Nutrients* **2021**, *13*(5), 1400. Doi: 10.3390/nu13051400.
67. Chen, X.; Hu, X.; Fang, J.; Sun, X.; Zhu, F.; Sun, Y.; Wang, Y. Association of oral microbiota profile with sugar-sweetened beverages consumption in school-aged children. *Int. J. Food. Sci. Nutr.* **2021**, *73*(1), 1–11.

68. Abranches, J.; Zeng, L.; Kajfasz, J.K.; Palmer, S.R.; Chakraborty, B.; Wen, Z.T.; Richards, V.P.; Brady, L.J.; Lemos, J.A.; Biology of oral *Streptococci*. *Microbiol. Spectr.* **2018**, *6*(5), Doi: 10.1128/microbiolspec.GPP3-0042-2018.
69. Kameda, M.; Abiko, Y.; Washio, J.; Tanner, A.C. R.; Kressirer, C.A.; Mizoguchi, I.; Takahashi, N. Sugar metabolism of *Scardovia wiggsiae*, a novel caries-associated bacterium. *Front. Microbiol.* **2020**, *11*, 479. Doi: 10.3389/fmicb.2020.00479.
70. Fluitman, K.S.; van den Broek, T.; Reinders, I.; Wijnhoven, H.A.H.; Nieuwdorp, M.; Visser, M.; IJzerman, R.G.; Keijser, B.J.F. The Effect of Dietary Advice Aimed at Increasing Protein Intake on Oral Health and Oral Microbiota in Older Adults: A Randomized Controlled Trial. *Nutrients* **2023**, *15*(21), 4567. Doi: 10.3390/nu15214567.
71. Cattaneo, C.; Riso, P.; Laureati, M.; Gargari, G.; Pagliarini, E. Exploring Associations between Interindividual Differences in Taste Perception, Oral Microbiota Composition, and Reported Food Intake. *Nutrients* **2019**, 11, 1167. Doi: 10.3390/nu11051167.
72. Cato, L.E.; McKay, A.K.A.; L'Heureux, J.E.; Vanhatalo, A.; Jones, A.M.; Askew, C.D.; Slater, G.J.; Burke, L.M. Low Carbohydrate, High Fat Diet Alters the Oral Microbiome without Negating the Nitrite Response to Beetroot Juice Supplementation. *Nutrients* **2023**, *15*(24), 5123. Doi: 10.3390/nu15245123.
73. Iwasaki, M.; Manz, M.; Moynihan, P.; Yoshihara, A.; Muramatsu, K.; Watanabe, R.; Miyazaki, H. Relationship between saturated fatty acids and periodontal disease. *J. Dent. Res.* **2011**, *90*, 861–867.
74. Serhan, C.N.; Chiang, N.; Dalli, J. The resolution code of acute inflammation: Novel pro-resolving lipid mediators in resolution. *In Seminars in Immunology* **2015**, 27(3), 200–215.
75. Guo, Y.; Li, Z.; Chen, F.; Chai, Y. Polyphenols in Oral Health: Homeostasis Maintenance, Disease Prevention, and Therapeutic Applications. *Nutrients* **2023**, *15*(20), 4384.
76. Goyal, D.; Sharma, S.; Mahmood, A. Inhibition of dextransucrase activity in *Streptococcus mutans* by plant phenolics. *Indian J. Biochem. Biophys.* **2013**, *50*, 48–53.
77. Goyal, D.; Ahmad, S.; Mahmood, A.; Chander Sharma, S. Interactions of dextransucrase purified from *Streptococcus mutans* 890 with plant polyphenols. *Biochem. Biophys. Rep.* **2021**, *26*, 100980
78. Goto, I.; Saga, S.; Ichitani, M.; Kimijima, M.; Narisawa, N. Investigation of components in roasted green tea that inhibit Streptococcus mutans biofilm formation. *Foods* **2023**, *12*, 2502.
79. Shahzad, M.; Millhouse, E.; Culshaw, S.; Edwards, C.A.; Ramage, G.; Combet, E. Selected dietary (poly)phenols inhibit periodontal pathogen growth and biofilm formation. *Food Funct.* **2015**, *6*, 719–729.
80. Jekabsone, A.; Sile, I.; Cochis, A.; Makrecka-Kuka, M.; Laucaityte, G.; Makarova, E.; Rimondini, L.; Bernotiene, R.; Raudone, L.; Vedlugaite, E. Investigation of antibacterial and anti-inflammatory activities of proanthocyanidins from *Pelargonium sidoides* DC root extract. *Nutrients* **2019**, *11*, 2829.
81. Khalil, M.A.; El-Sabbagh, M.S.; El Naggar, E.B.; El-Erian, R.H. Antibacterial activity of *Salvadora persica* against oral pathogenic bacterial isolates. *Niger. J. Clin. Pract.* **2019**, *22*, 1378–1387.
82. Ben Lagha, A.; Howell, A.; Grenier, D. Highbush blueberry proanthocyanidins alleviate *Porphyromonas gingivalis*-induced deleterious effects on oral mucosal cells. *Anaerobe* **2020**, *65*, 102266.
83. Ullah, H.; Minno, A.D.; Filippis, A.; Sommella, E.; Buccato, D.G.; Lellis, L.F.; El-Seedi, H.R.; Khalifa, S.A.M.; Piccinocchi, R.; Galdiero, M.; Campiglia, P. In vitro antimicrobial and antibiofilm properties and bioaccessibility after oral digestion of chemically characterized extracts obtained from *Cistus* × *incanus* L., *Scutellaria lateriflora* L., and their combination. *Foods* **2023**, *12*, 1826.
84. Liu, T.; Shen, H.; Wang, F.; Zhou, X.; Zhao, P.; Yang, Y.; Guo, Y. Thinned-young apple polyphenols inhibit halitosis-related bacteria through damage to the cell membrane. *Front. Microbiol.* **2021**, *12*, 745100.
85. Yacoub, A. Karmally, W. Nutrition in oral health. Nutrition in Lifestyle Medicine, Academic Press. **2017**, 193–209.
86. Doel, J.J.; Hector, M.P.; Amirtham, C.V.; Al-Anzan, L.A.; Benjamin N.; Allaker, R.P. Protective Effect of Salivary Nitrate and Microbial Nitrate Reductase Activity Against Caries. *Eur. J. Oral. Sci.* **2004**, *112*(5), 424–428.
87. Schairer, D.O.; Chouake, J.S.; Nosanchuk, J.D.; Friedman, A.J. The potential of nitric oxide releasing therapies as antimicrobial agents. *Virulence* **2012**, *3*(3), 271–9. Doi: 10.4161/viru.20328.
88. González-Soltero, R.; Bailén, M.; de Lucas, B.; Ramírez-Goercke, MI.; Pareja-Galeano, H.; Larrosa, M. Role of Oral and Gut Microbiota in Dietary Nitrate Metabolism and its Impact on Sports Performance. *Nutrients* **2020**, *12*(12), 3611. Doi: 10.3390/nu12123611.
89. Rosier, B.T.; Buetas, E.; Moya-Gonzalvez, E.M.; Artacho, A.; Mira, A. Nitrate as a potential prebiotic for the oral microbiome. *Sci. Rep.* **2020**, *10*(1), 12895.
90. Štšepetova, J.; Truu, J.; Runnel, R.; Nõmmela, R.; Saag, M.; Olak, J.; Mändar, R. Impact of polyols on Oral microbiome of Estonian schoolchildren. *BMC Oral Health* **2019**, *19*(1), 1–10.

91. Thorild, I.; Lindau, B.; Twetman, S. Effect of maternal use of chewing gums containing xylitol, chlorhexidine or fluoride on mutans streptococci colonization in the mothers' infant children. *OHPD* **2003**, *1*(1).
92. Lin, H.K.; Fang, C.E.; Huang, M.S.; Cheng, H.C.; Huang, T.W.; Chang, H.T.; Tam, K.W. Effect of maternal use of chewing gums containing xylitol on transmission of mutans streptococci in children: A meta-analysis of randomized controlled trials. *Int. J. Paediatr. Dent.* **2016**, 26(1), 35–44.
93. Machiulskiene, V.; Nyvad, B.; Baelum, V. Caries preventive effect of sugar-substituted chewing gum. *Community Dent. Oral Epidemiol.* **2001**, *29*(4), 278–288.
94. Van der Velden, U.; Kuzmanova, D.; Chapple, I.L.C. Micronutritional approaches to periodontal therapy. *J. Clin. Periodontol.* **2011**, *38*, 142–158.
95. Dommisch, H.; Kuzmanova, D.; Jönsson, D.; Grant, M.; and Chapple, I. Effect of micronutrient malnutrition on periodontal disease and periodontal therapy. *Periodontol.* **2018**, *78*, 129–153.
96. Cagetti, MG.; Wolf, TG.; Tennert, C.; Camoni, N.; Lingström, P.; Campus, G. The Role of Vitamins in Oral Health. A Systematic Review and Meta-Analysis. *Int. J Environ. Res. Public Health* **2020**, *17*(3), 938. Doi: 10.3390/ijerph17030938.
97. Chapple, I.L.; Bouchard, P.; Cagetti, M.G.; Campus, G.; Carra, M.C.; Cocco, F.; Nibali, L.; Hujoel, P.; Laine, M.L.; Lingström, P.; and Manton, D.J. Interaction of lifestyle, behaviour or systemic diseases with dental caries and periodontal diseases: consensus report of group 2 of the joint EFP/ORCA workshop on the boundaries between caries and periodontal diseases. *J. Clin. Periodontol.* **2017**, *44*, S39–S51.
98. Parthasarathy, P.; Priya, V.; Gayathri, R. Relationship between vitamin D and dental caries- review. *J. Pharm. Sci. Res.* **2016**, *8*, 459–460.
99. Nanci A. Ten Cate's oral histology: development, structure, and function. 9th ed., Elsevier Inc., USA: Elsevier, **2016**.
100. Glijer, B.; Peterfy, C.; Tenenhouse, A. The effect of vitamin D deficiency on secretion of saliva by rat parotid gland in vivo. *J. Physiol.* **1985**, *363*(1), 323–334.
101. Botelho, J.; Machado, V.; Proença, L.; Delgado, A.S.; Mendes, J.J. Vitamin D Deficiency and Oral Health: A Comprehensive Review. *Nutrients* **2020**, *12*(5), 1471. Doi: 10.3390/nu12051471.
102. Laky, M.; Bertl, K.; Haririan, H.; Andrukhov, O.; Seemann, R.; Volf, I.; Rausch-Fan, X. Serum levels of 25-hydroxyvitamin D are associated with periodontal disease. *Clin. Oral Investig.* **2017**, *21*, 1553–1558.
103. Grenier, D.; Morin, MP.; Fournier-Larente, J.; Chen, H. Vitamin D inhibits the growth of and virulence factor gene expression by *Porphyromonas gingivalis* and blocks activation of the nuclear factor kappa B transcriptionn factor in monocytes. *J. Periodontal Res.* **2016**, *51*, 359–365.
104. Nishida, M. Calcium and the risk for periodontal disease. *Clinical Calcium*, **2003**, *13*(5), 618–622.
105. Garcia, M.N.; Hildebolt, C.F.; Miley, D.D.; Dixon, D.A.; Couture, R.A.; Anderson Spearie, C.L.; Civitelli, R. One-year effects of vitamin D and calcium supplementation on chronic periodontitis. *J. Periodontol.* **2011**, *82*(1), 25–32.
106. Perayil, J.; Menon, K.S.; Kurup, S.; Thomas, A.E.; Fenol, A.; Vyloppillil, R.; Megha, S. Influence of vitamin D & calcium supplementation in the management of periodontitis. *J. Clin. Diagnostic Res.* **2015**, *9*(6), 35.
107. Poleník, P. Zinc in etiology of periodontal disease, *Med. Hypotheses.* **1993**, *40*(3), 182–185.
108. Johansson, I.; Fagernas, C. Effect of iron-deficiency anaemia on saliva secretion rate and composition in the rat. *Arch. Oral Biol.* **1994**, *39*(1), 51–56.
109. Kumari, S. Serum copper levels in periapical inflammation patients compare with healthy individuals. *Sch. Acad. J. Biosci.* **2015**, *3*(2B), 178–181.

5 Vegan Diet and Influence on Oral Microflora

Kirti Garg, Divya Sharma, and Reena Gupta
Faculty of Life Sciences, Department of Biotechnology,
Himachal Pradesh University, Shimla, Himachal Pradesh, India

5.1 INTRODUCTION

A nutrient-enriched, plant-based, or vegetarian diet aims to minimize consumption of processed foods, fats, and animal-based products (including dairy and eggs) while maximizing the consumption of plant-based food items. It encourages a diet limited in fat by eating lots of vegetables, fruits, beans, peas, pulses, lentils, legumes, seeds, and nuts [1]. Prominent authorities in the domain hold varying opinions regarding the components of the perfect vegetarian diet. A team of scientists recommended consuming restricted amounts of animal products, such as skim milk and egg whites, to treat various ailments. Nonetheless, a team of cardiologists advised against consuming any dietary item derived from animals, as well as almonds and soybeans [2]. The terms "vegan" and "plant-based" are occasionally used interchangeably. Diets that are strictly vegan due to moral or religious convictions may or may not be advantageous. Instead of assuming that a patient's diet is healthy, it is crucial to understand the precise meaning of associated diets and to find out the specifics of the patient's diet. There are various kinds of plant-based diets (Figure 5.1) [3]:

- *Vegan diet*: Sometimes known as a total vegetarian diet, abstains from all animal-based items, including dairy, eggs, meat, fish, and fowl. It doesn't limit fat or added sugar intake or demand the intake of whole foods.
- *Lacto-vegetarian diet*: This diet consists of dairy products and excludes meat, eggs, fish, and poultry.
- *Ovo-vegetarian diet*: This diet includes the consumption of eggs but abstains from milk products, meat, fish, and poultry.
- *Lacto-ovo-vegetarian diet*: This diet includes eating both eggs and milk products but abstains from meat, fish, and fowl.
- *Mediterranean diet*: A plant-based, whole foods diet that permits small quantities of milk products, eggs, chicken, and red meat one or two times a month. Olive oil and fish oil are recommended. [2].

Microorganisms, including bacteria, fungi, viruses, protozoa, and archaea are extensively present in the mouth of human beings. The oral bacterial communities have been demonstrated to be the most complex in the human body, followed by the colon communities [4]. Using advanced DNA sequencing methods alongside non-culture-based techniques to analyze the composition of the oral microbiome enables a much more comprehensive investigation than was previously possible. However, due to the genetic variability of bacterial species and the significant influence of surroundings on phenotype, species-level identification might not be sufficient [5].

Under a microscope, Antonie van Leeuwenhoek made the initial discovery and observation of bacteria in dental plaque. More than 250 species are thought to be present in the oral microbiome of humans at this time [6]. Among these are a number of important pathogens, including *Aggregatibacter actinomycetemcomitans*, *Tannerella forsythia*, *Porphyromonas gingivalis*, and *Streptococcus mutans*, which have been linked to the development of periodontal disease [4]. The imbalance of the commensal

 DOI: 10.1201/9781003478638-5

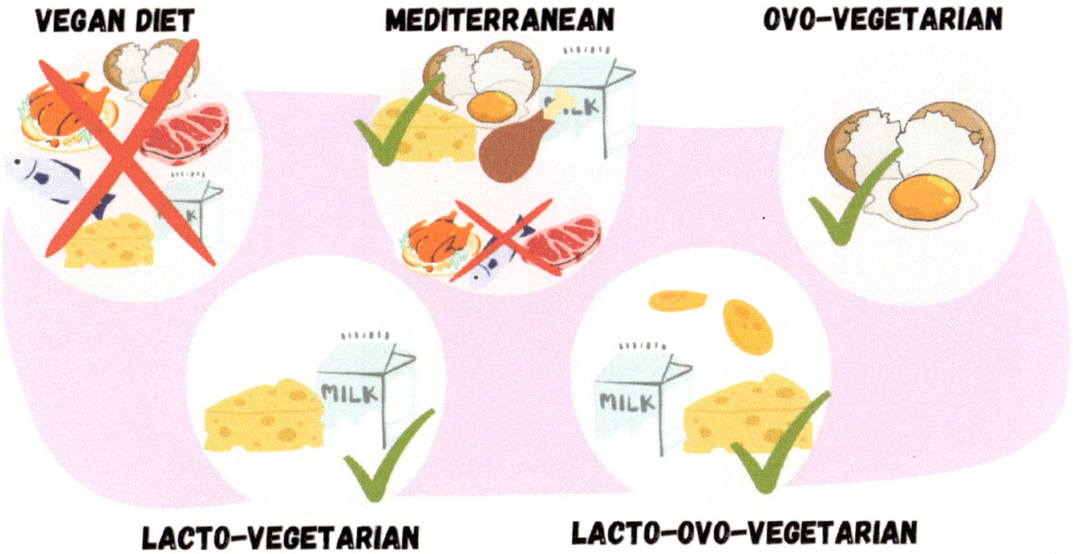

FIGURE 5.1 Types of plant-based diets.

oral microbiota promotes the onset and advancement of diseases and may impact the host's immune system [7]. Periodontal health is significantly affected by a number of nutrients [8]. Research suggests that a person's initial inflammatory state can be affected by the modulation of pro- and anti-inflammatory cascades by macro- and micronutrients [9]. The human oral microbiota receives nutrition from diet, which also acts as a selective pressure that favors the survival and growth of organisms, making them most suited to exploit particular food resources obtained from the host [10]. Over the past few decades, there has been a noticeable shift towards Western eating habits that are characterized by staple foods like processed grains, high-sugary dairy products, farmed animal meats, and refined vegetable oils. This shift has been linked to detrimental changes in oral microbiota, for example, an enhanced representation of organisms that produce acid and those that are acid tolerant [11]. This chapter aims at deriving a relationship between the vegan diet (or plant-based diet) and oral microbiota and potentially highlighting its implications for oral health and the overall well-being of humans. This majorly involves having knowledge of the microbiota in the oral cavity and the nutritional aspects of the vegan diet that can affect the oral microbiota. This study investigates how adherence to the vegan diet can affect the long-term composition of the human oral microbiota and comparatively analyzes the impact of other dietary patterns.

5.2 VEGAN NUTRITION

Plants are rich in phytochemicals, such as lycopene in tomatoes; polyphenols in grapes and nuts; indole-3-carbinol in vegetables like sprouts, cabbage, and cauliflower; and isoflavones in legumes. These compounds often lack any additional nutritional value, but they might influence the body's metabolic pathways, which can have a number of positive effects on health [12, 13]. Moreover, a vegan diet may result in calorie restriction and nutritional deficiencies of proteins, fats, vitamins, and minerals if it is not properly managed [14].

5.2.1 Macronutrients

Indigestible carbohydrates, commonly referred to as "dietary fiber," include non-starchy polysaccharides, resistant starch, indigestible oligosaccharides, and lignin [15]. Due to their resistance to digestion in the small intestine, they make their way into the large intestine, where after being fermented, they create bacterial metabolites like short-chain fatty acids (SCFAs) [16]. Dietary fiber appears to promote human health in a number of areas, including immunity, intestinal health, body weight management, cardiovascular disease, and prevention of colorectal cancer [16]. Prebiotics, specifically oligosaccharides derived from natural sources (like human milk) or synthetic sources (like fructo- and galacto-oligosaccharides), polyphenols and their derivatives, thiosulphates, and carotenoids, have been shown to have a number of positive impacts [17]. Some of these impacts are elevation in the number of bifidobacteria and lactobacilli, increased calcium absorption, reductions in protein fermentation, enhancement in gut immunity, generation of advantageous metabolites, and modifications in gut barrier permeability [18].

Omega-3 (n-3) polyunsaturated fatty acids are derived from vegetable oils, walnuts, cereals, chia seeds, linseeds, rapeseeds, camelina, hemp, and canola [19, 20]. Additional great suppliers of essential fatty acids include meals supplemented with docosahexaenoic acid (DHA) and microalgae supplements [21]. Vegetable proteins that have been concentrated or refined, such as gluten and soy, have high digestibility, whereas most vegetable products have low digestibility [22, 23]. Generally speaking, plant foods may supply essential amino acids needed for human nutrition if they are eaten in the right proportions. However, some amino acids may not be present in specific plants, such as methionine in legumes and lysine in cereals, rice, and maize [23]. Legumes are a source of protein that vegans typically consume in appropriate quantities. Research has suggested that legumes may protect against colon, prostate, and stomach cancer [14].

5.2.2 Micronutrients

For the primary metabolic processes, vitamins are typically needed in small amounts [24]. A vegan diet obviously provides vitamins from plant sources, such as carotenoids and vitamin C (*L*-ascorbic acid). Carotenoids, which include provitamin A and β-carotene, are precursors of vitamin A and are abundant in carrots. Liposoluble vitamin E is present in substantial concentrations in polyunsaturated vegetable oils. Many plant-based diets also include selenium (Se), a trace element that is crucial for the synthesis of glutathione peroxidase [25, 26]. Vitamin B12, a water-soluble vitamin, is primarily present in animal products. It is an essential vitamin for the nervous system and hematopoiesis, whose severe deficiency can be caused by inadequate nutrition or changes in absorption [27, 28]. This can lead to a number of medical conditions, including stroke, megaloblastic anemia, Parkinson's disease, vascular dementia, Alzheimer's disease, and cognitive impairment [29]. The following foods are good sources of vitamin B12: asparagus, broccoli, sprouts, some varieties of mushrooms, tempeh, tea leaves, algae (*Enteromorpha* spp.), microalgae, and cyanobacteria [30].

Bone health is greatly dependent on vitamin D, which is linked to both bone mineralization and calcium absorption. Because sufficient sun exposure is the primary factor determining its levels, supplementation may not be required, particularly for those who reside in low-latitude areas [18, 31]. While vitamin D2, also called ergocalciferol, is created by the action of ultraviolet (UV) radiation, vitamin D3, called cholecalciferol, can come from either plants or animals. UV-treated mushrooms can provide a significant amount of vitamin D [32]. Cereals and non-dairy milk alternatives such as rice, almond, and oat beverages are alternative sources of vitamin D [33]. There are many calcium-containing plant sources available; nevertheless, the bioavailability of calcium is negatively correlated with oxalate levels and, to a lesser degree, with phytate and fiber in vegetables [22]. Many green-leafy vegetables, tahini, tofu, and fortified products like cereals, rice, soy, almond drinks, coconut drinks, unsweetened cranberries, and low-sodium tomatoes [34] are examples of high-calcium foods. Every day, vegans can eat just as much iron as non-vegans. But compared to non-vegans, vegans have lower blood levels of iron and ferritin. Iron-rich foods include beans, legumes, whole grains, cereals, seeds, fruits, nuts, and green vegetables [35–37]. Anemia due to iron deficiency in vegans can be caused by low plasma zinc levels. Wholemeal bread, peas, corn, almonds, carrots, whole grains, soybeans, radish, cabbage, and legumes are examples of plant foods

TABLE 5.1
Association between Macronutrients, Micronutrients, and Periodontal Health

Nutrients	Plant-Based Dietary Sources	Importance in Oral and Periodontal Health	Reference
		Macronutrients	
Proteins	Pulses, cereals, some vegetables, legumes	• Protein enriched, neutral to alkaline environment during the progression of periodontal disease encourages the proliferation of periodontal pathogens.	[38, 39]
Fats (lipids)	• Saturated fatty acids (like butter, cheese, palm oil). • Monounsaturated fatty acids (like olive oil). • Polyunsaturated fatty acids (grape seed oil, sunflower oil, soya oil, nuts).	• A correlation that is positive between the development of periodontal lesions and saturated fatty acids. • Omega-3 fatty acids have been found in several trials to offer substantial advantages in periodontal therapy.	[40, 41]
Carbohydrates	• Lesser quantities (fruits, vegetables, whole grain, legumes). • Higher quantities (refined sugar, sugar beverages, white wheat flour).	• Carb-enriched food may lower the risk of periodontitis and gingivitis. • Consumption of high-glycemic meals may increase gingival and periodontal inflammation and bleeding. • Consuming a lot of processed carbs increases the chance of developing dental caries.	[42–44]
		Micronutrients	
Vitamin A	Carrots, capsicum, sweet potatoes, broccoli, green leafy vegetables	• Insignificant improvement in periodontal health.	[45]
B vitamins	*B1*: Oats, potatoes *B2*: Bananas, dairy, green beans *B3*: Mushrooms, nuts *B5*: Avocados, broccoli *B6*: Vegetables, nuts, bananas *B7*: Leafy vegetables, peanuts *B9*: Cereals, leafy vegetables *B12*: Fortified products	• Improve postsurgical healing by supplementation.	[46]
Vitamin C	Citrus fruits, peppers, grapefruits, vegetables, kiwis	• Deficiency can cause scurvy (gingival bleeding and inflammation). • Improved results of periodontal therapy.	[44, 47]
Vitamin D	Mushrooms, milk	• Delayed postsurgical healing due to deficiencies. • Reduced alveolar bone loss. • Accelerate osseointegration.	[48, 49]
Vitamin E	Green vegetables, vegetable oils, cereals, nuts	• Gingival bleeding due to deficiencies. • Reduce periodontal inflammation. • Hinder periodontitis progression.	[45, 50]
Vitamin K	Green vegetables, kale, cabbage, spinach, mustard	• Gingival bleeding due to deficiency. • Reduce periodontal inflammation.	[51, 52]

high in zinc. The impacts of both macronutrients and micronutrients on periodontal health have been well studied (Table 5.1).

5.3 MODELS OF VEGAN NUTRITION

Plant-based nutrition entered the formal scientific domain in the latter half of the twentieth century. According to ecological statistics, people who lived in African, Asian, and Mediterranean regions generally had longer lifespans and lower incidence of non-communicable diseases linked to nutrition [53, 54]. Plant foods make up a substantial portion of classical Asian and Mediterranean diets, which suggests that eating a plant-based diet has health benefits beyond just providing enough nutrients. This signaled the beginning of what is known as a paradigm shift, the substitution of a world view concept by others. Joan Sabaté from Loma Linda University was the first to document this paradigm shift towards vegetarian meals [55, 56]. Sabaté provided three graphical models to describe the phases of this paradigm change. The models show the anticipated health hazards and advantages of populations following a vegetarian or meat-based diet [57].

Model 1

According to the first model, there was a belief in the 1960s and even into the 1970s that people who followed a vegetarian diet were more likely to experience vitamin deficiencies than people who followed a meat-based diet. It is significant to highlight that the model did not account for the possible advantages of a vegan diet, such as preventing the development of degenerative and chronic disorders (Figure 5.2A). Surprisingly, this evaluation did not take into account the reality that individuals in certain Asian nations have been eating a vegan diet for centuries without experiencing vitamin deficiencies. This evaluation was predicated on the age-old finding that diets primarily consisted of plant foods in developing nations with high rates of malnutrition. However, poverty is mostly to blame for this kind of malnutrition since it causes an imbalanced diet [58].

Model 2

Various nutritional epidemiology studies conducted in the 1980s and 1990s provided evidence of the advantages of plant-based diets, including vegetarianism [59, 60]. The greater quantity and diversity of plant-based foods and their different constituents, as well as the absence of meat, have been specifically linked to decreased risks of several chronic and degenerative diseases (such as obesity, diabetes, heart disease, and a few cancers), overall mortality, and an increased lifespan. Plant-based diets were consistently found to have protective effects, while eating more meat was associated with harmful impacts. It seemed that eating plant-based foods could prevent disease more effectively than eating meat. The second model, which captures the consensus view of the final decades of the twentieth century, shows how this knowledge did not fully persuade the larger scientific community. A new feature of the model in Figure 5.2B was a slight rise in disease risk linked to a vegetarian diet, which was caused by obesity in inactive vegetarian individuals who consumed high amounts of ready-to-eat food items, which are typically high in monosaccharides, low-quality fat, and/or salt [55]. When a population consumes a diet high in meat, the risk of disease increases significantly due to the rise in adult and pediatric obesity, especially in those with wealthy lifestyles [55, 61].

Model 3

A new paradigm about vegetarian meals has arisen, based on progressively strong scientific facts (Figure 5.2C). The foundation of model 3 was the growing body of clinical, epidemiologic, and fundamental scientific evidence demonstrating the health benefits of a vegan diet [62, 63], especially those linked to the growing awareness of the diverse advantages of phytochemicals as antioxidants [64, 65]. Phytochemicals, as the name suggests, are only produced by plants and are found in trace levels in animal products. However, the complete role of phytochemicals in metabolism is not known yet. Many of the functions of plants are made possible by the synthesis of phytochemicals, which human bodies use for a variety of purposes, according to findings from fundamental science, clinical science, tissue-culture studies, and other sources [66]. These bioactive plant chemicals are partially responsible for the decrease in the incidence of cardiovascular disorders and some types of cancer in people who follow a vegan diet [67–70].

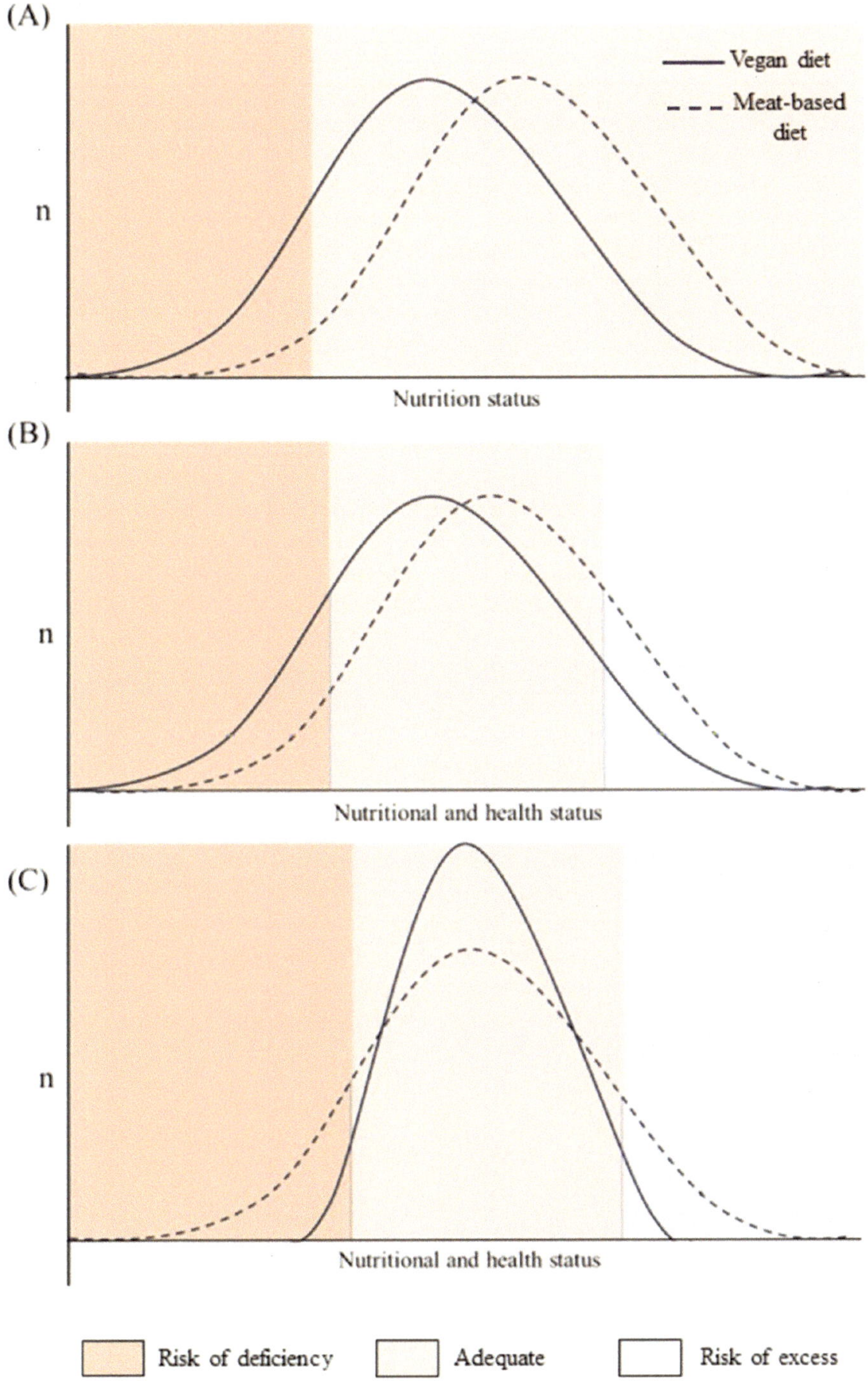

FIGURE 5.2 Models of vegan nutrition (where n = number of individuals in a population following a diet): (A) Early model on the suitability of vegetarian diets, which was popular until the 1960s. (B) Population health model based on studies in the 1980s and 1990s focused on the advantages and dangers of vegetarian and meat-based diets. (C) Optimal model of nutrition that is currently in use.

5.4 HEALTH BENEFITS OF A VEGAN DIET

Vegan diets or plant-based diets are typically lower in calories, cholesterol, saturated fat, omega-3 (n-3) fatty acids, Ca, Zn, vitamin D, and vitamin B12 and are richer in dietary fiber, folic acid, Mg, Fe, vitamin C, vitamin E, and phytochemicals. Vegetarians generally have a lower risk of type 2 diabetes, obesity, cardiovascular disease (CVD), cancer, and other lifestyle-related disorders [71]. A vegan diet seems to be beneficial for minimizing the consumption of dietary components linked to numerous chronic diseases and for enhancing the intake of phytochemicals and preventive elements [72]. In accordance with the Food and Agriculture Organization and World Health Organization evidence criteria, a high fruit and vegetable intake was associated with a reduced risk of cancer and CVD [73].

Case study: A 63-year-old man with a history of high blood pressure and cholesterol was diagnosed with type 2 diabetes. He was already taking atorvastatin, amlodipine, hydrochlorothiazide, and lisinopril. His doctor then advised him to follow a low-sodium vegetarian diet devoid of refined carbohydrates and animal products, as well as take glipizide, metformin, protamine, and insulin. It was advised that he must consume legumes, veggies, and beans every day, in addition to a small number of seeds and nuts. Biometric results (such as diabetes, lipid profile, and blood pressure) showed a considerable improvement after the man adhered to the rigorous diet and exercised every day. He was entirely weaned off of insulin, glipizide, amlodipine, and hydrochlorothiazide. His lisinopril dosage was lowered, and metformin by itself was sufficient to manage his diabetes. Numerous studies support the advantages of a plant-based diet [3] (Figure 5.3).

5.4.1 Obesity

A vegan or plant-based diet is quite successful for weight loss, according to a *Nutrition Reviews* study by authors Berkow and Barnard, who reviewed data from 87 reports [74]. According to their assessment, vegans lose weight at the rate of about one pound per week and are not dependent on exercise. The authors said that compared to meat-based diets, which may result in fewer calories burned since food is mostly stored as fat, a vegan diet increased the number of calories burned after meals [74]. According to a study conducted by a group of scientists, plant-based diets may be more nutrient dense and beneficial for controlling weight than diets containing meat. They found that vegetarians were slenderer than their peers who were eating meat in their study. It was also shown that vegetarians consumed less total fat and more Mg, K, Fe, riboflavin, folate, and thiamine [75].

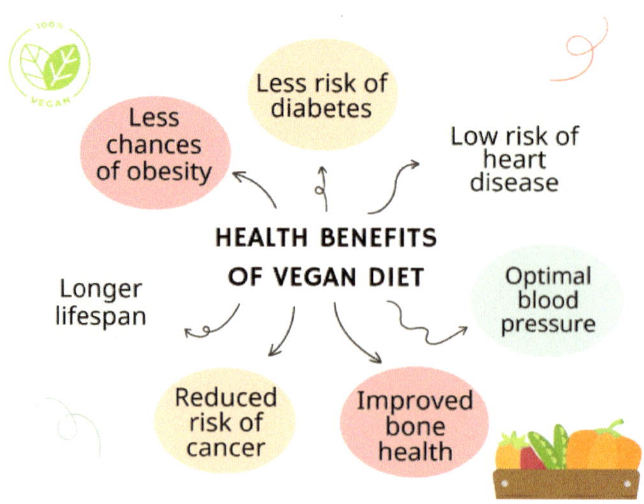

FIGURE 5.3 Impact of a vegan diet on disease prevention and human health.

In 2009, some researchers examined data from the National Health and Nutrition Examination Survey, conducted between 1999 and 2004, to examine the relationships between obesity and a meat-based diet. They demonstrated a favorable correlation between eating meat and obesity through the use of logistic and linear regression analysis [76]. Another study stated that a vegan diet is linked with a lower body mass index (BMI) and a reduced incidence of obesity in children as well as adults [77]. A 2-point lesser BMI was the outcome of a decreased weight difference of 3.3 kg in females and 7.6 kg in males, according to studies on adult vegan diets. Vegan children are also thinner than non-vegan youngsters, and this difference in BMI increases as the child reaches puberty. Research examining food groupings, dietary patterns, and the risk of overweight suggests that a vegan diet would be a reasonable strategy to avoid childhood obesity. Diets based primarily on plants have low energy density and high levels of fiber, water, and complex carbohydrates, which may boost energy expenditure during resting states and promote satiety [77].

5.4.2 Diabetes

When it comes to managing and preventing diabetes, vegans may be superior to those who are not. Vegans are about half as likely as non-vegans to acquire diabetes, according to the Adventist Health Studies [78]. The results of a controlled, randomized clinical trial analyzing a vegan diet low in calories with a diet tailored by the American Diabetes Association guidelines were published in 2006 [79]. Individuals following a vegan diet with minimal fat saw a 1.23-point decrease in their HbA1C, whereas those following the American Diabetes Association diet saw a 0.38-point decrease. Furthermore, compared to 26% of individuals following the American Diabetes Association diet, 43% of those following a vegan diet with minimal fat were able to cut back on their medicines. According to a study published in 2008 [80], non-vegans had a 74% higher risk of developing diabetes than the vegan population over a period of 17 years. According to a 2009 study, which included over 60,000 men and women, the incidence of diabetes was 2.9% for those following a vegan diet, while it was 7.6% for those who did not [81]. A vegan, low-fat diet that contains either no meat or very little meat, may help prevent or treat diabetes by lowering insulin resistance and increasing insulin sensitivity.

5.4.3 Heart Disease

According to one study, African American vegetarians had a considerably lower BMI and blood fats than lacto-ovo-vegetarians [82]. In a similar way, vegans and omnivores in Latin America had reduced blood lipid levels, with vegans having the lowest levels recorded [83]. In this study, vegans had lower low-density lipoprotein (LDL) cholesterol levels in plasma than omnivores, by 32% and 44%, respectively. The much-decreased mean BMI seen among vegans may be a key protective factor for decreasing blood cholesterol levels and lowering the risk of heart attack or stroke because obesity is a prominent risk factor for CVD. When it comes to vegetable and fruit consumption, vegans consume significantly more than omnivores [84, 85]. An increased intake of fruits and vegetables, which are abundant in fiber, antioxidants, folic acid, and several phytochemicals, has been linked to lowered blood lipid levels [86] and a decreased risk of stroke, heart disease, and mortality [87]. Additionally, almonds, soy, and whole grains are consumed in higher amounts by vegans [88], all of which have strong cardioprotective benefits [89, 90].

5.4.4 High Blood Pressure

The Dietary Guidelines Advisory Committee (DGAC) conducted a study in 2010 to find papers that looked at how dietary patterns affected adult blood pressure. Lowered diastolic and systolic blood pressures have been linked to vegan diets in this study. A sodium-limited vegetarian Japanese diet was found to significantly lower systolic blood pressure in a randomized crossover experiment [91].

5.4.5 Mortality

In 2010, the DGAC conducted a review of the literature to ascertain the impact of vegan diets on adult mortality rates, CVD, and stroke. Reducing red meat consumption may be the main way that vegan diets reduce mortality [92]. The advantages of limiting red meat consumption, which is linked to a higher risk of CVD and all-cause mortality, have been shown in numerous studies [93]. Increased lifespan has been linked to a low meat intake [94]. A meta-analysis of the mortality from CVD in vegans and non-vegans was conducted in 2012. Studies that provided estimated comparative risks, associating 95% confidence intervals, were the only ones that were included. Analysis was done on seven trials totaling 124,706 participants. The death rate from ischemic heart disease was 29% lower among vegans than among non-vegans [95].

5.4.6 Bone Health

Many factors other than the amounts of protein and calcium consumed play a role in defining the health of bones. Studies have indicated that foods like soy, fruits, and vegetables, as well as nutrients like vitamin K, vitamin D, magnesium, and potassium, have an impact on bone health [96]. Vegetarian diets are good sources of these essential macro- and micronutrients. For the health of bones, the acid-base balance must be maintained. Bone resorption is triggered by a decrease in extracellular pH, as calcium in bones serves as a buffer against this drop. Thus, a diet high in acid causes an increase in the excretion of calcium in the urine [97]. On the other hand, a vegan diet high in fruits and veggies benefits both men's and women's calcium economy, which is one of the markers of bone metabolism. Fruits and vegetables have a high K and Mg content, which produces an alkaline ash that prevents bone resorption [98]. Higher bone mineral density (BMD) of the lumbar spine and femoral neck in premenopausal women is linked to greater K intakes. The findings of two sizable prospective studies support the hypothesis that intake of vitamin K and the risk of hip fracture are related. The middle-aged women who consumed the highest amount of vitamin K had the least incidence of hip fractures in the Nurses' Health Study. Green leafy vegetables are the main source of vitamin K; a serving per day reduced the risk of hip fracture by 45% when compared to a serving per week [99]. Aged males and females in the top quartile of consumption of vitamin K had a 65% lower incidence of hip fracture than those in the lower quartile, according to the Framingham Heart Study [100]. Vegans typically consume a lot of fruits and vegetables, but they also frequently consume a lot of soy products, including tofu. It has been proposed that soy isoflavones may help postmenopausal women maintain healthy bones [101].

5.4.7 Cancer

Non-vegans were significantly more likely than vegans to develop colorectal and prostate cancer, according to data from the Adventist Health Study [102]. Numerous dietary components that prevent cancer are offered by a vegan diet. The mean BMI of vegans is significantly lower than that of non-vegans, suggesting that it may be a significant preventive factor in reducing the risk of developing cancer [71]. While frequent consumption of beans offers some defense against prostate and stomach cancer; fruits and vegetables are said to be preventive against cancer of the mouth, esophagus, stomach, and to a lesser extent, other locations. Furthermore, it has been demonstrated that a diet rich in fiber, carotenoids, ascorbic acid, flavonoids, some vitamins, and various phytochemicals protects against a variety of cancers; allium-containing vegetables like garlic and onion protect against colorectal and stomach cancer, respectively. Tomatoes and other lycopene-rich foods are recognized to offer protection from prostate cancer [103]. It is well known that a complex variety of phytochemicals found in fruits and vegetables have strong antiproliferative and antioxidant properties, as well as combined and mutually beneficial effects [104]. The phytochemicals disrupt a number of cellular functions that aid in the development of cancer. These mechanisms involve the inhibition of angiogenesis; blocking of the activation of nuclear factor-jB; induction of phase 2 enzymes, cell-cycle arrest, and apoptosis; inhibiting DNA adduct formation; signal transduction pathways, phase 1 enzymes, and the expression of oncogenes [104]. The risks of liver, esophageal, colorectal, and lung cancers were greater in individuals in the highest quintile of red meat eating than in those in the lowest quintile [105]. These risks ranged from 20%–60%. Furthermore, it was demonstrated earlier that eating eggs raises the probability of developing pancreatic cancer [106]. Even though vegans completely avoid eating eggs and red meat, they eat more legumes than omnivores. The Adventist Health Study found a negative correlation between this protein source and the risk of colon cancer [90]. According to published research, eating legumes might also help to moderately lower the possibility of developing prostate cancer [107].

5.5 IMPACT OF A VEGAN DIET ON THE ORAL MICROBIOTA

5.5.1 Oral Microbiome

The intestinal microbiota is the most complex microbiota in the human body, followed by the oral microbiota. The variety of microorganisms that are housed in the mouth, including viruses, bacteria, fungi, archaea, and protozoa (Figure 5.4), contribute to its complexity [108, 109]. The human oral microbiota database shows that the oral microbiome is composed of over 700 different microbial species, of which only 54% are known to be cultured and identified, 14% are cultured but not named, and 32% are uncultivated [110]. Various properties of the saliva, the soft tissue surfaces of the tongue and the oral mucosa, and the hard tissue surfaces of the teeth determine how complicated the oral environment remains. For the bacteria that live there, each component of the mouth creates a distinct habitat with the right nutrition and conditions [111]. Significant progress has been achieved in the last several decades in our understanding of the composition of the oral microbiota. The 16S rRNA sequencing method is one of the current methodologies that does not require culturing and

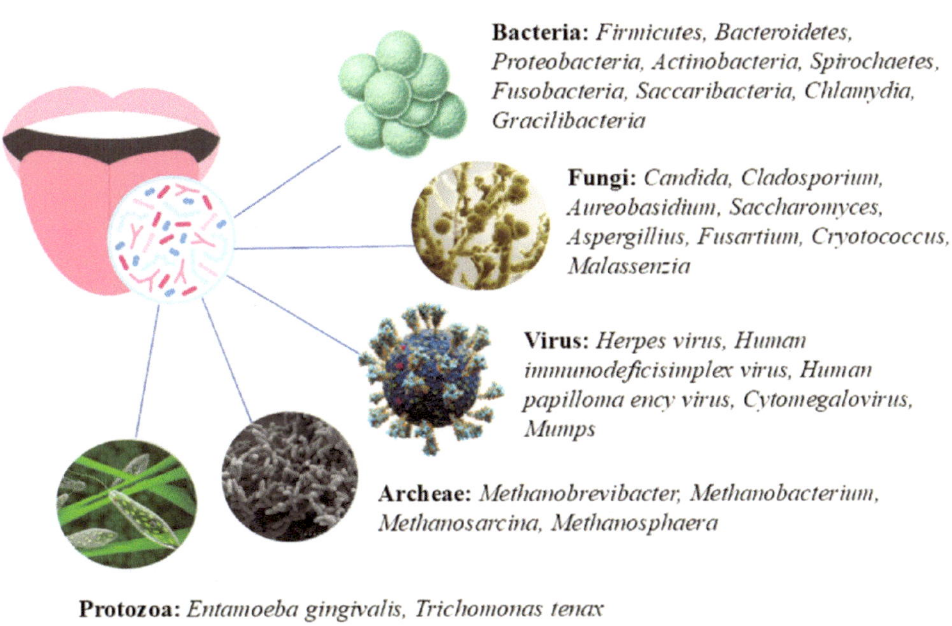

FIGURE 5.4 Composition of oral microbiota.

provides better knowledge of the composition of the oral microbiome [112]. The characteristics and genomic data of the currently recognized oral cavity bacteria are listed and thoroughly described in the Human Oral Microbiome Database.

5.5.1.1 Bacteria

Bacteria make up the majority of the oral microbiome. Most (94%) of the taxa found in the oral microbiota are classified into six major phyla: Bacteroidetes, Firmicutes, Actinobacteria, Proteobacteria, Fusobacteria, and Spirochaetes. The remaining 6% of the species are found in the phyla such as Synergistetes, Saccharibacteria, Gracilibacteria, Chloroflexi, Chlamydia, Chlorobi, and Tenericutes [113]. The gingival sulcus contains the majority of the isolated and known taxa, with much fewer being found on the tonsils, tooth surface, tongue, soft palate, and hard palate. It is important to remember that few species are site-specific, even though there is overlap among species isolated in various oral environments [114]. Salivary macromolecules serve as the main nutritional source for the local oral microbiota while also acting as a barrier against external stimuli that may compromise the overall health of the body, including the oral cavity. The salivary biofilm is made up of microorganisms along with some inorganic substances found in saliva, as well as the salivary microbiota. Salivary proteins (including glycoproteins) are absorbed onto the surface of teeth as soon as they erupt, creating a layer that allows germs to stick to the tooth surface. This layer protects against the effects of food, medication use, etc., and thus offers a favorable substrate for bacterial proliferation and survival [115].

5.5.1.2 Fungi

Fungi are essential components of a healthy dental microbiome, despite what the general public believes. They only turn into opportunistic infections in specific situations, such as when the microbiota balance is upset or when immunocompromised individuals are involved [116]. It has been noted that up to 101 different species of fungi could be found in the microbiome of healthy people. The most prevalent species are those of *Candida*, which are followed in frequency by *Aureobasidium, Cladosporium, Saccharomyces, Fusarium, Aspergillus,* and *Cryptococcus* [113].

5.5.1.3 Archaea

Archaea makes up only a small portion of the oral microbiome, among which Thermoplasmatales, *Methanobacterium*, *Methanobrevibacter*, *Methanosphaera*, and *Methanosarcina* are the methanogens that have been observed. They were discovered to be more common in people who had periodontitis [117]. Saliva from 98 healthy individuals revealed about 3621 different bacterial taxa, with the two most common phyla being Bacteroidetes (such as *Prevotella*) and Firmicutes (for example, *Streptococcus* and *Veillonella*). Because of the quick turnover of tissue, the oral mucosa contains few bacteria arranged in a monolayer on the palate and cheeks [118]. On the other hand, a biofilm made up of many layers of bacteria is what makes the tongue's surface unique. Exogenous foreign substances are a constant source of exposure for the oral microbial environment and are important factors in determining colonization and microbe resistance that comprise the oral microbiota [119].

The ordered biofilm that is both physically and functionally oriented onto the surface of teeth is created over time through a sequence of ordered stages [120]. As a result, teeth have unique surfaces that promote bacterial adherence and the development of biofilm or bacterial plaque [121]. According to one study, Firmicutes and Actinobacteria (such as *Actinomyces* and *Corynebacterium*) that produce acid and could survive in low pH settings are mostly responsible for caries in the supragingival plaque of healthy people [121]. The properties of the biofilm change as it travels down the root from the tooth surface to the subgingival area. As there is less saliva and more serum present, the environment becomes more anaerobic, and the temperature and pH rise to extreme levels [122].

5.5.2 Impact of Food Texture and Diet on the Oral Microbiome

There has been much discussion on the true impact of food on the oral microbiota. This idea is based on historical observations of the Neolithic period and the period of industrial revolutions. The key components of the human diet, such as the ratio of macronutrients (for example, glycemic load; composition of fatty acids; levels of Na, K, and other micronutrients; pH of the diet; and amount of fiber) were in fact dramatically changed by the foods and processing methods that emerged during these periods [123]. The emergence of agriculture during the Mesolithic–Neolithic transition, approximately 10,000 years back, significantly altered the dietary composition [124]. The primary foods of the pre-agricultural era were raw wild fruits, nuts, vegetables, and tubers. Cereals, legumes, dairy products, and meat were incorporated into the diets of human beings during the shift from hunter-gatherer to agrarian lifestyles [125]. The emergence of periodontal and dental diseases was linked to this shift due to modifications in the microbiota of the mouth. Despite variations in food across time, it has been noted that the microbial compositions of samples of plaque from the Neolithic era are similar with respect to the amounts and species of microorganisms present today [118]. Oral microorganisms associated with caries are more prevalent in samples collected from agriculture-related populations from the Neolithic era onwards than they are in samples from hunter-gatherer populations, in terms of both proportion and species. One group of scientists has studied the relationship between food and periodontitis in 10,000 participants of the National Health and Nutrition Examination Survey (NHANES) through cross-sectional research. According to that study, periodontal disease was found to be less common in people who followed a diet high in fruits, veggies, nuts, tea, and water and low in fermentable carbs, proteins, fatty acids, and sugar-filled beverages. This could be attributed to the oral microenvironment's decreased expression of periodontal microorganisms [126].

The industrial revolution approximately 200 years ago led to the growth of food processing methods, which worsened changes in the pre-Neolithic microbiota of the mouth. The nutritious qualities of cereals were greatly weakened by the removal of the outer bran layer and the separation of starchy elements (flour) when the development of mechanical mills made it possible to refine carbs [127]. Cereal processing has an impact on various dietary additions, including texture, bioavailability of starch, and total amount of nutrients. Consequently, the introduction of milk products, refined polysaccharides, alcohol, and vegetable oils into the diet of humans during the shift from hunter-gatherer to agrarian led to the development of these staples of the contemporary diet [125]. The adoption of these foods into the regular diet has been associated with a reduction in dental health as well as a number of distinct diseases, such as obesity, CVD, inflammatory bowel disease, and rheumatic disorders. Several studies have defined the relationship between the oral microbiome and the occurrence of various diseases (Table 5.2). Numerous studies conducted on animals have demonstrated the connection between food texture and oral health [128]. Saliva as well as crevicular fluid are the primary food sources for oral microorganisms [129]. These body fluids serve as nutrient sources for oral microbes, which seek to catabolize the substrates in these fluids to gain nutrition sources [118]. The relationship between the composition of oral microbiota and nutrition is not well known yet. As evidenced by some of the human studies, no variation in the growth rates of microbes in the oral cavity were seen in the presence or absence of food [130]. This theory was supported by a study that involved 161 participants and found that over 98% of them had a large baseline microbiota, with no changes identified based on the patients' style of diet (ovo-lacto-vegetarian, omnivorous, or vegan) [131]. However, in another study, slight variations in salivary metabolomic profiles were seen according to the kind of diet followed, such as ovo-lacto-vegetarian, omnivorous, or vegan. This could also be explained by the impact of nutrition on the gut microbiome, which then affected the oral microbiome [132].

TABLE 5.2
Examples of Oral Microbiome Organisms Associated with Specific Diseases

Disease	Associated Organism	Reference
Periodontitis	*Genera*: *Prevotella, Fusobacterium* *Species*: *Porphyromonas gingivalis, Treponema denticola, Tannerella forsythia,* *Filifactor alocis, Parvimonas micra*, and *Aggregatibacter actinomycetemcomitans*	[121, 133–135]
Dental caries	*Genera*: *Neisseria, Selenomonas, Propionibacterium* *Species*: *Streptococcus mutans, Lactobacillus* spp., and *Candida albicans*	[136, 137]
Oral cancer	*Species*: *Capnocytophaga gingivalis, Prevotella melaninogenica, Streptococcus mitis, Peptostreptococcus stomatis, Streptococcus salivarius, Streptococcus gordonii, Gemella haemolysans, Gemella morbillorum, Johnsonella ignava*, and *Streptococcus parasanguinis*	[138, 139]
Esophageal cancer	*Species*: *Tannerella forsythia* and *Porphyromonas gingivalis*	[140]
Colorectal cancer	*Genera*: *Lactobacillus, Rothia* *Species*: *Fusobacterium nucleatum*	[141, 142]
Pancreatic cancer	*Genera*: *Leptotrichia* *Species*: *Porphyromonas gingivalis* and *Aggregatibacter actinomycetemcomitans*	[143, 144]
Cystic fibrosis	*Streptococcus oralis, S. mitis, S. gordonii*, and *S. sanguinis*	[145]
Cardiovascular disease	*Species*: *Campylobacter rectus, Porphyromonas gingivalis, Porphyromonas endodontalis, Prevotella intermedia*, and *Prevotella nigrescens*	[146]
Rheumatoid arthritis	*Genera*: *Veillonella, Atopobium, Prevotella, Leptotrichia* *Species*: *Rothia mucilaginosa, Rothia dentocariosa, Lactobacillus salivarius*, and *Cryptobacterium curtum*	[147, 148]
Alzheimer's disease	*Phyla*: Spirochaetes *Species*: *Porphyromonas gingivalis*	[149, 150]
Diabetes	*Genera*: *Aggregatibacter, Neisseria, Gemella, Eikenella, Selenomonas, Actinomyces, Capnocytophaga, Fusobacterium, Veillonella*, and *Streptococcus*	[151]

5.6 DIFFERENCES IN DIETARY HABITS ON THE BASIS OF THE SALIVARY METABOLOME

There are several different ways through which microorganisms might enter the salivary flow, including exposure to the environment such as breathing in air or water vapor, eating, transdermal absorption, blood transfusion, gastrointestinal reflux, and oro-nasal tissues. Dietary differentiation is attributed to varying concentrations of several compounds, including urea, formate, and uridine. While the source of some molecules remains unknown, the physiological metabolism of humans might be responsible for others [131]. Though it is obvious that the taxonomic makeup of saliva is unrelated to diet and most likely doesn't contribute to metabolomic variations of saliva samples, an additional hypothesis is that the metabolism of oral microbiota as well as food intake play a role. Along with creatine, histidine, creatinine, and trimethylamine-*N*-oxide, urea (a by-product of protein digestion) is also a well-known indicator of animal protein intake [152–154]. In fact, the primary cause of alkaline pH of omnivores' saliva is urea. In addition to the action of chewing extremely fibrous food, the acidic saliva of non-omnivores, with special reference to vegetarians, has been suggested as the primary cause of the high occurrence of dental lesions [155]. Omnivores showed noticeably greater levels of formate, which could be produced by oral bacteria when they break down amino acids.

5.7 IMPACT OF VEGAN FOOD COMPONENTS ON THE HUMAN GUT MICROBIOTA

Polymers (such as proteins, fibers, lipids, and polyphenols) are frequently incorporated into the primary metabolic pathways of the gut microbiota. In addition to providing nutrition for microorganisms, ovo-lacto-vegetarian, omnivore, and vegetarian dietary patterns also have their own microbiome, which could have varying effects on the variety and quantity of the gut bacteria [156]. As compared to the gut microbiome of omnivores, the profiles of vegans may not vary and both comprise a higher abundance of good bacteria. In contrast, as non-vegan diets are typically associated with elevated amounts of fecal bile acids, the human gut microbiome seems to be changed more in omnivores than in vegetarians and is made up of bile-tolerant, possibly dangerous microbes [157]. Furthermore, animal-based diets might contribute to the spread of antibiotic resistance within the gut microbiota, as the prevalence and amount of antimicrobial resistance genes are observed to be much lower in the gut microbes of vegans compared to those of omnivores [158]. Contrary to *Bacteroides* spp., which are typically involved in multiple infections and provide antimicrobial resistance to a variety of antibiotics, *Prevotella* spp. have been found to have effective anti-inflammatory properties in certain diseases [159], such as inflammatory arthritis [160] and multiple sclerosis [161]. They may also be beneficial commensals for the human host. People with plant-based diets, such as those in Asian, African, and South American populations, have gut microbiomes dominated by *Prevotella* spp., whereas people in Western communities who eat diets high in animal protein and saturated fats have gut microbiomes driven by *Bacteroides* [162]. The quantity, type, and uniformity of bacterial species within the gut bacterial population could be influenced by dietary fiber. Therefore, indigestible carbohydrate diets high in whole grain and wheat bran are linked to an increase in *Lactobacillus* and *Bifidobacterium* species; on the other hand, lactic acid bacteria such as *Ruminococcus* species, *Roseburia* species, and *Eubacterium rectale* might also increase in response to resistant whole-grain barley and starch. Due to their ability to exclusively metabolize saccharolytic sugars, lactobacilli and bifidobacteria have been linked to a protective function in the human gut barrier, preventing the infiltration and proliferation of bacterial pathogens [16, 159].

Prevention of gut bacterial translocation, improvement of gut barrier function, prevention of obesity and inflammation, intestinal homeostasis, and overall metabolism have all been linked to *Akkermansia muciniphila*, a mucin-degrading bacterium of the intestinal microbiota that might comprise 3%–5% of the total microbial community in healthy subjects [163]. Additional research on animals has demonstrated that a pure membrane protein from the pasteurized bacterium *A. muciniphila* enhanced metabolism in diabetic and obese mice. An exploratory investigation including overweight and obese human volunteers demonstrated the advantageous effects of *A. muciniphila* supplementation [164]. It has been demonstrated that fermentable dietary fiber acts as a substrate for the metabolism of gut bacteria. Some metabolites, like single-chain fatty acids (SCFAs), are by-products of the metabolism of bacteria. The primary SCFAs are butyrate, which is vital for immunoregulation and tissue barrier maintenance and provides substrates for energy to epithelial cells in the gut. Acetate and propionate were utilized as glucose, sterol, and lipid metabolism substrates [15, 165]. They most likely have intestinal anti-inflammatory properties. Additionally, they play a role in a number of other critical physiological processes, such as lowering the pH and circulating cholesterol in the colon, enhancing insulin sensitivity and glucose tolerance, inhibiting the growth of appearing Enterobacteriaceae pathogens (such as *Salmonella* species and adherent-invasive *Escherichia coli*), promoting the absorption of water and sodium, supplying energy to epithelial cells of the colon, preventing the proliferation of cancer cells by hindering numerous mechanisms, and preventing obesity caused by high-fat diets (by promoting fat oxidation) [166, 167]. Consequently, SCFAs reinforce the mucosal barrier, lower body weight, enhance blood lipid profiles and glucose homeostasis, and act as a protective factor against a number of diseases (Figure 5.5), such as immunological diseases, inflammatory bowel disorder, and type 2 diabetes [159, 162, 168].

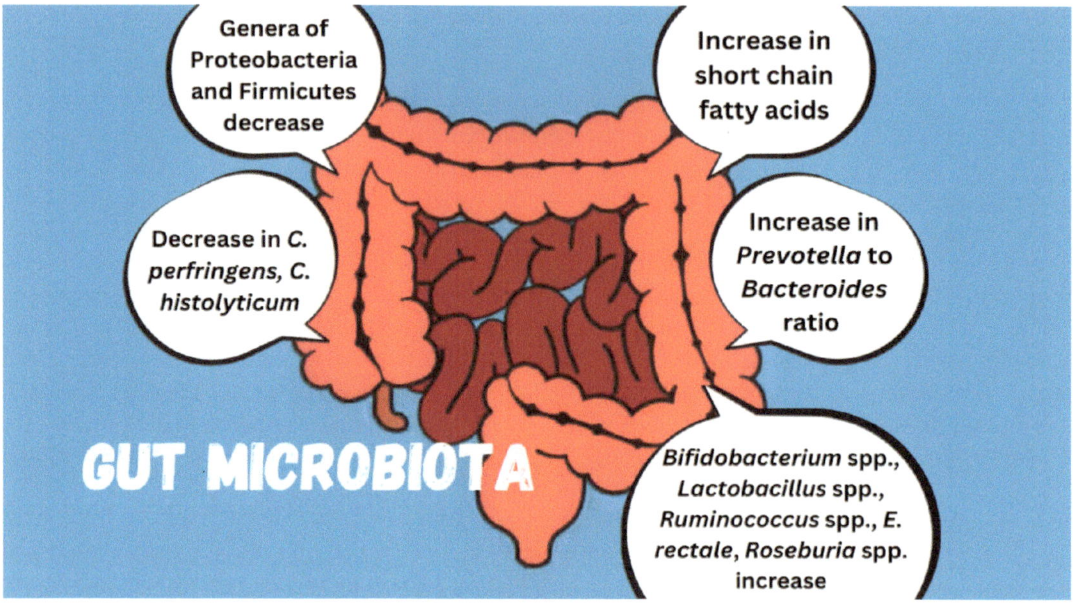

FIGURE 5.5 Effects of a vegan diet on gut microbiota.

In addition to fiber, the high concentration of polyphenols found in vegetarian diets boosts the number of *Lactobacillus* and *Bifidobacterium* species, which have antibacterial as well as anti-inflammatory properties and provide protection from CVD [159]. The majority of these chemicals, which include lignans, stilbenes, phenolic acids, secoiridoids, and flavonoids, show variations in structure. A minor amount might be absorbed in the small intestine, but they finally enter the colon, where they are metabolized by colonic bacteria, which affects their bioactivity [169]. Good sources of polyphenols include fruits (including sweet potatoes, grapes, mangoes, blueberries, and citrus fruits), vegetables, microalgae, medicinal plants, herbs, cereals, seeds, and drinks (like coffee, red wine, cocoa, and tea). Consuming polyphenols from fruit, seeds, tea, and wine is likely responsible for a reduction in potentially harmful *Clostridium histolyticum* and *Clostridium perfringens* [168]. In a different study, drinking red wine was linked to an increase in *Bifidobacteria* and species from genera like *Bacteroides*, *Enterococcus*, and *Prevotella* while inhibiting unfavorable bacteria like *Clostridium* species. This suggests that red wine polyphenols might have prebiotic benefits and lower cholesterol and C-reactive protein (CRP) [170]. Vegan diets are diets with minimal fat that are high in mono- and polyunsaturated fats, which changed the composition of microorganisms in the intestine by raising the ratio of Bacteroidetes to Firmicutes. Conversely, animal saturated fats reduce *Bifidobacterium* species and enhance the number of genera, like those of Proteobacteria and Firmicutes, which could cause inflammation and eventually cause metabolic disorders [159]. Substantial evidence suggests that eating a diet high in animal fats could play a significant role in the etiology of CVD by raising levels of LDL and total blood cholesterol [168]. It has been reported that vegans have a lower protein-energy status than omnivores. The gut microbes could synthesize certain micronutrients, such as vitamin K and B-complex vitamins like cobalamin, biotin, folate, pantothenic acid, nicotinic acid, riboflavin, pyridoxine, and thiamine, which are all involved in the metabolism of bacteria [171].

5.8 DIETARY RECOMMENDATIONS FOR AN OPTIMAL VEGAN DIET

1. *Fortified vitamin B12*: Vegans should routinely eat foods fortified with vitamin B12 (sources include fortified soy and rice beverages, morning cereals, meat substitutes, and nutritional yeast) or consume vitamin B12 supplements, one daily, to prevent vitamin B12

insufficiency. Seaweed, green vegetables, and fermented soy products are also some good sources of vitamin B12 [21].

2. *Fortified calcium*: Regularly ingesting calcium-fortified plant items, together with the typical vegan calcium sources (tofu, green vegetables, tahini, *etc.*), is recommended to guarantee appropriate calcium intake. Ready-to-eat cereals, rice drinks, soy beverages, and orange and apple juices are among these fortified foods with calcium. Calcium in milk has a similar bioavailability to calcium carbonate found in soy beverages and calcium citrate malate found in orange or apple beverages. Soy milk enriched with the mineral tricalcium phosphate, has a slightly lower calcium bioavailability than cow's milk [172, 173].
3. *Fortified vitamin D*: Vegans must regularly eat items fortified with vitamin D, such as rice milk, soy milk, breakfast cereals, and orange beverages, to maintain an acceptable level of the vitamin, particularly during the winter season. In the event that fortified foods are not accessible, 5–10 μg of vitamin D per day would need to be supplemented (most importantly for elderly individuals) [21].
4. *Food enriched with omega-3 fatty acids*: Vegetarians should regularly eat plant foods that are naturally enriched with omega-3 (n-3) fatty acids (sources include walnuts, hemp seed–based beverages, canola oil, soy products, and ground flaxseeds). Furthermore, it is advised that vegans eat foods like cereal bars and some soy milks that have been fortified with docosahexaenoic acid (DHA), an omega-3 fatty acid. People who need omega-3 fatty acids (like nursing and pregnant women) benefit from taking supplements that are high in DHA-enriched microalgae [21].
5. *Zinc-enriched food*: A vegan must eat foods rich in zinc (sources include pulses, whole grains, and soy products) to ensure an adequate intake of the mineral due to the high phytate content of the usual vegan diet. Vegans might also benefit from fortified cereals that are ready to eat and other meals enriched with zinc [174].

5.9 FUTURE OF VEGAN NUTRITION

Vegan nutrition has a bright future; however, there are an expanding number of food scandals, most of which involve meals that come from animals. This results in a decline in trust in these goods. Conversely, more people are becoming aware of the many long-term advantages of leading a vegetarian lifestyle. The notion of the "new nutrition science" [175] and the new scientific field of nutrition ecology both include these benefits. The pressing necessity to bring sustainability to the food chain and all human endeavors requires significant consideration [58]. The future of the vegan diet is bright for the following three specific reasons: (1) Moral, religious, ethical, and spiritual factors: People no longer desire for animals to live in misery to sate the appetite of non-vegans. (2) Economic, ecological, and political factors: People do not want climate change or the basis of life on Earth to be destroyed [176, 177]. Avoiding excess and waste is necessary. It is essential to develop a new prudential principle. (3) Health and hygiene concerns: Individuals no longer wish to endure avoidable, non-transmissible diseases linked to poor nutrition. Living a vegan lifestyle can significantly lower the risk of modern diseases [178].

5.10 CONCLUSION

In addition to avoiding or restricting animal products, added fats, and processed food (especially carbohydrates), the advice for patients who wish to follow a vegan diet may include consuming a range of fruits and vegetables, which may include beans, nuts, legumes, whole grains, and seeds. Individuals who choose to switch to a vegan diet gain a lot from it, including the potential to cut back on the amount of medications they take for diabetes, high blood pressure, various chronic diseases, lipid imbalance disorders, weight loss, cancer, and other heart diseases. A vegan diet is a lifestyle that is customized for each person rather than an all-or-nothing strategy. Health care teams should be aware that, even with a minimal risk of deficiencies, a motivated patient following a rigorous plant-based diet may require monitoring for certain nutrient deficits, as mentioned above. In the future, health care will evolve towards a paradigm in which eating more fruits and vegetables is the primary means of disease prevention and treatment rather than medication or surgery.

Vegetarian diets and their constituents have an impact on the metabolism of the oral microbiota and bacterial makeup by elevating the number of beneficial microbes. Furthermore, an improved understanding of the variety and nature of microbes in the oral cavity can result in novel medical methods and aid in the explanation of disease vulnerability. Many studies have demonstrated the significance of diet for the common oral dysbiosis associated with periodontal disease. Dietary factors can support the development of a microenvironment that is favorable for the growth and proliferation of certain disease-causing pathogens. In conclusion, it is believed that dietary patterns and associated consumption of supplements and pro- and prebiotics, along with the removal of biofilm through non-surgical therapies, induce several benefits that enhance periodontal health and decrease inflammation. Our understanding of the makeup and roles played by the oral microbiota in the maintenance of good health and disease prevention has significantly increased because of the application of recently discovered molecular techniques. Still, a lot has to be done. It is extremely difficult to research and comprehend how microbes interact with one another. Ultimately, the association between the microbiome and the human host is the primary cause of many disorders. Rarely have differences in how humans react to their microbiome been examined, and the impact of environmental variables adds even more complexity. Large volumes of new data will be available due to technological advancements in DNA and RNA sequencing, but processing and understanding the data will be very difficult.

REFERENCES

1. Blaney, D.; Diehl, H. *The optimal diet: the official chip cookbook: recipes to reverse and prevent obesity, high blood pressure, high cholesterol, diabetes, heart disease*, **2011**, Autumn House Pub.
2. Ornish, D.; Brown, S. E.; Billings, J. H.; Scherwitz, L. W.; Armstrong, W. T.; Ports, T. A.; Brand, R. J. Can lifestyle changes reverse coronary heart disease? The lifestyle heart trial. *The Lancet* **1990**, *336*(8708), 129–133. http://dx.doi.org/10.1016/0140-6736(90)91656-U
3. Tuso, P. J.; Ismail, M. H.; Ha, B. P.; Bartolotto, C. Nutritional update for physicians: plant-based diets. *The Permanente Journal* **2013**, *17*(2), 61. https://doi.org/10.7812%2FTPP%2F12-085
4. Wade, W. G. The oral microbiome in health and disease. *Pharmacological research* **2013**, *69*(1), 137–143. https://doi.org/10.1016/j.phrs.2012.11.006
5. Willis, J. R.; Gabaldón, T. The human oral microbiome in health and disease: from sequences to ecosystems. *Microorganisms* **2020**, *8*(2), 308. https://doi.org/10.3390/microorganisms8020308
6. Santonocito, S.; Giudice, A.; Polizzi, A.; Troiano, G.; Merlo, E. M.; Sclafani, R.; Isola, G. A cross-talk between diet and the oral microbiome: balance of nutrition on inflammation and immune system's response during periodontitis. *Nutrients* **2022**, *14*(12), 2426. https://doi.org/10.3390/nu14122426
7. Wade, W. G. Detection and culture of novel oral Bacteria. *Oral Microbial Ecology-Current Research and New Perspectives* **2013**; Jakubovics, NS, Palmer, RJ, Jr., Eds., Caister Academic Press, U.K.
8. Rosan, B.; Lamont, R. J. Dental plaque formation. *Microbes and Infection* **2000**, *2*(13), 1599–1607. https://doi.org/10.1016/S1286-4579(00)01316-2
9. Grosso, G.; Laudisio, D.; Frias-Toral, E.; Barrea, L.; Muscogiuri, G.; Savastano, S.; Colao, A. Anti-inflammatory nutrients and obesity-associated metabolic-inflammation: state of the art and future direction. *Nutrients* **2022**, *14*(6), 1137. https://doi.org/10.3390/nu14061137
10. Baker, J. L. Edlund, A; Exploiting the oral microbiome to prevent tooth decay: has evolution already provided the best tools? *Frontiers in Microbiology* **2019**, *9*, 3323. https://doi.org/10.3389/fmicb.2018.03323
11. Alhassani, A. A.; Hu, F. B.; Rimm, E. B.; Li, Y.; Rosner, B. A.; Willett, W. C.; Joshipura, K. J.; Dietary flavonoid intake and risk of periodontitis. *Journal of Periodontology* **2020**, *91*(8), 1057–1066. https://doi.org/10.1002/JPER.19-0463
12. Verneau, F.; La Barbera, F.; Furno, M. The role of health information in consumers' willingness to pay for canned crushed tomatoes enriched with Lycopene. *Nutrients* **2019**, *11*(9), 2173. https://doi.org/10.3390/nu11092173

13. Rajaram, S; Jones, J.; Lee, G. J. Plant-based dietary patterns, plant foods, and age-related cognitive decline. *Advances in Nutrition* **2019**, *10*(4), 422–S436.
14. Sakkas, H.; Bozidis, P.; Touzios, C.; Kolios, D.; Athanasiou, G.; Athanasopoulou, E.; Gartzonika, C. Nutritional status and the influence of the vegan diet on the gut microbiota and human health. *Medicina* **2020**, *56*(2), 88. https://doi.org/10.3390/medicina56020088.
15. Rinninella, E.; Cintoni, M.; Raoul, P.; Lopetuso, L. R.; Scaldaferri, F.; Pulcini, G.; Mele, M. C. Food components and dietary habits: keys for a healthy gut microbiota composition. *Nutrients* **2019**, *11*(10), 2393. https://doi.org/10.3390/nu11102393
16. Slavin, J. Fiber and prebiotics: mechanisms and health benefits. *Nutrients* **2013**, *5*(4), 1417–1435. https://doi.org/10.3390/nu5041417
17. Miqdady, M.; Al Mistarihi, J.; Azaz, A.; Rawat, D. Prebiotics in the infant microbiome: the past, present, and future. *Pediatric Gastroenterology, Hepatology & Nutrition* **2020**, *23*(1), 1. https://doi.org/10.5223/pghn.2020.23.1.1
18. Carlson, J. L.; Erickson, J. M.; Lloyd, B. B.; Slavin, J. L. Health effects and sources of prebiotic dietary fiber. *Current Developments in Nutrition* **2018**, *2*(3), 5.
19. Satija, A.; Bhupathiraju, S. N.; Spiegelman, D.; Chiuve, S. E.; Manson, J. E.; Willett, W.; Hu, F. B. Healthful and unhealthful plant-based diets and the risk of coronary heart disease in US adults. *Journal of the American College of Cardiology* **2017**, *70*(4), 411–422. https://doi.org/10.1016/j.jacc.2017.05.047
20. Thumann, T. A.; Pferschy-Wenzig, E. M.; Moissl-Eichinger, C.; Bauer, R. The role of gut microbiota for the activity of medicinal plants traditionally used in the European Union for gastrointestinal disorders. *Journal of Ethnopharmacology* **2019**, *245*, 112153. https://doi.org/10.1016/j.jep.2019.112153
21. Craig, W. J. Health effects of vegan diets. *The American Journal of Clinical Nutrition* **2009**, *89*(5), 1627–1633. https://doi.org/10.3945/ajcn.2009.26736N
22. Agnoli, C.; Baroni, L.; Bertini, I.; Ciappellano, S.; Fabbri, A.; Papa, M.; Pellegrini, N.; Sbarbati, R.; Scarino, M. L.; Siani, V.; Sieri, S.; Position paper on vegetarian diets from the working group of the Italian Society of Human Nutrition. *Nutrition, Metabolism and Cardiovascular Diseases* **2017**, *27*(12), 1037–1052. https://doi.org/10.1016/j.numecd.2017.10.020
23. Lemale, J.; Mas, E.; Jung, C.; Bellaiche, M.; Tounian, P.; Hepatology, F. S. P. Vegan diet in children and adolescents. Recommendations from the French-speaking Pediatric Hepatology, Gastroenterology and Nutrition Group (GFHGNP). *Archives de pédiatrie* **2019**, *26*(7), 442–450. https://doi.org/10.1016/j.arcped.2019.09.001
24. Chandra-Hioe, M. V.; Lee, C.; Arcot, J. What is the cobalamin status among vegetarians and vegans in Australia? *International Journal of Food Sciences and Nutrition* **2019**, *70*(7), 875–886. https://doi.org/10.1080/09637486.2019.1580681
25. Olza, J.; Aranceta-Bartrina, J.; González-Gross, M.; Ortega, R. M.; Serra-Majem, L.; Varela-Moreiras, G.; Gil, Á. Reported dietary intake and food sources of zinc, selenium, and vitamins A, E and C in the Spanish population: findings from the ANIBES study. *Nutrients* **2017**, *9*(7), 697. https://doi.org/10.3390/nu9070697
26. Irimie, A. I; Braicu, C.; Pasca, S.; Magdo, L.; Gulei, D.; Cojocneanu, R.; Berindan-Neagoe, I. Role of key micronutrients from nutrigenetic and nutrigenomic perspectives in cancer prevention. *Medicina* **2019**, *55*(6), 283. https://doi.org/10.3390/medicina55060283
27. Allès, B.; Baudry, J.; Méjean, C.; Touvier, M.; Péneau, S.; Hercberg, S.; Kesse-Guyot, E. Comparison of sociodemographic and nutritional characteristics between self-reported vegetarians, vegans, and meat-eaters from the NutriNet-Santé study. *Nutrients* **2017**, *9*(9), 1023. https://doi.org/10.3390/nu9091023
28. Tong, T. Y.; Key, T. J.; Gaitskell, K.; Green, T. J.; Guo, W.; Sanders, T. A.; Bradbury, K. E. Hematological parameters and prevalence of anemia in white and British Indian vegetarians and nonvegetarians in the UK Biobank. *The American Journal of Clinical Nutrition* **2019**, *110*(2), 461–472.
29. Medawar, E.; Huhn, S.; Villringer, A.; Veronica Witte, A. The effects of plant-based diets on the body and the brain: a systematic review. *Translational Psychiatry* **2019**, *9*(1), 226. https://doi.org/10.1038/s41398-019-0552-0
30. Woo, K. S.; Kwok, T. C.; Celermajer, D. S. Vegan diet, subnormal vitamin B-12 status and cardiovascular health. *Nutrients* **2014**, *6*(8), 3259–3273. https://doi.org/10.3390/nu6083259
31. Xie, L.; Wang, B.; Cui, X.; Tang, Q.; Cai, W.; Shen, X. Young adult vegetarians in Shanghai have comparable bone health to omnivores despite lower serum 25 (OH) vitamin D in vegans: a cross-sectional study. *Asia Pacific Journal of Clinical Nutrition* **2019**, *28*(2), 383–388.
32. Melina, V.; Craig, W.; Levin, S. Position of the academy of nutrition and dietetics: vegetarian diets. *Journal of the Academy of Nutrition and Dietetics* **2016**, *116*(12), 1970–1980. https://doi.org/10.1016/j.jand.2016.09.025

33. Veronese, N.; Reginster, J. Y. The effects of calorie restriction, intermittent fasting and vegetarian diets on bone health. *Aging Clinical and Experimental Research* **2019**, *31*, 753–758. https://doi.org/10.1007/s40520-019-01174-x
34. Ferruzzi, M. G.; Tanprasertsuk, J.; Kris-Etherton, P.; Weaver, C. M.; Johnson, E. J. Perspective: the role of beverages as a source of nutrients and phytonutrients. *Advances in Nutrition* **2020**, *11*(3), 507–523.
35. Wilson, A. K.; Ball, M. J. Nutrient intake and iron status of Australian male vegetarians. *European journal of clinical nutrition* **1999**, *53*(3), 189–194. https://doi.org/10.1038/sj.ejcn.1600696
36. Haider, L. M.; Schwingshackl, L.; Hoffmann, G.; Ekmekcioglu, C. The effect of vegetarian diets on iron status in adults: A systematic review and meta-analysis. *Critical Reviews in Food Science and Nutrition* **2018**, *58*(8), 1359–1374. https://doi.org/10.1080/10408398.2016.1259210
37. Gallego-Narbón, A; Zapatera, B; Vaquero, M. P. Physiological and dietary determinants of iron status in Spanish vegetarians. *Nutrients* **2019**, *11*(8), 1734. https://doi.org/10.3390/nu11081734
38. Takahashi, N.; Washio, J.; Mayanagi, G. Metabolomics of supragingival plaque and oral bacteria. *Journal of Dental Research* **2010**, *89*(12), 1383–1388. https://doi.org/10.1177/0022034510377792
39. Richter, C. K.; Skulas-Ray, A. C.; Champagne, C. M.; Kris-Etherton, P. M. Plant protein and animal proteins: do they differentially affect cardiovascular disease risk?. *Advances in Nutrition* **2015**, *6*(6), 712–728.
40. Iwasaki, M.; Manz, M. C.; Moynihan, P.; Yoshihara, A.; Muramatsu, K.; Watanabe, R.; Miyazaki, H. Relationship between saturated fatty acids and periodontal disease. *Journal of Dental Research* **2011**, *90*(7), 861–867. https://doi.org/10.1177/0022034511405384
41. Stein, C.; Cunha-Cruz, J.; Hugo, F. N. Is dietary pattern a mediator of the relationship between socioeconomic status and dental caries? *Clinical Oral Investigations* **2021**, *25*, 5441–5447. https://doi.org/10.1007/s00784-021-03852-5
42. Chapple, I. L.; Bouchard, P.; Cagetti, M. G.; Campus, G.; Carra, M. C.; Cocco, F.; Schulte, A. G. Interaction of lifestyle, behaviour or systemic diseases with dental caries and periodontal diseases: consensus report of group 2 of the joint EFP/ORCA workshop on the boundaries between caries and periodontal diseases. *Journal of Clinical Periodontology* **2017**, *44*, 39–51. https://doi.org/10.1111/jcpe.12685
43. Woelber, J. P.; Bremer, K.; Vach, K.; König, D.; Hellwig, E.; Ratka-Krüger, P.; Tennert, C. An oral health optimized diet can reduce gingival and periodontal inflammation in humans-a randomized controlled pilot study. *BMC Oral Health* **2017**, *17*(1), 1–8. https://doi.org/10.1186/s12903-016-0257-1
44. Woelber, J. P.; Tennert, C. Diet and periodontal diseases. *The Impact of Nutrition and Diet on Oral Health* **2020**, *28*, 125–133. https://doi.org/10.1159/000455380
45. Dodington, D. W.; Fritz, P. C.; Sullivan, P. J.; Ward, W. E.; Higher intakes of fruits and vegetables, β-carotene, vitamin C, α-tocopherol, EPA, and DHA are positively associated with periodontal healing after nonsurgical periodontal therapy in nonsmokers but not in smokers. *The Journal of Nutrition* **2015**, *145*(11), 2512–2519.
46. Neiva, R. F.; Al-Shammari, K.; Nociti Jr, F. H.; Soehren, S.; Wang, H. L. Effects of vitamin-B complex supplementation on periodontal wound healing. *Journal of Periodontology* **2005**, *76*(7), 1084–1091. https://doi.org/10.1902/jop.2005.76.7.1084
47. Sulaiman, A. E.; Shehadeh, R. M. Assessment of total antioxidant capacity and the use of vitamin C in the treatment of non-smokers with chronic periodontitis. *Journal of Periodontology* **2010**, *81*(11), 1547–1554. https://doi.org/10.1902/jop.2010.100173
48. Stein, S. H.; Tipton, D. A. Vitamin D and its impact on oral health-an update. *Journal of the Tennessee Dental Association* **2011**, *91*(2), 30.
49. Antonoglou, G. N.; Knuuttila, M.; Niemelä, O.; Raunio, T.; Karttunen, R.; Vainio, O.; Tervonen, T. Low serum level of 1, 25 (OH) 2D is associated with chronic periodontitis. *Journal of Periodontal Research* **2015**, *50*(2), 274–280. https://doi.org/10.1111/jre.12207
50. Iwasaki, M.; Moynihan, P.; Manz, M. C.; Taylor, G. W.; Yoshihara, A.; Muramatsu, K.; Miyazaki, H. Dietary antioxidants and periodontal disease in community-based older Japanese: a 2-year follow-up study. *Public Health Nutrition* **2013**, *16*(2), 330–338. https://doi.org/10.1017/S1368980012002637
51. Coutu, D. L.; Wu, J. H.; Monette, A.; Rivard, G. E.; Blostein, M. D.; Galipeau, J. Periostin, a member of a novel family of vitamin K-dependent proteins, is expressed by mesenchymal stromal cells. *Journal of Biological Chemistry* **2008**, *283*(26), 17991–18001. https://doi.org/10.1074/jbc.M708029200
52. Aral, K.; Alkan, B. A.; Saraymen, R.; Yay, A.; Şen, A.; Önder, G. Ö. Therapeutic effects of systemic vitamin k2 and vitamin d3 on gingival inflammation and alveolar bone in rats with experimentally induced periodontitis. *Journal of Periodontology* **2015**, *86*(5), 666–673. https://doi.org/10.1902/jop.2015.140467
53. Keys, A. *Seven countries: a multivariate analysis of death and coronary heart disease*, **1980**. Harvard University Press.

54. Trowell, H. C.; Burkitt, D. P. (Eds.) *Western diseases, their emergence and prevention,* **1981**. Harvard University Press.
55. Sabaté, J. The contribution of vegetarian diets to health and disease: a paradigm shift? *The American Journal of Clinical Nutrition* **2003**, *78*(3), 502–507.
56. Sabaté, J. The public health risk-to-benefit ratio of vegetarian diets: changing paradigms. *Vegetarian nutrition*, **2001**, CRC Press, 19–30.
57. Leitzmann, C. Vegetarian nutrition: past, present, future. *The American Journal of Clinical Nutrition* **2014**, *100*(1), 496–502.
58. Leitzmann, C. Nutrition ecology: the contribution of vegetarian diets. *The American Journal of Clinical Nutrition* **2003**, *78*(3), 657–659.
59. Key, T. J.; Appleby, P. N.; Spencer, E. A.; Travis, R. C.; Roddam, A. W.; Allen, N. E. Mortality in British vegetarians: results from the European prospective investigation into cancer and nutrition. *The American Journal of Clinical Nutrition* **2009**, *89*(5), 1613–1619.
60. Tonstad, S.; Stewart, K.; Oda, K.; Batech, M.; Herring, R. P.; Fraser, G. E. Vegetarian diets and incidence of diabetes in the Adventist Health Study-2. *Nutrition, Metabolism and Cardiovascular Diseases* **2013**, *23*(4), 292–299. https://doi.org/10.1016/j.numecd.2011.07.004
61. Ezzati, M.; Riboli, E. Behavioral and dietary risk factors for noncommunicable diseases. *New England Journal of Medicine* **2013**, *369*(10), 954–964. 10.1056/NEJMra1203528
62. Boeing, H.; Bechthold, A.; Bub, A.; Ellinger, S.; Haller, D.; Kroke, A.; Watzl, B. Critical review: vegetables and fruit in the prevention of chronic diseases. *European Journal of Nutrition* **2012**, *51*, 637–663. https://doi.org/10.1007/s00394-012-0380-y
63. Marsh, K.; Zeuschner, C.; Saunders, A. Health implications of a vegetarian diet: a review. *American Journal of Lifestyle Medicine* **2012**, *6*(3), 250–267. https://doi.org/10.1177/1559827611425762
64. González-Vallinas, M.; González-Castejón, M.; Rodríguez-Casado, A.; Ramírez de Molina, A. Dietary phytochemicals in cancer prevention and therapy: a complementary approach with promising perspectives. *Nutrition reviews*, **2013**, *71*(9), 585–599. https://doi.org/10.1111/nure.12051
65. Nile, S. H.; Park, S. W. Edible berries: Bioactive components and their effect on human health. *Nutrition* **2014**, *30*(2), 134–144.
66. Watzl, B.; Leitzmann, C. *Bioaktive Substanzen in Lebensmitteln,* **2005**. Georg Thieme Verlag.
67. Pettersen, B. J.; Anousheh, R.; Fan, J.; Jaceldo-Siegl, K.; Fraser, G. E. Vegetarian diets and blood pressure among white subjects: results from the Adventist Health Study-2 (AHS-2). *Public Health Nutrition* **2012**, *15*(10), 1909–1916. https://doi.org/10.1017/S1368980011003454
68. Crowe, F. L.; Appleby, P. N.; Travis, R. C.; Key, T. J. Risk of hospitalization or death from ischemic heart disease among British vegetarians and nonvegetarians: results from the EPIC-Oxford cohort study. *The American Journal of Clinical Nutrition* **2013**, *97*(3), 597–603.
69. Orlich, M. J.; Singh, P. N.; Sabaté, J.; Jaceldo-Siegl, K.; Fan, J.; Knutsen, S.; Fraser, G. E. Vegetarian dietary patterns and mortality in Adventist Health Study 2. *JAMA Internal Medicine* **2013**, *173*(13), 1230–1238. 10.1001/jamainternmed.2013.6473
70. Tantamango-Bartley, Y.; Jaceldo-Siegl, K.; Fan, J.; Fraser, G. Vegetarian diets and the incidence of cancer in a low-risk population. *Cancer Epidemiology, Biomarkers & Prevention* **2013**, *22*(2), 286–294. https://doi.org/10.1158/1055-9965.EPI-12-1060
71. Davey, G. K.; Spencer, E. A.; Appleby, P. N.; Allen, N. E.; Knox, K. H.; Key, T. J. EPIC–Oxford: lifestyle characteristics and nutrient intakes in a cohort of 33 883 meat-eaters and 31 546 non-meat-eaters in the UK. *Public Health Nutrition* **2003**, *6*(3), 259–268. https://doi.org/10.1079/PHN2002430
72. Dewell, A; Weidner, G.; Sumner, M. D.; Chi, C. S.; Ornish, D. A very-low-fat vegan diet increases intake of protective dietary factors and decreases intake of pathogenic dietary factors. *Journal of the American Dietetic Association* **2008**, *108*(2), 347–356. https://doi.org/10.1016/j.jada.2007.10.044
73. Ströhle, A.; Waldmann, A.; Wolters, M.; Hahn, A. Vegetarian nutrition: preventive potential and possible risks part 1: Plant foods. *Wiener Klinische Wochenschrift* **2006**, *118*, 580–593. https://doi.org/10.1007/s00508-006-0706-y
74. Berkow, S. E.; Barnard, N. Vegetarian diets and weight status. *Nutrition Reviews* **2006**, *64*(4), 175–188. http://doi.org/10.1111/j.1753-4887.2006.tb00200.x
75. Farmer, B.; Larson, B. T.; Fulgoni III, V. L.; Rainville, A. J.; Liepa, G. U. A vegetarian dietary pattern as a nutrient-dense approach to weight management: an analysis of the national health and nutrition examination survey 1999–2004. *Journal of the American Dietetic Association* **2011**, *111*(6), 819–827. https://doi.org/10.1016/j.jada.2011.03.012
76. Wang, Y.; Beydoun, M. A. Meat consumption is associated with obesity and central obesity among US adults. *International Journal of Obesity* **2009**, *33*(6), 621–628. https://doi.org/10.1038/ijo.2009.45

77. Sabaté, J.; Wien, M. Vegetarian diets and childhood obesity prevention. *The American Journal of Clinical Nutrition* **2010**, *91*(5), 1525S–1529S. http://dx.doi.org/10.3945/ajcn.2010.28701F
78. Snowdon, D. A.; Phillips, R. L. Does a vegetarian diet reduce the occurrence of diabetes?. *American Journal of Public Health* **1985**, *75*(5), 507–512. http://dx.doi.org/10.2105/AJPH.75.5.507
79. Barnard, N. D.; Cohen, J.; Jenkins, D. J.; Turner-McGrievy, G.; Gloede, L.; Jaster, B.; Talpers, S. A low-fat vegan diet improves glycemic control and cardiovascular risk factors in a randomized clinical trial in individuals with type 2 diabetes. *Diabetes Care* **2006**, *29*(8), 1777–1783. https://doi.org/10.2337/dc06-0606
80. Vang, A.; Singh, P. N.; Lee, J. W.; Haddad, E. H.; Brinegar, C. H. Meats, processed meats, obesity, weight gain and occurrence of diabetes among adults: findings from Adventist Health Studies. *Annals of Nutrition and Metabolism* **2008**, *52*(2), 96–104. http://dx.doi.org/10.1159/000121365
81. Tonstad, S.; Butler, T.; Yan, R.; Fraser, G. E. Type of vegetarian diet, body weight, and prevalence of type 2 diabetes. *Diabetes Care* **2009**, *32*(5), 791–796 http://dx.doi.org/10.2337/dc08-1886
82. Toohey, M. L.; Harris, M. A.; Williams, D.; Foster, G.; Schmidt, W. D.; Melby, C. L. Cardiovascular disease risk factors are lower in African-American vegans compared to lacto-ovo-vegetarians. *Journal of the American College of Nutrition* **1998**, *17*(5), 425–434. https://doi.org/10.1080/07315724.1998.10718789
83. De Biase, S. G.; Fernandes, S. F. C.; Gianini, R. J.; Duarte, J. L. G. Vegetarian diet and cholesterol and triglycerides levels. *Arquivos brasileiros de cardiologia* **2007**, *88*, 35–39. https://doi.org/10.1590/S0066-782X2007000100006
84. Keinan-Boker, L.; Peeters, P. H. M.; Mulligan, A. A.; Navarro, C.; Slimani, N.; Mattisson, I.; Riboli, E. Soy product consumption in 10 European countries: the European Prospective Investigation into Cancer and Nutrition (EPIC) study. *Public Health Nutrition* **2002**, *5*(6b), 1217–1226. https://doi.org/10.1079/PHN2002400
85. Larsson, C. L.; Johansson, G. K. Young Swedish vegans have different sources of nutrients than young omnivores. *Journal of the American Dietetic Association* **2005**, *105*(9), 1438–1441. https://doi.org/10.1016/j.jada.2005.06.026
86. Djoussé, L.; Arnett, D. K.; Coon, H.; Province, M. A.; Moore, L. L.; Ellison, R. C. Fruit and vegetable consumption and LDL cholesterol: the national heart, lung, and blood institute family heart study. *The American Journal of Clinical Nutrition* **2004**, *79*(2), 213–217.
87. Bazzano, L. A.; Serdula, M. K.; Liu, S. Dietary intake of fruits and vegetables and risk of cardiovascular disease. *Current Atherosclerosis Reports* **2003**, *5*(6), 492–499. https://doi.org/10.1007/s11883-003-0040-z
88. Larsson, C. L.; Johansson, G. K. Dietary intake and nutritional status of young vegans and omnivores in Sweden. *The American Journal of Clinical Nutrition* **2002**, *76*(1), 100–106.
89. Kelly, J. H.; Sabaté, J.; Nuts and coronary heart disease: an epidemiological perspective. *British Journal of Nutrition* **2006**, *96*(S2), S61–S67. https://doi.org/10.1017/BJN20061865
90. Mellen, P. B.; Walsh, T. F.; Herrington, D. M. Whole grain intake and cardiovascular disease: a meta-analysis. *Nutrition, Metabolism and Cardiovascular Diseases* **2008**, *18*(4), 283–290. https://doi.org/10.1016/j.numecd.2006.12.008.
91. Takahashi, Y.; Sasaki, S.; Okubo, S.; Hayashi, M.; Tsugane, S. Blood pressure change in a free-living population-based dietary modification study in Japan. *Journal of Hypertension* **2006**, *24*(3), 451–458. http://dx.doi.org/10.1097/01.hjh.0000209980.36359.16
92. Singh, P. N.; Sabaté, J.; Fraser, G. E. Does low meat consumption increase life expectancy in humans? *The American Journal of Clinical Nutrition* **2003**, *78*(3), 526S–532S.
93. Campbell II, T. M. *The China study: the most comprehensive study of nutrition ever conducted and the startling implications for diet, weight loss and long-term health*, **2004**. BenBella Books, Inc.
94. Sinha, R.; Cross, A. J.; Graubard, B. I.; Leitzmann, M. F.; Schatzkin, A. Meat intake and mortality: a prospective study of over half a million people. *Archives of Internal Medicine* **2009**, *169*(6), 562–571. http://dx.doi.org/10.1001/archinternmed.2009.6
95. Huang, T.; Yang, B; Zheng, J.; Li, G.; Wahlqvist, M. L.; Li, D. Cardiovascular disease mortality and cancer incidence in vegetarians: a meta-analysis and systematic review. *Annals of Nutrition and Metabolism* **2012**, *60*(4), 233–240. http://dx.doi.org/10.1159/000337301
96. Lanham-New SA. Importance of calcium, vitamin D and vitamin K for osteoporosis prevention and treatment. *Proc Nutr Soc* **2008**, *67*, 163–76.
97. Buclin, T.; Cosma, M.; Appenzeller, M.; Jacquet, A. F.; Decosterd, L. A.; Biollaz, J.; Burckhardt, P. Diet acids and alkalis influence calcium retention in bone. *Osteoporosis International* **2001**, *12*, 493–499. https://doi.org/10.1007/s001980170095

98. Tucker, K. L; Hannan, M. T; Kiel, D. P. The acid-base hypothesis: diet and bone in the Framingham Osteoporosis Study. *European Journal of Nutrition* **2001**, *40*, 231–237 https://doi.org/10.1007/s394-001-8350-8.
99. Feskanich, D.; Weber, P.; Willett, W. C.; Rockett, H.; Booth, S. L.; Colditz, G. A. Vitamin K intake and hip fractures in women: a prospective study. *The American Journal of Clinical Nutrition* **1999**, *69*, 74–9.
100. Booth, S. L.; Tucker, K. L.; Chen, H.; Hannan, M. T.; Gagnon, D. R.; Cupples, L. A.; Kiel, D. P. Dietary vitamin K intakes are associated with hip fracture but not with bone mineral density in elderly men and women. *The American Journal of Clinical Nutrition* **2000**, *71*(5), 1201–1208.
101. Cassidy, A.; Albertazzi, P.; Nielsen, I. L.; Hall, W.; Williamson, G.; Tetens, I.; Branca, F. Critical review of health effects of soyabean phyto-oestrogens in post-menopausal women. *Proceedings of the Nutrition Society* **2006**, *65*(1), 76–92. https://doi.org/10.1079/PNS2005476
102. Fraser, G. E. Associations between diet and cancer, ischemic heart disease, and all-cause mortality in non-Hispanic white California Seventh-day Adventists. *The American Journal of Clinical Nutrition* **1999**, *70*(3), 532–538.
103. Liu, R. H. Potential synergy of phytochemicals in cancer prevention: mechanism of action. *The Journal of Nutrition* **2004**, *134*(12), 3479–3485.
104. Liu, R. H. Health benefits of fruit and vegetables are from additive and synergistic combinations of phytochemicals. *The American Journal of Clinical Nutrition* **2003**, *78*(3), 517S–520S.
105. Cross, A. J.; Leitzmann, M. F.; Gail, M. H. Hollenbeck, A. R.; Schatzkin, A.; Sinha, R. A prospective study of red and processed meat intake in relation to cancer risk. *PLoS Medicine* **2007**, *4*(12), e325. https://doi.org/10.1371/journal.pmed.0040325
106. Chan, J. M.; Wang, F.; Holly, E. A. Pancreatic cancer, animal protein and dietary fat in a population-based study, San Francisco Bay Area, California. *Cancer Causes & Control* **2007**, *18*, 1153–1167. https://doi.org/10.1007/s10552-007-9054-0
107. Park, S. Y.; Murphy, S. P.; Wilkens, L. R.; Henderson, B. E.; Kolonel, L. N. Legume and isoflavone intake and prostate cancer risk: The Multiethnic Cohort Study. *International Journal of Cancer* **2008**, *123*(4), 927–932. https://doi.org/10.1002/ijc.23594
108. Krüger, W.; Vielreicher, S.; Kapitan, M.; Jacobsen, I. D.; Niemiec, M. J. Fungal-bacterial interactions in health and disease. *Pathogens* **2019**, *8*(2), 70. https://doi.org/10.3390/pathogens8020070
109. Li, Y.; Peng, X.; Zhou, X.; Ren, B.; Xiao, L.; Li, Y.; Guo, Q. Basic biology of oral microbes. *Atlas of Oral Microbiology: From Healthy Microflora to Disease,* **2020**, 1–24. https://doi.org/10.1007/978-981-15-7899-1_1
110. de Andrade, D. R.; Silva, P. A.; Colombo, A. P. V.; Silva-Boghossian, C. M. Subgingival microbiota in overweight and obese young adults with no destructive periodontal disease. *Journal of Periodontology* **2021**, *92*(10), 1410–1419. https://doi.org/10.1002/JPER.20-0187
111. Sonnenburg, J. L. Fischbach, M. A. Community health care: therapeutic opportunities in the human microbiome. *Science Translational Medicine* **2011**, *3*(78), 78ps12–78ps12. https://doi.org/10.1126/scitranslmed.3001626
112. Krishnan, K.; Chen, T.; Paster, B. J. A practical guide to the oral microbiome and its relation to health and disease. *Oral Diseases* **2017**, *23*(3), 276–286. https://doi.org/10.1111/odi.12509
113. Brun, A.; Nuzzo, A; Prouvost, B; Diallo, D; Hamdan, S; Meseguer, E; Range, H. Oral microbiota and atherothrombotic carotid plaque vulnerability in periodontitis patients. A cross-sectional study. *Journal of Periodontal Research* **2021**, *56*(2), 339–350. https://doi.org/10.1111/jre.12826
114. Aas, J. A.; Paster, B. J.; Stokes, L. N.; Olsen, I.; Dewhirst, F. E. Defining the normal bacterial flora of the oral cavity. *Journal of Clinical Microbiology* **2005**, *43*(11), 5721–5732. https://doi.org/10.1128/jcm.43.11.5721-5732.2005
115. Marsh, P. D.; Do, T.; Beighton, D.; Devine, D. A. Influence of saliva on the oral microbiota. *Periodontology* **2016**, *70*(1), 80–92. https://doi.org/10.1111/prd.12098
116. Krom, B. P.; Kidwai, S.; Ten Cate, J. M. Candida and other fungal species: forgotten players of healthy oral microbiota. *Journal of Dental Research* **2014**, *93*(5), 445–451. https://doi.org/10.1177/0022034514521814
117. Sedghi, L.; DiMassa, V.; Harrington, A.; Lynch, S. V.; Kapila, Y. L. The oral microbiome: Role of key organisms and complex networks in oral health and disease. *Periodontology* **2021**, *87*(1), 107–131. https://doi.org/10.1111/prd.12393
118. Adler, C. J.; Dobney, K.; Weyrich, L. S.; Kaidonis, J.; Walker, A. W.; Haak, W.; Cooper, A. Sequencing ancient calcified dental plaque shows changes in oral microbiota with dietary shifts of the Neolithic and Industrial revolutions. *Nature Genetics* **2013**, *45*(4), 450–455. https://doi.org/10.1038/ng.2536

119. Amaliya, Timmerman, M. F.; Abbas, F.; Loos, B. G.; Van der Weijden, G. A. Van Winkelhoff, A. J. Van der Velden, U. Java project on periodontal diseases: the relationship between vitamin C and the severity of periodontitis. *Journal of clinical periodontology* **2007**, *34*(4), 299–304. https://doi.org/10.1111/j.1600-051X.2007.01053.x
120. Marsh, P. D. Dental plaque as a biofilm and a microbial community–implications for health and disease. In *BMC Oral health* **2006** (Vol. 6, No. 1, pp. 1–7). BioMed Central. https://doi.org/10.1186/1472-6831-6-S1-S14
121. Costalonga, M.; Herzberg, M. C. The oral microbiome and the immunobiology of periodontal disease and caries. *Immunology Letters* **2014**, *162*(2), 22–38. https://doi.org/10.1016/j.imlet.2014.08.017
122. Paster, B. J.; Boches, S. K.; Galvin, J. L.; Ericson, R. E.; Lau, C. N.; Levanos, V. A.; Dewhirst, F. E. Bacterial diversity in human subgingival plaque. *Journal of Bacteriology* **2001**, *183*(12), 3770–3783. https://doi.org/10.1128/jb.183.12.3770-3783.2001
123. Halvorsrud, K.; Lewney, J.; Craig, D.; Moynihan, P. J. Effects of starch on oral health: systematic review to inform WHO guideline. *Journal of Dental Research* **2019**, *98*(1), 46–53. https://doi.org/10.1177/0022034518788283
124. Lycett, S. J. Cultural evolutionary approaches to artifact variation over time and space: basis, progress, and prospects. *Journal of Archaeological Science* **2015**, *56*, 21–31. https://doi.org/10.1016/j.jas.2015.01.004
125. Zhang, C.; Liu, K.; Hou, J. Extending the vitamin D pathway to vitamin D3 and CYP27A1 in periodontal ligament cells. *Journal of Periodontology* **2021**, *92*(7), 44–53. https://doi.org/10.1002/JPER.20-0225
126. Wright, D. M.; McKenna, G.; Nugent, A.; Winning, L.; Linden, G. J.; Woodside, J. V. Association between diet and periodontitis: a cross-sectional study of 10,000 NHANES participants. *The American Journal of Clinical Nutrition* **2020**, *112*(6), 1485–1491.
127. Li, E.; Dhital, S.; Hasjim, J. Effects of grain milling on starch structures and flour/starch properties. *Starch-Stärke* **2014**, *66*(1-2), 15–27. https://doi.org/10.1002/star.201200224
128. Crittenden, A. N.; Sorrentino, J.; Moonie, S. A.; Peterson, M.; Mabulla, A.; Ungar, P. S. Oral health in transition: The Hadza foragers of Tanzania. *PloS one* **2017**, *12*(3), e0172197. https://doi.org/10.1371/journal.pone.0172197
129. Loesche, W. J.; Gibbons, R. J. Influence of nutrition on the ecology and cariogenicity of the oral microflora. *The Science of Nutrition and its Application in Clinical Dentistry,* **1966**; *Saunders*: *Philadelphia, PA, USA*, 311.
130. Schmitt, D. P.; Alcalay, L.; Allensworth, M.; Allik, J.; Ault, L.; Austers, I.; ZupanÈiÈ, A. Patterns and universals of adult romantic attachment across 62 cultural regions: Are models of self and of other pancultural constructs? *Journal of Cross-Cultural Psychology* **2004**, *35*(4), 367–402. https://doi.org/10.1177/0022022104266105
131. De Filippis, F.; Vannini, L.; La Storia, A.; Laghi, L.; Piombino, P.; Stellato, G.; Ercolini, D. The same microbiota and a potentially discriminant metabolome in the saliva of omnivore, ovo-lacto-vegetarian and vegan individuals. *PloS one* **2014**, *9*(11), e112373. https://doi.org/10.1371/journal.pone.0112373
132. Wade, W. G. Resilience of the oral microbiome. *Periodontology* **2021**, *86*(1), 113–122. https://doi.org/10.1111/prd.12365
133. Jorth, P.; Turner, K. H.; Gumus, P.; Nizam, N.; Buduneli, N.; Whiteley, M. Metatranscriptomics of the human oral microbiome during health and disease. *MBio* **2014**, *5*(2), 10–1128. https://doi.org/10.1128/mbio.01012-14
134. Griffen, A. L.; Beall, C. J.; Campbell, J. H.; Firestone, N. D.; Kumar, P. S.; Yang, Z. K.; Leys, E. J. Distinct and complex bacterial profiles in human periodontitis and health revealed by 16S pyrosequencing. *The ISME Journal* **2012**, *6*(6), 1176–1185. https://doi.org/10.1038/ismej.2011.191
135. Liu, B.; Faller, L. L.; Klitgord, N; Mazumdar, V; Ghodsi, M; Sommer, D. D; Amar, S. Deep sequencing of the oral microbiome reveals signatures of periodontal disease. *PloS one* **2012**, *7*(6), e37919. https://doi.org/10.1371/journal.pone.0037919
136. Koo, H. Bowen, W. H. Candida albicans and Streptococcus mutans: a potential synergistic alliance to cause virulent tooth decay in children. *Future Microbiology* **2014**, *9*(12), 1295–1297. https://doi.org/10.2217/fmb.14.92
137. Gross, E. L.; Leys, E. J.; Gasparovich, S. R.; Firestone, N. D.; Schwartzbaum, J. A.; Janies, D. A; Griffen, A. L. Bacterial 16S sequence analysis of severe caries in young permanent teeth. *Journal of Clinical Microbiology* **2010**, *48*(11), 4121–4128. https://doi.org/10.1128/jcm.01232-10
138. Pushalkar, S.; Ji, X.; Li, Y.; Estilo, C.; Yegnanarayana, R.; Singh, B.; Saxena, D. Comparison of oral microbiota in tumor and non-tumor tissues of patients with oral squamous cell carcinoma. *BMC Microbiology* **2012**, *12*(1), 1–15. https://doi.org/10.1186/1471-2180-12-144

139. Wang, L.; Ganly, I. The oral microbiome and oral cancer. *Clinics in Laboratory Medicine* **2014**, *34*(4), 711–719. https://doi.org/10.1016/j.cll.2014.08.004
140. Peters, B. A.; Wu, J.; Pei, Z.; Yang, L.; Purdue, M. P.; Freedman, N. D.; Ahn, J. Oral microbiome composition reflects prospective risk for esophageal cancers. *Cancer Research* **2017**, *77*(23), 6777–6787.
141. Kato, I.; Vasquez, A. A.; Moyerbrailean, G.; Land, S.; Sun, J.; Lin, H. S.; Ram, J. L. Oral microbiome and history of smoking and colorectal cancer. *Journal of Epidemiological Research* **2016**, *2*(2), 92. https://doi.org/10.5430%2Fjer.v2n2p92
142. Broecker, F.; Russo, G.; Klumpp, J.; Moelling, K. Stable core virome despite variable microbiome after fecal transfer. *Gut Microbes* **2017**, *8*(3), 214–220. https://doi.org/10.1080/19490976.2016.1265196
143. Torres, P. J.; Fletcher, E. M.; Gibbons, S. M.; Bouvet, M.; Doran, K. S.; Kelley, S. T. Characterization of the salivary microbiome in patients with pancreatic cancer. *Peer J* **2015**, *3*, e1373. https://doi.org/10.7717/peerj.1373
144. Fan, X.; Alekseyenko, A. V.; Wu, J.; Peters, B. A.; Jacobs, E. J.; Gapstur, S. M.; Ahn, J. Human oral microbiome and prospective risk for pancreatic cancer: a population-based nested case-control study. *Gut* **2018**, *67*(1), 120–127. http://dx.doi.org/10.1136/gutjnl-2016-312580
145. Whiley, R. A.; Fleming, E. V.; Makhija, R.; Waite, R. D. Environment and colonisation sequence are key parameters driving cooperation and competition between Pseudomonas aeruginosa cystic fibrosis strains and oral commensal *streptococci*. *PLoS One* **2015**, *10*(2), e0115513. https://doi.org/10.1371/journal.pone.0115513
146. Chhibber-Goel, J.; Singhal, V.; Bhowmik, D.; Vivek, R.; Parakh, N.; Bhargava, B.; Sharma, A. Linkages between oral commensal bacteria and atherosclerotic plaques in coronary artery disease patients. *NPJ Biofilms and Microbiomes* **2016**, *2*(1), 7. https://doi.org/10.1038/s41522-016-0009-7
147. Zhang, X.; Zhang, D.; Jia, H.; Feng, Q.; Wang, D.; Liang, D.; Wang, J. The oral and gut microbiomes are perturbed in rheumatoid arthritis and partly normalized after treatment. *Nature Medicine* **2015**, *21*(8), 895–905. https://doi.org/10.1038/nm.3914
148. Roszyk, E.; Puszczewicz, M. Role of human microbiome and selected bacterial infections in the pathogenesis of rheumatoid arthritis. *Reumatologia/Rheumatology* **2017**, *55*(5), 242–250: https://doi.org/10.5114/reum.2017.71641.
149. Aguayo, S.; Schuh, C. M. A. P.; Vicente, B.; Aguayo, L. G. Association between Alzheimer's disease and oral and gut microbiota: are pore forming proteins the missing link?. *Journal of Alzheimer's Disease* **2018**, *65*(1), 29–46.
150. Dominy, S. S.; Lynch, C.; Ermini, F.; Benedyk, M.; Marczyk, A.; Konradi, A.; Potempa, J. *Porphyromonas gingivalis* in Alzheimer's disease brains: Evidence for disease causation and treatment with small-molecule inhibitors. *Science Advances* **2019**, *5*(1), eaau3333. https://doi.org/10.1126/sciadv.aau3333
151. Casarin, R. C. V.; Barbagallo, A.; Meulman, T.; Santos, V. R.; Sallum, E. A.; Nociti, F. H.; Gonçalves, R. B. Subgingival biodiversity in subjects with uncontrolled type-2 diabetes and chronic periodontitis. *Journal of Periodontal Research* **2013**, *48*(1), 30–36. https://doi.org/10.1111/j.1600-0765.2012.01498.x
152. de Almeida, P. D. V.; Gregio, A. M.; Machado, M. A.; De Lima, A. A.; Azevedo, L. R. Saliva composition and functions: a comprehensive review. *J Contemp Dent Pract* **2008**, *9*(3), 72–80.
153. Hedrick, V. E.; Dietrich, A. M.; Estabrooks, P. A.; Savla, J.; Serrano, E. Davy, B. M. Dietary biomarkers: advances, limitations and future directions. *Nutrition Journal* **2012**, *11*, 1–14. doi:10.1186/1475-2891-11-109.
154. O'Gorman, A.; Gibbons, H.; Brennan, L. Metabolomics in the identification of biomarkers of dietary intake. *Computational and Structural Biotechnology Journal* **2013**, *4*(5), e201301004. doi:10.5936/csbj.201301004.
155. Laffranchi, L.; Zotti, F.; Bonetti, S.; Dalessandri, D. O. M. E. N. I. C. O.; Fontana, P. Oral implications of the vegan diet: observational study. *Minerva Stomatologica* **2010**, *59*(11-12), 583–591.
156. Milanović, V.; Osimani, A.; Cardinali, F.; Litta-Mulondo, A.; Vignaroli, C.; Citterio, B.; Clementi, F. Erythromycin-resistant lactic acid bacteria in the healthy gut of vegans, ovo-lacto vegetarians and omnivores. *PloS one* **2019**, *14*(8), e0220549. https://doi.org/10.1371/journal.pone.0220549
157. De Angelis, M.; Ferrocino, I.; Calabrese, F. M.; De Filippis, F.; Cavallo, N.; Siragusa, S.; Cocolin, L. Diet influences the functions of the human intestinal microbiome. *Scientific Reports* **2020**, *10*(1), 4247.
158. Losasso, C.; Di Cesare, A.; Mastrorilli, E.; Patuzzi, I.; Cibin, V.; Eckert, E. M.; Corno, G. Assessing antimicrobial resistance gene load in vegan, vegetarian and omnivore human gut microbiota. *International Journal of Antimicrobial Agents* **2018**, *52*(5), 702–705.
159. Tomova, A. Bukovsky, I.; Rembert, E.; Yonas, W.; Alwarith, J.; Barnard, N. D.; Kahleova, H. The effects of vegetarian and vegan diets on gut microbiota. *Frontiers in Nutrition* **2019**, *6*, 47. https://doi.org/10.3389/fnut.2019.00047

160. Marietta, E. V.; Murray, J. A.; Luckey, D. H.; Jeraldo, P. R.; Lamba, A.; Patel, R.; Taneja, V. Human gut-derived *Prevotella histicola* suppresses inflammatory arthritis in humanized mice. *Arthritis & Rheumatology (Hoboken, NJ)* **2016**, *68*(12), 2878. https://doi.org/10.1002%2Fart.39785
161. Mangalam, A. K.; Murray, J. Microbial monotherapy with *Prevotella histicola* for patients with multiple sclerosis. *Expert Review of Neurotherapeutics* **2019**, *19*(1), 45–53. https://doi.org/10.1080/14737175.2019.1555473
162. Hjorth, M. F.; Blædel, T.; Bendtsen, L. Q.; Lorenzen, J. K.; Holm, J. B.; Kiilerich, P.; Astrup, A. Prevotella-to-Bacteroides ratio predicts body weight and fat loss success on 24-week diets varying in macronutrient composition and dietary fiber: results from a post-hoc analysis. *International Journal of Obesity* **2019**, *43*(1), 149–157.
163. Hu, F. B. Plant-based foods and prevention of cardiovascular disease: an overview. *Am J Clin Nutr* **2003**;78(suppl):544S–51S. https://doi.org/10.1038/s41366-018-0093-2
164. Everard, A.; Belzer, C.; Geurts, L.; Ouwerkerk, J. P.; Druart, C.; Bindels, L. B.; Cani, P. D. Cross-talk between Akkermansia muciniphila and intestinal epithelium controls diet-induced obesity. *Proceedings of the national academy of sciences* **2013**, *110*(22), 9066–9071. https://doi.org/10.1073/pnas.1219451110
165. Depommier, C.; Everard, A.; Druart, C.; Plovier, H.; Van Hul, M.; Vieira-Silva, S.; Cani, P. D. Supplementation with *Akkermansia muciniphila* in overweight and obese human volunteers: a proof-of-concept exploratory study. *Nature Medicine* **2019**, *25*(7), 1096–1103. https://doi.org/10.1038/s41591-019-0495-2
166. Tang, W. W.; Kitai, T.; Hazen, S. L. Gut microbiota in cardiovascular health and disease. *Circulation Research* **2017**, *120*(7), 1183–1196. https://doi.org/10.1161/CIRCRESAHA.117.309715
167. Klement, R. J.; Pazienza, V. Impact of different types of diet on gut microbiota profiles and cancer prevention and treatment. *Medicina* **2019**, *55*(4), 84. https://doi.org/10.3390/medicina55040084
168. McBurney, M. I.; Davis, C.; Fraser, C. M.; Schneeman, B. O.; Huttenhower, C.; Verbeke, K.; Latulippe, M. E. Establishing what constitutes a healthy human gut microbiome: state of the science, regulatory considerations, and future directions. *The Journal of Nutrition* **2019**, *149*(11), 1882–1895.
169. Singh, R. K.; Chang, H. W.; Yan, D. I.; Lee, K. M.; Ucmak, D.; Wong, K.; Liao, W. Influence of diet on the gut microbiome and implications for human health. *Journal of Translational Medicine* **2017**, *15*(1), 1–17. https://doi.org/10.1186/s12967-017-1175-y
170. Rowland, I.; Gibson, G.; Heinken, A.; Scott, K.; Swann, J.; Thiele, I.; Tuohy, K. Gut microbiota functions: metabolism of nutrients and other food components. *European Journal of Nutrition* **2018**, *57*, 1–24. https://doi.org/10.1007/s00394-017-1445-8
171. Queipo-Ortuño, M. I.; Boto-Ordóñez, M.; Murri, M.; Gomez-Zumaquero, J. M.; Clemente-Postigo, M.; Estruch, R.; Tinahones, F. J. Influence of red wine polyphenols and ethanol on the gut microbiota ecology and biochemical biomarkers. *The American Journal of Clinical Nutrition* **2012**, *95*(6), 1323–1334.
172. Gluba-Brzózka, A.; Franczyk, B.; Rysz, J. Vegetarian diet in chronic kidney disease—A friend or foe. *Nutrients* **2017**, *9*(4), 374. https://doi.org/10.3390/nu9040374
173. Andon, M. B.; Peacock, M.; Kanerva, R. L.; De Castro, J. A. Calcium absorption from apple and orange juice fortified with calcium citrate malate (CCM). *Journal of the American College of Nutrition* **1996**, *15*(3), 313–316. https://doi.org/10.1080/07315724.1996.10718604
174. Zhao, Y.; Martin, B. R.; Weaver, C. M. Calcium bioavailability of calcium carbonate fortified soymilk is equivalent to cow's milk in young women. *The Journal of Nutrition* **2006**, *135*(10), 2379–2382.
175. Messina, V.; Melina, V.; Mangels, A. R. A new food guide for North American vegetarians. *Canadian Journal of Dietetic Practice and Research* **2003**, *64*(2), 82–86. https://doi.org/10.3148/64.2.2003.82
176. Cannon, G.; Leitzmann, C. The new nutrition science project. *Public Health Nutrition* **2005**, *8*(6a), 673–694. https://doi.org/10.1079/PHN2005819
177. Macdiarmid, J. I.; Kyle, J.; Horgan, G. W.; Loe, J.; Fyfe, C.; Johnstone, A.; McNeill, G. Sustainable diets for the future: can we contribute to reducing greenhouse gas emissions by eating a healthy diet? *The American Journal of Clinical Nutrition* **2012**, *96*(3), 632–639.
178. Marlow, H. J.; Hayes, W. K.; Soret, S.; Carter, R. L.; Schwab, E. R.; Sabate, J. Diet and the environment: does what you eat matter? *The American Journal of Clinical Nutrition* **2009**, *89*(5), 1699–1703.

6 Oral Microbiome and Disease Susceptibility

Wamik Azmi, Vikrant, Anupam Sharma, and Ishan Sen
Department of Biotechnology, Himachal Pradesh University, Shimla, Himachal Pradesh, India

6.1 INTRODUCTION

The human microflora is a vibrant population of microorganisms such as bacteria, archaea, viruses, fungi, and protists. Varying human body parts function as distinct microbiomes, including diverse microbial communities with vastly different energy supplies and habitats [1]. Even within a single bodily location, the makeup of the microbiome differs among individuals with different health statuses, related to different lifestyles or other considerations [2]. An ounce of saliva fluid carries around 108 microbes, and 700 distinctive communities of prokaryotes have been recognized by a variety of studies [3]. Oral microflora in good health consists of an array of about 100–200 prominent bacteria. The mouth, as a microbiome, is the residence of several prominent colonies of microorganisms [4, 5]. Multiple studies have described the saliva, buccal, tongue, gums, teeth surface palate, both supragingival and subgingival plaque, tonsils, and throat microbiomes [1, 3] as exhibiting comprehensive likeness along with only small-scale differences, including higher levels of *Corynebacterium* in the two kinds of plaque and of Firmicutes in buccal mucosa and saliva in comparison with plaque [6, 7]. The mouth is having access point to the stomach and gastrointestinal systems, and the mouth also seems to have implications for the oral microbiome in other systemic disorders. These disorders can be impacted by a number of universally occurring pathogens, and all of these disorders are related to changes in the microflora, including dental caries, gingivitis, and periodontitis [8]. It's interesting to note that links between oral bacteria and health outside of the mouth may be found, since there may be oral microbial linkages with gastrointestinal malignancies, immunological, cardiometabolic, and respiratory problems as well as obstetric issues [9–11]. The impact of external variables on the initial system may lead to a resilient state in a microbial community. As a result, a person who practices poor dental hygiene will develop gingivitis by accumulating a biofilm on the surfaces of the buccal cavity. It will be challenging for microbial community linked to illness to revert to a microbiome linked to health, especially if mechanical therapy techniques (such as laser therapy, piezoelectric scaling and sonic scaling) don't disturb the biofilm. When disclosed to comparable environmental variables, people's susceptibilities to illnesses connected with dental plaque differ, as resilience has frequently been seen. Future studies should focus heavily on the robustness of residents of oral microbiome like archaeal, viral, fungal, and protozoal populations [3]. This shows that the oral microflora includes indicators useful in the recognition of some of the systemic disorders [1]. In fact, an increasing number of studies have discovered connections between changes in the oral microbiome and a lot of disorders [8]. For example, dental caries—a frequent chronic illness in children—occurs for multifarious reasons but normally results from regular sugar consumption being processed by a particular bacterial environment in the oral cavity, which raises acidity and demineralizes teeth [12]. A particular bacterial ecology triggers inflammatory reactions in the host, which results in the breakdown of periodontal tissue, the development of pockets, and thence the loss of teeth in individuals with periodontitis, a chronic condition. Similarly, gingival lesions linked to herpes and candidiasis can be caused by non-plaque-associated bacteria, viruses, and fungi, and there is growing evidence that a particular microecosystem in mouth cancer

DOI: 10.1201/9781003478638-6

characterizes malignant tissue. Hundreds of initiatives are currently under way worldwide, demonstrating the great increase in the research of the human microbiome; yet, the oral microflora has not received the same degree of attention as its gut cousin. Oral and systemic human health greatly depend on this microbial community [13].

New opportunities are created by next-generation sequencing technology in the field of large-scale metagenomic research across an array of populations. This is the reason to describe the morphology composition and functions of the oral microflora and in certain circumstances, its functional activities and effects on human health [8]. Therefore, knowing the factors that form and develop the oral microbiota is essential for comprehending both dental and general systemic health. Research endeavoring to delineate the makeup of the oral microflora across heterogeneous human populations is quickly advancing, as is the search for variance based on an increasing array of characteristics, especially those pertaining to health and illness [13]. The interaction of the oral microbiome with external variables will be covered in this chapter, along with disorders induced due to modifications in these interactions and the basic principles that govern the makeup of the bacterial population.

6.2 COMPOSITION OF THE HUMAN ORAL MICROBIOME

The core microbiome and the variable microbiome together to form the human microbiome. The core microbiome refers to those dominant microorganism species that inhabit different body parts in the normal state and the core microbiome is common to all human beings. On the other hand, the variable microbiome includes those microorganisms whose presence is a response caused by genotypic variability and specificity in lifestyle and is so named because it varies from individual to individual. The buccal cavity accommodates a diverse assortment of bacteria. It is constantly in contact with the environment and is subject to its effects [14]. Distinct niches like the tongue, cheek, palates, mouth floor, teeth, saliva, and throat in the buccal cavity furnish unique habitats for the populating of different microorganisms. Thus, the microbial environment of the buccal cavity is biologically rich and varies from person to person. Bacteria inhabit discrete niches in the buccal cavity due to the existence of unique adhesins on their surfaces that help them attach to corresponding receptors that exist on the surface of the buccal cavity [4]. The typical microbiome is composed of bacteria, fungi, viruses, archaea, and protozoa. At present, there are only a few research reports on the fungal microflora, and studies regarding bacterial microflora are also limited for the oral microbiota [15]. Although the taxonomic composition of the oral microbiota varies among people, there exist significant functional commonalities among the bacterial communities. This suggests that there is a significant overlap in the gathering and analysis of oral bacterial species [3].

Through community profiling examinations, almost 300 distinct species of bacteria have been identified in a single human at a particular time interval. These results have also demonstrated the remarkable stability of these species' proportions across time, indicating that they could be helpful for human identification [3, 16]. Significantly more new species of bacteria and archaea have been discovered inside the oral microflora by studies conducted. More than half of the total bacterial life that exists on Earth belongs to the candidate phyla radiation group in the bacteria domain, and its members seem to be widely distributed. Because of this, they become dependent on other bacteria to develop, either by invading other bacteria's cells or by growing near together in a process known as application [3, 17]. Three prospective phyla in the candidate phyla radiation group are Candidatus Saccharibacteria (formerly identified as Candidate Division TM7), Candidatus Gracilibacteria (GN02), and Candidatus Absconditabacteria (SR1) have been discovered in the oral cavity. Representatives of Saccharibacteria and Actinomyces species have been cultured together [18]. The oral cavity, with 392 taxa having at least one reference genome and around 1500 genomes in total, is one of the most investigated microbiomes to date. It contains over 700 species of identified prokaryotes. The distribution of all these species is across 185 genera and 12 phyla, of which about 54% have official names, 14% are unidentified but are farmed, and 32% are only known

TABLE 6.1
Major Bacteria That Inhabit the Oral Microbiota

Gram-negative bacteria	Rods	*Desulfobacter*, *Leptotrichia*, *Prevotella*, *Desulfovibrio*, *Eikenella*, *Fusobacterium*, *Selemonas*, *Simonsiella*, *Treponema*, *Wolinella*, *Campylobacter*, and *Capnocytophaga*
	Cocci	*Neisseria*, *Veillonella*, *Moraxella*
Gram-positive bacteria	Rods	*Propionibacterium*, *Actinomyces*, *Corynebacterium*, *Eubacterium*, *Rothia*, *Bifidobacterium*, and *Lactobacillus*
	Cocci	*Peptostreptococcus*, *Stomatococcus*, *Abiotrophia*, and *Streptococcus*

Source: Reference [21].

as uncultivated phylotypes [19]. SR1, Synergistetes, Firmicutes, Bacteroidetes, Fusobacteria, Actinobacteria, Proteobacteria, Chlamydiae, Chloroflexi, Spirochaetes, Saccharibacteria (TM7), and Gracilibacteria (GN02) are the 12 phyla [20]. In healthy mouths, the oral microbial population is preserved at the genus level, and major bacterial species present in the mouth are listed in Table 6.1. Despite the commonalities, the microbiome's diversity varies depending on the individual and the place [21]. The tongue is home to complex microflora, including anaerobes, due to its many papillae and sparse anaerobic locations. The palate and buccal mucosae are the regions with limited microbial diversity [22].

Oral microbiomes are dynamic with the host and can exhibit significant and quick changes in composition and activity both geographically and temporally. Numerous factors, including the diet's temporal frequency of host, the bacteria's reaction to pH changes, interactions between the bacteria, and, over a longer period, gene mutations and horizontal gene transfer that give the strain new properties, all contribute to these multiplex, non-equilibrium dynamics [23]. The bacteria in our mouth cavity have a symbiotic relationship based on reciprocal advantages. By preventing the pathogenic species from adhering to the mucosa, the commensal populations keep them in control and not harmed. Bacteria turn pathogenic when they pass the commensal barrier and start causing infection [21].

Since the oral microbiome is composed of many extremely diverse niches formed at different surfaces inside the mouth, it may need the unique study methodology of all the niches of the human body. Microorganisms use nutrition metabolism and biofilm development to build their tiny niches, and several variables, including variations in the attainability of nutrients and oxygen and saliva's capacity to moderate pH, all contribute to their proliferation [24]. Even while the same creatures form the majority of oral cavity niches, some may exist in varying amounts. Because saliva may control pH and food availability, saliva flow is a critical influence in regulating the makeup of the microbiome at each niche in the oral habitat. Other important elements may be the availability of oxygen and the type of surface [8]. For the comparison of these niches to one another, some researchers investigated each of them concurrently [7]. While some have focused on specific locations with an emphasis on localized dysbiosis in disease states, others have employed an oral rinse technique to get a comprehensive picture of the oral cavity. Research on certain oral niches usually tries to investigate an illness that is pertinent to that location. Because there might be a link between swallowed germs and pneumonia, the tongue microbiome of older people in Japan was investigated [8], and poorer oral hygiene samples had higher concentrations of bacteria linked to pneumonia. Additionally, the tongue was considered as a prospective component of diagnostic instruments that would use the microbiomes of the entire gastrointestinal system to identify pancreatic cancer [25]. It has been demonstrated that sickness samples from primary Sjögren's syndrome patients have higher Firmicutes/Proteobacteria ratios and higher abundances of 19 genera in comparison to healthy controls. It was proposed that germs linked to the condition might potentially be stored in the buccal mucosa [26].

In addition to a study of subgingival plaque and buccal mucosa demonstrating differences between periodontitis samples and healthy controls, numerous efforts have characterized the changes in subgingival and supragingival plaque related to disease progression, with many of the same organisms affected in both sites, though they also displayed unique species colonization [27]. Researchers looked into the palatine tonsils of human immunodeficiency virus (HIV)-positive individuals to learn more about the oral and systemic issues associated with the virus. They found that whereas the bacteriome was not much altered, the mycobiome was, because of its relatively constant microbial diversity. The microbiome of the mouth cavity may be easily identified from the microbiomes of other human body habitats [8].

6.3 ACQUISITION OF THE ORAL MICROBIOME

The acquisition of the oral microbiome, the diverse community of microorganisms residing in the mouth, begins at birth and evolves throughout life. Initial colonization occurs during delivery, with contact with the mother's microbiome, and continues through interactions with caregivers, the environment, and dietary factors. Vertical transmission of the oral microbiome from mother to child through the placenta initiates at birth, as described in Figure 6.1. The mode of delivery, whether vaginal or caesarian, significantly influences the initial microbial encounter for the newborn, determining whether the newborn is exposed to microorganisms from the vagina or skin [28]. This, in turn, impacts the diversity of the oral microbiome, with vaginally born infants exhibiting higher taxonomic diversity at 3 months of age [29]. Notably, the birth mode's impact may persist, as infants delivered via caesarian section acquire *Streptococcus mutans* almost 1 year earlier than vaginally born infants [30]. The method of feeding, whether breastfeeding or using infant formula, also influences the infant's oral microbiome. Breastfed infants at 3 months of age carry oral lactobacilli with antimicrobial properties not found in formula-fed infants [31, 32]. In addition to vertical transmission, horizontal transmission of oral microbiota among siblings and individuals sharing the same environment contributes to oral microbiome diversity [33, 34].

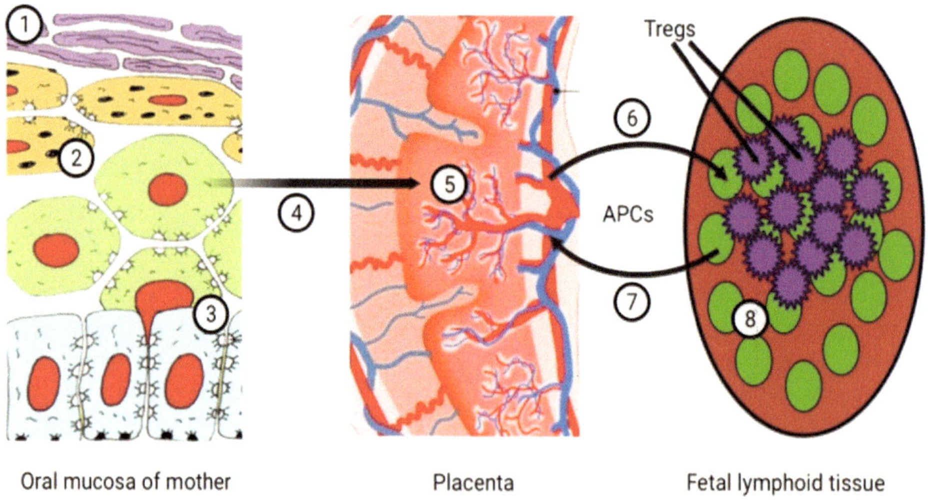

FIGURE 6.1 The significance of the placental microbiome in the development of maternal oral microbiome tolerance. (1) The oral microbiome of the mother; microbial cells attach to the oral epithelium. (2) Loosening of epithelial cell contacts due to pregnancy hormones; oral bacterial penetration. (3) Bacterial cells reach lamina propria and vascular tissues of the oral mucosa. (4) Transfer of bacterial cells to the placenta via bloodstream. (5) The oral bacteria of the mother are trapped in placental tissue. (6) Fetal antigen-presenting cells (APCs) acquire the alloantigen information in the placenta. (7) Fetal APCs return to fetal peripheral lymphoid organs. (8) Fetal regulatory T cells (Tregs) that recognise the mother's microbiome as "safe" are generated in fetal lymphoid tissue.)

6.4 FACTORS AFFECTING THE ORAL MICROBIOME

In the realm of clinical significance, the oral cavity stands out as one of the most pivotal microbial habitats, harboring a distinctly personalized consortium of microorganisms, encompassing bacteria, archaea, fungi, and viruses. Within the framework of the broader concept of the "human microbiome," denoting the collective assembly of microbes existing in symbiosis or commensalism with the human host, the term "oral microbiome" aptly characterized the specific microorganisms inhabiting the human mouth, as delineated by Dewhirst and coworkers [4]. The oral microbiome not only exerts a profound influence on the manifestation of the two most prevalent human oral maladies, namely, dental caries and periodontal diseases, but also emerges as a substantial risk factor for various systemic health conditions. These encompass, among others, tumor development [35], diabetes mellitus [36], cardiovascular diseases [37], bacteremia [38], and adverse pregnancy outcomes, including preterm birth and low birth weight in infants [39, 40]. Acknowledging the widely accepted perspective that oral microorganisms contribute to diseases primarily through synergistic or cooperative mechanisms, the intricacies of interspecies interactions within the oral community assume paramount importance in determining the pathogenic potential of oral microbiota. It is within this context that a comprehensive understanding of the complex interplay among these microorganisms necessitates an exploration of their endogenous residents. Consequently, a plethora of research endeavors has been undertaken to unravel the rich tapestry of oral microbial diversity. Following the establishment within the oral cavity, oral microorganisms embark on a journey of ecological succession, gradually giving rise to stable microbial communities. Yet, this delicate ecological equilibrium is subject to the influence of a myriad of both intrinsic and extrinsic factors. Among these factors are drug utilization, the lifestyle embraced by the host, environmental conditions, and the overall health status of the host, collectively contributing to variations in susceptibility to diseases. These factors exert their impact by intricately shaping the composition, structure, and metabolic functions of oral microorganisms.

6.4.1 Antibiotics Use

Antibiotics and antifungals can alter the oral microbiome by eliminating specific types of bacteria. This may lead to dysbiosis and an increased risk of oral health issues. Antibiotics have a significant impact on the makeup and activities of the oral microbiota in addition to causing certain metabolic changes when used in conjunction with other medications, as shown by a study [41]. In line with previous studies [42], a detailed study found that amoxicillin therapy decreased Shannon's diversity index when compared to the group which was untreated, and this decline continued for 6 months [43]. Remarkably, after 1 week, the antibiotic-treated group's relative abundance of *Actinobacteria* dropped by 60%, and after 6 months, it was still 50% below baseline. There were notable declines in *Fusobacteria* and Bacteroidetes as well. Remarkably, Proteobacteria became more abundant at the 1-month mark, while *Aggregatibacter paraphrophilus* continued to be present at the 3- and 6-month marks. Contrary to this, the control group did not contain any functioning taxonomic units that were able to last for 6 months [43]. Amoxicillin use was also linked to lower concentrations of *Neisseria, Streptococcus*, and *Veillonella* in the oral microbiota, as shown by research conducted [41, 44]. Apart from amoxicillin, other medications that have different working mechanisms also affect the oral microbiome, and at the same time the genes associated with antibiotic resistance also increase in number [15, 41]. According to recent research, some oral fungi exhibit resistance to the widely used antifungal azoles [45]. Furthermore, studies on *Candida albicans* obtained from patients having periodontal disease showed increased expression of genes linked to resistance to fluconazole and ketoconazole, indicating the critical role of drug-resistant strains of *Candida* in both chronic and acute periodontal infections [42]. Antibiotics certainly affect the oral microbiome, but their effects are far less severe than those observed in the gut microbiome, and it is important to remember this for further studies [15, 43]. This nuanced understanding underscores the need for judicious antibiotic use to minimize disruptions to the delicate nature of the oral microbial community.

6.4.2 Nutrition and Diet

The oral microbiome thrives on the nutrients derived from the diet, including carbohydrates, proteins, and fats. Carbohydrates, especially sugars and starches, serve as the main energy source for various microorganisms. Sugars, in particular, contribute to the production of acids by certain bacteria, leading to a drop in pH and creating an environment conducive to the development of dental caries. While there is a common assumption that the human diet significantly influences the oral microbiome due to the mouth being the entry point for food, recent findings suggest that oral bacteria primarily derive their nutrients from saliva and gingival crevicular fluid. *Streptococci*, particularly *Streptococcus oralis*, play a pivotal role in breaking down sugar residues from glycoproteins. Additionally, various oral bacteria, including proteolytic gram-negative anaerobes like *Porphyromonas* and *Prevotella* species, and aminopeptidase-producing organisms like *Parvimonas micra*, collaborate to metabolize proteins and peptides derived from glycan degradation [46]. This collaborative effort among oral bacteria underscores the complex and interdependent nature of their nutritional utilization mechanisms.

Research on newborns found that the infants who were fed formula had higher levels of *Actinomyces* and *Prevotella*, whereas breastfed infants had a higher amount of *Streptococcus*. It is interesting to note that babies fed on a diet of breast milk and solid food had less occurrence of oral candidiasis [47]. Another investigation linked the distinctive oral microorganism profiles in breastfed and formula-fed infants with the production of hydrogen peroxide, a compound which has antibacterial properties [48]. Dietary modification in adults was observed to have an influence on the oral microbiota and was also able to potentially mitigate related diseases [49]. Specific dietary elements, such as piscine monounsaturated fatty acids, medium-chain fatty acids, polyunsaturated fatty acids, and fibers, have been shown to have an association with the structure and community diversity of the oral microbiota [50]. On the other hand, there was a correlation between the prevalence of specific oral bacteria and the consumption of sugar and processed carbohydrates. For instance, the use of carbonated beverages was favorably correlated with the quantity of *Bacteroidetes*, *Gamma proteobacteria*, *Fusobacterium*, and *Veillonella* in the oral microbiome [51]. It has been established that diets with high sugar foster the growth of bacteria linked with dental caries, hence raising the risk of dental caries. Specific impacts on the composition of the salivary microbiome were seen when sugar alcohols such as erythritol, xylitol, and sorbitol were used as substitutes [52]. However, excessive intake of alcohol has been linked to oral microbiome imbalance [53]. Some also argue that moderate consumption of red wine might not alter the overall diversity and stability of representative human salivary microbiota and may even exert antibacterial effects [54]. Further research is required to validate the potential benefits or drawbacks of alcohol consumption on oral health. In essence, the relationship between the diet and oral microbiome is intricate, involving a dynamic interplay influenced by various dietary components and their impact on microbial composition and function. Understanding these connections provides valuable information that can help promote oral health through dietary choices.

6.4.3 Oral Hygiene

Undoubtedly, the practice of oral hygiene significantly shapes the modern oral microbiome by maintaining dental plaque in an immature state with elevated levels of early plaque-forming bacteria, particularly *Streptococci*. Archaeological evidence, found in preserved calculus, indicates a higher proportion of anaerobes in these plaques due to the disruption of plaque maturation by contemporary oral hygiene practices. Standard toothpaste formulations, including abrasives, detergents, and flavoring, often incorporate fluoride to enhance enamel resistance to acid-induced demineralization. Fluoride also exhibits antibacterial properties, reducing acid production and hindering enzyme activity. While fluoride's impact on the oral microbiome is challenging to study due to its widespread use, *in vitro* biofilm models show a reduction in acid production without affecting bacterial numbers. To optimize plaque removal, various chemical agents are added to dentifrices, such as chlorhexidine, cetylpyridinium chloride, triclosan, zinc citrate, delmopinol, amine and stannous fluorides, and essential oils. Scientific writings [55] suggested clinical benefits associated with their use. Detergents in toothpaste formulations, like sodium lauryl sulfate and triclosan, contribute antimicrobial properties similar to active ingredients. Despite laboratory potency, many active agents show limited efficacy *in vivo*. Salivary flow challenges retention, as binding to oral surfaces can deactivate agents. Substantivity, indicating the retention of anti-plaque activity when bound, is crucial. The natural occurrence of oral bacteria as biofilms presents a major hurdle, with biofilm-associated bacteria exhibiting significantly greater resistance to antimicrobials than planktonic cells. For instance, chlorhexidine, effective post-professional cleaning, faces limitations against established plaque due to its cationic nature, strong protein interaction, and poor penetration. The intricate interplay between oral hygiene practices, toothpaste formulations, and active agents underscores challenges in maintaining optimal oral health [3]. Recognizing the constraints posed by biofilm formation and salivary flow provides valuable insights for developing more effective strategies to control the oral microbiome.

6.5 ORAL MICROBIOME IN HEALTH AND DISEASE

To maintain oral health, the commensal oral microbiota is crucial. One such way by which commensals support oral health is their resistance to pathogen colonization, whereby they outcompete pathogenic species for colonization substrate, limiting the chances of exogenous pathogen integration [2]. On the other hand, dysbiosis of the oral microbiome causes states of oral illness, and several factors could lead to this kind of dysbiosis, such as the host's food, inflammatory reactions, systemic disturbances such as neurological disorders, adverse pregnancy outcomes, respiratory tract infections, heart-related diseases, and diabetes (Figure 6.2). These factors include drinking alcohol, smoking cigarettes, and using tobacco [56, 57]. Dental caries occurs as the result of a high

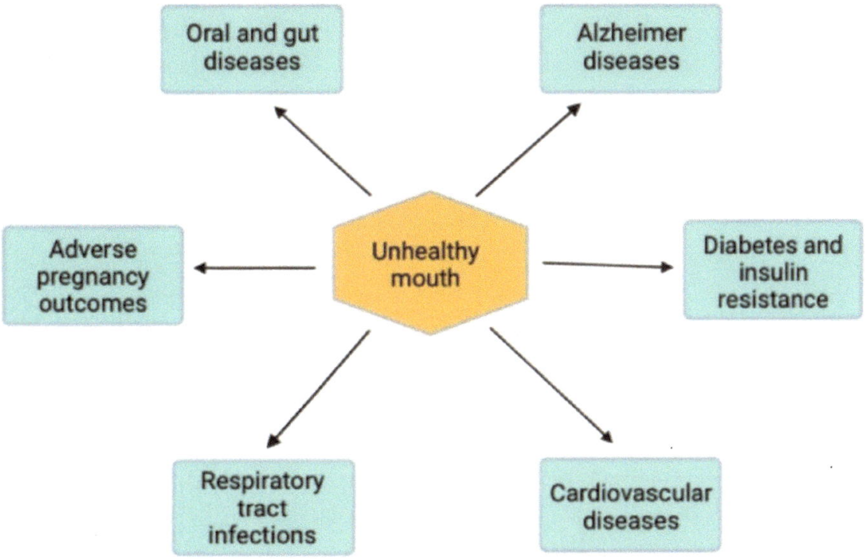

FIGURE 6.2 Different systemic disorders due to a disturbed oral microbiome.

consumption of carbohydrates in the diet, which causes bacteria to produce more acid and decreases saliva's buffering ability. It also lowers salivary pH and increases the development of the biofilm exopolysaccharide matrix, which traps and concentrates acids on enamel surfaces, and promotes the proliferation of acidogenic and aciduric species, such as *Lactobacillus* and *Streptococcus mutans* through induction positive feedback loops [2]. Dysbiosis of subgingival microbial populations is the cause of periodontal disease. This dysbiosis negatively impacts the host's immune system, causing uncontrollably high levels of inflammation in the gingival and periodontal tissues and impeding immunological subversion and tissue healing [58]. The species of the "red complex," which are *Porphyromonas gingivalis*, *Treponema denticola*, and *Tannerella forsythia*, along with other microorganisms such as the classes Clostridia, Negativicutes, Erysipelotrichia, and many more, are associated with periodontitis.

6.5.1 Oral Microbiome in Systemic Diseases

Oral microbiome dysbiosis results in systemic disease along with local (periodontal) sickness, as mentioned in Table 6.2. The three major ways through which oral infection and systemic pathology are linked are as follows: The infection spreads from the oral cavity because of transient bacteremia, microbial toxins circulate inside the body, and adverse immune responses to oral bacteria produce systemic inflammation [2, 59]. Approximately 0.75–1.5 L of saliva is produced by the oral cavity every day, containing bacteria from the tongue, cheeks, and dental and gingival plaque. The saliva is often swallowed either alone or in combination with food or drink; therefore, oral bacteria are often transmitted to the gastrointestinal tract. A healthy lower gastrointestinal tract is not a suitable environment for the majority of oral microbes to thrive in; yet, patients with inflammatory bowel disease, HIV, colon cancer, and liver cirrhosis have high levels of oral-associated microbes in their guts, and these diseases are frequently linked to disruptions in the gut microbiota, which may indicate that oral bacteria have a greater capacity for ectopic colonization when immunological dysregulation is present [60]. Oral microbiome dysbiosis is directly connected with esophageal, gastric, pancreatic, lung, hematologic, prostate, and breast cancers [61], and the cause of this connection is thought to be the production of carcinogenic substances such as acetaldehyde by ethanol-metabolizing taxa, nitrosamine by nitrate-reducing taxa, or the presence of cancer-causing viruses, e.g., the Epstein-Barr virus and cytomegalovirus. One study [62] found that the oral microbiome is similarly and strongly linked with breast cancer and non-malignant breast diseases as these illnesses have an inverse relationship with various genera, particularly some periodontal pathogens. Periodontitis also has frequently been associated with rheumatoid arthritis (RA), and *P. gingivalis* is once again implicated in this connection because it produces gingipains and peptidyl arginine deiminase (PAD), which facilitates protein citrullination, a significant RA trigger [3, 63].

Dysbiosis in the oral microbiota has also been connected with neurological disorders, and the most thorough investigation in this regard is the connection of *P. gingivalis* and Alzheimer's disease [64]. The gingipain load in the brain and other *P. gingivalis*-specific virulence factors, such as *P. gingivalis* lipopolysaccharide, have been correlated with Alzheimer's disease diagnosis. Gingipain cleaves the tau protein present in the brain and makes it misfolded and insoluble, and gingipain also increases neuronal degeneration [64]. Additionally, a higher incidence of oral squamous cell carcinoma is linked to periodontitis. Kamarajan and colleagues investigated in their study that *T. denticola*, *P. gingivalis*, and *Fusobacterium nucleatum* have little influence on cell proliferation and apoptosis, but they do improve cell motility, invasiveness, stemness, and oral tumorigenesis. The relationship between the oral microbiome and systemic health is bidirectional, which means that systemic disease may also exert an effect on the oral microbiome. For example, patients with type 2 diabetes and hyperglycemia related to this disease have higher chances of periodontitis due to the vascular complications of diabetes and an increased level of proinflammatory molecules [2].

TABLE 6.2
Oral and Systemic Diseases Associated with Oral Microbiome Dysbiosis

Disease	Causative Organisms (Dysbiosis in the Oral Microbiome)
Periodontitis	• An increased population of (red complex) *P. gingivalis*, *T. denticola*, *T. forsythia*, and archaeal methanogens. • Decreased population of Proteobacteria.
Caries	• An increased population of *Streptococcus mutans* and *Lactobacillus*. • Decreased population of non-mutans *Streptococci*.
Alzheimer's disease	• High population of spirochaetes and *Porphyromonas gingivalis*.
Cardiovascular disease	• An increased population of *Porphyromonas gingivalis, Porphyromonas endodontics, Prevotella intermedia, Prevotella nigrescens*, and *Campylobacter rectus*.
Cystic fibrosis	• An increased population of *Streptococci*. • Low or high (depending on environmental conditions) population of *Streptococcus oralis*.
Rheumatoid arthritis	• High population of *Veillonella, Atopobium, Prevotella, Leptotrichia, Lactobacillus salivarius, Cryptobacterium curtum, Rothia mucilaginosa*, and *Rothia dentocariosa*. • Low population of *Porphyromonas gingivalis, Haemophilus, Neisseria*, and *Rothia aeria*.
Diabetes	• An increased population of *Aggregatibacter, Neisseria*, and *Gemella*. • Decreased population of *Porphyromonas, Filifactor*, and *Eubacterium*.
Pancreatic cancer	• *Leptotrichia* (low early, high later). • *Porphyromonas gingivalis* and *Aggregatibacter actinomycetemcomitans* (high early, low later).
Esophageal cancer	• High population of *Tannerella forsythia* and *Porphyromonas gingivalis*. • Low population of *Neisseria* and *Streptococcus pneumoniae*.
Colorectal cancer	• High population of *Lactobacillus, Rothia*, and *Fusobacterium nucleatum*.

Source: Reference [8].

6.6 CONCLUSION

The human oral cavity is home to a large and diversified microbial life, making it the second largest and most diverse microbiome after the gut microbiota. Its relevance in human health and illness has recently been a significant investigation subject. Wide varieties of microorganisms are in continuous contact with the environment and are vulnerable to its impacts. As research advances, there are more findings that show how important the oral microbiota is for a number of health problems, including oral cavity illnesses. Microbiome dysbiosis is etiopathogenetically connected to a variety of illnesses affecting a wide range of organs and systems. In this view, the oral microbiome plays a major role in microbiome-related disorders and through postulated interactions with pulmonary, hepatic, mucosal, and cutaneous microbiomes. This, in turn, offers up new pathways for microbiome-based diagnostics and therapies, which benefit from the oral cavity's simple accessibility for microbiome monitoring and modification. Nonetheless, many hurdles are ahead. This chapter has presented the primary role of oral microbiome in health and diseases. The oral microbiome is critical to health because it may cause both oral and systemic disorders making it a fascinating and rapidly increasing area of research. Health is balanced because the ecosystem is formed by microorganisms living in biofilms throughout the buccal cavity. When there are imbalances in this homeostasis state, pathogens can develop and cause infection. Research in the field of microbiomes is quite new, a lot of research is being conducted, and the findings of several investigations are inconsistent. This might be affected by several factors like utilization of procedures, methods of standardization, and sample size. To discover consistent patterns of actual results, it is necessary to study large sample sizes from diverse locations of the microbiome. The human microbiome's initial research stage is the identification of the microbiome in health, followed by knowing the function of the microbiome in the change of functional and metabolic pathways associated with sick states. This will aid in identifying additional biomarkers and developing targeted medicines and personalized medicine for improved patient management in clinical practice.

REFERENCES

1. Baker, J. L.; Mark Welch, J. L.; Kauffman, K. M.; McLean, J. S.; He, X. The oral microbiome: diversity, biogeography and human health. *Nature Reviews Microbiology* **2023**, 1–16.
2. Sedghi, L.; DiMassa, V.; Harrington, A.; Lynch, S. V.; Kapila, Y. L. The oral microbiome: Role of key organisms and complex network in oral health and disease. *Periodontal 2000* **2021**, *87*(1), 107–131. https://doi.org/10.1111/prd.12393
3. Wade, W. G. Resilience of the oral microbiome. *Periodontology 2000* **2021**, *86*(1), 113–122. https://doi.org/10.1111/prd.12365
4. Dewhirst, F. E.; Chen, T.; Izard, J.; Paster, B. J.; Tanner, A. C. R.; Yu, W. H.; Lakshmanan, A.; Wade, W. G. The human oral microbiome. *Journal of Bacteriology* **2010**, *192*(19), 5002–5017. https://doi.org/10.1128/JB.00542-10
5. Huttenhower, C.; Gevers, D.; Knight, R.; Abubucker, S.; Badger, J. H.; Chinwalla, A. T.; Creasy, H. H.; Earl, A. M.; Fitzgerald, M. G.; Fulton, R. S.; Giglio, M. G.; Hallsworth-Pepin, K.; Lobos, E. A.; Madupu, R.; Magrini, V.; Martin, J. C.; Mitreva, M.; Muzny, D. M.; Sodergren, E. J.; White, O. Structure, function and diversity of the healthy human microbiome. *Nature* **2012**, *486*(7402), 207–214. https://doi.org/10.1038/nature11234
6. Ursell, L. K.; Metcalf, J. L.; Parfrey, L. W.; Knight, R. Defining the human microbiome. *Nutrition Reviews* **2012**, *70*(SUPPL. 1). https://doi.org/10.1111/j.1753-4887.2012.00493.x
7. Xu, X.; He, J.; Xue, J.; Wang, Y.; Li, K.; Zhang, K.; Guo, Q.; Liu, X.; Zhou, Y.; Cheng, L.; Li, M.; Li, Y.; Li, Y.; Shi, W.; Zhou, X. Oral cavity contains distinct niches with dynamic microbial communities. *Environmental Microbiology* **2015**, *17*(3), 699–710. https://doi.org/10.1111/1462-2920.12502
8. Willis, J. R.; Gabaldón, T. The human oral microbiome in health and disease: From sequences to ecosystems. *Microorganisms* **2020**, *8*(2). https://doi.org/10.3390/microorganisms8020308
9. Beck, J. D.; Slade, G.; Offenbacher, S. Oral disease, cardiovascular disease and systemic inflammation. *Periodontology* **2000**, *23*, 110–120.
10. Seymour, G. J.; Ford, P. J.; Cullinan, M. P.; Leishman, S.; Yamazaki, K. *REVIEW Relationship between periodontal infections and systemic disease* **2007**. http://www.who.int/cardiovascular_diseases/en/
11. Rubinstein, M. R.; Wang, X.; Liu, W.; Hao, Y.; Cai, G.; Han, Y. W. Fusobacterium nucleatum Promotes Colorectal Carcinogenesis by Modulating E-Cadherin/β-Catenin Signaling via its FadA Adhesin. *Cell Host and Microbe* **2013**, *14*(2), 195–206. https://doi.org/10.1016/j.chom.2013.07.012
12. Takahashi, N.; Nyvad, B.; The role of bacteria in the caries process: Ecological perspectives. In *Journal of Dental Research* **2011**, *90*(3), 294–303. https://doi.org/10.1177/0022034510379602
13. Gomez, A.; Espinoza, J. L.; Harkins, D. M.; Leong, P.; Saffery, R.; Bockmann, M.; Torralba, M.; Kuelbs, C.; Kodukula, R.; Inman, J.; Hughes, T.; Craig, J. M.; Highlander, S. K.; Jones, M. B.; Dupont, C. L.; Nelson, K. E. Host Genetic Control of the Oral Microbiome in Health and Disease. *Cell Host and Microbe* **2017**, *22*(3), 269–278.e3. https://doi.org/10.1016/j.chom.2017.08.013
14. Demmitt, B. A.; Corley, R. P.; Huibregtse, B. M.; Keller, M. C.; Hewitt, J. K.; McQueen, M. B.; Knight, R.; McDermott, I.; Krauter, K. S. Genetic influences on the human oral microbiome. *BMC Genomics* **2017**, *18*(1). https://doi.org/10.1186/s12864-017-4008-8
15. Zaura, E.; Nicu, E. A.; Krom, B. P.; Keijser, B. J. F. Acquiring and maintaining a normal oral microbiome: Current perspective. In *Frontiers in Cellular and Infection Microbiology* **2014**, *4*. Frontiers Media S.A. https://doi.org/10.3389/fcimb.2014.00085
16. Leake, S. L.; Pagni, M.; Falquet, L.; Taroni, F.; Greub, G. The salivary microbiome for differentiating individuals: proof of principle. *Microbes and Infection* **2016**, *18*(6), 399–405. https://doi.org/10.1016/j.micinf.2016.03.011
17. He, X.; McLean, J. S.; Edlund, A.; Yooseph, S.; Hall, A. P.; Liu, S. Y.; Dorrestein, P. C.; Esquenazi, E.; Hunter, R. C.; Cheng, G.; Nelson, K. E.; Lux, R.; Shi, W. Cultivation of a human-associated TM7 phylotype reveals a reduced genome and epibiotic parasitic lifestyle. *Proceedings of the National Academy of Sciences of the United States of America* **2015**, *112*(1), 244–249. https://doi.org/10.1073/pnas.1419038112

18. Hug, L. A.; Baker, B. J.; Anantharaman, K.; Brown, C. T.; Probst, A. J.; Castelle, C. J.; Butterfield, C. N.; Hernsdorf, A. W.; Amano, Y.; Ise, K.; Suzuki, Y.; Dudek, N.; Relman, D. A.; Finstad, K. M.; Amundson, R.; Thomas, B. C.; Banfield, J. F. A new view of the tree of life. *Nature Microbiology* **2016**, *1*(5). https://doi.org/10.1038/nmicrobiol.2016.48
19. Zhao, H.; Chu, M.; Huang, Z.; Yang, X.; Ran, S.; Hu, B.; Zhang, C.; Liang, J. Variations in oral microbiota associated with oral cancer. *Scientific Reports* **2017**, *7*(1). https://doi.org/10.1038/s41598-017-11779-9
20. Perera, M.; Al-Hebshi, N. N.; Speicher, D. J.; Perera, I.; Johnson, N. W. Emerging role of bacteria in oral carcinogenesis: A review with special reference to perio-pathogenic bacteria. *Journal of Oral Microbiology* **2016**, *8*(1). https://doi.org/10.3402/jom.v8.32762
21. Deo, P. N.; Deshmukh, R. Oral microbiome: Unveiling the fundamentals. *Journal of Oral and Maxillofacial Pathology* **2019**, *23*(1) 122–128. https://doi.org/10.4103/jomfp.JOMFP_304_18
22. Sultan, A. S.; Kong, E. F.; Rizk, A. M.; Jabra-Rizk, M. A. The oral microbiome: A Lesson in coexistence. *PLoS Pathogens* **2018**, *14*(1). https://doi.org/10.1371/journal.ppat.1006719
23. McLean, J. S. Advancements toward a systems level understanding of the human oral microbiome. *Frontiers in Cellular and Infection Microbiology* **2014**, *4*. https://doi.org/10.3389/fcimb.2014.00098
24. Laubichler, M. D.; Renn, J. Extended evolution: A conceptual framework for integrating regulatory networks and niche construction. *Journal of Experimental Zoology Part B: Molecular and Developmental Evolution* **2015**, *324*(7), 565–577. https://doi.org/10.1002/jez.b.22631
25. Lu, H.; Ren, Z.; Li, A.; Li, J.; Xu, S.; Zhang, H.; Jiang, J.; Yang, J.; Luo, Q.; Zhou, K.; Zheng, S.; Li, L. Tongue coating microbiome data distinguish patients with pancreatic head cancer from healthy controls. *Journal of Oral Microbiology* **2019**, *11*(1). https://doi.org/10.1080/20002297.2018.1563409
26. Van der Meulen, T. A.; Harmsen, H. J. M.; Bootsma, H.; Liefers, S. C.; Vila, A. V.; Zhernakova, A.; Fu, J.; Wijmenga, C.; Spijkervet, F. K. L.; Kroese, F. G. M.; Vissink, A. Dysbiosis of the buccal mucosa microbiome in primary Sjögren's syndrome patients. *Rheumatology (United Kingdom* **2018**, *57*(12), 2225–2234. https://doi.org/10.1093/rheumatology/key215
27. Wei, Y.; Shi, M.; Zhen, M.; Wang, C.; Hu, W.; Nie, Y.; Wu, X. Comparison of subgingival and buccal mucosa microbiome in chronic and aggressive periodontitis: A pilot study. *Frontiers in Cellular and Infection Microbiology* **2019**, *9*(MAR). https://doi.org/10.3389/fcimb.2019.00053
28. Dominguez-Bello, M. G.; Costello, E. K.; Contreras, M.; Magris, M.; Hidalgo, G.; Fierer, N.; Knight, R. Delivery mode shapes the acquisition and structure of the initial microbiota across multiple body habitats in newborns. Proceedings of the National Academy of Sciences of the United States of America **2010**, *107*(26), 11971–11975. https://doi.org/10.1073/pnas.1002601107
29. Lif Holgerson, P.; Harnevik, L.; Hernell, O.; Tanner, A. C.; Johansson, I. Mode of birth delivery affects oral microbiota in infants. *Journal of Dental Research* **2011**, *90*(10), 1183–1188. https://doi.org/10.1177/0022034511418973
30. Li, Y.; Caufield, P. W.; Dasanayake, A. P.; Wiener, H. W.; Vermund, S. H. Mode of delivery and other maternal factors influence the acquisition of Streptococcus mutans in infants. *Journal of Dental Research* **2005**, *84*(9), 806–811. https://doi.org/10.1177/154405910508400905
31. Holgerson, P. L.; Vestman, N. R.; Claesson, R.; Ohman, C.; Domellöf, M.; Tanner, A. C.; Hernell, O.; Johansson, I. Oral microbial profile discriminates breast-fed from formula-fed infants. *Journal of Pediatric Gastroenterology and Nutrition* **2013**, *56*(2), 127–136. https://doi.org/10.1097/MPG.0b013e31826f2bc6
32. Romani Vestman, N.; Timby, N.; Holgerson, P. L.; Kressirer, C. A.; Claesson, R.; Domellof, M.; Ohman, C.; Tanner, A. C.; Hernell, O.; Johansson, I. Characterization and in vitroproperties of oral lactobacilli in breastfed infants. *BMC Microbiology* **2013**, *13*, 193. https://doi.org/10.1186/1471-2180-13-193
33. Baca, P.; Castillo, A. M.; Liebana, M. J.; Castillo, F.; Martin-Platero, A.; Liebana, J. Horizontal transmission of Streptococcus mutans in schoolchildren. *Medicina oral, patologia oral y cirugiabucal* **2012**, *17*(3), e495–e500. https://doi.org/10.4317/medoral.17592
34. Stahringer, S. S.; Clemente, J. C.; Corley, R. P.; Hewitt, J.; Knights, D.; Walters, W. A.; Knight, R.; Krauter, K. S. Nurture trumps nature in a longitudinal survey of salivary bacterial communities in twins from early adolescence to early adulthood. *Genome Research* **2012**, *22*(11), 2146–2152. https://doi.org/10.1101/gr.140608.112
35. Farrell, J. J.; Zhang, L.; Zhou, H.; Chia, D.; Elashoff, D.; Akin, D.; Paster, B. J.; Joshipura, K.; Wong, D. T. Variations of oral microbiota are associated with pancreatic diseases including pancreatic cancer. *Gut* **2012**, *61*(4), 582–588. https://doi.org/10.1136/gutjnl-2011-300784
36. Loe H. Periodontal disease. The sixth complication of diabetes mellitus. *Diabetes Care* **1993**, *16*(1), 329–334.

37. Figuero, E.; Sanchez-Beltran, M.; Cuesta-Frechoso, S.; Tejerina, J. M.; del Castro, J. A.; Gutiérrez, J. M.; Herrera, D.; Sanz, M. Detection of periodontal bacteria in atheromatous plaque by nested polymerase chain reaction. *Journal of Periodontology* **2011**, *82*(10), 1469–1477. https://doi.org/10.1902/jop.2011.100719
38. Bahrani-Mougeot, F. K.; Paster, B. J.; Coleman, S.; Ashar, J.; Barbuto, S.; Lockhart, P. B.; Diverse and novel oral bacterial species in blood following dental procedures. *Journal of Clinical Microbiology* **2008**, *46*(6), 2129–2132. https://doi.org/10.1128/JCM.02004-07
39. Mitchell-Lewis, D.; Engebretson, S. P.; Chen, J.; Lamster, I. B.; Papapanou, P. N. Periodontal infections and pre-term birth: early findings from a cohort of young minority women in New York. *European Journal of Oral Sciences* **2001**, *109*(1), 34–39. https://doi.org/10.1034/j.1600-0722.2001.00966.x
40. Offenbacher, S.; Lin, D.; Strauss, R.; McKaig, R.; Irving, J.; Barros, S. P.; Moss, K.; Barrow, D. A.; Hefti, A.; Beck, J. D. Effects of periodontal therapy during pregnancy on periodontal status, biologic parameters, and pregnancy outcomes: a pilot study. *Journal of Periodontology* **2006**, *77*(12), 2011–2024. https://doi.org/10.1902/jop.2006.060047
41. Moraes, L. C.; Lang, P. M.; Arcanjo, R. A.; Rampelotto, P. H.; Fatturi-Parolo, C. C.; Ferreira, M. B. C.; Montagner, F. Microbial ecology and predicted metabolic pathways in various oral environments from patients with acute endodontic infections. *International Endodontic Journal* **2020**, *53*(12), 1603–1617. https://doi.org/10.1111/iej.13389
42. Monroy-Perez, E.; Rodriguez-Bedolla, R. M.; Garzon, J.; Vaca-Paniagua, F.; Arturo-Rojas Jimenez, E.; Paniagua-Contreras, G. L. Marked virulence and azole resistance in Candida albicans isolated from patients with periodontal disease. *Microbial Pathogenesis* **2020**, *148*, 104436. https://doi.org/10.1016/j.micpath.2020.104436
43. Menon, R. K.; Gomez, A.; Brandt, B. W.; Leung, Y. Y.; Gopinath, D.; Watt, R. M.; Crielaard, W.; Nelson, K. E.; Botelho, M. G. Long-term impact of oral surgery with or without amoxicillin on the oral microbiome-A prospective cohort study. *Scientific Reports* **2019**, *9*(1), 18761. https://doi.org/10.1038/s41598-019-55056-3
44. Larsson Wexell, C.; Ryberg, H.; Sjoberg Andersson, W. A.; Blomqvist, S.; Colin, P.; Van Bocxlaer, J.; Dahlen, G. Antimicrobial Effect of a Single Dose of Amoxicillin on the Oral Microbiota. *Clinical Implant Dentistry and Related Research* **2016**, *18*(4), 699–706. https://doi.org/10.1111/cid.12357
45. Gao, J.; Wang, H.; Li, Z.; Wong, A. H.; Wang, Y. Z.; Guo, Y.; Lin, X.; Zeng, G.; Liu, H.; Wang, Y.; Wang, J. *Candida albicans* gains azole resistance by altering sphingolipid composition. *Nature communications* **2018**, *9*(1), 4495. https://doi.org/10.1038/s41467-018-06944-1
46. Byers, H. L.; Tarelli, E.; Homer, K. A.; Hambley, H.; Beighton, D. Growth of *Viridans streptococci* on human serum alpha1-acid glycoprotein. *Journal of Dental Research* **1999**, *78*(7), 1370–1380. https://doi.org/10.1177/00220345990780071201
47. Oba, P. M.; Holscher, H. D.; Mathai, R. A.; Kim, J.; Swanson, K. S. Diet Influences the Oral Microbiota of Infants during the First Six Months of Life. *Nutrients* **2020**, *12*(11), 3400. https://doi.org/10.3390/nu12113400
48. Sweeney, E.; Alshehri, S.; Cowley, D.; Liley, H.; Bansal, N.; Charles, B.; Shaw, P.; Duley, J.; Knox, C. The effect of breastmilk and saliva combinations on the in vitro growth of oral pathogenic and commensal microorganisms. *Scientific Reports* **2018**, *8*. https://doi.org/10.1038/s41598-018-33519-3
49. Anderson, A. C.; Rothballer, M.; Altenburger, M. J.; Woelber, J. P.; Karygianni, L.; Vach, K.; Hellwig, E.; Al-Ahmad, A. Long-Term Fluctuation of Oral Biofilm Microbiota following Different Dietary Phases. *Applied and Environmental Microbiology* **2020**, *86*(20), e01421–20. https://doi.org/10.1128/AEM.01421-20
50. Hansen, T. H.; Kern, T.; Bak, E. G.; Kashani, A.; Allin, K. H.; Nielsen, T.; Hansen, T.; Pedersen, O. Impact of a vegan diet on the human salivary microbiota. *Scientific Reports* **2018**, *8*(1), 5847. https://doi.org/10.1038/s41598-018-24207-3
51. Chumponsuk, T.; Gruneck, L.; Gentekaki, E.; Jitprasertwong, P.; Kullawong, N.; Nakayama, J.; Popluechai, S. The salivary microbiota of Thai adults with metabolic disorders and association with diet. *Archives of Oral Biology* **2021**, *122*, 105036. https://doi.org/10.1016/j.archoralbio.2020.105036
52. Stsepetova, J.; Truu, J.; Runnel, R.; Nommela, R.; Saag, M.; Olak, J.; Nolvak, H.; Preem, J. K.; Oopkaup, K.; Krjutskov, K.; Honkala, E.; Honkala, S.; Makinen, K.; Makinen, P. L.; Vahlberg, T.; Vermeiren, J.; Bosscher, D.; de Cock, P.; Mandar, R. Impact of polyols on Oral microbiome of Estonian schoolchildren. *BMC Oral Health* **2019**, *19*(1), 60. https://doi.org/10.1186/s12903-019-0747-z
53. Yussof, A.; Yoon, P.; Krkljes, C.; Schweinberg, S.; Cottrell, J.; Chu, T.; Chang, S. L. A meta-analysis of the effect of binge drinking on the oral microbiome and its relation to Alzheimer's disease. *Scientific Reports* **2020**, *10*(1), 19872. https://doi.org/10.1038/s41598-020-76784-x

54. Barroso, E.; Martin, V.; Martínez-Cuesta, M. C.; Pelaez, C.; Requena, T. Stability of saliva microbiota during moderate consumption of red wine. *Archives of Oral Biology* **2015**, *60*(12), 1763–1768. https://doi.org/10.1016/j.archoralbio.2015.09.015
55. Serrano, J.; Escribano, M.; Roldan, S.; Martin, C.; Herrera, D. Efficacy of adjunctive anti-plaque chemical agents in managing gingivitis: a systematic review and meta-analysis. *Journal of Clinical Periodontology* **2015**, *42*(Suppl 16), S106–S138. https://doi.org/10.1111/jcpe.12331
56. Xiao, E.; Mattos, M.; Vieira, G.H.A.; Chen, S.; Correa, J.D.; Wu, Y.; Albiero, M.L.; Bittinger, K.; Graves, D.T. Diabetes enhances IL-17 expression and alters the oral microbiome to increase its pathogenicity. *Cell Host and Microbe* **2017**, *22*(1), 120–128. https://doi.org/10.1016/j.chom.2017.06.014.
57. Wang, Y.; Wang, S.; Wu, C.; Chen, X.; Duan, Z.; Xu, Q.; Jiang; W.; Xu; L.; Wang; T.; Su; L.; Wang, Y.; Chen, Y.; Zhang, J.; Huang, Y.; Tong, S.; Zhou, C.; Deng, S.; Qin, N. Oral Microbiome Alterations Associated with Early Childhood Caries Highlight the Importance of Carbohydrate Metabolic Activities. *Systems* **2019**, *4*(6), e00450–19. https://doi.org/10.1128/mSystems.00450-19.
58. Hajishengallis G. Periodontitis: from microbial immune subversion to systemic inflammation. *Nature Reviews Immunology* **2015**, *15*(1), 30–44. https://doi.org/10.1038/nri3785.
59. Li, X.; Kolltveit, K. M.; Tronstad, L.; Olsen, I. Systemic diseases caused by oral infection. *Clinical Microbiology Reviews* **2000**, *13*(4), 547–558. https://doi.org/10.1128/CMR.13.4.547.
60. Atarashi, K.; Suda, W.; Luo, C.; Kawaguchi, T.; Motoo, I.; Narushima, S.; Kiguchi, Y.; Yasuma, K.; Watanabe, E.; Tanoue, T.; Thaiss, C. A.; Sato, M.; Toyooka, K.; Said, H. S.; Yamagami, H.; Rice, S. A.; Gevers, D.; Johnson, R. C.; Segre, J. A.; Chen, K.; Honda, K. Ectopic colonization of oral bacteria in the intestine drives T_H1 cell induction and inflammation. *Science (New York, N.Y.)* **2017**, *358*(6361), 359–365. https://doi.org/10.1126/science.aan4526
61. Michaud, D. S.; Fu, Z.; Shi, J.; Chung, M. Periodontal Disease, Tooth Loss, and Cancer Risk. *Epidemiologic Reviews* **2017**, *39*(1), 49–58. https://doi.org/10.1093/epirev/mxx006
62. Wu, Z.; Byrd, D. A.; Wan, Y.; Ansong, D.; Clegg-Lamptey, J. N.; Wiafe-Addai, B.; Edusei, L.; Adjei, E.; Titiloye, N.; Dedey, F.; Aitpillah, F.; Oppong, J.; Vanderpuye, V.; Osei-Bonsu, E.; Dagnall, C. L.; Jones, K.; Hutchinson, A.; Hicks, B. D.; Ahearn, T. U.; Shi, J.; Vogtmann, E. The oral microbiome and breast cancer and nonmalignant breast disease, and its relationship with the fecal microbiome in the Ghana Breast Health Study. *International Journal of Cancer* **2022**, *151*(8), 1248–1260. https://doi.org/10.1002/ijc.34145
63. Kobayashi, T.; Murasawa, A.; Komatsu, Y.; Yokoyama, T.; Ishida, K.; Abe, A.; Yamamoto, K.; Yoshie, H. Serum cytokine and periodontal profiles in relation to disease activity of rheumatoid arthritis in Japanese adults. *Journal of Periodontology* **2010**, *81*(5), 650–657. https://doi.org/10.1902/jop.2010.090688
64. Dominy, S. S.; Lynch, C.; Ermini, F.; Benedyk, M.; Marczyk, A.; Konradi, A.; Nguyen, M.; Haditsch, U.; Raha, D.; Griffin, C.; Holsinger, L. J.; Arastu-Kapur, S.; Kaba, S.; Lee, A.; Ryder, M. I.; Potempa, B.; Mydel, P.; Hellvard, A.; Adamowicz, K.; Hasturk, H. Potempa *Porphyromonas gingivalis* in Alzheimer's disease brains: Evidence for disease causation and treatment with small-molecule inhibitors. *Science Advances* **2019**, *5*(1), eaau3333. https://doi.org/10.1126/sciadv.aau3333

7 Oral Microbiome and Acute Infections

Dimple, Ruby, Ajay Balda, Anil Kumar, Arup Giri
Department of Zoology, Baba Mastnath University, Rohtak, Haryana, India

Rajesh Kumar
Department of Biosciences, Himachal Pradesh University, Shimla, Himachal Pradesh, India

7.1 INTRODUCTION

The human oral cavity is a dynamic ecosystem teeming with microbial life, collectively referred to as the oral microbiome. This intricate microbial community composed of bacteria, viruses, fungi, and archaea interacts intricately with the host's immune system and plays a pivotal role in maintaining oral health and homeostasis [1, 2]. Over the past decades, research has increasingly focused on unraveling the complexities of the oral microbiome and its impact on various aspects of health and disease, including acute infections [3]. The relationship between the oral microbiome and acute infections is multifaceted and encompasses a broad spectrum of interactions. From influencing immune responses to directly interacting with pathogens, the oral microbiome exerts significant effects on susceptibility to viral, bacterial, and fungal infections [4]. Understanding these interactions is crucial for developing targeted interventions that can modulate the oral microbiome to promote health and mitigate disease risks.

This chapter embarks on a journey to explore the intricate interplay between the oral microbiome and acute infections. It begins by providing an overview of the composition of the oral microbiome, highlighting the diverse microbial species that inhabit this ecosystem and their dynamic relationships with each other and the host. Additionally, it delves into the factors that influence the composition of the oral microbiome, including environmental factors, diet, hygiene practices, and host genetics.

Building upon this foundation, the chapter delves into the fascinating realm of the oral microbiome's interactions with the host immune system. It examines how the oral microbiome shapes immune responses, both locally within the oral cavity and systemically throughout the body. Furthermore, the chapter explores the immunomodulatory effects of oral microbiota, shedding light on their ability to modulate inflammatory pathways and immune cell functions. Moving forward, this chapter shifts the focus to the role of the oral microbiome in viral infections, a topic of increasing significance in light of recent global health challenges. It investigates the mechanisms by which oral microbes interact with viruses, influencing viral susceptibility and disease severity. From respiratory viruses to emerging viral threats, the chapter examines the implications of oral microbiome dynamics on viral infections and discusses potential therapeutic strategies.

The chapter also delves into the realm of acute oral diseases, including dental caries, periodontal diseases, mucosal infections, and halitosis. It unravels the intricate connections between the oral microbiome composition and the onset, progression, and severity of these diseases, emphasizing the role of microbial dysbiosis in disease pathogenesis. Expanding the perspective, the chapter explores the linkages between the oral microbiome and acute systemic diseases, encompassing a wide range of conditions affecting various organ systems. From gastrointestinal disorders to cardiovascular diseases, the chapter investigates how the oral microbiome influences systemic health and disease outcomes.

In the realm of therapeutic interventions, the chapter highlights emerging strategies aimed at harnessing the potential of the oral microbiome. Probiotics, phage therapy, and gene therapies

DOI: 10.1201/9781003478638-7

targeting oral microbiome imbalances represent promising avenues for promoting oral health and combating acute infections. This chapter examines the scientific rationale behind these approaches and discusses their potential applications in clinical practice.

As the chapter navigates through these topics, it also reflects on the future directions and research challenges in the field of oral microbiomes and acute infections. Advancements in technology, such as high-throughput sequencing and multi-omics approaches, offer unprecedented opportunities to deepen our understanding of microbial communities and their interactions. However, translating scientific discoveries into clinical practice remains a formidable challenge, requiring interdisciplinary collaboration and innovative solutions. Moreover, this chapter serves as a comprehensive exploration of the intricate relationship between the oral microbiome and acute infections. By elucidating the mechanisms underlying these interactions and exploring novel therapeutic avenues, this chapter aims to contribute to the advancement of oral health care and the development of strategies to combat acute infections effectively.

7.2 ORAL MICROBIOMES: AN OVERVIEW

The oral microbiome is a complex and diverse ecosystem of microorganisms that inhabit the human oral cavity. It encompasses a wide array of bacteria, fungi, viruses, and other microbes, forming a dynamic community that plays a crucial role in maintaining oral health and overall well-being. Understanding the composition of oral microbiomes, the dynamic relationship between microbiota and host, and the factors influencing their composition is essential in comprehending their impact on human health and disease.

7.2.1 Compositions of Oral Microbiomes

The oral microbiome comprises a vast number of microbial species, with bacteria being the predominant group. These bacteria belong to various phyla, including Firmicutes, Bacteroidetes, Actinobacteria, Proteobacteria, and Fusobacteria, among others. Each phylum consists of numerous genera and species, creating a diverse microbial landscape within the oral cavity [1].

Furthermore, the oral microbiome is not uniform throughout the oral cavity but varies across different oral sites such as the teeth, gums, tongue, cheeks, and palate. For instance, the microbial community on tooth surfaces differs from that on the tongue's dorsal surface or the gingival crevices. This site-specific variation contributes to the overall diversity and complexity of oral microbiomes.

7.2.2 Dynamic Relationship between Microbiota and Host

The interaction between the oral microbiota and the host is dynamic and intricate, involving both cooperative and competitive relationships. Commensal bacteria in the oral cavity coexist with the host in a mutually beneficial manner, contributing to processes such as digestion, immune system development, and defense against pathogens [5].

However, this relationship can become dysregulated under certain conditions, leading to the overgrowth of pathogenic microbes and the development of oral diseases. Factors such as diet, oral hygiene practices, host genetics, immune status, and environmental influences can affect the balance of microbial communities and their interaction with the host.

7.2.3 Factors Influencing the Composition of Microbiomes

Several factors influence the composition and diversity of oral microbiomes. These include but are not limited to:

- *Host genetics*: Genetic variations in the host can influence the types of microbes that colonize the oral cavity and their abundance. Certain genetic traits may predispose individuals to specific oral microbiome compositions, affecting their susceptibility to oral diseases [6].

- *Diet and nutrition*: The type of diet consumed can significantly impact the oral microbiome. For example, diets high in sugar and carbohydrates can promote the growth of cariogenic bacteria associated with dental caries, while diets rich in fiber and nutrients can support a more diverse and balanced microbiome [7].
- *Oral hygiene practices*: Regular brushing, flossing, and oral rinsing play a crucial role in maintaining a healthy oral microbiome by reducing plaque accumulation and controlling microbial populations. Poor oral hygiene habits can lead to microbial dysbiosis and oral diseases.
- *Environmental exposures*: Environmental factors such as tobacco use, alcohol consumption, exposure to pollutants, and microbial transmission from close contacts can influence the oral microbiome's composition and stability [8].
- *Host's immune response*: The host's immune system plays a vital role in shaping the oral microbiome and maintaining microbial homeostasis. Immune deficiencies or hyperactive immune responses can disrupt the microbial balance and contribute to oral health issues [9].

The oral microbiome is a complex and dynamic ecosystem influenced by various factors, including microbial interactions, host factors, and environmental exposures. Understanding the compositions, dynamics, and influencing factors of oral microbiomes is essential for elucidating their role in health and disease and developing targeted strategies for maintaining oral and systemic well-being.

7.3 ORAL MICROBIOMES AND IMMUNE RESPONSE

7.3.1 Interactions between Microbes and Immune Response

The interactions between oral microbes and the immune response are highly complex and play a pivotal role in maintaining oral health and preventing infections. The oral cavity constitutes a dynamic environment hosting a diverse microbial community, including bacteria, viruses, fungi, and various microorganisms [10]. While oral colonization does not invariably lead to oral diseases, it can lead to oral infections, particularly when host immunity is compromised or there is a high risk of systemic infection [11]. The immune system in the oral mucosa effectively maintains a delicate balance between tolerance to commensal microorganisms and the ability to respond rapidly to potential pathogens. The immune system recognizes specific molecular patterns associated with microbes, known as pathogen-associated molecular patterns (PAMPs), as a part of the innate immune response. Pattern recognition receptors (PRRs), such as toll-like receptors (TLRs) present in immune cells, play a crucial role in identifying these PAMPs [12].

The oral mucosa harbors a significant population of commensal microorganisms, constituting the normal oral flora. These microorganisms contribute to the maintenance of oral health by competing with potential pathogens for resources and niche space [13]. The immune system has evolved mechanisms to recognize and tolerate these commensal microbes while effectively preventing their excessive growth and invasion. The oral mucosa is protected by two distinct immune systems, namely, the systemic and secretory [14]. Host defenses, particularly in the context of oral *Candida* species infections, rely on the prompt activation of the acute inflammatory response (innate immunity), followed by sequential and incremental stimulation of specific immune responses mediated by T cells (cellular immunity) or B cells (humoral immunity), thereby contributing to immune homeostasis [15].

Saliva exerts a significant impact on the oral microbiota and encompasses various antimicrobial components, including enzymes like lysozyme, antimicrobial peptides, and immunoglobulins. These constituents collectively play a pivotal role in averting microbial overgrowth and infections [16]. Notably, immunoglobulin A (IgA) plays a key role in mucosal immunity and is present in saliva [17]. A prime example is secretory immunoglobulin A (SIgA), found in mucous secretions within salivary glands, which functions as an initial defense mechanism that prevents the incursion of pathogens such as bacteria and viruses by inhibiting the pathogen attachment and settlement to the mucous membrane through binding and aggregating pathogens [18]. The flow of saliva aids in flushing away microbes while providing first-line innate defense and contributing significantly to the overall maintenance of oral homeostasis between the host and resident microbial communities [19].

Cells lining the oral cavity, including epithelial cells, serve as formidable physical barriers. These cells express PRRs, such as TLRs, enabling them to initiate innate immune responses upon detecting microbial invasion [20]. Antigen-presenting cells (APCs), such as dendritic cells, macrophages, or mast cells, capture and present microbial antigens to T cells [21]. This process initiates adaptive immune responses. The T cells, including helper T cells (CD4+) and cytotoxic T cells (CD8+), are critical for coordinating immune responses [22]. They help to regulate the activity of other immune cells and contribute to the elimination of infected cells. However, certain oral microbes have mechanisms to evade the immune system or modulate immune responses to their advantage in establishing persistent colonization. This may involve strategies to inhibit phagocytosis, interfere with cytokine signaling, or promote anti-inflammatory responses [23].

In conditions like dysbiosis, an imbalance in the oral microbial community is associated with various oral diseases, including periodontal disease and dental caries [24]. Chronic inflammatory conditions in the oral cavity may result from deregulated immune responses or failure to control microbial challenges effectively, potentially leading to tissue damage or bone loss [25]. For instance, *Porphyromonas gingivalis* has been implicated in inducing dysbiosis by compromising innate host defenses while simultaneously promoting inflammatory responses in phagocytic cells [26]. Resident immune cells, including neutrophils and macrophages, play a crucial role in responding to pathogens in the gingival tissues [27]. In the event of infection or tissue damage, these immune cells are activated and actively participate in the immune response. Notably, neutrophils play a central role in the innate immune response, which is recruited to sites of infection and can engulf and destroy microbes through phagocytosis.

Microbes organized into biofilms play a crucial role in the formation of dental plaque on tooth surfaces, potentially leading to the development of oral diseases such as cavities and periodontal diseases [28]. This process directly relates to the phenotypic resistance of these biofilm-based microbes to antimicrobials and is probably linked to their capacity to modulate host immunity, thereby evading protective responses [29]. The immune response is involved in the surveillance and control of dental plaque, with immune cells like neutrophils interacting actively with bacteria to maintain homeostasis [30].

7.3.2 Immunomodulatory Effect of the Oral Microbiota

The oral microbiota plays a crucial role in maintaining oral health and has been increasingly recognized for its potential impact on the overall immune system. The immunomodulatory effects of the oral microbiota refer to its ability to influence and regulate the immune response. Commensal bacteria exhibit both proinflammatory and anti-inflammatory activities, which is important for maintaining host-microbe homeostasis at heavily colonized sites [31]. Certain immunomodulatory commensal microbes can regulate the activities, development, and deployment of host immune cells, providing subtle effects on immune responses and overall immune status. For instance, specific members of the oral streptococci, particularly *Streptococcus mitis/oralis*, have demonstrated suppression of CXCL8 secretion by cultured epithelial cells, which in turn, inhibit the stimulation of TNF-α, leading to the activation of NF-κB [32]. Similarly, a study on *Candida* species infection demonstrated an increment in the proportion of regulatory T cells (Tregs) through the TLR2/

MyD88 pathway, which contributes to an elevated fungal burden, disrupting mucous membrane immunity and fostering the induction of commensal tolerance [33].

Emerging research suggests that the composition and balance of the oral microbiota influence systemic immune responses beyond the oral cavity. Alterations in the oral microbiota composition have been linked to inflammatory conditions, such as periodontal disease, diabetes, and other systematic diseases. For example, a study delved into the connection between *P. gingivalis* and inflammatory bowel disease (IBD), revealing that *P. gingivalis* aggravated colitis by the activation of CD4+ T cells and up-regulation of the Th17/Treg ratio *via* the JAK-STAT signaling pathway, thereby contributing to the development of bowel disease [34]. Lipopolysaccharides (LPS) and cell extracts from *Fusobacterium nucleatum* have been shown to elevate levels of inflammatory cytokines and chemokines, such as CCL20, creating a proinflammatory microenvironment. *Fusobacterium nucleatum* further enhances the abundance of CD11b+ myeloid cells, myeloid-derived suppressor cells (MDSCs), tumor-associated neutrophils (TANs), and tumor-associated macrophages (TAMs), thereby promoting the neoplastic progression [35].

The modulation of dendritic cells (DCs) in conditions like diabetes contributes to the loss of periodontal bone through an increased abundance of Th1 or Th17 lymphocytes or a reduction in regulatory T cells [36]. These Th1 and Th17 cells, which are known regulators of bone resorption, produce cytokines like RANKL and IL17, which are found in elevated levels in the case of periodontitis patients with diabetes, intensifying the pathogenicity of the oral flora and leading to the recruitment of neutrophils [37]. Studies have demonstrated that blocking the formation of advanced glycation end products (AGEs) can reduce levels of inflammatory cytokines, specifically tumor necrosis factor (TNF)-α, interleukin (IL)-1β, IL-6, and IL-18 [38]. Additionally, this intervention has been linked to decreased matrix metalloproteinase expression and a reduction in bone loss in the gums.

The oral and gut microbiota share a dynamic interconnection, and alterations in the oral microbiota have the potential to influence the gut immune system, thereby impacting overall immune homeostasis. A research study by Atarashi [39] revealed that ectopic gut colonization by oral *Klebsiella* species led to an increased abundance of Th1 cells. This resulted in the secretion of IL-18 by TLR4-activated colonic epithelial cells, leading to the amplification of the Th1 cell response. The oral microbiota can communicate with the host's immune system through various signaling pathways. This communication can involve the recognition of microbial patterns by host pattern recognition receptors (PRRs), which then trigger immune responses [40]. For instance, the stimulation of macrophages *via* TLR signaling tends to prime them towards anti-inflammatory profiles, resulting in lower levels of inflammatory cytokines, such as IL-6, reduced production of reactive oxygen species, and exposure to β-glucans promotes cell viability [41].

7.3.3 Implication for Acute Infection Susceptibility

The oral microbiota can have implications for an individual's susceptibility to acute infections. Disruptions in the balance of the oral microbiota, known as dysbiosis, can impact the local immune response, potentially increasing susceptibility to infections. Factors such as being immunecompromised, as well as the use of antibiotics and antifungals, can increase the likelihood of opportunistic bacterial and fungal pathogens causing infections in the oral cavity [42]. Moreover, inadequate oral hygiene and conditions such as periodontal disease can lead to inflammation and compromise the integrity of the oral tissues. This may create opportunities for pathogens to enter the bloodstream, potentially increasing the risk of systemic infections [43].

The oral cavity is intricately linked to the respiratory tract, and the aspiration of oral bacteria into the lungs can contribute to respiratory infections, such as pneumonia, especially in individuals with compromised immune function or underlying respiratory conditions [44]. For instance, a study indicated that oral bacteria, including periodontal pathogens, can disrupt the alveolar and bronchial epithelial barriers, exacerbating the progression of inflammation in the lower airways, which may extend to systemic organs, such as the lungs, through the bloodstream, leading to lung dysbiosis [45].

Furthermore, oral infections can also mediate complications in pregnancy, including preterm birth and low birth weight of infants, which occurs through the translocation of oral microbiomes to the fetoplacental unit, inducing fetal or maternal response that results in preterm birth [46]. The process may also involve the systemic dissemination of prostaglandins and inflammatory mediators such as IL-6, IL-8, TNF-α, and PGE2 on the fetal placental unit, as well as the impact of the oral microbiome reservoir of lipopolysaccharides on the fetal placental unit.

Persistent inflammation can contribute to the susceptibility to infections beyond the oral cavity, such as chronic periodontitis or trauma, which can lead to several clinical consequences as inflammatory mediators may not only drive the destruction of tooth-supporting structures, leading to tooth and alveolar bone loss, but also may increase the risk of hypertension, hyperglycemia, type 2 diabetes, chronic kidney disease, depression, neurodegenerative diseases, cardiovascular disease, and various types of cancer [47, 48]. Some studies also indicate that alterations in the composition of the oral microbiota are linked to sensitization and development of allergic reactions, extending beyond the oral health domain, including conditions such as asthma [49], and peanut allergies [50].

Dysbiosis and inflammation in the oral cavity may also lead to implications in neurodegenerative conditions, notably Alzheimer's disease. A few have shown a positive correlation between chronic periodontitis and Alzheimer's disease. Investigators have found that chronic periodontitis triggers the release of inflammatory mediators, such as C-reactive protein and interleukin-6, which have the potential to disrupt the blood-brain barrier, thereby contributing to neuroinflammation [51, 52].

7.4 ORAL MICROBIOMES AND VIRAL INFECTIONS

7.4.1 Exploration of the Role of Oral Microbiomes in Viral Infections

The oral microbiome plays a crucial role in maintaining oral health and influencing overall well-being. While the majority of research has been focused on the role of the oral microbiome in bacterial infections, there is emerging evidence suggesting its potential involvement in viral infections as well. The oral cavity acts as a gateway for various viruses, and some, including herpes simplex virus (HSV), human papillomavirus (HPV), and cytomegalovirus (CMV), can establish latent infections in the oral cavity. Some studies have detected these viruses in both clinically healthy gingival sulci and periodontal pockets [53], which may be attributed to a symbiotic relationship between microbes in these crevices and the viruses [54]. The composition and balance of the oral microbiome may influence the ability of viruses to establish and reactivate infections [5]. Disruptions in the balance of the oral microbiome may impact the defense mechanisms of the local immunity, potentially rendering the host susceptible to viral infections.

The oral microbiota can indirectly influence the colonization of non-indigenous microbes and subsequent infection by triggering host responses, such as mucosal immune defenses [55]. This dynamic interaction can alter the susceptibility and affect the outcomes of viral infections. The microbiota can prevent, suppress, or aggravate viral infections through various mechanisms, overcoming physical barriers and interacting with factors like tissue specificities, body temperature, mucus barriers, and epithelial secretions like IgA and defensins [56]. The presence of certain microorganisms in the oral microbiome may influence the transmission and replication of viruses [57]. Several studies have suggested that the interactions between the oral microbiome and viral infections, particularly involving specific bacteria, can either enhance or inhibit viral replication. For instance, a study indicated that oral or nasal inoculation of mice with various strains of *Lactobacillus*, such as *Lactobacillus gasseri*, *Lactobacillus rhamnosus*, and *Lactobacillus brevis* induced protective immunity against influenza virus infections [58, 59].

Viruses can be detected in saliva, and the oral microbiome plays a role in shaping the composition of saliva. Alterations in the oral microbiome have the potential to impact the viral load present in saliva, thereby influencing the transmission of certain viruses. For instance, a study indicated that the oral microbiome, containing specific oral microbes like *Streptococcus oralis* and *Streptococcus mitis* showed an increment in the viral load, in addition to affecting the neuraminidase (NA) activity of the influenza virus, hence, driving virus-oral microecological imbalance [60]. Emerging studies have also indicated that the oral cavity may act as a reservoir for SARS-CoV-2, potentially serving as an entry point to both the respiratory and gastrointestinal tracts, which may influence the infection as well as clinical symptoms [61].

7.4.2 Interactions between Oral Microbes and Viruses

The interaction between oral microbes and viruses is a dynamic process. The oral cavity provides a portal of entry for many viruses, and their ability to establish infection depends on interactions with host cells and the local microenvironment. Some oral bacteria may enhance the ability of viruses to infect host cells by facilitating attachment and entry. On the other hand, certain bacteria may produce substances that inhibit viral growth. For instance, a study indicated that *Prevotella* species overexpress certain proteins that appear to promote viral infections [62]. Similarly, the culture supernatants of NA-producing oral bacteria, such as *Streptococcus oralis* and *Streptococcus mitis*, have been observed to promote the release of influenza virus and facilitate cell-to-cell spread of the infection [63]. These bacteria also contribute to increased viral M1 protein expression levels and activation of cellular extracellular signal-regulated kinases (ERK). The enhancement in the proportion of neuraminidase-producing oral bacteria may elevate the risk of viral infections.

Previous research indicates that viruses have the potential to induce alterations in both local microbial composition and quantity. For example, a study observed an increased abundance of *Streptococcus*, *Actinomyces*, and *Rothia* in the oropharynx of flu patients experiencing bacterial complications [64]. While another study revealed a decreased presence of *Fusobacterium periodonticum* in individuals affected by SARS-CoV-2 infection [65]. Furthermore, *in vitro* investigations have revealed that coinfection of *P. gingivalis* and influenza virus can significantly lead to apoptosis and necrosis in epithelial cells of lungs over a short timeframe [66]. This process involves the activation of the Bcl2/Bax/Caspase3 signaling pathway [67]. Cytotoxic products of *P. gingivalis*, such as lipopolysaccharide, were shown to induce the release of inflammatory cytokines such as TNF-α, IL-1β, and IL-6, from host cells [68], which in turn contribute to damage in lung epithelial cells.

In addition to virus-microbe interactions, the interactions of herpes viruses, such as Epstein-Barr virus (EBV) and human cytomegalovirus (HCMV), with periodontopathic bacteria have the potential to contribute to the pathogenesis of periodontal diseases. A study also highlighted that the microbial synergy between periodontopathic bacteria, specifically *Porphyromonas gingivalis* and EBV, results in the production of butyric acid, which induces the reactivation of EBV, leading to the release of inflammatory cytokines and compromising local host defenses [69]. The elevated production of inflammatory cytokines due to the synergistic effects of EBV and periodontopathic bacteria contribute to the progression of periodontal disease and other EBV-related diseases. Moreover, certain microbes have also been identified to secrete proteases such as trypsin-like proteases (TLP) and others, which activate the infectivity of viral infections like influenza virus by proteolytically cleaving the hemagglutinin [70].

7.4.3 Implications for Viral Susceptibility and Severity

The oral microbiome may influence the establishment of viral infections, potentially impacting the local immune response and leading to increased susceptibility or severity of viral infections. Studies suggest that oral dysbiosis or an imbalance in the oral microbiome could result in decreased microbial communities during infections such as hepatitis C (HCV), human immunodeficiency virus (HIV), and influenza [71]. This dysbiosis may contribute to elevated severity symptoms, possibly attributed to a simultaneous increase of proinflammatory cytokines like IL-6, TNF-α, and IL-1β [72]. In addition, studies have indicated that alternations in the oral bacterial community may favor the severity of SARS-CoV-2 infection, facilitating enhanced local inflammation, promoting viral replication, contributing to the presence of oral myocytes, and diminishing mucosal sIgA response [73].

The viral infections can trigger the proliferation of a specific pathobiont, resulting in secondary infections [74]. For instance, an increased abundance of species like *Streptococcus* up-regulates adhesion receptors, facilitating viral entry during respiratory viral infections [75]. Similarly, a decrease in immunomodulatory organisms, such as *Streptococcus*, *Lactobacillus*, and *Fusobacterium*, can make individuals highly vulnerable to viral infections, leading to disease severity, increased viral replication, and progression of disease. Moreover, the enrichment of opportunistic pathogens in the oral cavity in individuals may induce alterations in angiotensin-converting enzyme 2 (ACE-2) expression levels, immunological profiling, oxidative stress, and mitochondrial dysfunction, which may promote higher susceptibility to viral infections, particularly in aged populations, by increasing viral replicability and the likelihood of coinfection [76, 77].

The oral microbiome is also responsible for increasing the abundance of specific species like *Prevotella* and *Veillonella*, which produce proteins that can enhance SARS-CoV-2 infection and inflammation and intensify the severity of viral infection, whichis associated with an increased risk of death in older individuals [78]. Additionally, in individuals with acute HIV infection, the dysbiosis of oral microbiota can lead to the reduction and dysfunction of innate and adaptive immune cells, particularly CD4+ T lymphocytes, which increases the vulnerability to HIV-related diseases and comorbidities associated with HIV that are not acquired immunodeficiency syndrome [79].

7.5 ORAL MICROBIOMES AND ACUTE ORAL DISEASES

The oral microbiome refers to the microorganisms (bacteria, fungi, viruses) found in the human oral cavity. Antonie van Leeuwenhoek first discovered it; he recognized the oral microbiome through a microscope. He was known as the father of microbiology andmade vast contributions to the identification of both protists and bacteria. The oral microbiome is the overall genome of microorganisms that inhabit the oral cavity. Everyone comprises a common type of core microbiome. However, each person has a variable microbiome that differs based on lifestyle and physiological variations. The oral microbiome plays a crucial role in overall health; it may lead to systemic and oral diseases (Figure 7.1). Microbesreside within biofilms that are found throughout the oral cavity, leading to healthy, balanced conditions [1]. However, sometimes, in homeostasis, there are some imbalances that let microbes emerge and spread diseases; this condition is known as dysbiosis. It was the first step to identifying the microbiome health; subsequently, it helps to understand how the microbiome affects the metabolic and functional pathways associated with disease states.

It has been evident in recent years that disease diagnosis may be achieved through analyzing the human oral microbiota; mouth (dental caries, periodontitis, halitosis) and systemic disease (cardiovascular, gastrointestinal, respiratory, nervous system related) development and progression are greatly impacted by oral microbial dysbiosis [80].

Two major oral pathologies originate from oral microbiota dysbiosis: caries and periodontitis. However, the characteristics of these two disorders differ [26].

7.5.1 Dental Caries

Caries is a very common infectious disease. It is the destruction of the hard tissues of the tooth. In addition, presence of cariogenic bacteria in tooth plaque caused demineralization of teeth. Consequently, any bacteria found in plaque that can convert sugar into lactic acid may play a role in causing the disease [81].

Salivary dysbiosis is actually the loss of bacterial variety, which occurs because of a disordered supragingival biofilm linked to an excessive sugar intake and poor dental hygiene. As a consequence of this imbalance, the composition of the biofilm changes, which causes an increase in gram-positive bacteria and medium acidification [82]. *Actinomyces*, *Lactobacillus*, and *Streptococcus mutans* are

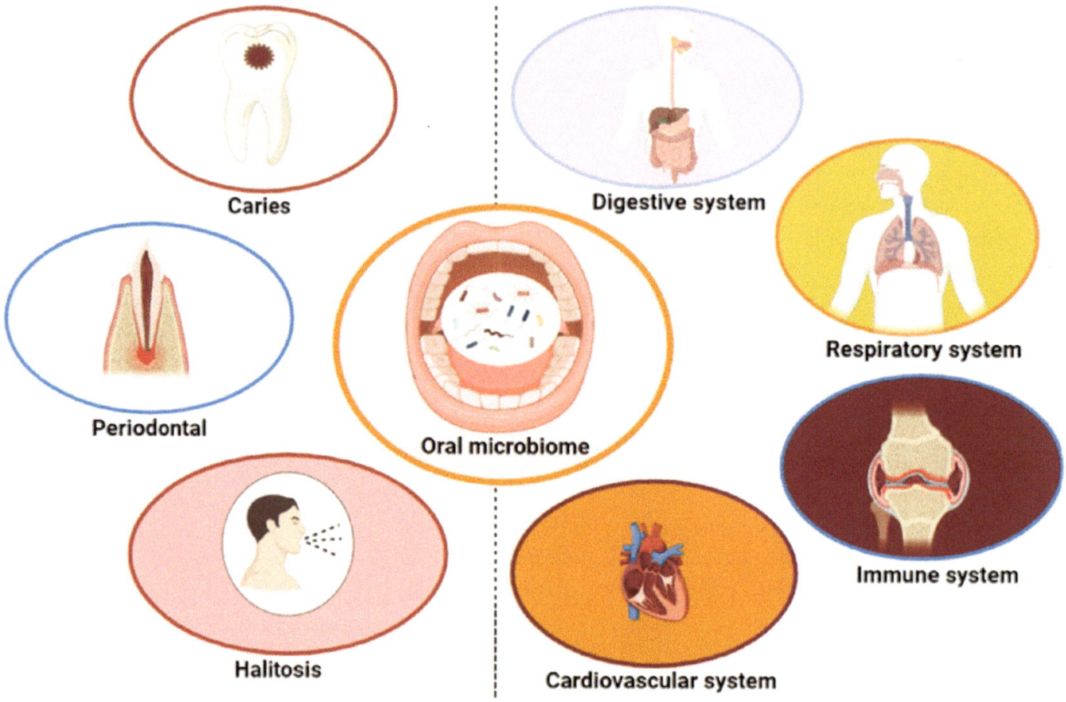

FIGURE 7.1 The role of the human oral microbiota on oral health conditions.

the primary pathogens that have grown. There are also higher concentrations of *Bifidobacterium* species, *Scardovia* species, and *Candida albicans* [83, 84]. While beneficial bacteria prefer to live at neutral pH levels, there has been a decrease in their number and metabolic activity along with an increase in acidification. Interestingly, many studies have suggested the involvement of the oral microbiome in the onset and progression of dental caries (Table 7.1).

A study reported by [85] stated a comparison of profiling the oral microbial between caries-free children and children with severe early childhood caries (SECC). They identified 379 different types of bacteria from every child and discovered a number of taxa, such as *Porphyromonas* and *Streptococcus*. It was revealed that *Actinomyces* showed a substantial correlation with SECC, and thus it may be considered as a useful biomarker for dental caries in primary dentition.

Johansson [86] investigated the study in which comparisons between Romanian adolescents with dental caries and Swedish caries active and caries free programs were attempted. The biofilm samples were analyzed by quantitative polymerase chain reaction (qPCR). It was revealed that *S. mutans* and *S. sobrinus* were typically found in Swedish adolescents involved in dental care programs. However, they were associated with caries-ridden Romanian teenagers with restricted access to dental care. In Swedish teenagers with active dental caries, *Actinomyces*, *Selenomonas*, *Prevotella*, and *Capnocytophaga* were commonly found to be colonized. A study demonstrated by Wang [87] utilized single-molecule real-time DNA analysis to compare and analyze saliva from 20 children of Han Chinese descent who had caries and 21 children who did not. It was found that *Prevotella* spp., *Lactobacillus* spp., *Dialister* spp., and *Filifactor* spp. could be associated with the onset and development of dental caries.

A study by Richards et al. [88] stated the oral microbiota association in the onset and progression of dental caries; analysis through Human Oral Microbe Identification Using Next-Generation Sequencing (HOMINGS) was used to analyze the microbiomes of specific supragingival dental plaques in children. They found that *Streptococcus sanguinis* and a majority of acidogenic/aciduric

TABLE 7.1
The Studies Demonstrated the Role of Oral Microbial Species in the Onset of Dental Caries

Infected Tissue	Microbial Species	Role in Caries	References
Dentine	*Porphyromonas*, *Streptococcus*, and *Actinomyces*	Initiation and progression	[85]
Dentine	*Actinomyces*, *Selenomonas*, *Prevotella*, and *Capnocytophaga*	Initiation and progression	[86]
Dentine	*Prevotella* spp., *Lactobacillus* spp., *Dialister* spp., and *Filifactor* spp.	Initiation and progression	[87]
Dentine	*Lactobacillus salivarius*, *Scardovia wiggsiae*, *Parascardovia denticolens*, and *Streptococcus mutans*	Onset and progression	[88]
Dentine	*Scardovia wiggsiae* in the *Actinomyces/Bifidobacterium* family	—	[89]
Dentine and enamel	*Streptococcus sanguinis*, *Streptococcus mutans*, *Scardovia wiggsiae*, and *Actinomyces*	Biofilm formation	[90]

organisms, including *Lactobacillus salivarius*, *Scardovia wiggsiae*, *Parascardovia denticolens*, and *Streptococcus mutans*, were detected in those with caries active with dentin carious lesions. Another study showed the role of the oral microbiome in the assay of the selection of cariogenic species in diseases. Through multiple assays, it was found that the caries association of *Scardovia wiggsiae* in the *Actinomyces/Bifidobacterium* family, both the cariogenic characteristics of acid generation and acid tolerance, are shared by a number of *Actinomyces* species [89]. Dental caries originates from a dysbiosis of biofilms associated with teeth and often penetrate the enamel to the dentin, which is distinct in composition and structure. It was revealed that there was an increased expression of *Streptococcus sanguinis*, *Streptococcus mutans*, *Scardovia wiggsiae*, and *Actinomyces* [90].

7.5.2 Periodontitis

The oral microbiome plays a crucial role in the onset and progression of acute oral infections such as periodontitis. Periodontitis is a disease of the oral cavity spurred on by periodontal dysbiosis and an adverse immuno-inflammatory response in the host. This causes the teeth to gradually and permanently deteriorate resulting in the periodontium, the attachment system with gum, and eventually loss of teeth. The most common type of acute infection in periodontitis is aggressive periodontitis. It is caused by a weakened immune system and an acute loss of periodontal attachment linked to highly pathogenic bacteria. It is divided into two categories: localized aggressive periodontitis (LAP) and generalized aggressive periodontitis (GAP). Another type of periodontal disease called acute necrotizing ulcerative gingivitis is characterized by very painful, bleeding gums and severe halitosis. *Porphyromonas gingivalis*, *Treponema denticola*, and *Tannerella forsythia*, collectively referred to as the "red complex," are widely believed to be the main pathogenic organisms associated with periodontitis [3]. It is characterized by the increased number of concentrations of gram-negative anaerobic bacteria (*Treponema gingivalis*, *Tannerella forsythia*, *Fusobacterium nucleatum*, *Prevotella intermedia*, *Dialister* spp., and *Selenomonas* spp.) [12].

New studies indicate that periodontitis could have been initiated and developed predominately by oral microorganisms. The oral microbiota exhibits the characteristics of biofilm and the development of dental plaques on tooth surfaces (Table 7.2).

Oral dysbiosis is a significant hallmark during periodontitis progression. Oral microorganisms like *Streptococcus oralis*, *Streptococcus mutants*, *Streptococcus intermedius*, *Streptococcus mitis*,

TABLE 7.2
The Impact of Oral Microbial Species Diversity on Periodontal Diseases

Periodontal Diseases	Microbial Species	Techniques	References
Localized aggressive periodontitis	*Fusobacterium nucleatum*	16S rRNA sequencing	[92]
Periodontitis	—	16S rRNA sequencing	[93]
Interdental biofilm disruption, periodontitis	*Tannerella forsythensis, Campylobacter rectus, T. denticola, Campylobacter concisus, and Eikenella corrodens*	—	[94]
Aggressive periodontitis	*Acteroidetes, Bacteriodetes, Peptostreptococcaceae, Veillonella, Treponema, Filifactor,* and *fretibacterium*	MiSeq sequencing of 16S rRNA gene	[95]

Veillonella parvula, and *Pseudomonas fluorescence* are responsible for an increase in virulence that drives development of oral dysbiosis [91].

Topcuoglu [92] collected samples from 84 participants by sequencing 16S rRNA, including those with generalized aggressive periodontitis, generalized chronic periodontitis, peri-implantitis, and localized aggressive periodontitis. The genes were chosen from ten species. The red complex bacteria were ultimately shown to be the most common, present in all groups at extremely high levels. High amounts of *Fusobacterium nucleatum* have been identified in every sample.

A study assessed, through analysis of DNA and amplified 16S rRNA genes,the oral microbiome of 16 systemically healthy participants with the following oral statuses: dental health, edentulism, periodontitis, and established caries. Healthy participants had the highest diversity, with 29 bacterial species present in considerably higher abundance than those with the other conditions [93]. The least common species were *Aggregatibacter actinomycetemcomitans*, *P. gingivalis*, and *Actinomyces viscosus*, while *F. nucleatum* was the most common one. With advancing molecular identification techniques, there was a significant increase in *Tannerella forsythensis*, *Campylobacter rectus*, *T. denticola*, *Campylobacter concisus*, and *Eikenella corrodens* [94] among the tested samples.

A study showed the comparisons of subgingival and buccal mucosal microbiome diversity in aggressive periodontitis. Through MiSeq sequencing involving 16S rRNA gene, 32 buccal swab samples and 113 subgingival samples were obtained, which included 12 from participants with aggressive periodontitis and 9 from periodontal healthy participants. It was revealed that the buccal mucosal and subgingival microbiomes differed from periodontal samples of the healthy participants. The most predominately found were *Acteroidetes*, *Bacteriodetes*, *Peptostreptococcaceae*, *Veillonella*, *Treponema*, *Filifactor*, and *Fretibacterium* [95].

7.5.3 Oral Mucosa

The oral mucosa is a living tissue of the mouth cavity, is a doorway to the body, and provides first-line protection against infections, airborne allergens, and exogenous chemicals in the form of a physical and microbiological-immune barrier. The oral mucosa is a very significant indicator of an individual's health. It is composed of two primary parts: a microbiological-immune barrier that maintains homeostasis in the internal environment and a physical barrier made up of cell-cell junctions and stratified epithelial cells. A number of variables, such as immune system components, saliva, proteins, and microorganisms have been implicated in the breakdown of the oral epithelial barrier. Mucosal structural and functional changes lead to several systemic diseases and oral diseases [96]. The impact of the oral microbiota in acute mucosal diseases is part of very few studies. It also has been noticed that the coexisting oral microbiota has a significant impact on the colonization, resistance, and induction of primary immunity of pathogenic bacteria. Oral dysbiosis can cause damage to oral health mucosal epithelial defenses, accelerating the onset of mucositis and ulcer diseases [97].

The dysbiosis of the mucosal and salivary microbiota is associated with recurrent aphthous stomatitis (RAS). Through analysis of 16S rRNA gene sequencing, rare species have been shown to be more common in the mucosal and salivary microbiota of RAS patients. RAS risk has been associated with decreased levels of *Streptococcus salivarius* and higher levels of *Acinetobacter johnsonii* in the mucosa [98]. Amer [99] analyzed the microbiome of oral leukoplakia in 36 patients compared to healthy mucosal tissue. Through illuminative gene sequencing analysis, it was revealed that there was an increased abundance of Fusobacteria and decreased levels of Firmicutes; also, an abundance of the acetaldehyde producing microbes like *Rothia mucilaginosa*. This microbe is generally located on oral leukoplakia at lingual sites.

Another study compared the microbiomes in ulcerative and normal mucosa of patients with recurrent RAS disease. It was revealed that Firmicutes, Proteobacteria, and Bacteroidetes were the most abundant phyla in the microbiomes. It was evident that *Escherichia coli* is a very common intestinal bacterium; *E. coli* colonization could be a cause of RAS and controlling *E. coli* colonization could thus help cure RAS [100].

7.5.4 Halitosis

Halitosis is a disorder in which an unpleasant smell comes from the breath. The illness may include both non-oral and oral diseases. The causes of halitosis can be complex and includeimpacted food or debris, deep carious lesions, exposed necrotic tooth pulps, pericoronitis, mucosal ulcerations, healing (mucosal) wounds, tongue coating, periodontal disease, and conditions that lower salivary flow rate [101]. Those with halitosis had a higher volatile sulfur compound (VSC) level, which the oral microbiota could have mediated [80]. Numerous studies have demonstrated that halitosis is due to oral microbiota dysbiosis (Table 7.3).

One study described the generation of volatile compounds in intra-oral halitosis (IOH), which are primarily created by anaerobic bacteria. These foul-smelling substances comprise aromatic compounds, VSCs, amines, short-chain fatty or organic aldehydes, ketones, alcohols, aliphatic molecules, and acids. *Actinomyces*, *Bacteroides*, *Dialister*, *Eubacterium*, *Fusobacterium*, *Leptotrichia*, *Peptostreptococcus*, *Porphyromonas*, *Prevotella*, *Selenomonas*, *Solobacterium*, *Tannerella*, and *Veillonella* species are the oral bacteria most closely linked to halitosis. In addition to causing periodontitis, the majority of halitosis-causing bacteria could also influence the growth of the digestive tract and oral cancer [102]. *Peptostreptococcus*, *Alloprevotella*, *Eubacterium nodatum*, and *Stomato baculum* were observed to have significantly beneficial associations with the total number of VSCs [103].

A study was conducted to investigate the salivary microorganisms and metabolites in people with and without halitosis. Out of the 52 individuals, 22 were categorized as having halitosis based on the volatile sulfur analysis of compounds using breath samples. The differences in microorganisms and metabolites between the saliva of the control and halitosis groups were examined using metabolomics and 16S rRNA gene amplicon sequencing techniques. *Megasphaera*, *Alloprevotella*, and *Prevotella* all have significantly higher relative abundances in halitosis; these microbes may be related to the putrescine and cadaverine pathways that cause halitosis [104]. Foo [105] demonstrated the composition of oral microbiota, various oral habitats, and the metabolic pathway associated with the IOH-associated compounds. Microorganisms such as *F. nucleatum*, *P. gingivalis*, *P. intermedia*, *P. nigrescens*, and *T. denticola* may contribute to the formation of sulfur compounds through methionine and cysteine metabolism. Aromatic chemical production also involves *P. intermedia*, *P. gingivalis*, and *F. nucleatum*. The root cause of IOH is a change in the composition of the oral microbiome. It was employed through 16S rRNA gene sequencing, which revealed that a higher abundance of *Microbacterium*, *Megasphaera*, *Proteobacteria*, *Campylobacter*, and Spirochaetales [106] was often associated with IOH.

TABLE 7.3
The Possible Link between Oral Microorganisms and the Onset of Oral Halitosis through Research Studies

Microbes	Compounds	References
Dialister spp., *Eubacterium* spp., *Fusobacterium* spp., *Leptotrichia* spp., *Peptostreptococcus* spp., *Porphyromonas* spp., *Prevotella* spp., *Selenomonas* spp., *Solobacterium* spp., *Tannerella forsythia*, and *Veillonella* spp.	Aromatic compounds, VSCs, amines, short-chain fatty or organic aldehydes, ketones, alcohols, aliphatic molecules	[102]
Prevotella, *Alloprevotella*, *Leptotrichia*, *Peptostreptococcus*, and *Stomatobaculum*	VSCs	[104]
Megasphaera, *Alloprevotella*, and *Prevotella*	VSCs	[104]
F. nucleatum, *P. gingivalis*, *P. intermedia*, *P. nigrescens*, and *T. denticola*	Sulfur and aromatic compounds	[105]
Microbacterium, *Megasphaera*, *Proteobacteria*, Campylobacter, and *Spirochaetales*	—	[106]

Abbreviations: IOH, intra-oral halitosis; VSCs, volatile sulfur compounds.

7.6 ORAL MICROBIOME AND ACUTE SYSTEMIC INFECTIONS

The gut and mouth microbiomes are reliant on one another and have the ability to control both pathological and physiological processes. The oral-gut barrier keeps the microbiomes of the mouth and the gut effectively separated even though the two are continuous areas joined by the gastrointestinal tract. However, when the mouth-gut barrier is compromised, the oral microbiota can move to the intestinal mucosa. Due to its high vascularity and role as the main intake route for almost all ingested substances, the mouth cavity offers lots of chances to influence activity at other body regions. Therefore, it should come as no surprise that the oral microbiota has been linked to several systemic diseases in addition to oral cavity disorders. The digestive system and lungs may both be directly accessed through the mouth [3]. It is widely understood by several research studies that oral microbiome dysbiosis, a change from commensal to a pathogenic oral microbiome, can either cause or exacerbate other chronic systemic disorders (Table 7.4).

7.6.1 Gastrointestinal Diseases

Oral microbiomes have been associated with an increasing variety of gastrointestinal disorders. Out of all the diseases found, inflammatory bowel disease is one of the oldest. In recent years, there has been increasing evidence to support the links between oral microbiomes and gastrointestinal cancer (colorectal and pancreatic cancer).

7.6.2 Inflammatory Bowel Disease

Inflammatory bowel disease (IBD) is a chronic inflammatory disease of the gastrointestinal tract, which includes Crohn's disease and ulcerative colitis. Based on recent research, the oral microbiota may contribute to the onset and development of IBD. Dysbiosis of the oral and gut microbiota, with a decrease in bacterial diversity, an increase in Proteobacteria and Bacteroidetes phylum abundance, and a decrease in Firmicutes phylum abundance, are characteristics of individuals with IBD [107].

TABLE 7.4
Depicting How Oral Microbial Diversity Dysbiosis Impacts Various Kinds of Systemic Diseases

DiseaseType	Microbial Species Increased in Number	Microbial Species Decreased in Number	Infected Tissue	References
Inflammatory bowel disease	*Klebsiella* spp.	Gastrointestinal disease	Intestinal	[39]
Inflammatory bowel disease associated with periodontitis	*Eikenellacorrodens*, *Prevotella* genus, and *Veillonella* genus	*Streptococcus* genus, *Haemophilus* genus, *Neisseria* genus, and *Gemella* genus	Intestinal	[108]
Inflammatory bowel disease	*Campylobacter* genus, Streptococcaceae, Enterobacteriaceae, and *Veillonellaceae*	Porphyromonadaceae and Neisseriaceae families	Intestinal	[109]
Pancreatic cancer	*Porphyromonas gingivalis* and *Aggregatibacter actinomycetemcomitans*	—	Pancreas	[110]
Acute appendicitis	*P. stomatitis*, *E. corrodens*, and *F. nucleatum*	—	Intestinal	[111]
Respiratory disorders				
Aspiration pneumonia	*Candida albicans*, *Staphylococcus aureus*, methicillin-resistant *S. aureus*, and *Pseudomonas aeruginosa*	—	Lungs	[115]
Pneumonia	*Pseudomonas aeruginosa, K. pneumoniae*, enterococci, and *Enterobacter* spp.	—	Lungs	[113]
COVID-19	*Capnocytophaga* and *Veillonella*	—	Lungs	[60]
COVID-19	*Prevotella* and *Veillonella*	—	Lungs	[78]
Nervous system				
Early onset Alzheimer's diseases	*Porphyromonas gingivalis*-LPS	—	CNS	[118]
Immune system				
RA-associated periodontitis	*Aggregatibacter actinomycetemcomitans* and *Porphyromonas gingivalis*	—	—	[120]
SADs	*Selenomonas* and *Veillonella*	*Streptococcus*	—	[121]
Cardiovascular system				
Endocarditis	*Haemophilus parainfluenzae*, *Aggregatibacter* spp. (*A. actinomycetemcomitans*, *A. aphrophilus*, *A. paraphrophilus*, and *A. segnis*), *Cardiobacterium* spp. (*C. hominis* and *C. valvaram*), *Eikenellacorrodens*, and *Kingella* spp. (*K. kingae* and *K. denitrificans*)	—	Endocardium, heart valves, atria, and ventricle walls	[123]

(*Continued*)

TABLE 7.4 (*Continued*)
Depicting How Oral Microbial Diversity Dysbiosis Impacts Various Kinds of Systemic Diseases

DiseaseType	Microbial Species Increased in Number	Microbial Species Decreased in Number	Infected Tissue	References
Post-infarct cardiac rupture	*Porphyromonas gingivalis*	—	Endocardium, heart valves, atria, and ventricle walls	[124]
Infective endocarditis	*Streptococcus sanguinis*	—	Endocardium, heart valves, atria, and ventricle walls	[125]
Acute CAD	*Aggregatibacter actinomycetemcomitans*, *Porphyromonas gingivalis*, *Porphyromonas endodontalis*, *Prevotella intermedia*, *Tannerella forsythia*, *Campylobacter rectus*, and *Fusobacterium nucleatum*	—	Coronary artery	[126]

Abbreviations: CAD, coronary artery disease; CNS, central nervous system; COVID-19, coronavirus disease 2019; CRC, colorectal cancer; IBD, inflammatory bowel disease; LPS, lipopolysaccharides; RA, rheumatoid arthritis; and SADs, systemic autoimmune disorders.

According to another recent study [39], strains of salivary microbiota-derived *Klebsiella* spp. colonize the gut and have the ability to cause chronic intestinal inflammation strongly. Various *Klebsiella* strains, when the salivary microbiome colonizes the gut via gnotobiotic methods, strongly stimulate the Th1 cells. These *Klebsiella* strains cause severe gut inflammation in genetically predisposed hosts, are resistant to several medications, and tend to colonize when the intestinal microbiota is dysbiosis. The potential intestinal pathobionts that might exacerbate intestinal illness may be stored in the oral cavity.

A study examined periodontal bacteria interlinked with periodontitis IBD; compared to healthy people, IBD patients had higher levels of *Eikenella corrodens*, *Prevotella* genus, *Veillonella* genus, and lower levels of *Streptococcus* genus, *Haemophilus* genus, *Neisseria* genus, and *Gemella* genus [108].

Another study provided an approach for identifying at-risk populations and clearly showed salivary dysbiosis and various ecotypes in IBD communities. The study significantly provided a clinically helpful method for monitoring IBD because saliva could be sampled readily. It was reported a decline level of Porphyromonadaceae and Neisseriaceae families and increase in the *Campylobacter* genus and Streptoccocaceace, Enterobacteriaceae, and Veillonellaceae families in saliva [109].

7.6.3 Other Gastrointestinal Diseases

A particular kind of highly fatal gastrointestinal cancer is pancreatic cancer. Antibodies for specific oral pathogens in the bloodstream and history of periodontal disease have been related to an increased risk of pancreatic cancer. A large sample size of case-control research investigated the association between oral microbiota and the risk of pancreatic cancer in the future. The presence of the oral infections *Porphyromonas gingivalis* and *Aggregatibacter actinomycetemcomitans* has been associated with an increased probability of pancreatic cancer in the direct evaluation of the genomic-based microbiome in oral samples. There is a correlation between the *Leptotrichia* genus and a reduced risk of pancreatic cancer. These oral bacteria may also function as easily accessible, non-invasive indicators for the risk of pancreatic cancer in the future, assisting in determining the presence of individuals who are at a high risk of developing the illness [110].

A study stated the role of the oral microbiome in acute appendicitis conditions. Out of 49 patients, 28 suffered from acute appendicitis and 21 were healthy children. Using 16S rDNA gene sequencing analysis, it was found that Firmicutes and *Streptococcus* were identified abundantly in children. *Peptostreptococcus stomatis, E. corrodens*, and *F. nucleatum* were identified in each group. As a result, it could be observed that an oral cavity could act as a significant source of acute appendicitis [111].

The intestinal colonization of oral microbiota may be explained by the diminished intestinal barrier functions seen in colorectal tumors, which are similar to IBD. A meta-analysis revealed an association between periodontitis and a higher risk of colorectal cancer (CRC). There was little evidence connecting the onset of CRC to oral dysbiosis and the mouth-gut microbiome [112].

7.6.4 Respiratory Disorders

It has been demonstrated that the onset and progression of several respiratory conditions, including lung cancer, cystic fibrosis lung disease, asthma, pneumonia, chronic obstructive pulmonary disease (COPD), and COVID-19 could be significantly influenced by the oral microbiome. Most of these diseases are chronic in behavior. The oral microbiota exhibits characteristics of the traditional biofilm and produces dental plaque on tooth surfaces, which starts the caries process. As the primary source of the lung microbiome, microbiota colonizing in the oral micro-ecosystem has been linked to the emergence and progression of respiratory disorders. As inhaled into the lower respiratory tract, native oral bacteria do not really cause respiratory infections and inflammation; rather, they are within the environment of the oral microbiome as a reservoir of respiratory pathogens [113]. Recent studies have also demonstrated the role of the oral microbiome in respiratory diseases primarily in the early phase and progression of respiratory diseases.

7.6.5 Pneumonia

Pneumonia is a very common infection that inflames the air sacs in one or both lungs. Dental plaque is recognized as a potential source of common respiratory infections that cause pneumonia. It has been known that oral bacteria could cause aspiration pneumonia and that the number of oral bacteria increased the risk of aspiration pneumonia [114]. A study reported that the pathogenicity of microbes in oral microbiota is interrelated to pneumonia risk. The prevalence of oral microbes *Candida albicans, Staphylococcus aureus*, methicillin-resistant *S. aureus*, and *Pseudomonas aeruginosa* in aspiration pneumonia [115] was associated with the pneumonia in the patients.

Gram-positive cocci, like *Staphylococcus aureus* and *S. pneumoniae*, were present in the pathogenic microbial spectrum of hospital-associated pneumonia (HAP). Still, there were more gram-negative bacteria, including *Pseudomonas aeruginosa, K. pneumoniae, enterococci*, and *Enterobacter* species. These are potential respiratory pathogens that develop pneumonia [113].

7.6.6 Coronavirus Disease 2019 (COVID-19)

Severe acute respiratory syndrome coronavirus 2 (SARS-CoV-2) is a new coronavirus that is the cause of COVID-19, which has spread around the world. In the past couple of years, numerous studies have been published that described the correlation of oral microbiota in Covid-19 disease. Bao [60] described that some potential oral microbiome pathogens closely accompanied SARS-CoV-2 affected patients, in addition to outrageously co-infecting the lungs. The species of *Capnocytophaga*, *Veillonella*, and other oral pathogens were also found.

Another study also demonstrated the role of oral microbiome dysbiosis-related symptom severity in the local inflammatory response of COVID-19 patients. The findings showed that COVID-19 patients had oral dysbiosis in contrast with matched controls, with reduced species richness and a considerably lower alpha-diversity value in COVID-19 participants. However, oral dysbiosis and symptom severity were associated with COVID-19, and elevated local inflammation was also observed [73]. Inflammation-based oral microbiota dysbiosis led to prolonged duration of COVID-19 symptoms. It was found that higher levels of microbiota caused inflammation, including species that generate LPS and belonged to the genera *Prevotella* and *Veillonella* [78].

7.6.7 Nervous System

It has been established that disorders of the neurological system and oral microbiomes are related. Seeing nervous system disorders from a different angle is interesting. There is not much research that explains how oral microbiota and acute infections affect the neural system. According to several studies examining the oral microbiota and its links to the brain, oral microorganisms may also have an impact on the progression of non-proliferative disorders. However, very few studies described the interrelation between the oral microbiome and acute infection(s) of the nervous system. Still unclear are the paths and processes that provide information regarding the interactions between the brain and oral microbiome in neuropsychiatric disorders (NPDs). The brain is impacted by oral microbiota. The theoretical basis for an oral-microbiota-brain axis (OMBA) possibly has been the relationships between the central nervous system (CNS) and the presence of oral microbiota [116].

Another study demonstrated the role of microbiota dysbiosis in the early stage of Parkinson's disease. The 48 patients had early Parkinson's disease. The results showed significantly altered microbial taxa; specifically, alterations in bacterial, yeast, and phage taxa were observed. Interestingly, a significant association between the microbiota and a wide range of functional measures, including those that indicate balance, cognition, and the severity of the disease was recorded [117]. A mouse model study demonstrated the impact of *Porphyromonas gingivalis* (Pg) and liposaccharides (LPSs) induced systemic microbiota dysfunctions in the early onset and progression of Alzheimer's disease (AD). The findings showed that in an AD mouse model, the continuous brain exposure to Pg-LPS caused sarcopenia and heart damage without worsening cognitive impairment of the animals [118].

7.6.8 Immune System

The functioning of the human immune system is firmly related to oral microbiomes, which is the reason they are connected to the onset and development of diseases of the human immune system, such as rheumatoid arthritis, and thus impact the performance of the immune system in several system disorders including infection with HIV. An increasing number of research studies have shown that dysbiosis of the oral microbiota has a role in controlling the onset and development of several autoimmune disorders. A number of variables, including host genetics, food choices, stress, smoking, the use of antibiotics, tissue damage, and infection can cause oral microbiota dysbiosis. Through a number of processes, the disruption of the oral microbiome plays a key role in both initiating and perpetuating autoimmune disorders [119]. Many studies have discussed the role of oral microbiota diversity in autoimmunity disorders; however, very few have discussed the relationship between acute infections of autoimmune illnesses and alterations in the oral microbiota.

According to a study, periodontal disease and the bacteria that cause it could potentially be the origin of rheumatoid arthritis. Some periodontal bacteria, such as *Aggregatibacter actinomycetemcomitans* and *Porphyromonas gingivalis*, might be involved in the emergence of RA autoantibodies. The potential link between the onset of RA and higher rate of periodontitis [120]. A study showed that the oral microbiome plays a role in the onset and course of systemic autoimmune disorders (SADs), including rheumatoid arthritis, Sjögren's syndrome (SS), and systemic lupus erythematosus (SLE). The study revealed that as compared to healthy controls, individuals with SAD had different bacterial populations and compositions in their oral samples; three SADs had considerably higher levels of *Selenomonas* and *Veillonella*, whereas the SADs had significantly lower levels of *Streptococcus* in control groups. The deregulation of the oral microbiome was a major factor in the development and promotion of autoimmune disorders [121].

7.6.9 Cardiovascular System

Cardiovascular disease (CVD) pathogenesis has been implicated in the oral microbiota. It has been observed that disruption of the oral microbiota increases cardiovascular inflammation and CVD. The correlation between the gut microbiome and CVD is widely known; however, the extent of the disturbance (dysbiosis) of the oral microbiome remains mostly unclear. Heart failure, peripheral artery disease, coronary artery disease, infective endocarditis, and rheumatic heart disease have all been related to the oral microbiota. It is yet unclear how exactly oral dysbiosis contributes to the etiology of CVD. However, there is growing evidence from basic research and clinical studies that oral dysbiosis induced periodontal disease and increased the onset of cardiovascular disease [122].

Endocarditis is an acute condition in which inflammation occurs in the inner heart valves. The HACEK group of bacteria includes *Haemophilus parainfluenzae*, *Aggregatibacter* spp. (*A. actinomycetemcomitans*, *A. aphrophilus*, *A. paraphrophilus*, and *A. segnis*), *Cardiobacterium* spp. (*C. hominis* and *C. valvarum*), *Eikenellacorrodens*, and *Kingella* spp. (*K. kingae* and *K. denitrificans*). These showed potential pathogenicity in endocarditis [123].

Another study demonstrated that the oral periodontal pathogen (*Porphyromonas gingivalis*) promoted post-infarct cardiac rupture. The post-infarct cardiac rupture is an acute condition. The finding showed a close link between periodontal disease and atherosclerosis. *Porphyromonas gingivalis* entered the myocardium, encouraging apoptosis and the myocardium's Matrix metalloproteinase-9 (MMP-9) activity, which eventually resulted in cardiac rupture [124].

The severe cardiovascular condition known as infective endocarditis (IE) is caused by inflammation of the endocardium, heart valves, atria, and ventricle walls. It is known that the commensal bacteria *Streptococcus sanguinis*, which is prevalent in periodontitis, is the source of IE [125]. In the past, various studies investigated the relationship between acute coronary artery disease (CAD) and salivary levels. The studies revealed the level of subgingival microbes *Aggregatibacter actinomycetemcomitans*, *Porphyromonas gingivalis*, *Porphyromonas endodontalis*, *Prevotella intermedia*, *Tannerella forsythia*, *Campylobacter rectus*, and *Fusobacterium nucleatum* [126]. The IE is often brought through bacteria that enter the circulation and embed in the blood vessels or heart valves. The study provided insight into the connection between IE and oral microbiota [127].

7.7 ORAL MICROBIOMES AND EMERGING THERAPIES

The oral microbiome is a complex and dynamic ecosystem within the human mouth. It is home to a very diverse community of microorganisms that plays a vital role in maintaining oral health. Understanding the delicate balance of this microbial community is key to understanding oral health and its implications for overall well-being. As the advantages of scientific research continue, novel therapies are emerging to target and manipulate the oral microbiome for improved health outcomes. The exploration of these emerging therapies, such as the use of probiotics to introduce beneficial microorganisms, phage therapy to selectively combat harmful bacteria, and gene therapies to modulate microbial imbalances has been the need of the time. Additionally, the concept of combining probiotics and prebiotics, and personalized approaches based on individual microbial profiles, are shaping the frontier of oral health interventions. In this part, we focus on how, by examining these innovative approaches, we can explore the potential to revolutionize oral care and open new areas for exploration in healthcare and hence establish a more balanced oral microbiome.

7.7.1 Probiotics and Their Role in Maintaining a Healthy Oral Microbiome

Probiotics are live microorganisms that provide health benefits when ingested in sufficient quantities, constituting specifically tailored microorganisms to manipulate the oral microbiome. Probiotics contribute to oral health by inhibiting the growth of harmful bacteria, promoting the growth of beneficial bacteria, and modulating the host's immune response. The mechanism of action for the probiotics in the oral cavity includes the production of bacteriocins and stimulation of the local immune response. The probiotics also work by changing the composition of the oral cavity by competing with pathogenic bacteria for nutrients and mucosal receptors. The probiotics influence the production of antibodies and cytokines by up-regulating the T lymphocytes. *Lactobacillus* has been seen to influence the production of proinflammatory and anti-inflammatory responses. Moreover, they colonize the oral mucosal surfaces by adhering to receptors located on the cell surface [128].

The microorganisms present in the mouth form dental plaque consisting of bacterial cells, bacterial metabolites/toxins, and food debris [3]. Consumption of probiotic lactobacilli decreases the potential for pathogenic bacteria to bind with the oral tissue to form biofilm. It also increases the acidity of the dental plaque, increasing the chances of survival for acid-tolerant bacteria such as *S. mutans* and *Viridans streptococci*, increasing the pathogenicity of the dental biofilm [129]. A study revealed that the *Lactobacillus acidophilus*, *L. rhamnosus*, *Bifidobacterium longum*, and *S. boulardii* intervention for 3 weeks in children significantly reduced plaque accumulation [130]. The administration of *Streptococcus salivarius* M18 to children for 3 months significantly reduced dental plaque without affecting non-target microorganisms. The colonization rate was low for *S. salivarius* M18, but in colonized children, it showed anti-caries activity [131].

A study by Sarmento [132] studied the effect of probiotics on the composition of saliva in children and found a significant decrease in the density of *Aggregatibacter actinomycetemcomitans* and *Porphyromonas gingivalis* after treatment with product with *L. casei*. A 4-week use of *Bifidobacterium animalis* subsp. *lactis* DN-173010 supplementation in yogurt led to a positive impact on gingival inflammation at a 4-week stage [133]. Similar results were seen in the studies by Deshmukh [134] and Nadkerny [135], where a 2-week probiotic administration led to a reduction in gingival inflammation and plaque formation in treatment groups compared to the control. A randomized placebo-controlled clinical trial [136] evaluated the effect of probiotic tablets containing *Lactobacillus rhamnosus* and *Bifidobacterium lactis* on the oral microbiome, dental plaque, and gingival health. Therefore, probiotic supplements may serve as a simple adjunct to standard oral care for promoting the oral health in adolescents. Another study [137] discovered the presence of different strains of lactic acid bacteria, such as *Lactobacillus fermentum*, *Weissella confusa*, and *Lactobacillus delbrueckii*. The best probiotic characteristics for oral health were shown by a combination of *L. fermentum* N 2, TC 3-11 and *L. delbrueckii* subsp. *lactis* VG 2.

In a study by Boyeena [138], the impact of probiotic combinations was compared to the application of tetracycline fibers directly into the periodontal pocket filling plaque removal. The use of probiotics led to a notable reduction in bleeding on probing and pocket depth. Shah [139] studied the use of *L. brevis* with and without doxycycline in treating patients with aggressive periodontitis after plaque removal. Patients on *L. brevis* and doxycycline use for 2 weeks showed a significant reduction in pocket depth compared to other patients. The impact of different probiotics such as *L. rhamnosus* SP1 strain [140], *Streptococcus salivarius* M18 [141], *Streptococcus uberis* KJ2TM, *Streptococcus oralis* KJ3TM, and *Streptococcus rattus* JH145TM [142] was studied on dental caries in children, and all these studies led to a significant reduction in dental caries and prevented demineralization of enamel.

7.7.2 Phage Therapy for Targeting Harmful Oral Bacteria

Phage therapy is a promising alternative approach to antibiotics for oral pathogens, as it plays a key role in the natural balance in the prey-predator relationship with bacteria. Bacteriophages are viruses that can specifically target and kill bacteria. They are highly efficient against biofilms, starin-specific,and easy to isolate and manipulate, making them ideal candidate(s) for therapeutic use. They do not boost antibiotic resistance and have been successfully used to treat oral infectious diseases caused by Actinobacteria [143], Firmiculates [144], and Fusobacteria [145]. There is a large gap in phage therapy use in humans, but it is considered an alternative to antibiotics [146].

The studies carried out in the field of phage therapy have been based on the evolutionary relationship between the bacteriophage and formed biofilm. They have been proven effective even against multidrug-resistant bacteria and the destruction of the biofilm [147, 148]. Only lytic bacteriophages are used in phage therapy, and it is considered a novel method for the prevention and treatment of oral infectious diseases. The extracellular polysaccharides of the bacterial cell membrane are altered in *S. mutans* bacteria by the T4 RNA ligase 1 bacteriophage [149]. It was also demonstrated that the bacteriophages belonging to the Siphoviridae family can lead to a decrease in biofilm formation by 70% in the case of *F. nucleatum* infection by damaging their membrane structures [150]. However, there have been studies to confirm the feasibility of the phages. Unfortunately, they are not sufficient and more research is needed in this area to make it a reality.

7.7.3 Gene Therapies Targeting Oral Microbiome Imbalances

Gene therapy is used to manipulate the expression of specific genes related to specific health issues. In the context of the oral microbiome, it aims to modulate the activity of genes associated with microbial imbalance. Gene therapy can help in designing target-specific pathways to modulate microbial dysbiosis for restoring a healthier balance of microorganisms in the oral cavity. In nature, horizontal gene transfer is seen in the case of microorganisms through conjugation, transduction, and transformation. Conjugation is the direct transfer of genes between two live cells using conjugation plasmids and transposons. In transduction, genomic DNA from the host is packaged in bacteriophage heads and is introduced to the recipient. While in transformation, DNA fragments are picked from the environment.

Horizontal gene transfer has been studied in different bacteria from the oral microbiome. *In vitro* studies have shown the transfer of genes by conjugation and transformation between different species of bacteria. A study by Warburton [151] showed the transfer of transposons of doxycycline resistance during the treatment of periodontitis using doxycycline therapy. Another study by Roberts [152] showed the transfer of genes between bacterial stains enables them to colonize the oral biofilm. Studies have also shown an abundance of lysogenic bacteriophages in saliva [153]. Bacteriophages isolated from the saliva were lysogenic and were specific to *Enterococcus faecalis* [65] and *Aggregatibacter actinomycetemcomitans* [154]. Based on these studies, new emerging products are being developed. Research in the field of gene therapy for oral microbiomes has been gaining importance, but still more research and clinical trials are needed for clinical applications.

7.8 FUTURE DIRECTIONS AND RESEARCH CHALLENGES

Advances in oral microbiome research have opened new avenues for understanding the intricate interplay between oral microbes and human health. As we delve deeper into this field, several key areas of interest emerge as focal points for future research and clinical development, presenting both opportunities and challenges in advancing our knowledge and therapeutic approaches.

7.8.1 Advances in Oral Microbiome Research

The rapid evolution of sequencing technologies, such as metagenomics and single-cell sequencing, has revolutionized our ability to characterize the oral microbiome with unprecedented depth and resolution. These advances have enabled researchers to identify previously uncultivable microbes, unravel microbial community dynamics, and elucidate functional pathways within the oral ecosystem. Moreover, interdisciplinary collaborations between microbiologists, immunologists, geneticists, and clinicians have facilitated a holistic understanding of how oral microbiomes influence systemic health and disease outcomes. The integration of omics data, including genomics, transcriptomics, proteomics, and metabolomics, has provided comprehensive insights into microbial-host interactions and biomolecular mechanisms underlying oral diseases. However, the development of *in vitro* and *in vivo* models that recapitulate the oral microbiome's complexity and dynamics has accelerated translational research, enabling the testing of novel therapeutic interventions and microbiome-targeted therapies.

7.8.2 Future Research and Clinical Development

Looking ahead, several key areas merit attention in the realm of oral microbiome research and clinical development, which include:

- *Personalized oral microbiome medicine*: Advancing towards precision medicine approaches, leveraging microbiome data, host genetics, and clinical phenotypes to tailor preventive and therapeutic strategies for individuals based on their unique oral microbiome profiles.
- *Microbiome-based diagnostics*: Harnessing microbial biomarkers and molecular signatures to develop non-invasive diagnostic tools for early detection, risk stratification, and monitoring of oral diseases, including periodontitis, caries, oral cancers, and systemic conditions linked to oral health.
- *Microbiome engineering and therapeutics*: Exploring innovative strategies such as targeted antimicrobials, microbial consortia engineering, probiotics, phage therapy, and gene editing technologies to modulate the oral microbiome towards a healthier state and mitigate dysbiosis-associated diseases.
- *Bioinformatics and computational modeling*: Leveraging advanced bioinformatics tools, machine learning algorithms, and systems biology approaches to analyze large-scale microbiome datasets, predict microbial interactions, identify microbial biomarkers, and optimize therapeutic interventions.
- *Longitudinal cohort studies*: Conducting longitudinal studies encompassing diverse populations, age groups, and geographic regions to elucidate temporal dynamics, developmental trajectories, and environmental influences on the oral microbiome, paving the way for targeted interventions at different life stages.
- *Ethical and regulatory considerations*: Addressing ethical implications, privacy concerns, and regulatory frameworks surrounding microbiome research, including data sharing, informed consent, intellectual property rights, and equitable access to microbiome-based therapies.

The future of oral microbiome research and clinical development holds immense promise in transforming oral healthcare paradigms, from diagnostics and risk assessment to personalized therapeutics and preventive interventions. By embracing multidisciplinary collaborations, technological innovations, and ethical frameworks, it may be possible to navigate the challenges and seize the opportunities presented by the dynamic and complex world of oral microbiomes.

7.9 CONCLUSION

The exploration of the oral microbiome and its intricate relationship with human health has unveiled a fascinating landscape of microbial diversity, interactions, and implications for disease susceptibility and therapeutic interventions. Various aspects of the oral microbiome, from its compositions and dynamics to its role in immune response modulation, viral infections, oral diseases, systemic conditions, and emerging therapies, have been described. The oral microbiome is not merely a collection of microbes inhabiting the oral cavity but rather a dynamic ecosystem shaped by complex interactions between microbial communities and the host environment. Comprising diverse bacterial, fungal, viral, and archaeal species, the oral microbiome exhibits site-specific variations and temporal dynamics influenced by factors such as diet, genetics, hygiene practices, immune responses, and environmental exposures. Understanding the composition and dynamics of oral microbiomes is crucial for deciphering their functional roles in health and disease. The symbiotic relationship between commensal microbes and the host immune system plays a pivotal role in maintaining oral homeostasis and defending against pathogens. However, dysbiosis of the oral microbiome, characterized by microbial imbalance and dysregulated immune responses, can contribute to the onset and progression of various acute oral diseases, including caries, periodontal diseases, mucosal infections, and halitosis. Moreover, emerging evidence underscores the intricate connections between the oral microbiome and systemic health, with implications for the gastrointestinal, respiratory, nervous, and immune systems and CVD. Understanding the mechanisms underlying these associations offers novel opportunities for developing microbiome-targeted therapies, including probiotics, phage therapy, gene therapies, and personalized interventions tailored to individuals' unique oral microbiome profiles. As we navigate the future of oral microbiome research and clinical practice, several challenges and opportunities lie ahead. Advances in sequencing technologies, omics data integration, bioinformatics tools, and computational modeling will empower precision medicine approaches, microbiome-based diagnostics, therapeutic innovations, and longitudinal cohort studies. Ethical considerations, regulatory frameworks, and equitable access to microbiome-based therapies will also shape the ethical landscape of microbiome research and healthcare delivery. By embracing multidisciplinary collaborations, technological advancements, ethical frameworks, and patient-centered approaches, we can harness the potential of the oral microbiome to revolutionize healthcare paradigms and improve patient outcomes in the realm of acute infections and beyond.

REFERENCES

1. Deo, P. N.; Deshmukh, R. Oral Microbiome: Unveiling the Fundamentals. *J. Oral Maxillofac. Pathol.* **2019**, *23*(1), 122–128. doi: 10.4103/JOMFP.JOMFP_304_18
2. Sedghi, L.; DiMassa, V.; Harrington, A.; Lynch, S. V.; Kapila, Y. L. The Oral Microbiome: Role of Key Organisms and Complex Networks in Oral Health and Disease. *Periodontol.* **2021**, *87* (1), 107–131. doi: 10.1111/PRD.12393
3. Willis, J. R.; Gabaldón, T. The Human Oral Microbiome in Health and Disease: From Sequences to Ecosystems. *Microorg.* **2020**, *8*(2), 308. doi:10.3390/MICROORGANISMS8020308
4. Santacroce, L.; Passarelli, P. C.; Azzolino, D.; Bottalico, L.; Charitos, I. A.; Cazzolla, A. P.; Colella, M.; Topi, S.; Godoy, F. G.; D'Addona, A. Oral Microbiota in Human Health and Disease: A Perspective. *Exp. Biol. Med. (Maywood).* **2023**, *248*(15), 1288–1301. doi: 10.1177/15353702231187645
5. Li, N.; Ma, W. T.; Pang, M.; Fan, Q. L.; Hua, J. L. The Commensal Microbiota and Viral Infection: A Comprehensive Review. *Front. Immunol.* **2019**, *10*(JULY), 443190. doi: 10.3389/FIMMU.2019.01551/BIBTEX
6. Gomez, A.; Espinoza, J. L.; Harkins, D. M.; Leong, P.; Saffery, R.; Bockmann, M.; Torralba, M.; Kuelbs, C.; Kodukula, R.; Inman, J.; et al. Host Genetic Control of the Oral Microbiome in Health and Disease. *Cell Host Microbe.* **2017**, *22*(3), 269–278.e3. doi: 10.1016/j.chom.2017.08.013
7. Santonocito, S.; Giudice, A.; Polizzi, A.; Troiano, G.; Merlo, E. M.; Sclafani, R.; Grosso, G.; Isola, G. A Cross-Talk between Diet and the Oral Microbiome: Balance of Nutrition on Inflammation and Immune System's Response during Periodontitis. *Nutr.* **2022**, *14*(12), 2426. doi: 10.3390/NU14122426

8. Thomas, A. M.; Gleber-Netto, F. O.; Fernandes, G. R.; Amorim, M.; Barbosa, L. F.; Francisco, A. L. N.; Guerra De Andrade, A.; Setubal, J. C.; Kowalski, L. P.; Nunes, D. N.; et al. Alcohol and Tobacco Consumption Affects Bacterial Richness in Oral Cavity Mucosa Biofilms. *BMC Microbiol.* **2014**, *14*(1), 1–12. doi: 10.1186/S12866-014-0250-2/FIGURES/5
9. Li, X.; Liu, Y.; Yang, X.; Li, C.; Song, Z. The Oral Microbiota: Community Composition, Influencing Factors, Pathogenesis, and Interventions. *Front. Microbiol.* **2022**, *13*, 895537. doi: 10.3389/FMICB.2022.895537/BIBTEX
10. Idris, A.; Hasnain, S. Z.; Huat, L. Z.; Koh, D. Human Diseases, Immunity and the Oral Microbiota—Insights Gained from Metagenomic Studies. *Oral Sci. Int.* **2017**, *14*(2), 27–32. doi: 10.1016/S1348-8643(16)30024-6
11. Černáková, L.; Rodrigues, C. F. Microbial Interactions and Immunity Response in Oral Candida Species. *Future Microbiol.* **2020**, *15*(17), 1653–1677. doi: 10.2217/FMB-2020-0113
12. Thomas, C.; Minty, M.; Vinel, A.; Canceill, T.; Loubières, P.; Burcelin, R.; Kaddech, M.; Blasco-Baque, V.; Laurencin-Dalicieux, S. Oral Microbiota: A Major Player in the Diagnosis of Systemic Diseases. *Diagnostics* **2021**, *11*(8), 1376. doi: 10.3390/DIAGNOSTICS11081376
13. Pisano, M.; Giordano, F.; Sangiovanni, G.; Capuano, N.; Acerra, A.; D'Ambrosio, F. The Interaction between the Oral Microbiome and Systemic Diseases: A Narrative Review. *Microbiol. Res.* **2023**, *14*(4), 1862–1878. doi: 10.3390/MICROBIOLRES14040127
14. Russell, M. W.; Moldoveanu, Z.; Ogra, P. L.; Mestecky, J. Mucosal Immunity in COVID-19: A Neglected but Critical Aspect of SARS-CoV-2 Infection. *Front. Immunol.* **2020**, *11*, 611337. doi: 10.3389/FIMMU.2020.611337/BIBTEX
15. Dineshshankar, J.; Sivakumar, M.; Karthikeyan, M.; Udayakumar, P.; Shanmugam, K. T.; Kesavan, G. Immunology of Oral Candidiasis. *J. Pharm. Bioallied Sci.* **2014**, *6*(SUPPL. 1). doi: 10.4103/0975-7406.137251
16. Vila, T.; Rizk, A. M.; Sultan, A. S.; Jabra-Rizk, M. A. The Power of Saliva: Antimicrobial and Beyond. *PLOS Pathog.* **2019**, *15*(11), e1008058. doi: 10.1371/JOURNAL.PPAT.1008058
17. de Sousa-Pereira, P.; Woof, J. M. IgA: Structure, Function, and Developability. *Antibodies* **2019**, *8*(4), 57. doi: 10.3390/ANTIB8040057
18. Tada, A.; Senpuku, H. The Impact of Oral Health on Respiratory Viral Infection. *Dent. J.* **2021**, *9*(4), 43. doi: 10.3390/DJ9040043
19. Marsh, P. D.; Do, T.; Beighton, D.; Devine, D. A. Influence of Saliva on the Oral Microbiota. *Periodontol.* **2016**, *70*(1), 80–92. doi: 10.1111/PRD.12098
20. Koren, N.; Zubeidat, K.; Saba, Y.; Horev, Y.; Barel, O.; Wilharm, A.; Heyman, O.; Wald, S.; Eli-berchoer, L.; Shapiro, H.; et al. Maturation of the Neonatal Oral Mucosa Involves Unique Epithelium-Microbiota Interactions. *Cell Host Microbe* **2021**, *29*(2), 197–209.e5. doi: 10.1016/j.chom.2020.12.006
21. El-Howati, A.; Thornhill, M. H.; Colley, H. E.; Murdoch, C. Immune Mechanisms in Oral Lichen Planus. *Oral Dis.* **2023**, *29*(4), 1400–1415. doi: 10.1111/ODI.14142
22. Enomoto, A.; Sato, E.; Yasuda, T.; Isomura, T.; Nagao, T.; Chikazu, D. Intraepithelial CD8+ Lymphocytes as a Predictive Diagnostic Biomarker for the Remission of Oral Lichen Planus. *Hum. Pathol.* **2018**, *74*, 43–53. doi: 10.1016/J.HUMPATH.2017.12.008
23. Devine, D. A.; Marsh, P. D.; Meade, J. Modulation of Host Responses by Oral Commensal Bacteria. *J. Oral Microbiol.* **2015**, *7*(1), 1–4. doi: 10.3402/JOM.V7.26941
24. O'Donnell, L. E.; Millhouse, E.; Sherry, L.; Kean, R.; Malcolm, J.; Nile, C. J.; Ramage, G. Polymicrobial Candida Biofilms: Friends and Foe in the Oral Cavity. *FEMS Yeast Res.* **2015**, *15*(7), 77. doi: 10.1093/FEMSYR/FOV077
25. Liu, Y.; Qv, W.; Ma, Y.; Zhang, Y.; Ding, C.; Chu, M.; Chen, F. The Interplay between Oral Microbes and Immune Responses. *Front. Microbiol.* **2022**, *13*, 1009018. doi: 10.3389/FMICB.2022.1009018/BIBTEX
26. Lamont, R. J.; Koo, H.; Hajishengallis, G. The Oral Microbiota: Dynamic Communities and Host Interactions. *Nat. Rev. Microbiol.* **2018**, *16*(12), 745–759. doi: 10.1038/s41579-018-0089-x
27. Valm, A. M. The Structure of Dental Plaque Microbial Communities in the Transition from Health to Dental Caries and Periodontal Disease. *J. Mol. Biol.* **2019**, *431*(16), 2957–2969. doi: 10.1016/J.JMB.2019.05.016
28. Bertolini, M.; Costa, R. C.; Barão, V. A. R.; Cunha Villar, C.; Retamal-Valdes, B.; Feres, M.; Silva Souza, J. G. Oral Microorganisms and Biofilms: New Insights to Defeat the Main Etiologic Factor of Oral Diseases. *Microorg.* **2022**, *10*(12), 2413. doi: 10.3390/MICROORGANISMS10122413

29. Ebersole, J. L.; Dawson, D.; Emecen-Huja, P.; Nagarajan, R.; Howard, K.; Grady, M. E.; Thompson, K.; Peyyala, R.; Al-Attar, A.; Lethbridge, K.; et al. The Periodontal War: Microbes and Immunity. *Periodontol* **2017**, *75*(1), 52–115. doi: 10.1111/PRD.12222
30. Chang, A. M.; Liu, Q.; Hajjar, A. M.; Greer, A.; McLean, J. S.; Darveau, R. P. Toll-like Receptor-2 and -4 Responses Regulate Neutrophil Infiltration into the Junctional Epithelium and Significantly Contribute to the Composition of the Oral Microbiota. *J. Periodontol.* **2019**, *90*(10), 1202–1212. doi: 10.1002/JPER.18-0719
31. Hathaway-Schrader, J. D.; Aartun, J. D.; Poulides, N. A.; Kuhn, M. B.; McCormick, B. E.; Chew, M. E.; Huang, E.; Darveau, R. P.; Westwater, C.; Novince, C. M. Commensal Oral Microbiota Induces Osteoimmunomodulatory Effects Separate from Systemic Microbiome in Mice. *JCI Insight* **2022**, *7*(4). doi: 10.1172/JCI.INSIGHT.140738
32. Myers, S.; Do, T.; Meade, J. L.; Tugnait, A.; Vernon, J. J.; Pistolic, J.; Hancock, R. E. W.; Marsh, P. D.; Trivedi, H. M.; Chen, D.; et al. Immunomodulatory Streptococci That Inhibit CXCL8 Secretion and NFκB Activation Are Common Members of the Oral Microbiota. *J. Med. Microbiol.* **2021**, *70*(3), 001329. doi: 10.1099/JMM.0.001329/CITE/REFWORKS
33. Pandiyan, P.; Bhaskaran, N.; Zou, M.; Schneider, E.; Jayaraman, S.; Huehn, J. Microbiome Dependent Regulation of Tregs and Th17 Cells in Mucosa. *Front. Immunol.* **2019**, *10*(MAR), 440566. doi: 10.3389/FIMMU.2019.00426/BIBTEX
34. Jia, L.; Wu, R.; Han, N.; Fu, J.; Luo, Z.; Guo, L.; Su, Y.; Du, J.; Liu, Y. Porphyromonas Gingivalis and Lactobacillus Rhamnosus GG Regulate the Th17/Treg Balance in Colitis via TLR4 and TLR2. *Clin. Transl. Immunol.* **2020**, *9*(11), e1213. doi: 10.1002/CTI2.1213
35. Gholizadeh, P.; Eslami, H.; Kafil, H. S. Carcinogenesis Mechanisms of Fusobacterium Nucleatum. *Biomed. Pharmacother.* **2017**, *89*, 918–925. doi: 10.1016/J.BIOPHA.2017.02.102
36. Graves, D. T.; Ding, Z.; Yang, Y. The Impact of Diabetes on Periodontal Diseases. *Periodontol.* **2020**, *82*(1), 214–224. doi: 10.1111/PRD.12318
37. Huang, N.; Dong, H.; Luo, Y.; Shao, B. Th17 Cells in Periodontitis and Its Regulation by A20. *Front. Immunol.* **2021**, *12*, 742925. doi: 10.3389/FIMMU.2021.742925/BIBTEX
38. Wu, Y. Y.; Xiao, E.; Graves, D. T. Diabetes Mellitus Related Bone Metabolism and Periodontal Disease. *Int. J. Oral Sci.* **2015**, *7*(2), 63–72. doi: 10.1038/ijos.2015.2
39. Atarashi, K.; Suda, W.; Luo, C.; Kawaguchi, T.; Motoo, I.; Narushima, S.; Kiguchi, Y.; Yasuma, K.; Watanabe, E.; Tanoue, T.; et al. Ectopic Colonization of Oral Bacteria in the Intestine Drives TH1 Cell Induction and Inflammation. *Science* **2017**, *358*(6361), 359–365. doi: 10.1126/SCIENCE.AAN4526
40. Kohil, A.; Abdalla, W.; Ibrahim, W. N.; Al-Harbi, K. M.; Al-Haidose, A.; Al-Asmakh, M.; Abdallah, A. M. The Immunomodulatory Role of Microbiota in Rheumatic Heart Disease: What Do We Know and What Can We Learn from Other Rheumatic Diseases? *Med.* **2023**, *59*(9), 1629. doi: 10.3390/MEDICINA59091629
41. Moreno, C. M.; Boeree, E.; Freitas, C. M. T.; Weber, K. S. Immunomodulatory Role of Oral Microbiota in Inflammatory Diseases and Allergic Conditions. *Front. Allergy* **2023**, *4*, 1067483. doi: 10.3389/FALGY.2023.1067483/BIBTEX
42. Cannon, R. D. Oral Fungal Infections: Past, Present, and Future. *Front. Oral Heal.* **2022**, *3*, 838639. doi: 10.3389/FROH.2022.838639/BIBTEX
43. Martínez-García, M.; Hernández-Lemus, E. Periodontal Inflammation and Systemic Diseases: An Overview. *Front. Physiol.* **2021**, *12*, 709438. doi: 10.3389/FPHYS.2021.709438/BIBTEX
44. Pathak, J. L.; Yan, Y.; Zhang, Q.; Wang, L.; Ge, L. The Role of Oral Microbiome in Respiratory Health and Diseases. *Respir. Med.* **2021**, *185*, 106475. doi: 10.1016/J.RMED.2021.106475
45. Imai, K.; Iinuma, T.; Sato, S. Relationship between the Oral Cavity and Respiratory Diseases: Aspiration of Oral Bacteria Possibly Contributes to the Progression of Lower Airway Inflammation. *Jpn. Dent. Sci. Rev.* **2021**, *57*, 224–230. doi: 10.1016/J.JDSR.2021.10.003
46. Saadaoui, M.; Singh, P.; Al Khodor, S. Oral Microbiome and Pregnancy: A Bidirectional Relationship. *J. Reprod. Immunol.* **2021**, *145*, 103293. doi: 10.1016/J.JRI.2021.103293
47. Furman, D.; Campisi, J.; Verdin, E.; Carrera-Bastos, P.; Targ, S.; Franceschi, C.; Ferrucci, L.; Gilroy, D. W.; Fasano, A.; Miller, G. W.; et al. Chronic Inflammation in the Etiology of Disease across the Life Span. *Nat. Med.* **2019**, *25*(12), 1822–1832. doi: 10.1038/s41591-019-0675-0
48. Konkel, J. E.; O'Boyle, C.; Krishnan, S. Distal Consequences of Oral Inflammation. *Front. Immunol.* **2019**, *10*(JUN), 431939. doi: 10.3389/FIMMU.2019.01403/BIBTEX
49. Cai, Y.; Zhao, Y.; Kang, Y.; Yang, Y. Future Prospect of Oral Microbiota Influencing Allergy/Asthma Development. *Front. Microbiol.* **2022**, *13*, 875664. doi: 10.3389/FMICB.2022.875664/BIBTEX

50. Mariman, R.; Reefman, E.; Tielen, F.; Persoon-Deen, C.; van de Mark, K.; Worms, N.; Koning, F.; Nagelkerken, L. Lactobacillus Plantarum NCIMB8826 Ameliorates Inflammation of Colon and Skin in Human APOC1 Transgenic Mice. *Benef. Microbes* **2016**, *7*(2), 215–226. doi: 10.3920/BM2015.0074
51. Furutama, D.; Matsuda, S.; Yamawaki, Y.; Hatano, S.; Okanobu, A.; Memida, T.; Oue, H.; Fujita, T.; Ouhara, K.; Kajiya, M.; et al. IL-6 Induced by Periodontal Inflammation Causes Neuroinflammation and Disrupts the Blood–Brain Barrier. *Brain Sci.* **2020**, *10*(10), 679. doi: 10.3390/BRAINSCI10100679
52. Teixeira, F. B.; Saito, M. T.; Matheus, F. C.; Prediger, R. D.; Yamada, E. S.; Maia, C. S. F.; Lima, R. R. Periodontitis and Alzheimer's Disease: A Possible Comorbidity between Oral Chronic Inflammatory Condition and Neuroinflammation. *Front. Aging Neurosci.* **2017**, *9*(OCT), 263378. doi: 10.3389/FNAGI.2017.00327/BIBTEX
53. Pallos, D.; Ruivo, G. F.; Ferrari-Junior, S. H.; Pannuti, C. S.; Perozini, C.; Sarmento, D. J. S.; Palmieri, M.; Souza, A. C. M. F.; Tozetto-Mendoza, T. R.; Doglio, A.; et al. Periodontal Disease and Detection of Human Herpesviruses in Saliva and Gingival Crevicular Fluid of Chronic Kidney Disease Patients. *J. Periodontol.* **2020**, *91*(9), 1139–1147. doi: 10.1002/JPER.19-0583
54. Chen, C.; Feng, P.; Slots, J. Herpesvirus-Bacteria Synergistic Interaction in Periodontitis. *Periodontol.* **2020**, *82*(1), 42–64. doi: 10.1111/PRD.12311.
55. Libertucci, J.; Young, V. B. The Role of the Microbiota in Infectious Diseases. *Nat. Microbiol.* **2018**, *4*(1), 35–45. doi: 10.1038/s41564-018-0278-4
56. Baghbani, T.; Nikzad, H.; Azadbakht, J.; Izadpanah, F.; Haddad Kashani, H. Dual and Mutual Interaction between Microbiota and Viral Infections: A Possible Treat for COVID-19. *Microb. Cell Factories* **2020**, *19*(1), 1–25. doi: 10.1186/S12934-020-01483-1
57. Bourgeois, D.; Gonçalves, L. S.; Lima-Junior, J. da C.; Carrouel, F. Editorial: The Oral Microbiome Is a Key Factor in Oral and Systemic Health. *Front. Microbiol.* **2022**, *13*, 855668. doi: 10.3389/FMICB.2022.855668/BIBTEX
58. Zelaya, H.; Tada, A.; Vizoso-Pinto, M. G.; Salva, S.; Kanmani, P.; Agüero, G.; Alvarez, S.; Kitazawa, H.; Villena, J. Nasal Priming with Immunobiotic Lactobacillus Rhamnosus Modulates Inflammation–Coagulation Interactions and Reduces Influenza Virus-Associated Pulmonary Damage. *Inflamm. Res.* **2015**, *64*(8), 589–602. doi: 10.1007/S00011-015-0837-6
59. Khan, R.; Petersen, F. C.; Shekhar, S. Commensal Bacteria: An Emerging Player in Defense against Respiratory Pathogens. *Front. Immunol.* **2019**, *10*(MAY), 452323. doi: 10.3389/FIMMU.2019.01203/BIBTEX
60. Bao, L.; Zhang, C.; Dong, J.; Zhao, L.; Li, Y.; Sun, J. Oral Microbiome and SARS-CoV-2: Beware of Lung Co-Infection. *Front. Microbiol.* **2020**, *11*, 546497. doi: 10.3389/FMICB.2020.01840/BIBTEX
61. Xiang, Z.; Koo, H.; Chen, Q.; Zhou, X.; Liu, Y.; Simon-Soro, A. Potential Implications of SARS-CoV-2 Oral Infection in the Host Microbiota. *J. Oral Microbiol.* **2021**, *13*(1), 1853451. doi: 10.1080/20002297.2020.1853451
62. Khan, A. A.; Khan, Z. COVID-2019-Associated Overexpressed Prevotella Proteins Mediated Host–Pathogen Interactions and Their Role in Coronavirus Outbreak. *Bioinformatics* **2020**, *36*(13), 4065–4069. doi: 10.1093/BIOINFORMATICS/BTAA285
63. Kamio, N.; Imai, K.; Shimizu, K.; Cueno, M. E.; Tamura, M.; Saito, Y.; Ochiai, K. Neuraminidase-Producing Oral Mitis Group Streptococci Potentially Contribute to Influenza Viral Infection and Reduction in Antiviral Efficacy of Zanamivir. *Cell. Mol. Life Sci.* **2015**, *72* (2), 357–366. doi: 10.1007/S00018-014-1669-1/METRICS
64. Lu, H. F.; Li, A.; Zhang, T.; Ren, Z. G.; He, K. X.; Zhang, H.; Yang, J. Z.; Luo, Q. X.; Zhou, K.; Chen, C. L.; et al. Disordered Oropharyngeal Microbial Communities in H7N9 Patients with or without Secondary Bacterial Lung Infection. *Emerg. Microbes Infect.* **2017**, *6*(1). doi: 10.1038/EMI.2017.101
65. Moore, S. C.; Penrice-Randal, R.; Alruwaili, M.; Randle, N.; Armstrong, S.; Hartley, C.; Haldenby, S.; Dong, X.; Alrezaihi, A.; Almsaud, M.; et al. Amplicon-Based Detection and Sequencing of SARS-CoV-2 in Nasopharyngeal Swabs from Patients With COVID-19 and Identification of Deletions in the Viral Genome That Encode Proteins Involved in Interferon Antagonism. *Viruses* **2020**, *12*(10), 1164. doi: 10.3390/V12101164
66. Connolly, E.; Millhouse, E.; Doyle, R.; Culshaw, S.; Ramage, G.; Moran, G. P. The Porphyromonas Gingivalis Hemagglutinins HagB and HagC Are Major Mediators of Adhesion and Biofilm Formation. *Mol. Oral Microbiol.* **2017**, *32*(1), 35–47. doi: 10.1111/OMI.12151
67. Chen, Y.; Zhou, R.; Yi, Z.; Li, Y.; Fu, Y.; Zhang, Y.; Li, P.; Li, X.; Pan, Y. Porphyromonas Gingivalis Induced Inflammatory Responses and Promoted Apoptosis in Lung Epithelial Cells Infected with H1N1 via the Bcl-2/Bax/Caspase-3 Signaling Pathway. *Mol. Med. Rep.* **2018**, *18*(1), 97–104. doi: 10.3892/MMR.2018.8983

68. Ke, X.; Lei, L.; Li, H.; Li, H.; Yan, F. Manipulation of Necroptosis by Porphyromonas Gingivalis in Periodontitis Development. *Mol. Immunol.* **2016**, *77*, 8–13. doi: 10.1016/J.MOLIMM.2016.07.010
69. Tonoyan, L.; Vincent-Bugnas, S.; Olivieri, C. V.; Doglio, A. New Viral Facets in Oral Diseases: The EBV Paradox. *Int. J. Mol. Sci.* **2019**, *20*(23), 5861. doi: 10.3390/IJMS20235861
70. Kawamoto, M.; Tanaka, H.; Sakurai, A.; Otagiri, H.; Karasawa, I.; Yamada, S. I.; Kurita, H. Exploration of Correlation of Oral Hygiene and Condition with Influenza Infection. *PLOS One* **2021**, *16*(8), e0254981. doi: 10.1371/JOURNAL.PONE.0254981
71. Yildiz, S.; Mazel-Sanchez, B.; Kandasamy, M.; Manicassamy, B.; Schmolke, M. Influenza A Virus Infection Impacts Systemic Microbiota Dynamics and Causes Quantitative Enteric Dysbiosis. *Microbiome* **2018**, *6*(1), 1–17. doi: 10.1186/S40168-017-0386-Z/FIGURES/5
72. Ramos-Sevillano, E.; Wade, W. G.; Mann, A.; Gilbert, A.; Lambkin-Williams, R.; Killingley, B.; Nguyen-Van-Tam, J. S.; Tang, C. M. The Effect of Influenza Virus on the Human Oropharyngeal Microbiome. *Clin. Infect. Dis.* **2019**, *68*(12), 1993–2002. doi: 10.1093/CID/CIY821
73. Soffritti, I.; D'Accolti, M.; Fabbri, C.; Passaro, A.; Manfredini, R.; Zuliani, G.; Libanore, M.; Franchi, M.; Contini, C.; Caselli, E. Oral Microbiome Dysbiosis is Associated with Symptoms Severity and Local Immune/Inflammatory Response in COVID-19 Patients: A Cross-Sectional Study. *Front. Microbiol.* **2021**, *12*, 687513. doi: 10.3389/FMICB.2021.687513/BIBTEX
74. Wu, Y.; Cheng, X.; Jiang, G.; Tang, H.; Ming, S.; Tang, L.; Lu, J.; Guo, C.; Shan, H.; Huang, X. Altered Oral and Gut Microbiota and Its Association with SARS-CoV-2 Viral Load in COVID-19 Patients during Hospitalization. *NPJ Biofilms Microbiomes* **2021**, *7*(1), 1–9. doi: 10.1038/s41522-021-00232-5
75. Man, W. H.; De Steenhuijsen Piters, W. A. A.; Bogaert, D. The Microbiota of the Respiratory Tract: Gatekeeper to Respiratory Health. *Nat. Rev. Microbiol.* **2017**, *15*(5), 259–270. doi: 10.1038/nrmicro.2017.14
76. Farshbafnadi, M.; Kamali Zonouzi, S.; Sabahi, M.; Dolatshahi, M.; Aarabi, M. H. Aging & COVID-19 Susceptibility, Disease Severity, and Clinical Outcomes: The Role of Entangled Risk Factors. *Exp. Gerontol.* **2021**, *154*, 111507. doi: 10.1016/J.EXGER.2021.111507
77. Gupta, A.; Bhanushali, S.; Sanap, A.; Shekatkar, M.; Kharat, A.; Raut, C.; Bhonde, R.; Shouche, Y.; Kheur, S.; Sharma, A. Oral Dysbiosis and Its Linkage with SARS-CoV-2 Infection. *Microbiol. Res.***2022**, *261*, 127055. doi: 10.1016/J.MICRES.2022.127055
78. Haran, J. P.; Bradley, E.; Zeamer, A. L.; Cincotta, L.; Salive, M. C.; Dutta, P.; Mutaawe, S.; Anya, O.; Meza-Segura, M.; Moormann, A. M.; et al. Inflammation-Type Dysbiosis of the Oral Microbiome Associates with the Duration of COVID-19 Symptoms and Long COVID. *JCI Insight* **2021**, *6*(20). doi: 10.1172/JCI.INSIGHT.152346
79. Li, S.; Su, B.; He, Q. S.; Wu, H.; Zhang, T. Alterations in the Oral Microbiome in HIV Infection: Causes, Effects and Potential Interventions. *Chin. Med. J. (Engl).* **2021**, *134*(23), 2788–2798. doi: 10.1097/CM9.0000000000001825
80. Lee, Y. H.; Chung, S. W.; Auh, Q. S.; Hong, S. J.; Lee, Y. A.; Jung, J.; Lee, G. J.; Park, H. J.; Shin, S. Il; Hong, J. Y. Progress in Oral Microbiome Related to Oral and Systemic Diseases: An Update. *Diagnostics* **2021**, *11*(7), 1283. doi: 10.3390/DIAGNOSTICS11071283
81. Scannapieco, F. A. The Oral Microbiome: Its Role in Health and in Oral and Systemic Infections. *Clin. Microbiol. Newsl.* **2013**, *35*(20), 163–169. doi: 10.1016/J.CLINMICNEWS.2013.09.003
82. Marsh, P. D.; Zaura, E. Dental Biofilm: Ecological Interactions in Health and Disease. *J. Clin. Periodontol.* **2017**, *44*, S12–S22. doi: 10.1111/JCPE.12679
83. Hajishengallis, G. Periodontitis: From Microbial Immune Subversion to Systemic Inflammation. *Nat. Rev. Immunol.* **2015**, *15*(1), 30–44. doi: 10.1038/nri3785
84. Mira, A.; Simon-Soro, A.; Curtis, M. A. Role of Microbial Communities in the Pathogenesis of Periodontal Diseases and Caries. *J. Clin. Periodontol.* **2017**, *44*, S23–S38. doi: 10.1111/JCPE.12671
85. Ma, C.; Chen, F.; Zhang, Y.; Sun, X.; Tong, P.; Si, Y.; Zheng, S. Comparison of Oral Microbial Profiles between Children with Severe Early Childhood Caries and Caries-Free Children Using the Human Oral Microbe Identification Microarray. *PLOS One* **2015**, *10*(3), e0122075. doi: 10.1371/JOURNAL.PONE.0122075
86. Johansson, I.; Witkowska, E.; Kaveh, B.; Lif Holgerson, P.; Tanner, A. C. R. The Microbiome in Populations with a Low and High Prevalence of Caries. *J. Dental Res.* **2015**, *95*(1), 80–86. doi: 10.1177/0022034515609554
87. Wang, Y.; Zhang, J.; Chen, X.; Jiang, W.; Wang, S.; Xu, L.; Tu, Y.; Zheng, P.; Wang, Y.; Lin, X.; et al. Profiling of Oral Microbiota in Early Childhood Caries Using Single-Molecule Real-Time Sequencing. *Front. Microbiol.* **2017**, *8*(NOV), 303364. doi: 10.3389/FMICB.2017.02244/BIBTEX

88. Richards, V. P.; Alvarez, A. J.; Luce, A. R.; Bedenbaugh, M.; Mitchell, M. L.; Burne, R. A.; Nascimento, M. M. Microbiomes of Site Specific Dental Plaques from Children with Different Caries Status. *Infect. Immun.* **2017**, *85*(8). doi: 10.1128/IAI.00106-17/SUPPL_FILE/ZII999092088S1.PDF
89. Tanner, A. C. R.; Kressirer, C. A.; Rothmiller, S.; Johansson, I.; Chalmers, N. I. The Caries Microbiome: Implications for Reversing Dysbiosis. *Adv. Dent. Res.* **2018**, *29*(1), 78–85. doi: 10.1177/0022034517736496
90. Kressirer, C. A.; Chen, T.; Lake Harriman, K.; Frias-Lopez, J.; Dewhirst, F. E.; Tavares, M. A.; Tanner, A. C. R. Functional Profiles of Coronal and Dentin Caries in Children. *J. Oral Microbiol.* **2018**, *10*(1). doi: 10.1080/20002297.2018.1495976
91. Yost, S.; Duran-Pinedo, A. E.; Teles, R.; Krishnan, K.; Frias-Lopez, J. Functional Signatures of Oral Dysbiosis during Periodontitis Progression Revealed by Microbial Metatranscriptome Analysis. *Genome Med.* **2015**, *7*(1). doi: 10.1186/S13073-015-0153-3
92. Topcuoglu, N.; Kulekci, G.16S RRNA Based Microarray Analysis of Ten Periodontal Bacteria in Patients with Different Forms of Periodontitis. *Anaerobe* **2015**, *35*, 35–40. doi: 10.1016/J.ANAEROBE.2015.01.011
93. Pozhitkov, A. E.; Leroux, B. G.; Randolph, T. W.; Beikler, T.; Flemmig, T. F.; Noble, P. A. Towards Microbiome Transplant as a Therapy for Periodontitis: An Exploratory Study of Periodontitis Microbial Signature Contrasted by Oral Health, Caries and Edentulism. *BMC Oral Health* **2015**, *15*(1), 1–11. doi: 10.1186/S12903-015-0109-4/FIGURES/5
94. Carrouel, F.; Viennot, S.; Santamaria, J.; Veber, P.; Bourgeois, D. Quantitative Molecular Detection of 19 Major Pathogens in the Interdental Biofilm of Periodontally Healthy Young Adults. *Front. Microbiol.* **2016**, *7*(JUN), 192219. doi: 10.3389/FMICB.2016.00840/BIBTEX
95. Wei, Y.; Shi, M.; Zhen, M.; Wang, C.; Hu, W.; Nie, Y.; Wu, X. Comparison of Subgingival and Buccal Mucosa Microbiome in Chronic and Aggressive Periodontitis: A Pilot Study. *Front. Cell. Infect. Microbiol.* **2019**, *9*(MAR), 445517. doi: 10.3389/FCIMB.2019.00053/BIBTEX
96. Şenel, S. An Overview of Physical, Microbiological and Immune Barriers of Oral Mucosa. *Int. J. Mol. Sci.* **2021**, *22*(15), 7821. doi: 10.3390/IJMS22157821/S1
97. Min, Z.; Yang, L.; Hu, Y.; Huang, R. Oral Microbiota Dysbiosis Accelerates the Development and Onset of Mucositis and Oral Ulcers. *Front. Microbiol.* **2023**, *14*, 1061032. doi: 10.3389/FMICB.2023.1061032/BIBTEX
98. Kim, Y. J.; Choi, Y. S.; Baek, K. J.; Yoon, S. H.; Park, H. K.; Choi, Y. Mucosal and Salivary Microbiota Associated with Recurrent Aphthous Stomatitis. *BMC Microbiol.* **2016**, *16*(1), 1–10. doi: 10.1186/S12866-016-0673-Z/FIGURES/4
99. Amer, A.; Galvin, S.; Healy, C. M.; Moran, G. P. The Microbiome of Potentially Malignant Oral Leukoplakia Exhibits Enrichment for Fusobacterium, Leptotrichia, Campylobacter, and Rothia Species. *Front. Microbiol.* **2017**, *8*(DEC.), 302760. doi: 10.3389/FMICB.2017.02391/BIBTEX
100. Yang, Z.; Cui, Q.; An, R.; Wang, J.; Song, X.; Shen, Y.; Wang, M.; Xu, H. Comparison of Microbiomes in Ulcerative and Normal Mucosa of Recurrent Aphthous Stomatitis (RAS)-Affected Patients. *BMC Oral Health* **2020**, *20*(1), 1–8. doi: 10.1186/S12903-020-01115-5/FIGURES/4
101. Li, Z.; Li, J.; Fu, R.; Liu, J.; Wen, X.; Zhang, L. Halitosis: Etiology, Prevention, and the Role of Microbiota. *Clin. Oral Investig.* **2023**, *27* (11), 6383–6393. doi: 10.1007/S00784-023-05292-9/METRICS
102. Hampelska, K.; Jaworska, M. M.; Babalska, Z. Ł.; Karpiński, T. M. The Role of Oral Microbiota in Intra-Oral Halitosis. *J. Clin. Med.* **2020**, *9* (8), 2484. doi: 10.3390/JCM9082484
103. Ye, W.; Zhang, Y.; He, M.; Zhu, C.; Feng, X. P. Relationship of Tongue Coating Microbiome on Volatile Sulfur Compounds in Healthy and Halitosis Adults. *J. Breath Res.* **2019**, *14*(1), 016005. 10.1088/1752-7163/AB47B4
104. Jo, J. K.; Seo, S. H.; Park, S. E.; Kim, H. W.; Kim, E. J.; Na, C. S.; Cho, K. M.; Kwon, S. J.; Moon, Y. H.; Son, H. S. Identification of Salivary Microorganisms and Metabolites Associated with Halitosis. *Metab.* 2**021**, *11*(6), 362. doi: 10.3390/METABO11060362
105. Foo, L. H.; Balan, P.; Pang, L. M.; Laine, M. L.; Seneviratne, C. J. Role of the Oral Microbiome, Metabolic Pathways, and Novel Diagnostic Tools in Intra-Oral Halitosis: A Comprehensive Update. *Crit. Rev. Microbiol.* **2021**, *47* (3), 359–375. doi: 10.1080/1040841X.2021.1888867
106. Huang, Z.; Cheng, Y. Oral Microbiota Characteristics in Intra-Oral Halitosis Patients and Innovative Therapeutic Approaches. **2022**, 1–19.
107. Muhvić-Urek, M.; Tomac-Stojmenović, M.; Mijandrusic-Sincic, B. Oral Pathology in Inflammatory Bowel Disease. *World J. Gastroenterol.* **2016**, *22*(25), 5655–5667. doi: 10.3748/WJG.V22.I25.5655
108. Schmidt, J.; Weigert, M.; Leuschner, C.; Hartmann, H.; Raddatz, D.; Haak, R.; Mausberg, R. F.; Kottmann, T.; Schmalz, G.; Ziebolz, D. Active Matrix Metalloproteinase-8 and Periodontal Bacteria—Interlink between Periodontitis and Inflammatory Bowel Disease? *J. Periodontol.* **2018**, *89*(6), 699–707. doi: 10.1002/JPER.17-0486

109. Xun, Z.; Zhang, Q.; Xu, T.; Chen, N.; Chen, F. Dysbiosis and Ecotypes of the Salivary Microbiome Associated with Inflammatory Bowel Diseases and the Assistance in Diagnosis of Diseases Using Oral Bacterial Profiles. *Front. Microbiol.* **2018**, *9*(MAY), 323562. doi: 10.3389/FMICB.2018.01136/BIBTEX
110. Fan, X.; Alekseyenko, A. V.; Wu, J.; Peters, B. A.; Jacobs, E. J.; Gapstur, S. M.; Purdue, M. P.; Abnet, C. C.; Stolzenberg-Solomon, R.; Miller, G.; et al. Human Oral Microbiome and Prospective Risk for Pancreatic Cancer: A Population-Based Nested Case-Control Study. *Gut* **2018**, *67*(1), 120–127. doi: 10.1136/GUTJNL-2016-312580
111. Blod, C.; Schlichting, N.; Schülin, S.; Suttkus, A.; Peukert, N.; Stingu, C. S.; Hirsch, C.; Elger, W.; Lacher, M.; Bühligen, U.; et al. The Oral Microbiome—the Relevant Reservoir for Acute Pediatric Appendicitis? *Int. J. Colorectal Dis.* **2018**, *33*(2), 209–218. doi: 10.1007/S00384-017-2948-8/METRICS
112. Li, W.; Xu, J.; Zhang, R.; Li, Y.; Wang, J.; Zhang, X.; Lin, L. Is Periodontal Disease a Risk Indicator for Colorectal Cancer? A Systematic Review and Meta-Analysis. *J. Clin. Periodontol.* **2021**, *48*(3), 336–347. doi: 10.1111/JCPE.13402
113. Dong, J.; Li, W.; Wang, Q.; Chen, J.; Zu, Y.; Zhou, X.; Guo, Q. Relationships Between Oral Microecosystem and Respiratory Diseases. *Front. Mol. Biosci.* **2022**, *8*, 718222. doi: 10.3389/FMOLB.2021.718222/BIBTEX
114. Nishizawa, T.; Niikura, Y.; Akasaka, K.; Watanabe, M.; Kurai, D.; Amano, M.; Ishii, H.; Matsushima, H.; Yamashita, N.; Takizawa, H. Pilot Study for Risk Assessment of Aspiration Pneumonia Based on Oral Bacteria Levels and Serum Biomarkers. *BMC Infect. Dis.* **2019**, *19*(1), 1–8. doi: 10.1186/S12879-019-4327-2/FIGURES/3
115. Khadka, S.; Khan, S.; King, A.; Goldberg, L. R.; Crocombe, L.; Bettiol, S. Poor Oral Hygiene, Oral Microorganisms and Aspiration Pneumonia Risk in Older People in Residential Aged Care: A Systematic Review. *Age Ageing* **2021**, *50*(1), 81–87. doi: 10.1093/AGEING/AFAA102
116. Bowland, G. B.; Weyrich, L. S. The Oral-Microbiome-Brain Axis and Neuropsychiatric Disorders: An Anthropological Perspective. *Front. Psychiatry* **2022**, *13*, 810008. doi: 10.3389/FPSYT.2022.810008/BIBTEX
117. Mihaila, D.; Donegan, J.; Barns, S.; LaRocca, D.; Du, Q.; Zheng, D.; Vidal, M.; Neville, C.; Uhlig, R.; Middleton, F. A. The Oral Microbiome of Early Stage Parkinson's Disease and Its Relationship with Functional Measures of Motor and Non-Motor Function. *PLoS One* **2019**, *14*(6), e0218252. doi: 10.1371/JOURNAL.PONE.0218252
118. Hayashi, K.; Hasegawa, Y.; Takemoto, Y.; Cao, C.; Takeya, H.; Komohara, Y.; Mukasa, A.; Kim-Mitsuyama, S. Continuous Intracerebroventricular Injection of Porphyromonas Gingivalis Lipopolysaccharide Induces Systemic Organ Dysfunction in a Mouse Model of Alzheimer's Disease. *Exp. Gerontol.* **2019**, *120*, 1–5. doi: 10.1016/J.EXGER.2019.02.007
119. Huang, X.; Huang, X.; Huang, Y.; Zheng, J.; Lu, Y.; Mai, Z.; Zhao, X.; Cui, L.; Huang, S. The Oral Microbiome in Autoimmune Diseases: Friend or Foe? *J. Transl. Med.* **2023**, *21*(1), 1–24. doi: 10.1186/S12967-023-03995-X
120. Cheng, Z.; Meade, J.; Mankia, K.; Emery, P.; Devine, D. A. Periodontal Disease and Periodontal Bacteria as Triggers for Rheumatoid Arthritis. *Best Pract. Res. Clin. Rheumatol.* **2017**, *31*(1), 19–30. doi: 10.1016/J.BERH.2017.08.001
121. Gao, L.; Cheng, Z.; Zhu, F.; Bi, C.; Shi, Q.; Chen, X. The Oral Microbiome and Its Role in Systemic Autoimmune Diseases: A Systematic Review of Big Data Analysis. *Front. Big Data* **2022**, *5*, 927520. doi: 10.3389/FDATA.2022.927520/BIBTEX
122. Tonelli, A.; Lumngwena, E. N.; Ntusi, N. A. B. The Oral Microbiome in the Pathophysiology of Cardiovascular Disease. *Nat. Rev. Cardiol.* **2023**, *20*(6), 386–403. doi: 10.1038/s41569-022-00825-3
123. Revest, M.; Egmann, G.; Cattoir, V.; Tattevin, P. HACEK Endocarditis: State-of-the-Art. *Expert Rev. Anti. Infect. Ther.* **2016**, *14*(5), 523–530. doi: 10.1586/14787210.2016.1164032
124. Shiheido, Y.; Maejima, Y.; Suzuki, J. ichi; Aoyama, N.; Kaneko, M.; Watanabe, R.; Sakamaki, Y.; Wakayama, K.; Ikeda, Y.; Akazawa, H.; et al. Porphyromonas Gingivalis, a Periodontal Pathogen, Enhances Myocardial Vulnerability, Thereby Promoting Post-Infarct Cardiac Rupture. *J. Mol. Cell. Cardiol.* **2016**, *99*, 123–137. doi: 10.1016/J.YJMCC.2016.03.017
125. Zhu, B.; Macleod, L. C.; Kitten, T.; Xu, P. *Streptococcus sanguinis* Biofilm Formation & Interaction with Oral Pathogens. *Future Microbiol.* **2018**, *13*(8), 915–932. doi: 10.2217/FMB-2018-0043/ASSET/IMAGES/LARGE/FIGURE3.JPEG
126. Liljestrand, J. M.; Paju, S.; Pietiäinen, M.; Buhlin, K.; Persson, G. R.; Nieminen, M. S.; Sinisalo, J.; Mäntylä, P.; Pussinen, P. J. Immunologic Burden Links Periodontitis to Acute Coronary Syndrome. *Atherosclerosis* **2018**, *268*, 177–184. doi: 10.1016/J.ATHEROSCLEROSIS.2017.12.007

127. Del Giudice, C.; Vaia, E.; Liccardo, D.; Marzano, F.; Valletta, A.; Spagnuolo, G.; Ferrara, N.; Rengo, C.; Cannavo, A.; Rengo, G. Infective Endocarditis: A Focus on Oral Microbiota. *Microorg.* **2021**, *9*(6), 1218. doi: 10.3390/MICROORGANISMS9061218
128. Janczarek, M.; Bachanek, T.; Mazur, E.; Chałas, R. The Role of Probiotics in Prevention of Oral Diseases. *Postepy Hig. Med. Dosw. (Online)* **2016**, *70*(0), 850–857. doi: 10.5604/17322693.1214381
129. Marttinen, A.; Haukioja, A.; Karjalainen, S.; Nylund, L.; Satokari, R.; Öhman, C.; Holgerson, P.; Twetman, S.; Söderling, E. Short-Term Consumption of Probiotic Lactobacilli Has No Effect on Acid Production of Supragingival Plaque. *Clin. Oral Investig.* **2012**, *16*(3), 797–803. doi: 10.1007/S00784-011-0584-1/METRICS
130. Thakkar, P.; Imranulla, M.; Naveen Kumar, P.; Prashant, G.; Sakeenabi, B.; Sushanth, V. Effect of Probiotic Mouth Rinse on Dental Plaque Accumulation: A Randomized Controlled Trial. *Dent. Med. Res.* **2013**, *1*(1), 7. doi: 10.4103/2348-1471.124988
131. Burton, J. P.; Drummond, B. K.; Chilcott, C. N.; Tagg, J. R.; Thomson, W. M.; Hale, J. D. F.; Wescombe, P. A. Influence of the Probiotic Streptococcus Salivarius Strain M18 on Indices of Dental Health in Children: A Randomized Double-Blind, Placebo-Controlled Trial. *J. Med. Microbiol.* **2013**, *62* (PART6), 875–884. doi: 10.1099/JMM.0.056663-0/CITE/REFWORKS
132. Sarmento, É. G.; Cesar, D. E.; Martins, M. L.; de Oliveira Góis, E. G.; Furtado Martins, E. M.; da Rocha Campos, A. N.; Del'Duca, A.; de Oliveira Martins, A. D. Effect of Probiotic Bacteria in Composition of Children's Saliva. *Food Res. Int.* **2019**, *116*, 1282–1288. doi: 10.1016/J.FOODRES.2018.10.017
133. Kuru, B. E.; Laleman, I.; Yalnızoğlu, T.; Kuru, L.; Teughels, W. The Influence of a Bifidobacterium Animalis Probiotic on Gingival Health: A Randomized Controlled Clinical Trial. *J. Periodontol.* **2017**, *88*(11), 1115–1123. doi: 10.1902/JOP.2017.170213
134. Deshmukh, M. A.; Dodamani, A. S.; Karibasappa, G.; Khairnar, M. R.; Naik, R. G.; Jadhav, H. C. Comparative Evaluation of the Efficacy of Probiotic, Herbal and Chlorhexidine Mouthwash on Gingival Health: A Randomized Clinical Trial. *J. Clin. Diagn. Res.* **2017**, *11* (3), ZC13–ZC16. doi: 10.7860/JCDR/2017/23891.9462
135. Nadkerny, P.; Ravishankar, P.; Pramod, V.; Agarwal, L.; Bhandari, S. A Comparative Evaluation of the Efficacy of Probiotic and Chlorhexidine Mouth Rinses on Clinical Inflammatory Parameters of Gingivitis: A Randomized Controlled Clinical Study. *J. Indian Soc. Periodontol.* **2015**, *19*(6), 633. doi: 10.4103/0972-124X.168491
136. Alanzi, A.; Honkala, S.; Honkala, E.; Varghese, A.; Tolvanen, M.; Söderling, E. Effect of Lactobacillus Rhamnosus and Bifidobacterium Lactis on Gingival Health, Dental Plaque, and Periodontopathogens in Adolescents: A Randomised Placebo-Controlled Clinical Trial. *Benef. Microbes* **2018**, *9*(4), 593–602. doi: 10.3920/BM2017.0139
137. Atanasov, N.; Evstatieva, Y.; Nikolova, D. Probiotic Potential of Lactic Acid Bacterial Strains Isolated from Human Oral Microbiome. *Microbiol. Res.* **2023**, *14*(1), 262–278. doi: 10.3390/MICROBIOLRES14010021
138. Boyeena, L.; Koduganti, R.; Panthula, V.; Jammula, S. Comparison of Efficacy of Probiotics versus Tetracycline Fibers as Adjuvants to Scaling and Root Planting. *J. Indian Soc. Periodontol.* **2019**, *23* (6), 539. doi: 10.4103/JISP.JISP_590_18
139. Shah, M. P.; Gujjari, S. K.; Chandrasekhar, V. S. Evaluation of the Effect of Probiotic (Inersan®) Alone, Combination of Probiotic with Doxycycline and Doxycycline Alone on Aggressive Periodontitis - a Clinical and Microbiological Study. *J. Clin. Diagn. Res.* **2013**, *7* (3), 595–600. doi: 10.7860/JCDR/2013/5225.2834
140. Rodríguez, G.; Ruiz, B.; Faleiros, S.; Vistoso, A.; Marró, M. L.; Sánchez, J.; Urzúa, I.; Cabello, R. Probiotic Compared with Standard Milk for High-Caries Children: A Cluster Randomized Trial. *J. Dent. Res.* **2016**, *95* (4), 402–407. doi: 10.1177/0022034515623935
141. Di Pierro, F.; Zanvit, A.; Nobili, P.; Risso, P.; Fornaini, C. Cariogram Outcome after 90 Days of Oral Treatment with Streptococcus Salivarius M18 in Children at High Risk for Dental Caries: Results of a Randomized, Controlled Study. *Clin. Cosmet. Investig. Dent.* **2015**, *7*, 107–113. doi: 10.2147/CCIDE.S93066
142. Hedayati-Hajikand, T.; Lundberg, U.; Eldh, C.; Twetman, S. Effect of Probiotic Chewing Tablets on Early Childhood Caries – A Randomized Controlled Trial. *BMC Oral Health* **2015**, *15* (1). doi: 10.1186/S12903-015-0096-5
143. Hatfull, G. F. Actinobacteriophages: Genomics, Dynamics, and Applications. *Annu. Rev. Virol.* **2020**, *7*, 37–61. doi: 10.1146/ANNUREV-VIROLOGY-122019-070009/1
144. Wolfoviz-Zilberman, A.; Kraitman, R.; Hazan, R.; Friedman, M.; Houri-Haddad, Y.; Beyth, N. Phage Targeting *Streptococcus mutans* In Vitro and In Vivo as a Caries-Preventive Modality. *Antibiotics* **2021**, *10*(8). doi: 10.3390/ANTIBIOTICS10081015

145. Machuca, P.; Daille, L.; Vinés, E.; Berrocal, L.; Bittner, M. Isolation of a Novel Bacteriophage Specific for the Periodontal Pathogen *Fusobacterium nucleatum*. *Appl. Environ. Microbiol.* **2010**, *76*(21), 7243–7250. doi: 10.1128/AEM.01135-10
146. Shlezinger, M.; Khalifa, L.; Houri-Haddad, Y.; Coppenhagen-Glazer, S.; Resch, G.; Que, Y.-A.; Beyth, S.; Dorfman, E.; Hazan, R.; Beyth, N. Phage Therapy: A New Horizon in the Antibacterial Treatment of Oral Pathogens. *Curr. Top. Med. Chem.* **2017**, *17*(10), 1199–1211. doi: 10.2174/1568026616666160930145649
147. Baker, J. L.; Bor, B.; Agnello, M.; Shi, W.; He, X. Ecology of the Oral Microbiome: Beyond Bacteria. *Trends Microbiol.* **2017**, *25*(5), 362–374. doi: 10.1016/J.TIM.2016.12.012
148. Torres-Barceló, C. The Disparate Effects of Bacteriophages on Antibiotic-Resistant Bacteria. *Emerg. Microbes Infect.* **2018**, *7*(1). doi: 10.1038/S41426-018-0169-Z
149. Chen,J.;Chen,Z.;Yuan,K.;Huang,Z.;Mao,M.RecombinantBacteriophageT4RnlI ImpactsStreptococcus Mutans Biofilm Formation. *J. Oral Microbiol.* **2021**, *13*(1). doi: 10.1080/20002297.2020.1860398
150. Kabwe, M.; Brown, T. L.; Dashper, S.; Speirs, L.; Ku, H.; Petrovski, S.; Chan, H. T.; Lock, P.; Tucci, J. Genomic, Morphological and Functional Characterisation of Novel Bacteriophage FNU1 Capable of Disrupting Fusobacterium Nucleatum Biofilms. *Sci. Rep.* **2019**, *9*(1). doi: 10.1038/S41598-019-45549-6
151. Warburton, P. J.; Palmer, R. M.; Munson, M. A.; Wade, W. G. Demonstration of in Vivo Transfer of Doxycycline Resistance Mediated by a Novel Transposon. *J. Antimicrob. Chemother.* **2007**, *60*(5), 973–980. doi: 10.1093/JAC/DKM331
152. Roberts, A. P.; Pratten, J.; Wilson, M.; Mullany, P. Transfer of a Conjugative Transposon, Tn5397 in a Model Oral Biofilm. *FEMS Microbiol. Lett.* **1999**, *177* (1), 63–66. doi: 10.1111/J.1574-6968.1999.TB13714.X
153. Pride, D. T.; Salzman, J.; Haynes, M.; Rohwer, F.; Davis-Long, C.; White, R. A.; Loomer, P.; Armitage, G. C.; Relman, D. A. Evidence of a Robust Resident Bacteriophage Population Revealed through Analysis of the Human Salivary Virome. *ISME J.* **2012**, *6*(5), 915–926. doi: 10.1038/ISMEJ.2011.169
154. Willi, K.; Sandmeier, H.; Kulik, E. M.; Meyer, J. Transduction of Antibiotic Resistance Markers among *Actinobacillus actinomycetemcomitans* Strains by Temperate Bacteriophages Aa Phi 23. *Cell. Mol. Life Sci.* **1997**, *53*(11–12), 904–910. doi: 10.1007/S000180050109

8 Oral Microbiome and Chronic Infections

Sukhraj Kaur, Anmol Narang, and Rahul Sharma
Department of Microbiology, Guru Nanak Dev University, Amritsar, Punjab, India

8.1 INTRODUCTION

The human microbiome is a group of microorganisms that includes bacteria, fungi, protozoa, viruses, archaea, and their genes existing in a commensal relationship with human cells and that also have complex interactions with one another. Studies highlighting the differences in microbes at various body sites, and between the states of health and disease, are as old as microbiology itself. Antonie van Leewenhoek in the 1680s [1] studied the diversity of his oral and fecal microbiota. He compared the differences of microbiota between these two habitats and between samples of individuals with various states of health and disease. Modern metagenomic studies have further yielded qualitative and quantitative data about the human microbiome. An average adult human body consists of about 30 trillion human cells and 38 trillion bacteria [2]. The highest diversity of the microbiome is present in the colon, which contains 10^{11} bacterial cells/g, 10^9 virus-like particles (VLP)/g, and 10^5–10^6 fungal cells/g of fecal matter. Fungi account for only 0.1% of the total 9.9 million genes of eukaryotic origin in a human gut microbial reference catalog [3]. The three predominant fungal phyla isolated from the human gut are Ascomycota, Basidiomycota, and Zygomycota [4]. Archaebacteria accounts for 1.2% of the total human microbiome, and the most prevalent archaea in the human gut are Methanobacteriales and Methanomassiliicoccales [5]. The human virome, which is believed to shape the diversity and functional aspects of the bacteriome, is not well characterized. It is, therefore, known as the "viral dark matter". Virome includes phages that infect bacteria and archaea, and viruses that infect human cells. The predominant core human phageome belongs to the order Caudovirales (Myoviridae, Podoviridae, and Siphoviridae families), which have double-strand DNA, and the Microviridae family, which has single-stranded DNA [6].

8.1.1 Oral Bacteriome

The "core microbiome" in humans is confined to a bacterial microbiome, i.e., a bacteriome [7]. The majority of the oral microbiome is found in oral biofilms, and lesser numbers are present as planktonic bacteria (Table 8.1). A biofilm is an organized community of bacterial cells held together in an extracellular polymeric matrix (EPM) and adhered to an inert or live surface [8, 9]. Planktonic bacteria typically consisting of facultative anaerobes and saccharolytic aerobes act as "early colonizers" that primarily consume salivary mucins and oral glycoproteins and adhere to the surface of oral biofilms with the help of specific surface adhesins [10]. The surface-attached bacteria alter their metabolic and gene expression profiles after initial colonization to create and secrete EPMs, which are composed of polysaccharides, proteins, lipids, and extracellular DNA [11, 12]. The distribution/abundance of different oral bacterial biofilm species varies in relation to the three diverse habitat zones in the oral cavity, i.e., dental plaque, tongue dorsum, and keratinized gingiva [13].

DOI: 10.1201/9781003478638-8

TABLE 8.1
Major Bacteria Found in the Oral Cavity of a Healthy Person

Gram Positive		Gram Negative	
Cocci	**Rods**	**Cocci**	**Rods**
Streptococcus	*Actinomycetes*	*Moraxella*	*Capnocytophaga*
Stomatococcus	*Bifidobacterium*	*Neisseria*	*Desulfobacter*
Peptostreptococcus	*Corynebacterium*	*Veillonella*	*Desulfovibrio*
	Eubacterium		*Fusobacterium*
	Propionibacterium		*Prevotella*
			Leptotrichia

8.1.2 Virobiome/Virome

The oral virome in humans is highly conserved [14]. It varies with host gender, age. and hygiene. and it is highly personalized [15]. Bacteriophages, which have a very stable lytic/lysogenic cycle, make up the great majority of viruses in the oral cavity [16, 17]. Some of these bacteriophages have been linked to *Veillonella* and *Streptococcus* species, two of the most prevalent commensal bacterial genera in the oral cavity [15]. The three main bacteriophage families detected in the oral cavity of healthy individuals were Siphoviridae, Myoviridae, and Podoviridae, all of which belong to the Caudovirales order [16, 17]. Myoviridae and Podoviridae are mainly lytic viruses that quickly destroy their bacterial hosts, but Siphoviridae are mainly lysogenic viruses that create a dynamic equilibrium with their linked host species [18]. The oral viral community has only been investigated in a few numbers of studies [19, 20]. Therefore, additional research is still required to determine the virome's contribution to the oral microbiome.

8.1.3 Mycobiome

Fungi/yeast species are also found in the human oral cavity [21]. The "basal oral mycobiome" was initially identified by Ghannoum [22]. The *Candida* genera present in 70% of healthy persons [22, 23] are the most prevalent commensal fungi and constituents of the basal oral mycobiome [24]. It is followed by *C. glabrata*, *C. parapsilosis*, *C. tropicalis*, *C. krusei*, *C. stellatoidea*, *C. kefyr*, *C. khmerensis*, and *C. metapsilosis* [25]. Apart from *Candida* spp., other present species belong to the genera *Cladosporium*, *Aureobasidium*, *Sacchromycetales*, *Aspergillus*, *Fusarium*, and *Cryptoccoccus*.

8.1.4 Protozoome

The presence of protozoans as oral commensals is well known. In 1928, *Entamoeba gingivalis* and *Trichomonas tenax* as common oral parasites were described for the first time [26]. However, subsequent research conducted revealed that these oral species were generally present in the oral cavity of healthy individuals as commensals [27, 28]. However, their numbers increase in people with poor oral hygiene that can lead to their association with pathogenic oral bacteria such as *Porphyromonas gingivalis*, *Treponema denticola*, and *Eubacterium nodatum* and thereby induce oral pathological state [29].

8.1.5 Archaeome

Five different methanogenic archaea genera, including *Methanobrevibacter*, *Methanosphaera*, *Methanosarcina*, *Thermoplasma*, and *Methanobacterium*, have been found in the oral cavity of healthy humans [30]. *Methanobrevibacter oralis*, *Methanobacterium congelense/curvum*, and *Methanosarcina mazzei* are the primary archaea species detected in healthy individuals. *Methanobrevibacter oralis* predominate over other species with a prevalence of 40% [31, 32]. Although they are part of the oral microbiome, archaea are thought to be less common and diverse than oral bacteria [33].

8.2 ESTABLISHMENT OF THE ORAL MICROBIOME

The oral microbiome resides in close proximity to the host cells and exhibits interspecies as well as host-microbiome interactions [34]. The interspecific interactions are categorized into three classes: synergistic, signaling, and antagonistic interactions as shown in Figure 8.1.

8.2.1 Synergism

8.2.1.1 Coaggregation

Oral microbes have a propensity to adhere to other microbes, which promotes the development of multispecies biofilms [34]. Coadhesion is a process that profoundly alters the gene expression of both of the involved cells [35] and promotes the formation of biofilms. For instance, a yeast like *C. albicans* can coaggregate with oral *Streptococci*, producing a synergistic relationship in which the bacteria increase the invasive qualities of *C. albicans* and the yeast increases *Streptococci*'s ability to form biofilms [36, 37]. These coaggregated cells secrete EPM around them that prevents molecules from moving freely, limiting the diffusion of nutrients and oxygen, resulting in nutritional gradients throughout the biofilm [11]. Further, these nutritional gradients also support nutritional

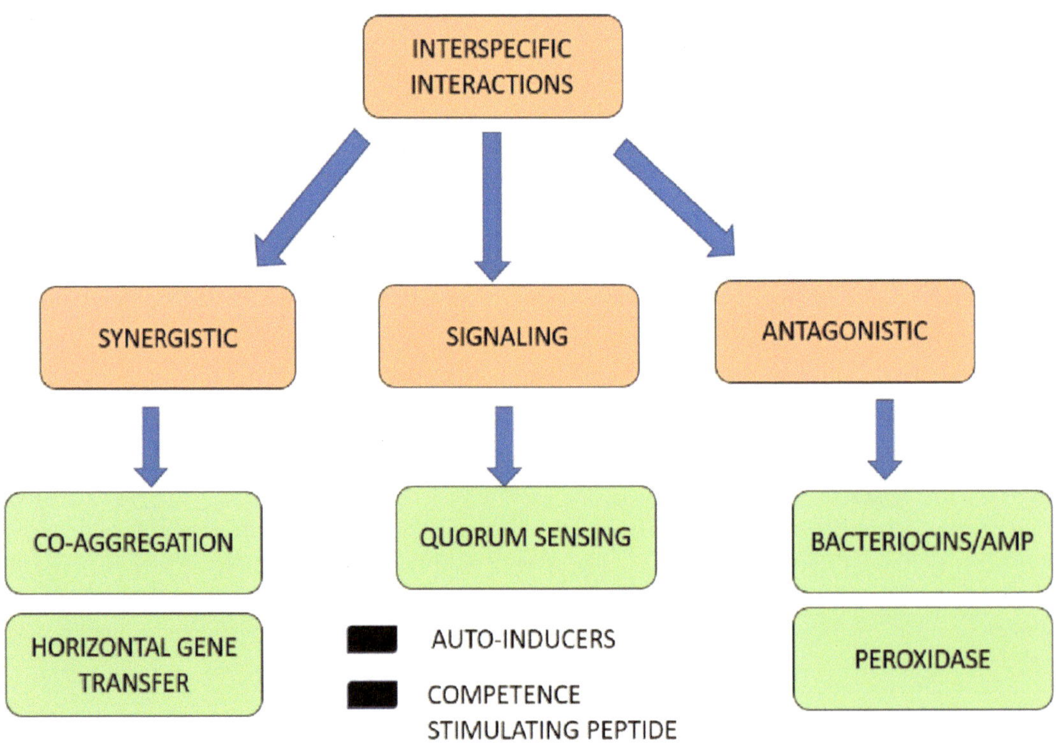

FIGURE 8.1 Types of interspecies interactions and various mediators that play a role in the establishment of the oral microbiome.

interactions when the metabolic end products of one microbe become the food supply for another. Another useful effect of coaggregation is the defense of facultative and oxygen-tolerant anaerobes in anoxic niches created by nearby oxygen-consuming organisms inside the biofilm [34].

8.2.1.2 Horizontal Gene Transfer

Horizontal gene transfer frequently occurs in the oral microbiota that helps bacteria to access new genes and develop new traits such as virulence factors and antibiotic resistance [38]. For instance, it has been discovered that *Streptococcus mutans* cells undergo transformations 10–600 times greater in biofilms than in their planktonic stage [38]. Similarly, Tribble et al. (2012) reported that the *Porphyromonas gingivalis* strains have a natural capacity for transformation, which is crucial to the species' overall survival and persistence in the oral cavity [39]. *Streptococcus pneumoniae* and other oral streptococci, such as *S. oralis* and *S. mitis*, share portions of the *pbp2x* gene by the process of conjugation, which increases their resistance to penicillin [40].

8.2.2 Signaling

8.2.2.1 Autoinducer

A class of chemicals called autoinducers can cause phenotypic modifications in bacteria. These compounds are produced by a variety of bacterial species and build up in the extracellular environment until a detection threshold is achieved, at which point downstream signaling and effector responses emerge [41]. Prokaryotes secrete autoinducer-2 (AI-2), which is produced by the enzyme LuxS and derived from 4,5-dihydroxypentane-2,3-dione. In the oral cavity, AI-2 appears to control a number of characteristics of the bacteria such as pathogenicity and biofilm formation [42, 43]. For instance, pure AI-2 derived from *Veillonella tobetsuensis* leads to reduction in the production of biofilm by *S. gordonii*. Similarly, knocking down the AI-2 gene from *S. oralis* prevents *Actinomyces naeslundii* to form biofilms [44, 45]. Two different species of the same genus responded extremely differently to AI-2 generated by a strain of *Fusobacterium nucleatum* [46].

8.2.2.2 Competence-Stimulating Peptide

Competence-stimulating peptide (CSP) is a 17-amino-acid peptide pheromone encoded by the *comC* gene. CSPs secreted by *S. pneumoniae* attach to and activate the transmembrane histidine kinase receptor ComD, which in turn activates the synthesis of proteins involved in the formation of biofilms, cell competence, the production of bacteriocins, stress resistance, and autolysis [34, 47]. A study showed that CSP might be counterintuitively implicated in releasing environmental (eDNA) in *S. mutans* biofilms leading to stable biofilms [48]. However, in another study, contrasting results were obtained where *S. gordonii* mutants lacking the *comC* gene produced more eDNA and more stable biofilms in the presence of *C. albicans* [49].

8.2.3 Antagonistic Interactions

Antagonistic interactions are common among microbiota in order to gain survival advantage in an ecological niche. These are mediated by various antimicrobial compounds such as antimicrobial peptides (AMPs) and hydrogen peroxide. AMPs produced by bacteria are also known as bacteriocins. These are mostly cationic peptides containing 12–100 amino acids that are released extracellularly for the inhibition of the target microbial growth [50]. Hydrogen peroxide can also be produced by some microbes. The harmful effects of hydrogen peroxide on bacterial cells result from the production of hydroxyl radicals in the presence of Fe(II). This causes the hydroxyl radicals to interact with other molecules, causing cellular harm by degrading the iron-sulfur clusters on enzymes, which renders them inactive, or by oxidizing macromolecules such as DNA and proteins [11]. *In vitro* and under aerobic conditions, the growth of a variety of gram-positive bacteria is inhibited by several oral streptococci species that can produce hydrogen peroxide at certain levels [51].

8.3 LOCALIZED DISEASES RELATED TO ORAL HEALTH

8.3.1 Dental Caries

Dental caries is the most common chronic oral disease and is caused by the bacterial infection of the tooth enamel and dentin mostly by oral microbiota. It affects an estimated 2 billion people worldwide including 514 million children worldwide [52]. In children, severe early childhood caries (ECC) can adversely affect primary dentition [53] and can cause caries in permanent teeth that is difficult to treat. Caries is linked to frequent consumption of dietary sugars that results in the replacement of the normal oral microbiota to acidogenic and cariogenic biofilm-forming microbiota [54]. The acidogenic bacteria in turn produce organic acids that cause demineralization of the tooth enamel resulting in the formation of dental caries [55]. Formation of caries not only causes tooth destruction but also leads to pulp and periapical infection [56].

8.3.1.1 Role of Oral Dysbiosis in the Onset of Dental Caries

If sugar consumption is low, the production of organic acids is less and infrequent and the saliva is able to neutralize small amounts of organic acids, thereby protecting demineralization and acid-mediated erosion of tooth enamel. However, in the case of a large intake of fermentable sugar, the growth of acidogenic and aciduric microorganisms is favored. This further facilitates the production of EPS and acidic metabolism, which changes the diversity of the oral microbiota [57]. As the EPS matrix grows, it wraps around bacterial cells and supports the formation of microscopic colonies that lead to plaque formation on the tooth enamel [58]. Within the plague, the diverse microbial communities produce consistent amounts of organic acids lowering the oral pH to less than 5.0, which ultimately leads to erosion of tooth enamel and dental caries.

Streptococcus mutans (SM) has been identified as the primary bacteria responsible for the initiation of dental caries. It is evidenced by various experiments that have shown that the implantation of SM in germ-free animals fed a sucrose-rich diet causes dental caries. SM possesses various characteristic properties that increase its cariogenic propensity as compared to other *Streptococcal* species. SM is highly acidogenic and aciduric. Secondly, it has the ability to produce polymeric glucose, known as glucan, from simple sugars that initiate the process of biofilm formation. It has surface antigens I/II, which promote bacterial adhesion to the tooth surface and to other bacteria. SM could be predominantly isolated from the initial stages of caries, where it constitutes about 30% of the total flora [59–61]. On the other hand, SM is less frequently present at the progressive stage of caries, where lactobacilli, *Prevotellae*, and *Bifidobacterium* spp. are more prevalent. On the contrary, certain metagenomic studies have contradicted the notion that dental caries are primarily dominated by SM. Instead, these studies reveal a significant presence of other bacterial genera, such as *S. gordonii*, *Leptotrichia buccalis*, and *Veillonella parvula*, in individuals with caries [62]. Additionally, strains of *Fusobacterium* spp. play a crucial role in the maturation of dental biofilm, facilitating coaggregation with other gram-negative bacteria like Bacteroidetes and Spirochaetes [63].

Various molecular techniques and metagenomic studies have shown that the samples of supragingival plaque, saliva, and dentin of caries-infected children, adults, and elderly people show the presence of SM along with other bacteria such as *Veillonella*, *Actinomyces*, *Granulicatella*, *Leptotrichia*, *Thiomonas*, *Bifidobacterium*, and *Prevotella* [64–67]. Adult patients with dental caries also show the presence of *Streptococcus*, *Lactobacillus*, *Actinomycetes*, *Propionibacterium*, and *Veillonella* [62]. In elderly people with root caries, 16S rRNA sequencing studies have shown the presence of *Atopobium*, *Olsenella*, *Pseudoramibacter*, *Propionibacterium*, and *Selenomonas* [68]. In addition, it has been seen that even at same stage of caries development, there is variation in predominant bacteria between primary and permanent dentition. For example, the bacteria most frequently detected from cavity surfaces are *S. salivarius*, *S. parasanguis*, *Corynebacterium*, and *Actinomyces gerencseriae* in primary dentition and *S. salivarius*,

S. parasanguis, *Camphylobacter*, and *Selenomonas* in permanent dentition [68]. There is variation in the type of species of *Prevotella* between healthy and caries-active salivary samples. For example, the percentage of *P. histicola* significantly ($p<0.05$) increased, but *P. shahii* significantly ($p<0.1$) decreased in the caries-active samples compared to the samples of the healthy individuals [69]. In a carrier active host, genes related to DNA uptake, acid production, and stress response were highly expressed [62]. The cariogenic microbiota dominated by *S. mutans* and non-mutans *Streptococcus*, *Actinomyces*, *Bifidobacterium*, *Lactobacillus*, and *Scardovia* spp. and the synergistic effect of these bacteria will promote EPS generation and microenvironment acidification [70].

8.3.2 Periodontitis

Periodontitis is the common inflammatory condition affecting the gums, and if left untreated, it results in the gradual deterioration of the crucial structural components supporting the teeth. These structural components include the cementum, periodontal ligament, alveolar bone, and gingival tissue [71]. Globally, it affects 20%–50% of the population, with predictions indicating a potential increase in the future [72]. Gingivitis, an initial inflammatory gum disease, is characterized by redness and bleeding of the gingiva. If neglected, gingivitis can progress to periodontitis [73]. The onset involves the disruption of bacterial homeostasis, leading to the accumulation of plaque between the gums [74]. Subsequently, specific symptoms of periodontitis including inflammation, bleeding, and attachment loss of the tooth due to resorption of alveolar bone become apparent.

8.3.2.1 Role of Oral Dysbiosis in the Onset of Periodontitis

The microbiome of the oral cavity consists of heterogenous microbial communities due to distinct environmental conditions in different sites of the oral cavity [57]. Various factors such as a change in nutrient availability and pH may contribute to dysbiosis in the oral cavity [51, 75]. Taking medications such as steroids or those for cancer therapy can also increase periodontitis [76].

Porphyromonas gingivalis is one of the main causative bacterial agents of periodontitis that drive the onset of dysbiosis and the consequent development of chronic periodontal disease [77]. The other dysbiotic species involved in chronic periodontitis include bacterial genera such as *Neisseria* spp., *Serratia marcescens*, *Pseudomonas* spp., and *Hafnia alvei* as identified by DNA-DNA hybridization technique [78]. In another study, 16S rRNA gene sequencing showed that periodontitis patients exhibited an overabundance of various bacterial genera such as *Tannerella*, *Prevotella*, *Filifactor*, *Mycoplasma*, *Phocaeicola*, *Johnsonella*, *Desulfobulbus*, and *Mogibacterium* [79].

The dysbiotic microflora in the microbial plaque induces the host-mediated inflammatory immune response that acts as a primary etiological factor in periodontitis [80]. Microbial plaque formed on the teeth and gingiva surface influences the initial host inflammation response, leading residential cells to produce cytokines such as TNF-α, IL-1β, and IL-6 [81]. *Porphyromonas gingivalis* is shown to induce high levels of proinflammatory cytokines, such as IL-1*β* and IL-6 by peripheral $CD4^+$ T helper cells [82].

Proinflammatory cytokines along with various enzymes that are upregulated during dysbiosis result in localized inflammation in the periodontal tissue, leading to the destruction of the tissue. The ongoing destruction of periodontal tissue results in the resorption of bone tissue and the degradation of the extracellular matrix [80]. Matrix degradation is attributed to the up-regulation of the expression of metalloproteinases, a family of Zn^{2+} and Ca^{2+}-dependent enzymes. While these enzymes are expressed in low amounts in healthy periodontal tissue and play roles in physiological functions like tissue development and turnover, they also contribute to pathological processes. Specifically, they are involved in the degradation of gingival and periodontal ligament collagen during connective tissue catabolism in periodontitis [83].

8.3.3 Halitosis

Halitosis refers to chronic unpleasant breath and air odor that comes from the mouth. It is derived from the Latin words "halitus" (breathable air) and "osis" (pathological change). A meta-analysis of various studies reported that the prevalence of halitosis was 31.8% with the occurrence of halitosis higher in low-middle income countries (39.8%) as compared to developed countries [84].

8.3.3.1 Role of Oral Dysbiosis in the Onset of Halitosis

The origin of bad breath in halitosis can be intra-oral or extra-oral. Intra-oral halitosis is due to the factors andconditions within the oral cavity, and it accounts for 80%–90% of cases [72]. Extra-oral halitosis is due to factors not related to the oral cavity. Other factors such as gastrointestinal, respiratory, or hepatic diseases can cause the bad odor in the mouth. Intra-oral halitosis can be due to various oral diseases such as gingivitis, periodontitis, and in the case of healthy individuals, poor oral hygiene or the tongue coating. Dietary habits, diabetes, smoking, alcohol consumption, and stress also have an effect on halitosis [85–87]. The tongue coating contains food debris, desquamated epithelial cells, and blood cells, creating a perfect environment for nourishment of bacteria [88]. These bacteria break down organic substate such as glucose, mucins, peptides, and protein present in saliva and thus produce odor compounds such as sulfur compounds, aromatic compounds, nitrogen-containing compounds, short-chain fatty acids, alcohol and phenyl compounds, and ketones [89–91]. Among them, volatile sulfur compounds (VSCs) are responsible for bad breath, with hydrogen sulfide and methyl mercaptan accounting for 90% of VSCs [92, 93]. A study using 16S rRNA analysis of tongue flora found that the dominant bacteria in intra-oral halitosis patients include *Atopium parvulum*, *Solobacterium moorei*, *Atopobium parvulum*, and *Eubacterium sulci*. On the other hand, *S. salivarius*, *Rothia mucilaginosa*, and an uncharacterized cultivable *Eubacterium* species were abundant in healthy controls [94]. In healthy individuals, in the absence of the habit of proper tongue cleaning, the anaerobic environment of the tongue may change to gram-negative bacteria. Gram-negative bacteria such as *Porphyromonas gingivalis*, *Treponema denticola*, *and Tannerella forsythia* produce hydrogen sulfide by metabolism of sulfur-containing amino acid compounds [95–97]. Gram-positive bacteria also promote the production of VSCs by gram-negative bacteria, as they are known to secrete certain enzymes that cleave sugar chains from glycoprotein thereby increasing the bioavailability of proteins for gram-negative bacteria [98].

8.4 SYSTEMIC DISEASES RELATED TO ORAL HEALTH

Growing evidence indicates that the oral microbiome may potentially affect the onset and progression of some of the chronic systemic diseases. Dysbiosis of oral microorganisms affects the inflammatory microenvironment of tissues and organs, which in turn may affect the normal physiology of the organs and can lead to systemic diseases. Oral biofilms close to the periodontal vasculature can easily find their way to the bloodstream and thereafter spread to remote organs like the heart and liver. Moreover, during oral inflammation such as in cases of periodontitis or gingivitis, bacterial transit into organs is promoted due to increased permeability of the blood vessels. Some of the systemic diseases in which the role of the oral microbiome has been explored are cardiovascular diseases, Alzheimer's disease, diabetes mellitus (DM), and rheumatoid arthritis (Figure 8.2).

8.4.1 Systemic Disorders Due to the Inflammatory Response of the Oral Microbiota

8.4.1.1 Diabetes Mellitus

DM is a medical condition arising due to insufficient production of insulin in the case of type-I and defective response of insulin in type-II. India has earned the notorious title as "Capital of diabetes" as it harbours world's largest number of diabetic patients (72 million in 2017).

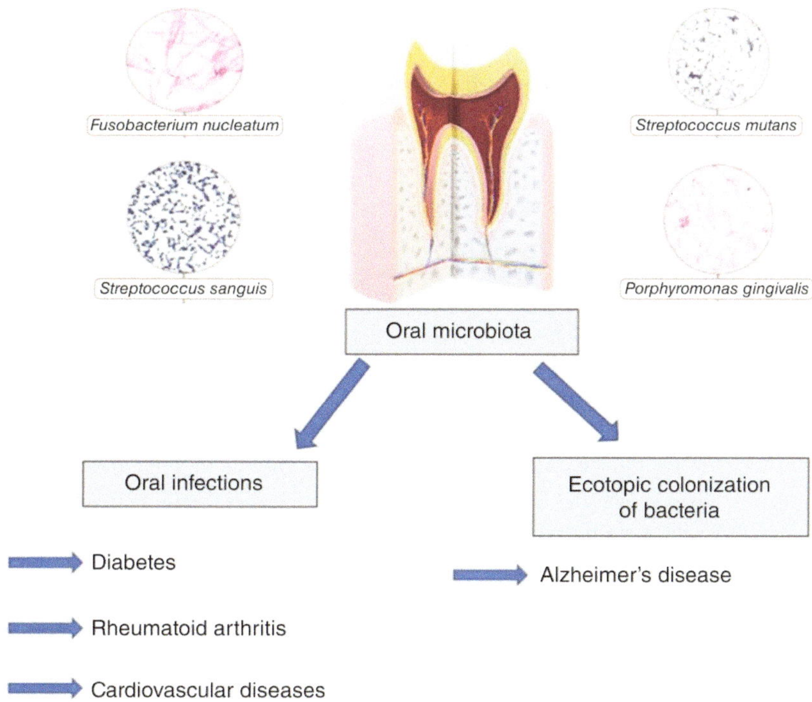

FIGURE 8.2 Schematic representation of the involvement of the oral microbiota in the development of systemic diseases through the inflammatory response and ectopic colonization of bacteria in different tissues of the body.

8.4.1.2 Role of the Oral Microbiome in the Onset of DM

Oral microorganisms can influence the onset and progression of DM by influencing the systemic immune homeostasis [99]. A direct correlation between the numbers of periodontal pathogenic bacteria such as *P. gingivalis* and *A. actinomycetemcomitans* and glycemic index/DM risk has been observed [100, 101]. Another study witnessed a substantial decrease in biological and phylogenetic diversity of the oral microbiome in DM patients as compared to healthy subjects [47, 102]. Patients with type 2 DM had much lower levels of the phylum Actinobacteria as compared to controls [103]. Conversely, a different study found that DM patients had reduced *Alloprevotella* abundances and greater *Actinomyces* and *Selenomonas* abundances as compared to non-DM persons [104]. Oral microbial profiles of peridontitic patients who were diabetic, or nondiabetic were compared. Patients without diabetes mellitus (DM) who have periodontitis show oral microbial profiles resembling those of diabetic patients with the disease, which also present unique characteristics [105, 106]. When compared to non-DM patients with periodontitis, the subgingival microbiome in DM patients with periodontitis reveals a greater abundance of Actinobacteria and Proteobacteria. *Actinomyces* and *Aggregatibacter*, *Tannerella forsythia*, *Capnocytophaga* [105–107], but the abundance of *P. gingivalis*, and *T. denticola* remained same. In contrast, another study did not find any major differences in the oral microbiome of periodontitis patients with and without DM [108]. DM patients with varying glycemic levels did not differ in their levels of known periodontopathogens, such as *P. gingivalis*, *T. forsythia*, and *T. denticola*, which are symptomatic of periodontitis [109].

The interaction of the microflora with the immune system of the surrounding periodontal tissues can lead to the production of inflammatory cytokines. For instance, in periodontitis, the excess response of mononuclear macrophages to periodontal bacteria results in increased local levels of inflammatory cytokines such as IL-1, IL-6, and TNF-α [110, 111]. The endotoxin of *P. gingivalis* is known to activate the toll-like receptor (TLR) 2/4 signaling pathway via binding to CD14 molecules on macrophage surfaces (Figure 8.3). The activated TLR 2/4 binds to the N-terminal death domain of MyD88 (myeloid differentiation primary response gene 88) and Toll/IL-1receptor (TIR) homology domain leading to the activation of IL-1 receptor-associated kinase (IRAK). Thereafter, the activated IRAK interacts with components associated with the TNF receptor to initiate the NF-κB, mitogen-activated protein kinase (MAPK), and c-Jun N-terminal kinase (JNK) signal transduction pathways eventually leading to a persistent inflammatory condition. NF-κB is also known to inhibit the glucose transporter 4 protein (GLUT4), one of the crucial proteins in the insulin signaling system [112]. The elevated level of free fatty acids in the blood due to TNF stimulation of lipid breakdown in fat cells also decreases insulin sensitivity. Thus, insulin resistance is a consequence of chronic inflammation that further exacerbates systemic inflammatory response and results in altered blood glucose metabolism.

Secondly, dysbiosis of the oral microbiome may cause increased numbers of oral gram-negative bacteria. Certain genera of gram-negative bacteria and their metabolites have the potential to enter the periodontal tissue through the damaged and loose gingival sulcus epithelium and spread to the surrounding tissues, leading to the enhanced production of reactive oxygen species (ROS) in inflammatory cells and vascular endothelial cells. A dysbiotic oral microbiome can also down-regulate the levels of antioxidant enzymes that are an essential component of host immune protection and microbial eradication. Thus, in the absence of a compensatory endogenous antioxidant response, oxidative stress generated due to ROS can lead to damage of pancreatic β cells that produce insulin leading to progression of DM [113]. ROS is known to damage insulin receptors and thereby negatively impact their downstream signaling pathways, resulting in a failure to respond adequately to insulin levels resulting in insulin resistance. Increased oxidative stress inhibits tyrosine phosphorylation, which further activates a number of serine/threonine kinase cascades via the NF-κB activation enzymes such as p38 MAPK, JNK/SAPK, and protein kinase C isoenzymes leading to disruption in the insulin transduction pathway (Figure 8.3) [114].

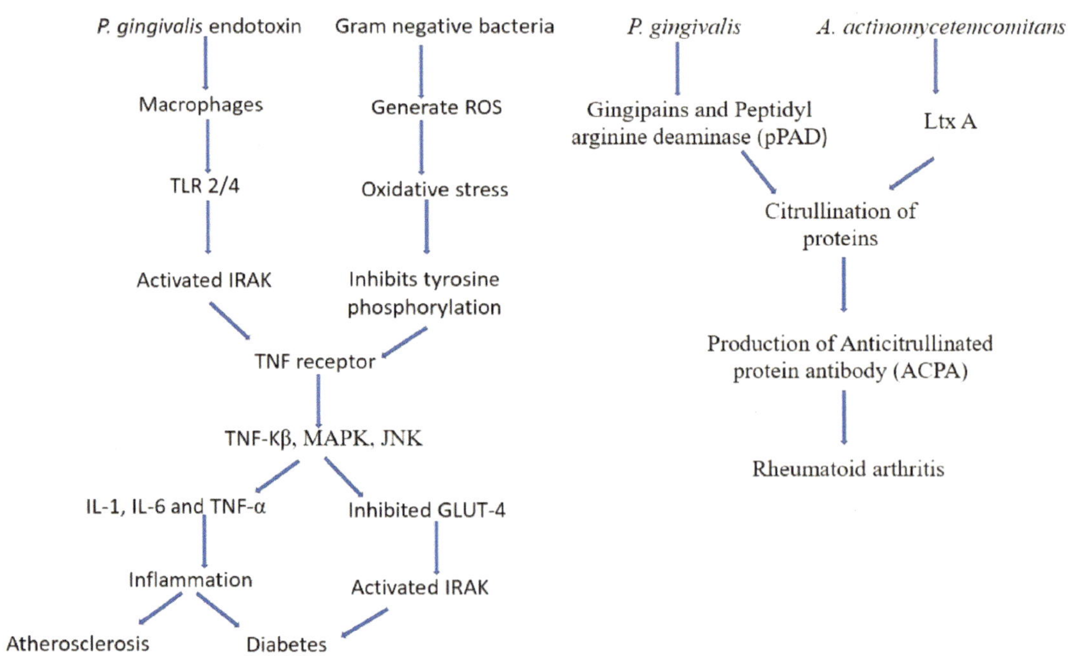

FIGURE 8.3 Inflammatory response pathways induced by oral microbiota that lead to systemic disorders. *Abbreviations*: GLUT4, glucose transporter 4 protein; IRAK, IL-1 receptor-associated kinase; JNK, c-Jun N-terminal kinase; MAPK, mitogen-activated protein kinase; TLR, toll-like receptor; TNF, tumor necrosis factor.

8.4.2 Rheumatoid Arthritis

Rheumatoid arthritis (RA) is a multifactorial, symmetric polyarticular autoimmune disease with uncertain etiology that mostly affects the joints but can also appear extra-articular. The primary cause of the RA pathology arises due to the overproduction of autoantibodies that leads to deposition of antigen-antibody complexes in the joints and kidneys leading to a type III hypersensitivity reaction and inflammation. The annual incidence of RA varies between 0.5%–1% and depends on gender and race/ethnicity.

8.4.2.1 Role of the Oral Microbiome in the Onset of RA

The connection between oral bacteria and RA has undergone new understanding in light of recent research findings. The immunoglobulins known as anti-citrullinated protein antibodies (ACPA) are very specific in RA patients [115]. These antibodies are thought to have a key role in the connection between periodontitis and RA. Citrullination or deamination is a posttranslational modification of the protein due to an enzyme known as peptidyl arginine deiminase (PAD) that converts arginine to citrulline amino acid. This alters the host protein metabolic characteristics and immunology, leading to the production of novel antigenic determinants, which trigger the body to make autoantibodies that are directed against these epitopes. PAD homologs have been found in *P. gingivalis* and in many different microbes in the gut and environment. Gingipains, a cysteine-like protease and PAD of *P. gingivalis* can together cause citrullination of the host protein resulting in autoantibody responses [116]. The arginine at the C-terminus of polypeptides is split by gingipain, and the liberated arginine can then be citrullinated by PAD. Anti-PAD antibody titers in humans are incredibly low and difficult to detect in serum.

In contrast to *P. gingivalis*, which produces bacterial PAD by self-synthesis, *A. actinomycetemcomitans* mostly induces endogenous PAD activity, which in turn causes the synthesis of autoantigens, the development of ACPA and rheumatoid factor, and the onset of RA. Similarly, *A. actinomycetemcomitans* secretes leukotoxin (LtxA), which causes high levels of protein citrullination in neutrophils and the development of self-antigens [117]. This causes an imbalance in neutrophil PAD activity. Furthermore, LtxA can alter the shape of neutrophils, mimic the synthesis of extracellular neutrophil traps, lead to neutrophil lysis, and finally release a significant amount of citrullinated antigen. The oral microbiota has a great sensitivity and can be used as a fingerprint to track the progression of the disease and determine the prognosis for RA.

8.4.3 Cardiovascular Diseases

Cardiovascular diseases are diseases of the heart and vascular system including a microbial infection of the heart, known as endocarditis, and coronary artery disease (CAD). A key pathogenic factor in CAD is atherosclerosis, which is the hardening and loss of elasticity of the arteries that eventually leads to loss of blood flow. An estimated 7.4 million people died in 2012 due to CAD arising from atherosclerosis out of the total 17.5 million people that died from cardiovascular diseases. Lipid deposition on the vascular endothelium of the big and middle arteries results in atherosclerotic plaques that limit the blood flow in arteries thereby leading to myocardial infarction. However, atherosclerosis is an inflammation-mediated event that is initiated in response to endothelium injury caused by the retention of lipoproteins. Activated endothelial cells recruit macrophages that further increase oxidative stress which oxidizes the deposited lipoproteins. Macrophages also secrete proinflammatory cytokines thereby recruiting other leukocytes such as T cells, B cells, and dendritic cells [118].

8.4.3.1 Role of the Oral Microbiome in the Onset of Cardiovascular Disease

It is being slowly recognized that the inflammatory immune responses induced by oral and gut microbiota can exacerbate inflammation in the body that in turn affects the cardiovascular health.

The oral microbiota also regulates the diversity of gut microbiota due to the transfer of oral microbiota into the gut through saliva secretions. Oral microbial dysbiosis ultimately manifests as systemic inflammation, immunoreaction, oxidative stress, and thrombosis. A meta-analysis of 63 studies demonstrated that five oral commensal bacteria, i.e., *Campylobacter rectus*, *P. endodontalis*, *P. gingivalis*, *Prevotella intermedia*, and *Pr. nigrescens*, were present in the atherosclerotic plaques in CAD patients and their relative abundances were high when compared to those of non-cardiac patients [119]. These bacteria secrete various metabolites, and 36 of the bacteria-derived metabolites are specifically associated with the plaques that can further cause inflammation. Additionally, a nearly twofold increased incidence of angiographically verified stable CAD was linked to the presence of *A. actinomycetemcomitans* in the subgingival region [120]. *Streptococcus mutans* was the most abundant bacteria isolated from diseased heart valves and atheromatous plaques [121]. The virulence proteins, toxins, and extracellular enzymes secreted by the dysbiotic oral flora in the gingival crevice activates the immune system resulting in inflammation. The levels of proinflammatory cytokines such as TNF-α, IL-1, IL-6, and prostaglandins increased when periodontal bacteria infected epithelial cells. These cytokines further cause atherosclerosis to develop. For example, *P. gingivalis* was specifically shown to promote atherosclerosis in mice by increasing the oxidative stress and inflammation [122].

Epithelial cells produced IL-6, IFN, or TNF in response to *P. gingivalis* in the early stages of infection, which led to the destruction of nearby tissue. The degraded gingival tissues and activated endothelial cells, bacteria, and their virulence factors (like gingipains, lipopolysaccharide, or fimbriae) entered the bloodstream and caused an inflammatory response under the control of proinflammatory mediators (like IL-8 and MCP-1), growth factors, differentiation factors, cell-adhesion molecules, and toll-like receptors.

8.4.4 Systemic Disorders Due to Ectopic Colonization of the Oral Microbiota

8.4.4.1 Alzheimer's Disease

Alzheimer's disease (AD) is characterized by dementia and loss of mental faculties due to the buildup of extracellular β-amyloid (Aβ) protein and intracellular neuronal protein called tau. The amyloid precursor protein and presenilin genes, both of which are connected to Aβ metabolism, become mutated, leading to AD. Clinically, AD can be of two types. Early onset AD, also known as familial AD, occurs before the age of 60, and second type is sporadic AD that has a late onset after the age of 60 years. With more than 15 million cases worldwide, sporadic AD is much more prevalent than familial illness. The disease's varied nature, coupled with aging and a complicated interplay of genetic and environmental risk factors, makes the origin of the sporadic form of the disease uncertain.

8.4.4.2 Role of the Oral Microbiome in the Onset of AD

The link between oral microorganisms and AD is recently being explored. It was shown that gingipains, a virulence factor of *P. gingivalis*, is neurotoxic *in vivo* and *in vitro* and damage tau protein, which is important for normal brain function (Figure 8.4) [123]. Studies have shown that *P. gingivalis*–induced periodontitis increased extracellular deposition of Aβ and the levels of IL1β and TNF-α, which impacted the cognitive function of mice [124]. In another investigation, oral administration of *P. gingivalis* or gingipains to ten C57BL/6 mice for 22 weeks resulted in experimentally induced chronic periodontitis and, concomitantly, the mice displayed a neurodegenerative disorder resembling AD due to accumulation of extracellular protein Aβ1-42 [125]. *Porphyromonas gingivalis* can colonize mice's brains and boost the buildup of Aβ1-42 [123]. Inhibitors of gingipains are known to decrease the amount of *P. gingivalis* in brain tissue, preventing the formation of Aβ1-42, reduced neuroinflammation and hippocampus neuronal injury [123].

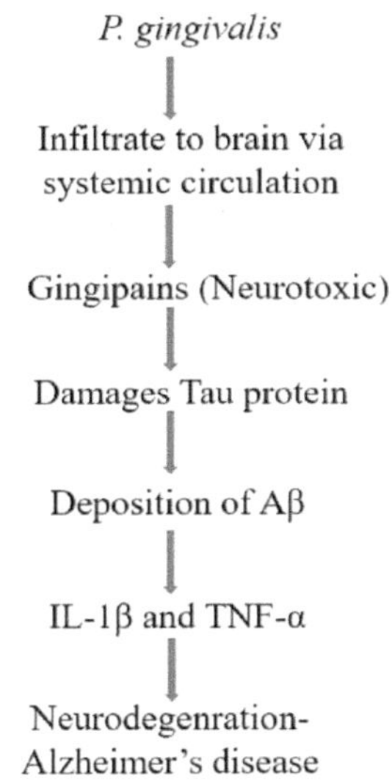

FIGURE 8.4 Diagrammatic representation of systemic disorders due to ectopic colonization of the oral microbiota.

There are two ways in which oral bacteria can impact brain cells. First periodontal bacteria can induce proinflammatory cytokines that infiltrate the brain tissue via the systemic circulation. The blood-brain barrier is believed to become more permeable with aging, a condition that often precedes the development of AD. Genetically engineered mice (5XFAD mice) with mutations in the amyloid precursor protein gene exhibited enhanced blood-brain barrier permeability and senile plaque accumulation [126].

The second mechanism involves periodontal microbes or their by-products that directly infiltrate the central nervous system through peripheral nerves such as the trigeminal and/or glossopharyngeal nerve. Oral *Treponema* spp. have been isolated from the hippocampus in the brain [127] and the trigeminal ganglion [128]. Rather than protecting the brain, the blood-brain barrier could serve as a gateway for bacteria to reach the brain. Additionally, AD and chronic periodontitis have a mutually reinforcing association. AD patients often have decreased mobility and poor dental hygiene, which encourages the accumulation of periodontal tissue inflammation and eventual tooth loss. Further research is required to clarify the association and mechanism of the interaction between periodontitis and AD.

8.5 CONCLUSION AND FUTURE OUTLOOK

The oral microbiome is a significant component of the human microbiome and is both systemic and community specific. However, when compared to the gut microbiome, research on the oral microbiome is still in its early stages. The development of metagenomics and high-throughput sequencing technologies has helped to understand the link between the oral microbiome and systemic diseases. These studies indicate that the composition and function of oral bacteria is altered in a diseased state and may further change the equilibrium between human health and disease state. When dysbiosis takes place, harmful host-pathogen interactions can cause the oralome to change from a healthy condition to a pathogenic state. Thus, identifying the "signature" microbes implicated in the health and disease will further help to devise microbiome-based diagnostic methods for various diseases and to develop therapeutic interventions for the prevention and treatment of oral and human systemic diseases.

REFERENCES

1. Martin, J. J.; Schatz, V.; Schatz, A. Contribution to the microbiology of the mouth by Antony van Leeuwenhoek (24 October 1632-26 August 1732). *The Journal of Nihon University School of Dentistry* **1972**, *14*(3), 106–112. doi: 10.2334/josnusd1959.14.106
2. Sender, R.; Fuchs, S.; Milo, R. Revised estimates for the number of human and bacteria cells in the body. *PLoS Biology* **2016**, 14(8). doi: 10.1371/journal.pbio.1002533
3. Li, J.; Jia, H.; Cai, X.; Zhong, H.; Feng, Q.; Sunagawa, S.; Arumugam, M.; Kultima, J.R.; Prifti, E.; Nielsen, T.; Juncker, A.S. An integrated catalog of reference genes in the human gut microbiome. *Nature Biotechnology* **2014**, *32*(8), 834–841. doi: 10.1038/nbt.2942
4. Hallen-Adams, H.E.; Suhr, M.J. Fungi in the healthy human gastrointestinal tract. *Virulence* **2017**, *8*(3), 352–358. doi: 10.1080/21505594.2016.1247140
5. Chibani, C.M.; Mahnert, A.; Borrel, G. A catalogue of 1,167 genomes from the human gut archaeome. *Nature Microbiology* **2022**, *7*, 48–61.
6. Manrique, P.; Bolduc, B.; Walk, S.T.; van der Oost, J.; de Vos, W.M.; Young, M.J. Healthy human gut phageome. *Proceedings of the National Academy of Sciences* **2016**, *113*(37), 10400–10405. doi: 10.1073/pnas.1601060113.
7. Bandara, H. M. H. N., Panduwawala, C. P., Samaranayake, L. P. Biodiversity of the human oral mycobiome in health and disease. *Oral Diseases* **2019**, *25*(2), 363–371. doi: 10.1111/odi.12899.
8. Jamal, M.; Ahmad, W.; Andleeb, S.; Jalil, F.; Imran, M.; Nawaz, M. A.; Kamil, M. A. Bacterial biofilm and associated infections. *Journal of the Chinese Medical Association* **2018**, *81*(1), 7–11. doi: 10.1016/j.jcma.2017.07.012.
9. Saini, R.; Saini, S.; Sharma, S. Biofilm: A dental microbial infection. *Journal of Natural Science, Biology, and Medicine* **2011**, *2*(1), 71. doi: 10.4103/0976-9668.82317.
10. Velsko, I. M.; Fellows Yates, J. A.; Aron, F.; Hagan, R. W.; Frantz, L. A.; Loe, L.; Larson, G. Microbial differences between dental plaque and historic dental calculus are related to oral biofilm maturation stage. *Microbiome* **2019**, *7*(1), 1–20. doi: 10.1186/s40168-019-0717-3.
11. Jakubovics, N. S. Intermicrobial interactions as a driver for community composition and stratification of oral biofilms. *Journal of Molecular Biology* **2015**, *427*(23), 3662–3675. doi: 10.1016/j.jmb.2015.09.022.
12. Davey, M. E.; Costerton, J. W. Molecular genetics analyses of biofilm formation in oral isolates. *Periodontology 2000* **2006**, *42*(1), 13–26. doi: 10.1111/j.1600-0757.2006.00052.x.
13. Eren, A. M.; Borisy, G. G.; Huse, S. M.; Mark Welch, J. L. Oligotyping analysis of the human oral microbiome. *Proceedings of the National Academy of Sciences* **2014**, *111*(28), E2875–E288. doi: 10.1073/pnas.1409644111.
14. Robles-Sikisaka, R.; Ly, M.; Boehm, T.; Naidu, M.; Salzman, J.; Pride, D. T. Association between living environment and human oral viral ecology. *The ISME Journal* **2013**, *7*(9), 1710–1724. doi: 10.1038/ismej.2013.63
15. Pride, D. T.; Salzman, J.; Haynes, M.; Rohwer, F.; Davis-Long, C.; White, R. A.; Relman, D. A. Evidence of a robust resident bacteriophage population revealed through analysis of the human salivary virome. *The ISME Journal* **2012**, *6*(5), 915–926. doi: 10.1038/ismej.2011.169.
16. Ly, M.; Abeles, S. R.; Boehm, T. K.; Robles-Sikisaka, R.; Naidu, M.; Santiago-Rodriguez, T.; Pride, D. T. Altered oral viral ecology in association with periodontal disease. *mBio* **2014**, *5*(3). doi: 10.1128/mbio.01133-14
17. Perez-Brocal, V.; Moya, A. The analysis of the oral DNA virome reveals which viruses are widespread and rare among healthy young adults in Valencia (Spain). *PloS one* **2018**, *13*(2). doi: 10.1371/journal.pone.0191867.
18. Baker, J. L.; Bor, B.; Agnello, M.; Shi, W.; He, X. Ecology of the oral microbiome: beyond bacteria. *Trends in Microbiology* **2017**, *25(5)*, 362–374. doi: 10.1016/j.tim.2016.12.012.
19. Gao, L.; Kang, M.; Zhang, M. J.; Reza Sailani, M.; Kuraji, R.; Martinez, A.; Zhan, L. Polymicrobial periodontal disease triggers a wide radius of effect and unique virome. *NPJ Biofilms and Microbiomes* **2020**, *6*(1), 10. doi: 10.1038/s41522-020-0120-7.
20. De la Cruz Pena, M. J.; Martinez-Hernandez, F.; Garcia-Heredia, I.; Lluesma Gomez, M.; Fornas, O.; Martinez-Garcia, M. Deciphering the human virome with single-virus genomics and metagenomics. *Viruses* **2018**, *10*(3), 113. doi: 10.3390/v10030113.
21. Ward, T. L.; Dominguez-Bello, M. G.; Heisel, T.; Al-Ghalith, G.; Knights, D.; Gale, C. A. Development of the human mycobiome over the first month of life and across body sites. *MSystems* **2018**, *3*(3), doi: 10.1128/msystems.00140-17.

22. Ghannoum, M. A.; Jurevic, R. J.; Mukherjee, P. K.; Cui, F.; Sikaroodi, M.; Naqvi, A.; Gillevet, P. M. Characterization of the oral fungal microbiome (mycobiome) in healthy individuals. *PLoS Pathogens* **2010**, *6*(1). doi: 10.1371/journal.ppat.1000713.
23. Peters, B. A.; Wu, J.; Hayes, R. B.; Ahn, J. The oral fungal mycobiome: characteristics and relation to periodontitis in a pilot study. *BMC Microbiology* **2017**, *17*, 1–11. doi: 10.1186/s12866-017-1064-9.
24. Naglik, J. R.; Tang, S. X.; Moyes, D. L. Oral colonization of fungi. *Current Fungal Infection Reports* **2013**, *7*, 152–159. doi: 10.1007/s12281-013-0129-y.
25. He, J.; Li, Y.; Cao, Y.; Xue, J.; Zhou, X. The oral microbiome diversity and its relation to human diseases. *Folia Microbiologica* **2015**, *60*, 69–80. doi: 10.1007/s12223-014-0342-2.
26. Kofoid, C. A.; The protozoa of the human mouth. presidential address, American Society of Parasitologists, New York, December 29, 1928. *The Journal of Parasitology* **1929**, *15*(3), 151–174. doi: 10.2307/3271673.
27. Maybodi, F. R.; Ardakani, A. H.; Bafghi, A. F.; Ardakani, A. H.; Zafarbakhsh, A. The effect of non-surgical periodontal therapy on *Trichomonas tenax* and *Entamoeba gingivalis* in patients with chronic periodontitis. *Journal of Dentistry* **2016**, *17*(3), 171.
28. Nagarajan, M.; Prabhu, V. R.; Kamalakkannan, R. Metagenomics: implications in oral health and disease. In Nagarajan, M. Ed. *Metagenomics*, Academic Press. **2018**, 179–195. doi: 10.1016/B978-0-08-102268-9.00009-4.
29. Dubar, M.; Zaffino, M.-L.; Remen, T.; Thilly, N.; Cunat, L.; Machouart, M.C.; Bisson, C. Protozoans in subgingival biofilm: clinical and bacterial associated factors and impact of scaling and root planing treatment. *Journal of Oral Microbiology* **2020**, *12*(1), 1693222. doi: 10.1080/20002297.2019.1693222.
30. Nguyen-Hieu, T.; Khelaifia, S.; Aboudharam, G.; Drancourt, M. Methanogenic archaea in subgingival sites: a review. *Apmis* **2013**, *121*(6), 467–477. doi: 10.1111/apm.12015.
31. Sogodogo, E.; Doumbo, O.; Aboudharam, G.; Kouriba, B.; Diawara, O.; Koita, H.; Drancourt, M. First characterization of methanogens in oral cavity in Malian patients with oral cavity pathologies. *BMC Oral Health* **2019**, *19*, 1–6. doi: 10.1186/s12903-019-0929-8.
32. Mosaddad, S. A.; Tahmasebi, E.; Yazdanian, A.; Rezvani, M. B.; Seifalian, A.; Yazdanian, M.; Tebyanian, H. Oral microbial biofilms: an update. *European Journal of Clinical Microbiology & Infectious Diseases* **2019**, *38*, 2005–2019. doi: 10.1007/s10096-019-03641-9.
33. Belmok, A.; de Cena, J.; Kyaw, C.; Dame-Teixeira, N. The oral archaeome: a scoping review. *Journal of Dental Research* **2020**, *99*(6), 630–643. doi: 10.1177/0022034520910435
34. Marsh, P., Zaura, E. Dental biofilm: ecological interactions in health and disease. *Journal of Clinical Periodontology* **2017**, *44*, S12–S22. doi: 10.1111/jcpe.12679.
35. Wright, C. J.; Burns, L. H.; Jack, A. A.; Back, C. R.; Dutton, L. C.; Nobbs, A. H.; Jenkinson, H. F. Microbial interactions in building of communities. *Molecular Oral Microbiology* **2013**, *28*(2), 83–101. doi: 10.1111/omi.12012.
36. Xu, H.; Jenkinson, H. F.; Dongari-Bagtzoglou, A. Innocent until proven guilty: mechanisms and roles of *Streptococcus–Candida* interactions in oral health and disease. *Molecular Oral Microbiology* **2014**, *29*(3), 99–116. doi: 10.1111/omi.12049.
37. Diaz, P. I. Microbial diversity and interactions in subgingival biofilm communities. *Periodontal Disease* **2012**, *15*, 17–40. doi: 10.1159/000329669.
38. Li, Y.H.; Lau, P. C.; Lee, J. H.; Ellen, R. P.; Cvitkovitch, D. G. Natural genetic transformation of *Streptococcus mutans* growing in biofilms. *Journal of Bacteriology* **2001**, *183*(3), 897–908. doi: 10.1128/JB.183.3.897-908.2001.
39. Tribble, G. D.; Rigney, T. W.; Dao, D.H. V.; Wong, C. T.; Kerr, J. E.; Taylor, B. E.; Kaplan, H. B. Natural competence is a major mechanism for horizontal DNA transfer in the oral pathogen *Porphyromonas gingivalis*. *MBio* **2012**, *3*(1), doi: 10.1128/mbio.00231-11.
40. Chi, F.; Nolte, O.; Bergmann, C.; Ip, M.; Hakenbeck, R. Crossing the barrier: evolution and spread of a major class of mosaic pbp2x in *Streptococcus pneumoniae*, *S. mitis* and *S. oralis*. *International Journal of Medical Microbiology* **2007**, 297(8), 503–512. doi: 10.1016/j.ijmm.2007.02.009.
41. Pereira, C. S.; Thompson, J. A.; Xavier, K. B. AI-2-mediated signalling in bacteria. *FEMS Microbiology Reviews* **2013**, *37*(2), 156–181. doi: 10.1111/j.1574-6976.2012.00345.x.
42. Tourneroche, A.; Lami, R.; Hubas, C.; Blanchet, E.; Vallet, M.; Escoubeyrou, K.; Prado, S. Bacterial–fungal interactions in the kelp endomicrobiota drive autoinducer-2 quorum sensing. *Frontiers in Microbiology* **2019**, *10*, 1693. doi: 10.3389/fmicb.2019.01693.
43. Ryu, E.J.; Sim, J.; Sim, J.; Lee, J.; Choi, B.K. D-Galactose as an autoinducer 2 inhibitor to control the biofilm formation of periodontopathogens. *Journal of Microbiology* **2016**, *54*, 632–637. doi: 10.1007/s12275-016-6345-8.

44. Mashima, I.; Nakazawa, F. Role of an autoinducer-2-like molecule from *Veillonella tobetsuensis* in *Streptococcus gordonii* biofilm formation. *Journal of Oral Biosciences* **2017**, *59*(3), 152–156. doi: 10.1016/j.job.2017.06.002.
45. Rickard, A. H.; Palmer Jr, R. J.; Blehert, D. S.; Campagna, S. R.; Semmelhack, M. F.; Egland, P. G.; Kolenbrander, P. E. Autoinducer 2: a concentration-dependent signal for mutualistic bacterial biofilm growth. *Molecular Microbiology* **2006**, *60*(6), 1446–1456. doi: 10.1111/j.1365-2958.2006.05202.x.
46. Jang, Y.J. Autoinducer 2 of *Fusobacterium nucleatum* as a target molecule to inhibit biofilm formation of periodontopathogens. *Archives of oral biology* **2013**, *58*(1),17–27. doi: 10.1016/j.archoralbio.2012.04.016.
47. Yang, Y.; Lin, J.; Harrington, A.; Cornilescu, G.; Lau, G. W.; Tal-Gan, Y. Designing cyclic competence-stimulating peptide (CSP) analogs with pan-group quorum-sensing inhibition activity in *Streptococcus pneumoniae. Proceedings of the National Academy of Sciences* **2020**, *117*(3), 1689–1699. doi: 10.1073/pnas.1915812117.
48. Nagasawa, R.; Yamamoto, T.; Utada, A. S.; Nomura, N.; Obana, N. Competence-stimulating-peptide-dependent localized cell death and extracellular DNA production in *Streptococcus mutans* biofilms. *Applied and Environmental Microbiology* **2020**, *86*(23), e02080–02020. doi: 10.1128/aem.02080-20
49. Jack, A. A.; Daniels, D. E.; Jepson, M. A.; Vickerman, M. M.; Lamont, R. J.; Jenkinson, H. F.; Nobbs, A. H. *Streptococcus gordonii* comCDE (competence) operon modulates biofilm formation with *Candida albicans. Microbiology* **2015**, *161*(2), 411. 10.1099/mic.0.000010
50. Radaic, A.; de Jesus, M. B; Kapila, Y. L. Bacterial anti-microbial peptides and nano-sized drug delivery systems: The state of the art toward improved bacteriocins. *Journal of Controlled Release* **2020**, *321*, 100–118. doi: 10.1016/j.jconrel.2020.02.001.
51. Liu, B.; Faller, L. L.; Klitgord, N.; Mazumdar, V.; Ghodsi, M.; Sommer, D. D.; Li, S. Deep sequencing of the oral microbiome reveals signatures of periodontal disease. *PloS one* **2012**, *7*(6), e37919. doi: 10.1371/journal.pone.0037919.
52. World Health Organization. Global oral health status report: towards universal health coverage for oral health by 2030. WHO Geneva **2022**.
53. Hughes, C. V.; Dahlan, M.; Papadopolou, E.; Loo, C. Y.; Pradhan, N. S.; Lu, S. C.; Tanner, A. C. Aciduric microbiota and *mutans streptococci* in severe and recurrent severe early childhood caries. *Pediatric Dentistry* **2012**, *34*(2), 16E–23E.
54. Takahashi, N.; Nyvad, B.; The role of bacteria in the caries process: ecological perspectives. *Journal of Dental Research* **2011**, *90*(3), 294–303. doi: 10.1177/0022034510379602.
55. Kidd, E. A. Clinical threshold for carious tissue removal. *Dental Clinics* **2010**, *54*(3), 541–549. doi: 10.1016/j.cden.2010.03.001.
56. Balakrishnan, M.; Simmonds, R. S.; Tagg, J. R. Dental caries is a preventable infectious disease. *Australian Dental Journal* **2000**, *45*(4), 235–245. doi: 10.1111/j.1834-7819.2000.tb00257.x.
57. Bowen, W. H.; Burne, R. A.; Wu, H.; Koo, H. Oral biofilms: pathogens, matrix, and polymicrobial interactions in microenvironments. *Trends in Microbiology* **2018**, *26*(3), 229–242. doi: 10.1016/j.tim.2017.09.008.
58. Flemming, H.C.; Wingender, J.; Szewzyk, U.; Steinberg, P.; Rice, S. A.; Kjelleberg, S. Biofilms: an emergent form of bacterial life. *Nature Reviews Microbiology* **2016**, *14*(9), 563–575. doi: 10.1038/nrmicro.2016.94.
59. Harper, D. S.; Loesche, W. J. Growth and acid tolerance of human dental plaque bacteria. *Archives of Oral Biology* **1984**, *29*(10), 843–848. doi: 10.1016/0003-9969(84)90015-3.
60. Milnes, A. R.; Bowden, G. H. W. The microflora associated with developing lesions of nursing caries. *Caries Research* **1985**, *19*(4), 289–297. 10.1159/000260858.
61. Boue, D.; Armau, E.; Tiraby, G. A bacteriological study of rampant caries in children. *Journal of Dental Research*, **1987**, *66*(1), 23–28. doi: 10.1177/00220345870660010501.
62. Belda-Ferre, P.; Alcaraz, L. D.; Cabrera-Rubio, R.; Romero, H.; Simon-Soro, A.; Pignatelli, M., & Mira, A. The oral metagenome in health and disease. *The ISME Journal* **2012**, *6*(1), 46–56. doi: 10.1038/ismej.2011.85.
63. Benitez-Paez, A.; Belda-Ferre, P.; Simon-Soro, A.; Mira, A. Microbiota diversity and gene expression dynamics in human oral biofilms. *BMC Genomics* **2014**, *15*(1), 1–13. doi: 10.1186/1471-2164-15-311.
64. Becker, M. R.; Paster, B. J.; Leys, E. J.; Moeschberger, M. L.; Kenyon, S. G.; Galvin, J. L. Molecular analysis of bacterial species associated with childhood caries. *Journal of Clinical Microbiology* **2002**, *40*(3), 1001–1009. doi: 10.1128/JCM.40.3.1001-1009.2002.
65. Kanasi, E.; Dewhirst, F.; Chalmers, N.; Kent, R Jr.; Moore, A.; Hughes, C.; Tanner, A. Clonal analysis of the microbiota of severe early childhood caries. *Caries Research* **2010**, *44*(5), 485–497. doi: 10.1159/000320158.

66. Ling, Z.; Kong, J.; Jia, P.; Wei, C.; Wang, Y.; Pan, Z.; Xiang, C. Analysis of oral microbiota in children with dental caries by PCR-DGGE and barcoded pyrosequencing. *Microbial Ecology* **2010**, *60*, 677–690. doi: 10.1007/s00248-010-9712-8.
67. Tanner, A.; Kent Jr, R.; Holgerson, P. L.; Hughes, C.; Loo, C.; Kanasi, E.; Johansson, I.; Microbiota of severe early childhood caries before and after therapy. *Journal of Dental Research* **2011**, *90*(11), 1298–1305. doi: 10.1177/0022034511421201.
68. Preza, D.; Olsen, I.; Aas, J. A.; Willumsen, T.; Grinde, B.; Paster, B. J. Bacterial profiles of root caries in elderly patients. *Journal of Clinical Microbiology* **2008**, *46*(6), 2015–2021. doi: 10.1128/jcm.02411-07.
69. Yang, F.; Zeng, X.; Ning, K.; Liu, K.L.; Lo, C.C.; Wang, W.; Chang, X. Saliva microbiomes distinguish caries-active from healthy human populations. *The ISME Journal* **2012**, *6*(1), 1–10. doi: 10.1038/ismej.2011.71.
70. Lamont, R. J.; Koo, H.; Hajishengallis, G. The oral microbiota: dynamic communities and host interactions. *Nature Reviews Microbiology* **2018**, *16*(12), 745–759. doi: 10.1038/s41579-018-0089-x.
71. Papapanou, P. N.; Sanz, M.; Buduneli, N.; Dietrich, T.; Feres, M.; Fine, D. H.; Graziani, F. Periodontitis: Consensus report of workgroup 2 of the 2017 World Workshop on the classification of periodontal and peri-implant diseases and conditions. *Journal of Periodontology* **2018**, *89*, S173–S182. doi: 10.1111/jcpe.12946.
72. Nazir, M. A.; Almas, K.; Majeed, M. I. The prevalence of halitosis (oral malodor) and associated factors among dental students and interns, Lahore, Pakistan. *European Journal of Dentistry* **2017**, *11*(04), 480–485. doi: 10.4103/ejd.ejd_142_17.
73. Zhang, Y.; Ding, Y.; Guo, Q. Probiotic species in the management of periodontal diseases: An overview. *Frontiers in Cellular and Infection Microbiology* **2022**, *12*, 308. doi: 10.3389/fcimb.2022.806463.
74. Van Dyke, T. E. Pro-resolving mediators in the regulation of periodontal disease. *Molecular Aspects of Medicine* **2017**, *58*, 21–36. doi: 10.1016/j.mam.2017.04.006.
75. Willis, J. R.; Gabaldón, T. The human oral microbiome in health and disease: from sequences to ecosystems. *Microorganisms* **2020**, *8*(2), 308. doi: 10.3390/microorganisms8020308.
76. Kim, J.; Amar, S. Periodontal disease and systemic conditions: a bidirectional relationship. *Odontology* **2006**, *94*, 10–21. doi: 10.1007/s10266-006-0060-6.
77. Mysak, J.; Podzimek, S.; Sommerova, P.; Lyuya-Mi, Y.; Bartova, J.; Janatova, T.; Duskova, J. *Porphyromonas gingivalis*: major periodontopathic pathogen overview. *Journal of Immunology Research* **2014**. doi: 10.1155/2014/476068.
78. Colombo, A. P. V.; MagalhNes, C. B.; Hartenbach, F. A. R. R.; do Souto, R. M., da Silva-Boghossian, C. M. Periodontal-disease-associated biofilm: A reservoir for pathogens of medical importance. *Microbial Pathogenesis* **2016**, *94*, 27–34. doi: 10.1016/j.micpath.2015.09.009.
79. Chen, C.; Hemme, C.; Beleno, J.; Shi, Z. J.; Ning, D.; Qin, Y.; Zhou, J. Oral microbiota of periodontal health and disease and their changes after nonsurgical periodontal therapy. *The ISME journal* **2018**, *12*(5), 1210–1224. doi: 10.1038/s41396-017-0037-1.
80. Cekici, A.; Kantarci, A.; Hasturk, H.; Van Dyke, T. E. Inflammatory and immune pathways in the pathogenesis of periodontal disease. *Periodontology* **2014**, *64*(1), 57–80. doi: 10.1111/prd.12002.
81. Cheng, W. C.; van Asten, S. D.; Burns, L. A.; Evans, H. G.; Walter, G. J.; Hashim, A.; Taams, L. S. Periodontitis-associated pathogens *Porphyromonas gingivalis* and *A. actinomycetemcomitans* activate human CD14+ monocytes leading to enhanced Th17/IL-17 responses. *European Journal of Immunology* **2016**, *46*(9), 2211–2221. doi: 10.1002/eji.201545871.
82. Gonzales, J. R.; Groeger, S.; Johansson, A.; Meyle, J. T helper cells from aggressive periodontitis patients produce higher levels of interleukin-1 and interleukin-6 in interaction with *Porphyromonas gingivalis*. *Clinical Oral Investigations* **2014**, *18*, 1835–1843. doi: 10.1007/s00784-013-1162-5.
83. Checchi, V.; Maravic, T.; Bellini, P.; Generali, L.; Consolo, U.; Breschi, L.; Mazzoni, A. The role of matrix metalloproteinases in periodontal disease. *International Journal of Environmental Research and Public Health* **2020**, *17*(14), 4923. doi: 10.3390/ijerph17144923.
84. Silva, M. F.; Leite, F. R.; Ferreira, L. B.; Pola, N. M.; Scannapieco, F. A.; Demarco, F. F.; Nascimento, G. G. Estimated prevalence of halitosis: a systematic review and meta-regression analysis. *Clinical Oral Investigations* **2018**, *22*, 47–55. doi: 10.1007/s00784-017-2164-5.
85. De Lima, P. O.; Nani, B. D.; Rolim, G. S.; Groppo, F. C.; Franz-Montan, M.; De Moraes, A. B. A.; Marcondes, F. K. Effects of academic stress on the levels of oral volatile sulfur compounds, halitosis-related bacteria and stress biomarkers of healthy female undergraduate students. *Journal of Breath Research* **2020**, *14*(3), 036005. doi: 10.1088/1752-7163/ab944d.
86. Hampelska, K.; Jaworska, M. M.; Babalska, Z. Ł.; Karpiński, T. M. The role of oral microbiota in intra-oral halitosis. *Journal of Clinical Medicine* **2020**, *9*(8), 2484. doi: 10.3390/jcm9082484.
87. Wu, J.; Cannon, R.; Ji, P.; Farella, M.; Mei, L. Halitosis: prevalence, risk factors, sources, measurement and treatment–a review of the literature. *Australian Dental Journal* **2020**, *65*(1), 4–11. 10.1111/adj.12725.

88. Kamaraj, D.; Bhushan, K. S.; Vandana, K. An evaluation of microbial profile in halitosis with tongue coating using PCR (polymerase chain reaction)-a clinical and microbiological study. *Journal of Clinical and Diagnostic Research* **2014**, *8*(1), 263. doi: 10.7860/JCDR/2014/6213.3856.
89. McNamara, T. F.; Alexander, J. F.; Lee, M. The role of microorganisms in the production of oral malodor. *Oral Surgery, Oral Medicine, Oral Pathology* **1972**, *34*(1), 41–48. doi: 10.1016/0030-4220(72)90271-X.
90. Persson, S.; Claesson, R.; Carlsson, J.; The capacity of subgingival microbiotas to produce volatile sulfur compounds in human serum. *Oral Microbiology and Immunology* **1989**, *4*(3), 169–172. 10.1111/j.1399-302x.1989.tb00246.x.
91. Thorn, R. M. S.; Greenman, J. Microbial volatile compounds in health and disease conditions. *Journal of Breath Research* **2012**, *6*(2), 024001. doi: 10.1088/1752-7155/6/2/024001.
92. Payne, D.; Gordon, J. J.; Nisbet, S.; Karwal, R.; Bosma, M. L. A randomised clinical trial to assess control of oral malodour by a novel dentifrice containing 0.1% w/wo-cymen-5-ol, 0.6% w/w zinc chloride. *International Dental Journal* **2011**, *61*, 60–66. doi: 10.1111/j.1875-595x.2011.00051.x.
93. Van den Velde, S.; van Steenberghe, D.; Van Hee, P.; Quirynen, M. Detection of odorous compounds in breath. *Journal of Dental Research* **2009**, *88*(3), 285–289. doi: 10.1177/0022034508329741.
94. Kazor, C.; Mitchell, P.; Lee, A.; Stokes, L.; Loesche, W.; Dewhirst, F.; Paster, B. Diversity of bacterial populations on the tongue dorsa of patients with halitosis and healthy patients. *Journal of Clinical Microbiology* **2003**, *41*(2), 558–563. doi: 10.1128/jcm.41.2.558-563.2003.
95. Bodet, C.; Chandad, F.; Grenier, D. Pathogenic potential of *Porphyromonas gingivalis*, *Treponema denticola* and *Tannerella forsythia*, the red bacterial complex associated with periodontitis. *Pathologie-biologie* **2006**, *55*(3-4), 154–162. doi: 10.1016/j.patbio.2006.07.045.
96. Chen, W.; Kajiya, M.; Giro, G.; Ouhara, K.; Mackler, H. E.; Mawardi, H.; Kawai, T. Bacteria-derived hydrogen sulfide promotes IL-8 production from epithelial cells. *Biochemical and Biophysical Research Communications*, **2010**, *391*(1), 645–650. doi: 10.1016/j.bbrc.2009.11.113.
97. Nakamura, S.; Shioya, K.; Hiraoka, B. Y.; Suzuki, N.; Hoshino, T.; Fujiwara, T.; Yoshida, A. *Porphyromonas gingivalis* hydrogen sulfide enhances methyl mercaptan-induced pathogenicity in mouse abscess formation. *Microbiology* **2018**, *164*(4), 529–539. doi: 10.1099/mic.0.000640.
98. Phillips, L.; Chu, L.; Kolodrubetz, D. Multiple enzymes can make hydrogen sulfide from cysteine in *Treponema denticola*. *Anaerobe* **2020**, *64*, 102231. doi: 10.1016/j.anaerobe.2020.102231.
99. Ruff, W. E.; Greiling, T. M.; Kriegel, M. A. Host–microbiota interactions in immune-mediated diseases. *Nature Reviews Microbiology* **2020**, *18*(9), 521–538. doi: 10.1038/s41579-020-0367-2.
100. Makiura, N.; Ojima, M.; Kou, Y.; Furuta, N.; Okahashi, N.; Shizukuishi, S.; Amano, A.; Relationship of *Porphyromonas gingivalis* with glycemic level in patients with type 2 diabetes following periodontal treatment. *Oral Microbiology and Immunology* **2008**, *23*(4), 348–351. 10.1111/j.1399-302x.2007.00426.x.
101. Aemaimanan, P.; Amimanan, P.; Taweechaisupapong, S. Quantification of key periodontal pathogens in insulin-dependent type 2 diabetic and non-diabetic patients with generalized chronic periodontitis. *Anaerobe*, **2013**, *22*, 64–68. doi: 10.1016/j.anaerobe.2013.06.010.
102. Saeb. A.T.M.; Al-Rubeaan. K.A.; Aldosary. K. Relative reduction of biological and phylogenetic diversity of the oral microbiota of diabetes and pre-diabetes patients. *Microbial Pathogenesis* **2019**, 128, 215–229. doi: 10.1016/j.micpath.2019.01.009.
103. Chen. B.; Wang. Z.; Wang. J. The oral microbiome profile and biomarker in Chinese type 2 diabetes mellitus patients. *Endocrine*. **2020**, 68, 564–572. doi: 10.1007/s12020-020-02269-6.
104. Ogawa. T.; Honda-Ogawa. M.; Ikebe. K. Characterizations of oral microbiota in elderly nursing home residents with diabetes. *Journal of Oral Science* **2017**, 59, 549–555. doi: 10.2334/josnusd.16-0722.
105. Casarin, R.C.; Barbagallo, A.; Meulman, T. Subgingival biodiversity in subjects with uncontrolled type-2 diabetes and chronic periodontitis. *Journal of Periodontal Research*. **2013**, 48, 30–36. doi: 10.1111/j.1600-0765.2012.01498.x.
106. Farina. R.; Severi. M.; Carrieri. A. Whole metagenomic shotgun sequencing of the subgingival microbiome of diabetics and non-diabetics with different periodontal conditions. *Archives of Oral Biology* **2019**, *104*, 13–23. doi: 10.1016/j.archoralbio.2019.05.025.
107. Zhou, M.; Rong, R.; Munro, D. Investigation of the effect of type 2 diabetes mellitus on subgingival plaque microbiota by high-throughput 16S rDNA pyrosequencing. *PLoS One* **2013**, 8, 61516. doi: 10.1371/journal.pone.0061516.
108. Balmasova, I.P.; Olekhnovich, E.I.; Klimina, K.M. Drift of the subgingival periodontal microbiome during chronic periodontitis in Type 2 diabetes mellitus patients. *Pathogens* **2021**, *10*, 504. doi: 10.3390/pathogens10050504.

109. Longo, P.L.; Dabdoub, S.; Kumar, P. Glycaemic status affects the subgingival microbiome of diabetic patients. *Journal of Clinical Periodontology* **2018**, 45, 932–940. doi: 10.1111/jcpe.12908.
110. Liu, Y.; Duan, D.; Ma, R.; Ding, Y.; Xu, Y.; Zhou, X.; Xu, X.; The combined use of salivary biomarkers and clinical parameters to predict the outcome of scaling and root planing: A cohort study. *Journal of Clinical Periodontology* **2020**, *47*(11), 1379–1390. doi: 10.1111/jcpe.13367.
111. Isola, G.; Giudice, A. L.; Polizzi, A.; Alibrandi, A.; Murabito, P.; Indelicato, F. Identification of the different salivary interleukin-6 profiles in patients with periodontitis: a cross-sectional study. *Archives of Oral Biology* **2021**, *122*, 104997. doi: 10.1016/j.archoralbio.2020.104997.
112. Mattera, M. S. d. L. C.; Chiba, F. Y.; Lopes, F. L.; Tsosura, T. V. S.; Peres, M. A.; Brito, V. G. B.; Dos Santos, R. M.; et al. Effect of maternal periodontitis on GLUT4 and inflammatory pathway in adult offspring. *Journal of Periodontology* **2019**, *90*(8), 884–893. doi: 10.1002/jper.18-0568.
113. Maritim, A.; Sanders, A.; Watkins Iii, J. Diabetes, oxidative stress, and antioxidants: a review. *Journal of Biochemical and Molecular Toxicology* **2003**, *17*(1), 24–38. 10.1002/jbt.10058.
114. Yaribeygi, H.; Sathyapalan, T.; Atkin, S. L.; Sahebkar, A. Molecular mechanisms linking oxidative stress and diabetes mellitus. *Oxidative Medicine and Cellular longevity* **2020**. doi: 10.1155/2020/8609213.
115. Sakkas, L. I.; Daoussis, D; Liossis, S.N.; Bogdanos, D. P. The infectious basis of ACPA-positive rheumatoid arthritis. *Frontiers in Microbiology* **2017**, *8*, 1853. doi: 10.3389/fmicb.2017.01853.
116. Munoz-Atienza, E.; Flak, M.; Sirr, J.; Paramonov, N.; Aduse-Opoku, J.; Pitzalis, C.; Curtis, M.; The *P. gingivalis* autocitrullinome is not a target for ACPA in early rheumatoid arthritis. *Journal of Dental Research* **2020**, *99*(4), 456–462. doi: 10.1177/0022034519898144.
117. Gomez-Banuelos, E.; Mukherjee, A.; Darrah, E.; Andrade, F.; Rheumatoid arthritis-associated mechanisms of *Porphyromonas gingivalis* and *Aggregatibacter actinomycetemcomitans*. *Journal of Clinical Medicine* **2019**, *8*(9), 1309. doi: 10.3390/jcm8091309.
118. Linton, M. F.; Yancey, P. G.; Davies, S. S.; Jerome, W. G.; Linton, E. F.; Song, W. L.; Vickers, K. C. The role of lipids and lipoproteins in atherosclerosis. In: Feingold KR, Anawalt B, Blackman MR, et al., editors. Endotext. South Dartmouth (MA). https://www.ncbi.nlm.nih.gov/books/NBK343489/
119. Chhibber-Goel, J.; Singhal, V.; Bhowmik, D. Linkages between oral commensal bacteria and atherosclerotic plaques in coronary artery disease patients. *NPJ Biofilms Microbiomes* **2016**, 2, 7. doi: 10.1038/s41522-016-0009-7.
120. Mantyla, P.; Buhlin, K.; Paju, S.; Persson, G. R.; Nieminen, M. S.; Sinisalo, J.; Pussinen, P. J. Subgingival *Aggregatibacter actinomycetemcomitans* associates with the risk of coronary artery disease. *Journal of Clinical Periodontology* **2013**, *40*(6), 583–590. doi: 10.1111/jcpe.12098.
121. Kozarov, E.; Sweier, D.; Shelburne, C.; Progulske-Fox, A.; Lopatin, D. Detection of bacterial DNA in atheromatous plaques by quantitative PCR. *Microbes and Infection* **2006**, *8*(3), 687–693. doi: 10.1016/j.micinf.2005.09.004.
122. Xie, M.; Tang, Q.; Nie, J.; Zhang, C.; Zhou, X.; Yu, S.; Chen, L. BMAL1-downregulation aggravates *Porphyromonas gingivalis*-induced atherosclerosis by encouraging oxidative stress. *Circulation Research* **2020**, *126*(6), e15–e29. doi: 10.1161/circresaha.119.315502.
123. Stephan, A.; Phillips, A. A case for a non-transgenic animal model of Alzheimer's disease. *Genes, Brain and Behavior* **2005**, *4*(3), 157–172. doi: 10.1111/j.1601-183x.2004.00113.x.
124. Dominy, S. S.; Lynch, C.; Ermini, F.; Benedyk, M.; Marczyk, A.; Konradi, A.; Griffin, C. *Porphyromonas gingivalis* in Alzheimer's disease brains: Evidence for disease causation and treatment with small-molecule inhibitors. *Science advances* **2019**, *5*(1), eaau3333. doi: 10.1126/sciadv.aau3333.
125. Zeng, F.; Liu, Y.; Huang, W.; Qing, H.; Kadowaki, T.; Kashiwazaki, H.; Wu, Z. Receptor for advanced glycation end products up-regulation in cerebral endothelial cells mediates cerebrovascular-related amyloid β accumulation after *Porphyromonas gingivalis* infection. *Journal of Neurochemistry* **2021**, *158*(3), 724–736. doi: 10.1111/jnc.15073.
126. Kook, S.; Lee, K.; Kim, Y.; Cha, M.; Kang, S.; Baik, S.; Mook-Jung, I. High-dose of vitamin C supplementation reduces amyloid plaque burden and ameliorates pathological changes in the brain of 5XFAD mice. *Cell death & Disease* **2014**, *5*(2), e1083–e1083. doi: 10.1038/cddis.2014.26.
127. Riviere, G.R.; Riviere; K.H.; Smith, K.S. Molecular and immunological evidence of oral Treponema in the human brain and their association with Alzheimer's disease. *Oral Microbiology and Immunology* **2002**, *17*(2), 113–118. doi: 10.1046/j.0902-0055.2001.00100.x.
128. Horowitz, M.; Horowitz, M.; Ochs, M.; Carrau, R.; Kassam, A. Trigeminal neuralgia and glossopharyngeal neuralgia: two orofacial pain syndromes encountered by dentists. *The Journal of the American Dental Association* **2004**, *135*(10), 1427–1433. doi: 10.14219/jada.archive.2004.0052.

9 Oral Microbiome and Drug Resistance

Shamsher S. Kanwar, Vicky, Vandna Kumari, Sohini Sharma, and Sandipan Acharjee
Department of Biotechnology, Himachal Pradesh University, Shimla, Himachal Pradesh, India

9.1 INTRODUCTION

The microbiota breaks down potentially harmful food ingredients, boosts the immune system, and synthesizes amino acids, vitamins, and other chemicals. The word "microbiome" was given by American molecular scientist Joshua Lederberg to refer to the commensal, symbiotic, and pathogenic microorganisms that live inside human bodies as parasites or in commensalism [1]. The bacteria found in the human buccal cavity are collectively known as the oral microbiome or oral microflora [2]. Antonie van Leeuwenhoek, recognized as the father of microbiology, pioneered the discovery of the oral microbiome using a microscope of his own creation [3]. The buccal cavity itself constitutes a miniature ecosystem forming a microhabitat environment, including gums, teeth, and the lining of the oral cavity, housing both breathed and consumed microorganisms. Over 700 distinct species may be found in this complex habitat, including viruses, archaea, bacteria, fungi, and protozoa. The bulk of the microorganisms in this environment are bacteria, which are also the primary components of this particular community. The Human Oral Microbiome Database has been expanded, revealing 774 different species of oral bacteria that have been found and studied. Among these species, approximately 58% have undergone cultivation and have official taxonomic names assigned, 16% have been cultivated but are yet to be officially named, and the remaining 26% have not been successfully cultivated. These species primarily fall into six major phyla: Proteobacteria, Firmicutes, Fusobacteria, Actinobacteria, Bacteroidetes, and Spirochaetes [4]. The main bacterial genera that are present in a healthy mouth cavity include gram-negative bacteria, such as *Campylobacter*, *Moraxella*, *Neisseria*, *Veillonella*, *Prevotella*, *Desulfovibrio*, *Eikenella*, *Fusobacterium*, *Haemophilus*, *Capnocytophaga*, *Leptotrichia*, *Selenomonas*, *Simonsiella*, *Treponema*, *Desulfobacter*, and *Wolinella*, and gram-positive bacteria, such as *Actinomyces*, *Bifidobacterium*, *Corynebacterium*, *Eubacterium*, *Lactobacillus*, *Pseudoramibacter*, *Propionibacterium*, *Rothia*, *Abiotrophia*, *Peptostreptococcus*, *Streptococcus*, and *Stomatococcus*. Within this miniature ecosystem, some microbial species are considered beneficial or friendly, contributing to oral health, while others are regarded as potentially harmful or pathogenic, posing risks to oral health and systemic well-being [5]. The emphasis on the positive impacts of the microbiome of humans has largely centered on the digestive tracts, but numerous benefits likely extend to the oral microbiome as well [6]. The oral microbiota has the capacity to control blood pressure by releasing nitric oxide during nitrate reduction, which highlights the systemic effects that it may have [7]. The microbiome makeup is influenced by modern lifestyle factors that change the relative proportions of its members, such as dietary choices, sugar intake, tobacco use, oral hygiene practices, use of antibiotics, other antimicrobials, and immunization [8]. An imbalance in the oral microbiome could lead to conditions like cavities, periodontal diseases, or candidiasis. Oral illnesses such as periodontitis and caries begin when these microorganisms go out of balance or lose their homeostasis [9,10].

DOI: 10.1201/9781003478638-9

9.2 THE ORAL MICROBIOME'S DEVELOPMENT

Usually, the womb is considered infection-free [11]. A study has indicated that up to 70% of pregnant women's intrauterine environments have been colonized by oral bacteria, especially in the amniotic fluid [12]. During delivery, the baby is exposed to the mother's uterine and vaginal microbiota; following birth, it is also exposed to airborne bacteria. Despite the high risk of contamination, a newborn's buccal cavity is often sterile. From the first meal onward, the mouth is continuously colonized by bacteria, initiating the process of creating a resident oral microflora. One of the most frequently discovered cultivable microbes is *Fusobacterium nucleatum*. The resident microflora of any surface is acquired through the sequential transfer of microorganisms to the possible colonization site. The principal route of transmission is through saliva, while there are additional passive routes as well, including through bacteria in milk, water, and the environment. Colonization begins either right after birth or at birth. Pioneer species include *Streptococcus salivarius*, *Streptococcus mitis*, *Streptococcus sanguinis*, and *Streptococcus gordonii*, which are the first species to settle in a new location after birth [13]. Once tooth eruption starts, these germs might reside on non-shedding surfaces. As each tooth erupts, more surfaces become available for colonization. Gingival fissures develop as a result of periodontal microbial colonization. Different microbial colonies could establish themselves in various areas of the tooth, such as pits and fissures, where plaque deposition occurs. This process leads to high species diversity and microbial succession. Several factors contribute to the formation and evolution of communities over time, such as specific cell-to-cell binding that determines the early communities' composition, organisms' selective adherence to tooth surfaces or epithelium, and their interactions modifying the local environment, initiating the path towards oral diseases [14].

9.3 DRUG RESISTANCE IN THE ORAL MICROBIOME

American scientists Selman Waksman and colleagues first used the word "antibiotic" to describe substances produced by germs that inhibit the development of other microorganisms. Although the practice of antibiotics in medicine and public health began in the middle of the twentieth century, ancient Greece, Egypt, China, and many other parts of the world were among the societies that used microorganisms to treat microbial illnesses. The present era of antibiotics began in 1928 when Alexander Fleming accidentally came across penicillin in the culture filtrate of the fungi *Penicillium notatum* [15]. Antibiotics have ever since evolved the face of contemporary medicine, thus saving lives of many people across the globe. The United States claimed to have discovered methicillin-resistant *Staphylococcus aureus* (MRSA) in 1968, whereas the United Kingdom reported its first case in 1962. Unfortunately, resistance to almost every antibiotic ever created has developed over time. There has been a global surge in infections initiated by multidrug-resistant bacteria, posing an increasing threat of diseases that have not been curable since the twenty-first century began [16]. The oral microbiome, which is the body's second-largest microbial community followed by the guts, is found in the mouth cavity and serves as the body's interface with the external environment during activities like eating [17]. It is crucial to characterize the oral microbiome's potential for antibiotic resistance genes in humans. Throughout human history, antimicrobial resistance genes have been discovered in the oral microbiome. These genes have been detected in humans of all ages, from newborns to adults. Furthermore, it is well recognized that the oral microbiome is a place where resistance may be intrinsic and acquired through horizontal gene transfer. As per the 2019 World Health Organization (WHO) report, the incidence of antimicrobial resistance has caused 700,000 fatalities and projections suggested a potential surge to 20 million deaths by 2050, resulting in an economic cost exceeding US $2.9 trillion. Consequently, this issue has become a matter of great concern, posing a serious risk to the economy and our well-being.

9.4 THE MECHANISM OF DRUG ACTION

The antibacterial activity of drugs is commonly categorized into five mechanisms: interference with metabolic pathways, disruption of bacterial membrane function, suppression of bacterial protein synthesis, obstruction of bacterial nucleic acid creation, and disruption of bacterial cell wall synthesis (Figure 9.1).

9.4.1 Antibiotics Targeting the Cell Wall

Bacterial cell walls are made of peptidoglycan, which is made up of prolonged sugar polymers. [18]. Peptidoglycan maintains the structure of cells and is necessary for their division and proliferation [15]. Peptidoglycan comprises a polysaccharide chain linked by β-1-4 bonds and alternating between *N*-acetylmuramic acid (Mur*N*Ac) and *N*-acetylglucosamine (Glc*N*Ac) units. *N*-acetylglucosamine (Glc*N*Ac) and Mur*N*Ac are two examples of substrates that are used in the three-partitioned process that creates the peptidoglycan layer in the cytoplasm, periplasm, and membrane of each bacterium. This synthesis, which begins in the cytoplasm and extends to the inner cell membrane, begins with the synthesis of nucleotide precursors. Mur enzymes primarily generate a shared pool of precursors in the cytoplasm [19]. MurA and MurB operate sequentially to catalyze the synthesis of the first precursor, UDP-Mur*N*Ac. MurA forms UDP-GlcNAc-enolpyruvate by transferring enolpyruvate from phosphoenolpyruvate (PEP) to the UDP-Glc*N*Ac. MurA has no human counterpart (Figure 9.2). It is extensively conserved in bacteria and is necessary for cell survival. UDP-Glc*N*Ac-enolpyruvate is later reduced to UDP-Mur*N*Ac by MurB (Table 9.1). The UDP-Mur*N*Ac-pentapeptide is created when the adensonine triphosphate (ATP)-dependent Mur ligases MurB, MurA, MurD, MurC, MurI, MurF, and MurE

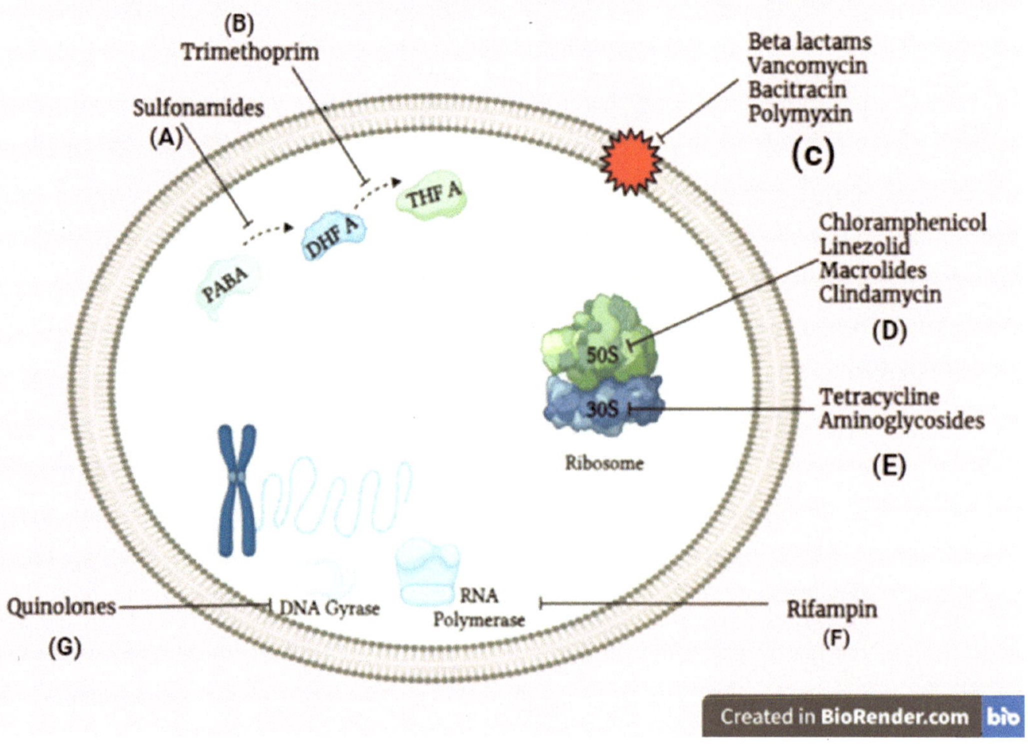

FIGURE 9.1 The mechanisms of action of the common antibiotics. (Created by Bio.render.com.)

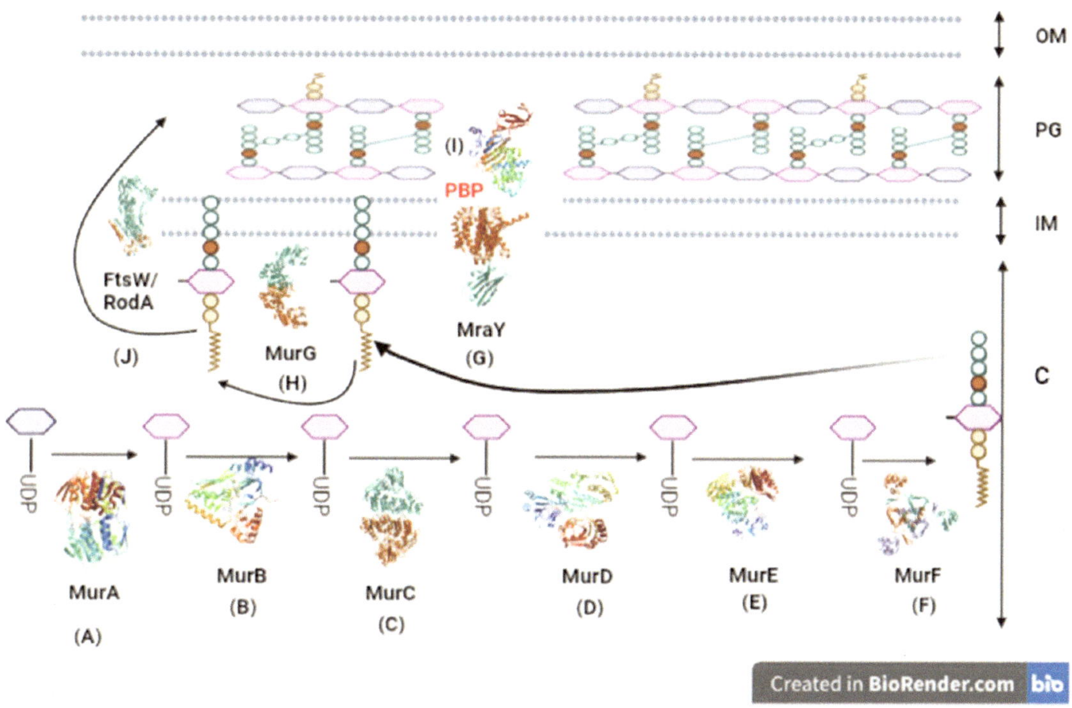

FIGURE 9.2 Peptidoglycan synthesis biosynthetic pathway.

are consecutively attached to the five amino acid residues to UDP-Mur*N*Ac [20]. Targeting a Mur enzyme represents a modern therapeutic strategy to combat drug-resistant bacteria [21].

The pentapeptide side chain structure of the lactyl side chain of Mur*N*Ac units is generally expressed as *L*-Ala-γ-*D*-Glu-X-*D*-Ala-*D*-Ala, where X can represent either *L*-Lys (gram-positive bacteria) or meso-diaminopimelic acid (DAP, gram-negative bacteria). Transpeptidation is the process by which peptidoglycan cross-linking is accomplished. It involves the *D*-Ala residue at position 4 and the ε-amino group of the Lys/DAP residue at position 5. This is the last phase in the biosynthetic route and is hindered by β-lactam antibiotics, whereas the vancomycin group of glycopeptide antibiotics targets the transglycosylation step that comes before it (Figure 9.3). Transpeptidases

TABLE 9.1
Inhibitors of Mur Enzymes

Mur Enzymes	Inhibitors	Reference
MurA	Fosfomycin, cyclic disulfide, purine analog, pyrazolopyrimidine	[21]
MurB	4-Thiazolidinones	[21]
MurC	Phosphinate and *L*-alanine analogs (β-alanine, β-CN-*L*-alanine, *L*-vinylglycine, *L*-cysteine, β-cyano-*L*-alanine, β-fluoro-*L*-alanine)	[21]
MurD	Phosphinate, analogs of *D*-glutamic acid (*DL*-homocysteic acid, *D*-erythro-4-methylglutamic acid, and *DL*-erythro-3-methylglutamic acid)	[21]
MurE	Analogs of phosphonate, A2pma: (2 S, 3 R, 6 S)3-Hydroxy-A2pm, *N*-Fluoro-A2pm	[21]
MurF	Aminoalkylphosphinates, 2,4,6-trisubstituted 1,3,5-triazines, sulfonamides, diarylquinolones, hydroxylamines, phosphorylated hydroxylamines, thiazolylaminopyrimidines	[21]

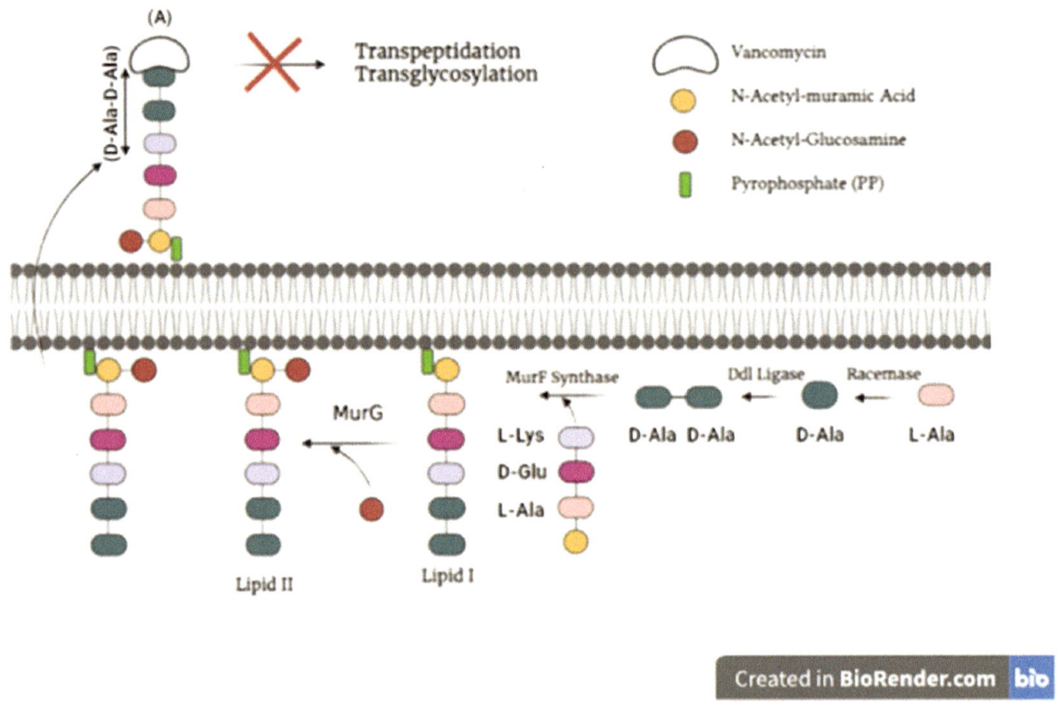

FIGURE 9.3 Mechanisms of vancomycin resistance in bacterial cell wall synthesis.

serve as formidable targets primarily because their active sites are accessible from the periplasmic side in gram-negative membranes and from the external environment in gram-positive bacteria. β-Lactam antibiotics work as bactericidal agents by covalently attaching to essential penicillin-binding proteins (PBPs) which disrupt the formation of the bacterial cell wall. Each species of bacterium has a unique set of PBPs which can include anywhere from three to eight different enzymes. These crucial PBPs are occasionally referred to as the antibiotic's death site [22]. These enzymes are involved in the final stages of peptidoglycan cross-linking in gram-positive as well as gram-negative bacteria [23]. Glycine residues are used by PBPs to assist the cross-linking of the peptide chain's *D*-alanyl-alanine region. The cell wall is strengthened by this process of cross-linking. However, the formation of the cell wall is interfered with by β-lactams and glycopeptides.

9.4.2 Action of β -Lactam Antibiotics

It is widely acknowledged that β-lactam antibiotics, like penicillin, kill bacteria by inhibiting their enzymatic activity and preventing their cell walls from being cross-linked [24]. The main focus of β-lactam agents is the PBPs. The β-lactam ring of β-lactam antibiotics imitate the peptide chain's *D*-alanyl-*D*-alanine portion, which is usually bound by PBPs. The β-lactam ring interaction prevents PBPs from being used in the manufacture of new peptidoglycans (Table 9.2). Bacterial lysis is the outcome of this peptidoglycan layer breakdown rRNA within the 50S subunit [25].

9.4.3 Glycopeptides

Glycopeptides attach to the *D*-alanyl-*D*-alanine region of the peptide side chain within the precursor peptidoglycan subunit (Figure 9.3). Vancomycin is a large drug molecule that prevents this *D*-alanyl component from binding to the PBPs (Table 9.2), thereby impeding the formation of cell walls [23, 26].

TABLE 9.2
Modes of Action and Drug Resistance of Commonly Used Antibiotics

Antibiotic Classes	Mode of Action of Antibiotic	Mode of Antibiotic Resistance	Examples	References
B-Lactams	Biosynthesis of peptidoglycan	Target alteration, efflux and hydrolysis	Penicillin's (ampicillin), monobactams (aztreonam), cephalosporins (cephamycin) and penems (meropenems)	[25]
Aminoglycosides	Translation	Target modification, efflux, phosphorylation, acetylation, and nucleotidylation	Gentamycin, streptomycin, and spectinomycin	[34]
Glycopeptides	Peptidoglycan biosynthesis	Peptidoglycan biosynthesis reprogramming	Vancomycin and teicoplanin	[35,36]
Tetracyclines	Translation	Changed target, outflow, and monooxygenation	Minocycline and tigecycline	[37,38]
Macrolides	Translation	Altered target, efflux, glycosylation, acetylation, phosphorylation and hydrolysis	Erythromycin and azithromycin	[39,40]
Lincosamides	Translation	Target alteration, efflux, and nucleotidylation	Clindamycin	[27]
Streptogramins	Translation	Changed target, efflux, C-O lyase (type B streptogramins), and acetylation (type A streptogramins)	Synercid	[27]
Oxazolidinones	Translation	Target modification and efflux	Linezolid	[41]
Phenicols	Translation	Changed target, efflux, and acetylation	Chloramphenicol	[27]
Quinolones	DNA replication	Altered target, efflux, and acetylation	Ciprofloxacin	[42]
Dihydropyrimidine	Metabolism	Target modification and efflux	Trimethoprim	[43]
Sulfonamides	Metabolism	Target modification and efflux	Sulfamethoxazole	[44]
Rifamycins	Transcription	Ribosylation, efflux, and changed target (Adenosine diphosphate, i.e. ADP)	Rifampin	[43]
Lipopeptides	Cell membrane	Altered target	Daptomycin	[45]
Cationic peptidase	Cell membrane	Target modification and efflux	Colistin	[22]

9.4.4 Inhibitors of Protein Biosynthesis

Transcribing bacterial DNA information into messenger RNA (mRNA) is the first step in the transcription process. Subsequently, proteins specified in mRNA are synthesized by the ribosome, a macromolecular structure in a process known as translation. The ribosomes along with cytoplasmic factors facilitate protein production. There are inhibitors that target either the 30S or 50S component of the bacterial ribosome, which is composed of the 70S ribosome with two ribonucleoprotein subunits [27]. The primary site of action for these inhibitors is the bacterial ribosome. For medication to be effective, it needs to cross the cytoplasmic membrane through an energy-dependent active bacterial transport mechanism requiring oxygen along with an active proton motive force. Due to this mechanism, aminoglycosides are effective against aerobic bacteria but have minimal impact on anaerobic ones. They demonstrate synergistic effects with antibiotics that impede cell wall synthesis, such as β-lactams and glycopeptides, enhancing aminoglycoside penetration into the cell at lower doses. Aminoglycosides create hydrogen bonds with 16S rRNA of the 30S ribosomal subunit located proximal to the acceptor (A) site resulting in misreading thereby premature termination of mRNA translation (Table 9.2).

9.4.4.1 Tetracyclines

The first tetracycline in the world, chlortetracycline, was introduced in the year 1948 following an extensive screening process that involved isolating organisms from soil specimens collected across different regions, known to produce antibiotics [28]. The tetracyclines recognized as broad-spectrum antibiotics exhibited potent efficacy against numerous gram-positive and gram-negative bacteria such as *mycoplasmas*, *chlamydia*, and specific protozoa. These antibiotics drove their name from their structure comprising four fused cyclic rings. The chemical components of various derivatives linked to this fundamental ring structure have undergone only minor alteration. Tetracyclines such as doxycycline, tetracycline hydrochloride, chlortetracycline, and minocycline function by binding to the 30S bacterial ribosome obstructing aminoacyl tRNA from reaching the A site on the mRNA-ribosome complex thereby impeding bacterial protein synthesis [29]. Moreover, tetracyclines have the capacity to alter the cytoplasmic membrane, permitting nucleotides and other substances to escape from the bacterial cell [30]. Tetracycline serves as the parent compound for semi-synthetic antibiotics like doxycycline and minocycline. In comparison to tetracycline hydrochloride, these two compounds exhibit superior lipid solubility, offering advantages such as a reduced dosage administration, prolonged half-life in serum, increased penetration into tissue fluids, and fewer gastrointestinal adverse effects [11].

9.4.5 Inhibitors of the 50S Subunit

9.4.5.1 Chloramphenicol

Chloramphenicol prevents tRNA from attaching to the A site on ribosomes (Table 9.2), which obstructs the production of proteins (Figure 9.4). This action occurs through interaction with conserved regions located inside the peptidyl transferase cavity of the 23S rRNA within the 50S subunit [29].

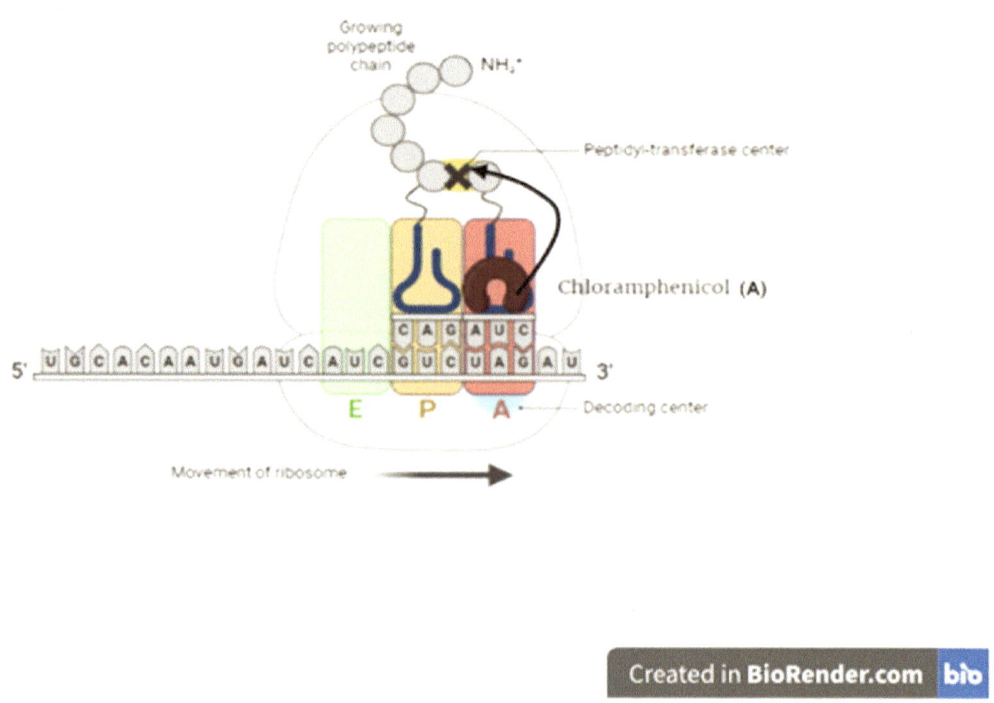

FIGURE 9.4 The mode of action of chloramphenicol in a bacterial cell.

9.4.5.2 Macrolides

The first known macrolide, erythromycin, has been used for more than 40 years in medicine. The structures of the semi-synthetic macrolides azithromycin and clarithromycin are similar to those of erythromycin. Azithromycin is distinct from erythromycin due to the inclusion of a methyl-substituted nitrogen atom to the lactone ring which makes it a member of the recently discovered azalides subclass of macrolides. In comparison to erythromycin, this modification results in a molecule that is more acid stable, has a serum half-life longer than usual, increases penetration within tissues, and has a higher efficacy against gram-negative pathogens [28]. These substances interact with conserved regions in the peptidyl transferase center of the 23S rRNA present in the 50S ribosomal subunit to affect translocation, which is the first stage of protein synthesis (Table 9.2). As a result, some peptide chains break off too soon [29]. It was only lately that the true process was revealed. Macrolide antibiotics, which have a lactone ring with 12 to 16 members that have been replaced with one or more sugars, can prevent the elongation of developing polypeptides by binding to the ribosomal exit tunnel opening and triggering a drop-off event that accumulates toxic peptidyltransferase RNA. Streptogramin B, lincosamides, and macrolides all have comparable modes of action [31].

9.4.5.3 Oxazolidinones

Within this class of oxazolidinones antibiotics, linezolid is a member of a recently approved synthetic group. Oxazolidinones impede the process of protein synthesis in two ways: first, by binding to the 50S subunit's 23S rRNA (Figure 9.5), they prevent protein synthesis; second, they reduce 70S inhibition and interact with peptidyl-tRNA [32].

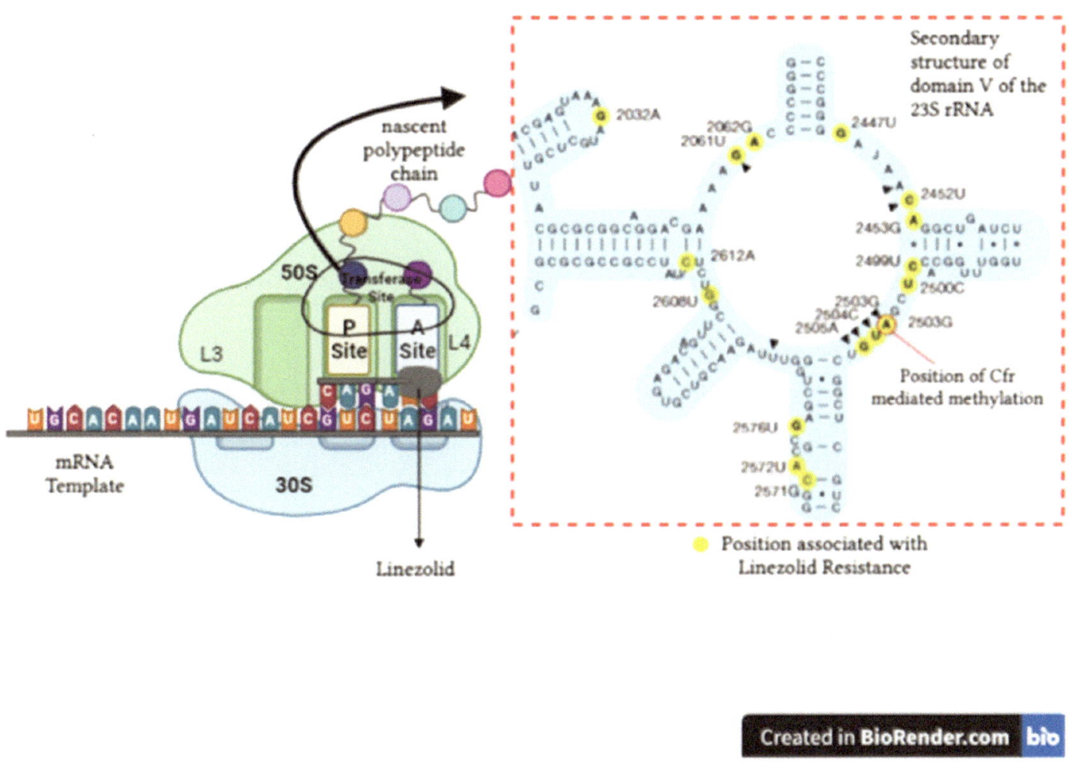

FIGURE 9.5 The mode of action of oxazolidinones.

9.4.6 Inhibitors of DNA Replication

9.4.6.1 Quinolones

The fluoroquinolones block the bacterial DNA gyrase, which nicks double-stranded DNA and adds negative supercoils before closing the nicked ends. The DNA gyrase consists of two A subunits and two B subunits; the A subunit nicks the DNA, while the B subunit adds negative supercoils. The fluoroquinolones bind tightly to the A subunit and disrupt its ability to cut and reseal DNA strands (Table 9.2). Topoisomerase IV nicks and separates daughter DNA strands following replication of DNA, the susceptible target of activity in gram-positive bacteria. Its ability to bind to this enzyme more strongly may increase its effectiveness against gram-positive bacteria. Mammalian cells contain topoisomerase II which has a relatively lower affinity for fluoroquinolones and is less harmful to cells than DNA gyrase or topoisomerase IV [23].

9.4.7 Inhibitors of Folic Acid Metabolism

9.4.7.1 Trimethoprim and Sulfonamides

The folate metabolic pathway is responsible for producing essential cellular precursors and features a crucial point dihydrofolate reductase (DHFR) common to both prokaryotes and eukaryotes. Targeting the DHFR enzyme, methotrexate is utilized in anticancer treatments, trimethoprim is used in antibacterial applications, and pyrimethamine is used in antiprotozoal applications. Dihydropteroate synthase (DHPS), which is exclusive to prokaryotes since they are unable to obtain folate through food, is another target of antifolate. It has been shown to be the main target of the oldest class of antibiotics, known as sulfonamides, which work in concert with DHFR inhibitors. Studies have shown that the majority of DHPS enzymes are capable of using sulfa medications metabolically to form other products that likely block enzymes downstream that need the generated dihydropteroate. Each of these medications hinders specific stages within the folic acid metabolism process. Trimethoprim and sulfa medications work in tandem to reduce the rate of resistance mutation when they operate at different stages on the same biosynthetic pathway. DHPS is competitively inhibited by sulfonamides, which have a higher affinity for the enzyme than *p*-aminobenzoic acid, which is its natural substrate (Figure 9.1). Trimethoprim and other substances function in a later step of folic acid production (Table 9.2) by blocking the dihydrofolate reductase enzyme [33].

9.5 CAUSES OF DRUG RESISTANCE IN THE ORAL MICROBIOME

Bacteria constantly change and adjust to their surroundings to survive as living things [46]. One of the major problems facing modern medicine is the natural growth of antibiotic resistance in microorganisms. Antibiotic resistance has complex origins, influenced by various factors. These include the overuse and misuse of antibiotics, incorrect diagnosis and prescription procedures, lowered patient sensitivity resulting in self-medication, subpar health care facilities, unhygienic living conditions, and the extensive application of antibiotics in agriculture [47]. In dentistry, antibiotics like penicillins, tetracyclines, macrolides, and metronidazole are commonly employed for treating conditions such as dental caries, periodontal diseases, and oral inflammation. These antibiotics have been playing a crucial role in managing oral infections and maintaining oral hygiene.

9.5.1 Origin of Resistance

Some antimicrobial drugs do not reliably lead to consistent susceptibility or resistance in specific bacterial species or groups. Within similar bacterial groups, resistance might vary significantly.

Susceptibility and resistance are typically assessed by determining the minimum inhibitory concentration (MIC), which is the lowest concentration of a drug that prevents bacterial growth. Susceptibility is characterized by a range of mean MICs for a particular drug among bacteria of the same kind. A species is considered inherently resistant to a medicine if its average MIC is in the resistant portion of this range. Moreover, bacteria can obtain resistance genes from closely related organisms, resulting in different degrees of resistance based on the species and the particular genes obtained [48,49]. The development of antibiotic resistance mainly occurs through two mechanisms: natural resistance [50] and the procurement of genes that provide resistance [51] via horizontal gene transfer.

9.5.2 The Mechanisms behind Antibiotic Resistance

Antibiotic resistance exists in two primary forms: natural and acquired/adaptive. Natural resistance can be innate (present within organisms) or mediated (where genes for resistance are naturally present in bacteria but activated to resistant levels after antibiotic exposure).

9.5.2.1 Adaptive Resistance

Bacteria acquire resistance genes through the main mechanisms of genetic material transfer: transformation, transposition, and conjugation, which are collectively known as horizontal gene transfer. This involves the transmission of efflux pumps encoded genes using plasmids, as *TetK* and *TetL* in *S. aureus* [52]. Furthermore, bacteria may also develop resistance through mutations in their chromosomal DNA. This acquisition can be either transient or permanent. The most frequent means by which bacteria obtain foreign genetic material is by the transfer of resistance genes via plasmids; bacteriophage transmission is quite uncommon. Some bacteria, like *Acinetobacter* spp., demonstrate natural competence, allowing them to directly absorb genetic material from their surroundings. Internally, insertion sequences and integrins facilitate the movement of genetic material, and various stressors [such as starvation, ultraviolet (UV) radiation, and chemicals] commonly lead to genetic mutations like substitutions or deletions [48, 51]. It also involved enzymes capable of modifying the antibiotic or its target [53].

9.5.2.2 Horizontal Gene Transfer

The human oral microbiome is not only a potential source of antibiotic resistance but also a significant site for horizontal gene transfer (HGT) [54]. Research on the metagenome of modern dental plaque has revealed a significant number of genes likely associated with resistance to various antibiotics and toxic substances [54,55]. A major phenotypic benefit of bacterial cells is their ability to withstand antimicrobial agents that are frequently present in the oral environment. Numerous oral cavity bacteria have been found to carry genes for acquired antibiotic resistance [16]. The HGT processes, such as bacteriophage transduction, plasmid conjugation, and free DNA transformation, enable the movement of antibiotic resistance genes between different bacterial populations [56,57]. Natural selection favors the employment of one of the three HGT processes more than the others, yet all three are extensively utilized in nature. Conjugative plasmids are frequently used by enterobacteria to exchange genetic information, although streptococci have the ability to become naturally competent and therefore participate efficiently in transformation [58]. Conjugative and mobilizable elements are the main players in HGT. The latter depends on the conjugation capabilities of coresident elements, such as conjugative plasmids or transposons, to transfer to other hosts, whereas the former contains all the genetic information required to independently move from one bacterium to another. Certain elements have the ability to relocate to other locations within the genome, but they could not be transferred to a new host by themselves. However, if they integrate into a conjugative element, they can be transferred to new hosts. These include mobile introns and transposons. Many gram-negative and gram-positive bacteria release membrane vesicles from their surfaces, which can contain proteins, polysaccharides, and, crucially, DNA that contributes to microbial adaptation. Other capable bacteria may use this DNA as a transformation substrate. The significance of oral biofilm extracellular DNA (eDNA) lies in its release by membrane vesicles from *Streptococcus mutans* into the growing biofilm, serving as a crucial source of genetic material [43]. It has been shown that HGT takes place between a variety of bacteria that live in the mouth cavity of humans. Multiple bacterial species have been observed to experience both conjugation and transformation [16]. However, because of the difficulties associated with identifying recombination events outside of the laboratory, it remains uncertain to what extent transformation or transduction contributes to the spread of resistance genes in the environment or clinical settings. Certain environments are believed to be hotspots for genetic exchange due to the high density of phages, bacteria, and plasmids present. These environments, including sewage and wastewater treatment plants, hospital effluents, aquaculture settings, agricultural sites, and abattoir waste, are considered ideal locations for such exchange events.

9.5.2.2.1 Transformation

Bacteria have the potential to acquire antibiotic-resistant genes through the process of transformation. The process occurs in both gram-positive and gram-negative bacteria. Competitive bacteria are those that have the ability to absorb DNA from their surroundings. During this procedure, the recipient bacteria absorb the bare DNA and either transpose or recombine it into the host genome. The bacterial diet and environmental stressors like antibiotics or chemicals that damage DNA are the main elements that determine competence [23]. The evolution of *Streptococcus* and *Neisseria* strains that are resistant to antibiotics appeared to have been significantly influenced by transformation [1]. Extracellular DNA accessible to competent bacteria is generally required for transformation. A regulatory link between competence development and the production of extracellular DNA has been observed, where the release of chromosomal DNA occurs as part of a lytic process controlled by the competence system known as fratricide [59]. The regulatory collaboration of competence acquisition and DNA release rewards the population-wide decision to focus energy on competence development by providing the substrate for absorption at the same time. Since extracellular DNA generated during clone growth may vary due to mutations, there will be sufficient genetic diversity to potentially foster the emergence of new phenotypic traits under appropriate selection pressures.

The homologous extracellular DNA released by a competent population of pneumococci would facilitate straightforward integration through homologous recombination. The lack of complex diversity that comes with homologous extracellular DNA, such as genes for novel or alternate metabolic pathways or antimicrobial resistance genes, may be an evolutionary disadvantage. As a result, integrating heterologous DNA from ecologically comparable and non-clonal species is most likely how diversification occurs. In fact, bacterial communities linked to humans, such as oral biofilm, provide this variety. *Streptococcus* is a well-known genus of oral biofilm, and communication development is widespread within this genus. The unique biofilm environment of oral streptococci is caused by their high cell density and variety of species. One of the most widespread human-associated communities found to date, the mature oral biofilm community is formed when oral streptococci coaggregate with other bacterial species.

9.5.2.2.2 Transduction

In the process of transduction, bacterial DNA is used to package the phage particles rather than the actual phage. There are two types of transductions: specialized, in which DNA adjacent to the phage insertion site is packaged, and generalized, where any fragment of bacterial DNA can be incorporated into the phage head. Transduction is believed to play a significant role in the evolution of *S. aureus* resistance. Despite the lack of relevant *in vitro* models or verified instances of transduction in the oral cavity, there is now considerable evidence that bacteriophages are widespread in the oral cavity. Bacteriophages found in saliva are capable of lysogenizing specific bacterial pathogens, such as *Enterococcus faecalis* and *Aggregatibacter actinomycetemcomitans* [46, 56].

9.5.2.2.3 Conjugation

The direct, DNase-insensitive exchange between live donors and recipients of cells is known as conjugation. Conjugative plasmids and conjugative transposons utilize this type of transfer [60,61]. Mobile genetic elements (MGEs), known as conjugative transposons, are able to conjugatively transfer across a wide host range of bacterial cells, frequently belonging to distinct genera. The Tn*916* family of conjugative transposons, which typically, but not always, encode tetracycline resistance given by Tet(M), is the most thoroughly researched group of conjugative transposons. Tetracycline and minocycline resistance is often mediated by the tet(M) protein, which is a ribosomal protein that binds reversibly to the 23S rRNA subunit of the ribosome. These binding blocks tetracycline from attaching, thereby either preventing protein synthesis or removing a bound tetracycline molecule before it can bind to the ribosome [62].

Tn*916* represents a foundational element for a large family of MGEs, many of which utilize it to encode additional resistance genes, including Tn*2009*, Tn*6009*, Tn*1545*, and Tn*6003*. The *erm*(B) gene, which confers resistance to streptogramins, lincosamides, and macrolides, is carried by Tn*2009*. Tn*6009* encodes resistance to both inorganic and organic mercury, while Tn*1545* and Tn*6003* provide resistance to kanamycin [52, 63]. Interestingly, the first antiseptic resistance gene was recently identified, linked to a Tn*916*-like element known as Tn*6078*. This antiseptic resistance gene, *qrg*, is flanked by two copies of the IS*1216* insertion sequence. It could successfully transform a competent oral streptococcus to cetyltrimethylammonium bromide resistance [9].

9.5.2.3 Natural Resistance

There are two types of natural resistance: intrinsic resistance, which is inherent to the species, and induced resistance, which arises when normally occurring genes in bacteria are activated to a higher level of resistance following exposure to antibiotics. For instance, metronidazole is ineffective against some oral bacteria, such as many streptococci, because they lack the nitroreductases needed to convert the drug into its active form. A bacterial species' intrinsic resistance was a universal feature that was unrelated to horizontal gene transfer and unaffected by previous antibiotic exposure [64,65]. The primary bacterial mechanisms associated with intrinsic resistance included decreased permeability of the outer membrane, particularly in gram-negative bacteria, focusing on lipopolysaccharide and the inherent function of efflux pumps. Additionally, induced resistance commonly involved multidrug efflux pumps as a prevailing mechanism [8]. A microorganism's innate resistance to an antibiotic is referred to as intrinsic resistance. This resistance develops spontaneously.

9.5.2.3.1 Outer Membrane

The gram-negative bacteria generally exhibit higher resistance to lipophilic and amphiphilic inhibitors compared to gram-positive bacteria [66]. A variety of chemicals, including detergents, free fatty acids, dyes, antibiotics, and other chemotherapeutic drugs, are included in this category of inhibitors [24]. The inherent resistance observed in gram-negative bacteria is frequently linked solely to the existence of the outer membrane barrier. Since gram-positive bacteria do not have an outer barrier, medication penetration is typically less restricted. This barrier indeed plays a role in resistance since the restricted porin channels impede the entry of even small hydrophilic solutes and the reduced fluidity within the lipopolysaccharide leaflet decreases the speed of transmembrane diffusion for lipophilic solutes [66]. The bacteria that lack a cell wall, such as *Mycoplasma* and similar species, naturally exhibit resistance to all drugs aimed at targeting the cell wall, including β-lactams and glycopeptides [67]. Enterococci, whether susceptible or resistant to vancomycin, contain distinct complexes comprised of peptidoglycans and pentapeptides situated outside their cellular membrane. These glycopeptide antibiotics bind to this particular structure. The pentapeptides are composed of tripeptide precursors with attached dipeptides that differ between vancomycin-susceptible and vancomycin-resistant enterococci. For vancomycin-susceptible strains, the terminal amino acid is *D*-alanine (*D*-Ala-*D*-Ala), while for vancomycin-resistant strains, it's *D*-serine (*D*-Ala-*D*-Ser) or *D*-lactate (*D*-Ala-*D*-Lac). Substituting *D*-alanine with *D*-serine alters the bonds with vancomycin, weakening its efficacy. Replacing *D*-alanine with *D*-lactate reduces the number of bonds with vancomycin. Strains with *D*-Ala-*D*-Ser dipeptides are susceptible to teicoplanin but show low vancomycin resistance. Conversely, strains with *D*-Ala-*D*-Lac dipeptides are highly resistant to vancomycin. These structures play a role in cell wall synthesis and inhibit vancomycin by blocking its binding to the pentapeptide molecule, consequently halting cell wall construction. In the latter scenario, the antibiotic's attachment to the peptidoglycan precursor is restricted, potentially enabling cell wall synthesis to continue even in the presence of vancomycin.

Mechanisms conferring resistance to chlorhexidine (CHX) include the activation of multidrug efflux pumps and modifications to the bacterial cell membrane in the oral microbiome such as *Pseudomonas stutzeri* [68]. In *Pseudomonas stutzeri*, resistance to CHX is associated with specific alterations in the bacterium's outer membrane components. Changes in outer membrane proteins and modifications to LPS profiles play critical roles in this resistance. These alterations reduce the permeability of the membrane to CHX, limiting its entry into the bacterial cell and thereby diminishing its antimicrobial efficacy. Additionally, the activation of multidrug efflux pumps, which actively transport CHX out of the cell, further enhances resistance by reducing intracellular CHX concentrations. Together, these mechanisms highlight the complex adaptations employed by *Pseudomonas stutzeri* to survive in the presence of CHX.

9.5.2.3.2 Porins Channels

When chemicals enter bacteria with large outer membranes, they usually do so through porin channels. Specifically, in gram-negative bacteria, these porin channels typically facilitate the entry of hydrophilic molecules [69]. The peptidoglycan polymer is quite near to the cell surface in gram-positive bacteria, which makes it easy for the antibiotic to enter the cell. Gram-negative bacteria on the other hand have an outer membrane made of lipopolysaccharides which serve as a barrier against hydrophilic channels that are larger than a particular molecular weight and often hydrophobic molecules. Changes in the number of available porins and mutations that alter the selectivity of the porin channel are the two main ways that changes in porins might limit drug absorption [70]. Enterobacteriaceae members developed resistance by diminishing the quantity of porins and, in some cases, ceasing the production of specific porins altogether [71]. Mutations leading to alterations in the porin channel have been observed in *Enterobacter aerogenes*, resulting in resistance to imipenems and particular cephalosporins. Similarly, in *Neisseria gonorrhoeae*, these mutations confer resistance to β-lactams and tetracycline [24, 53].

9.5.2.3.3 Biofilm

It has been estimated that there are about 700 distinct kinds of bacteria in the human mouth. Most of them are connected to tooth plaque and have created a very well-structured biofilm [72]. Another common event in bacterial colonization is the development of a biofilm within the bacterial community. These biofilms might be dominated by a single organism (e.g., *Pseudomonas aeruginosa* in the lungs) or encompass various organisms, as observed in the biofilm community of the normal gut flora. In pathogenic organisms, biofilm formation serves as a shield, safeguarding bacteria from host immune system assaults and offering resistance against antimicrobial agents [71]. In the oral cavity, most commensals acquire their niche via biofilm formation, quorum sensing, biofilm-structuring, and acquiring nutrition from the host gums as their exquisite source of nutrients. More than one variety of organisms could be often found, even species from non-commensal nature as well as pathogenic organisms associate to form biofilm resulting is dental calculus or tartar. These organisms expel mostly lipopolysaccharides, which results in the tensile and rigid structure of the biofilm layers with microbial growth and metabolism to evade antagonistic organisms or host immune defenses.

9.5.2.3.3.1 Resistance Mechanism of Biofilm The biofilm's stable structure and the close proximity of bacterial cells create an optimal environment for horizontal gene transfer, promoting the dissemination of antibiotic resistance genes among the biofilm's members. Antibiotic resistance develops gradually as a result of the physiological and structural traits of biofilm-forming bacteria that increase their resistance to antibiotics. Bacteria that survive and thrive within a biofilm exhibit significant antibiotic resistance [73]. The mechanisms of antibiotic resistance include capsule, cell membrane modification, efflux mechanism, plasmids/enzyme-mediated resistance, genetic adaptation, quorum sensing, metabolic factors, persistence bacteria, and stress response (Table 9.3).

TABLE 9.3
Mechanisms of Antibiotic Resistance in Biofilm-Forming Bacteria

Mechanisms	Description	References
Capsule	The capsule is made up of glycoproteins and polysaccharides that are affected by environmental factors. The bacteria acquire resistance to antibiotics and other antimicrobial treatments as a result of the capsule or glycocalyx layer accumulating antibacterial chemicals, which can make up to 25% of the bacteria's weight. It traps antimicrobial molecules, leading to the degradation of drugs by bacterial exoenzymes, thereby contributing to antibiotic resistance. The thickness of the capsule in both gram-positive and gram negative bacteria has a range of 0.2–1.0 μm.	[73–75]
Cell membrane modification	The alterations in cell membrane permeability and porin protein expression prevent antibiotics from entering bacterial cells within the biofilm, leading to increased resistance.	[73–75]
Efflux mechanism	The efflux pumps are a group of transport proteins in bacteria that actively remove toxic substances, including antibiotics, preventing their entry into the cell and contributing to multidrug resistance.	[73, 75,76]
Plasmids/enzyme-mediated resistance	The biofilm-forming bacteria have plasmids that include genes for proteins and enzymes, which provide resistance to antibiotics and heavy metals. These plasmids frequently include many resistance genes that can propagate resistance across bacteria through horizontal gene transfer.	[73, 75, 77]
Genetic adaptation	The continuous exposure to antibiotics results in the horizontal transfer of resistance genes (HGT) and long-term genetic adaptations, leading to the emergence of drug-resistant bacterial strains.	[75, 77]
Quorum sensing	The bacteria communicate via quorum sensing, regulating activities like biofilm formation and defense against antimicrobials. This process enhances virulence and adaptation to changing environments.	[73, 78]
Metabolic factors	The bacterial metabolism and growth are impacted by variations in oxygen and nutrition availability inside the biofilm. This results in diverse populations where dormant cells live longer and are thus less vulnerable to antibiotics.	[78]
Persistence bacteria	The persister cells, which tolerate antimicrobial agents, are difficult to detect and contribute to chronic infections. The reduced ATP levels decrease antibiotic target activity, leading to antibiotic tolerance.	[79]
Stress response	The bacterial stress responses, triggered by environmental changes such as nutrient deprivation, extreme temperatures, and pH fluctuations, enhance stress tolerance and aid in cellular damage repair, contributing to resistance.	[79]

Abbreviations: ATP, adenosine triphosphate; HGT, horizontal gene transfer.

9.5.2.3.4 Efflux Pumps

Antibiotics are exported by membrane proteins called efflux pumps (particularly macrolides, tetracyclines, and fluoroquinolones except polymyxin) from the cell, thereby maintaining low intracellular concentrations of these drugs [51, 80]. Efflux pumps reside in the cytoplasmic membrane, as opposed to porins, which are found in the external membrane.

Both gram-positive and gram-negative bacteria have the capability to possess single-drug and/or multidrug efflux pumps. Efflux pumps differ in both their specificity for certain substances and their mechanisms of action. Inducible multidrug efflux pumps are accountable for the inherent antibiotic resistance found in numerous organisms. Mutations affecting the regulatory components governing efflux pump production can enhance antibiotic resistance. The seven primary families of distinct multidrug efflux proteins have been identified in bacteria: the proteobacterial antimicrobial compound efflux (PACE) family, the *p*-aminobenzoyl-glutamate transporter (AbgT) family, the small multidrug resistance (SMR) family, the multidrug and toxic compound extrusion (MATE) family, the resistance-nodulation-division (RND) family, the major facilitator superfamily (MFS), and the ATP-binding cassette (ABC) superfamily (Figure 9.6). They have the capability to expel a wide range of unrelated antibiotics, such as fluoroquinolones, macrolides, and tetracyclines, significantly contributing to the rise of organisms resistant to multiple drugs [81].

The oral microbiome predominantly includes bacteria such as *S. aureus*, *P. aeruginosa*, *Neisseria gonorrhoeae*, *Enterococcus faecalis*, *Lactococcus lactis*, *and Streptococcus* species (including *Streptococcus mutans* and *Streptococcus salivarius*). Among these, a recently identified tetracycline resistance determinants, *tet*AB(46), has been found in *Streptococcus australis*, a bacterium commonly isolated from the human oral cavity. This resistance mechanism is mediated by a predicted heterodimeric ATP-binding cassette (ABC) transporter, *tet*AB(46), which functions to expel tetracycline from the bacterial cell, reducing its intracellular concentration and thereby conferring resistance [82]. The *tet*AB(46) is linked to multidrug resistance (MDR)-type ABC transporters present in both gram-positive and gram-negative bacteria, but it specifically provides resistance only to tetracyclines. Genes closely related to *tet*AB(46) have been identified in the genomes of other tetracycline-resistant oral streptococci, including multiple strains of *Streptococcus parasanguinis* and another strain of *Streptococcus australis*. The well-known bacteria *Enterococcus* is responsible for infections in the urinary system, pelvic and intraabdominal regions, surgical sites, and oral cavity. ABC transporters are highly expressed in and may be potential target molecules for testing the resistance of fecal enterococci biofilms to alkaline environments [83]. Endogenous transporters in *Staphylococcus aureus* aid in safeguarding the bacteria from antibiotics and contribute to protection against environmental toxins. Exposure of *Staphylococcus aureus* to pyocyanin from *Pseudomonas aeruginosa* increases the efflux pump overexpression of four types: NorA, NorB, NorC, and tet38. The NorA efflux pump in *Staphylococcus aureus* confers resistance to hydrophilic fluoroquinolones, including norfloxacin and ciprofloxacin, along with some other antibiotics [84]. In *P. aeruginosa*, the MexAB-OprM efflux pump is typically activated by drug presence. However, mutations in its regulator (*mex*R) result in heightened expression of MexAB-OprM (Table 9.4), resulting in heightened resistance to β-lactam antibiotics [85,86].

The oral microbiome features bacteria with significant efflux pump mechanisms, such as *Streptococcus australis* with the TetAB(46) pump for tetracycline resistance and *Enterococcus* with ABC transporter efflux pumps. The MexAB-OprM efflux pump in *Pseudomonas aeruginosa* is usually triggered in response to the presence of drugs. *Staphylococcus aureus* utilizes multiple efflux pumps, including NorA, NorB, and NorC, to resist antibiotics and toxins. These pumps play a crucial role in bacterial survival and resistance, impacting treatment strategies for oral infections.

9.5.2.3.5 Drug Inactivation Mechanism

The synthesis of enzymes that deactivate antibiotics by adding certain chemical components to the compound or by dissolving the molecule itself is a very efficient strategy used by bacteria to

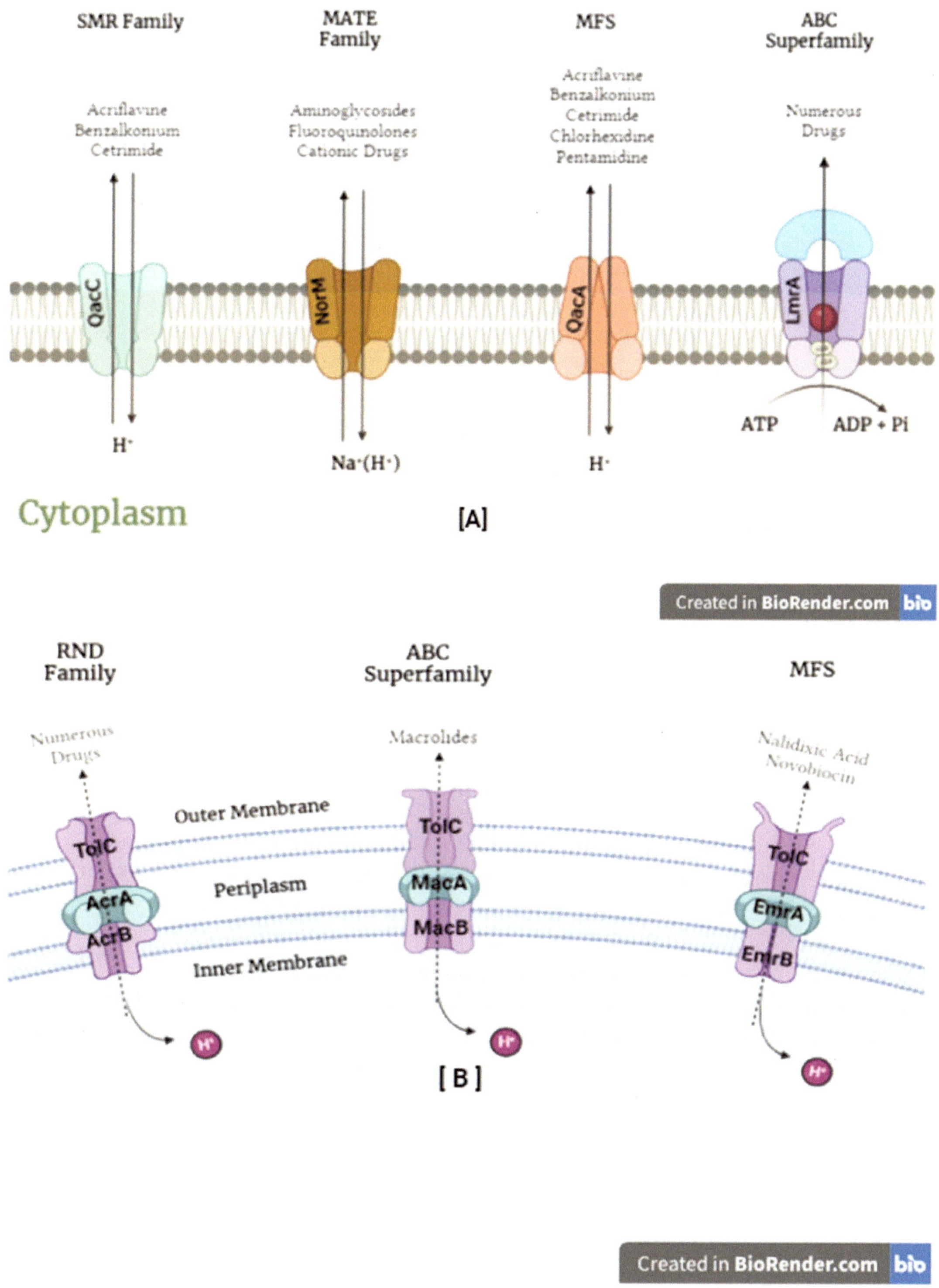

FIGURE 9.6 Efflux pumps of gram-positive and gram-negative bacteria.

TABLE 9.4
The Key Details of Various Efflux Pumps and Their Role in Antibiotic Resistance in Different Bacteria

Aspects	Details	References
Antibiotics export mechanism	Antibiotics are exported by membrane proteins called efflux pumps, maintaining low intracellular concentrations. Efflux pumps expel macrolides, tetracyclines, and fluoroquinolones, except polymyxin.	[51, 80]
Efflux pump families	Seven primary families: ABC superfamily, RND family, MFS family, SMR family, MATE family, PACE family, and AbgT family.	[81]
Function of efflux pumps	Efflux pumps contribute to inherent antibiotic resistance. Mutations in regulatory components can enhance resistance.	[87]
Oral microbiome bacteria	Includes *Enterococcus faecalis*, *Lactococcus lactis*, *Streptococcus* species, *Staphylococcus aureus*, *Pseudomonas aeruginosa*, *Neisseria gonorrhoeae*, and *Stenotrophomonas maltophilia*.	[5]
Streptococcus australis	Contains the TetAB(46) efflux pump for tetracycline resistance, associated with ABC transporters.	[82]
Enterococcus	Known for infections and resistance in various parts of the body; ABC transporters are key in biofilm resistance to alkaline environments.	[83]
Staphylococcus aureus	Contributes to survival by using the efflux pumps NorA, NorB, NorC, and Tet38 to increase resistance to toxins and antibiotics.	[84]
Pseudomonas aeruginosa	MexAB-OprM efflux pump is typically triggered by drugs; mutations in mexR lead to increased MexAB-OprM expression, enhancing resistance to β-lactam antibiotics.	[85, 86]
Acinetobacter baumannii	The RND family of efflux pumps is a significant contributor to antibiotic resistance in a variety of bacteria: the AdeABC efflux pump in *Acinetobacter baumannii* and the AcrAB-TolC efflux pump in *E. coli*.	[87]
Neisseria gonorrhoeae	The MATE family of efflux pumps plays a significant role in antibiotic resistance. The NorM efflux pump is found in *Neisseria gonorrhoeae*, a bacterium that can inhabit the oral cavity.	[88]

Abbreviations: ABC, adenosine triphosphate (ATP)-binding cassette; AbgT, *p*-aminobenzoyl-glutamate transporter; MATE, multidrug and toxic compound extrusion; MFS, major facilitator superfamily; PACE, proteobacterial antimicrobial compound efflux; RND, resistance nodulation division; SMR, small multidrug resistance.

combat the presence of antibiotics (Figure 9.7). This renders the antibiotics inactive in their interaction with their target [89]. Bacteria use several effective strategies to combat antibiotics such as a decrease in membrane permeability, overexpression of efflux pumps, enzyme inactivation, and target alteration.

9.5.2.3.5.1 Chemical Modifications to the Antibiotic The process of modifying antibiotics is frequently employed to nullify their effectiveness, notably with aminoglycoside antibiotics such as gentamycin, kanamycin, chloramphenicol, streptomycin, and β-lactams. A range of aminoglycoside modification enzymes (AMEs) like *N*-acetyl transferases (AACs), *O*-phosphotransferases (APHs), and *O*-adenyl transferases (ANTs) are capable of acetylating, phosphorylating, or adenylating the aminoglycoside antibiotic. These enzymes were first identified in *Streptomyces* species in the 1970s and exhibit comparable metabolic processes to those observed in clinical strains of antibiotic-resistant bacteria [51, 69, 90]. Numerous classes of modifying enzymes have been identified, commonly catalyzing biochemical reactions: for example, for phosphorylation (aminoglycosides and chloramphenicol), adenylation (aminoglycosides and lincosamides), and acetylation (aminoglycosides, chloramphenicol, and streptogramins) [67].

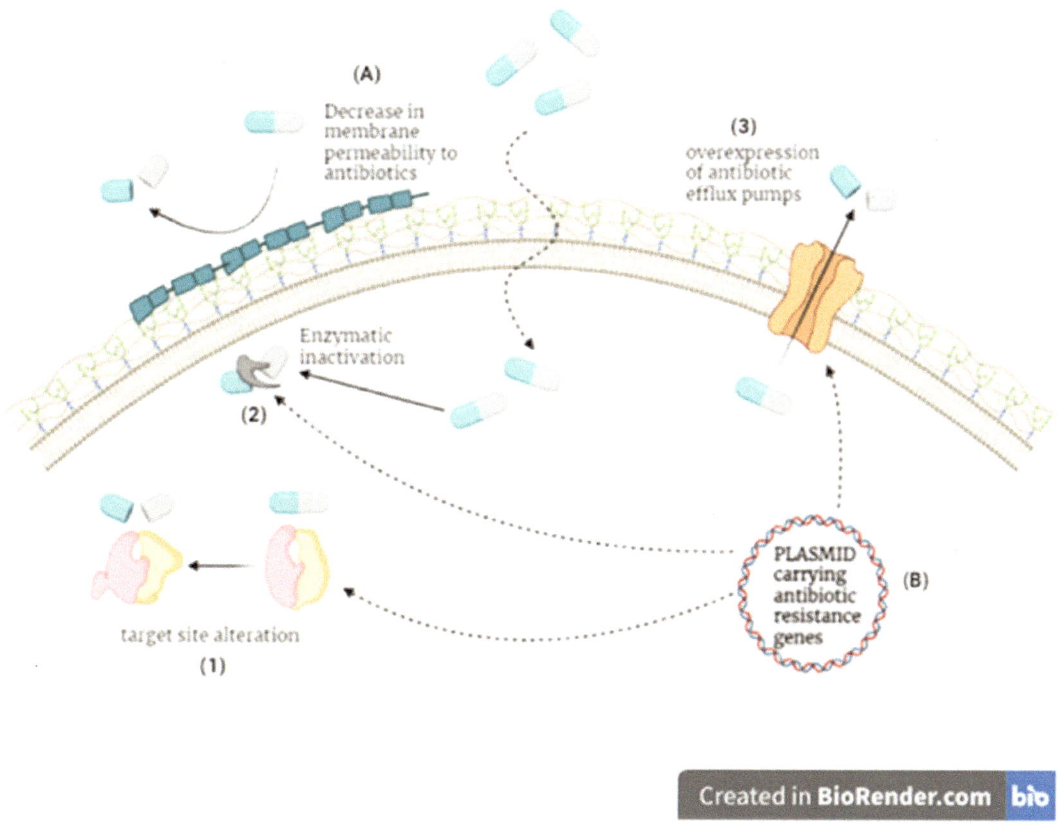

FIGURE 9.7 Drug inactivation mechanisms present in bacteria.

9.5.2.3.5.2 Destruction of the Antibiotic Compound The β-lactamase-producing bacteria in the oral cavity include *Streptococcus* species, *Neisseria* species, *Haemophilus* species, *Prevotella* species, and *Fusobacterium* species. The β-lactamases possess the ability to hydrolyze a wide range of β-lactams that contain ester and amide bonds, including penicillins, cephalosporins, monobactams, and carbapenems. Around 300 β-lactamases have been identified to date [23]. The predominant mechanism leading to resistance against penicillin and cephalosporin involves β-lactamase-mediated breakdown of the β-lactam ring, rendering the antibiotic inactive. Depending on the specific drug, these enzymes are often termed as penicillinases or cephalosporinases. Some of these enzymes exhibit activity against both classes of antibiotics. Gram-positive bacteria generally produce a substantial quantity of β-lactamase, which is released extracellularly. In a mixed bacterial population, such as the subgingival microbiota within a periodontal pocket [63], the synthesis of β-lactamase by a small number of gram-positive bacteria could protect other bacteria that were typically susceptible to penicillins. Conversely, in gram-negative bacteria, β-lactamases are found in relatively smaller amounts and are located in the periplasmic space between the inner and outer cell membranes. Positioned close to the outer surface membrane where cell-wall synthesis enzymes are situated, these β-lactamases offer effective defenses for microbes. Consequently, even a low production of β-lactamase within the outer membrane of gram-negative bacteria could often maintain the internal antibiotic concentration below inhibitory levels. β-Lactamases might exist as chromosomal or plasmid-encoded, and they could be either constitutively present or inducible.

9.6 INTRINSIC RESISTOME: EVOLUTIONARILY ANCIENT PHENOTYPES IN BACTERIAL SPECIES

The group of chromosomal genes linked to intrinsic resistance within a species of bacteria is known as species-specific bacterial intrinsic resistance [91]. It is not the consequence of horizontal gene transfer and is present in different strains of the same bacterial species regardless of previous exposure to antibiotics. It is noteworthy that the most prominent intrinsically resistant bacteria come from non-clinical settings, where there is a far smaller chance of strong antibiotic-selective pressure than in clinical settings. This suggested that the main physiological role of the elements associated with this feature in these bacteria's native habitats is not to confer resistance to antibiotics that are now used in medicine. All strains of a bacterial species possess an intrinsic resistome, which is an evolutionary ancient feature that has not been gained as a result of modern antibiotic therapy [92]. *Pseudomonas aeruginosa* is thought to have inherent resistance to antibiotics due to the activation of multidrug efflux pumps and decreased permeability of its bacterial envelope. Nonetheless, this assumption lacks empirical confirmation since a comprehensive exploration of other genetic elements contributing to intrinsic resistance hasn't been extensively conducted. Understanding the genetic networks responsible for antibiotic resistance in bacterial pathogens is therefore pivotal in mitigating and forecasting resistance, a task that has definitely garnered theoretical attention [59].

9.7 CONCLUSION

The oral microbiome is crucial to health as it can cause both oral and systemic diseases. The available knowledge on the impact of dentists' antibiotic prescriptions on the selection of resistant microorganisms remains limited. However, studies suggest an increasing resistance in periodontal microbiota to antibiotics over the last decade. Among the commonly prescribed antibiotics in dentistry are β-lactams, often prescribed to treat oral infections and as prophylaxis for bacterial endocarditis or postsurgery infections. Notably, certain practices, such as immediate amoxicillin use after implant placement, lack scientific support and should be avoided to mitigate the risk of bacterial resistance. Periodontitis, being an infectious disease, has prompted discussions about systemic antibiotic therapy as a treatment option. It is crucial to acknowledge that the unique characteristics associated with periodontitis genesis may significantly influence the efficacy of antimicrobial medications in periodontal therapy. Extrapolating outcomes of antibiotic therapy from other human illnesses to periodontics should be avoided. Antimicrobial susceptibility tests often evaluate planktonic cultures, yet subgingival bacteria within biofilms exhibit higher resistance to antimicrobial therapy. Targeting biofilms is quite challenging as it might require higher antibiotic doses. Therefore, compelling evidence for antibiotic use in periodontal treatment is best derived from *in vivo* studies. The controlled clinical trials, conducted with increased rigor by various researchers, have helped alleviate several concerns regarding the appropriate systemic antibiotic use in periodontics. Dentists must be concerned that indiscriminate or nonspecific antibiotic usage in patients often leads to increased side effects and bacterial drug resistance. Nonetheless, when used appropriately, systemic antibiotics are crucial therapeutics for managing oral infections and periodontal disorders.

REFERENCES

1. McLean, J. S. Advancements toward a systems level understanding of the human oral microbiome. *Front Cell Infect Microbiol* **2014**, *4*, 98.
2. Sommer, M.O.; Dantas, G.; Church, G.M. Functional characterization of the antibiotic resistance reservoir in the human microflora. *Science* **2009**, *325*, 1128–1131.
3. Kilian M. The oral microbiome – friend or foe. *Eur. J. Oral Sci.* **2018**, *126*, 5–12.
4. Xie, G. Community and gene composition of a human dental plaque microbiota obtained by metagenomic sequencing. *Mol Oral Microbiol.* **2010**, *25*, 391–405.

5. Spellberg, B.:Blaser, M.; Guidos, R.; Boucher, H.W.; Bradley, J.S.; Eisenstein, B.; Gerding, D.; Lynfield, R.; Reller, L.B.; Rex, J.; Schwarz.; Septimus, E.; Tenover, F.C.; Gilbert, D.N. For the Infectious Diseases Society of America. Combating antimicrobial resistance: policy recommendations to save lives. *Clin. Infect. Dis.* **2011**, *52*, S397–S428.
6. Zhang, Y.; Wang, X. Human oral microbiota and its modulation for oral health. *Biomed. Pharmacother.* **2018**, *99*, 883–893.
7. Reynolds, P.E. Structure, biochemistry and mechanism of action of glycopeptide antibiotics. *Eur. J. Clin. Microbiol. Infect. Dis.* **1989**, *8*, 943–950.
8. Gotsman, I.; Lotan, C.; Soskolne, W.A.; Rassovsky, S.; Pugatsch, T.; Lapidus, L.; Novikov, Y.; Masrawa, S.; Stabholz, A. Periodontal destruction is associated with coronary artery disease and periodontal infection with acute coronary syndrome. *J. Periodontol.* **2007**, *78*, 849–858.
9. Costalonga, M.; Herzberg, M.C. The oral microbiome and the immunobiology of periodontal disease and caries. *Immunol. Let.* **2014**, *162*, 22–38.
10. Song, X.; Sun, J.; Mikalsen, T.; Roberts, A. P.; and Sundsfjord, A. Characterisation of the plasmidome within *Enterococcus faecalis* isolated from marginal periodontitis patients in Norway. *PLOS ONE* **2013**, *8*.
11. Batabyal, B.; Chakraborty, S.; Biswas, S.; Bébéar, C.M.; and Pereyre, S. Role of the oral microflora in Mechanisms of drug resistance in *Mycoplasma pneumoniae. Curr. Drug Targets.* **2005**, 5, 263–271.
12. Van A, H.; Van D, P.; and Coenye, T. Molecular mechanisms of antimicrobial tolerance and resistance in bacterial and fungal biofilms. *Trends Microbiol.* **2014**, *22*(6), 326–333.
13. Baker, J. L.; and Edlund, A. Exploiting the oral microbiome to prevent tooth decay: has evolution already provided the best tools? *Front. Microbiol.* **2019**, *9*, 33–23
14. Könönen, E. Development of oral bacterial flora in young children. *Ann. Med.* **2000**, *32*, 107–12.
15. Stevens, R. H.; Porras, O. D.; and Delisle, A. L. Bacteriophages induced from lysogenic root canal isolates of *Enterococcus faecalis. Oral Microbiol. Immunol.* **2009**, *24*, 278–284.
16. Ciric, L.; Ellatif, M.; Sharma, P.; Patel, R.; Song, X.; Mullany, P.; et al. Tn916- like elements from human, oral, commensal *streptococci* possess a variety of antibiotic and antiseptic resistance genes. *Int J. Antimicrob. Agents* **2012**, *39*, 360–361.
17. Barza, M.; Brown, RB.; Shanks, C.; Gamble, C.; Weinstein, L. Relation between lipophilicity and pharmacological behavior of minocycline, doxycycline, tetracycline and oxytetracycline in dogs. *Antimicrob. Agents Chemother.* **1975**, *8*, 713–20.
18. Kahne, D.; Leimkuhler, C.; Lu, W.; Walsh, C. Glycopeptide and lipoglycopeptide antibiotics. *Chem. Rev.* **2005**, *105*, 425–48.
19. Amera, G. M.; Canarutto, G.; Jain, M.; Wubie, G.; Adal, M.; Singh, R. P.; Muthukumaran, J.; Singh, A. K.; Piazza, S. Delving into the Structural and Functional Characteristics of Mur Family Proteins in *Acinetobacter baumannii* strain AYE: A comprehensive Review. *Preprints.* **2024**, 2024021040. https://doi.org/10.20944/preprints202402.1040.v1
20. Silver, L. L. Viable screening targets related to the bacterial cell wall. *Ann New York Acad Sci.* **2013**, *1277*(1), 29–53.
21. El Zoeiby, A.; Sanschagrin, F.; and Levesque, R. C. Structure and function of the Mur enzymes: development of novel inhibitors. *Molecular Microbiology* **2003**, *47*(1), 1–12.
22. Higgins, P.G.; Fluit, A.C.; Schmitz, F.J. Fluoroquinolones: Structure and target sites. *Curr. Drug Targets* **2003**, *4*, 181–90.
23. Claverys, J. P.; Prudhomme, M.; and Martin, B. Induction of competence regulons as a general response to stress in gram-positive bacteria. *Annu. Rev. Microbiol.* **2006**, *60*, 451–475.
24. Blázquez, J.; Couce, A.; Rodríguez, B.J. Antimicrobials as promoters of genetic variation. *Curr. Opin. Microbiol.* **2012**, *15*, 561–569.
25. Bush, K.; Bradford, PA.; β-lactams and β-lactamase inhibitors: an overview. *Cold Spring Harb Perspect Med.* **2016**, 6.
26. Roberts, A. P.; and Mullany, P.; Tn916-like genetic elements: a diverse group of modular mobile elements conferring antibiotic resistance. *FEMS Microbiol. Rev.* **2011**, *35*, 856–871.
27. Wright, G.D. The antibiotic resistome: the nexus of chemical and genetic diversity. *Nat. Rev. Microbiol.* **2007**, *5*, 175–186.
28. Hoepelman, I.M.; Schneider, M.M. Azithromycin: the first of the tissue-selective azalides. *Int. J. Antimicrob. Agents.* **1995**, *5*, 145–167.
29. walker and Walker, J. B. Streptomycin biosynthesis and metabolism. Enzymatic phosphorylation of dihydrostreptobiosamine moieties of dihydrostreptomycin-(streptidino) phosphate and dihydrostreptomycin by Streptomyces extracts. *J. Biol. Chem.* **1970**, *245*, 6683–6689.

30. Poole, K. Multidrug efflux pumps and antimicrobial resistance in *P. aeruginosa* and related organisms, *J. Mol. Microbiol. Biotechnol.* **2001**, *3*, 225–264.
31. Raizada, M.K.; Joe, B.; Bryan, N.S.; Chang, E.B.; Dewhirst, F.E.; Borisy, G.G.; Galis, Z.S.; Henderson, W.; Jose, P.A.; Ketchum, C.J.; Lampe, J.W.; Pepine, C.J.; Pluznick, J.L.; Raj, D.; Seals D.R.; Gioscia.; Ryan, R.A.; Tang, W.H.W.; Oh, Y. S. Report of the National Heart, Lung, and Blood Institute Working Group on the Role of Microbiota in Blood Pressure Regulation: current status and future directions. *Hypertension.* **2017**.
32. Bozdogan, B.; and Appelbaum, P.C. Oxazolidinones: Activity, mode of action, and mechanism of resistance. *Int. J. Antimicrob. Agent.* **2004**, *23*, 113–9.
33. Sreeja, M.K.; Gowrishankar, N.L.; Adisha, S.; Divya, K.C. Antibiotic resistance reasons and the most common resistant pathogens - a review. *Res. J. Pharm. Technol.* **2017**, *10*, 1886–1890.
34. Zhou, Y.; Yong, Y.; Zhu, C.; Yang, H.; and Fang, B. Exogenous D-ribose promotes gentamicin treatment of several drug-resistant Salmonella. *Front Microbiol.* **2022**, *13*, 1053330.
35. Selim, S. Mechanisms of gram-positive vancomycin resistance. *Biomed Rep.* **2022**, *16*(1), 1–6.
36. Goldstein, B. P.; Rita R.; and Francesco P. Teicoplanin. *Glycopeptide Antibiotics.* CRC Press**. 2020**. 273–308.
37. Asadi, A.; Abdi, M.; Kouhsari, E.; Panahi, P.; Sholeh, M.; Sadeghifard, N.; Gholami, M. Minocycline, focus on mechanisms of resistance, antibacterial activity, and clinical effectiveness: Back to the future. *J Global Antimicrob Resist.* **2020**, *22*, 161–174.
38. Korczak, L.; Majewski, P.; Iwaniuk, D.; Sacha, P.; Matulewicz, M.; Wieczorek, P.; Tryniszewska, E.; Molecular mechanisms of tigecycline-resistance among Enterobacterales. *Front Cell Infect Microbiol.* **2024**, *14*, 1289396.
39. Mahfouz, A. A.; Said, H. S.; Elfeky, S. M.; and Shaaban, M. I. Inhibition of erythromycin and erythromycin-induced resistance among *Staphylococcus aureus* clinical isolates. *Antibiotics.* **2023**, *12*(3), 503.
40. Firth, A.; and Praveen P. Azithromycin: the first broad-spectrum therapeutic. *Eur J Med Chem.* **2020**, *207*, 112739.
41. Manoharan, M.; Sawant, A. R.; Prashanth, K.; and Sistla, S. Multiple mechanisms of linezolid resistance in *Staphylococcus haemolyticus* detected by whole-genome sequencing. *J Med Microbiol.* **2023**, *72*(7), 001737.
42. Shariati, A.; Arshadi, M.; Khosrojerdi, M. A.; Abedinzadeh, M.; Ganjalishahi, M.; Maleki, A.; Khoshnood, S. The resistance mechanisms of bacteria against ciprofloxacin and new approaches for enhancing the efficacy of this antibiotic. *Front Public Health.* **2022**, *10*, 1025633.
43. Davies, J.; and Davies, D. Origins and evolution of antibiotic resistance. *Microbiol. Mol. Biol. Rev.* **2010**, *74*, 417–433.
44. Song, X.; Su, R.; Wang, Y.; Zhang, Y.; Gao, B.; Wang, Y.; Li, Q. Visible light-driven chlorite activation process for enhanced sulfamethoxazole antibiotics degradation, antimicrobial resistance reduction and biotoxicity elimination. *Chem Eng J.* **2023**, *452*, 139103.
45. Nguyen, A. H.; Hood, K. S.; Mileykovskaya, E.; Miller, W. R.; Tran, T. T. Bacterial cell membranes and their role in daptomycin resistance: A review. *Frontiers in Molecular Biosciences* **2022**, *9*, 1035574.
46. Wardal, E.; Gawryszewska, I.; Hryniewicz, W.; and Sadowy, E. Abundance and diversity of plasmid-associated genes among clinical isolates of *Enterococcus faecalis. Plasmid.* **2013**, 70, 329–342.
47. Zarco, M.F.; Vess, T.J.; Ginsburg, G.S. The oral microbiome in health and disease and the potential impact on personalized dental medicine. *Oral Dis.* **2012**, *18*, 109–20.
48. Coculescu, B.I. Antimicrobial resistance induced by genetic changes. *J. Med. Life.* **2009**, *2*, 114–123.
49. Pato, M.L. Tetracycline inhibits propagation of deoxyribonucleic acid replication and alters membrane properties. *Antimicrob. Agents Chemother.* **1977**, *11*, 318–323.
50. Martinez, J.L.; Baquero, F. Mutation frequencies and antibiotic resistance. *Antimicrob. Agents Chemother.* **2000**, *44*, 1771–1777.
51. Gill, M.J.; Simjee, S.; Hattawi, K.; *et al.* **1998** *Gonococcal* resistance to β-lactams and tetracycline involves mutation in loop 3 of the porin encoded at the penB locus. *Antimicrob. Agents Chemother.* 1998, *42*, 2799–2803.
52. Rogers, H. J.; Perkins, H. R.; and Ward, J. B. Microbial Cell Walls and Membranes, Chapman & Hall, *London* **1980**.
53. Vannuff, P.; and Cocito, C. Mechanism of action of streptogramins and macrolides. *Drugs.* **1996**, *51*, 1, 20–30.
54. Sowmya, Y. A review on the human oral microflora. *Res. Rev.* **2016**, *4*, 1–5.

55. Wintersdorff, C. J.; Penders, J.; van Niekerk, J. M.; Mills, N. D.; Majumder, S.; van Alphen, L. B.; *et al.* Dissemination of antimicrobial resistance in microbial ecosystems through horizontal gene transfer. *Front. Microbiol.* **2016**, *7*, 173.
56. Willi, K.; Sandmeier, H.; Kulik, E. M.; and Meyer, J. Transduction of antibiotic resistance markers among *Actinobacillus actinomycetemcomitans* strains by temperate bacteriophages Aa phi 23. *Cell. Mol. Life Sci.* **1997**, *53*, 904–910.
57. Hu, Y.; Gao, G.F.; and Zhu, B. The antibiotic resistome: gene flow in environments, animals and human beings. *Front. Med.* **2017**, *11*, 161–168.
58. Moscoso, M.; and Claverys, J. P. Release of DNA into the medium by competent *Streptococcus pneumoniae:* kinetics, mechanism and stability of the liberated DNA. *Mol. Microbiol.* **2004**, *54*, 783–794.
59. Martin, B.; Quentin, Y.; Fichant, G.; and Claverys, J. P. Independent evolution of competence regulatory cascades in *streptococci*? *Trends Microbiol.* **2006**, *14*, 339–345.
60. Kenzaka, T.; Tani, K.; and Nasu, M. High-frequency phage-mediated gene transfer in freshwater environments determined at single-cell level. *ISME J.* **2010**, *4*, 648–659.
61. Luria, S. E.; and Delbrück, M. Mutations of bacteria from virus sensitivity to virus resistance. *Genetics.* **1943**, *28*, 491–511.
62. Roberts, A. P.; and Mullany, P. Genetic basis of horizontal gene transfer among oral bacteria. *Periodontol.* **2006**, *42*, 36–46.
63. Walker, C.B. The acquisition of antimicrobial resistance in the periodontal microflora. *Periodontol.* **2000**, *10*, 79–88.
64. Cox, G.; Wright, G.D. Intrinsic antibiotic resistance: mechanisms, origins, challenges and solutions. *Int. J. Med. Microbiol.* **2013**, *303*, 287–292.
65. Martinez, J.L. General principles of antibiotic resistance in bacteria. *Drug Discov. Today.* **2014**, *11*, 33–39.
66. Patil, S.; Rao, R.S.; Amrutha, N.; Sanketh, D.S. Oral microbial flora in health World *Dent. J.* **2013**, *4*, 262–6.
67. Benveniste, R.; and Davies, J. Aminoglycoside antibiotic-inactivating enzymes in *actinomycetes* similar to those present in clinical isolates of antibiotic-resistant bacteria. *Proc. Natl. Acad. Sci. U.S.A.* **1973**. *70*, 2276–2280. doi: 10.1073/pnas.70.8.2276.
68. Cieplik, F.; Jakubovics, N. S.; Buchalla, W.; Maisch, T.; Hellwig, E.; and Al-Ahmad, A. Resistance toward chlorhexidine in oral bacteria–is there cause for concern?. *Front Microbiol.* **2019**, *10*, 587.
69. Blair, J.M.; Richmond, G.E.; Piddock, L.J. Multidrug efflux pumpsin Gram-negativebacteria and their role in antibiotic resistance. *Future Microbiol.* **2014**, 1165–1177.
70. Kumar, A.; Schweizer, H.P. Bacterial resistance to antibiotics: active efflux and reduced uptake. *Adv. Drug Deliver Rev.* **2005**, 1486–1513.
71. Marsh, P.D. Role of the oral microflora in health. *Microbial Ecol. Health.* **2012**.
72. Marsh, P. D. Dental plaque: biological significance of a biofilm and community life-style. *J Clin Periodontol.* **2005**, *32*, 7–15.
73. Donlan, R. M.; Biofilms: microbial life on surfaces. *Emerg Infect Dis.* **2002**, *8*(9), 881.
74. Singh, S.; Singh, S. K.; Chowdhury, I.; and Singh, R. Understanding the mechanism of bacterial biofilms resistance to antimicrobial agents. *Open Microbiol J.* **2017**, *11*, 53.
75. Stewart, P. S. Antimicrobial tolerance in biofilms. *Microbiol. Spectrum.* **2015**, *3*(3), 10–1128.
76. Soto, S. M. Role of efflux pumps in the antibiotic resistance of bacteria embedded in a biofilm. *Virulence.* **2013**, *4*(3), 223–229.
77. Roberts, A. P.; and Mullany, P. Oral biofilms: a reservoir of transferable, bacterial, antimicrobial resistance. *Expert Rev Anti Infect Ther.* **2010**, *8*(12), 1441–1450.
78. Li, Y. H.; and Tian, X. Quorum sensing and bacterial social interactions in biofilms. *Sensor.* **2012**, *12*(3), 2519–2538.
79. Wimpenny, J.; Manz, W.; and Szewzyk, U. Heterogeneity in biofilms. *FEMS Microbiol Rev.* **2000**, *24*(5), 661–671.
80. Lambert, P.A. Bacterial resistance to antibiotics: Modified target sites. *Adv Drug Deliv Rev.* **2005**, *57*, 1471–85.
81. Zack, K. M.; Sorenson, T.; and Joshi, S. G. Types and Mechanisms of Efflux Pump Systems and the Potential of Efflux Pump Inhibitors in the Restoration of Antimicrobial Susceptibility, with a Special Reference to *Acinetobacter baumannii. Pathogens.* **2024**, *13*(3), 197.
82. Warburton, P. J.; Ciric, L.; Lerner, A.; Seville, L. A.; Roberts, A. P.; Mullany, P.; and Allan, E. TetAB (46), a predicted heterodimeric ABC transporter conferring tetracycline resistance in Streptococcus australis isolated from the oral cavity. *J Antimicrob Chemother.* **2013**, *68*(1), 17–22.

83. Jiang, W.; Zhang, Y.; Yan, J.; He, Z.; and Chen, W. Differences of protein expression in *enterococcus faecalis* biofilm during resistance to environmental pressures. *Technol Health Care.* **2024**, (Preprint), 1–13.
84. Truong-Bolduc, Q. C.; Yonker, L. M.; Wang, Y.; Lawton, B. G.; and Hooper, D. C. NorA efflux pump mediates *Staphylococcus aureus* response to *Pseudomonas aeruginosa* pyocyanin toxicity. *Antimicrobial Agents and Chemotherapy* **2024**, *68*(2), e01001–23.
85. Köhler, T.; Epp, S.F.; Curty, L.K.; and Pechère, J.C.; Characterization of MexT, the 87 regulator of the MexE-MexF-OprN multidrug efflux system of *Pseudomonas aeruginosa*, *J. Bacteriol.* **1999**, *181*, 6300–6305.
86. Poulsen, S.M.; Kofoed, C.; and Vester, B. Inhibition of the ribosomal peptidyl transferase reaction by the mycarose moiety of the antibiotics carbomycin, spiramycin and tylosin. *J. Mol. Biol.* **2000**, *304*, 471–481.
87. Abdi, S. N.; Ghotaslou, R.; Ganbarov, K.; Mobed, A.; Tanomand, A.; Yousefi, M.; and Kafil, H. S. *Acinetobacter baumannii* efflux pumps and antibiotic resistance. *Infect Drug Resist.* **2020**, 423–434.
88. Li, X. Z.; Plésiat, P.; and Nikaido, H. The challenge of efflux-mediated antibiotic resistance in Gram-negative bacteria. *Clin Microbiol Rev.* **2015**, *28*(2), 337–418.
89. Lambert, P.A. Mechanisms of antibiotic resistance in *Pseudomonas aeruginosa*. *J R Soc Med.* **2002**, *95*, 41, 22–6.
90. Zhang, J.S.; Chu, C.H.; and Yu, O.Y. Oral Microbiome and Dental Caries Development. *Dent J.* **2022**, *10*, 184.
91. Yoneyama, H.; and Katsumata, R. Antibiotic resistance in bacteria and its future for novel antibiotic development. *Biosci Biotechnol Biochem.* **2006**, *70*, 1060–75.
92. Liao, S.; Klein, M.I.; Heim, K.P.; Fan, Y.; Bitoun, J.P.; Ahn, S.J.; *et al. Streptococcus mutans* extracellular DNA is upregulated during growth in biofilms, actively released via membrane vesicles, and influenced by components of the protein secretion machinery. *J Bacteriol.* **2014**, *196*, 2355–2366.

10 Oral Microbiome and Mouth Cancer

Shamsher S. Kanwar, Ekta Arya, and Duni Chand
Department of Biotechnology, Himachal Pradesh University, Shimla, Himachal Pradesh, India

10.1 INTRODUCTION

The human oral cavity, the entrance to the food and respiratory tract, has a complex environment, comprising bacteria, microeukaryotes, archaea, and viruses. The oral microbiota also interfaces with the immune system of the human host and has an important role in not only the health of the oral cavity but also systemic health. As constituents of the diet, a variety of organic, inorganic, aliphatic, and aromatic compounds are routinely exposed to the native microbial flora of the oral cavity. The consequence of such interactions may promote biofilm formation, microbial colonization and invasive factors, tissue inflammation, chemical deposition and/or absorption, and possible mutagenesis of the oral microorganism. The quest of the native oral cavity microorganism for survival under varying biochemical dietary environments may prompt some of these microorganisms to cause growth aberration in the oral squamous tissue cells lining the oral cavity. Each year, a substantial amount of new invasive cancer cases are diagnosed. Oral cancer is one of the most common of these malignancies. Although there are differences in the microbiome between a healthy oral cavity and oral squamous cell carcinoma (OSCC), the specific pathogenic bacteria or bacterial combinations accountable for OSCC remain unidentified.

Oral cancer, also known as mouth cancer, represents the presence of malignant growth within the oral cavity, including the tongue, lips, mouth floor, gums, palate, and inner cheek lining. The main feature of this condition is the presence of squamous cell carcinoma, originating from the lining of the mouth and throat composed of squamous cells. It develops as persistent sores, lumps, or patches in the oral region that do not heal, as well as changes in the color or texture of oral tissues, difficulties swallowing or chewing, enduring discomfort, numbness, and bleeding. Tobacco use (both smoking and non-smoking), excessive alcohol consumption, human papillomavirus (HPV) infection, prolonged sun exposure (especially for lip cancer), poor oral hygiene, and a low-fruit and vegetable diet are all risk factors for mouth cancer.

The oral microbiome, the complex collection of microorganisms that live within the mouth, plays a critical role in sustaining oral health and impacting systemic health. The oral microbiome exhibits diversity in its makeup owing to the ongoing interaction of the oral cavity with the external environment. Factors such as temperature variations, dietary choices, pH levels, and feeding behaviors play significant roles in shaping the oral microbiome [1]. The American Cancer Society reports that around 48,000 Americans develop oral cancer and 8500 people die of the disease annually. India records more than 100,000 cases of oral cancer every year. Lip cancer, a frequent malignant tumor, accounts for about 25%–30% of all oral malignancies. The high rate of oral and lip cancer is a major worldwide health concern, with significant geographical differences in incidence and mortality rates. Despite its extensive impact, there are still not a lot of detailed statistics on the worldwide prevalence and mortality rates for this malignancy [2].

This chapter attempts to thoroughly investigate the complex association between the oral microbiome and OSCC, also known as oral cancer. It studies the complex interactions between varied microbial communities in the oral cavity and various environmental conditions, host features, and carcinogenic processes. It seeks to understand the processes behind OSCC initiation and

DOI: 10.1201/9781003478638-10

development by studying how the oral microbiome supports biofilm formation, microbial colonization, and inflammatory responses within the oral cavity. Furthermore, this investigation will look into how dietary substrates, oral hygiene habits, and lifestyle factors influence the makeup and diversity of the oral microbiome, as well as the potential consequences of OSCC risk.

Oral cancer, can occur in a wide range of sites in the mouth cavity and the surrounding areas of the head and neck. More than 90% of oral cavity malignancies are squamous cell carcinomas. Normally, the lining of the throat and mouth is made up of squamous cells, which have a flat appearance and resemble fish scales when viewed under a microscope. Squamous cell carcinoma develops when some of these squamous cells acquire mutation, resulting in unusual growth. These cancers are typically detected among individuals over the age of 64, however they may additionally impact younger individuals as well. Approximately 20% (1 in 5) of instances involve people under the age of 55 [3].

10.2 DIFFERENT TYPES OF ORAL CANCER

Oral cancer can occur in various forms, impacting distinct areas such as the mouth, lips, tongue, throat, and adjacent areas. Several prevalent forms of oral cancer are reviewed in this section (Figure 10.1).

10.2.1 Oral Squamous Cell Carcinoma

The development of OSCC is a multistep process and often begins as a small ulcer or lesion in the mouth that does not heal, but it can grow into a larger mass or tumor (Figure 10.2). Tobacco use (both smoking and non-smoking), excessive alcohol intake, human papillomavirus (HPV) infection and prolonged sun exposure on the lips are all risk factors for OSCC. Early identification and therapy are critical for improving outcomes for patients with OSCC. The number of cases of OSCC has increased significantly in several countries, particularly among younger populations. The question of whether age impacts the prognosis of OSCC patients remains under discussion, with differences between younger and older individuals continuing to be a matter of controversy [4]. Tobacco, alcohol and HPV can cause normal oral mucosal epithelial cells to undergo epithelial-mesenchymal transitions (EMTs). EMT allows epithelial cells to acquire mesenchymal features such as increased ECM component production, motility, invasion capability, angiogenesis, lymphogenesis and apoptosis suppression [5].

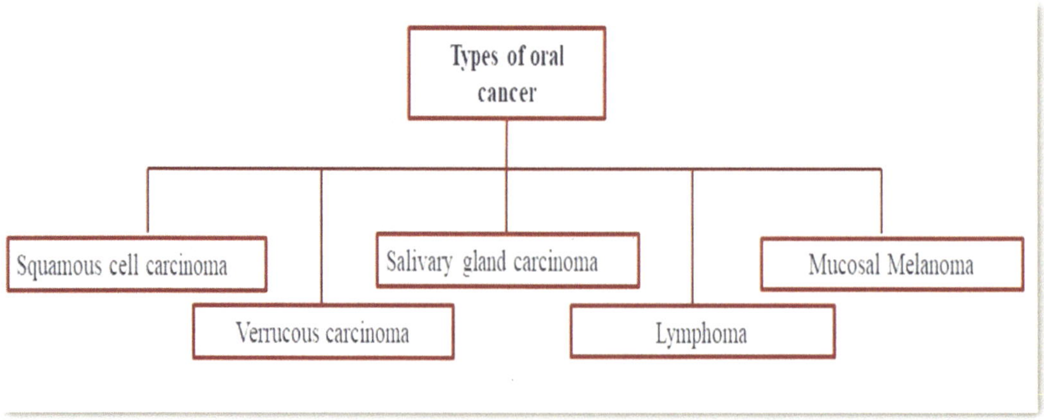

FIGURE 10.1 Schematic representation of different types of oral cancer.

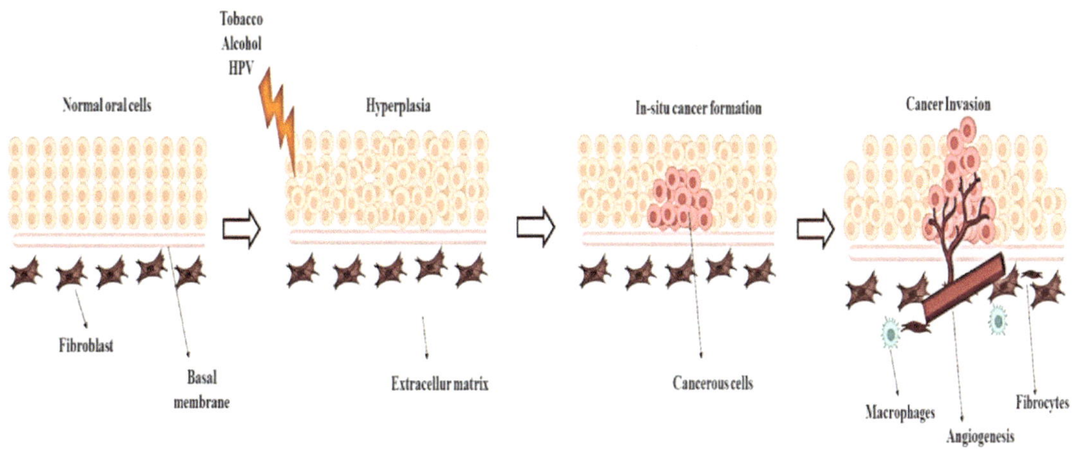

FIGURE 10.2 Schematic representation of the multistep process of oral cancer. (From Ref. [3].)

10.2.2 Verrucous Carcinoma

Verrucous squamous cell carcinoma (VSCC), also known as Ackerman's tumor, is a well-differentiated, low-grade subtype of squamous cell carcinoma. It is distinguished by its outward development, which resembles warts and a large expanding infiltration edge beneath the neighboring normal epithelial basement membrane. The tumor is made up of bland squamous epithelial cells that lack mitotic activity [6]. As a form of squamous cell carcinoma (SCC), it is strongly associated with tobacco and alcohol abuse. It constitutes approximately 5% of oral cavity malignancies. Although verrucous carcinoma can appear in different areas, it most commonly affects the oral cavity, specifically the buccal mucosa, gingiva, and palate. It can also appear in the genital area (penis and vulva), the soles of the feet (plantar verrucous carcinoma), and other mucosal sites.

VSCC is a highly differentiated subtype of SCC characterized by epithelial features lacking malignant cytology. Consequently, there is surface maturation, minimal pleomorphism, and generally few mitotic figures, although occasional basal mitoses may be observed. The surface displays pronounced epithelial hyperplasia, manifesting as numerous finger-like projections of well-differentiated squamous epithelium accompanied by significant keratosis (including hyperkeratosis and church-spire keratosis), parakeratosis, and parakeratotic crypting/keratin plugging. However, it is the thickened, club-shaped, or elephant foot-like broad pushing border of stromal infiltration beneath the adjacent unaffected epithelium basement membrane that is crucial for diagnosis. This downward extension of the epithelium forms a cup or "arms" at the periphery while maintaining an intact basement membrane.

10.2.3 Minor Salivary Gland Carcinoma

Salivary gland cancers (SGCs) are infrequent, accounting for approximately 1%–6% of all head and neck neoplasms. They are diverse in respect to their genesis and pathophysiology. This disease consists of different malignancies that can develop in the minor salivary glands and metastasize to the mouth and neck area. Some examples are adenoid cystic carcinoma, polymorphous low-grade adenocarcinoma, and mucoepidermoid carcinoma. SGCs are classified as those that originate in the major or minor salivary glands. Major SGCs are most abundant in the parotid glands, followed by the submandibular and sublingual glands. Minor SGCs, which account for 9%–23% of all salivary gland cancers, are most commonly found in the oral cavity's hard palate.

10.2.4 Lymphoma

Lymphomas refer to oral tumors that form in lymph tissue, which is part of the immune system. The tonsils and the tongue both contain lymphoid tissue. Lymphomas are malignancies that develop from lymphocyte cell lines. They constitute approximately 14% of all malignant neoplasms impacting the head and neck. They are differentiated into non-Hodgkin's lymphomas (NHLs) and Hodgkin's lymphomas (HLs) which vary in clinical characteristics, histology, and prognosis. NHLs are more frequently found in extranodal sites, with an incidence rate of 10%–35%. Extranodal onset is a rare subtype of HL, comprising only 5% of cases. In contrast, HL typically presents as a localized disease in lymph nodes, notably in the mediastinum and neck. Both HL and NHL can result in both nodal and extranodal involvement [7]

10.2.5 Mucosal Melanoma

Primary oral mucosal melanoma (OMM) is a rare malignant tumor that arises from the malignant transformation and clonal growth of melanocytes in the mouth cavity. It accounts for only 0.5% of all oral malignant tumors. OMM makes up 2% of all mucosal and cutaneous melanomas and 25% of all head and neck mucosal melanomas (HNMMs). HNMMs are most commonly found in the sinonasal tract. Asian communities have the highest frequency of OMM worldwide, particularly among males. OMM is distinguished by its high aggressiveness, which leads to a poorer prognosis than cutaneous melanomas (CM). Furthermore, OMM is distinct from cutaneous melanomas in terms of genesis and biological behavior. Unlike CM, OMM is not related to sun exposure, which is the key risk factor for CM.

10.3 RISK FACTORS AND ETIOLOGY

Oral cancer (OC) is very common in India, Pakistan, Brazil, France, Afghanistan, Bangladesh, Sri Lanka, Bhutan, Nepal, Iran, and the Maldives. It is either the leading or second most common cancer in these countries [8]. Various risk factors or causal agents for OC have been identified such as alcohol consumption, tobacco use, HPV infection, syphilis, dietary deficiencies, oro-dental factors, and certain viruses.

10.3.1 Chemical Factors

10.3.1.1 Tobacco

Abundant evidence suggests that tobacco, whether smoked, chewed, or consumed as betel quid, exerts a carcinogenic effect on the oral cavity. Smoking remains the predominant method of tobacco use. Hookah, also referred to as chillum, enjoys popularity in several Asian countries, notably India, with cigarettes, cigars, pipes, and bidis being common forms. In certain regions of India, the practice of Mizoram "smoke on the water", a unique water preparation containing the extracts of tobacco smoke, is used and is locally known as tuibur (tobacco smoke-infused water), which turns cognac in color and has a pungent smell. This practice represents a unique form of tobacco consumption.

10.3.1.2 Alcohol

Although, there remains some debate regarding whether alcohol alone can directly cause cancer, numerous studies have highlighted alcohol as a major risk factor for OC. Alcohol may exert an additive effect, potentially facilitating the entry of carcinogens into exposed cells and altering the metabolism of oral mucosal cells. Research indicates that individuals who consume more than 170 grams of whisky daily have a tenfold higher risk of OC compared to light drinkers.

10.3.2 Biological Factors

10.3.2.1 Viruses

Viruses can hijack their host's cellular machinery, altering DNA and chromosomal architecture and driving cell proliferation. HPV and herpes simplex virus (HSV) are responsible for OC. HPV has been found in approximately 23.5% of OC cases, and the most common HPV in head and neck SCC is HPV-16 which accounts for 90%–95% of all HPV-positive cases, followed by HPV-18, HPV-31, and HPV-33. Some studies, however, have revealed that HPV-positive OC cases have a better prognosis and disease-specific survival rates.

10.3.2.2 Syphilis

There have been studies indicating a serological positivity rate of 6%–19% for syphilis among patients with tongue cancer. The biological mechanism linking syphilis to oral cancer remains unclear. Chronic inflammation, which is characteristic of syphilis, has been implicated in the development of certain cancers.

10.3.2.3 Candida

Candida infection may play a role in the initiation of OC indicating nodular leukoplakia, a heightened risk of dysplasia, and malignant transformation. Laboratory studies have shown that infection with *Candida albicans* can induce squamous metaplasia and increased proliferation in the epithelium of chick embryos, which raises the possibility of a link between *Candida* infection and the development of oral cancer.

10.3.3 Dental Hygiene and Related Factors

Poor oral hygiene and prolonged discomfort from sharp teeth have been considered as potential contributors showing an inverse relationship with the incidences of OC. This is thought to occur through the accumulation of bacteria and toxins in the mouth, leading to chronic inflammation and tissue damage, which can promote the development of cancerous lesions. In addition, poor oral hygiene and dental sepsis are thought to exacerbate tobacco's carcinogenic effects.

10.3.4 Nutritional Factors

Dietary factors associated with OC development include antioxidant and micronutrient deficiencies impairing the body's ability to combat oxidative stress and inflammation, which are key factors in cancer initiation and progression. Some studies suggest that a diet high in fruits and vegetables, which contain important nutrients and protective chemicals, may lower the incidence of OC [9].

10.4 CURRENT DIAGNOSTIC AND TREATMENT APPROACHES

Standard diagnostic procedures in clinical practice commonly include vital staining, oral cytology, and optical imaging technology (Figure 10.3).

10.4.1 Vital Staining

Vital staining, a conventional tissue-staining method, employs toluidine blue, a metachromatic dye having high sensitivity and modest specificity to detect cells with elevated DNA content as well as dysplastic or malignant cells exhibiting abnormal DNA. Lugol's iodine staining, which results in a reddish or mahogany color, is an additional method for distinguishing the oral mucosa. It reacts with glycogen in normal oral tissue, but malignant lesions do not stain with Lugol's iodine, giving them a pale look when compared to healthy tissue nearby. When paired with toluidine blue staining,

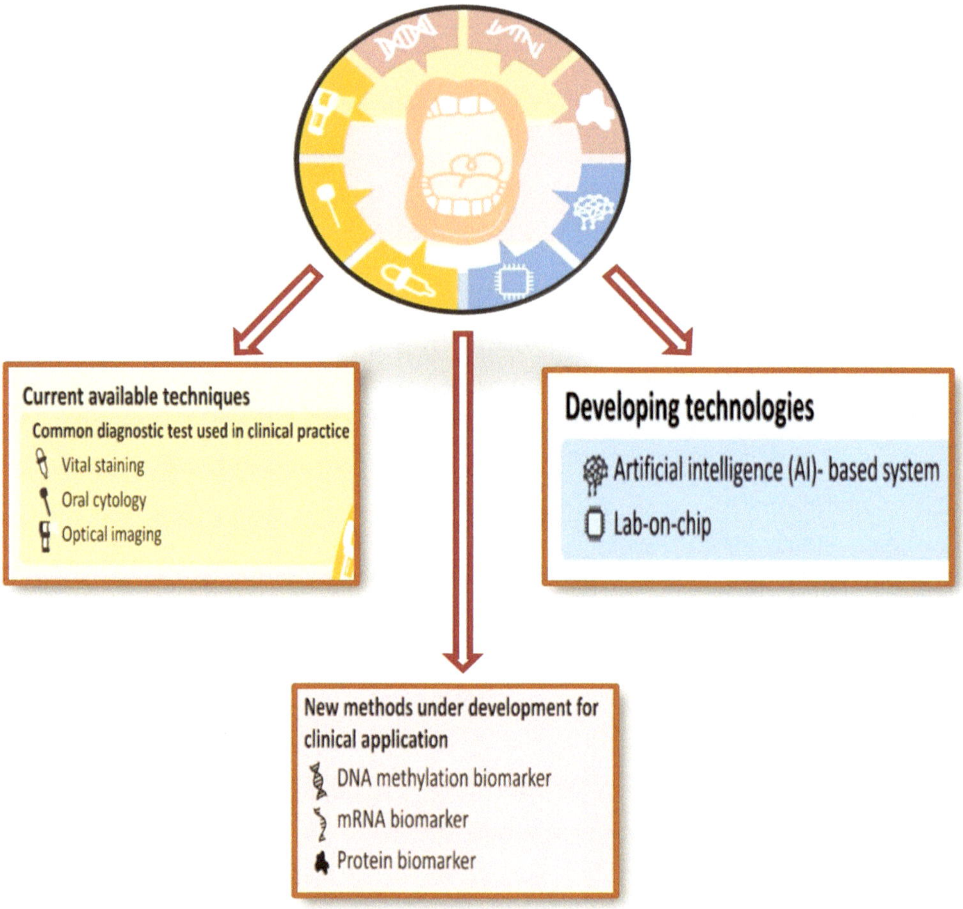

FIGURE 10.3 Schematic representation outlining both traditional early diagnostic methods and advancing technologies for oral cancer detection. (From Ref. [26].)

this method helps distinguish between normal and pathological epithelium, which is highly useful for finding biopsy sites before treatment, especially when there are many lesions spread across a large area.

10.4.2 Oral Cytology

Exfoliative oral cytology is a conventional approach to collecting oral mucosal cells via scraping, washing, and brushing with a tongue blade or brush. After collection, the specimens are preserved, stained, and inspected for morphology under a microscope by an experienced pathologist. It was first used in 1963 to analyze human oral mucosal lesions. This approach is simple, minimally invasive, and somewhat painless. However, studies have shown that it has a low sensitivity in diagnosing oral cancer, which could be attributed to insufficient or non-representative sampling, an increased risk of procedural errors, or the examiner's subjective interpretation.

10.4.3 Optical Imaging

Clinicians have long utilized light-based technology to detect oral potentially malignant disorders (OPMDs), such as oral cancer. Chemiluminescence or fluorescent light is employed as an intraoral detector to identify lesions on the oral mucosa. The color of the reflected light from the oral epithelium is then used to evaluate its condition.

10.4.4 Autofluorescence-Based Method

Autofluorescence is the phenomenon where light of a particular wavelength interacts with cells, causing excitation and subsequent emission of light at different wavelengths. Tissues emit autofluorescence because of the presence of naturally occurring fluorophores like collagen, tryptophan, elastin, keratin, hemoglobin, and nicotinamide adenine dinucleotide plus hydrogen (NADH). These fluorophores exhibit varying concentrations in OPMDs and cancerous lesions, leading to alterations in the natural light scattering and absorption characteristics of the tissues.

10.4.5 Chemiluminescence-Based Method

Chemiluminescence is the emission of visible light waves resulting from the excitation of electrons through a chemical process, followed by their return to their ground state. Light photons are emitted as electronic potential energy transitions occur within molecules. This method relies on reflectance, which assesses a surface's capacity to reflect incident light. For years, this technique has been employed to inspect the oral mucosa and detect OPMDs or malignant tumors [10].

10.4.6 Multispectral Fluorescence- and Reflectance-Based Methods

Multispectral fluorescence and reflectance methods are used to identify oral cancer. These methods involve evaluating fluorescence emissions and reflectance signals at multiple wavelengths to provide detailed information about the oral tissues. These approaches, which combine fluorescence and reflectance measurements, seek to improve the sensitivity and specificity of oral cancer detection, allowing for more precise diagnosis and early intervention.

10.5 EMERGING METHODS AND TECHNOLOGIES FOR CLINICAL APPLICATIONS

10.5.1 DNA Methylation Biomarkers (*ZNF582/PAX1*)

ZNF582 belongs to the extensive group of transcription regulators that produce Krüppel-associated box zinc finger proteins (KRAB-ZFPs) and acts as a tumor suppressor in nasopharyngeal cancer (NPC) [11]. *ZNF582* regulates the transcription and expression levels of adhesion molecules such as Nectin-3 and NRXN3, which inhibits metastasis of NPC via hypermethylation. Its tumor suppressor activity includes esophageal and anal carcinomas, and its methylation shows promise as a biomarker for cancer detection [12].

PAX1, which is distinguished by its paired-box domain, serves an important developmental role in embryogenesis and also as a tumor-suppressor gene. Elevated *PAX1* methylation contributes to carcinogenesis, particularly in oral and cervical carcinomas [13].

10.5.2 mRNA Biomarker (OAZ1/SAT and DUSP1)

Ornithine decarboxylase antizyme (OAZ1) undergoes activation through a polyamine-dependent mechanism and is associated with DNA repair, a factor that aligns with the metastatic potential of human OSCC cell lines [14]. SAT serves as the primary acetyltransferase in the catabolic pathway of polyamine metabolism. Research indicates that SAT expression levels are notably elevated in both prostate and oral malignancies [15]. It facilitates the acetylation of spermidine and spermine, playing a crucial role in regulating intracellular polyamine levels and their efflux from cells. DUSP1 belongs to type I cysteine-based protein tyrosine phosphatase and participates in various signaling pathways. Salivary DUSP1 mRNA levels are notably elevated in OSCC patients as compared to normal controls [16].

The National Cancer Institute's Early Detection Research Network: Biomarker Reference Laboratories (NCI-EDRN-BRL) conducted a case study in five cohorts aimed at comparing the effectiveness of seven mRNAs (OAZ1, IL-1 β, IL-8, SAT, H3F3A, DUSP1, and SP100P) and three proteins (IL-8, M2BP, and IL-1 β) in differentiating between OSCC patients and healthy individuals and the expression levels of all seven mRNAs (particularly, IL-8 and SAT) and three proteins were higher in OSCC patients compared to controls. The validation study further affirmed that these biomarkers effectively differentiate OSCC patients from healthy controls [16].

10.6 PROTEIN BIOMARKER

10.6.1 CD44

CD44, a cell surface glycoprotein, plays an important role in numerous cellular regulatory processes such as migration and cell proliferation. It serves as a biomarker for tumor initiation and stem cell association, showing overexpression in oral cancer. This overexpression contributes to tumor initiation in multiple cancer types. CD44 is mainly expressed in the basal and suprabasal areas of the normal epithelium. CD44 expression spreads into the superficial layers as dysplasia proceeds, indicating its participation in the early phases of carcinogenesis. Furthermore, recurrent tumors have a significantly higher concentration of CD44 on their surfaces than initial tumors. In advanced oral malignancies, overexpressed metalloproteinases cleave CD44 and release it as a soluble form (solCD44).

Several studies have looked into the efficacy of identifying CD44 and total protein levels. A hospital-based case-control study included 150 cases and 150 frequency-matched normal controls regardless of demographic or risk factors (such as age, gender, race, ethnicity, and tobacco and alcohol use) to determine whether it correlates with mouth cancer. A CD44 level of ≥ 5.33 ng/mL was highly linked with case status [17].

10.6.2 S100A7

S100A7, a calcium-binding protein belonging to the calcium-modulated multigenic S100 family, is typically localized in the upper, well-differentiated spinous layer normal epithelium. However, in high-risk oral lesions linked with cancer development, there is a notable increase in the expression of S100A7. This overexpression suggests its potential role as a biomarker for the progression of oral lesions towards malignancy [18].

In a study aimed at evaluating protein biomarkers for diagnosing high-risk oral dysplastic lesions prone to cancer progression, researchers conducted immunohistochemistry on samples from 110 individuals with biopsy-proven oral dysplasia. These patients' clinical outcomes were meticulously documented. The study also investigated the correlation between these biomarkers and both p16 expression and HPV 16/18 status. Among the array of protein biomarkers analyzed cytoplasmic S100A7 overexpression emerged as a standout marker significantly linked to development of cancer in dysplastic lesions. Patients exhibiting cytoplasmic S100A7 overexpression had notably shorter oral cancer-free survival (OCFS) durations, averaging 68.6 months, compared to those with weak or absent S100A7 expression in cytoplasmic immunostaining, who had a mean OCFS of 122.8 months. These findings underscore the potential of cytoplasmic S100A7 as a prognostic indicator for oral cancer progression in dysplastic lesions [18].

10.7 DEVELOPING TECHNOLOGIES

10.7.1 Artificial Intelligence-Based System

The ability of artificial intelligence (AI) to manage large amounts of data and find patterns allows it to help health care personnel make more accurate and timely diagnoses. In the instance of an oral cancer diagnosis, AI algorithms have been created to evaluate photographs and scans of oral tissues, assisting in the identification of precancerous or potentially malignant tumors. AI technologies can help improve patient outcomes by swiftly spotting irregularities that the human eye may overlook. AI consists of two primary subsets: machine learning and deep learning. Machine learning (ML) employs algorithms and statistical models to complete tasks without explicit directions. It includes supervised and unsupervised learning. Supervised learning is the process of training a model on a labeled dataset such that the algorithm can make predictions by connecting input data with output labels. Artificial neural networks (ANN), decision trees, support vector machines (SVMs), and a variety of other classifiers are all examples of supervised learning techniques [19]. Unsupervised learning, on the other hand, entails training models using unlabeled data to detect latent patterns or structures in the dataset. Unsupervised learning methods often include clustering algorithms, dimensionality reduction techniques, and association rule learning. Deep learning, a subclass of machine learning, focuses on neural networks with numerous layers (hence the name "deep"). Deep learning algorithms, which are inspired by the structure and function of the human brain, may learn data representations automatically by layering abstractions. Deep learning architectures include convolutional neural networks (CNNs) for image identification, recurrent neural networks (RNNs) for sequential data processing, and deep belief networks (DBNs).

The integration of AI-powered technology into clinical decision making or diagnostic systems represents a valuable approach to lesion discrimination, oral cancer screening, and predictive modeling [20]. A multicenter study highlighted the creation of a cost-effective deep learning–supported smartphone-based oral cancer probe designed specifically for populations at higher risk living in rural areas with limited infrastructure [21]. This smartphone-based probe, featuring an imaging system and guided by OSCC risk factors, provides guidance for oral mucosa screening. In a study, researchers developed a tablet-based mobile microscope tool for an oral cancer screening. This device, utilizing an iPad Mini equipped with collection optics, Bluetooth-controlled motors, and LED illumination was employed to scan slide specimens and take high-resolution images of stained brush biopsy samples. Results revealed agreement between cytology, histology, and image evaluation by a remote pathologist, showing the potential of this device to improve screening efficiency in health care settings with limited access to specialists [22]. In the area of oral cancer screening, AI has demonstrated its ability to enable remote health care interactions, particularly benefiting low- and middle-income countries. Using an algorithm, intraoral lesion images are classified as "suspicious" or "not suspicious," achieving specificity, sensitivity, positive predictive value, and negative predictive values in the range of 81%–95%. [23].

AI technology has been instrumental in developing prediction models, to unveil a personalized prediction model accessible at web.opmd-risk.com. This model utilizes a machine learning algorithm derived from data on 266 patients with OPMDs to forecast cancer risk. Impressively, the model can differentiate between high -and low-risk lesions with notable sensitivity and specificity. Furthermore, it exhibits the capability to anticipate the likelihood of future oral cancer development, offering valuable insights for clinical decision making and patient management [24].

10.7.2 Lab-on-a-Chip

Lab-on-a-Chip (LOC) is an innovative microelectromechanical system capable of detecting protein and ribonucleic acid biomarkers. It leverages microfluidics engineering to integrate multiple laboratory processes, including specimen preparation, fluid separation (such as serum, saliva, blood, or urine), signal amplification, and detection onto a single chip in an automated fashion. Because of its simplicity and automation, LOC can serve as a point-of-care device, removing the need for specialist laboratory personnel and making it suitable for usage in rural locations or large-scale screenings. Several molecular diagnostics, biochemical methods, and immunoassays have been effectively used on LOC systems, including nucleic acid and protein assays, as well as cell sorting [25].

In another study, 1 mL of saliva was collected and placed into an LOC device. The device removed salivary lymphocytic cells with antibodies coupled to magnetic beads, whereas precancerous and cancerous cells were captured and concentrated with antibodies specific to dysplasia and cancer markers [e.g., epithelial cell adhesion molecules (EpCAMs)] [26]. Research has shown that a combination of various salivary biomarkers, including thioredoxin and IL-8 proteins, as well as SAT, ODZ, IL-8, and IL-1b mRNAs, could identify oral cancer with high accuracy [27]. These advantages lead to a significant reduction in reagent usage, exposure to infectious agents, the need for hazardous materials, and the risk of specimen contamination. Furthermore, LOC improves reproducibility at a reasonable cost. It has enormous potential for improving oral cancer diagnosis; however, it is still in the early stages of development. Rapid identification of various expression profiles, proving predictive value, and optimizing specimen-collecting processes are all challenges that must be addressed before they can be widely adopted.

10.8 THE ORAL MICROBIOME: BASICS AND DIVERSITY

A human's oral microbiome, the microorganisms that live in the human mouth, changes dramatically throughout the lifespan. It begins with the uptake of bacteria at birth and develops as we age. The earliest settlers are *Streptococcus* species, such as *S. salivarius* on the tongue and in saliva, *S. mitis* on the inner cheek, and *S. sanguinis* on the teeth. These pioneer germs alter the environment, establishing conditions for other bacteria to thrive.

As we develop teeth, new bacteria enter the system, including *Prevotella melaninogenica*, *Fusobacterium nucleatum*, *Veillonella*, *Neisseria*, and others. The formation of teeth generates new habitats such as the gingival crevice, which is the space between the gums and the teeth. Over time, the microbial community settles into a stable state known as the climax community, in which various bacteria interact and contribute to the environment. However, interruptions can occur, resulting in dental disorders such as periodontitis. According to the "keystone pathogen" idea, certain low-abundance infections can cause such disorders by disrupting the balance of good bacteria [28].

10.9 INFLUENCE OF DIFFERENT FACTORS ON ORAL MICROBIOME COMPOSITION

Biofilm formation in the oral cavity is influenced by various factors such as the host's age, environmental conditions (e.g., pH, temperature, redox potential, oxygen levels, and nutrition), and lifestyle choices (e.g., dietary habits, tobacco and alcohol consumption, and oral hygiene practices), which all impact the composition of the oral microbiome. Additionally, saliva plays a crucial role as it transports nutrients, peptides, and partially dissolved carbohydrates necessary for biofilm formation and salivary glycoproteins control the adhesion of microorganisms to oral surfaces. Biofilm generation in dental plaque is significantly associated with the development of caries and periodontal disorders; however, its buildup is limited by oral epithelia. Furthermore, gingival crevicular fluid (GCF) contains iron, which may promote the pathogenic processes of oral bacteria associated with periodontal disease. Lactoperoxidase, in saliva, catalyzes the conversion of hydrogen peroxide into hypothiocyanite, which has antibacterial properties by blocking bacterial glycolysis. Another antibacterial salivary component is nitrite, which oral bacteria produce from dietary nitrates, which is then converted to nitric oxide, inhibiting the proliferation of cariogenic bacteria.

10.10 TECHNIQUES FOR STUDYING THE ORAL MICROBIOME

10.10.1 Culture Technique

Previously, bacterial taxa were identified using culture-dependent methods such as microscopy, biochemical and phenotypic assays, growth condition assessment, sugar usage, and antibiotic sensitivity testing. Despite advances, roughly 31% of known oral taxa are still uncultivable *in vitro*, highlighting the continued importance of bacterial culture in microbiology. However, except for antibiotic sensitivity testing, culture-dependent techniques for diagnostics are labor intensive, expensive, and less thorough than molecular DNA-based technologies. These molecular approaches eliminate the requirement for cultivation, providing a more complete understanding of microbial ecosystems.

10.10.2 Gel-Based Technologies

Culture-independent techniques have made it possible to conduct high-throughput microbial community analyses. Initially, techniques like temperature gradient gel electrophoresis and denaturing gradient gel electrophoresis enable community fingerprinting by separating DNA of similar sizes, which could then be sequenced for identification. Additionally, restriction fragment length polymorphism (RFLP) can be used to fragment homologous DNA sequences, with changes in the resulting fragment lengths serving as a genome-mapped tool. These technologies enable macro-level investigation, allowing the detection of substantial shifts or fluctuations in the population of specific microbial communities. These community-profiling approaches use polymerase chain reaction (PCR) to amplify regions of interest using specified primers, which can be further analyzed.

10.10.3 DNA Microarray

DNA microarrays provide a comprehensive approach to identifying the oral microbiome, enabling a better knowledge of oral health and illness. The process usually begins with the collection of samples, such as saliva, dental plaques, or swabs from various oral areas. DNA is then isolated and amplified from these samples to ensure that the amount present will be adequate for analysis. The next step is to create a microarray, which involves selecting probes that target certain microbial species or genes of interest within the oral microbiome. After hybridizing the amplified DNA with the microarray, fluorescence or other detection methods reveal the bound DNA fragments, allowing the identification of microbial species and their relative abundance. Data interpretation is also critical, with statistical analysis and bioinformatics tools used to identify patterns and comprehend the structure of the microbial community. Validation of data using other techniques such as quantitative PCR or next-generation sequencing ensures the accuracy of the result.

10.10.4 Sequencing

Sequencing of the microbial genome has become an important approach to characterizing the oral microbiome, providing remarkable depth and resolution in understanding its composition and functions. Initially, samples are taken from several oral locations, including saliva, tooth plaque, and mucosal swabs. Following that, DNA extraction is used to isolate genetic material from the microbial population. Library preparation involves fragmenting and adapting the isolated DNA for sequencing. High-throughput sequencing systems, such as Illumina and PacBio, produce massive volumes of sequence data from microbial DNA. These data are processed using bioinformatics, which assembles reads and identifies microbial species based on genetic signatures. Taxonomic profiling then categorizes the detected sequences, providing information about the richness and abundance of oral microbial species.

10.11 INTERPLAY BETWEEN THE ORAL MICROBIOME AND MOUTH CANCER

According to the Human Oral Microbiome Database (HOMD), the most common phyla in the human oral cavity are Firmicutes, Bacteroidetes, Proteobacteria, Actinobacteria, and Fusobacteria, a consensus backed by various research. In March 2020, HOMD listed 784 bacterial species and 1567 genomes in the human oral microbiome taxonomic hierarchy [29]. The majority are from the Firmicutes phylum, notably the Streptococcaceae family. *Streptococcus* is the most common genus in the oral cavity, along with *Haemophilus*, *Leptotrichia*, *Porphyromonas*, *Prevotella*, *Propionibacterium*, *Staphylococcus*, *Veillonella*, and *Treponema*. Some bacterial species are site-specific, whereas others can live in numerous areas at the same time as shown in Figure 10.4. The oral microbiota has moderate β-diversity, but high α-diversity compared to other body locations, with negligible variation in composition across people. Due to frequent contact with exogenous bacteria through drinking, eating, and breathing, differentiating indigenous from transient species in the oral cavity can be difficult. Persistent oral fungal infections like *Candida* spp. and high-risk human papillomavirus (HPV) and Epstein-Barr virus (EBV) infections can induce oncogenic

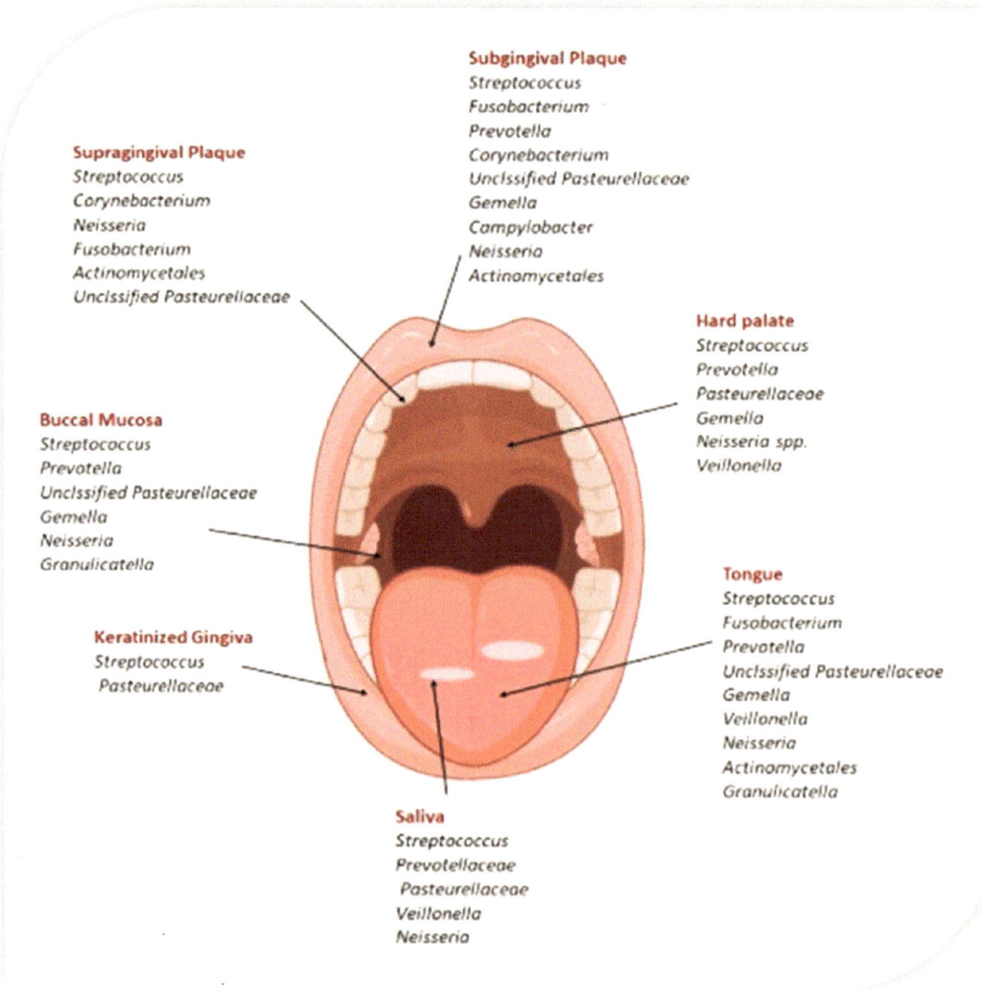

FIGURE 10.4 The most abundant bacterial species in the mouth cavity based on anatomical areas.

mutations and contribute to the development of OSCC. High-risk HPV DNA has even been detected in infants, though its long-term health implications are still unclear. Furthermore, the makeup of the oral microbiota varies with gender, age, and level of education.

OSCC has been linked to reduced levels of Firmicutes and increased levels of *Fusobacterium*. OSCC samples typically exhibit elevated quantities of *Peptostreptococcus*, *Fusobacterium*, *Prevotella* (especially *P. melaninogenica*), *Porphyromonas*, *Veillonella* (primarily *Veillonella parvula*), *Haemophilus*, *Rothia*, and *Streptococcus*.

10.12 SPECIFIC MICROBIAL SPECIES ASSOCIATED WITH MOUTH CANCER

In recent years, significant data has accumulated about two carcinogenic oral bacteria, *P. gingivalis* and *F. nucleatum*, that have been demonstrated both *in vitro* and in animal models. These bacteria have been shown to induce the generation of inflammatory cytokines, cell proliferation, and cellular invasion in OSCC in a variety of ways. *Porphyromonas gingivalis* increases the production of interleukins, TNF-α, and matrix metalloproteinases (MMPs), while blocking apoptosis and decreasing the *p53* tumor suppressor gene. Additionally, continuous exposure to *P. gingivalis* has been demonstrated to increase the invasiveness of OSCC (Figure 10.5). *Fusobacterium nucleatum*, on the other hand, boosts cell proliferation, increases interleukin production, and enhances MMP activity, all of which contribute to tumor invasion and metastasis. These actions cause increased transcriptional activity of proinflammatory cytokines and oncogenes (Figure 10.6). Both *F. nucleatum* and *P. gingivalis* produce endotoxins, such as lipopolysaccharides, which stimulate inflammation-associated cytokine production. This inflammation-associated cytokine production plays a crucial role in bacteria-induced inflammation and contributes to carcinogenesis [30].

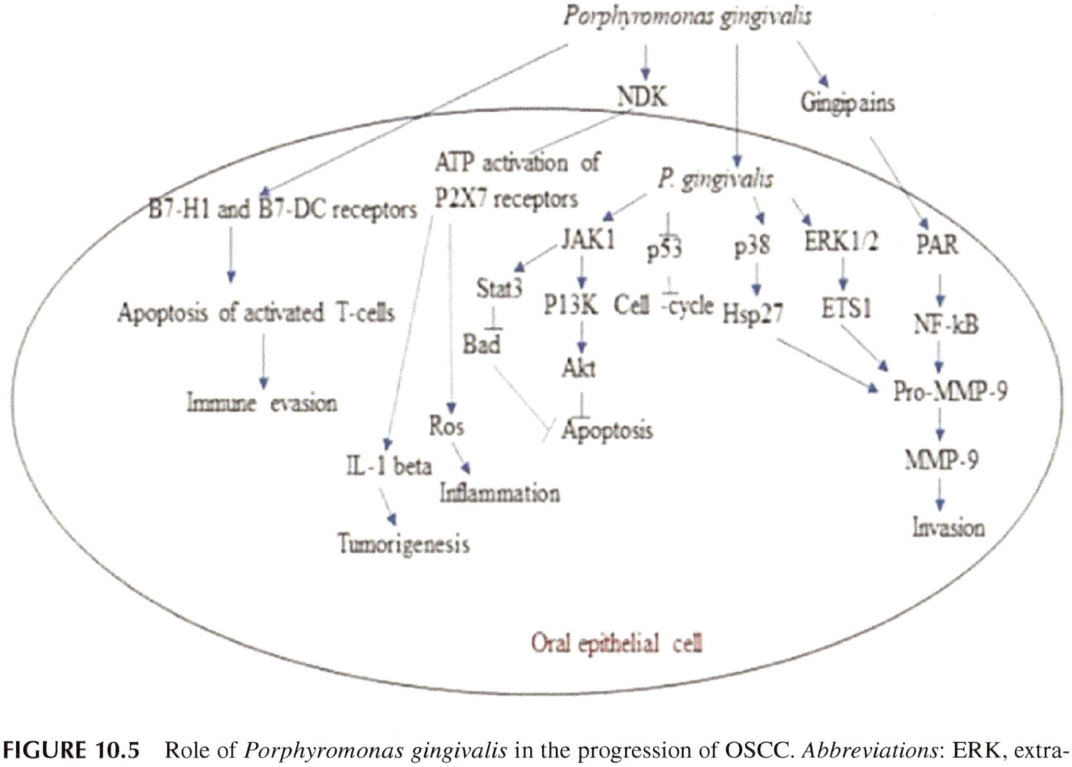

FIGURE 10.5 Role of *Porphyromonas gingivalis* in the progression of OSCC. *Abbreviations*: ERK, extracellular signal regulated kinase; ETS1, ETS Proto-Oncogene 1, Transcription Factor; IL, Interleukin; JAK, Janus kinase; NDK, nucleoside diphosphate kinase; ROS, reactive oxygen species; STAT, signal transducers and activators of transcription; PAR, Protease-activated receptors. (From Ref. [5].)

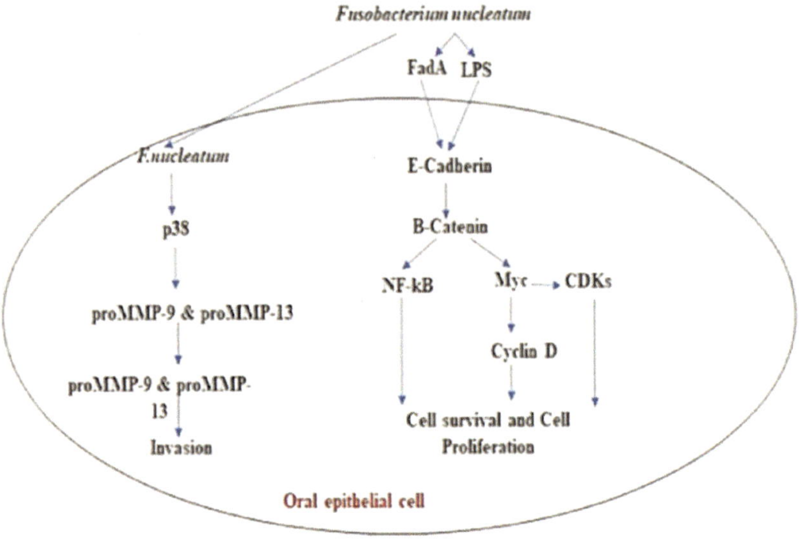

FIGURE 10.6 Role of *Fusobacterium nucleatum* in the progression of OSCC. *Abbreviations*: LPS, lipopolysacchardies; Myc, myelocytomatosis oncogene; CDKs, cyclin dependent kinases. (From Ref. [5].)

10.13 DYSBIOSIS AND ITS ROLE IN CANCER DEVELOPMENT

The studies have discovered links between oral cancer and habits like not brushing teeth often, infrequent dental checkups, and having lost or missing teeth. However, these links seem to be significant mainly in people who don't smoke or drink alcohol. Another study showed that periodontal diseases, which affect the gums and bone supporting the teeth, are associated with an increased risk of developing oral cancers. Additionally, research involving 51 patients with tongue cancer and 54 people without cancer found that long-term inflammation in the gums due to periodontal disease is a risk factor for cancer [31]. In another observational study with a large number of patients in the United States, researchers investigated how dental care influences the risk of oral cancer. They discovered that people who visited the dentist in the previous year had a 62% lower risk of oral cancer than those who had not had any dental checkups in the same period [32]. In a study, it was found that the oral microbiome's plays an important role in the development of oral cancer. They conducted a thorough examination of biological samples collected from 121 OC patients and 242 healthy controls who were age and gender matched. Multivariate analysis revealed substantial differences in the oral microbiota between people with poor dental hygiene, OC patients, and smokers. Notably, the microbiota changes identified in OC patients were more pronounced, particularly after tooth loss. Both findings highlight the importance of oral microbiome changes and tooth loss as substantial risk factors for oral cancer development, which are related to molecular and microenvironmental changes inside the oral cavity caused by both occurrences [33].

10.14 MECHANISMS OF MICROBIAL INVOLVEMENT IN CARCINOGENESIS

The complete understanding of the association between oral cancer and dysbiosis remains elusive. Various mechanisms have been proposed to elucidate how the oral microbiome influences cancer pathogenesis, including the stimulation of chronic inflammation by bacteria. This inflammatory process leads to the production of mediators that can initiate or facilitate mutagenesis, uncontrolled cell proliferation, angiogenesis, and cell degeneration, all of which are implicated in neurodegenerative disorders and cancer.

Moreover, bacteria can regulate cell proliferation by activating the nuclear factor κB (NF-κB) and inhibiting cell apoptosis, thus influencing the development of different cancer types. Furthermore, bacteria in the mouth cavity produce reactive oxygen species (ROS) and nitrogen oxygen species, as well as carcinogenic metabolites such as nitrosamines, which cause genetic damage in oral mucosal cells. Recent research has shown that microbiota and dysbiosis in the oral mucosa led to the accumulation of numerous epigenetic changes, which predispose individuals to neoplastic transformation [33]. Additionally, the integration of viral oncogenes into host genomes or the disruption of epithelial barrier integrity may promote genome instability, leading to irreversible cellular damage.

Dysbiosis and oral infection help to create a proinflammatory milieu, which is characterized by the release of inflammatory cytokines and matrix metalloproteinases, and also disrupts epithelial barriers leading to tumor formation and progression. Furthermore, oral dysbiosis is linked to a variety of epigenetic modifications, including changes in onco-miRs and DNA methylation patterns, both of which lead to tumor growth (Figure 10.7).

Chronic inflammatory mediators have the potential to disrupt normal cellular and tissue homeostasis, thereby contributing to or facilitating several processes such as cell proliferation, mutagenesis, angiogenesis, and oncogene activation. Certain anaerobic oral bacteria, including species like *Fusobacterium*, *Porphyromonas*, and *Prevotella*, are implicated in periodontal diseases, which in turn lead to chronic inflammation. Mediators for chronic inflammation have the potential to disrupt normal cellular and tissue homeostasis, thereby contributing to or facilitating several processes such as cell proliferation, angiogenesis, mutagenesis, and oncogene activation.

Periodontal bacteria influence the paracrine production of various proinflammatory mediators, including tumor necrosis factor-alpha (TNF-α), interleukins, and proteinases leading to the degradation of extracellular matrix (MMP 8, 9, and 13). Gram-negative bacteria release a proinflammatory lipopolysaccharide endotoxin which stimulates the production of IL1-β, IL-6, and TNF-α by binding to the toll-like receptor (TLR) of leukocytes. Particularly, inflammatory cytokines induce the overexpression of other proinflammatory proteins, thereby stimulating the release of prostaglandins (PG), phospholipase A2, and acute-phase proteins. The heightened production of inflammatory proteins, particularly of IL-6 and MMP-9, is predictive of a poorer prognosis and a more aggressive tumor. Bacteria produce a wide array of compounds, some of which have been implicated in promoting genetic damage and tumor growth. The production of acids, free radicals, sulfur compounds, and particularly ROS and reactive nitrogen species (RNS) can contribute to DNA damage. Nitric oxide (NO) and its derivatives, which are produced by nitric oxide synthase (NOS), dysregulation of which can cause the overproduction of NO and RNS, contribute to chronic inflammation and lead to cancer development.

10.15 FACTORS INFLUENCING THE ORAL MICROBIOTA IN MOUTH CANCER

Several factors can regulate the oral microbiota and potentially affect the development and progression of mouth cancer. These factors are detailed in this section.

10.15.1 Lifestyle Factors

10.15.1.1 Tobacco Use and Smoking

Smoking cigarettes, cigars, pipes, or using smokeless tobacco products drastically raises your chance of acquiring mouth cancer. Tobacco includes carcinogens, which can harm oral cells and cause malignant growth. Tobacco smoking also affects the oral microbiome, boosting pathogenic bacteria growth and raising inflammation, both of which can contribute to cancer development. The complex cocktail of carcinogenic substances present in tobacco smoke causes direct injury to the sensitive cells lining the mouth and throat, promoting mutations that lead to cancer. Furthermore, smoking undermines the body's natural repair systems, reducing its ability to repair DNA damage

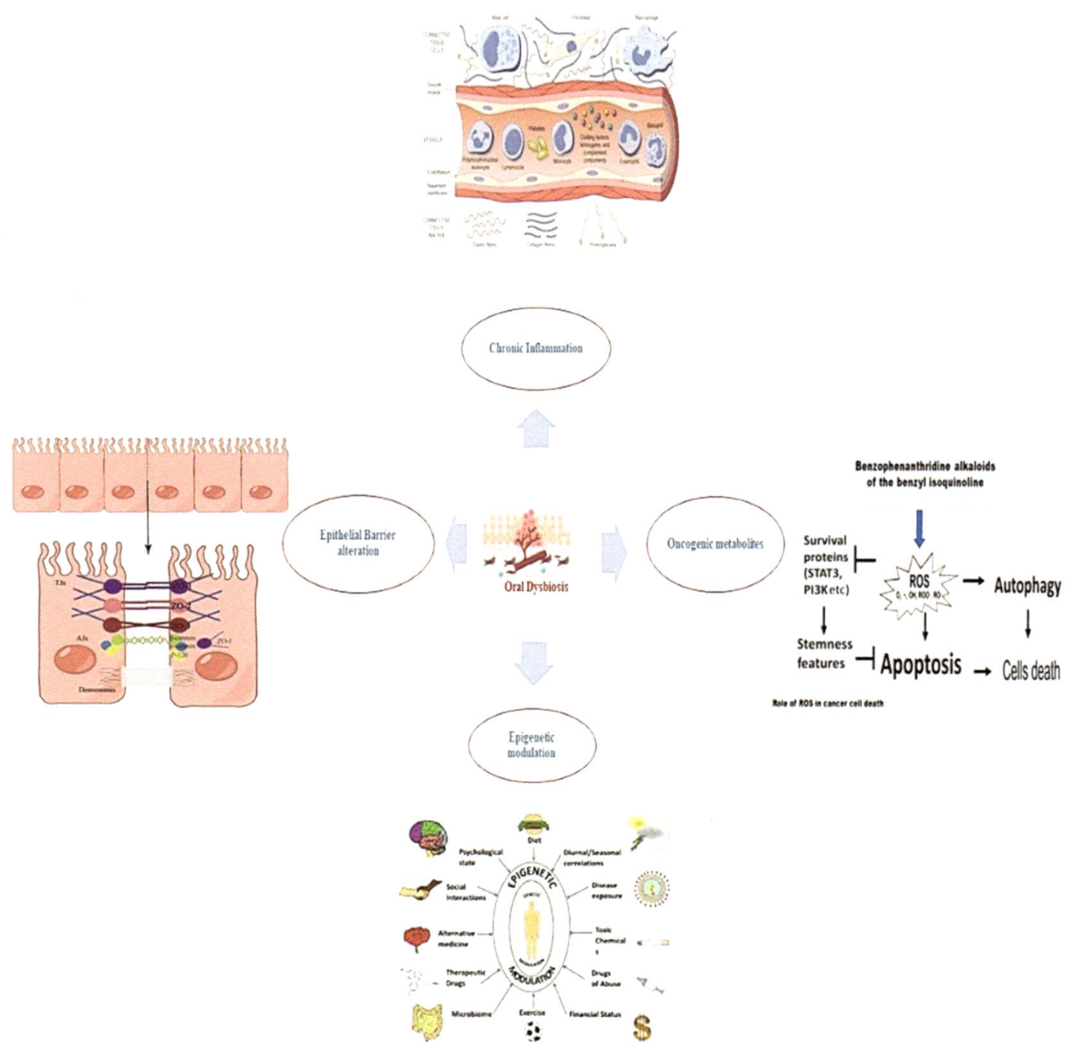

FIGURE 10.7 Oral microbiome dysbiosis is linked to oral cancer development via different mechanisms. (From Ref. [15].)

caused by these carcinogens. As if weakening the immune system's defenses against abnormal cells wasn't enough, smoking also causes persistent inflammation in the mouth cavity, producing an ideal habitat for cancer to grow and thrive. This persistent inflammation not only prepares the soil for tumor formation but also promotes the expansion of harmful microorganisms, which exacerbates the malignant process. The combination of smoking and strong alcohol intake only heightens the danger, intensifying the attack on oral tissues and upsetting the delicate balance of the oral microbiome.

10.15.1.2 Alcohol Consumption

Heavy alcohol intake is another significant risk factor for oral cancer. Alcohol can irritate the cells that line the mouth, increasing their susceptibility to carcinogens. Furthermore, alcohol consumption disturbs the equilibrium of the oral microbiome, allowing dangerous bacteria to flourish and boosting inflammation, both of which can promote cancer progression.

10.15.1.3 Dietary Habits

A diet high in sugar and processed foods and low in fruits and vegetables can be detrimental to oral health and raise the risk of oral cancer. Sugary diets encourage the growth of cariogenic (cavity-causing) bacteria in the mouth, and a lack of fruits and vegetables depletes the body of key nutrients and antioxidants that aid in oral health and cancer prevention.

10.15.1.4 Oral Hygiene Practices

Poor oral hygiene, such as infrequent brushing and flossing, can cause the buildup of dental plaque, which harbors germs that can contribute to oral disorders. Maintaining proper dental hygiene is critical for lowering the risk of oral cancer and improving overall oral health. Regular brushing, flossing, and dental checkups are critical in preventing the buildup of dental plaque, which contains germs that can lead to oral disorders such as cancer. Brushing twice a day with fluoride toothpaste and flossing every day eliminates food particles, plaque, and germs from the teeth and gums, lowering the risk of cavities, gum disease, and oral cancer. Furthermore, regular dental checkups and expert cleanings are critical for detecting and treating oral health issues early. Dentists can recognize indicators of gum disease.

10.15.1.5 Systemic Conditions and Medications

Systemic diseases and drugs have a complex impact on oral health and the risk of oral cancer. Notably, oral symptoms of systemic disorders can be markers of underlying inflammation or immunological dysregulation, emphasizing the significance of a thorough oral health assessment in those with systemic conditions. Close coordination between health care professionals and dentists is critical for controlling systemic illnesses, minimizing medication side effects, and addressing oral health issues to reduce the risk of oral cancer and preserve overall dental well-being.

10.15.1.6 Patient Education and Management Strategies

Caregivers in oral health care should shift their primary focus from illness management to actively maintaining a healthy microbiota. It is critical to incorporate the fundamentals of oral flora and its clinical consequences into the training curriculum for both undergraduate and postgraduate students in community dentistry. These trained experts can then teach patients how to make informed lifestyle choices, use effective plaque management procedures, and build good oral hygiene practices to maintain healthy oral microbiota. Initiatives like tobacco control and diabetes management can also be incorporated into microbiome-centric community oral health care programs. Currently, the majority of microbiome-based treatments are geared towards individuals or clinical settings. However, there is great promise for tackling public health issues through broader microbiome-based interventions. Incorporating probiotics or prebiotics into meals to support or supplement them is one such attempt with the potential for widespread benefit.

10.16 CONCLUSION AND FUTURE PERSPECTIVE

Understanding how the oral microbiota functions in mouth cancer is similar to solving a complex puzzle with profound consequences. This complex ecosystem of bacteria, fungi, and viruses in the human mouth is essential for oral health, but if disrupted, it can contribute to the development and spread of mouth cancer. The recent oral microbiology investigations that employ sophisticated sequencing, microfluidics, or microscopy techniques confront substantial challenges, including sample size limitations, differences in DNA extraction methods, various bioinformatics pipelines, and sequencing system selection. These limitations are most likely due to the high costs connected with such research, as well as the early stages of development of methodological and analytical processes. However, the application of second-generation sequencing techniques and robust statistical analysis is extending to larger cohorts, potentially providing insights into host-microbiome interactions. Furthermore, it is critical to understand that the majority of oral microbiome research has concentrated on bacterial diversity, while the richness and relevance of fungal and viral inhabitants in the oral cavity have been underestimated. Given the extensive fungal diversity observed in oral health conditions, their interaction with oral bacteria, and the proposed role of bacteriophages in controlling oral microbial communities, future research should focus on understanding the involvement of viral and fungal components in both health and disease contexts.

REFERENCES

1. Ali, J.; Sabiha, B.; Jan, H. U.; Haider, S. A.; Khan, A. A.; Ali, S. S. Genetic etiology of oral cancer. *Oral Oncology* **2017**, *70*, 23–28.
2. Almangush, A.; Mäkitie, A. A.; Triantafyllou, A.; de Bree, R.; Strojan, P.; Rinaldo, A.; Leivo, I. Staging and grading of oral squamous cell carcinoma: An update. *Oral Oncology* **2020**, *107*, 104799.
3. Chamoli, A.; Gosavi, A. S.; Shirwadkar, U. P.; Wangdale, K. V.; Behera, S. K.; Kurrey, N. K.; Mandoli, A. Overview of oral cavity squamous cell carcinoma: Risk factors, mechanisms, and diagnostics. *Oral Oncology* **2021**, *121*, 105451.
4. Chang, C.L.; Ho, S.C.; Su, Y.F.; Juan, Y.C.; Huang, C.Y.; Chao, A.S.; Hsu, Z.S.; Chang, C.F.; Fwu, C.W.; Chang, T.C. DNA methylation marker for the triage of hrHPV positive women in cervical cancer screening: Real-world evidence in Taiwan. *Gynecologic Oncology* **2021**, *161*(2), 429–435.
5. Chattopadhyay, I.; Verma, M.; Panda, M. Role of oral microbiome signatures in diagnosis and prognosis of oral cancer. *Technology in Cancer Research & Treatment* **2019**, *18*, 1533033819867354.
6. De Almeida Geraldino, R.; Rezende, L.; da-Silva, C.Q.; Almeida, J.C.F. Remote diagnosis of traumatic dental injuries using digital photographs captured via a mobile phone. *Dental Traumatology* **2017**, *33*, 350–357.
7. Elashoff, D.; Zhou, H.; Reiss, J.; Wang, J.; Xiao, H.; Henson, B.; Hu, S.; Arellano, M.; Sinha, U.; Le, A.; et al. Prevalidation of salivary biomarkers for oral cancer detection. *Cancer Epidemiology, Biomarkers & Prevention* **2012**, *21*, 664–672.
8. Gau, V. Wong, D. Oral fluid nanosensor test (OFNASET) with advanced electrochemical-based molecular analysis platform. *Annals of New York Academy of Science* **2007**, *1098*, 401–410.
9. Holmes, L. Jr.; desVignes-Kendrick, M.; Slomka, J.; Mahabir, S.; Beeravolu, S.; Emani, S.R. Is dental care utilization associated with oral cavity cancer in a large sample of community-based United States residents. *Community Dent Oral Epidemiol.* **2009**, *37*, 134–142.
10. Huang, J.; Wang, G.; Tang, J.; Zhuang, W.; Wang, L.P.; Liou, Y.L.; Liu, Y.Z.; Zhou, H.H.; Zhu, Y.S. DNA methylation status of PAX1 and ZNF582 in esophageal squamous cell carcinoma. *International Journal of Environmental Research and Public Health* **2017**, *14*, 216.
11. Ilhan, B.; Guneri, P.; Wilder-Smith, P. The contribution of artificial intelligence to reducing the diagnostic delay in oral cancer. *Oral Oncology* **2021**, *116*, 105254.
12. Ilhan, B.; Lin, K.; Guneri, P.; Wilder-Smith, P. Improving oral cancer outcomes with imaging and artificial intelligence. *Journal of Dental Research* **2020**, *99*(3), 241–248.
13. Kaur, J.; Matta, A.; Kak, I.; Srivastava, G.; Assi, J.; Leong, I.; Witterick, I.; Colgan, T.J.; Macmillan, C.; Siu, K.W.; et al. S100A7 overexpression is a predictive marker for high risk of malignant transformation in oral dysplasia. *International Journal of Cancer* **2014**, *134*, 1379–1388.
14. Krishnan, K.; Chen, T.; Paster, B. J. A practical guide to the oral microbiome and its relation to health and disease. *Oral Diseases* **2017**, *23*(3), 276–286.

15. La Rosa, G. R. M.; Gattuso, G.; Pedullà, E.; Rapisarda, E.; Nicolosi, D.; Salmeri, M. Association of oral dysbiosis with oral cancer development. *Oncology Letters* **2020**, *19*(4), 3045–3058.
16. Martin, J.L.; Gottehrer, N.; Zalesin, H.; Hoff, P.T.; Shaw, M.; Clarkson, J.H.; Haan, P.; Vartanian, M.; McLeod, T.; Swanick, S.M. Evaluation of salivary transcriptome markers for the early detection of oral squamous cell cancer in a prospective blinded trial. *Compendium of Continuing Education in Dentistry* **2015**, *36*, 365–373.
17. Oral Cavity & Oropharyngeal Cancer Key Statistics **2021** | American Cancer Society.
18. Pandya, D.; Nagarajappa, A.K.; Reddy, S.; Bhasin, M. Lab-on-a-chip — Oral cancer diagnosis at your door step. *Journal of International Oral Health* **2015**, *7*, 122–128.
19. Pereira, L.H.; Reis, I.M.; Reategui, E.P.; Gordon, C.; Saint-Victor, S.; Duncan, R.; Gomez, C.; Bayers, S.; Fisher, P.; Perez, A.; et al. Risk stratification system for oral cancer screening. *Cancer Prevention Research* **2016**, *9*, 445–455.
20. Ram, H.; Sarkar, J.; Kumar, H.; Konwar, R.; Bhatt, M. L. B.; Mohammad, S. Oral cancer: Risk factors and molecular pathogenesis. *Journal of Maxillofacial and Oral Surgery* **2011**, *10*, 132–137.
21. Salehiniya, H. Raei, M. Oral cavity and lip cancer in the world: an epidemiological review. *Biomedical Research and Therapy* **2020**, *7*(8), 3898–3905.
22. Sharma, N.; Bhatia, S.; Sodhi, A. S.; Batra, N. Oral microbiome and health. *AIMS Microbiology* **2018**, *4*(1), 42.
23. Skandarajah, A.; Sunny, S.P.; Gurpur, P.; Reber, C.D.; D'Ambrosio, M.V.; Raghavan, N.; James, B.L.; Ramanjinappa, R.D.; Suresh, A.; Kandasarma, U.; et al. Mobile microscopy as a screening tool for oral cancer in India: A pilot study. *PLoS One* **2017**, *12*, e0188440.
24. Song, B.; Sunny, S.; Uthoff, R.D.; Patrick, S.; Suresh, A.; Kolur, T.; Keerthi, G.; Anbarani, A.; Wilder-Smith, P.; Kuriakose, M.A.; et al. Automatic classification of dual-modality, smartphone-based oral dysplasia, and malignancy images using deep learning. *Biomedical Optics Express* **2018**, *9*, 5318–5329.
25. Sorrentino, A.; Ferragina, F.; Barca, I.; Arrotta, A.; Cristofaro, M. G. Extra-nodal lymphomas of the head and neck and oral cavity: a retrospective study. *Current Oncology* **2022**, *29*(10), 7189–7197.
26. Su, Y. F.; Chen, Y. J.; Tsai, F. T.; Li, W. C.; Hsu, M. L.; Wang, D. H.; Yang, C. C. Current insights into oral cancer diagnostics. *Diagnostics* **2021**, *11*(7), 1287.
27. Tezal, M.; Sullivan, M.A.; Reid, M.E.; Marshall, J.R.; Hyland, A.; Loree, T.; Lillis, C.; Hauck, L.; Wactawski-Wende, J.; & Scannapieco, F.A.; Chronic periodontitis and the risk of tongue cancer. *Archives of Otolaryngology - Head & Neck Surgery* **2007**, *133*, 450–454.
28. Thompson, L. D.; Verrucous squamous cell carcinoma. *Ear, Nose & Throat Journal* **2021**, *100*(5_ suppl), 540S–541S.
29. Tsuji, T.; Katsurano, M.; Ibaragi, S.; Shima, K.; Sasaki, A.; & Hu, G.F.; Ornithine decarboxylase antizyme upregulates DNA-dependent protein kinase and enhances the nonhomologous end-joining repair of DNA double-strand breaks in human oral cancer cells. *Biochemistry* **2007**, *46*, 8920–8932.
30. Tuominen, H. & Rautava, J.; Oral microbiota and cancer development. *Pathobiology* **2021**, *88*(2), 116–126.
31. Wang, X.; Yang, J.; Wei, C.; Zhou, G.; Wu, L.; Gao, Q.; He, X.; Shi, J.; Mei, Y.; Liu, Y.; et al. A personalized computational model predicts cancer risk level of oral potentially malignant disorders and its web application for promotion of non-invasive screening. *Journal of Oral Pathology & Medicine*. **2020**, *49*, 417–426.
32. Zhao, Y.; Hong, X.H.; Li, K.; Li, Y.Q.; Li, Y.Q.; He, S.W.; Zhang, P.P.; Li, J.Y.; Li, Q.; Liang, Y.L.; et al. ZNF582 hypermethylation promotes metastasis of nasopharyngeal carcinoma by regulating the transcription of adhesion molecules Nectin-3 and NRXN3. *Cancer Communications* **2020**, *40*, 721–737.
33. Ziober, B.L.; Mauk, M.G.; Falls, E.M.; Chen, Z.; Ziober, A.F.; & Bau, H.H.; Lab-on-a-chip for oral cancer screening and diagnosis. *Head & Neck* **2008**, *30*, 111–121.

11 Oral Microbiome and Malignancies

Shamsher S. Kanwar, Sohini Sharma, Pranjal Verma, and Shahzeb Khan
Department of Biotechnology, Himachal Pradesh University, Shimla, Himachal Pradesh, India

11.1 INTRODUCTION

The oral cavity serves as a primary entry point into the body and hosts the second largest and most diverse microbiome, comprising over 770 different bacterial species. This microbial diversity encompasses various regions within the oral cavity, including the teeth, soft and hard palates, buccal mucosa, and tongue. These diverse microbial communities collectively form the oral microbiome, which includes bacteria, fungi, and viruses. Bacteria constitute the predominant population of the oral microbiome [1].

Cancer is a debilitating, complex, and deeply frightening illness with the potential to impact any organ or tissue within the body [2]. The ongoing worldwide cancer epidemic persists as an important matter for public health, given that cancer stands firmly as the second most important factor for mortality in United States [3]. Numerous findings both from medical and statistical investigations unequivocally demonstrate the impact of lifestyle choices in humans on both the onset and prevention of cancer. A diet characterized by high levels of saturated fats and processed foods, leading to weight gain coupled with sedentary behavior, alcohol intake, and smoking has contributed to a rise in cancer incidence [4]. Infections, primarily induced by viruses, are believed to account for approximately 20% of all human tumors. Despite this, the direct role of microorganisms in the development of cancer has been mostly overlooked. Although epidemiological evidence associates bacterial infections with specific types of cancer, the prevailing assumption attributes tumor formation solely to the resulting inflammatory responses. However, it is essential to understand that numerous bacteria directly influence their host cells throughout their infection cycle. These manipulations can compromise host cell integrity and significantly contribute to the initiation of cancer [5].

Numerous investigations have additionally demonstrated the microbial community as the root cause of varied malignancies. Among these, the most prevalent malignancies include those linked with colorectal, lung, pancreatic, and oral cancer [6]. The findings from the meta-analysis revealed a heightened incidence of cancer due to periodontal bacterial infection [7]. For many years, the significant interplay between bacteria and tumors received little attention. However, the tumorigenic potential of *Helicobacter pylori* in gastric cancer (GC) was confirmed in the early 1990s. Subsequently, other associations between bacterial infections and various types of cancers have emerged: *Fusobacterium nucleatum* with colon cancer, *Salmonella typhi* with gallbladder cancer, *Chlamydia pneumoniae* with lung cancer, and *Propionibacterium acnes* with prostate cancer. These discoveries have shifted perspectives of scientists in this field, highlighting the crucial role of microorganisms in tumor development [1].

11.2 MALIGNANCIES CAUSED BY THE ORAL MICROBIOME

The fact that many oral microorganisms thought of as commensals in the mouth cavity are frequently linked to different tumors in distant organs is one feature of the oral microbiome that sets it apart from other body regions. There are two types of relationships between oral microbes and distant

 DOI: 10.1201/9781003478638-11

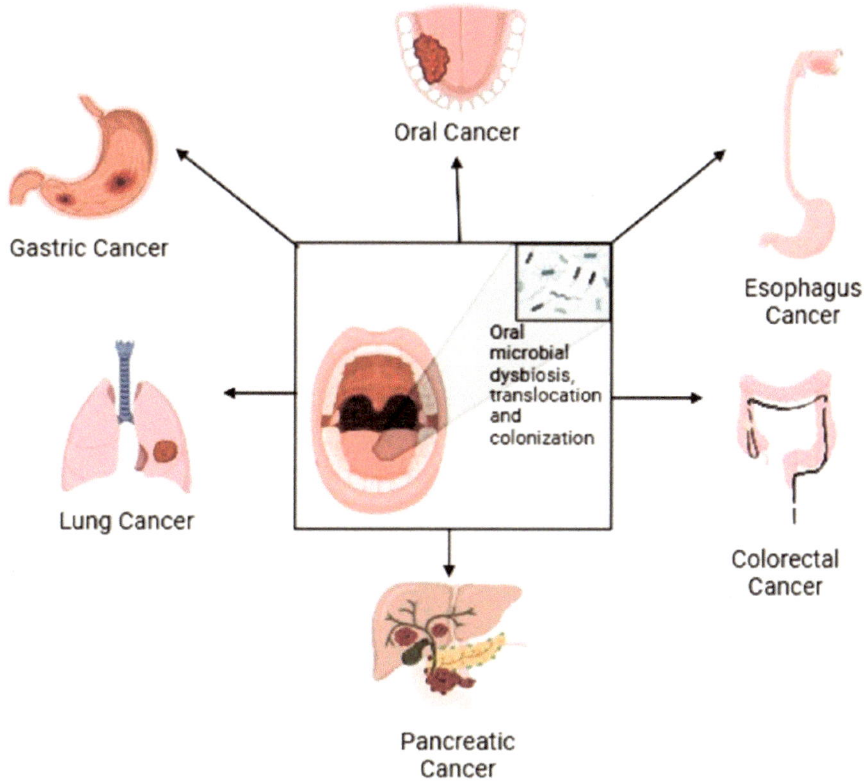

FIGURE 11.1 Different malignancies caused by oral microbial dysbiosis.

malignancies. The first is seen in some cancer types when the etiology of the tumor is not directly influenced by oral microorganisms, but there is a persistent alteration in the oral microbiome's composition that is linked to cancer. The second type of relationship is that of organisms implicated in carcinogenesis, as the correlation between *F. nucleatum* and colorectal cancer [8]. Microbial dysbiosis, colonization, and translocation of oral microbiota are essential for initiating various carcinogenic processes (Figure 11.1). Several mechanisms underlie oral microbiota–induced carcinogenesis, including heightened inflammatory responses, host immunosuppression, facilitation of malignant transformation, antiapoptotic activities, and secretion of carcinogens (Table 11.1). While considerable evidence links oral bacteria to cancers, the precise mechanisms of how oral microbiota influence carcinogenesis remain unclear. Before directly impacting carcinogenesis, certain microbial behaviors associated with tumorigenesis must first be observed. In oral in situ carcinogenesis, colonization, survival of microbiota, and subsequent microbial dysbiosis are prerequisites for the successful exertion of microbial oncogenic effects. For systemic distal carcinogenesis, the ability of microbiota to translocate from the oral cavity to other parts of the human body is a crucial factor [7].

11.2.1 Head and Neck Cancer

Head and neck cancer (HNC) is the seventh most common type of cancer globally, with over 660,000 newly diagnosed cases and approximately 325,000 fatalities reported each year [9]. Significant alterations in the abundance of oral microbiota, such as Bifidobacteriales, Lactobacillaceae, and Firmicutes have been observed in patients with oral cancer compared to those who are healthy [10]. Analysis of oral microbiome taxa revealed associations between certain bacterial groups

TABLE 11.1
Microbiological Dispersion in Various Cancer Types

S. No.	Types of Cancers	Elevated Levels of Bacteria	Decreased Levels of Associated Bacteria
1.	Head and neck cancer	Bifidobacteriales, Lactobacillaceae, Firmicutes, *P. gingivalis*, and *F. nucleatum*	*Prevotella nanceiensis*, *Capnocytophaga leadbetteri*, *Selenomonas sputigena*, *Prevotella nanceiensis*, *Capnocytophaga leadbetteri*, and *Selenomonas sputigena*
2.	Esophageal cancer	Firmicutes, Negativicutes, Selenomonadales, Prevotellaceae, *Prevotella*, and Veillonellaceae	*Proteobacteria*, *Betaproteobacteria*, *Neisseriales*, and Neisseriaceae
3.	Gastrointestinal cancer	*P. gingivalis*	
4.	Lung cancer	*Capnocytophaga* and *Veillonella*	*Streptococcus*
5.	Pancreatic cancer	*P. gingivalis*, *S. mitis*, and *A. actinomycetemcomitans*	*Fusobacteria* and *Leptotrichia*
6.	Colorectal cancer	*P. gingivalis* and *F. nucleatum*	*Bifidobacterium*, *Faecalibacterium*, and *Blautia*

and head and neck cancer risk. For instance, a greater abundance of the Actinobacteria phylum was associated with increased risk, while specific taxa within Actinobacteria and Proteobacteria phyla were associated with reduced risk. Notably, several species such as *Prevotella nanceiensis*, *Capnocytophaga leadbetteri*, and *Selenomonas sputigena* showed inverse associations with these carcinomas. Additionally, certain bacterial genera were associated with reduced risk of larynx cancer, while others were associated with reduced risk of pharynx or oral cavity cancer. These findings suggest potential roles of specific oral microbiome taxa in modulating the risk of different types of head and neck cancers [11]. Microbial examinations had also revealed that *Streptococcus salivarius*, *Streptococcus gordonii*, *Gemella haemolysans*, *Gemella morbillorum*, *Johnsonella ignava*, and *Streptococcus parasanguinis* were highly associated with the tumor site, while *Granulicatella adiacens* was prevalent at the non-tumor site. *Streptococcus intermedius* was present in 70% of both non-tumor and tumor sites [12].

Oral squamous cell carcinoma (OSCC), which originates from the oral cavity and lips, represents a prevalent form of malignancy within the head and neck region. According to statistics compiled by the Global Cancer Observatory (GCO), the yearly occurrence of OSCC reached 377,713 cases globally in 2020 [13]. Li et al. reported a substantial increase in *Fusobacterium* spp., in all cases of oral cancer [14]. Studies have also established synergistic relations between *P. gingivalis* and *F. nucleatum* to be associated with OCSS. These bacteria aid in the progression of cancer by altering cellular metabolism and increasing the proliferation of oral tumors [15].

11.2.2 Esophageal Cancer

A comparison of the oral microbiome in healthy individuals and patients diagnosed with esophageal cancer revealed notable differences. Specifically, individuals with esophageal cancer exhibited elevated levels of certain bacterial taxa including Firmicutes, Negativicutes, Selenomonadales, Prevotellaceae, *Prevotella*, and Veillonellaceae. Conversely, there was a decrease in the percentage of taxa such as Proteobacteria, Betaproteobacteria, Neisseriales, Neisseriaceae, and *Neisseria* in patients with esophageal cancer. These findings suggest distinct microbial compositions between healthy individuals and those afflicted with esophageal cancer, indicating a potential role for the oral microbiome in the development or progression of this malignancy [16].

11.2.3 Gastrointestinal Cancer

Several studies have determined the existence of *P. gingivalis* in individuals diagnosed with gastrointestinal (GI) cancers. A notable study focused on serum-positive samples revealed that the serum IgG antibody levels specific to *P. gingivalis* exhibited a positive correlation with the mortality rates associated with GI malignancies. This finding suggests a potential link between the presence of *P. gingivalis* and the progression or severity of GI cancers [1].

11.2.4 Lung Cancer

Oral bacteria have been identified as potential colonizers of the lungs via the pharynx, potentially contributing to chronic obstructive pulmonary disease (COPD), a condition closely associated with lung cancer. Some studies have revealed significantly elevated levels of *Capnocytophaga* and *Veillonella* in the saliva of individuals diagnosed with lung cancer (LC) compared to control samples. These findings underscore the intricate interplay between oral microbiota and lung health, justifying additional research into their function in lung-related diseases [17]. Examination of airway rinse samples from 39 lung cancer patients and 36 non-cancer patients demonstrated a heightened presence of *Streptococcus* and *Veillonella* in the lower airways of those afflicted with lung cancer. This study unveiled a correlation between *Streptococcus* and *Veillonella* and the increased activation of ERK and PI3K signaling pathways in the process of carcinogenesis [17].

11.2.5 Pancreatic Cancer

An increased prevalence of periodontopathogens such as *P. gingivalis*, *Porphyromonas*, *S. mitis*, and *Aggregatibacter actinomycetemcomitans* in the oral cavity has been associated with elevated pancreatic cancer risk [18]. The phylum Fusobacteria and its genus *Leptotrichia* were linked to a lower risk of pancreatic cancer. These findings suggest a potential role for specific oral pathogens and microbial profiles in the development or progression of pancreatic cancer, emphasizing the importance of further research into the oral microbiome's influence on pancreatic health and disease [19].

11.2.6 Colorectal Cancer

Metabolites generated by pathogenic bacteria, such as *P. gingivalis* and *F. nucleatum*, can heighten the risk of developing colorectal cancer (CRC) and potentially accelerate its progression [20]. The potential for *F. nucleatum* to function as a carcinogen is plausible, given its demonstrated capability to provoke inflammation and hinder local immune responses. The production of inflammatory cytokines and reactive oxygen species as a result of *F. nucleatum* infection could assist in the initiation of cancer by promoting mutations, genomic instability, and epigenetic changes. While research has indicated an enrichment of fusobacteria in tumors, direct evidence linking this colonization to an increased danger of malignant transformation is currently lacking. Further studies are needed to elucidate the precise mechanisms by which *F. nucleatum* may contribute to colorectal cancer development and progression [21].

11.3 ORAL MICROBIOTA AND SYSTEMIC HEALTH

The interactive oral microbiota collaborates with the host to convey information regarding immunity and metabolism through bidirectional communication across the oral cavity and systemic organs. As microbiome research progresses, investigations into the relationship between oral microbes and numerous chronic human conditions have expanded. These circumstances involve but are not restricted to inflammatory bowel disease, Alzheimer's disease, rheumatoid arthritis, and preterm birth [22]. Oral microorganisms have the ability to traverse oral mucous membranes and periodontal pockets, disseminating to various anatomical locations throughout the body and triggering systemic or localized infections. Persistent systemic inflammation induced by oral pathogens could exacerbate the progression of severe systemic diseases like diabetes or cardiovascular disease [23].

11.3.1 Digestive Disorders

Inflammatory bowel disease (IBD) and liver cirrhosis are the two most prevalent medical disorders linked to oral microbiome dysbiosis [22]. IBD is characterized as persistent inflammation of the intestines, stemming from interactions between the host and microbial communities within a genetically susceptible individual [24]. In the realm of IBD, two notable conditions of concern are primary sclerosing cholangitis (PSC) and CRC. PSC has been connected to an intestinal accumulation of oral bacteria, notably showing elevated levels of *Rothia mucilaginosa*, *Veillonella parvulas*, and *Streptococcus* spp. [25]. The potential connection between *P. gingivalis*, the primary pathogen responsible for periodontal diseases, and IBD has been suggested. Studies have shown that the presence of *P. gingivalis* can heighten the risk of colon epithelial loss and inflammatory cell infiltration during Dextran sulfate sodium induced IBD in mouse models [26].

Cirrhosis is characterized by the histological formation of regenerative nodules encircled by fibrous bands due to prolonged liver injury, resulting in portal hypertension and advanced liver disease [27]. Individuals with periodontitis exhibit elevated serum levels of antibodies to *P. gingivalis*. Clinical observations have documented a severely obese patient with cirrhosis linked to nonalcoholic steatohepatitis (NASH), who succumbed to sepsis resulting from a *P. gingivalis* infection. This case potentially underscores the significant role of *P. gingivalis* in advancing cirrhosis among NASH patients [28].

11.3.2 Cardiovascular Diseases

Research indicates a connection between periodontitis and cardiovascular disease (CVD), attributed to pathogenic oral bacteria present in atherothrombotic tissue in mice [29]. A meta-analysis has demonstrated that people with periodontitis face a moderate to heightened risk of ischemic stroke, CVD, and coronary heart disease development [30]. Anaerobic gram-negative bacteria, like *P. gingivalis* and *Prevotella intermedia*, and spirochetes, such as *Treponema denticola*, are commonly associated with periodontitis [31]. These bacteria, including *P. gingivalis*, thrive in low-oxygen environments resulting from inflammation, thereby exacerbating inflammation. Evidence suggests that treating periodontitis leads to a reduction in CVD markers such as C-reactive protein and circulating lipids [30]. This underscores the importance of addressing periodontal health in managing cardiovascular risk factors.

11.3.3 Adverse Pregnancy Outcomes

Preterm labor, early rupture of the membranes, pre-eclampsia, miscarriage, intrauterine growth restriction, low birth weight, stillbirth, and neonatal sepsis are just a few of the conditions that fall under the umbrella of adverse pregnancy outcomes (APO) [31]. Experiments conducted on pregnant mice showed that intravenous administration of *F. nucleatum* resulted in targeted colonization and multiplication of this microorganism within the fetal-placental unit. Conversely, injecting bacteria into the decidua, mimicking chorioamnionitis, ultimately led to premature birth and fetal demise. Likewise, the migration of *P. gingivalis* and *Campylobacter rectus* to placental tissues was associated with restricted fetal growth [32].

11.3.4 Rheumatoid Arthritis

The connection between periodontal disease (PD) and rheumatoid arthritis (RA) is primarily grounded in their shared pathophysiology, characterized by inflammatory destruction, and is supported by increasing epidemiological evidence [33]. Increasing evidence supports the link between periodontitis and rheumatoid arthritis, with *P. gingivalis* being a prominent suspect [34]. *Porphyromonas gingivalis* not only increases IL-17 production, a key cytokine in RA inflammation, but also promotes the generation of Th17 cells, further implicating its role in RA pathogenesis [22]. However, in mouse models, a combination of *P. gingivalis, F. nucleatum*, and *Aggregatibacter actinomycetemcomitans* shows less alveolar bone loss compared to *P. gingivalis* alone, though *F. nucleatum* and *A. actinomycetemcomitans* accelerate arthritis onset and progression [34].

11.3.5 Alzheimer's Disease

Dominy and coworkers noted a link between *P. gingivalis* and the advancement of Alzheimer's disease (AD), observing the presence of this bacterium in the brains of AD patients. Additionally, in some studies, *P. gingivalis* infection was found to facilitate brain colonization and the proliferation of amyloid plaque components in mice [22]. *Porphyromonas gingivalis* secretes various proteolytic enzymes, heme, and iron capture proteins, adhesins, and internalins, which are concentrated on outer membrane vesicles (OMVs). These OMVs have been observed to penetrate the blood-brain barrier, potentially exacerbating AD neuropathology by enhancing neuroinflammation, promoting plaque and tangle formation, disrupting iron homeostasis, and ultimately triggering ferroptosis, resulting in neuronal demise and neurodegeneration [35, 36].

11.3.6 Diabetes

Oral microbes, notably *P. gingivalis*, can influence diabetes onset and progression by disrupting systemic immune balance. *Porphyromonas gingivalis* endotoxin activates the toll-like receptor (TLR) 2/4 pathway in macrophages, leading to chronic inflammation via the c-Jun N-terminal kinase (JNK), mitogen-activated protein kinase (MAPK), and nuclear factor kappa-light-chain enhancer of activated B cells (NF-κB) pathways. NF-κB inhibits insulin signaling proteins like glucose transporter type 4 (GLUT4), while tumor necrosis factors (TNFs) promote lipid breakdown, reducing insulin sensitivity. Persistent inflammation induces insulin resistance, exacerbating systemic inflammatory imbalance and affecting glucose metabolism. Additionally, *P. gingivalis* endotoxin up-regulates immune-inflammatory and insulin signaling genes in pancreatic β-cells, while animal studies demonstrated a link between periodontitis-induced dysbiosis and insulin resistance in high-fat-diet–fed mice [23]. Subjects with diabetes exhibited significantly higher levels of Fusobacteria and Actinobacteria including *Pseudomonas* spp., and *Neisseria* spp., whereas the abundance of Proteobacteria was notably reduced [37, 38].

11.4 MECHANISM OF ORAL CANCER

Numerous methods of action such as bacterial activation of chronic inflammation, production of carcinogens, etc., have been discovered to clarify the impact of oral microbiota on cancer development (Figure 11.2). This process results in the creation of inflammatory mediators that can induce or contribute to mutagenesis, unchecked cell division, angiogenesis, and cell deterioration, which are the root causes of cancer and neurological diseases [39–42]. The initiation of persistent inflammation by microorganisms results in the release of inflammatory mediators that either induce or promote angiogenesis, mutagenesis, oncogene activation, and cell proliferation. Furthermore, by inhibiting cell death and activating nuclear factor κB (NF-κB), bacteria can control the growth of cells either encouraging or preventing the growth of various cancer forms [42–45]. Additionally, the introduction of viral oncogenes into host genomes or modifications of the integrity of the epithelial barrier may encourage genomic instability and permanent cellular damage [46]. It is interesting that an intricate relationship between inflammation, epithelial barriers, and microbiota may play a significant part in the carcinogenic process [47–50]. The relationship between oral mucosa dysbiosis

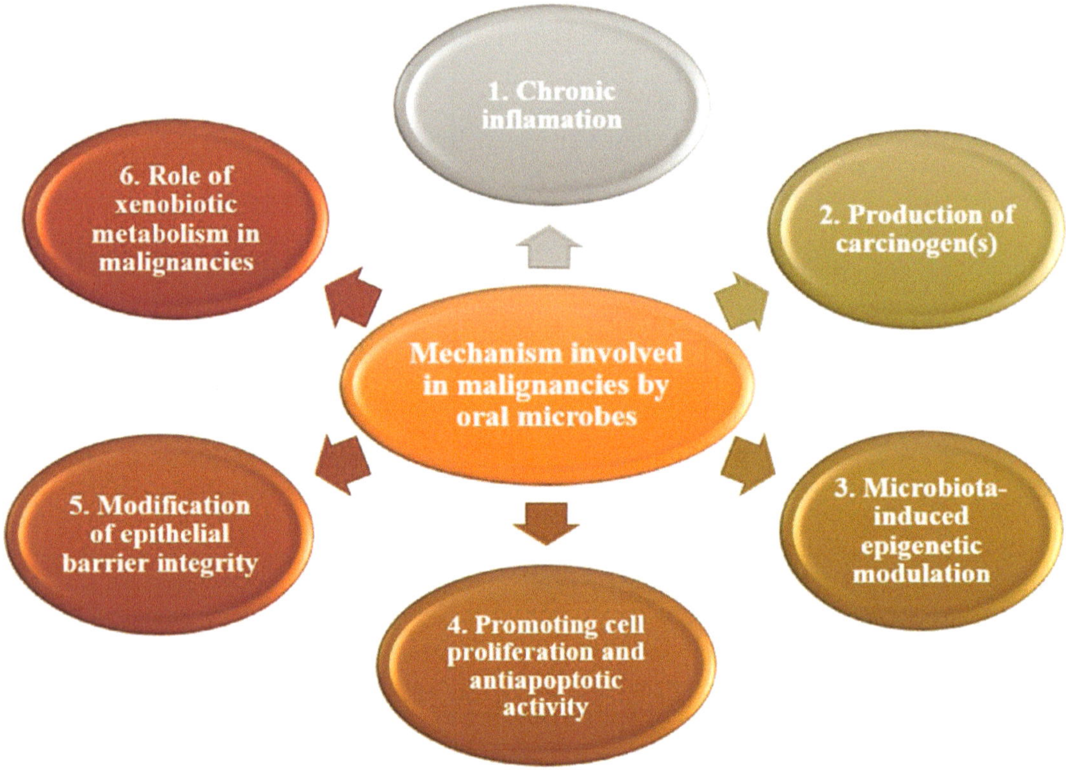

FIGURE 11.2 Mechanism involved in oral-microbiome–induced carcinogenesis.

and microbiota has been shown to result in the accumulation of several epigenetic changes that increase the risk of neoplastic transformation [51].

11.4.1 Chronic Inflammation

Around 25% of human cancer cases are linked to chronic inflammation, highlighting inflammation as a major hallmark of cancer [52]. Chronic inflammatory mediators could disrupt the normal balance of cells and tissues, potentially leading to or facilitating processes such as cell proliferation, angiogenesis, mutagenesis, and the activation of oncogenes [52].

Certain anaerobic oral bacteria, such as *Fusobacterium*, *Porphyromonas*, and *Prevotella* species, are linked to periodontal diseases and cause persistent inflammation. These bacterial cells secrete inflammatory mediators that can engage with various tissue cells, triggering widespread inflammatory responses. Periodontal bacteria affect the paracrine production of multiple proinflammatory mediators, including interleukins (IL-1, IL-6, IL-17, and IL-23), tumor necrosis factor-α (TNF-α), and extracellular matrix-degrading proteinases [53, 54]. Many types of cancer arise from the excessive production of inflammatory proteins, with the overexpression of IL-6 and MMP-9 in particular being associated with a worse prognosis and more aggressive tumor behavior [55–57].

Excessive expression of cytokines and other inflammatory agents disrupts numerous molecular pathways, such as those regulating cell growth and metabolism. For example, periodontal disease stemming from alterations in the oral microbiota can significantly impact receptor for advanced glycation end products (RAGE) protein expression, potentially leading to cancer [58]. Additionally, gram-negative bacteria secrete a lipopolysaccharide (LPS) endotoxin that stimulates the production of TNF-α, IL-1β, and IL-6 by binding to leukocytes' TLR receptors [59, 60]. Specifically, these proinflammatory cytokines cause additional proinflammatory proteins to be overexpressed, which in turn triggers the release of acute-phase proteins, prostaglandins (PG), and phospholipase A2 [61, 62]. According to studies, increased IL-1 facilitates the angiogenetic milieu, which aids in the tumor's metastasis [63–65]. Matrix metalloproteinase-9 (MMP-9) is concurrently induced by IL-1 [66–68].

Further research showed that the overexpression of IL-6 facilitates tumor spread by increasing levels of adhesion molecules, matrix-metalloproteinases, and adhesion molecules of endothelial leukocytes [69]. These findings underscore the unique role of interleukins, particularly IL-6, in neoplastic transformation [70]. In addition to interleukins, changes in TNF-α levels, influenced by alterations in the Wnt and NF-κB pathways, have been linked to tumor development [71, 72]. These results emphasize the crucial role of NF-κB in cancer. Although NF-κB acts as an immune-stimulating factor against cancer cells, its elevated protein levels could also contribute to cancer development by functioning as an oncogene in certain malignancies [73, 74]. There has been evidence to suggest that dysbiosis-associated cancer could be dependent on aberrant NF-κB activation [75]. Consequently, the immune system also plays a crucial role as it triggers the synthesis of NF-κB when pathogens are present [76].

The activation of certain inflammatory signaling pathways, such as the Wnt, JAK-STAT3 cascades, and NF-κB, by these cytokines is associated with carcinogenesis from a genetic perspective. This partially explains how the microbiota influences cancer development. Following infection with *P. gingivalis*, microarray data analysis revealed changes in key hub genes, including IL-6, signal transducer and activator of transcription 1 and C-X-C motif ligand. The majority of these alterations may serve as upstream regulators in cancer development [77]. Microbiome-related immunosuppression plays a key role in the development of oral cancer by altering immune cell function. *Porphyromonas gingivalis* interferes with CD8+ T-cell cytotoxicity by increasing PD-L1 expression in dendritic cells through the Akt-STAT3 pathway [78]. Through the GalNAc–Autophagy–TBC1D5–GLUT1 signaling axis, *F. nucleatum* causes cancer cells to create more lactate, which in turn promotes the development of M2-like tumor-associated macrophages and aids in immune evasion [79].

11.4.2 Production of Carcinogens

A variety of compounds produced by bacteria have been suggested to possess carcinogenic properties. For instance, sulfur compounds, acids, and reactive species such as nitric and oxygen species are generated through bacterial metabolism and may contribute to genetic damage that supports tumor development. Moreover, microorganisms involved in alcohol metabolism produce acetaldehyde a compound known to facilitate the neoplastic transformation [60].

It is widely recognized that alterations in nicotinamide adenine dinucleotide phosphate (NADPH) oxidase and nitric oxide synthase (NOS) activities result in the buildup of reactive oxygen species (ROS) and reactive nitrogen species (RNS). These molecules contribute to chronic inflammation, which may eventually lead to the onset of cancer [80, 81]. Additionally, hydrogen peroxide (H_2O_2) is generated by oral bacteria such as *Streptococcus gordonii*, *S. mitis*, *S. oligofermentans*, *S. oralis*, *S. sanguinis*, *Bifidobacterium adolescentis*, *Lactobacillus acidophilus*, *L. fermentum*, *L. jensenii*, and *L. minutus*, which are known for their peroxygenase activity [60, 82]. Other carcinogens produced by the oral bacteria include sulfides and nitrosamines. Bacteria, such as Firmicutes and Bacteroides species, can ferment an excess of protein in the host to produce nitrosamines and sulfides. These dangerous chemicals have the ability to damage DNA in onco-suppressor genes or oncogenes [83, 84].

Moreover, oral microbes like *Bifidobacterium*, *Lactobacillus*, *Lactococcus*, *Leuconostoc*, *Pediococcus*, *Peptostreptococcus stomatis*, and *Streptococcus* generate various acids such as isobutyric, isovaleric, isocaproic, lactic acid, acetic acid, and butyric acid. These compounds reduce the pH level of their local environment within the oral cavity [85–87]. These acids help to create the perfect tissue milieu that encourages the growth of cancer cells and their metastasis [88, 89]. Studies analyzing the microbiomes in cancerous tissues and normal mucosa of oral cancer patients have emphasized the crucial role and expression of superoxide dismutase (SOD) activity [90]. Inhibiting the harmful effects of the superoxide ion (O_2^-) requires the activity of SOD specifically. It has been proved that the presence of Fe^2+ in tumor samples reacts with H_2O_2 to produce hazardous reactive species that drive neoplastic transformation through DNA mutations that impact important genes involved in cell cycle regulation [91].

Research has found that tryptophanase activity in the microbiome may be a carcinogenetic mechanism in cancer patients, where the tumor is in the normal or tumorous site. Specifically, there appears to be a connection between the development of cancer and the increased conversion of *L*-tryptophan to secondary metabolites (pyruvate, ammonium, and indole). In this context, the role of the aryl hydrocarbon receptor (AHR) is essential [92]. Similar to this, patients with oral cancer exhibit elevated levels of other enzymes like glutamate dehydrogenase (GDH), which could contribute to alterations in the cellular redox state [93, 94]. Several oral bacteria, including *S. gordonii*, *S. oralis*, *S. salivarius*, *S. mitis*, and *S. sanguinis* [95], and yeasts (*Candida*) [96] have been implicated in the metabolism of alcohol into acetaldehyde, which has the potential to cause cancer [76, 96]. The production of acetaldehyde by these bacteria makes them all potentially harmful for the development of OSCC [97, 98]. A number of oral bacteria, such as *F. nucleatum*, *P. gingivalis*, *Aggregatibacter actinomycetemcomitans*, *Prevotella*, and *Intermedia*, are known to produce volatile sulfur compounds (VSCs), which include methyl mercaptan (CH_3SH), dimethyl sulfide $(CH_3)_2S$, and hydrogen sulfide (H_2S). The air within the mouth has the highest concentration of H_2S, whereas CH_3SH is the main compound in the gingival pockets [99, 100]. VSCs are harmful to tissues even at low concentrations and are involved in the development of chronic inflammation and diseases like periodontitis [101]. As a recognized genotoxic agent, H_2S can cause accumulating mutations or genomic instability. Certain H_2S-producing enzymes have been shown to overexpress in case of cancer cells, especially in ovarian and colon malignancies. Increased H_2S generation from overexpression of cystathionine-β-synthase affects tumor development and spread by triggering invasive signaling pathways, migration, and proliferation, in addition to this promoting tumor angiogenesis [102]. It has also been discovered that H_2S has dichotomous effects (stimulatory and inhibitory) on a number of gastrointestinal processes, including apoptosis, cancer, and inflammation [103]. While pure ethanol does not cause cancer, acetylcholine (ACH) hinders normal replication of DNA and forms DNA protein adducts that cause point mutations and chromosomal abnormalities. In addition to the breakdown of ingested alcohol by oral microbiota and mucosal alcohol dehydrogenase (ADH) and the alcoholic beverage itself playing a role in malignancy, Smedra presented an alternative theory that ACH can result from microbial dysbiosis in the oral cavity [104].

11.4.3 Microbiota-Induced Epigenetic Modulation

Interestingly, the environmental elements, such as lifestyle and diet choices and natural compounds, have been made clear to have a beneficial effect on both genetic and epigenetic changes, as well as inhibiting the onset of chronic degenerative diseases that predispose people to the emergence of numerous illnesses [105–107]. Specifically, the intake and consumption of food impact numerous cellular and molecular functions, influencing them in various manners [108–110]. Consuming particular foods and nutrients can alter inflammatory and regulatory pathways of the cell cycle to preserve ideal cell homeostasis and delay the commencement of disease [111, 112]. Conversely, aberrant food intake and/or absorption cause epigenetic alterations linked to several diseases, including cancer. However, it is not yet clear that the consumption and intake of nutrients play a role in the development of disease. Research has demonstrated how a person's choice of diet could affect their redox state, oral and gut microbes, status of DNA methylation, and the expression levels of microRNAs (miRNAs) [113–116].

Specifically, the current research indicates that epigenetic changes play a crucial role in the process of neoplastic transformation as well as in various other diseases [58, 117]. In this regard, numerous miRNAs, which are a type of non-coding RNA consisting of 20–22 nucleotides, have been associated with the initiation and progression of various cancers [118–124]. Furthermore, an abundance of reports has indicated the existence of a reciprocal link between host miRNAs and the human microbiotas [125–127]. Liu and colleagues found that fecal miRNAs from epithelial cells and Hopx-positive cells could infiltrate bacteria such as *F. nucleatum* and *Escherichia coli*. This interaction altered the makeup of the microbiota and influenced bacterial cell growth [128]. It is concluded by animal-based experiments that the fecal miRNAs come from sources other than intestinal cells. Numerous investigations demonstrate the fecal miRNAs could be obtained from diets and taken in by intestinal epithelia where they play a role in alteration of host gene expression levels [129, 130]. However, studies have also proved that milk-derived miRNAs played important roles; the majority of these miRNAs were seeded exosome-derived miRNAs [131–133]. It is possible for all these food-derived miRNAs to interact with gut and oral bacteria [134, 135]. Accordingly, exogenous miRNAs could function as tiny RNAs produced by bacteria to disrupt bacterial gene expression and affect the microbiome as a whole [136, 137]. Microbiota present in the mouth mucosa and intestines could alter the few specific miRNA expressions demonstrating the reciprocal link and mutual influence of miRNAs and microbiota [138]. The synthesis of various compounds that cause major alterations in host-cell metabolism and consequent modifications to gene and miRNA expressions is one of the methods by which the microbiota alters the host epithelial miRNAs' levels of expression [139–141].

11.4.4 Promoting Cell Proliferation and Antiapoptotic Activity

Through their antiapoptotic properties and promotion of cell growth, oral bacteria aid in the expansion of cancer. In human malignancies, cell cycle disruption is one of the major prevalent characteristics. *Porphyromonas gingivalis* promotes the development of OSCC by quickening the G1 stage of the cell cycle by up-regulating the production of cyclin D1 through the miR-21/PDCD4/AP-1 signaling pathway [142]. The intact film of *P. gingivalis*'s fimbrial structure and gingipain proteases stimulate the proliferation and death of cancer cells [143, 144].

In addition, *P. gingivalis* possesses the capacity to stop the apoptosis of gingival epithelial cells that is brought about by ATP-induced purinergic receptor P2X7 ligation. This receptor is essential for facilitating cell proliferation, neovascularization, metastasis, and the release of inflammatory cytokines. These bacteria have proven their abilities to release nucleoside diphosphate kinase (NDK), an antiapoptotic enzyme that modifies ATP/P2X7-signaling by cleaving ATP and inhibiting the pro-apoptotic P2X7 receptor's activation [145]. Additionally, *P. gingivalis* NDK secretion could control ROS triggered by ATP in the cytosol and mitochondria, in addition to the antioxidant glutathione response produced by the P2X7/NADPH-oxidase interactome [146]. Downregulating *p*27, which is linked to aberrant cell proliferation, in the presence of *F. nucleatum* resulted in cell cycle arrest in the S phase and promotes abnormal cell proliferation. In the interim, there was a down-regulation of the DNA repair proteins Ku70 and *p*53, which impaired their capacity to repair cells [147]. High putrescine levels are also linked to *F. nucleatum* and could cause cancer cells to proliferate malignantly [148]. *Staphylococcus aureus* could promote the growth of human oral keratinocytes by triggering the COX-2/PGE2 pathway [149].

Apoptosis, or programmed cell death, is how the body gets rid of necrotic cells from the body. This process is affected during pathological conditions and causes tumors to grow. Nucleoside-diphosphate-kinase released by *P. gingivalis* not only inhibits ATP-dependent apoptosis and reduces IL-1β production but it also phosphorylates heat-shock protein-27, thereby preventing epithelial cell death [150]. *Porphyromonas gingivalis* decreases the intrinsic mitochondrial apoptotic activity by phosphorylating the pro-apoptotic protein Bad and blocking the production of cytochrome-C via the JAK1/AKT/STAT3 pathway [151, 152]. In a study on *F. nucleatum* infection, there was also a notable rise in the expression of MYC, JAK1, and STAT3 [153]. *Treponema denticola* had the ability to directly induce cell death, control the cell cycle, and stimulate cell growth in CAL-27 cells [154]. It was also discovered that HPV E7 inhibits proptosis by enlisting TREM21 to ubiquitinate and degrade the IFI16 inflammasome [155].

The lipopolysaccharides (LPS) of *F. nucleatum* induce the production of inflammatory cytokines, such as TNF-α, IL-6, and IL-1β. The ongoing inflammatory process leads to tissue damage and the loss of periodontal attachment [156]. Multiple antiapoptotic pathways are modulated by infection with *F. nucleatum*; when bacteria activate TLRs, they trigger NF-κB signaling [157]. *Fusobacterium nucleatum* and cancer were directly related, and one important factor in this link is the fusobacterial adhesin/invasin FadA, which attaches to E-cadherin on cancer cells and triggers β-catenin signaling. This route led to increased wnt transcriptional activity, proinflammatory cytokine activation, oncogene activation, and encouragement of cancer cell proliferation [158]. One of *F. nucleatum's* primary virulence factors, FadA, modified the infiltration of macrophages and the methylation of the promoter of the cyclin-dependent kinase inhibitor 2A (CDKN2A) in cancer lesions [159]. Additionally, the LPS from *F. nucleatum* might activate β-catenin signaling, leading to an up-regulation of β-catenin expression and the oncogenes C-*myc* and *cyclin* D1 [160, 161]. Furthermore, *F. nucleatum* activated p38, which caused MMP-9 and MMP-13 to be secreted. These proteins are crucial for the invasion and spread of cancer cells [162].

11.4.5 Alterations in the Integrity of the Epithelial Barrier

Researchers described the connection between dysbiosis and carcinogenesis, as well as how it affects immune activation and homeostasis through altered epithelial barriers [47]. Dysfunction of the epithelium barrier and changes to the microenvironment might result from changes in anatomic structure, microbial composition, or mucus production [163]. The ensuing imbalance between epithelia and microbiota is a significant factor in infections as well as other microbial disorders, like malignancies [164, 165].

As previously indicated, changes in microbiological conditions that promote inflammation play a key role in the modification in the epithelial barrier. Cancer and microbiota are related to inflammatory reactions [166]. Furthermore, inflammation has the ability to alter the population of bacteria, promote microbial translocation, and stimulate the growth of particular bacteria [163, 167]. Certain microbial metabolites such as ROS and hydroxyl radicals, as well as toxins, e.g., cyto-lethal distending toxin (CDT), caused chromosomal damage that made epithelial cells become cancerous. Furthermore, bacteria use virulence genes, such as the AvrA virulence factor, to activate various types of signal transduction pathways. The transforming growth factor-β (TGF-β) is another factor that is induced by bacterial proteins and bacteriocins, causing aberrant cell proliferation [47, 168]. Moreover, TGF-β functions as an immunomodulatory element that suppresses dendritic cells (DCs); as a result, T-cells receptors function as factors that encourage tumor growth [169]. Remarkably, the dysregulated inflammatory microenvironment triggered by microbiota is linked to the process of TGF-β signaling in promoting cancer [47].

11.4.6 Role of Xenobiotic Metabolism in Malignancies

Alcohol consumption and tobacco usage are the two primary risk factors for oral cancer. It has been proved that chewing betel nuts with or without tobacco is independently linked to the emergence of OSCC. Although acetaldehyde, the by-product of ethanol metabolism, is highly carcinogenic, alcohol itself is not. Alcohol, betel nuts, and tobacco can change the oral microbiota. The microbiome of the oral mucosa contributes to the synthesis of the genotoxic metabolite acetaldehyde, which damages DNA by oxidizing ethanol and creating DNA adducts in oral epithelial cells. The oral microbiota is altered by chronic smoking, producing more ethanol-derived acetaldehyde. The oral microbiota may work in concert with basic risk factors like alcohol and tobacco use as a synergistic risk factor [170].

In betel nut chewers leukoplakia, erythroplakia, and oral submucous fibrosis are precursor lesions of oral cancer. Oral submucous fibrosis development has been linked to the microbiome. When used consistently, tannic acid, a component of betel nuts, prevents commensal bacteria from growing. There are also oscillations in the abundance of common taxa in the oral microbiome caused by cigarette smoking [171]. Alcohol use along with chewing betel nut may change the profiles of the oral microbiota. According to reports, there was a substantial difference in bacterial diversity between individuals who chewed betel nuts excessively and those who did not [172, 173]. Alcohol dehydrogenase, which is present in *S. salivarius*, *S. intermedius*, *S. mitis*, and nonpathogenic *Neisseria* subspecies, is in charge of converting ethanol into carcinogenic substances like acetaldehyde and hydroxyl ethyl radicals [174–176].

A reduced risk of head and neck squamous cell carcinomas (HNSCCs) was demonstrated by the overrepresentation of the commensal bacterial genera *Corynebacterium* and *Kingella*. The xenobiotic biodegradation and metabolic pathways that are able to metabolize many toxicants present in cigarette smoke are functionally connected to *Corynebacterium* and *Kingella*. Good dental health has been linked to higher abundances of *Neisseria*, *Abiotrophia*, *Actinomyces*, *Veillonella*, and *Capnocytophaga*. HNSCC was inversely correlated with *Prevotella nanceiensis*, *Capnocytophaga leadbetteri*, and *Selenomonas sputigena*. *Aggregatibacter actinomycetemcomitans*, *S. mutans*, *P. gingivalis*, and *T. forsythia* were all positively correlated with HNSCC risk. A lower risk of laryngeal cancer was linked to *Kingella dentificans* and *Streptococcus sanguinis*; a lower risk of pharynx cancer was linked to *Actinomyces oris* and *V. denticariosi*; and a reduced risk of mouth cancer was linked to *P. micra* and *Neisseria sicca* [11]. The pathways and microorganisms involved in a certain form of cancer are summarized in Table 11.2.

TABLE 11.2
Primary Mechanisms and Oral Bacteria Implicated in Cancer

S. No.	Mechanisms	Oral Bacteria Involved	Types of Cancers	References
1	Chronic inflammation	*Porphyromonas gingivalis*, *Fusobacterium nucleatum*, and *Prevotella* spp.	Oral squamous cell carcinoma, esophageal and oropharyngeal cancers	[44, 78, 143]
2	Production of carcinogens	*Streptococcus* spp., *Bifidobacterium adolescentis*, *Lactobacillus* spp., *Lactococcus* spp., *Leuconostoc* spp., *Pediococcus* spp., *Peptostreptococcus stomatis*	Oral squamous cell carcinoma, lung cancer, pancreatic cancer, ovarian and colorectal malignancies, etc.	[85, 86]
3	Microbiota-induced epigenetic modulation	*Fusobacterium nucleatum*, *Porphyromonas gingivalis*, and gut microbiota	OSCC, colon and pancreatic cancers, esophageal and oropharyngeal cancers, etc.	[113, 128]
4	Promoting cell proliferation and antiapoptotic activity	*Fusobacterium nucleatum*, *Porphyromonas gingivalis*, and *Streptococcus* spp.	Oral squamous cell carcinoma, esophageal and oropharyngeal cancers, and colorectal cancers	[148, 149]
5	Modification of epithelial barrier integrity	Change in the composition of *Porphyromonas* spp. And *Fusobacterium* spp.	Head and neck cancers, lung cancer, pancreatic cancer, etc.	[47, 165]
6	Role of xenobiotic metabolism in malignancies	*Streptococcus* spp., *Corynebacterium* spp., *Kingella* spp., *Neisseria* spp., *Actinomyecete* spp., and *Porphyromonas selemonas*	OSCC, HNSCC, laryngeal cancer, and pharynx cancer	[11]

11.5 ORAL MICROBIOME MALIGNANCIES: MANAGEMENT, PROGNOSIS, AND DIAGNOSIS

Tumors or malignant growths in the lips, throat, or mouth are examples of oral cancer. The most varied microbiota in the human body, the oral microbiota, has been involved in the progression of several other cancers [19]. The human microbiota has the ability to influence the response of cancer therapy through several means, which could have a favorable or negative effect on the result [177]. Although each person's prognosis for oral cancer is unique, it generally helps to identify how the illness is progressing and the chance of survival. Reduced long-term prognosis and higher local recurrence are the consequences of insufficient tumor clearance. A prognostic factor is an attribute of the cancer or a characteristic of the patient (like smoking habit), which the physician will consider making a prediction. It is believed that the patient, the surgeon, and the type of tumor all have an influence on the patient's chance of survival. Determining how these prognostic indicators combine and impact the prognosis of the patient is critical. The prognostic indicators would be helpful in precisely determining the biological aggressiveness of these malignancies and would allow customized treatment regimens for individual tumors [178].

11.5.1 Impact of Various Oral Microbes in Cancer Prognosis

Several investigations have established a correlation between certain oral bacteria and the prognosis of cancer [8]. For instance, it has been proven that *Fusobacterium* species are positively associated with a bad prognosis in cases of colorectal, pancreatic, and OSCC [179]. It was discovered in some studies that while *Malassezia* was more prevalent in patients with a good prognosis for OSCC, increased salivary carriage of the genus *Candida* was linked to a bad prognosis [180]. In the prognosis for nasopharyngeal cancer, reduced within-community diversity of the oral microbiome was linked to greater mortality, particularly in cases involving elderly patients [181]. *Bacteroides fragilis* and *F. nucleatum* were more prevalent in groups with poorer prognoses [182]. Patients who have less variety in their oral microbiome or who have positive *P. gingivalis* or high levels of *Fusobacterium* in their malignant tissue are significantly more probable to possess poor overall survival and disease-specific survival [178]. Studies have revealed that salivary flora in patients with esophageal squamous cell carcinoma are less diverse, with an increase in relative abundance of *Prevotella*, *Streptococcus*, and *Porphyromonas*. Furthermore, *P. gingivalis* has been found in esophageal squamous cell carcinoma lesions up to 61% of the time, although it is only found in nearby areas at 12% [183]. Therefore, the variety of oral microbiomes and specific flora is linked to tumor prognosis, and these biomarkers show promise for assessing tumor prognosis due to their ease of detection and assessment.

11.5.2 Prognosis of Most Prominent Oral Malignancy (OSCC)

Of the various kinds of malignancies, OSCC has a poor prognosis, showing a 59.2% disease-specific survival (DSS) rate and a 51.2% overall survival (OS) rate [184]. Numerous institutes have conducted considerable research on the survival rate of OSCC, and the variable rates of 47%–71% are revealed by the 5-year cumulative OS rate among patients who had surgery [185]. The most well-known assessable important variables linked to OSCC survival include tumor thickness, depth of invasion, and neck metastasis. The American Joint Committee on Cancer (AJCC) developed the pTNM staging system, which is a comprehensive prognostic marker that considers these crucial elements: primary tumor (pT), involvement of locoregional lymph nodes (N), and distance metastasis (M) [186]. There is a 50% decrease in patient survival for individuals with neck metastases, which is widely recognized as the most important prognostic marker in OSCC. Furthermore, prior research has demonstrated the impact of various other variables on patient survival, including drinking habits, smoking, locations, age, sex, postoperative radiation, cell differentiation, perineural invasion, lympho-vascular invasion, neck dissection, and recurrence [187]. An intriguing discovery, which needs confirmation through additional investigation, indicates that lesions of OSCC linked to a high concentration of *F. nucleatum* had a better prognosis when treated with medications like pembrolizumab [188]. To improve the patient's chances of survival, perioperatively expressing these data calls for careful monitoring and intensive treatment.

11.5.3 Modulation of the Oral Microbiome for Improved Cancer Prognosis

The microbiome has been linked to influencing the toxicity and effectiveness of cancer treatments, such as immunotherapy, chemotherapy, and radiotherapy. Patients with head and neck cancers are more likely to experience side effects from anticancer treatments, particularly from high-dose radiation. It appears that the radiation dosage administered to the oral mucosa has a critical role in the development of oral mucositis, adversely affecting these patients' quality of life and nutrition. Finding alterations in the oral microbiota may make it possible to modify it in a eubiotic way, which would alleviate that symptom. However, more complex problems, like osteoradionecrosis, can arise in the oral cavity and may take a long time to manifest following therapy. As a result, it is imperative to provide proper dental care before, during, and after the management phase. It may make it possible to lessen the frequency and importance of oral problems related to treatment. Furthermore, the intricate course of treatment that includes giving these individuals probiotics appears to be working well for them [189]. It has been demonstrated that certain probiotics, which contain live strains of bacteria from genera including *Bifidobacterium*, *Lactobacillus*, and *Streptococcus*, enhance alpha diversity and improve dental health [190]. A number of strategies for keeping the oral microbiome in good health have been suggested. Figure 11.3 basically showcases potential interventions aimed at restoring microbial balance, such as probiotics, prebiotics, antibiotics, and lifestyle modifications, which may help in reducing cancer risk or improving treatment outcomes. It underscores the importance of understanding and manipulating the oral microbiome as a novel approach in cancer management and prevention strategies.

Personalized therapies can be guided by an individual's oral microbiota analysis. Drugs that specifically target dangerous germs while preserving good ones are being investigated by researchers. Fecal Microbiota Transplantation (FMT) is used to develop microbiota-targeted therapies. Despite being mainly utilized for gut health, FMT may have an effect on the mouth microbiota. This field is still being researched [191]. Furthermore, preclinical evidence indicates that altering the microbiota may develop into a unique tactic for enhancing the effectiveness of immune-based cancer treatments [192].

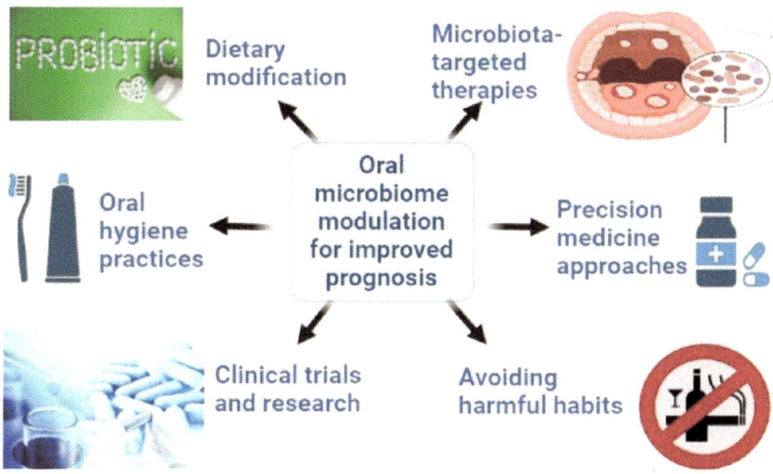

FIGURE 11.3 Modulation of oral microbiome for management of cancer outcomes.

11.5.4 Biomarkers Used for Various Malignancies

Biomarkers are intended to serve a variety of purposes, such as diagnosis, screening, prognostication, and therapy choice guidance. Furthermore, the confirmation of these biomarkers would make it easier to find novel treatment targets, which could lower the incidence of oral cancer globally by utilizing customized therapy [193]. Despite advances in oral oncology and therapy research, there have been no significant changes in OSCC survival rates worldwide in recent years. This is primarily owing to the diagnosis of OSCC in advanced stages. In this regard, the identification of biomarkers for oral microbial malignant transformation might offer a chance to identify high-risk individuals and, as a result, contribute to the global decrease in the incidence of oral cancer [194]. To help clinicians stratify high-risk patients or lesions, decide between active therapies and clinical follow-up, and minimize needless invasive procedures, a strong malignant transformation prediction biomarker should be able to guide clinical procedures [195].

The immunohistochemical biomarker that is most commonly reported was P53, a consequence of the tumor suppressor gene *TP53*. P53 is assumed to defend the genome and is essential for DNA repair, apoptosis, and cell cycle regulation [196]. Other biomarkers that can be used includes Ki-67, podoplanin, p16, hepatocyte growth factor receptor (HGFR/Met), aldehyde dehydrogenase 1 (ALDH1), minichromosome maintenance protein 2 (MCM2), cyclooxygenase-2 (COX-2), SMAD4, carbonic anhydrase 9 (CA9), human antigen R (HuR), insulin-like growth factor II mRNA-binding protein (IMP3), β-catenin, *c*-Jun, epidermal growth factor receptor (EGFR), ΔNp63, and cyclin D1 (CD1) [197].

It is essential to identify sensitive and readily detectable tumor markers because conventional treatments, also known as adjuvant therapy, have poor efficacy. This makes the relationship between the oral microbiota and tumors a potentially fruitful path for the discovery of novel biomarkers and treatment targets for a variety of cancer types. Even though there are numerous markers used to track the development and prognosis of tumors, further research is necessary to fully understand the significance of the oral microbiota in this situation [178]. Oral cancer needs more investigation as the majority of current research on the microbial impacts of cancer therapy focuses on digestive tract tumors. A growing body of research has demonstrated that bacteria and anticancer medicines are closely related, even though the exact mechanism by which these microbes influence the effectiveness of anticancer therapy is still unknown. Microbiota can be used as a target for treating OSCC as well as an adjuvant to improve therapeutic outcomes and reduce side effects [198].

11.6 CONCLUSION

Advancements in technology and interdisciplinary scientific approaches have made the study of microorganisms and cancer a widely discussed subject. A systematic assessment of the relationship between oral microbiota and the world's most frequent malignancies and several mechanisms by which oral bacteria cause these malignancies has been suggested. Numerous oral microbiota members have been linked to the onset and spread of cancer, both in the oral cavity and in other distant anatomical regions. Specifically, *F. nucleatum* and *P. gingivalis* carriage is linked to the onset of malignancy. Through systematic investigations, it has become increasingly obvious that the composition and activity of oral microorganisms may impact the onset and course of different cancers. The intricate interplay between microbial communities and host factors within the oral cavity might contribute to carcinogenesis through mechanisms such as inflammation, immune modulation, and the production of carcinogenic metabolites. Additionally, emerging evidence suggested that oral dysbiosis, characterized by imbalances in microbial populations, might serve as a potential biomarker for cancer risk assessment and prognosis. The microbiome has been linked to mediating resistance to anticancer therapy in addition to promoting the onset and spread of cancer. However, more research is necessary to elucidate the specific mechanisms underlying these associations and to explore the therapeutic propositions of targeting the oral microbiome in cancer prevention and treatment strategies. Overall, understanding the intricate relationship between the oral microbiome and malignancies holds great promise for advancing our knowledge of cancer etiology and improving clinical management approaches. In conclusion, this chapter elucidates the role of the oral microbiome in cancer development and progression, offering insights into personalized approaches for cancer management and improving patient outcomes.

REFERENCES

1. Nasiri, K.; Amiri Moghaddam, M.; Etajuri, E. A.; Badkoobeh, A.; Tavakol, O.; Rafinejad, M.; Mirhosseii, A. F.; Fathi, A. Periodontitis and progression of gastrointestinal cancer: Current knowledge and future perspective. *Clinical and Translational Oncology* **2023**, *25*(10), 2801–2811.
2. Yusuf, K.; Sampath, V.; Umar, S. Bacterial infections and cancer: Exploring this association and its implications for cancer patients. *International Journal of Molecular Sciences* **2023**, *24*(4), 3110.
3. Siegel, R. L.; Miller, K. D.; Fuchs, H. E.; Jemal, A. Cancer statistics, 2021. *CA: A Cancer Journal for Clinicians* **2021**, *71*(1), 7–33.
4. Marino, P.; Mininni, M.; Deiana, G.; Marino, G.; Divella, R.; Bochicchio, I.; Giuliano, A.; Lapadula, S.; Lettini, A. R.; Sanseverino, F. Healthy lifestyle and cancer risk: Modifiable risk factors to prevent cancer. *Nutrients* **2024**, *16*(6), 800.
5. van Elsland, D.; Neefjes, J. Bacterial infections and cancer. *EMBO Reports* **2018**, *19*(11), e46632.
6. Teles, F. R. F.; Alawi, F.; Castilho, R. M.; Wang, Y. Association or causation? Exploring the oral microbiome and cancer links. *Journal of Dental Research* **2020**, *99*(13), 1411–1424.
7. Irfan, M.; Delgado, R. Z. R.; Frias-Lopez, J. The oral microbiome and cancer. *Frontiers in Immunology* **2020**, *11*, 591088.
8. Sun, J.; Tang, Q.; Yu, S.; Xie, M.; Xie, Y.; Chen, G.; Chen, L. Role of the oral microbiota in cancer evolution and progression. *Cancer Medicine* **2020**, *9*(17), 6306–6321.
9. Gormley, M.; Creaney, G.; Schache, A.; Ingarfield, K.; Conway, D. I. Reviewing the epidemiology of head and neck cancer: Definitions, trends and risk factors. *British Dental Journal* **2022**, *233*(9), 780–786.
10. Lim, Y.; Totsika, M.; Morrison, M.; Punyadeera, C. Oral microbiome: A new biomarker reservoir for oral and oropharyngeal cancers. *Theranostics* **2017**, *7*(17), 4313.
11. Hayes, R. B.; Ahn, J.; Fan, X.; Peters, B. A.; Ma, Y.; Yang, L.; Agalliu, I.; Burk, R. D.; Galy, I.; Purdue, M. P.; Freedman, N. D.; Gapstur, S. M.; Pei, Z. Association of oral microbiome with risk for incident head and neck squamous cell cancer. *JAMA Oncology* **2018**, *4*(3), 358–365.
12. Wang, L.; Ganly, I. The oral microbiome and oral cancer. *Clinics in Laboratory Medicine* **2014**, *34*(4), 711–719.
13. Ali, K. Oral cancer-the fight must go on against all odds. *Evidence-Based Dentistry* **2022**, *23*, 4–5.
14. Li, Z.; Fu, R.; Wen, X.; Wang, Q.; Huang, X.; Zhang, L. The significant clinical correlation of the intratumor oral microbiome in oral squamous cell carcinoma based on tissue-derived sequencing. *Frontiers in Physiology* **2023**, *13*, 1089539.
15. Murugan, K.; Padmavathy, K.; Aswath, N.; Mallika, L. Taciturn role of oral microbiome in oral squamous cell carcinoma: The hidden truth–a systematic review. *Journal of Indian Academy of Oral Medicine and Radiology* **2023**, *35*(4), 599–604.
16. Zhou, J.; Shrestha, P.; Qiu, Z.; Harman, D. G.; Teoh, W. C.; Al-Sohaily, S.; Liem, H.; Turner, I.; Ho, V. Distinct microbiota dysbiosis in patients with non-erosive reflux disease and esophageal adenocarcinoma. *Journal of Clinical Medicine* **2020**, *9*(7), 2162.
17. Yan, X.; Yang, M.; Liu, J.; Gao, R.; Hu, J.; Li, J.; Zhang, L.; Shi, Y.; Guo, H.; Cheng, J.; Razi, M.; Pang, S.; Yu, X.; Hu, S. Discovery and validation of potential bacterial biomarkers for lung cancer. *American Journal of Cancer Research* **2015**, *5*(10), 3111.
18. Lan, Z.; Liu, W. J.; Cui, H.; Zou, K. L.; Chen, H.; Zhao, Y. Y.; Yu, G. T. The role of oral microbiota in cancer. *Frontiers in Microbiology* **2023**, *14*, 1253025.
19. Tuominen, H.; Rautava, J. Oral microbiota and cancer development. *Pathobiology* **2021**, *88*(2), 116–126.
20. Fan, X.; Alekseyenko, A. V.; Wu, J.; Peters, B. A.; Jacobs, E. J.; Gapstur, S. M.; Purdue, M. P.; Abnet, C. C.; Stolzenberg-Solomon, R.; Miller, G.; Ravel, J.; Hayes, R. B.; Ahn, J. Human oral microbiome and prospective risk for pancreatic cancer: A population-based nested case-control study. *Gut* **2018**, *67*(1), 120–127.

21. Kudra, A.; Muszyński, D.; Sobocki, B. K.; Atzeni, A.; Carbone, L.; Kaźmierczak-Siedlecka, K.; Polom, K.; Kalinowski, L. Insights into oral microbiome and colorectal cancer–on the way of searching new perspectives. *Frontiers in Cellular and Infection Microbiology* **2023**, *13*, 1159822.
22. McIlvanna, E.; Linden, G. J.; Craig, S. G.; Lundy, F. T.; James, J. A. *Fusobacterium nucleatum* and oral cancer: A critical review. *BMC Cancer* **2021**, *21*, 1–11.
23. Peng, X.; Cheng, L.; You, Y.; Tang, C.; Ren, B.; Li, Y.; Xu, X.; Zhou, X. Oral microbiota in human systematic diseases. *International Journal of Oral Science* **2022**, *14*(1), 14.
24. Sampaio-Maia, B.; Caldas, I. M.; Pereira, M. L.; Pérez-Mongiovi, D.; Araujo, R. The oral microbiome in health and its implication in oral and systemic diseases. *Advances in Applied Microbiology* **2016**, *97*, 171–210.
25. Fakhoury, M.; Negrulj, R.; Mooranian, A.; Al-Salami, H. Inflammatory bowel disease: Clinical aspects and treatments. *Journal of Inflammation Research* **2014**, 113–120.
26. Read, E.; Curtis, M. A.; Neves, J. F. The role of oral bacteria in inflammatory bowel disease. *Nature Reviews Gastroenterology & Hepatology* **2021**, *18*(10), 731–742.
27. Lee, Y. C.; Liu, C. Y.; Lee, C. L.; Zhang, R. H.; Huang, C. J.; Yen, T. L. The periodontopathic pathogen, *Porphyromonas gingivalis*, involves a gut inflammatory response and exacerbates inflammatory bowel disease. *Pathogens* **2022**, *11*(1), 84.
28. Schuppan, D.; Afdhal, N. H. Liver cirrhosis. *The Lancet* **2008**, *371*(9615), 838–851.
29. Lei, Y.; Li, S.; He, M.; Ao, Z.; Wang, J.; Wu, Q.; Wang, Q. Oral pathogenic bacteria and the oral–gut–liver axis: A new understanding of chronic liver diseases. *Diagnostics* **2023**, *13*(21), 3324.
30. Parvaneh, M.; Witting, P. K.; Ku, J.; Moradi, T.; Eroglu, E.; Freedman, B.; Sutherland, G. T.; McCorkindale, A.; Guennewig, B.; Choowong, P.; Bell-Anderson, K.; Cooney, G.; Thomas, S. R.; Eberhard, J. Periodontitis induces endothelial dysfunction in mice. *Scientific Reports* **2021**, *11*(1), 14993.
31. Mohammed, L. I.; Zakaria, Z. Z.; Benslimane, F. M.; Al-Asmakh, M. Exploring the role of oral microbiome dysbiosis in cardiometabolic syndrome and smoking. *Experimental Lung Research* **2024**, *50*(1), 65–84.
32. Belibasakis, G. N.; Belstrøm, D.; Eick, S.; Gursoy, U. K.; Johansson, A.; Könönen, E. Periodontal microbiology and microbial etiology of periodontal diseases: Historical concepts and contemporary perspectives. *Periodontology 2000* **2023**.
33. Bobetsis, Y. A.; Graziani, F.; Gürsoy, M.; Madianos, P. N. Periodontal disease and adverse pregnancy outcomes. *Periodontology 2000* **2020**, *83*(1), 154–174.
34. de Aquino, S. G.; Abdollahi-Roodsaz, S.; Koenders, M. I.; van de Loo, F. A.; Pruijn, G. J.; Marijnissen, R. J.; Walgreen, B.; Helsen, M. M.; van den Bersselaar L. A.; de Molon, R. S.; Campos, M. J. A.; Cunha, F. Q.; Cirelli, J. A.; van den Berg, W. B. Periodontal pathogens directly promote autoimmune experimental arthritis by inducing a TLR2-and IL-1–driven Th17 response. *The Journal of Immunology* **2014**, *192*(9), 4103–4111.
35. Ebbers, M.; Lübcke, P. M.; Volzke, J.; Kriebel, K.; Hieke, C.; Engelmann, R.; Lang, H.; Kreikemeyer, B.; Müller-Hilke, B. Interplay between *P. gingivalis, F. nucleatum* and *A. actinomycetemcomitans* in murine alveolar bone loss, arthritis onset and progression. *Scientific Reports* **2018**, *8*(1), 15129.
36. Liu, S.; Butler, C. A.; Ayton, S.; Reynolds, E. C.; Dashper, S. G. *Porphyromonas gingivalis* and the pathogenesis of Alzheimer's disease. *Critical Reviews in Microbiology* **2023**, 1–11.
37. Matsha, T. E.; Prince, Y.; Davids, S.; Chikte, U.; Erasmus, R. T.; Kengne, A. P.; Davison, G. M. Oral microbiome signatures in diabetes mellitus and periodontal disease. *Journal of Dental Research* **2020**, *99*(6), 658–665.
38. Teles, F.; Wang, Y.; Hajishengallis, G.; Hasturk, H.; Marchesan, J. T. Impact of systemic factors in shaping the periodontal microbiome. *Periodontology 2000* **2021**, *85*(1), 126–160.
39. Cali, F.; Cantone, M.; Cosentino, F. I. I.; Lanza, G.; Ruggeri, G.; Chiavetta, V.; Ferri, R. Interpreting genetic variants: Hints from a family cluster of Parkinson's disease. *Journal of Parkinson's Disease* **2019**, *9*(1), 203–206.
40. Pennisi, M.; Lanza, G.; Cantone, M.; Schepis, C.; Ferri, R.; Barone, R.; Bella, R. Unusual neurological presentation of nevoid basal cell carcinoma syndrome (Gorlin-Goltz Syndrome). *Journal of Clinical Neurology (Seoul, Korea)* **2017**, *13*(4), 439.
41. La Rosa, G. R. M.; Gattuso, G.; Pedullà, E.; Rapisarda, E.; Nicolosi, D.; Salmeri, M. Association of oral dysbiosis with oral cancer development. *Oncology Letters* **2020**, *19*(4), 3045–3058.
42. Zhang, Y.; Wang, X.; Li, H.; Ni, C.; Du, Z.; Yan, F. Human oral microbiota and its modulation for oral health. *Biomedicine & Pharmacotherapy* **2018**, *99*, 883–893.

43. Guarneri, C.; Bevelacqua, V.; Polesel, J.; Falzone, L.; Cannavò, P. S.; Spandidos, D. A.; Libra, M. NF-κB inhibition is associated with OPN/MMP-9 downregulation in cutaneous melanoma. *Oncology Reports* **2017**, *37*(2), 737–746.
44. Leonardi, G. C.; Falzone, L.; Salemi, R.; Zanghì, A.; Spandidos, D. A.; Mccubrey, J. A.; Libra, M. Cutaneous melanoma: From pathogenesis to therapy. *International Journal of Oncology* **2018**, *52*(4), 1071–1080.
45. Dolcet, X.; Llobet, D.; Pallares, J.; Matias-Guiu, X. NF-kB in development and progression of human cancer. *Virchows Archiv* **2005**, *446*, 475–482.
46. Pang, X.; Tang, Y. J.; Ren, X. H.; Chen, Q. M.; Tang, Y. L.; Liang, X. H. Microbiota, epithelium, inflammation, and TGF-β signaling: An intricate interaction in oncogenesis. *Frontiers in Microbiology* **2018**, *9*, 367461.
47. Couturier-Maillard, A.; Secher, T.; Rehman, A.; Normand, S.; De Arcangelis, A.; Haesler, R.; Chamaillard, M. NOD2-mediated dysbiosis predisposes mice to transmissible colitis and colorectal cancer. *The Journal of Clinical Investigation* **2013**, *123*(2), 700–711.
48. Hu, B.; Elinav, E.; Huber, S.; Strowig, T.; Hao, L.; Hafemann, A.; Flavell, R. A. Microbiota-induced activation of epithelial IL-6 signaling links inflammasome-driven inflammation with transmissible cancer. *Proceedings of the National Academy of Sciences* **2013**, *110*(24), 9862–9867.
49. Francescone, R.; Hou, V.; Grivennikov, S. I. Microbiome, inflammation, and cancer. *The Cancer Journal* **2014**, *20*(3), 181–189.
50. Ojesina, A. I.; Lichtenstein, L.; Freeman, S. S.; Pedamallu, C. S.; Imaz-Rosshandler, I.; Pugh, T. J.; Meyerson, M. Landscape of genomic alterations in cervical carcinomas. *Nature* **2014**, *506*(7488), 371–375.
51. Allen, J.; Sears, C. L. Impact of the gut microbiome on the genome and epigenome of colon epithelial cells: Contributions to colorectal cancer development. *Genome Medicine* **2019**, *11*, 1–18.
52. Multhoff, G.; Molls, M.; Radons, J. Chronic inflammation in cancer development. *Frontiers in Immunology* **2012**, *2*, 17950.
53. Szkaradkiewicz, A. K.; Karpiński, T. M. Microbiology of chronic periodontitis. *Journal of Biology and Earth Sciences* **2013**, *3*, 14–20.
54. Leonardi, R.; Talic, N. F.; Loreto, C.; MMP-13 (collagenase 3) immunolocalisation during initial orthodontic tooth movement in rats. *Acta Histochemica* **2007**, *109*(3), 215–220.
55. Salemi, R.; Falzone, L.; Madonna, G.; Polesel, J.; Libra, M. MMP-9 as a candidate marker of response to BRAF inhibitors in melanoma patients with BRAFV600E mutation detected in circulating-free DNA. *Frontiers in Pharmacology* **2018**, *9*, 409395.
56. Silva, E. M.; Mariano, V. S.; Pastrez, P. R. A.; Pinto, M. C.; Castro, A. G.; Syrjanen, K. J.; Longatto-Filho, A. High systemic IL-6 is associated with worse prognosis in patients with non-small cell lung cancer. *PloS One* **2017**, *12*(7), e0181125.
57. Falzone, L.; Salemi, R.; Travali, S.; Scalisi, A.; McCubrey, J. A.; Candido, S.; Libra, M. MMP-9 overexpression is associated with intragenic hypermethylation of MMP9 gene in melanoma. *Aging (Albany NY)* **2016**, *8*(5), 933.
58. Katz, J.; Wallet, S.; Cha, S. Periodontal disease and the oral-systemic connection: "Is it all the RAGE?". *Quintessence International* **2010**, *41*(3), 229–237.
59. Zhang, G.; Ghosh, S. Molecular mechanisms of NF-κB activation induced by bacterial lipopolysaccharide through toll-like receptors. *Journal of Endotoxin Research* **2000**, *6*(6), 453–457.
60. Karpiński, T. M. Role of oral microbiota in cancer development. *Microorganisms* **2019**, *7*(1), 20.
61. Hou, L. T.; Liu, C. M.; Liu, B. Y.; Lin, S. J.; Liao, C. S.; Rossomando, E. F. Interleukin-1β, clinical parameters and matched cellular-histopathologic changes of biopsied gingival tissue from periodontitis patients. *Journal of Periodontal Research* **2003**, *38*(3), 247–254.
62. Konopka, Ł.; Brzezińska-Błaszczyk, E. Cytokines in gingival crevicular fluid as potential diagnostic and prognostic markers of periodontitis. *Dental and Medical Problems* **2010**, *47*(2), 206–213.
63. Carmi, Y.; Dotan, S.; Rider, P.; Kaplanov, I.; White, M. R.; Baron, R.; Voronov, E. The role of IL-1β in the early tumor cell–induced angiogenic response. *The Journal of Immunology* **2013**, *190*(7), 3500–3509.
64. Jin, L.; Yuan, R. Q.; Fuchs, A.; Yao, Y.; Joseph, A.; Schwall, R.; Rosen, E. M.; Expression of interleukin-1β in human breast carcinoma. *Cancer: Interdisciplinary International Journal of the American Cancer Society* **1997**, *80*(3), 421–434.
65. Voronov, E.; Shouval, D. S.; Krelin, Y.; Cagnano, E.; Benharroch, D.; Iwakura, Y.; Apte, R. N. IL-1 is required for tumor invasiveness and angiogenesis. *Proceedings of the National Academy of Sciences* **2003**, *100*(5), 2645–2650.

66. Pannone, G.; Santoro, A.; Feola, A.; Bufo, P.; Papagerakis, P.; Lo Muzio, L.; Di Domenico, M. The role of E-cadherin down-regulation in oral cancer: CDH1 gene expression and epigenetic blockage. *Current Cancer Drug Targets* **2014**, *14*(2), 115–127.
67. Wong, S. H. M.; Fang, C. M.; Chuah, L. H.; Leong, C. O.; Ngai, S. C. E-cadherin: Its dysregulation in carcinogenesis and clinical implications. *Critical Reviews in Oncology/Hematology* **2018**, *121*, 11–22.
68. Kossakowska, A. E.; Edwards, D. R.; Prusinkiewicz, C.; Zhang, M. C.; Guo, D.; Urbanski, S. J.; Janowska-Wieczorek, A. Interleukin-6 regulation of matrix metalloproteinase (MMP-2 and MMP-9) and tissue inhibitor of metalloproteinase (TIMP-1) expression in malignant non-Hodgkin's lymphomas. *Blood, The Journal of the American Society of Hematology* **1999**, *94*(6), 2080–2089.
69. Natali, P.; Nicotra, M. R.; Cavaliere, R.; Bigotti, A.; Romano, G.; Temponi, M.; Ferrone, S. Differential expression of intercellular adhesion molecule 1 in primary and metastatic melanoma lesions. *Cancer Research*, *50*(4) **1990**, 1271–1278.
70. Haura, E. B.; Turkson, J.; Jove, R. Mechanisms of disease: Insights into the emerging role of signal transducers and activators of transcription in cancer. *Nature Clinical Practice Oncology* **2005**, *2*(6), 315–324.
71. Szlosarek, P.; Charles, K. A.; Balkwill, F. R. Tumour necrosis factor-α as a tumour promoter. *European Journal of Cancer* **2006**, *42*(6), 745–750.
72. Rivas, M. A.; Carnevale, R. P.; Proietti, C. J.; Rosemblit, C.; Beguelin, W.; Salatino, M.; Schillaci, R. TNFα acting on TNFR1 promotes breast cancer growth *via* p42/P44 MAPK, JNK, Akt and NF-κB-dependent pathways. *Experimental Cell Research* **2008**, *314*(3), 509–529.
73. Karin, M. NF-κB as a critical link between inflammation and cancer. *Cold Spring Harbor Perspectives in Biology* **2009**, *1*(5), a000141.
74. Hoesel, B.; Schmid, J. A. The complexity of NF-κB signaling in inflammation and cancer. *Molecular Cancer* **2013**, *12*, 1–15.
75. Vogelmann, R.; Amieva, M. R. The role of bacterial pathogens in cancer. *Current Opinion in Microbiology* **2007**, *10*(1), 76–81.
76. Agassandian, M.; Shurin, G. V. Bacterial infections and cancer development. *Infection and cancer: Bi-directorial Interactions* **2015**, 49–74.
77. Geng, F.; Wang, Q.; Li, C.; Liu, J.; Zhang, D.; Zhang, S.; Pan, Y. Identification of potential candidate genes of oral cancer in response to chronic infection with *Porphyromonas gingivalis* using bioinformatical analyses. *Frontiers in Oncology* **2019**, *9*, 91.
78. Ren, J.; Han, X.; Lohner, H.; Hoyle, R. G.; Li, J.; Liang, S.; Wang, H. *P. gingivalis* infection upregulates PD-L1 expression on dendritic cells, suppresses CD8+ T-cell responses, and aggravates oral cancer. *Cancer Immunology Research* **2023**, *11*(3), 290–305.
79. Sun, J.; Tang, Q.; Yu, S.; Xie, M.; Zheng, W.; Chen, G.; Chen, L. *F. nucleatum* facilitates oral squamous cell carcinoma progression via GLUT1-driven lactate production. *EBioMedicine* **2023**, *88*.
80. Hussain, S. P.; Hofseth, L. J.; Harris, C. C. Radical causes of cancer. *Nature Reviews Cancer* **2003**, *3*(4), 276–285.
81. Piao, J. Y.; Lee, H. G.; Kim, S. J.; Kim, D. H.; Han, H. J.; Ngo, H. K. C.; Surh, Y. J. Helicobacter pylori activates IL-6-STAT3 signaling in human gastric cancer cells: potential roles for reactive oxygen species. *Helicobacter* **2016**, *21*(5), 405–416.
82. Brauncajs, M.; Sakowska, D.; Krzeminski, Z. Production of hydrogen peroxide by lactobacilli colonising the human oral cavity. *Medycyna Doswiadczalna I Mikrobiologia* **2001**, *53*(4), 331–336.
83. Carbonero, F.; Benefiel, A. C.; Alizadeh-Ghamsari, A. H.; Gaskins, H. R. Microbial pathways in colonic sulfur metabolism and links with health and disease. *Frontiers in Physiology* **2012**, *3*, 33851.
84. Bhatt, A. P.; Redinbo, M. R.; Bultman, S. J. The role of the microbiome in cancer development and therapy. *CA: A Cancer Journal for Clinicians* **2017**, *67*(4), 326–344.
85. Karpiński, T. M.; Szkaradkiewicz, A. K. Characteristic of bacteriocines and their application. *Polish Journal of Microbiology* **2013**, *62*(3), 223–35.
86. Senneby, A.; Davies, J. R.; Svensäter, G.; Neilands, J. Acid tolerance properties of dental biofilms in vivo. *BMC Microbiology* **2017**, *17*, 1–8.
87. Downes, J.; Wade, W. G. *Peptostreptococcus stomatis* sp. nov.; isolated from the human oral cavity. *International Journal of Systematic and Evolutionary Microbiology* **2006**, *56*(4), 751–754.
88. Lunt, S. J.; Chaudary, N.; Hill, R. P. The tumor microenvironment and metastatic disease. *Clinical & Experimental Metastasis* **2009**, *26*, 19–34.
89. Mazzio, E. A.; Smith, B.; Soliman, K. F. Evaluation of endogenous acidic metabolic products associated with carbohydrate metabolism in tumor cells. *Cell Biology and Toxicology* **2010**, *26*, 177–188.

90. Yost, S.; Stashenko, P.; Choi, Y.; Kukuruzinska, M.; Genco, C. A.; Salama, A.; Frias-Lopez, J. Increased virulence of the oral microbiome in oral squamous cell carcinoma revealed by metatranscriptome analyses. *International Journal of Oral Science* **2018**, *10*(4), 32.
91. Franco, R.; Schoneveld, O.; Georgakilas, A. G.; Panayiotidis, M. I. Oxidative stress, DNA methylation and carcinogenesis. *Cancer letters* **2008**, *266*(1), 6–11.
92. Murray, I. A.; Patterson, A. D.; Perdew, G. H.; Aryl hydrocarbon receptor ligands in cancer: Friend and foe. *Nature Reviews Cancer* **2014**, *14*(12), 801–814.
93. Jin, L.; Li, D.; Alesi, G. N.; Fan, J.; Kang, H. B.; Lu, Z.; Kang, S. Glutamate dehydrogenase 1 signals through antioxidant glutathione peroxidase 1 to regulate redox homeostasis and tumor growth. *Cancer Cell* **2015**, *27*(2), 257–270.
94. Liu, G.; Zhu, J.; Yu, M.; Cai, C.; Zhou, Y.; Yu, M.; Zeng, B. Glutamate dehydrogenase is a novel prognostic marker and predicts metastases in colorectal cancer patients. *Journal of Translational Medicine* **2015**, *13*, 1–10.
95. Pavlova, S. I.; Jin, L.; Gasparovich, S. R.; Tao, L. Multiple alcohol dehydrogenases but no functional acetaldehyde dehydrogenase causing excessive acetaldehyde production from ethanol by oral streptococci. *Microbiology* **2013**, *159*(Pt_7), 1437–1446.
96. Marttila, E.; Bowyer, P.; Sanglard, D.; Uittamo, J.; Kaihovaara, P.; Salaspuro, M.; Rautemaa, R. Fermentative 2-carbon metabolism produces carcinogenic levels of acetaldehyde in C andida albicans. *Molecular Oral Microbiology* **2013**, *28*(4), 281–291.
97. Salaspuro, M. Microbial metabolism of ethanol and acetaldehyde and clinical consequences. *Addiction Biology* **1997**, *2*(1), 35–46.
98. Meurman, J. H.; Uittamo, J. Oral micro-organisms in the etiology of cancer. *Acta Odontologica Scandinavica* **2008**, *66*(6), 321–326.
99. Koczorowski, R.; Karpiński, T. M. Halitosis—Problem społeczny. Halitosis—A social problem. *Nowiny Lekarskie* **2001**, *70*, 657–664.
100. Koczorowski, R.; Karpiński, T. M.; Hofman, J. Badanie zależności między halitosis a chorobami przyzębia. *A study of the relationship between halitosis and periodontal diseases. Dent. Forum* **2004**, *30*, 51–56.
101. Milella, L. The negative effects of volatile sulphur compounds. *Journal of Veterinary Dentistry* **2015**, *32*(2), 99–102.
102. Hellmich, M. R.; Szabo, C. Hydrogen sulfide and cancer. *Chemistry, Biochemistry and Pharmacology of Hydrogen Sulfide* **2015**, 233–241.
103. Singh, S. B.; Lin, H. C. Hydrogen sulfide in physiology and diseases of the digestive tract. *Microorganisms* **2015**, *3*(4), 866–889.
104. Smędra, A.; Berent, J. The influence of the oral microbiome on oral cancer: A literature review and a new approach. *Biomolecules* **2023**, *13*(5), 815.
105. Lanza, G.; Centonze, S. S.; Destro, G.; Vella, V.; Bellomo, M.; Pennisi, M.; Ciavardelli, D. Shiatsu as an adjuvant therapy for depression in patients with Alzheimer's disease: A pilot study. *Complementary Therapies in Medicine* **2018**, *38*, 74–78.
106. Pennisi, M.; Lanza, G.; Cantone, M.; Ricceri, R.; Ferri, R.; D'Agate, C. C.; Bella, R. Cortical involvement in celiac disease before and after long-term gluten-free diet: A Transcranial Magnetic Stimulation study. *PLoS One* **2017**, *12*(5), e0177560.
107. Acquaviva, R.; Sorrenti, V.; Santangelo, R.; Cardile, V.; Tomasello, B.; Malfa, G.; Di Giacomo, C. Effects of an extract of *Celtis aetnensis* (Tornab.) Strobl twigs on human colon cancer cell cultures. *Oncology Reports* **2016**, *36*(4), 2298–2304.
108. Malfa, G. A.; Tomasello, B.; Acquaviva, R.; Genovese, C.; La Mantia, A.; Cammarata, F. P.; Di Giacomo, C. *Betula etnensis* Raf.(Betulaceae) extract induced HO-1 expression and ferroptosis cell death in human colon cancer cells. *International Journal of Molecular Sciences* **2019**, *20*(11), 2723.
109. Soldati, L.; Di Renzo, L.; Jirillo, E.; Ascierto, P. A.; Marincola, F. M.; De Lorenzo, A. The influence of diet on anti-cancer immune responsiveness. *Journal of Translational Medicine* **2018**, *16*, 1–18.
110. Farhud, D. D.; Yeganeh, M. Z. Nutrigenomics and nutrigenetics. *Iranian Journal of Public Health* **2010**, *39*(4), 1.
111. Martucci, M.; Ostan, R.; Biondi, F.; Bellavista, E.; Fabbri, C.; Bertarelli, C.; Santoro, A. Mediterranean diet and inflammation within the hormesis paradigm. *Nutrition Reviews* **2017**, *75*(6), 442–455.
112. Minihane, A. M.; Vinoy, S.; Russell, W. R.; Baka, A.; Roche, H. M.; Tuohy, K. M.; Calder, P. C. Low-grade inflammation, diet composition and health: Current research evidence and its translation. *British Journal of Nutrition* **2015**, *114*(7), 999–1012.

113. Murtaza, N.; Burke, L. M.; Vlahovich, N.; Charlesson, B.; O'Neill, H. M.; Ross, M. L.; Morrison, M. Analysis of the effects of dietary pattern on the oral microbiome of elite endurance athletes. *Nutrients* **2019**, *11*(3), 614.
114. Kadayifci, F. Z.; Zheng, S.; Pan, Y. X. Molecular mechanisms underlying the link between diet and DNA methylation. *International Journal of Molecular Sciences* **2018**, *19*(12), 4055.
115. Guillemin, G. J.; Essa, M. M.; Song, B. J.; Manivasagam, T. Dietary supplements/antioxidants: Impact on redox status in brain diseases. *Oxidative Medicine and Cellular Longevity* **2017**, *2017*.
116. Quintanilha, B. J.; Reis, B. Z.; Duarte, G. B. S.; Cozzolino, S. M.; Rogero, M. M. Nutrimiromics: Role of microRNAs and nutrition in modulating inflammation and chronic diseases. *Nutrients* **2017**, *9*(11), 1168.
117. Falzone, L.; Lupo, G.; La Rosa, G. R. M.; Crimi, S.; Anfuso, C. D.; Salemi, R.; Candido, S. Identification of novel microRNAs and their diagnostic and prognostic significance in oral cancer. *Cancers* **2019**, *11*(5), 610.
118. Aggarwal, B. B.; Vijayalekshmi, R. V.; Sung, B. Targeting inflammatory pathways for prevention and therapy of cancer: Short-term friend, long-term foe. *Clinical Cancer Research* **2009**, *15*(2), 425–430.
119. Falzone, L.; Romano, G. L.; Salemi, R.; Bucolo, C.; Tomasello, B.; Lupo, G.; Candido, S. Prognostic significance of deregulated microRNAs in uveal melanomas. *Molecular Medicine Reports* **2019**, *19*(4), 2599–2610.
120. Falzone, L.; Scola, L.; Zanghì, A.; Biondi, A.; Di Cataldo, A.; Libra, M.; Candido, S. Integrated analysis of colorectal cancer microRNA datasets: Identification of microRNAs associated with tumor development. *Aging (Albany NY)* **2018**, *10*(5), 1000.
121. Polo, A.; Crispo, A.; Cerino, P.; Falzone, L.; Candido, S.; Giudice, A.; Costantini, S. Environment and bladder cancer: Molecular analysis by interaction networks. *Oncotarget* **2017**, *8*(39), 65240.
122. McCubrey, J. A.; Fitzgerald, T. L.; Yang, L. V.; Lertpiriyapong, K.; Steelman, L. S.; Abrams, S. L.; Libra, M. Roles of GSK-3 and microRNAs on epithelial mesenchymal transition and cancer stem cells. *Oncotarget* **2017**, *8*(8), 14221.
123. Falzone, L.; Candido, S.; Salemi, R.; Basile, M. S.; Scalisi, A.; McCubrey, J. A.; Libra, M. Computational identification of microRNAs associated to both epithelial to mesenchymal transition and NGAL/MMP-9 pathways in bladder cancer. *Oncotarget* **2016**, *7*(45), 72758.
124. Hafsi, S.; Candido, S.; Maestro, R.; Falzone, L.; Soua, Z.; Bonavida, B.; Libra, M. Correlation between the overexpression of Yin Yang 1 and the expression levels of miRNAs in Burkitt's lymphoma: A computational study. *Oncology Letters* **2016**, *11*(2), 1021–1025.
125. Dalmasso, G.; Nguyen, H. T. T.; Yan, Y.; Laroui, H.; Charania, M. A.; Ayyadurai, S.; Merlin, D. Microbiota modulate host gene expression via microRNAs. *PloS o#ne* **2011**, *6*(4), e19293.
126. Peck, B. C.; Mah, A. T.; Pitman, W. A.; Ding, S.; Lund, P. K.; Sethupathy, P. Functional transcriptomics in diverse intestinal epithelial cell types reveals robust microRNA sensitivity in intestinal stem cells to microbial status. *Journal of Biological Chemistry* **2017**, *292*(7), 2586–2600.
127. Yuan, C.; Subramanian, S. microRNA-mediated tumor–microbiota metabolic interactions in colorectal cancer. *DNA and Cell Biology* **2019**, *38*(4), 281–285.
128. Liu, S.; Da Cunha, A. P.; Rezende, R. M.; Cialic, R.; Wei, Z.; Bry, L.; Weiner, H. L. The host shapes the gut microbiota via fecal microRNA. *Cell Host & Microbe* **2016**, *19*(1), 32–43.
129. Zhang, L.; Chen, T.; Yin, Y.; Zhang, C. Y.; Zhang, Y. L. Dietary microRNA—a novel functional component of food. *Advances in Nutrition* **2019**, *10*(4), 711–721.
130. Teodori, L.; Petrignani, I.; Giuliani, A.; Prattichizzo, F.; Gurău, F.; Matacchione, G.; Albertini, M. C. Inflamm-aging microRNAs may integrate signals from food and gut microbiota by modulating common signalling pathways. *Mechanisms of Ageing and Development* **2019**, *182*, 111127.
131. Lang, C.; Karunairetnam, S.; Lo, K. R.; Kralicek, A. V.; Crowhurst, R. N.; Gleave, A. P.; Ingram, J. R. Common variants of the plant microRNA-168a exhibit differing silencing efficacy for human low-density lipoprotein receptor adaptor protein 1 (LDLRAP1). *MicroRNA* **2019**, *8*(2), 166–170.
132. Golan-Gerstl, R.; Elbaum Shiff, Y.; Moshayoff, V.; Schecter, D.; Leshkowitz, D.; Reif, S. Characterization and biological function of milk-derived miRNAs. *Molecular Nutrition & Food Research* **2017**, *61*(10), 1700009.
133. Wagner, A. E.; Piegholdt, S.; Ferraro, M.; Pallauf, K.; Rimbach, G. Food derived microRNAs. *Food & Function* **2015**, *6*(3), 714–718.
134. Teng, Y.; Ren, Y. I.; Sayed, M.; Hu, X.; Lei, C.; Kumar, A.; Zhang, H. G. Plant-derived exosomal microRNAs shape the gut microbiota. *Cell Host & Microbe* **2018**, *24*(5), 637–652.
135. Adami, G. R.; Tangney, C. C.; Tang, J. L.; Zhou, Y.; Ghaffari, S.; Naqib, A.; Schwartz, J. L. Effects of green tea on miRNA and microbiome of oral epithelium. *Scientific Reports* **2018**, *8*(1), 5873.

136. Felden, B.; Gilot, D. Modulation of bacterial sRNAs activity by epigenetic modifications: Inputs from the eukaryotic miRNAs. *Genes* **2018**, *10*(1), 22.
137. Bloch, S.; Węgrzyn, A.; Węgrzyn, G.; Nejman-Faleńczyk, B. Small and smaller—sRNAs and microRNAs in the regulation of toxin gene expression in prokaryotic cells: A mini-review. *Toxins* **2017**, *9*(6), 181.
138. Yuan, C.; Steer, C. J.; Subramanian, S. Host–MicroRNA–microbiota interactions in colorectal cancer. *Genes* **2019**, *10*(4), 270.
139. Miro-Blanch, J.; Yanes, O. Epigenetic regulation at the interplay between gut microbiota and host metabolism. *Frontiers in Genetics* **2019**, *10*, 460718.
140. Patrignani, P.; Tacconelli, S.; Bruno, A. Gut microbiota, host gene expression, and aging. *Journal of Clinical Gastroenterology* **2014**, *48*, S28–S31.
141. Masotti, A. Interplays between gut microbiota and gene expression regulation by miRNAs. *Frontiers in Cellular and Infection Microbiology* **2012**, *2*, 137.
142. Chang, C.; Wang, H.; Liu, J.; Pan, C.; Zhang, D.; Li, X.; Pan, Y. *Porphyromonas gingivalis* infection promoted the proliferation of oral squamous cell carcinoma cells through the miR-21/PDCD4/AP-1 negative signaling pathway. *ACS Infectious Diseases* **2019**, *5*(8), 1336–1347.
143. Gao, S.; Liu, Y.; Duan, X.; Liu, K.; Mohammed, M.; Gu, Z.; Wang, H. *Porphyromonas gingivalis* infection exacerbates oesophageal cancer and promotes resistance to neoadjuvant chemotherapy. *British Journal of Cancer* **2021**, *125*(3), 433–444.
144. Zhou, Y.; Sztukowska, M.; Wang, Q.; Inaba, H.; Potempa, J.; Scott, D. A.; Lamont, R. J. Noncanonical activation of β-catenin by *Porphyromonas gingivalis*. *Infection and Immunity* **2015**, *83*(8), 3195–3203.
145. Yilmaz, Ö.; Yao, L.; Maeda, K.; Rose, T. M.; Lewis, E. L.; Duman, M.; Ojcius, D. M. ATP scavenging by the intracellular pathogen *Porphyromonas gingivalis* inhibits P2X7-mediated host-cell apoptosis. *Cellular Microbiology* **2008**, *10*(4), 863–875.
146. Choi, C. H.; Spooner, R.; DeGuzman, J.; Koutouzis, T.; Ojcius, D. M.; Yilmaz, Ö. *Porphyromonas gingivalis*-nucleoside-diphosphate-kinase inhibits ATP-induced reactive-oxygen-species via P 2 X 7 receptor/NADPH-oxidase signaling and contributes to persistence. *Cellular Microbiology* **2013**, *15*(6), 961–976.
147. Geng, F.; Zhang, Y.; Lu, Z.; Zhang, S.; Pan, Y. *Fusobacterium nucleatum* caused DNA damage and promoted cell proliferation by the Ku70/p53 pathway in oral cancer cells. *DNA and Cell Biology* **2020**, *39*(1), 144–151.
148. Ding, N.; Cheng, Y.; Liu, H.; Wu, Y.; Weng, Y.; Cui, H.; Cui, Y. *Fusobacterium nucleatum* infection induces malignant proliferation of esophageal squamous cell carcinoma cell by putrescine production. *Microbiology Spectrum* **2023**, *11*(2), e02759–22.
149. Wang, Y.; Liu, S.; Li, B.; Jiang, Y.; Zhou, X.; Chen, J.; Cheng, L. *Staphylococcus aureus* induces COX-2-dependent proliferation and malignant transformation in oral keratinocytes. *Journal of Oral Microbiology* **2019**, *11*(1), 1643205.
150. Lee, J.; Roberts, J. S.; Atanasova, K. R.; Chowdhury, N.; Yilmaz, Ö. A novel kinase function of a nucleoside-diphosphate-kinase homologue in *Porphyromonas gingivalis* is critical in subversion of host cell apoptosis by targeting heat-shock protein 27. *Cellular Microbiology* **2018**, *20*(5), e12825.
151. Mao, S.; Park, Y.; Hasegawa, Y.; Tribble, G. D.; James, C. E.; Handfield, M.; Lamont, R. J. Intrinsic apoptotic pathways of gingival epithelial cells modulated by *Porphyromonas gingivalis*. *Cellular Microbiology* **2007**, *9*(8), 1997–2007.
152. Yao, Á.; Jermanus, C.; Barbetta, B.; Choi, C.; Verbeke, P.; Ojcius, D. M.; Yilmaz, Ö. *Porphyromonas gingivalis* infection sequesters pro-apoptotic Bad through Akt in primary gingival epithelial cells. *Molecular Oral Microbiology* **2010**, *25*(2), 89–101.
153. Harrandah, A. M.; Chukkapalli, S. S.; Bhattacharyya, I.; Progulske-Fox, A.; Chan, E. K. Fusobacteria modulate oral carcinogenesis and promote cancer progression. *Journal of Oral Microbiology* **2021**, *13*(1), 1849493.
154. Peng, R. T.; Sun, Y.; Zhou, X. D.; Liu, S. Y.; Han, Q.; Cheng, L.; Peng, X. Treponema denticola promotes OSCC development via the TGF-β signaling pathway. *Journal of Dental Research* **2022**, *101*(6), 704–713.
155. Song, Y.; Wu, X.; Xu, Y.; Zhu, J.; Li, J.; Zou, Z.; Cheng, H. HPV E7 inhibits cell pyroptosis by promoting TRIM21-mediated degradation and ubiquitination of the IFI16 inflammasome. *International Journal of Biological Sciences* **2020**, *16*(15), 2924.
156. Baqui, A. A. M. A.; Meiller, T. F.; Chon, J. J.; Turng, B. F.; Falkler Jr, W. A. Granulocyte-macrophage colony-stimulating factor amplification of interleukin-1I and tumor necrosis factor alpha production in THP-1 human monocytic cells stimulated with lipopolysaccharide of oral microorganisms. *Clinical Diagnostic Laboratory Immunology* **1998**, *5*(3), 341–347.

157. Gallimidi, A. B.; Fischman, S.; Revach, B.; Bulvik, R.; Maliutina, A.; Rubinstein, A. M.; Elkin, M. Periodontal pathogens *Porphyromonas gingivalis* and *Fusobacterium nucleatum* promote tumor progression in an oral-specific chemical carcinogenesis model. *Oncotarget* **2015**, *6*(26), 22613.
158. Rubinstein, M. R.; Wang, X.; Liu, W.; Hao, Y.; Cai, G.; Han, Y. W. *Fusobacterium nucleatum* promotes colorectal carcinogenesis by modulating E-cadherin/β-catenin signaling via its FadA adhesin. *Cell Host & Microbe* **2013**, *14*(2), 195–206.
159. Park, H. E.; Kim, J. H.; Cho, N. Y.; Lee, H. S.; Kang, G. H. Intratumoral *Fusobacterium nucleatum* abundance correlates with macrophage infiltration and CDKN2A methylation in microsatellite-unstable colorectal carcinoma. *Virchows Archiv* **2017**, *471*, 329–336.
160. Chen, Y.; Peng, Y.; Yu, J.; Chen, T.; Wu, Y.; Shi, L.; Fu, X. Invasive *Fusobacterium nucleatum* activates beta-catenin signaling in colorectal cancer via a TLR4/P-PAK1 cascade. *Oncotarget* **2017**, *8*(19), 31802.
161. Wu, Y.; Wu, J.; Chen, T.; Li, Q.; Peng, W.; Li, H.; Fu, X. *Fusobacterium nucleatum* potentiates intestinal tumorigenesis in mice via a toll-like receptor 4/p21-activated kinase 1 cascade. *Digestive Diseases and Sciences* **2018**, *63*, 1210–1218.
162. Uitto, V. J.; Baillie, D.; Wu, Q.; Gendron, R.; Grenier, D.; Putnins, E. E.; Firth, J. D. *Fusobacterium nucleatum* increases collagenase 3 production and migration of epithelial cells. *Infection and Immunity* **2005**, *73*(2), 1171–1179.
163. Schwabe, R. F.; Jobin, C. The microbiome and cancer. *Nature Reviews Cancer* **2013**, *13*(11), 800–812.
164. Sellers, R. S.; Morton, D. The colon: From banal to brilliant. *Toxicologic Pathology* **2014**, 42(1), 67–81.
165. Taur, Y.; Pamer, E. G. Microbiome mediation of infections in the cancer setting. *Genome Medicine* **2016**, *8*, 1–7.
166. Elinav, E.; Nowarski, R.; Thaiss, C. A.; Hu, B.; Jin, C.; Flavell, R. A. Inflammation-induced cancer: Crosstalk between tumours, immune cells and microorganisms. *Nature Reviews Cancer* **2013**, *13*(11), 759–771.
167. Arthur, J. C.; Perez-Chanona, E.; Mühlbauer, M.; Tomkovich, S.; Uronis, J. M.; Fan, T. J.; Jobin, C. Intestinal inflammation targets cancer-inducing activity of the microbiota. *Science* **2012**, *338*(6103), 120–123.
168. White, R. A.; Malkoski, S. P.; Wang, X. J. TGFβ signaling in head and neck squamous cell carcinoma. *Oncogene* **2010**, *29*(40), 5437–5446.
169. Yang, L. TGFβ and cancer metastasis: An inflammation link. *Cancer and Metastasis Reviews* **2010**, *29*, 263–271.
170. Gaonkar, P. P.; Patankar, S. R.; Tripathi, N.; Sridharan, G. Oral bacterial flora and oral cancer: The possible link. *Journal of Oral and Maxillofacial Pathology* **2018**, *22*(2), 234–238.
171. Wu, J.; Peters, B. A.; Dominianni, C.; Zhang, Y.; Pei, Z.; Yang, L.; Ahn, J. Cigarette smoking and the oral microbiome in a large study of American adults. *The ISME Journal* **2016**, *10*(10), 2435–2446.
172. Hernandez, B. Y.; Zhu, X.; Goodman, M. T.; Gatewood, R.; Mendiola, P.; Quinata, K.; Paulino, Y. C. Betel nut chewing, oral premalignant lesions, and the oral microbiome. *PloS One* **2017**, *12*(2), e0172196.
173. Charlson, E. S.; Chen, J.; Custers-Allen, R.; Bittinger, K.; Li, H.; Sinha, R.; Collman, R. G. Disordered microbial communities in the upper respiratory tract of cigarette smokers. *PloS One* **2010**, *5*(12), e15216.
174. Kurkivuori, J.; Salaspuro, V.; Kaihovaara, P.; Kari, K.; Rautemaa, R.; Grönroos, L.; Salaspuro, M. Acetaldehyde production from ethanol by oral streptococci. *Oral Oncology* **2007**, *43*(2), 181–186.
175. Muto, M.; Hitomi, Y.; Ohtsu, A.; Shimada, H.; Kashiwase, Y.; Sasaki, H.; Esumi, H. Acetaldehyde production by non-pathogenic Neisseria in human oral microflora: Implications for carcinogenesis in upper aerodigestive tract. *International Journal of Cancer* **2000**, *88*(3), 342–350.
176. Sasaki, M.; Yamaura, C.; Ohara-Nemoto, Y.; Tajika, S.; Kodama, Y.; Ohya, T.; Kimura, S. Streptococcus anginosus infection in oral cancer and its infection route. *Oral Diseases* **2005**, *11*(3), 151–156.
177. Pignatelli, P.; Romei, F. M.; Bondi, D.; Giuliani, M.; Piattelli, A.; Curia, M. C. Microbiota and oral cancer as a complex and dynamic microenvironment: A narrative review from etiology to prognosis. *International Journal of Molecular Sciences* **2022**, *23*(15), 8323.
178. Li, S.; Wang, T.; Ren, Y.; Liu, Z.; Gao, J.; Guo, Z. Prognostic impact of oral microbiome on survival of malignancies: A systematic review and meta-analysis. *Systematic Reviews* **2024**, *13*(1), 41.
179. Neuzillet, C.; Marchais, M.; Vacher, S.; Hilmi, M.; Schnitzler, A.; Meseure, D.; Bieche, I. Prognostic value of intratumoral *Fusobacterium nucleatum* and association with immune-related gene expression in oral squamous cell carcinoma patients. *Scientific Reports* **2021**, *11*(1), 7870.
180. Mohamed, N.; Litlekalsøy, J.; Ahmed, I. A.; Martinsen, E. M. H.; Furriol, J.; Javier-Lopez, R.; Costea, D. E. Analysis of salivary mycobiome in a cohort of oral squamous cell carcinoma patients from Sudan identifies higher salivary carriage of malassezia as an independent and favorable predictor of overall survival. *Frontiers in Cellular and Infection Microbiology* **2021**, *11*, 673465.

181. Du, Y.; Feng, R.; Chang, E. T.; Debelius, J. W.; Yin, L.; Xu, M.; Ye, W. Influence of pre-treatment saliva microbial diversity and composition on nasopharyngeal carcinoma prognosis. *Frontiers in Cellular and Infection Microbiology* **2022**, *12*, 831409.
182. Wei, Z.; Cao, S.; Liu, S.; Yao, Z.; Sun, T.; Li, Y.; Zhou, Y. Could gut microbiota serve as prognostic biomarker associated with colorectal cancer patients' survival? A pilot study on relevant mechanism. *Oncotarget* **2016**, *7*(29), 46158.
183. Gao, S.; Li, S.; Ma, Z.; Liang, S.; Shan, T.; Zhang, M.; Feng, X. Presence of *Porphyromonas gingivalis* in esophagus and its association with the clinicopathological characteristics and survival in patients with esophageal cancer. *Infectious Agents and Cancer* **2016**, *11*, 1–9.
184. Kim, M. J.; Ahn, K. M. Prognostic factors of oral squamous cell carcinoma: The importance of recurrence and pTNM stage. *Maxillofacial Plastic and Reconstructive Surgery* **2024**, *46*(1), 1–10.
185. Sim, Y. C.; Hwang, J. H.; Ahn, K. M. Overall and disease-specific survival outcomes following primary surgery for oral squamous cell carcinoma: Analysis of consecutive 67 patients. *Journal of the Korean Association of Oral and Maxillofacial Surgeons* **2019**, *45*(2), 83.
186. Moeckelmann, N.; Ebrahimi, A.; Tou, Y. K.; Gupta, R.; Low, T. H. H.; Ashford, B.; Clark, J. R. Prognostic implications of the 8th edition American Joint Committee on Cancer (AJCC) staging system in oral cavity squamous cell carcinoma. *Oral Oncology* **2018**, *85*, 82–86.
187. Prieto, I.; Prieto, A.; Bravo, M.; Bascones, A. Prognostic factors for cancer of the oral cavity. *Quintessence International* **2005**, *36*(9).
188. Sami, A.; Elimairi, I.; Stanton, C.; Ross, R. P.; Ryan, C. A. The role of the microbiome in oral squamous cell carcinoma with insight into the microbiome–treatment axis. *International Journal of Molecular Sciences* **2020**, *21*(21), 8061
189. Makarewicz, J.; Kaźmierczak-Siedlecka, K.; Sobocki, B. K.; Dobrucki, I. T.; Kalinowski, L.; Stachowska, E. Anti-cancer management of head and neck cancers and oral microbiome—what can we clinically obtain?. *Frontiers in Cellular and Infection Microbiology* **2024**, *14*.
190. Willis, J. R.; Gabaldón, T. The human oral microbiome in health and disease: From sequences to ecosystems. *Microorganisms* **2020**, *8*(2), 308.
191. Wang, X. L.; Xu, H. W.; Liu, N. N. Oral microbiota: A new insight into cancer progression, diagnosis and treatment. *Phenomics* **2023**, *3*(5), 535–547.
192. Fessler, J.; Matson, V.; Gajewski, T. F. Exploring the emerging role of the microbiome in cancer immunotherapy. *Journal for Immunotherapy of Cancer* **2019**, *7*, 1–15.
193. Goossens, N.; Nakagawa, S.; Sun, X.; Hoshida, Y. Cancer biomarker discovery and validation. *Translational Cancer Research* **2015**, *4*(3), 256.
194. Warnakulasuriya, S. Potentially malignant disorders of the oral cavity. Textbook of Oral Cancer: Prevention, Diagnosis and Management **2020**, 141–158.
195. Villa, A.; Celentano, A.; Glurich, I.; Borgnakke, W. S.; Jensen, S. B.; Peterson, D. E.; Farah, C. S. World Workshop on Oral Medicine VII: Prognostic biomarkers in oral leukoplakia: A systematic review of longitudinal studies. *Oral Diseases* **2019**, *25*, 64–78.
196. Vogelstein, B.; Lane, D.; Levine, A. J. Surfing the p53 network. *Nature* **2000**, *408*(6810), 307–310.
197. Mello, F. W.; Melo, G.; Guerra, E. N. S.; Warnakulasuriya, S.; Garnis, C.; Rivero, E. R. C. Oral potentially malignant disorders: A scoping review of prognostic biomarkers. *Critical Reviews in Oncology/Hematology* **2020**, *153*, 102986.
198. Wang, J.; Gao, B. Mechanisms and potential clinical implications of oral microbiome in oral squamous cell carcinoma. *Current Oncology* **2023**, *31*(1), 168–182.

12 Role of Oral Microbiome in Systemic Infections

A Current Narrative

Abhilasha Sood, Kiranjeet Kaur, and Arpit Mehrotra
Department of Allied Health Sciences, Chitkara School of Health Sciences, Chitkara University, Rajpura, Punjab, India

12.1 INTRODUCTION

The microbial community exists in the human body showing a mutualistic beneficial niche. The oral cavity is a vivacious habitat hosting an array of microorganisms, playing an extensively important role in human health and disease [1, 2]. It serves as an important ecosystem wherein microorganisms such as bacteria, fungi, viruses, and protozoa coexist with human cells and organelles supporting their overall well-being [3] and engaging in mutualistic relationships while competing and interacting with the complex community existing there. The key taxa contributing to this intricate oral microbial landscape are found to include *Streptococcus*, *Actinomyces*, *Prevotella*, and *Porphyromonas*, among others [4, 5]. Hence, characterizing, understanding, and utilizing this intersection of the microbiome and oral ecosystem stands as a major key point, wherein the focus of upcoming research could act as a major contributing factor in uplifting the overall systemic health of the human as compared to the traditional inference, predominantly linked to oral health conditions like periodontal disease, dental caries, and gingivitis [6, 7].

Therefore, it stands as an important avenue to understand various pathways that could elaborate on the oral microbiome's impact on different organs and systems [8], as shown in Figure 12.1.

It is crucial to comprehend and embark on this journey of understanding oral-systemic link and the involvement of the oral microbiome in the overall health of the living system [9, 10]. A variety of activities such as poor oral hygiene, disease, and other factors are known to lead to this dysbiosis, which is disseminated in the form of health hazards through various routes into the bloodstream leading to injury and infection exerting its effects beyond the oral cavity [11]. Mounting evidence has suggested the effects of the oral microbiome on the immune system thereby influencing inflammatory and immunogenic pathways [12]. This interplay potentially impacts systemic health and could be a contributing factor in the exacerbation of clinical conditions like diabetes, respiratory infections, and cardiovascular and neurological conditions [13–15]. Therefore, this chapter aims to delve deeply into the multifaceted interplay of the oral microbiome and systemic health and disease, unveiling the current status and established relationship between them.

12.2 ORAL MICROBES AND DIGESTIVE SYSTEM DISEASES

The oral cavity acts as the primary gateway to the gastrointestinal system, affecting the digestion and absorption of important nutrients [3]. Increasing research has revealed a complex interplay between these two important facilities contributing to human health and disease. While the periodontal diseases are characterized by bacterial infections and inflammation in the teeth and gums, these have been extensively linked to digestive disorders like inflammatory bowel disease, emphasizing the importance of the inflammatory pathway and the oral microbial population [16, 17].

 DOI: 10.1201/9781003478638-12

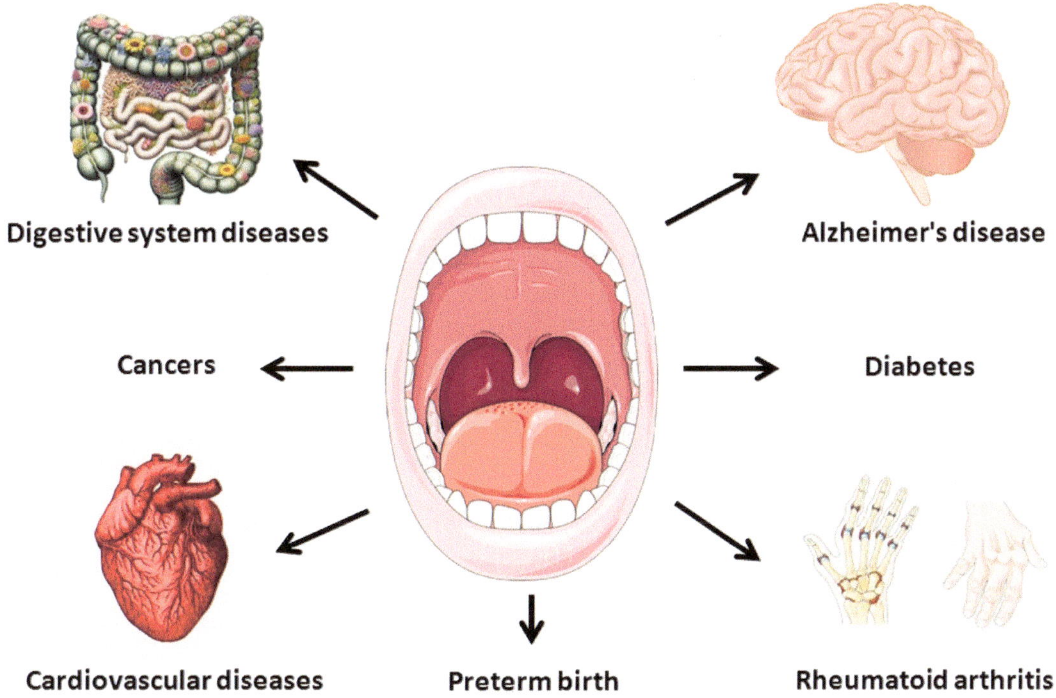

FIGURE 12.1 Role of the oral microbiome in mediating systemic infections.

The translocation of oral bacteria or their by-products into the gut is reportedly a major contributing factor in gastrointestinal health and disease exacerbation [18]. It can lead to dental erosion, enamel damage, and a change in the microbiome of the oral cavity. It is essential to highlight that *Helicobacter pylori*, which is a known inhabitant of the oral cavity, is recognized to cause gastric ulcers and gastritis and could serve as a potential source of gut infections [19, 20]. Moreover, there is emerging evidence pointing to the influence of the oral microbiome on colorectal cancer and its progression [21]. Hence, understanding this dynamic interplay is a promising prospect that could prove instrumental in mitigating the progression, management, and risk of gastrointestinal diseases and the development of preventive strategies [22, 23].

12.3 ORAL BACTERIA AND CANCERS

Viruses are well known to be associated with cancers, and with the recognition of *Helicobacter pylori* as a causative agent of gastric cancers, new avenues to explore the correlation between bacterial agents and cancers have been proposed [24, 25]. In this context, the role of the oral microbiome has gained considerable attention, and efforts are being made to see the impact of oral microbes in cancer. The oral cavity is known to be inhabited by more than 700 bacterial species that generally endure in a homeostatic mode with the host. These include *Porphyromonas gingivalis*, *Streptococcus pyogenes*, *Fusobacterium*, *Propionibacterium acnes*, *Streptococcus viridans*, and *Streptococcus salivarius* [26]. Different factors such as temperature, pH, oxygen requirement, and saliva dictate the type, location, and number of oral microbes in the cavity. However, dysbiosis of the oral microbiome resulting in either chronic inflammation and/or synthesis of metabolites leading to mutations could favor the initiation/progression of cancer [27]. Perhaps, the most likely carcinogenic link with oral bacteria is with oral squamous cell carcinoma [28]. This is associated with significantly higher levels of *Fusobacteria* and *Porphyromonas*, with concomitant decreased levels of *Streptococcus*, *Haemophilus*, and *Actinomyces* [29]. Interestingly, one characteristic feature that distinguishes the oral microbiome from others is that it is not only localized; rather it is associated with distant organs too. For instance, apart from oral squamous cell carcinoma, various other cancers such as esophageal cancer, pancreatic cancer, hepatic cancer, lung cancer, colorectal cancer, and others are linked with disturbances in the oral microbiome [30–32]. The increase in levels of key orally present species belonging to *Porphyromonas gingivalis* and *Fusobacterium* species was observed in cases of esophageal, pancreatic, and hepatic as well as in colorectal cancer [33]. However, the genus *Veillonella* was observed to be increased in cases of lung cancer [34].

Although the microbial communities differ depending on the type of cancer, *Porphyromonas* and *Fusobacterium* genera are widely reported to be associated with cancer progression [35, 36]. The former bacteria possess antiapoptotic properties and can suppress even chemically induced apoptosis. It is shown to activate the Jak1/Akt/Stat3 signaling that majorly controls intrinsic mitochondrial apoptosis pathways. It further leads to the inhibition of proapoptotic activity of Bad and Bcl2 [37, 38]. On the other hand, *F. nucleatum* infection targets kinases involved in cell cycle control and can elevate the cell proliferation and migration [39]. Both the microbes are involved in tumor invasion and metastasis. Recent studies have shown the data comparing the cancer patients with healthy ones, wherein evident differences between microbial architecture were observed. However, the finding on association between microbiota and risk of developing cancer is still limited [40]. Among the several cancers, oral, colorectal, pancreatic, and esophageal cancers are the most important cancers closely associated with the oral microbiome [41, 42]. Thus, it can be concluded that improved oral hygiene would be beneficial in effectively reducing the development or spread of cancer.

12.4 ORAL MICROBES AND CARDIOVASCULAR DISEASES

A complex relationship has been unveiled between the oral microbiome and the incidence of cardiovascular diseases (CVDs). Periodontal disease due to oral inflammatory conditions has shown use as a focal point of the oral-cardiovascular connection [43]. A significant correlation has been suggested between the manifestation of periodontitis into CVDs such as atherosclerosis, stroke, and coronary artery disease, highlighting it as an independent cardiovascular risk factor [44]. Periodontal pathogens have also been reported to trigger systemic inflammation leading to vascular endothelial dysfunctions and the formation of artherosclerotic plaque [45]. Oral bacteria associated with periodontal disease conditions such as *Porphyromonas gingivalis* and *Fusobacterium nucleatum* have shown implications in CVDs [46]. The bacteria can enter the bloodstream through the compromised periodontal tissue/oral mucosa, further infiltrating and impacting vascular endothelial cells, promoting inflammation, and exacerbating cardiovascular conditions. Moreover, similar oral health conditions like gum diseases have been related to increased inflammation, a potential risk factor for progression of atherosclerosis and destabilization of preexisting cardiovascular conditions [47].

Interestingly, recent studies have also inferred potential involvement of the oral microbiota community in the metabolism of a dietary compound known as nitrate, producing potentially toxic nitric oxide: a molecule affecting cardiac health, vascular function, and blood pressure regulation [48]. Thus, the balance of the oral microbial community is an important consideration in the understanding and management of oral health and might be a leading point in further management and mitigation of the risk/progression of cardiovascular diseases. Ongoing research on the taxonomic profiling of identified oral microbiome bacteria to the species level, following comparison with the Human Oral Microbiome Database, is a future hope in this direction. It might be able to suggest a species-specific connection between the oral microbiota and CVDs, further uncovering this intricate link [49].

12.5 ORAL MICROBES AND ALZHEIMER'S DISEASE

Alzheimer's disease (AD) is one of the most common forms of dementia, representing 60%–80% of all the dementia cases [50, 51]. In 2015, the World Health Organization also recognized dementia as a global public threat and came up with an action plan (2015–2025) to tackle the disease [52]. Around 47 million people are reported to be affected by AD worldwide, and the figure is projected to rise to 75 million in 2030 and to 132 million by 2050 [53]. In the United States alone, the expected number of cases is marked as 13.85 million by the year 2060 [54]. Looking at the rising number of dementia cases as well as the lack of adequate treatment options for AD, the key focus of research is to identify the potential risk factors for AD. Overall AD presents clinically with depression and declining cognition accompanied with two neuropathological lesions [55]. The lesions basically correspond to plaques in the brain that are composed of amyloid-beta and hyperphosphorylated tau protein coated neurofibrillary tangles. Recent studies have revealed a causal link between microbes and the pathogenesis of the brain, especially the central nervous system in AD. Researchers believe that the host immune response against pathogenic microbes, including the oral microbiome, can result in the formation of amyloid products [56, 57]. Different studies have proposed varied underlying mechanisms on how oral microbiome led to development and progression of AD. These possibilities include: (1) movement of microbial products from the oral cavity to the brain, (2) inflammation in the mouth eventually influencing inflammation in the brain, (3) the oral microbiome itself crosses the blood-brain barrier and enters into the brain, and (4) oral pathogenic microbial metabolites or their by-products migrate into the brain further inducing the translocation of proinflammatory markers [7, 58].

The three most common microbes linked with AD pathology include *Porphyromonas gingivalis*, *Treponema denticola*, and *Aggregatibacter actinomycetemcomitans*, while association of *Streptococcus mutans*, *Fusobacterium nucleatum*, and *Actinomycin meyeri* still lacks strong evidence [59]. *Porphyromonas gingivalis*, *T. forsythia*, and *T. denticola* constitute the three major oral microbes linked with the etiology of chronic periodontitis and are among the most prevalent microbes in AD. For instance, *P. gingivalis* has always been reported as a risk factor for cognitive impairment, development of Aβ plaques, and dementia [60]. In fact, researchers have detected the presence of *P. gingivalis* using quantitative polymerase chain reaction (qPCR) or monoclonal antibodies. Notably, the patients with AD showed higher levels of anti–*P. gingivalis* IgG compared to healthy subjects in tested blood samples [61]. The *in vivo* studies conducted in male C57BL/6J and SAMP8 mice using an intraperitoneal injection of lipopolysaccharide (LPS) obtained from *P. gingivalis* resulted in the reduced expression of neprilysin in the hippocampus, which is associated with increased Aβ deposition in AD [62]. Similarly, LPS from serotype-B of *A. actinomycetemcomitans*, has also led to increased secretion of the Aβ1-42 peptide in mixed hippocampal cells accompanied with high levels of proinflammatory cytokine production in microglial cells [59]. Notably, in animal studies conducted using oral inoculation of *T. denticola* into C57BL/6 mice, the neuronal apoptosis was also present along with Aβ1-42 in the hippocampus [63]. Previous research has provided evidence of both *in vitro* and *in vivo* studies that *S. mutans* can produce amyloid [64]. However, there is a lack of direct evidence of a causal link between AD and *S. mutans*. AD is a complex pathological condition with multiple contributing factors, and further investigations are required to strongly establish the influence of the oral microbial population on the advancement of AD.

12.6 ORAL MICROBES AND DIABETES

The recently explored association between roles of oral microbes in diabetes advancement has surfaced as a captivating area of research, disclosing possible bidirectional cross talk between the oral community and progression of diabetes mellitus [65]. It is well reported that affected individuals with diabetes are known to be more vulnerable to oral health complications, predominantly periodontal disease [66]. The existence of chronic hyperglycemic circumstances in diabetes leads to compromised functioning of immunological machinery, thereby making affected individuals even more exposed to infections, including those affecting the oral cavity [67]. Periodontal disease showing characteristic features of inflammation, with concomitant bacterial infection of the gums and its supporting structures, have a tendency to be more stern and common in individuals affected with diabetes [68]. On the contrary, such inflammatory responses elicited by periodontal disease may also contribute to insulin resistance and further deteriorate glycemic control in diabetic patients [69].

Additionally, certain specific oral pathogens allied with the progression of periodontal disease, such as *P. gingivalis* and *T. denticola*, have been found to be strongly correlated with the existence of insulin resistance and altered glucose metabolism [70]. These oral bacteria have been shown to directly influence systemic inflammation and thus impact insulin signaling pathways, simultaneously aggravating the progression of diabetes. On the other hand, diabetes may also manipulate the oral microbiome. For example, an enhanced level of glucose in the saliva and crevicular fluid generates a milieu favorable for the propagation of certain pathogenic oral bacteria, thereby altering the composition of the oral microbial population [71]. This dysbiosis within the oral microbiome in diabetic patients worsens the oral health conditions and further promotes a systemic inflammation response. Such an existence of a bidirectional association between diabetes and the oral microbiome emphasizes the significance of maintaining comprehensive oral care in diabetes management [72]. Recovery of oral hygiene and periodontal treatment has shown prospective benefits in glycemic control among diabetic patients, thus highlighting the probable impact of oral health interferences on systemic health outcomes. Exploring the complicated relationship between oral microbes and diabetes not only highlights the significance of oral health in diabetes management but also suggests possible approaches that could be employed for targeted interventions [73]. Hence, including oral health as a fundamental aspect of diabetes management may hold promise in improving overall health outcomes for individuals affected with diabetes mellitus.

12.7 ORAL MICROBES AND RHEUMATOID ARTHRITIS

The existence of a relationship between the oral microbial population and rheumatoid arthritis has been of considerable research interest over the past few years, revealing prospective links between oral health, particularly periodontal disease, and the concerted progress or exacerbation of such an autoimmune disorder [74]. Periodontal disease has been widely explored in relation to rheumatoid arthritis. Investigations have exposed a higher occurrence and severity of periodontal disease in individuals affected with rheumatoid arthritis as compared to the general population [75]. Interestingly, the existence of commonalities in inflammatory pathways between periodontal disease and rheumatoid arthritis has been recognized, suggesting a possible mechanistic link [76].

The oral microbial population, particularly the specific periodontal pathogens like *P. gingivalis, A. actinomycetemcomitans*, and *P. intermedia*, is found to be directly associated in the pathogenesis of rheumatoid arthritis [77, 78]. These oral bacteria harbor the enzyme peptidylarginine deiminase, which is capable of citrullinating or deiminating the host proteins. Such a citrullination mechanism is also a hallmark feature observed during rheumatoid arthritis, which leads to the elicited generation of autoantibodies against such citrullinated peptides and thus contributes to an ensuing autoimmune response as observed in rheumatoid arthritis [79]. Furthermore, the systemic propagation of oral pathogens or their subsequent by-products from the periodontal pockets into the main bloodstream has shown to activate or intensify the inflammatory responses observed in rheumatoid arthritis. This chronic inflammation coupled with periodontal disease might further propagate the systemic inflammatory mechanisms in individuals with rheumatoid arthritis, potentially impacting the disease severity and its subsequent progression [80].

Conversely, there's evidence suggesting that rheumatoid arthritis, as a systemic inflammatory condition, might influence the oral microbiome. Rheumatoid arthritis–related alterations in the immune response could potentially impact the oral microbial community, contributing to dysbiosis and exacerbating oral health issues [81]. Understanding the interplay between oral microbes and rheumatoid arthritis underscores the significance of comprehensive oral care in managing rheumatoid arthritis. Targeted periodontal treatment options and maintaining optimal oral hygiene may not only ease oral health concerns but also potentially impact the inflammatory burden associated with rheumatoid arthritis, highlighting the prospective for employing integrated approaches in managing such autoimmune conditions [82].

12.8 ORAL MICROBES AND PRETERM BIRTH

The presence of an intricate relationship between the oral microbial population and preterm birth has acquired significant consideration due to growing facts, suggesting the existence of a prospective relationship between compromised oral health (particularly periodontal disease) and unfavorable pregnancy outcomes, including preterm birth and low birth weight [83]. Periodontal disease has also been closely associated with systemic health issues, including impediments during pregnancy. Investigations have shown that pregnant females affected with periodontal disease may have the propensity of delivering prematurely or having infants with lower birth weight [84]. Certain specific periodontal pathogens like *P. gingivalis* and *F. nucleatum* have been shown to adversely affect pregnancy outcomes [85]. These bacteria and their by-products are known to enter the main bloodstream and trigger a systemic inflammatory response, which further releases proinflammatory cytokines and thus contributes to an interruption of normal pregnancy processes, including premature labor or impaired development of the fetus [86].

Moreover, such inflammatory responses associated with periodontal disease are known to contribute to a stimulation of the production of prostaglandins, which are known to be involved in inducing uterine contractions [87]. This suggests a possible direct mechanistic link between oral pathogens, inflammation, and the concomitant initiation of preterm labor. On the contrary, hormonal changes observed during pregnancy, principally the fluctuations in estrogen and progesterone levels, can directly influence the oral microbiome, potentially leading to an augmented risk of developing gingivitis or exacerbating the existing periodontal conditions [88]. Occurrence of such oral dysbiosis and its associated inflammation might contribute to the systemic inflammatory burden and, in turn, influence the pregnancy outcomes. Therefore, understanding and managing oral health during pregnancy, including regular dental checkups and maintenance of good oral hygiene, may hold promise in extenuating the risk of preterm birth. Integrated health care advancements that involve both obstetric care and dental interventions aim to manage oral health issues during pregnancy, potentially lowering the risk of unfavorable pregnancy outcomes associated with oral microbial imbalances [85].

12.9 CONCLUSIONS AND PERSPECTIVES

In conclusion, the prominent role of the oral microbiome in regulating systemic infections represents an appealing connection between oral and systemic health. This complicated association between the oral microbial population and systemic diseases like cardiovascular disorders, diabetes, rheumatoid arthritis, and preterm birth emphasizes the implications of oral health beyond the mouth. Unraveling the precise mechanisms through which oral microbes navigate systemic barriers, manipulate the immune system, and contribute to the commencement, succession, or exacerbation of systemic infections is highly desirable. Dysbiosis within the oral microbiome has been correlated with altered immune system–mediated responses, systemic inflammation, and propagation of oral pathogens or their by-products, thus impacting remote organs or systems. This bidirectional cross talk between oral health and systemic infections warrants further integrated health care investigations. Managing oral health is suggestive of circumventing the risk, development, or management of various systemic diseases. Strategies focusing on comprehensive oral care, regular dental examination, maintenance of oral hygiene, and targeted interventions to restore oral microbial equilibrium may offer preventative and therapeutic opportunities in the management of systemic infections.

Moreover, unrelenting research efforts are necessary to unravel the complexities of the oral microbiome's impact on systemic health comprehensively. Improvements in exploring these relationships may suggest ways for designing innovative diagnostic tools and targeted therapies, including personalized interventions that could exploit the capacity of the oral microbiome in improving overall health outcomes. Eventually, recognizing the deep influence of the oral microbiome on regulating systemic infections opens up opportunities for designing preventative health care approaches that will prioritize oral health as a fundamental constituent of systemic well-being, thus promising the development of futuristic approaches to managing both oral and systemic health.

CONFLICT OF INTEREST

None declared by the authors among themselves or with their parent institutions.

REFERENCES

1. Gendron, R.; Grenier, D.; Maheu-Robert, L.-F. The oral cavity as a reservoir of bacterial pathogens for focal infections. *Microbes and Infection* **2000**, *2*, 897–906.
2. He, H.; Hao, Y.; Fan, Y.; Li, B.; Cheng, L. The interaction between innate immunity and oral microbiota in oral diseases. *Expert Review of Clinical Immunology* **2023**, *19*, 405–415.
3. Kitamoto, S.; Nagao-Kitamoto, H.; Hein, R.; Schmidt, T.M.; Kamada, N. The bacterial connection between the oral cavity and the gut diseases. *Journal of Dental Research* **2020**, *99*, 1021–1029.
4. Pietrobon, G.; Tagliabue, M.; Stringa, L.M.; De Berardinis, R.; Chu, F.; Zocchi, J.; Carlotto, E.; Chiocca, S.; Ansarin, M. Leukoplakia in the oral cavity and oral microbiota: a comprehensive review. *Cancers* **2021**, *13*, 4439.
5. Verma, D.; Garg, P.K.; Dubey, A.K. Insights into the human oral microbiome. *Archives of Microbiology* **2018**, *200*, 525–540.
6. Willis, J.R.; Gabaldon, T. The human oral microbiome in health and disease: from sequences to ecosystems. *Microorganisms* **2020**, *8*, 308.
7. Sedghi, L.; DiMassa, V.; Harrington, A.; Lynch, S.V.; Kapila, Y.L. The oral microbiome: role of key organisms and complex networks in oral health and disease. *Periodontology 2000* **2021**, *87*, 107–131.
8. Sampaio-Maia, B.; Caldas, I.M.; Pereira, M.L.; Perez-Mongiovi, D.; Araujo, R. The oral microbiome in health and its implication in oral and systemic diseases. *Advances in Applied Microbiology* **2016**, *97*, 171–210.
9. Peng, X.; Cheng, L.; You, Y.; Tang, C.; Ren, B.; Li, Y.; Xu, X.; Zhou, X. Oral microbiota in human systematic diseases. *International Journal of Oral Science* **2022**, *14*, 14.
10. Sharma, B.; Kumar, A.; Sharma, U.; Pal, D.; Prashar, S. The potential role of gut microbiota in the pathogenesis of type 2 diabetes mellitus via epigenetics and inflammasome. *Endocrine, Metabolic & Immune Disorders-Drug Targets (Formerly Current Drug Targets-Immune, Endocrine & Metabolic Disorders)* **2022**, *22*, 1331–1343.
11. Radaic, A.; Kapila, Y.L. The oralome and its dysbiosis: new insights into oral microbiome-host interactions. *Computational and Structural Biotechnology Journal* **2021**, *19*, 1335–1360.
12. Costalonga, M.; Herzberg, M.C. The oral microbiome and the immunobiology of periodontal disease and caries. *Immunology Letters* **2014**, *162*, 22–38.
13. Hou, K.; Wu, Z.-X.; Chen, X.-Y.; Wang, J.-Q.; Zhang, D.; Xiao, C.; Zhu, D.; Koya, J.B.; Wei, L.; Li, J. Microbiota in health and diseases. *Signal Transduction and Targeted Therapy* **2022**, *7*, 1–28.
14. Qin, H.; Li, G.; Xu, X.; Zhang, C.; Zhong, W.; Xu, S.; Yin, Y.; Song, J. The role of oral microbiome in periodontitis under diabetes mellitus. *Journal of Oral Microbiology* **2022**, *14*, 2078031.
15. Nicholson, J.S.; Landry, K.S. Oral dysbiosis and neurodegenerative diseases: correlations and potential causations. *Microorganisms* **2022**, *10*, 1326.
16. Park, S.-Y.; Hwang, B.-O.; Lim, M.; Ok, S.-H.; Lee, S.-K.; Chun, K.-S.; Park, K.-K.; Hu, Y.; Chung, W.-Y.; Song, N.-Y. Oral-gut microbiome axis in gastrointestinal disease and cancer. *Cancers* **2021**, *13*, 2124.
17. Olsen, I.; Yamazaki, K. Can oral bacteria affect the microbiome of the gut? *Journal of Oral Microbiology* **2019**, *11*, 1586422.
18. Imai, J.; Kitamoto, S.; Kamada, N. The pathogenic oral-gut-liver axis: new understandings and clinical implications. *Expert Review of Clinical Immunology* **2021**, *17*, 727–736.

19. Ansari, S.A.; Khan, T.A.; Kazmi, S.U. Association of oral Helicobacter pylori with gastric complications. *Life Sciences* **2018**, *205*, 125–130.
20. Adler, I.; Muin, A.; Aguas, S.; Harada, L.; Diaz, M.; Lence, A.; Labbrozzi, M.; Muino, J.M.; Elsner, B.; Avagnina, A. Helicobacter pylori and oral pathology: relationship with the gastric infection. *World Journal of Gastroenterology: WJG* **2014**, *20*, 9922.
21. Flemer, B.; Warren, R.D.; Barrett, M.P.; Cisek, K.; Das, A.; Jeffery, I.B.; Hurley, E.; Micheal, O.R.; Shanahan, F.; Paul, W.T. The oral microbiota in colorectal cancer is distinctive and predictive. *Gut* **2018**, *67*, 1454–1463.
22. Koliarakis, I.; Messaritakis, I.; Nikolouzakis, T.K.; Hamilos, G.; Souglakos, J.; Tsiaoussis, J. Oral bacteria and intestinal dysbiosis in colorectal cancer. *International Journal of Molecular Sciences* **2019**, *20*, 4146.
23. Chowdhury, M.; Raj Chaudhary, N.; Kaur, P.; Goyal, A.; Sahu, S.K. Different Strategies Targeting Gut Microbiota for the Management of Several Disorders: a Sustainable Approach. *Infectious Disorders-Drug Targets (Formerly Current Drug Targets-Infectious Disorders)* **2024**, *24*, 45–56.
24. Alipour, M. Molecular mechanism of Helicobacter pylori-induced gastric cancer. *Journal of Gastrointestinal Cancer* **2021**, *52*, 23–30.
25. Polk, D.B.; Peek Jr, R.M. Helicobacter pylori: gastric cancer and beyond. *Nature Reviews Cancer* **2010**, *10*, 403–414.
26. Yamashita, Y.; Takeshita, T. The oral microbiome and human health. *Journal of Oral Science* **2018**, *59*, 201–206.
27. Whitmore, S.E.; Lamont, R.J. Oral bacteria and cancer. *PLoS Pathogens* **2014**, *10*, e1003933.
28. Givony, S. Oral squamous cell carcinoma (OSCC) an overview. *Journal of Medical Sciences.* **2020**, *8*, 67–74.
29. Chang, C.; Geng, F.; Shi, X.; Li, Y.; Zhang, X.; Zhao, X.; Pan, Y. The prevalence rate of periodontal pathogens and its association with oral squamous cell carcinoma. *Applied Microbiology and Biotechnology* **2019**, *103*, 1393–1404.
30. Irfan, M.; Delgado, R.Z.R.; Frias-Lopez, J. The oral microbiome and cancer. *Frontiers in Immunology* **2020**, *11*, 591088.
31. Peters, B.A.; Wu, J.; Pei, Z.; Yang, L.; Purdue, M.P.; Freedman, N.D.; Jacobs, E.J.; Gapstur, S.M.; Hayes, R.B.; Ahn, J. Oral microbiome composition reflects prospective risk for esophageal cancers. *Cancer Research* **2017**, *77*, 6777–6787.
32. Gholizadeh, P.; Eslami, H.; Yousefi, M.; Asgharzadeh, M.; Aghazadeh, M.; Kafil, H.S. Role of oral microbiome on oral cancers, a review. *Biomedicine & Pharmacotherapy* **2016**, *84*, 552–558.
33. Irfan, M.; Solbiati, J.; Duran-Pinedo, A.; Rocha, F.G.; Gibson Iii, F.C.; Frias-Lopez, J. A *Porphyromonas gingivalis* hypothetical protein controlled by the type IC CRISPR-Cas system is a novel adhesin important in virulence. *Msystems* **2023**, e01231–01223.
34. Zhang, Y.; Niu, Q.; Fan, W.; Huang, F.; He, H. Oral microbiota and gastrointestinal cancer. *OncoTargets and Therapy* **2019**, 4721–4728.
35. Harrandah, A.M.; Chukkapalli, S.S.; Bhattacharyya, I.; Progulske-Fox, A.; Chan, E.K.L. Fusobacteria modulate oral carcinogenesis and promote cancer progression. *Journal of Oral Microbiology* **2021**, *13*, 1849493.
36. Gallimidi, A.B.; Fischman, S.; Revach, B.; Bulvik, R.; Maliutina, A.; Rubinstein, A.M.; Nussbaum, G.; Elkin, M. Periodontal pathogens *Porphyromonas gingivalis* and *Fusobacterium nucleatum* promote tumor progression in an oral-specific chemical carcinogenesis model. *Oncotarget* **2015**, *6*, 22613.
37. Okuyama, K.; Yanamoto, S. Oral bacterial contributions to gingival carcinogenesis and progression. *Cancer Prevention Research* **2023**, *16*, 199–209.
38. Zhou, Y.; Luo, G.-H. *Porphyromonas gingivalis* and digestive system cancers. *World Journal of Clinical Cases* **2019**, *7*, 819.
39. Groeger, S.; Zhou, Y.; Ruf, S.; Meyle, J. Pathogenic mechanisms of Fusobacterium nucleatum on oral epithelial cells. *Frontiers in Oral Health* **2022**, *3*, 831607.
40. White, M.G.; Damania, A.; Alshenaifi, J.; Sahasrabhojane, P.; Peacock, O.; Losh, J.; Wong, M.C.; Lutter-Berkova, Z.; Chang, G.J.; Futreal, A. Young-onset Rectal Cancer: unique Tumoral Microbiome and Correlation with Response to Neoadjuvant Therapy. *Annals of Surgery* **2023**, *278*, 538–548.
41. Asili, P.; Mirahmad, M.; Rezaei, P.; Mahdavi, M.; Larijani, B.; Tavangar, S.M. The association of oral microbiome dysbiosis with gastrointestinal cancers and its diagnostic efficacy. *Journal of Gastrointestinal Cancer* **2023**, *54*, 1082–1101.

42. Zhang, Z.; Yang, J.; Feng, Q.; Chen, B.; Li, M.; Liang, C.; Li, M.; Zhang, L. Compositional and functional analysis of the microbiome in tissue and saliva of oral squamous cell carcinoma. *Frontiers in Microbiology* **2019**, *10*, 451189.
43. Shetty, B.; Fazal, I.; Khan, S.F.; Nambiar, M.; Prasad, R.; Raj, A. Association between cardiovascular diseases and periodontal disease: more than what meets the eye. *Drug Target Insights* **2023**, *17*, 31.
44. Carra, M.C.; Range, H.; Caligiuri, G.; Bouchard, P. Periodontitis and atherosclerotic cardiovascular disease: a critical appraisal. *Periodontology 2000* **2023**.
45. Czerniuk, M.R.; Surma, S.; Romańczyk, M.; Nowak, J.M.; Wojtowicz, A.; Filipiak, K.J. Unexpected Relationships: periodontal diseases: atherosclerosis-plaque destabilization? From the teeth to a coronary event. *Biology (Basel)* **2022**, *11*, doi: 10.3390/biology11020272.
46. Curia, M.C.; Pignatelli, P.; D'Antonio, D.L.; D'Ardes, D.; Olmastroni, E.; Scorpiglione, L.; Cipollone, F.; Catapano, A.L.; Piattelli, A.; Bucci, M.; et al. Oral Porphyromonas gingivalis and Fusobacterium nucleatum Abundance in Subjects in Primary and Secondary Cardiovascular Prevention, with or without Heterozygous Familial Hypercholesterolemia. *Biomedicines* **2022**, 10, doi: 10.3390/biomedicines10092144.
47. Gianos, E.; Jackson, E.A.; Tejpal, A.; Aspry, K.; O'Keefe, J.; Aggarwal, M.; Jain, A.; Itchhaporia, D.; Williams, K.; Batts, T.; et al. Oral health and atherosclerotic cardiovascular disease: a review. *Am. J. Prev. Cardiol.* **2021**, *7*, 100179, doi: 10.1016/j.ajpc.2021.100179.
48. Gonzalez-Soltero, R.; Bailen, M.; de Lucas, B.; Ramirez-Goercke, M.I.; Pareja-Galeano, H.; Larrosa, M. Role of oral and gut microbiota in dietary nitrate metabolism and its impact on sports performance. *Nutrients* **2020**, *12*, doi 10.3390/nu12123611.
49. Chopra, A.; Franco-Duarte, R.; Rajagopal, A.; Choowong, P.; Soares, P.; Rito, T.; Eberhard, J.; Jayasinghe, T.N. Exploring the presence of oral bacteria in non-oral sites of patients with cardiovascular diseases using whole metagenomic data. *Scientific Reports* **2024**, *14*, 1476.
50. Tahami Monfared, A.A.; Byrnes, M.J.; White, L.A.; Zhang, Q. Alzheimer's disease: epidemiology and clinical progression. *Neurology and Therapy* **2022**, *11*, 553–569.
51. Behl, T.; Kaur, D.; Sehgal, A.; Singh, S.; Sharma, N.; Zengin, G.; Andronie-Cioara, F.L.; Toma, M.M.; Bungau, S.; Bumbu, A.G. Role of monoamine oxidase activity in Alzheimer's disease: an insight into the therapeutic potential of inhibitors. *Molecules* **2021**, *26*, 3724.
52. Vradenburg, G. A pivotal moment in Alzheimer's disease and dementia: how global unity of purpose and action can beat the disease by 2025. *Expert Review of Neurotherapeutics* **2015**, *15*, 73–82.
53. Tiwari, S.; Atluri, V.; Kaushik, A.; Yndart, A.; Nair, M. Alzheimer's disease: pathogenesis, diagnostics, and therapeutics. *International Journal of Nanomedicine* **2019**, 5541–5554.
54. Roth, S.; Burnie, N.; Suridjan, I.; Yan, J.T.; Carboni, M. Current diagnostic pathways for Alzheimer's disease: a cross-sectional real-world study across six countries. *Journal of Alzheimer's Disease Reports* **2023**, *7*, 659–674.
55. Behl, T.; Kaur, G.; Sehgal, A.; Bhardwaj, S.; Singh, S.; Buhas, C.; Judea-Pusta, C.; Uivarosan, D.; Munteanu, M.A.; Bungau, S. Multifaceted role of matrix metalloproteinases in neurodegenerative diseases: pathophysiological and therapeutic perspectives. *International Journal of Molecular Sciences* **2021**, *22*, 1413.
56. Miller, A.L.; Bessho, S.; Grando, K.; Tukel, C. Microbiome or infections: amyloid-containing biofilms as a trigger for complex human diseases. *Frontiers in Immunology* **2021**, *12*, 638867.
57. Kozak, M.; Pawlik, A. The role of the oral microbiome in the development of diseases. *International Journal of Molecular Sciences* **2023**, *24*, 5231.
58. Sansores-Espana, L.D.; Melgar-Rodriguez, S.; Olivares-Sagredo, K.; Cafferata, E.A.; Martinez-Aguilar, V.M.; Vernal, R.; Paula-Lima, A.C.; Diaz-Zuniga, J. Oral-gut-brain axis in experimental models of periodontitis: associating gut dysbiosis with neurodegenerative diseases. *Frontiers in Aging* **2021**, *2*, 781582.
59. Diaz-Zuninga, J.; Munoz, Y.; Melgar-Rodriguez, S.; More, J.; Bruna, B.; Lobos, P.; Monasterio, G.; Vernal, R.; Paula-Lima, A. Serotype b of *Aggregatibacter actinomycetemcomitans* triggers pro-inflammatory responses and amyloid beta secretion in hippocampal cells: a novel link between periodontitis and Alzheimer's disease? *Journal of Oral Microbiology* **2019**, *11*, 1586423.
60. Singhrao, S.K.; Olsen, I. Assessing the role of Porphyromonas gingivalis in periodontitis to determine a causative relationship with Alzheimer's disease. *Journal of Oral Microbiology* **2019**, *11*, 1563405.
61. Noble, J.M.; Scarmeas, N.; Celenti, R.S.; Elkind, M.S.V.; Wright, C.B.; Schupf, N.; Papapanou, P.N. Serum IgG antibody levels to periodontal microbiota are associated with incident Alzheimer disease. *PloS one* **2014**, *9*, e114959.

62. Morikawa, T.; Uehara, O.; Paudel, D.; Yoshida, K.; Harada, F.; Hiraki, D.; Sato, J.; Matsuoka, H.; Kuramitsu, Y.; Michikawa, M. Systemic administration of lipopolysaccharide from *Porphyromonas gingivalis* decreases neprilysin expression in the mouse hippocampus. *In Vivo* **2023**, *37*, 163–172.
63. Wang, R.P.-H.; Ho, Y.-S.; Leung, W.K.; Goto, T.; Chang, R.C.-C. Systemic inflammation linking chronic periodontitis to cognitive decline. *Brain, Behavior, and Immunity* **2019**, *81*, 63–73.
64. di Cologna, N.M.; Samaddar, S.; Valle, C.A.; Vargas, J.; Aviles-Reyes, A.; Morales, J.; Ganguly, T.; Pileggi, R.; Brady, L.J.; Lemos, J. Amyloid aggregation of Streptococcus mutans Cnm influences its collagen-binding activity. *Applied and Environmental Microbiology* **2021**, *87*, e01149–01121.
65. Ahmad, R.; Haque, M. Oral health messiers: diabetes mellitus relevance. *Diabetes, Metabolic Syndrome and Obesity* **2021**, 3001–3015.
66. Verhulst, M.J.L.; Loos, B.G.; Gerdes, V.E.A.; Teeuw, W.J. Evaluating all potential oral complications of diabetes mellitus. *Frontiers in Endocrinology* **2019**, *10*, 56.
67. Dhar, A.K.; Anitha, K.; Bhatt, S. Immune system and diabetes mellitus. In *Biochemical Immunology of Diabetes and Associated Complications*; Elsevier: **2024**; pp. 19–47.
68. Liccardo, D.; Cannavo, A.; Spagnuolo, G.; Ferrara, N.; Cittadini, A.; Rengo, C.; Rengo, G. Periodontal disease: a risk factor for diabetes and cardiovascular disease. *International journal of Molecular Sciences* **2019**, *20*, 1414.
69. Rapone, B.; Ferrara, E.; Corsalini, M.; Qorri, E.; Converti, I.; Lorusso, F.; Delvecchio, M.; Gnoni, A.; Scacco, S.; Scarano, A. Inflammatory status and glycemic control level of patients with type 2 diabetes and periodontitis: a randomized clinical trial. *International Journal of Environmental Research and Public Health* **2021**, *18*, 3018.
70. Thouvenot, K.; Turpin, T.; Taile, J.; Clement, K.; Meilhac, O.; Gonthier, M.-P. Links between insulin resistance and periodontal bacteria: insights on molecular players and therapeutic potential of polyphenols. *Biomolecules* **2022**, *12*, 378.
71. Mosaddad, S.A.; Tahmasebi, E.; Yazdanian, A.; Rezvani, M.B.; Seifalian, A.; Yazdanian, M.; Tebyanian, H. Oral microbial biofilms: an update. *European Journal of Clinical Microbiology & Infectious Diseases* **2019**, *38*, 2005–2019.
72. Nyvad, B.; Takahashi, N. Integrated hypothesis of dental caries and periodontal diseases. *Journal of Oral Microbiology* **2020**, *12*, 1710953.
73. Lalla, E.; Papapanou, P.N. Diabetes mellitus and periodontitis: a tale of two common interrelated diseases. *Nature Reviews Endocrinology* **2011**, *7*, 738–748.
74. Correa, J.D.; Fernandes, G.R.; Calderaro, D.C.; Mendonca, S.M.S.; Silva, J.M.; Albiero, M.L.; Cunha, F.Q.; Xiao, E.; Ferreira, G.A.; Teixeira, A.L. Oral microbial dysbiosis linked to worsened periodontal condition in rheumatoid arthritis patients. *Scientific Reports* **2019**, *9*, 8379.
75. Xiao, F.; Li, C.; Lin, Y.; Peng, Z.; Xu, X.; Wen, Y.; Huang, N.; Zhang, P. Increased risk of periodontitis occurrence in patients with rheumatoid arthritis and its association with the levels of IL-1β and TNF-α in gingival crevicular fluid. *Annals of Palliative Medicine* **2021**, *10*, 9078087–9079087.
76. Hajishengallis, G.; Chavakis, T. Local and systemic mechanisms linking periodontal disease and inflammatory comorbidities. *Nature Reviews Immunology* **2021**, *21*, 426–440.
77. Persson, G.R. Rheumatoid arthritis and periodontitis-inflammatory and infectious connections. Review of the literature. *Journal of Oral Microbiology* **2012**, *4*, 11829.
78. Krutyholowa, A.; Strzelec, K.; Dziedzic, A.; Bereta, G.P.; Lazarz-Bartyzel, K.; Potempa, J.; Gawron, K. Host and bacterial factors linking periodontitis and rheumatoid arthritis. *Frontiers in Immunology* **2022**, *13*, 980805.
79. Tilvawala, R.; Nguyen, S.H.; Maurais, A.J.; Nemmara, V.V.; Nagar, M.; Salinger, A.J.; Nagpal, S.; Weerapana, E.; Thompson, P.R. The rheumatoid arthritis-associated citrullinome. *Cell Chemical Biology* **2018**, *25*, 691–704.e696.
80. Martinez-Garcia, M.; Hernandez-Lemus, E. Periodontal inflammation and systemic diseases: an overview. *Frontiers in physiology* **2021**, *12*, 709438.
81. Wilson, T.M.; Trent, B.; Kuhn, K.A.; Demoruelle, M.K. Microbial influences of mucosal immunity in rheumatoid arthritis. *Current Rheumatology Reports* **2020**, *22*, 1–8.
82. du Teil Espina, M.; Gabarrini, G.; Harmsen, H.J.M.; Westra, J.; van Winkelhoff, A.J.; van Dijl, J.M. Talk to your gut: the oral-gut microbiome axis and its immunomodulatory role in the etiology of rheumatoid arthritis. *FEMS Microbiology reviews* **2019**, *43*, 1–18.
83. Vidmar Šimic, M.; Maver, A.; Zimani, A.N.; Hočevar, K.; Peterlin, B.; Kovanda, A.; Premru-Sršen, T. Oral microbiome and preterm birth. *Frontiers in Medicine* **2023**, *10*, 1177990.

84. Puertas, A.; Magan-Fernandez, A.; Blanc, V.; Revelles, L.; O'Valle, F.; Pozo, E.; León, R.; Mesa, F. Association of periodontitis with preterm birth and low birth weight: a comprehensive review. *The Journal of Maternal-Fetal & Neonatal Medicine* **2018**, *31*, 597–602.
85. Cobb, C.M.; Kelly, P.J.; Williams, K.B.; Babbar, S.; Angolkar, M.; Derman, R.J. The oral microbiome and adverse pregnancy outcomes. *International Journal of Women's Health* **2017**, 551–559.
86. Chopra, A.; Radhakrishnan, R.; Sharma, M. Porphyromonas gingivalis and adverse pregnancy outcomes: a review on its intricate pathogenic mechanisms. *Critical Reviews in Microbiology* **2020**, *46*, 213–236.
87. Cekici, A.; Kantarci, A.; Hasturk, H.; Van Dyke, T.E. Inflammatory and immune pathways in the pathogenesis of periodontal disease. *Periodontology 2000* **2014**, *64*, 57–80.
88. Bayardo-Gonzalez, R.A.; Pena-Rodriguez, M.; Pereira-Suarez, A.L.; Rubio-Sanchez, A.X.; Garcia-Chagollan, M. Valenzuela-Orozco, D.N.; Lizarazo-Taborda, M.d.R.; Mora-Mora, J.; Vega-Magaria, N. Insights into estrogen impact in oral health & microbiome in COVID-19. *BMC Microbiology* **2024**, *24*, 32.

13 Role of Oral Bacteria in Cardiac Diseases

Gaytri Mahajan, Sandipan Acharjee, Sanjana Gupta, Hema Bhardwaj, and Reena Gupta
Department of Biotechnology, Himachal Pradesh University, Shimla, Himachal Pradesh, India

13.1 INTRODUCTION

The population of microbes that live in our body is referred to as the microbiome. The term "microbiome" was coined by Nobel laureate Joshua Lederberg to refer to the biological community that includes commensal, pathogenic, and symbiotic microorganisms [1]. The oral microflora or oral microbiome refers to the bacteria that are present in the human mouth cavity [2]. The oral microbiome was first identified by the Dutchman Antonie van Leeuwenhoek using a handmade microscope [3]. The oral microbiome is the collective genome of all the microorganisms that inhabit the oral cavity. After the gut, it is the second-largest microbial colony in humans. Compared to other body organs, they have an amazing spectrum of predicted protein activity [4, 5]. The human microbiome is composed of a core microbiome and a changeable microbiome. Everybody has the same basic microbiome, but because of differences in lifestyle and physiological composition, everyone also has a unique variable microbiome. The oral cavity's hard and soft structures, including the mucosa, are susceptible to bacterial invasion [6, 7]. The surfaces of the oral cavity are covered in a thick layer of bacteria known as bacterial biofilm [8]. The average body temperature of the human mouth cavity is 37°C, providing a stable environment for these bacteria. Saliva serves as a medium that transports nutrients to microorganisms while also keeping the bacteria hydrated [9]. The intricate association between oral health and overall well-being has long been a subject of fascination and concern within the field of medicine.

While the importance of maintaining good oral hygiene is well-recognized for preventing conditions such as gum disease and tooth decay, a growing body of research has shed light on a more surprising connection_the link between oral health and cardiac diseases. Cardiac diseases, like coronary artery disease, heart attacks, and stroke, are among the leading causes of mortality worldwide. In recent years, scientists have begun to unravel a complex and unexpected connection between the bacteria that inhabit our mouths and the health of our hearts [10]. Our mouth hosts a diverse and dynamic community of microorganisms, collectively known as the oral microbiome. While some of these bacteria are benign and even beneficial, others can become opportunistic pathogens, causing inflammation and infections in the oral cavity. As research advances, it is becoming increasingly evident that these oral pathogens can journey beyond the mouth, gaining access to the bloodstream and potentially impacting distant organs, including the heart.

One of the significant pathological processes in coronary heart disease is atherosclerosis. Atherosclerosis is closely related to alterations in vascular intimal cell function and vascular smooth muscle cell proliferation. In the vascular endothelium of middle and large arteries, lipid deposition occurs, which results in atherosclerotic plaques that are dispersed or flaky and narrow the arterial lumen, the primary pathological symptom of atherosclerosis [11]. Bacteria associated with periodontal disease induce inflammation, damage to vascular endothelial cells, and the formation of atherosclerotic plaques when they penetrate the bloodstream through periodontal tissue injury [12, 13]. These factors affect the immune system [14]. When a mouse's mouth is inoculated with *Porphyromonas gingivalis*, there will be a drop in the number of Firmicutes and an increase in Bacteroides, both of

DOI: 10.1201/9781003478638-13

which are linked to systemic inflammatory reactions and endotoxemia [15, 16]. According to reports, the emergence of atherosclerotic vascular disease is significantly correlated with an elevated amount of C-reactive protein (CRP) in the body due to an oral bacterial infection [17]. Infective endocarditis is associated with oral bacteria *Streptococcus* and *Lactobacillus* [18, 19]. *Streptococcus sanguinis* is an early dental plaque colonizer; it has the highest rate of detection in patient endocardium and is closely linked to the endocarditis inflammatory process [20]. The development of thrombus-like protrusions on the surface of the vascular endothelium may help in pathogenesis, which encourages bacterial adherence and ultimately results in infection. The animal model research of endocarditis significantly increases the virulence factor of *S. gordonii*, suggesting a potential relationship between oral bacteria virulence factor and endocardial infection process [21]. C-reactive protein, a marker of systemic inflammatory expansion, was significantly lower in patient serum after periodontitis treatment, as were some other serum inflammatory markers [22, 23]. As a result, the inflammatory components in the serum have a direct impact on the level of periodontal inflammation. The nexus between oral bacteria and cardiac diseases is an emerging field that bridges the domains of dentistry and cardiology. This chapter seeks to unravel the various mechanisms through which oral bacteria might contribute to the development of cardiac diseases and the risk factors that come into play. Throughout this chapter, we will navigate the oral microbiome and its locations, different virulence factors of oral bacteria, periodontitis, different cardiac diseases related to oral bacteria, inflammatory markers, and the epigenetic link between oral bacteria and cardiac diseases, shedding light on the importance of oral hygiene and regular dental checkups.

13.2 ORAL MICROBIOME

The human oral cavity represents a remarkable and intricate ecosystem, teeming with a diverse community of microorganisms collectively known as the oral microbiome. Within this complex microbial world, a wide array of microorganisms, including bacteria, viruses, fungi, and archaea, coexist in a dynamic equilibrium. While the oral microbiome includes a variety of microorganisms, bacteria predominantly occupy this microbial landscape, making them the primary focus of scientific inquiry [24]. The diversity observed within the oral microbiome is, in part, a reflection of the uniqueness of each individual's oral ecosystem [25]. Genetics, lifestyle, dietary choices, and environmental factors all contribute to shaping the composition of the oral microbiome, resulting in a personalized microbial community that varies from person to person. This individuality underscores the intricate nature of the oral microbiome and its adaptability to the specific conditions of each individual's oral environment. In biofilms, these microorganisms grow on the surfaces of teeth as complex, varied, and mutually reliant colonies that are closely packed against the tooth in the deeper layers and are more mobile in the superficial layers [26, 27]. According to cultural research, dental plaque contains more than 500 different bacteria species. However, 16S rDNA amplification molecular techniques provide an even more diverse picture of the subgingival bacterial ecology and suggest that a substantial portion of even this well-researched and well-known microbial environment remains uncharacterized [28, 29]. The bacterial count on the supragingival (above the gum line) surface of a single tooth can surpass 1×10^9 bacteria. A healthy shallow cleft beneath the gum has 1×10^3 bacteria, but a periodontal pocket has more than 1×10^8 [30, 31]. Every 48 hours, teeth cleanings can keep the biofilm mass at a level that doesn't compromise gingival health [32]. Within the oral microbiome, a significant portion of microorganisms is considered commensal, meaning they coexist harmoniously with their human host and offer various essential benefits. These beneficial bacteria play critical roles in maintaining overall health. They participate in digestion, aid in the synthesis of vitamins, and contribute to the protection of the host against harmful invaders. The commensal bacteria in the oral microbiome are, therefore, essential contributors to several physiological processes that support human well-being [33].

However, alongside these beneficial microorganisms, there are pathogenic bacteria that have the potential to cause oral health issues when their populations grow unchecked. These pathogenic bacteria can lead to dental problems such as cavities and periodontal disease, undermining the integrity of the oral cavity. In more severe cases, pathogenic oral bacteria can gain access to the bloodstream, potentially leading to systemic infections [34]. This duality of commensal and pathogenic microorganisms highlights the complex nature of the oral microbiome and the balance that must be maintained for optimal oral health. The oral microbiome is also characterized by intricate and dynamic interactions between different species of microorganisms [35]. Within this ecosystem, bacteria compete for space and essential nutrients, form symbiotic relationships, and produce substances that can influence the growth and behavior of other microorganisms. These interactions are not static but change over time in response to various factors, including diet, oral hygiene practices, and the individual's overall health. Understanding these interactions is critical to unraveling the complexities of the oral microbiome and its potential impact on systemic health [36, 37]. Moreover, the composition of the oral microbiome can vary in different regions of the mouth, including the tongue, teeth, cheeks, and gums. Each area offers a unique environment influenced by factors such as pH levels, moisture, and oxygen content, creating distinct niches for specific microorganisms to thrive. This spatial distribution within the oral cavity adds another layer of complexity to the oral microbiome's composition and function [38, 39]. The diversity of oral microorganisms is a critical aspect of understanding the oral microbiome's significance. It underscores the intricate nature of this microbial world and its potential influence on overall health. Ongoing research in this field aims to delve deeper into the complexities of the oral microbiome, identifying novel species, elucidating their specific roles, and exploring potential connections between oral health and broader systemic health [40]. This expanding knowledge is instrumental in shedding light on the hidden world of the oral microbiome and may lead to innovative approaches for promoting not only oral health but also the well-being of the entire body [41].

13.2.1 Locations of the Oral Microbiota

The oral microbiota is a dynamic and diverse community of microorganisms that colonize various locations within the oral cavity, contributing significantly to the complex ecosystem of the mouth. One primary location for these microorganisms is the tooth surfaces, where dental plaque forms, particularly along the gumline [42]. Plaque is a biofilm composed of bacteria and other microorganisms embedded in a matrix of extracellular polymers, and its presence is closely linked to dental health. Apart from teeth, the tongue's surface serves as another significant habitat for oral microbiota. The tongue is covered with papillae and fissures, creating a textured environment where different microbial communities can thrive. The interactions between these microorganisms on the tongue contribute to the overall balance of the oral microbiome. Furthermore, the buccal mucosa (inner lining of the cheeks), palate (roof of the mouth), and throat are additional locations where distinct microbial populations reside. Each of these sites offers a unique environment influenced by factors such as moisture, acidity, and nutrient availability, shaping the composition of the microbial communities present [43]. Understanding the distribution and dynamics of oral microbiota in these various locations is crucial for comprehending oral health and disease. The delicate balance between different microbial species, as well as their interactions with the host's immune system, plays a pivotal role in maintaining oral health or contributing to conditions like dental caries, periodontal diseases, and systemic health issues [44]. Researchers and health care professionals continually explore the intricate relationships within the oral microbiota to develop strategies for promoting oral health and preventing associated diseases.

13.2.2 Virulence Factors Unveiled in Oral Bacteria

The intricate world of oral microbiology continues to reveal its secrets as researchers delve into the virulence factors of oral bacteria. These factors are essential components that enhance the pathogenicity of bacteria and play a crucial role in the development and progression of oral diseases.

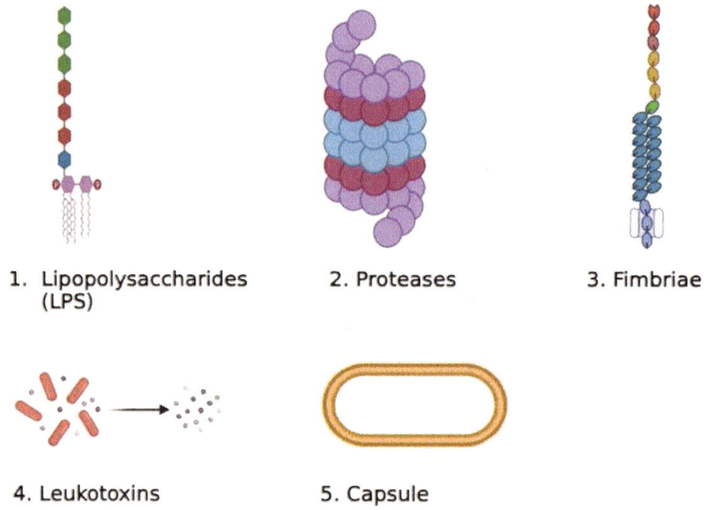

FIGURE 13.1 Various virulence factors of oral bacteria.

Unveiling these factors provides valuable insights into the complex interplay between bacteria and the host environment within the oral cavity. Figure 13.1 illustrates different virulence factors of oral bacteria.

13.2.2.1 Lipopolysaccharide

Lipopolysaccharide (LPS) constitutes the outer membrane of gram-negative bacteria. Toll-like receptors (TLRs) are the means by which the innate immune system identifies the LPS. Pathogen-associated microbial products (PAMPs) with preserved structures, like LPS, can be recognized by TLRs [43]. TLR-4 typically identifies bacterial LPS; however, *P. gingivalis* LPS is detected by TLR-2 due to a structural alteration [44]. TLR ligation initiates signal transduction pathways that promote acquired immunity specific to antigens and a fast innate inflammatory response [43]. When detecting bacterial LPS, CD14 functions as a coreceptor in conjunction with TLR-4. Patients with periodontal disease had higher serum levels of soluble CD14 and antibodies against LPS of periodontopathogens, which suggests that for these patients, prolonged exposure to bacterial LPSs produces systemic effects. Many proinflammatory molecules, such as prostaglandin E2 [44], IL-1, IL-6, and tumor necrosis factor-alpha (TNF-α), are secreted in response to CD14 activation. Inflammatory mediators, such as prostaglandins, histamine, bradykinin, and platelet activation factor, are then stimulated to release [45].

It is also known that bacterial LPS triggers the complement system, which is a collection of over 30 proteins involved in inflammation and tissue damage [46]. One of the first host immunological reactions to periodontal infection is complement system activation, and there exists a correlation between the degree of inflammation and the activation of the complement system [47]. Because of the proteolytic activity exhibited by cell surfaces, some oral bacteria are able to avoid being opsonized, which aids in the breakdown of complement system components.

13.2.2.2 Proteases

Proteases are classified based on their catalytic activity and their ability to cleave peptide bonds. Proteolytic enzymes can be secreted in significant quantities by *P. gingivalis*. Key virulence factors among *P. gingivalis*'s extracellular endopeptidases are cysteine proteases called gingipains. Although they are harmful to the host, their actions are necessary for the survival of the bacterium [48]. For *P. gingivalis* to remain attached to host tissues, gingipains facilitate the formation of fimbriae and binding of hemoglobin to collect host amino acids and heme, two essential sources of nourishment for the bacterium. These enzymes can break down a wide variety of proteins, such as collagens, which make up a significant amount of periodontal extracellular matrix proteins like laminin and fibronectin and connective tissue [49], which can cause periodontal tissue to break down. Additionally, Rgp and Kgp cleave and render inactive cytokines and immunoglobulins, impairing host defense systems [50]. This suppresses IL-6-mediated inflammatory reactions and facilitates the destruction of periodontal tissue [51].

Furthermore, gingipains were discovered to cleave IL-8 and IL-1b, among other cytokines. It has also been shown that Rgp and Kgp cause the coagulation cascade to be dysregulated and the kallikrein-kinin cascade to be activated. Furthermore, Rgp produces a chemotactic factor that attracts leukocytes to the periodontal tissues and increases vascular permeability [52]. By activating the kallikrein-kinin system in the coagulation intrinsic route, Kgp promotes alveolar bone resorption, which releases prostaglandin and bradykinin. Furthermore, gingipains have the ability to degrade the CD14 bacterial LPS receptor on the surface of T cells and monocytes, which reduces their sensitivity to LPS [53]. Additionally, it has been demonstrated that gingipains activate and induce matrix metalloproteinases (MMPs), which demolish periodontal tissue. These factors that increase tissue damage may also facilitate the entry of germs into the bloodstream [54]. Furthermore, animals given *P. gingivalis* gingipain mutant injections had higher levels of bacterial clearance. This could be because gingipains play a crucial role in finA maturation, which is necessary for bacterial invasion and adherence. Furthermore, the absence of gingipains will keep bacteria from getting heme, which is a nutrient supply needed for growth and development [55]. In a mouse model, gingipain immunization prevented *P. gingivalis* infection from spreading [56].

13.2.2.3 Fimbriae

The hair-like protein structures known as fimbriae are present on the cell surface of bacteria, especially on gram-negative bacteria. Fimbriae support bacterial adhesion to host cells, colonization, and resistance to host defenses [57]. The most researched of *P. gingivalis* proteins, fimA, is required for the initial adhesion and penetration of target cells, as well as their colonization of gingival tissues [58]. A *P. gingivalis* strain exhibiting loss of or a decrease in host cell adhesion or invasion was found to have mutations in the fimA genes [59]. Human endothelial, epithelial, fibroblastic, and periodontal ligament cells have all been demonstrated to be invaded by *P. gingivalis in vitro* [60]. Based on differences in nucleotide sequence, six genotypes of fimA from *P. gingivalis* (types I-V and Ib) have been categorized. Type II and IV fimA are thought to be more virulent and have been found to be more common in patients with periodontitis who also had cardiovascular disease, whereas type I fimA was primarily found in those who had good periodontal health. Additionally, *P. gingivalis* type IV showed higher hemagglutination activity compared to type II, which showed lower BAPNA (trypsin-like, arginine-specific activity) [61]. It has been demonstrated that type II fimA *P. gingivalis* impairs host cellular function, including adhesion, migration, and proliferation, because it can degrade paxillin, focal adhesion kinase, and integrin-related signaling molecules at much higher concentrations than other strains.

13.2.2.4 Leukotoxins

A class of exotoxins known as leukotoxins primarily damages leukocytes, especially polymorphonuclear cells (PMNs). *Aggregatibacter actinomycetemcomitans* leukotoxin creates holes in the lipid bilayer and, through a particular receptor on the cell membrane, is able to destroy human leukocytes. Human PMNs and macrophages are stimulated by leukotoxins to move cytoplasmic granules to the peripheral, whereupon they release their lysosomal contents and increase oxidative activity, causing inflammation and subsequent tissue damage to the host [62]. Additionally, leukotoxins stimulate endogenous nuclease, which increases the disintegration of chromosomal DNA and kills macrophages, PMNs, T lymphocytes, and natural killer (NK) cells. This process is known as programmed cell death, or apoptosis [63]. Crucially, leukotoxin aids in the ability of

A. actinomycetemcomitans to elude the host immune system and develop a local infection. In addition, leukotoxin serves as a bridge to link the bacteria to leukocytes, which permits the bacteria to silently proliferate throughout host tissues.

13.2.2.5 Capsules

Capsules prevent phagocytosis in addition to aiding in cell adhesion and the subsequent formation of biofilms. Capsular antigens have a role in cell recognition and are targets of the immune system. According to reports, *P. gingivalis* strains with capsules are more invasive and virulent than strains without [64].

13.3 PERIODONTITIS: AN OVERVIEW

Periodontal disease is a chronic inflammatory oral condition brought on by the localized infiltration and accumulation of oral bacteria. The kind of periodontal disease that is of interest is periodontitis, a disorder that damages the supporting tissues of teeth permanently. It is an inflammatory disease of the teeth's supporting structures that is chronic and progressive, ultimately resulting in their loss [65]. Plaque accumulation in predisposed individuals first causes gingivitis, which is a reversible inflammation of the gums and soft tissues surrounding the teeth, before periodontitis develops [66, 67]. Inflammatory mediators like tumor necrosis factor-α, interleukin-1β, interleukin-6, and C-reactive protein can be found in the bloodstream as early as the beginning of the disease. Gingivitis and chronic periodontitis are two serious infections of the gums that, if left untreated, can result in tooth loss. It has been estimated that 80% of adults worldwide have periodontal disease, with bacterial plaque, also known as biofilm, being the primary cause [68]. Biofilm is a sticky, colorless film that continuously forms on teeth. When gingivitis is left untreated, it can progress to periodontitis and cause red, swollen, and bleeding gums that sometimes happen when brushing teeth [69].

The bacteria in plaque biofilm produce toxins that irritate the gums, and over time, the film can grow and spread below the gum line (subgingival area). The body essentially turns on itself as a result of those toxins, causing a chronic inflammatory response that breaks down and destroys the tissues and bone supporting the teeth [70, 71]. During this process, the gums split off from the teeth, creating infected periodontal pockets—the spaces between the teeth and gums. The pockets get deeper, and more bone and gum tissue are lost as the condition worsens. The symptoms of this damaging process are often quite mild, though teeth may eventually come loose and require extraction. Despite the fact that the causes of chronic periodontitis are complex, there is evidence to suggest that certain gram-negative bacteria found in the biofilm of subgingival plaque are crucial to the disease's development. *Treponema denticola*, *Tannerella forsythia*, and *P. gingivalis* are the three subgingival plaque bacterial species for which there is strong evidence of roles in the onset and development of the disease, with *P. gingivalis* being identified as a major etiological agent [72]. Figure 13.2 depicts schematic visualization of healthy gums versus a periodontal-linked gum.

13.4 PERIODONTAL AND CARDIAC DISEASES

Periodontal disease, a chronic inflammatory condition affecting the supporting structures of the teeth, has garnered increased attention due to its potential impact on cardiovascular health. The link between periodontal disease and cardiovascular diseases has been the subject of extensive research, exploring the intricate connections between oral health and systemic well-being [73, 74]. Periodontal disease is characterized by inflammation of the gums (gingivitis) and, if left untreated, can progress to involve the surrounding bone (periodontitis). The oral cavity serves as a reservoir for diverse microbial communities, and in the presence of periodontal disease, the balance of these microorganisms is disrupted [75]. Chronic inflammation associated with periodontitis can lead to the release of inflammatory mediators and bacterial by-products into the bloodstream. Research has

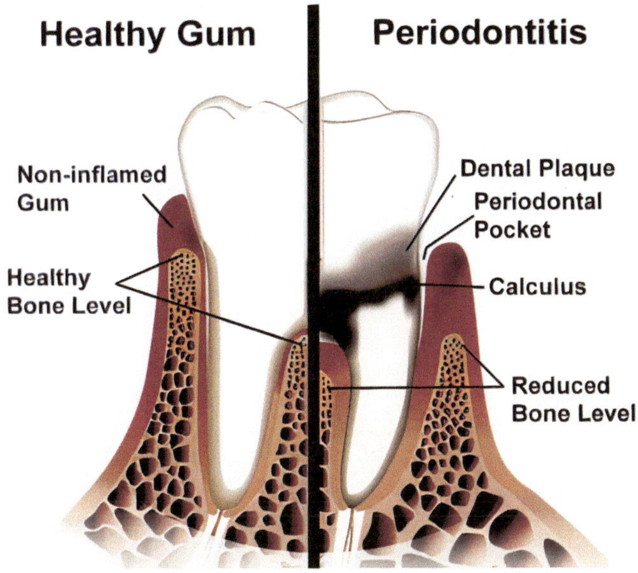

FIGURE 13.2 Schematic visualization of a healthy gum versus a periodontal-linked gum.

shown that these inflammatory mediators, such as cytokines and acute-phase proteins, can contribute to a systemic inflammatory burden [76]. This systemic inflammation is considered a potential factor in the development and progression of cardiovascular diseases. Moreover, specific periodontal pathogens, including *Porphyromonas gingivalis*, have been identified in atherosclerotic plaques, suggesting a potential role in the initiation or exacerbation of cardiovascular conditions. The mechanisms underlying the relationship between periodontal disease and cardiovascular health are multifaceted [77, 78]. Chronic inflammation may contribute to endothelial dysfunction, a critical step in atherosclerosis. Additionally, the systemic dissemination of periodontal pathogens may directly influence the vascular walls, promoting the formation of atherosclerotic lesions. Furthermore, the association between periodontal disease and cardiovascular health extends beyond inflammation. Shared risk factors, such as smoking, diabetes, and age, may contribute to both conditions, emphasizing the importance of a holistic approach to health management [79]. While the evidence supporting the link between periodontal disease and cardiovascular health is compelling, it is essential to acknowledge that correlation does not imply causation. Additional research is needed to elucidate the precise mechanisms and establish causative relationships definitively. In light of these findings, there is growing recognition of the importance of oral health in overall health maintenance [80]. Integrated health care approaches that consider the interplay between oral health and systemic conditions may hold promise in preventing and managing both periodontal disease and cardiovascular disorders. Figure 13.3 shows periodontitis and cardiac disease.

13.5 ATHEROSCLEROSIS: THE ROLE OF ORAL BACTERIA

Atherosclerosis is a complex and progressive cardiovascular condition characterized by the buildup of plaques within the arteries. These plaques consist of cholesterol, fatty substances, cellular waste, and other materials, ultimately leading to the narrowing and hardening of the arterial walls [81, 82]. While traditional risk factors such as high cholesterol, hypertension, and smoking are well-established contributors to atherosclerosis, emerging research suggests a potential link between oral bacteria, particularly those associated with periodontal disease, and the development or exacerbation of atherosclerosis. The development of coronary artery heart disease through the formation

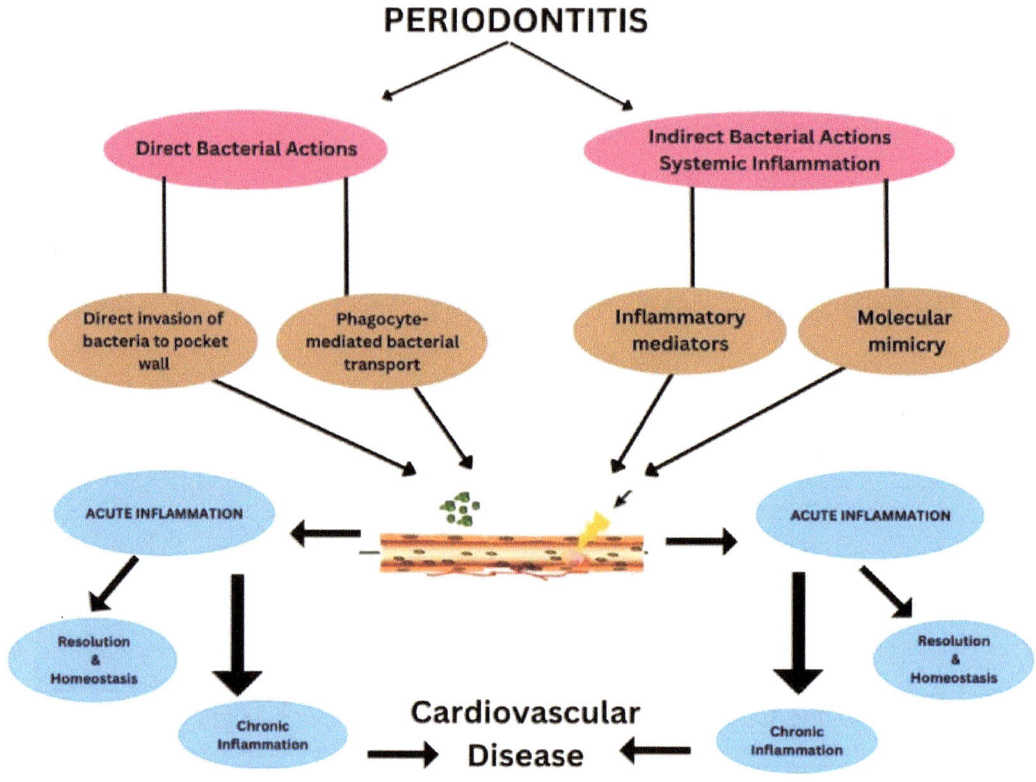

FIGURE 13.3 Periodontitis and cardiac disease.

of atherosclerotic plaques is termed atherogenesis. Key participants in this progression include immune cells such as T and B cells, along with monocytes and macrophages [83]. Atherosclerotic lesions in the arteries develop more quickly as a result of the secretion of inflammatory cytokines, metalloproteinases, and reactive oxygen species, which are released when inflammatory cells accumulate in the inner membrane of the artery [84]. Persistent inflammation promotes the formation of a necrotic core, composed of foam cells generated through the phagocytosis of oxidized lipid molecules, macrophages, and deceased smooth muscle cells [85]. The coronary artery's slender fibrous cap and extensive necrotic core constitute the fundamental factors behind plaque rupture and thrombus formation. Additionally, these signals induce the increased expression of chemoattractants [e.g., interleukin (IL)-8, thrombin] and adhesion molecules [e.g., intercellular adhesion molecule (ICAM)-1, vascular cell adhesion molecule (VCAM)-1, E-selectin, P-selectin] [86, 87]. Moreover, proinflammatory cytokines, such as IL-6 and IL-2, are released by migrating leukocytes along with proteases and reactive oxygen species (ROS) that help to degrade the extracellular matrix of endothelium [88]. Low-density lipoproteins (LDLs) accumulate beneath the intima layer of the endothelium, where macrophages phagocytize them, transforming them into foam cells. The formation of foam cells further amplifies the production of proinflammatory cytokines such as IL-1, IL-6, and TNF-α [2, 9]. Fibrosis and the development of fibrous plaque are brought on by the secretion of metalloproteinases (MMPs) by smooth muscle cells located between the media and intima layers of the arteries. MMPs stimulate myogenesis. Lipid streaks, which are caused by cholesterol deposits on the plaque over time, have the potential to obstruct vascular flow [88]. When plaque ruptures, thrombotic factors are released, which can lead to a thromboembolism by triggering coagulation when in contact with platelets. Figure 13.4 illustrates the association of periodontitis and acute cardiovascular diseases (ACVDs).

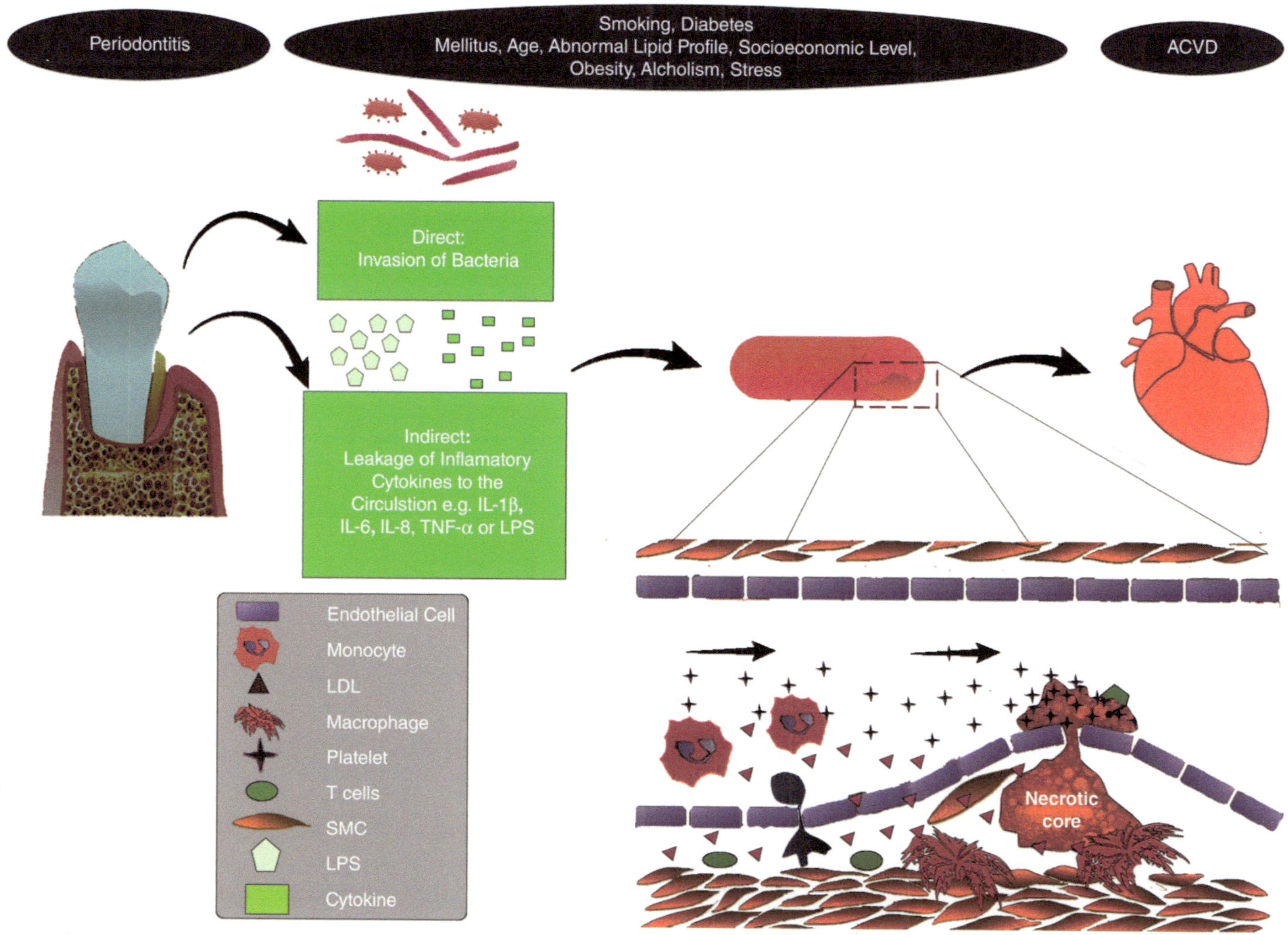

FIGURE 13.4 Association of periodontitis and acute cardiovascular diseases (ACVDs).

13.5.1 Pathogenesis of Atherosclerosis

Atherosclerosis begins with endothelial dysfunction, and its development is influenced by risk factors like smoking, hypertension, diabetes, abnormal cholesterol, and elevated inflammatory markers [89]. The process involves the buildup of oxidized LDL, triggering expression of adhesion molecules. Monocytes differentiate into macrophages, forming foam cells and fatty streaks. A T-leukocyte-induced immune response accelerates atherogenesis, leading to fibrous capsule formation and eventual plaque vulnerability [90]. Periodontitis, indicated by elevated gram-negative bacteria levels, is linked to increased risk of myocardial infarction (MI). Specific periodontal pathogens, including *P. gingivalis*, can enter the bloodstream, weaken epithelial barriers, and use intracellular tactics to evade host immunity. Cooperation between *F. nucleatum* and *T. forsythia* enhances *P. gingivalis* invasion [91]. Understanding these processes sheds light on the intricate connection between periodontal health and cardiovascular risks.

13.5.2 Mechanism Linking Periodontitis and Atherosclerosis

There are two possible explanations for how periodontitis affects ACVD. First, endothelial cells are directly invaded by periodontal pathogens through a direct mechanism. This polymerase chain reaction tests for atherosclerotic plaques corroborate this idea. *Prevotella intermedia*, *T. forsythia*, and *P. gingivalis* were among the other bacteria found in atherosclerotic lesions that developed in coronary arteries [92]. The exact mechanism by which the presence of bacteria intracellularly affects atherosclerosis is unknown, but certain pathogens, like *P. gingivalis*, may cause foam cells to form or remain within the cells, causing a secondary inflammatory state that eventually results in endothelial dysfunction. Figure 13.5 shows the mechanism of oral bacteria causing atherosclerosis [93].

The second theory, or indirect pathway, attributes periodontal disease-related increases in systemic inflammatory cytokine levels. Periodontal tissues release inflammatory mediators and bacterial byproducts into the systemic circulation. This pathway suggests that periodontal pathogens and their virulence factors, such as lipopolysaccharides, trigger a local immune response in the periodontal pocket, leading to the release of pro-inflammatory cytokines like IL-1β, IL-6, and TNF-α. These cytokines can then enter the bloodstream, contributing to a systemic inflammatory state. Atherosclerotic vascular disease and periodontal disease are linked to elevated levels of several cytokines, such as TNF-α, IL-1β, IL-6, IL-8, and monocyte chemoattractant protein-1, which are caused by a systemic inflammatory response [94]. Certain substances have the power to hasten the liver's production and release of intravascular plasma proteins, such as fibrinogen and CRP. Moreover, bacterial by-products such as LPS have the potential to enter the bloodstream and elicit a robust immune response [95]. The aforementioned factors may be the first to cause atherosclerosis by changing lipid metabolism, increasing oxidative stress and affecting endothelial cells. Other pathways are:

- *Inflammatory pathways*: Chronic inflammation is a hallmark of both periodontal disease and atherosclerosis. In the context of periodontal disease, the constant presence of bacterial biofilm triggers an immune response, resulting in the release of inflammatory mediators such as cytokines and chemokines. These inflammatory molecules can enter the bloodstream, contributing to systemic inflammation. In atherosclerosis, inflammation plays a crucial role in the initiation and progression of plaques [96]. The artery-lining endothelial cells are activated by a variety of stimuli, including signals related to inflammation. This activation makes it possible for immune cells, especially macrophages, to penetrate the arterial walls while causing fatty streaks and ultimately mature plaques to develop.
- *Systemic dissemination of oral pathogens*: Certain oral bacteria, most notably *Porphyromonas gingivalis*, have been found in atherosclerotic plaques. This implies that

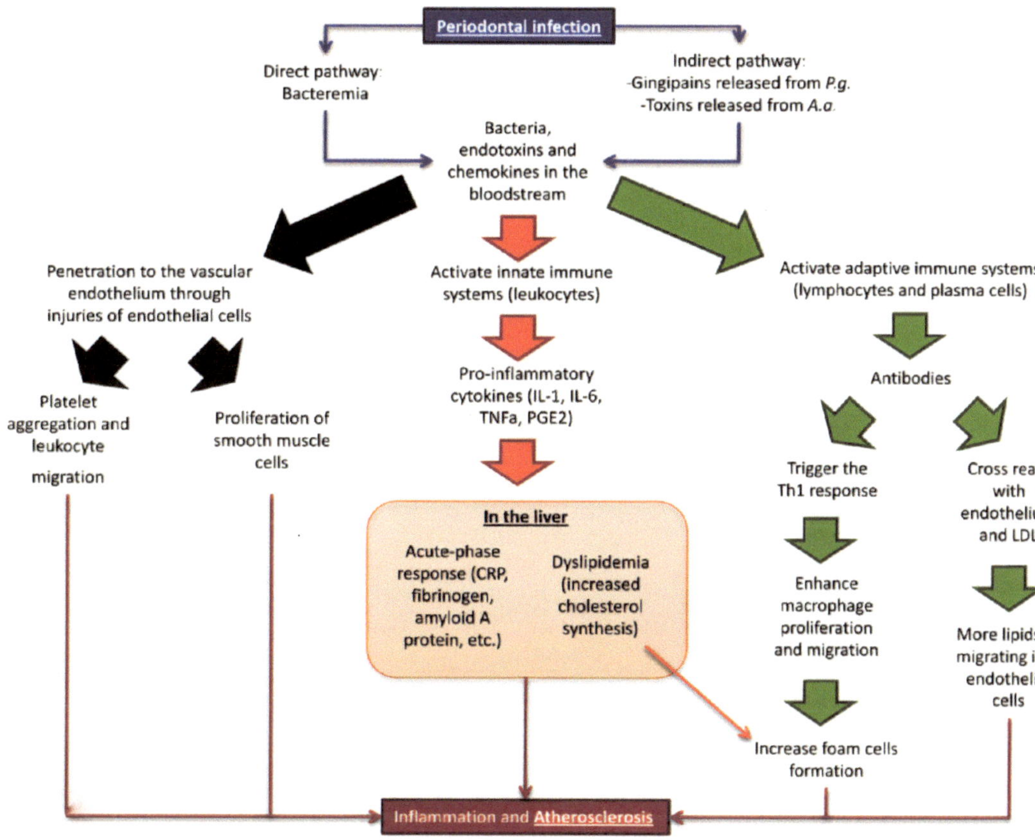

FIGURE 13.5 Mechanism of oral bacteria causing atherosclerosis.

oral pathogens that start in the mouth have the ability to get into the bloodstream and travel long distances, such as through the arteries. These bacteria may exacerbate the inflammation that is linked to atherosclerosis once they are inside the arterial walls.

- *Induction of endothelial dysfunction*: Atherosclerosis's pivotal early event is endothelial dysfunction. The endothelial cells that line the arteries may become less functional as a result of the long-term inflammation that oral bacteria cause. These cells' capacity to control vascular tone, encourage vasodilation, and inhibit the adhesion of immune cells and platelets—all crucial for preserving vascular health—is compromised by this dysfunction. [97].
- *Platelet activation and thrombosis*: Plaques may become unstable and burst in advanced stages of atherosclerosis, disclosing the tissue beneath them into the bloodstream. Blood clots form, and platelet activation is triggered by this event. Because oral bacteria are involved in systemic inflammation, they may increase the risk of a prothrombotic state and clot formation in the arteries [98]. Although there is strong evidence linking oral bacteria to atherosclerosis, more research is necessary to definitively establish the causal relationship. Understanding that atherosclerosis is a complex disorder impacted by a variety of environmental, lifestyle, and genetic variables is crucial. However, the possibility that oral bacteria contribute to the inflammatory environment linked to atherosclerosis highlights the significance of oral hygiene as a crucial element of cardiovascular health [99]. To prevent and treat diseases like atherosclerosis, integrated health care strategies that take into account oral and holistic wellness are still essential.

13.6 ASSOCIATION OF PERIODONTITIS WITH ACUTE CORONARY SYNDROME

Chronic inflammatory periodontitis, which affects the tissues that support teeth, is gaining more and more attention due to its possible effects on conditions other than oral health, especially when it comes to cardiovascular conditions like acute coronary syndrome (ACS) [100]. At the core of the periodontitis-ACS nexus lies chronic inflammation, a shared characteristic between these seemingly disparate conditions. The oral cavity serves as a central location for studying the systemic effects of periodontal disease because it is an evolving ecosystem supporting a variety of microbial populations. A major character in this story is the periodontal pathogen *Porphyromonas gingivalis* [101]. This bacterium, with its ability to manipulate host immune responses and exhibit various virulence factors, has been detected within atherosclerotic plaques, suggesting a potential direct role in the pathogenesis of ACS. The journey from oral health to cardiovascular events begins with the systemic dissemination of oral bacteria or their by-products.

The intricate interplay involves the entry of these microbial elements, particularly gram-negative bacteria associated with periodontitis, into the bloodstream [102]. This prepares the groundwork for a systemic inflammatory stress that may impact distant vascular beds and aid in the development and advancement of atherosclerosis. This periodontitis-induced persistent inflammatory state has the consequence of causing dysfunction in endothelial cells, an essential phase in the onset of ACS. The inflamed endothelial cells lining the coronary arteries exhibit altered function, leading to reduced vasodilation and increased permeability. This sets the groundwork for the formation of atherosclerotic plaques, where the inflammatory milieu, characterized by elevated levels of cytokines and acute-phase proteins, further fuels the atherogenic process [103, 104]. The microbial aspect adds layers of complexity to this association. Beyond the systemic dissemination of bacteria, specific periodontal pathogens possess unique virulence factors that influence cellular events. *Porphyromonas gingivalis*, for instance, has been found to enter host cells as a defensive tactic, altering cellular machinery and evading host immune responses [105]. Moreover, the interplay among distinct periodontal bacteria could potentially augment the infiltration of host cells, thereby augmenting the systemic consequences of periodontitis on cardiovascular well-being. The fact that diabetes, hypertension, and smoking are common risk factors highlights how closely periodontitis and ACS are related. Periodontopathic bacteria create oxidative stress and trigger an inflammatory and immune response [106]. Both the pathophysiology of chronic periodontitis and ACS involves oxidative stress. Oxidative stress damages both DNA and RNA. One of the most reliable indicators of DNA damage that has been demonstrated to be increased in periodontitis and ACS is 8-hydroxyguanosine [107]. Blood sample tests of fibrinogen alongside CRP, which both show elevated levels indicating their association with periodontitis, and salivary analyses for 8-hydroxyguanosine, malondialdehyde, carbonyl protein assay, and the overall antioxidant capacity on early morning samples gathered after a fasting session are two potential non-invasive diagnostic tools.

13.6.1 Factors Affecting Acute Coronary Syndrome via Periodontitis

13.6.1.1 Bacterial Association

Atherosclerosis has been shown to be facilitated by both acute and long-term systemic inflammation. Periodontal bacteria, such as *Porphyromonas gingivalis* and *Aggregatibacter actinomycetemcomitans,* exert direct or indirect effects by infiltrating various cells, including, leukocytes, smooth muscle cells, endothelial cells, and platelets. This influence may occur through the release of paracrine factors that impede cellular functions. These bacteria possess the capability to discharge their components into the bloodstream, thereby heightening the pro-atherogenic responses of endothelial cells [108]. Specifically, the outer layer of vesicles, known as gingipains, and the soluble bacterial elements from *Porphyromonas gingivalis*, as well as *Aggregatibacter actinomycetemcomitans*, have the potential to irritate and inflame endothelial cells [109]. Periodontitis increases a patient's vulnerability to bacterial colonization of endothelial cells and widespread propagation, which can lead to more serious complications and ultimately result in bacteremia. In experimental animal models, *P. gingivalis* infection has been shown to quicken the pathogenesis of ACVD by causing inflammatory processes and accumulation of lipids in vascular walls.

13.6.1.2 Inflammatory Factors

The ulcerated periodontium engages in the secretion and absorption of proinflammatory cytokines (TNF-α, IL-1, and IL-6) and other chemokines, leading to the synthesis of acute-phase proteins in the liver, including fibrinogen, amyloid A protein, and CRP. In response to prolonged bacteremia, the adaptive immune system generates antibodies specific to harmful periodontal pathogens and their pathogen-associated molecular patterns (PAMPs). This immune reaction is crucial as it can initiate a cross-reaction between endothelial cells and oxidized LDL, facilitating the translocation of lipids into the intima layer of the vessel wall [110]. The previously mentioned inflammatory cytokines and antibodies also elicit type 1 T-helper (Th1) responses, recruiting monocytes and macrophages to the atheroma location where oxidized LDLs are utilized. This process leads to the formation of foam cells, expediting the atherogenesis process. Unintentional persistence of moderate-to-severe periodontitis with subgingival disease caused by red complex microbes is linked to systemic inflammation [111, 112]. This case-control study found that patients with periodontitis had higher systemic levels of ICAM-1, myeloperoxidase (MPO), and E-selectin compared to periodontally healthy controls. Notably, MPO is recognized as a pathogenic factor in atherosclerosis because it often co-occurs with inflammation and oxidative stress. During these processes, it catalyzes the production of various ROS and functions as a defense against pathogens, including periodontal bacteria [113]. According to recent research, non-stimulated circulating neutrophils from patients with periodontitis were more responsive to the existence of periodontal bacteria in *ex vivo* models and produced significantly more ROS than the control group. Furthermore, several observational studies have shown a connection among these inflammatory proteins and cardiovascular events, atheroma plaque formation, and endothelial dysfunction.

13.6.1.3 Genetics

Genes have risk alleles related to periodontitis as well as to coronary artery disease (CAD). Molecule studies have linked inflammation, obesity, CAD, and periodontitis mechanistically through the integration of Antisense Noncoding RNA in the INK4 Locus (ANRIL) and (Vesicle-Associated Membrane Protein 3 (VAMP3) into the regulation network that connects immunological response to glucose and fatty acid metabolism. The disruption of these pathways by genetic factors may be a pathogenic common denominator between CAD and periodontitis [114]. Plasminogen (PLG) is an additional genetic risk factor shared by atherosclerosis and periodontitis [98, 115]. The disruption of these processes by genetic factors may be a pathogenic common denominator between CAD and periodontitis [116]. Numerous pathogenic bacteria have been found to bind human PLG on infectious PLG receptors, which causes the bacteria to become proteolytic organisms. Moreover, all the common risk genes for CAD and periodontitis that have been found belong to the TGF-β signaling pathway. TGF-β signaling may be the point of intersection for the various downstream problems of obesity-induced complications, such as periodontitis and CAD. Two greater candidate genes that are currently under investigation are *TGFBRAP1* and *FURIN* [117, 118].

13.7 ASSOCIATION OF PERIODONTITIS WITH STROKE

Numerous studies have clarified the link between periodontitis and stroke, especially in populations with a high risk of atherosclerosis. Notably, studies have linked periodontitis to increased carotid artery intima media thickness, which is a crucial indicator of atherosclerosis [119]. Figure 13.6 illustrates periodontitis and heart stroke. Investigations further revealed that a significant percentage (up to 31.3%) of people with periodontitis have unilateral carotid calcifications, which are suggestive

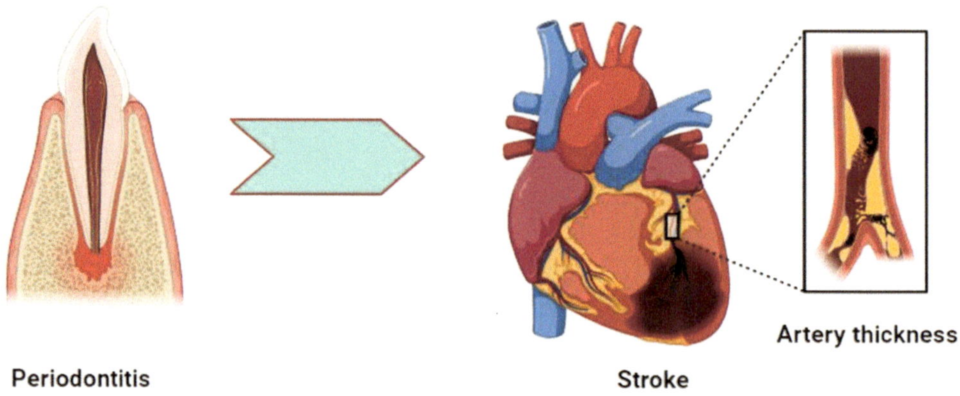

FIGURE 13.6 Association of periodontitis and heart stroke.

of subclinical carotid disease. Additionally, compared to controls, patients with periodontitis have larger amounts of flowing CRP and noticeably thicker carotid artery intima media [120]. The raised thickness of the carotid artery intima media is correlated with elevated CRP levels, highlighting the function of inflammation in this correlation. The association between dental health and risk for stroke is further supported by case-control research. The incidence of stroke and the periodontal index have been found to be strongly correlated, highlighting the possible influence of oral health on cerebrovascular events.

Once in the bloodstream, these substances contribute to systemic inflammation, increasing the progression of atherosclerosis, which is defined by the thickness and hardness of artery walls as a result of plaque accumulation. This systemic inflammation causes endothelial dysfunction, causing increased arterial thickness and plaque instability. Atherosclerosis increases the risk of thrombus development and subsequent cardiovascular complications such as heart attacks and strokes [121].

There are two main types of strokes: ischemic strokes and hemorrhagic strokes. A stroke is a medical condition caused by an abrupt brain hemorrhage or a blocked blood supply to the brain [122, 123]. Beyond these signs and symptoms, prolonged periodontitis has been linked to increased levels of tumor necrosis factor (TNF) and CRP, as well as the presence of lacunar infarcts. These findings highlight the systemic effect of periodontal disease on markers of inflammation and cerebral health. Ischemic strokes are caused by a blockage in the blood supply, which deprives the brain of oxygen and nutrients. Hemorrhagic strokes are caused by sudden bleeding within the brain, which damages brain cells [124]. Although the epidemiologic connection among periodontitis and stroke is still up for debate, it is known that periodontitis is associated with elevated inflammatory markers, which are independent indicators of stroke risk. After taking into consideration a range of risk factors, lifestyle factors, and socioeconomic characteristics, a meta-analysis confirmed that periodontal diseases and stroke are related in both potential and retrospective research. This thorough understanding underscores the need for comprehensive dental strategies that take into account both oral and systemic health by emphasizing the importance of improving periodontal health in reducing the risk of stroke and mitigating inflammatory markers [125].

13.7.1 Factors Resulting in Stroke Due to Periodontitis

The relationship between periodontitis and stroke is complex, involving various factors that contribute to cerebrovascular events. Through a number of pathways, periodontitis, a chronic inflammatory disease that affects the gums and the tissues that support the teeth, has been linked to the etiology of stroke. Here, we delve into the detailed factors that link periodontitis to an increased risk of stroke:

- *Inflammatory mediators*: Periodontitis is characterized by persistent inflammation. Systemic inflammation can be exacerbated by the inflammatory mediators generated in the bloodstream as a result of the oral bacterial burden [126]. A systemic inflammatory state is influenced by a rise in levels of proinflammatory cytokines, including TNF-α and interleukin-6 (IL-6). One of the main contributing factors to the onset and advancement of atherosclerosis, a disorder that can result in stroke, is this inflammation.
- *Atherosclerosis and vascular damage*: The accumulation of plaque in arteries is a vascular disease known as atherosclerosis. Periodontitis-related chronic inflammation can aggravate endothelial dysfunction, which can start and speed up atherosclerotic processes. The arteries delivering to the brain may narrow or block as atherosclerosis advances, raising the risk of an ischemic stroke—a kind of stroke brought on by decreased circulation to the brain.
- *Thromboembolism*: Thromboembolic events have been associated with a higher risk of periodontitis. Blood clots may form as a result of the long-term inflammation and vascular impairment brought on by periodontal disease [127]. An ischemic stroke can result from these clots blocking smaller blood vessels in the brain if they become loose and move there.
- *Microbial factors*: *Porphyromonas gingivalis* is one of the specific oral bacteria that have been found in atherosclerotic plaques. The development of plaque and vascular inflammation may be directly attributed to these bacteria or their by-products. These microbiological components may increase the risk of stroke and worsen systemic inflammation in the blood.
- *Systemic inflammatory response*: A spike in circulatory inflammatory markers is caused by the systemic inflammatory response that periodontitis elicits. For instance, elevated CRP levels have been linked to a higher risk of stroke as well as periodontitis. The general prothrombotic and atherogenic milieu may be exacerbated by this systemic inflammation [128].
- *Endothelial dysfunction*: Endothelial function may be compromised by the persistent inflammatory state linked to periodontitis. Atherosclerosis is largely influenced by endothelial dysfunction, which also raises the risk of hypertension, another stroke risk factor. The chain of events that culminates in stroke may be facilitated by the compromised vascular integrity.
- *Shared risk factors*: Stroke and periodontitis are associated with similar risk factors, including diabetes, hypertension, and smoking [129]. The combined effect of these risk factors and the inflammatory burden associated with periodontal disease can increase the risk of stroke overall.

Gaining knowledge of these variables helps to clarify the complex relationship between periodontitis and stroke. The evidence suggests that treating oral health may have consequences for lowering the risk of stroke while contributing to overall cardiovascular well-being, even though more research is required to definitively establish causation [130, 131]. When it comes to complete stroke prevention in people with periodontitis, integrated health care approaches that take into account both oral as well as systemic wellness are essential.

13.8 ASSOCIATION OF PERIODONTITIS WITH ATRIAL FIBRILLATION

Cardiovascular research has focused on the relationship between atrial fibrillation (AF), a type of cardiac disorder, and periodontitis, a chronic inflammatory disease that affects the tissues that support teeth. The complex interactions between inflammatory mechanisms and their possible effects on the cardiovascular system serve as the foundation for this relationship [132]. Recognizing periodontitis as a cause of systemic inflammation is essential to comprehending this relationship. Proinflammatory cytokines along with additional mediators are released into the bloodstream by this illness, resulting in an inflammatory response in the body that affects the heart and other organs [133]. Because atrial fibrillation can result in blood clots, it is a serious health risk. Atrial fibrillation is characterized by irregular and frequently rapid heartbeats in the upper chambers of the heart. There are multiple underlying mechanisms that explain the association between AF and periodontitis. The disturbance of the coagulation cascade, which is the series of actions that result in the formation of blood clots, is one important factor. The inflammatory reaction linked to periodontitis may encourage the formation of blood clots in the atria, raising the possibility of embolic events [134]. Moreover, studies, including those utilizing animal models, have shed light on the specific impact of periodontitis on the atrial myocardium, the heart muscle in the atria. This condition has been shown to induce increased immune activation within the atrial myocardium, leading to disruptions in its electrophysiological functions.

The consequences include a shorter atrial effective refractory period (AERP), which denotes the time during which the atria are not excitable after a previous activation. The shortening of AERPs has been implicated in the facilitation of atrial fibrillation, thereby increasing susceptibility to abnormal heart rhythms [135, 136]. Furthermore, the research indicates that periodontitis may contribute to the progression of atrial fibrillation inducibility. Animal studies have demonstrated an accelerated pace in the development and sustainment of AF under conditions of periodontitis. This suggests that the inflammatory milieu created by periodontitis may make it easier for atrial fibrillation to be triggered and sustained [138]. While the existing evidence highlights intriguing connections between periodontitis and atrial fibrillation, it is crucial to acknowledge the need for further research to establish a definitive cause-and-effect relationship [137]. Nonetheless, these findings underscore the potential significance of oral health, particularly in managing inflammatory burdens, as an integral component of overall cardiovascular well-being.

13.8.1 Factors Causing Atrial Fibrillation Due to Periodontitis

The main role of oral microorganisms and pathogenic bacteria linked to periodontitis, AF, and poor dental hygiene can be explained by two main host-response mechanisms: an autoimmune response to the pathogen and a systemic inflammatory reaction in response to it. Figure 13.7 depicts the systemic inflammatory and autoimmune response in atrial fibrillation.

- *Systemic inflammatory response*: Subgingival biofilm in a periodontal pocket allows periodontal pathogens to translocate through the bloodstream and increases serum biomarkers of inflammation and cytokines, such as CRP, TNF-α, and IL-6, which in turn causes inflammatory responses in the secondary and systemic circulation [139, 140]. CRP plays a major role in endothelial dysfunction through the instability of endothelial nitric oxide (NO)-synthase messenger RNA (mRNA) and down-regulation of endothelial NO synthase gene transcription in endothelial cells. One way to encourage endothelial cell apoptosis is to suppress NO synthesis. Atrial fibrillation (AF) related to inflammation is caused by TNF-α, which also causes abnormal calcium homeostasis and increases pulmonary vein arrhythmogenicity.
- *Autoimmune response*: Another host-response mechanism is the autoimmune response to heat-shock proteins (HSPs) produced by microorganisms, especially gram-negative bacteria [141]. HSPs prevent damage to cardiomyocytes and maintain proteostasis in response to stress in human cells. Because their amino acid sequences are similar, oral microorganisms especially periodontal pathogens and human HSP60 show structural homology. The two molecules then experience an immunological cross-reaction as a result, a process called "molecular mimicry." The antibodies generated against bacterial HSP60 cause an

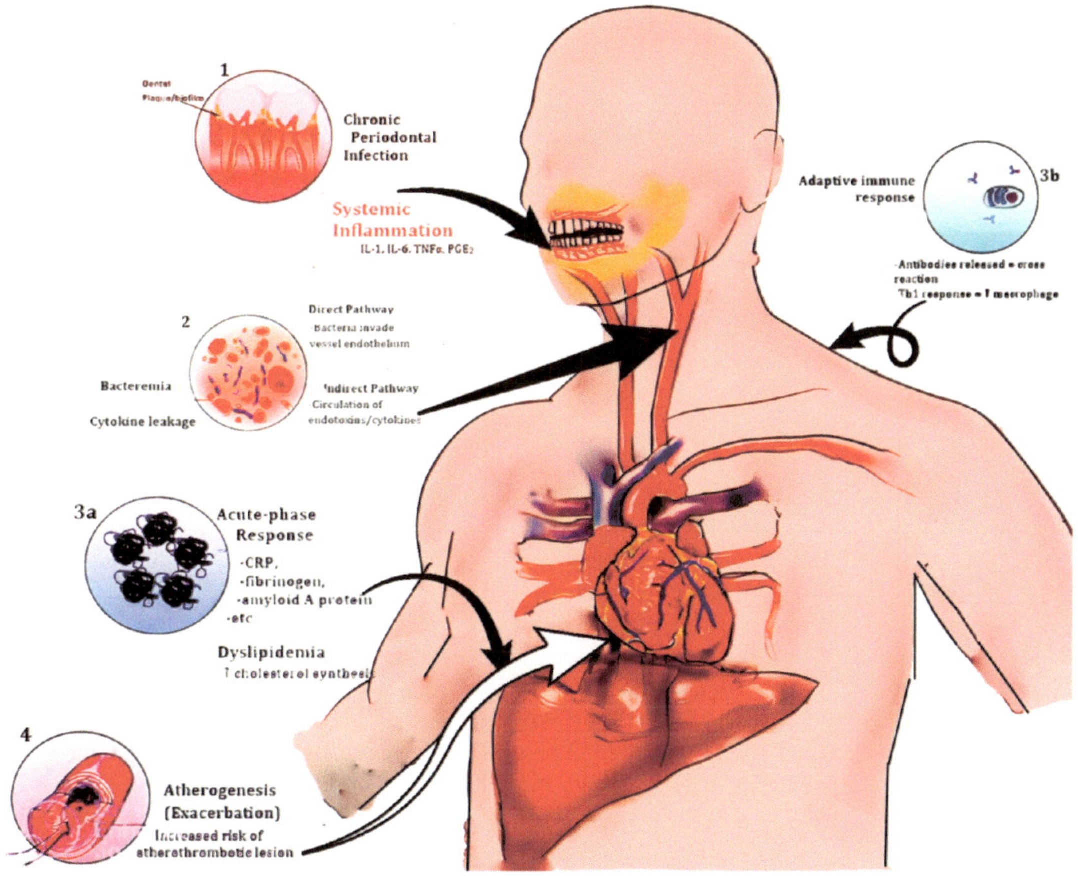

FIGURE 13.7 Systemic inflammatory and autoimmune response in atrial fibrillation.

autoimmune response, and these antibodies also mature into autoantibodies that comprehend human autologous HSP [142]. When confronted with GroEL, the HSP 60/65 homologous of Pg, which cross-reacts with the host's intrinsic HSPs, humans undergo a humoral and cellular immune response. In addition to the autoimmune response triggered by similar epitopes in host proteins and microbial HSPs, the cytotoxicity of some bacterial HSPs might also contribute to tissue deterioration. It has been demonstrated that patients with AF exhibit changed cardiac HSP60 expression and that people with periodontitis exhibit auto-reactivity to HSPs. [143, 144].

13.9 INFLAMMATORY MARKERS COMMON IN PERIODONTITIS AND CARDIAC DISEASES

Inflammation plays an important role in the development, progression, and presentation of CVDs. Proinflammatory cytokines and other inflammatory markers have been shown in clinical studies to mitigate endothelial nitrite oxidase synthase expression, boost endothelial nicotinamide adenine dinucleotide phosphate oxidase synthesis, as well as increase endothelial cell adhesion molecule expression. Patients with changes in these inflammatory parameters may be at a higher risk for CVD, because proinflammatory cytokines have been linked to an increased risk even when there is a presence of regulated low-density lipoprotein levels [144]. Significantly, research has identified an inverse relationship between the quantity of teeth and CVD mortality. Reduced baseline tooth counts are associated with diminished survival in CVD, and this connection seems to be impacted by increased levels of C-reactive protein [145].

Inflammatory markers that rise in both periodontitis and cardiovascular disease include intercellular adhesion molecule 1, fibrinogen, white blood cells, C-reactive protein, along with proinflammatory cytokines. Furthermore, the risk variables for both diseases are similar and include diabetes, obesity, stress, smoking, poor oral hygiene, and dropped physical activity. Using blood pressure tracking and oral health conditions, researchers came across an elevated likelihood of hypertension associated with periodontitis.

Preliminary findings of a study add to the evidence of an association among oral and cardiovascular wellness. They suggest that patients with periodontal disease who have higher levels of certain species of bacteria, including *F. nucleatum*, *A. actinomycetemcomitans*, and *P. intermedia* are more susceptible to hypertension than healthy controls [55, 146]. Whole bacteria, including *P. gingivalis* and infectious lipids emanating from a dysbiotic oralome, have been found in arterial plaques from CVD patients.

Patients with severe periodontitis and high levels of proinflammatory mediators like fibrinogen, C-reactive protein, and interleukin-1 had more substantial peripheral blood neutrophil levels than healthy controls. The chemoattractants leukotrienes A4 and B4 are among those that attract neutrophils to tissues. Because indicative human atherosclerotic plaques demonstrate leukotrienes A4 hydrolase, 5-lipoxygenase, and 5-lipoxygenase-activating protein, these leukotrienes are also involved in atherosclerosis [2, 147]. Numerous illnesses linked to oral health disruptions, such as cardiovascular disease, show changes in blood microbiota composition and circulating neutrophil phenotypes. Microbes in the oral cavity and their products can influence neutrophil activity. Because neutrophils are capable of entering the peripheral bloodstream, this has the potential to have a secondary impact on systemic health outcomes. Endotoxin from bacteria in the mouth can vary based on the communities of bacteria and the condition of oral health [145]. This can have a variety of impacts on cellular lipopolysaccharide acceptance mechanisms, which in turn can influence the periodontitis-associated neutrophil phenotype. Microorganisms and their products, such as lipopolysaccharides and proinflammatory cytokines, can activate the formation of neutrophil extracellular traps, which are comprised of a DNA scaffold enhanced by antimicrobial peptides and made up of activated platelets or endothelial cells, and interactions with these cells contribute to periodontitis-related tissue damage [148]. Parallel to this, it was discovered that neutrophil extracellular traps facilitated thrombosis and atherosclerosis.

13.10 THE INTERPLAY OF EPIGENETICS IN PERIODONTITIS AND CARDIOVASCULAR RISK

Epigenetics plays a significant part in the relationship of cardiovascular risk and periodontal disease [149]. The epithelial cells of the oral cavity are the first line of defense against infections and the existence of bacteria, and the by-products they produce may alter their epigenome, affecting the dynamics and capabilities of inflammatory cells by causing changes in the expression of genes and signaling pathways.

Therefore, DNA molecules do not contain the encoded information of the changes. Nevertheless, epigenetics determines a thorough remodeling of chromatin through subsequent chemical changes, leading to the activation of gene expression or repression [150]. Environmental factors like habits, tobacco use, and being exposed to radiation, heavy metals, as well as infectious agents all have a significant impact on epigenetic dynamics [151, 152]. DNA methylation, posttranscriptional histone modifications that affect the arrangement of chromatin, and non-coding RNA (e.g., micro-RNA)-mediated gene expression regulation are all important epigenetic mechanisms. Single nucleotide polymorphisms may engage with epigenetic processes that regulate gene expression, further complicating matters. Micro-RNA and epigenetic modifications are important in the occurrence of

CVDs, contribute to the progression of atherosclerosis, and serve a critical role in the underlying causes of these ailments and the susceptibility of atherosclerotic plaques [153]. The primary pathways through which epigenetic mechanisms operate were compiled in a recent review. The first pathway can alter human cells at the DNA level and deals with how external environmental factors affect human cells. The epigenetic initiator, epigenetic maintainer, and non-coding RNAs are then in charge of passing on these modifications to subsequent cell generations [154, 155]. Conversely, cytokines that are key players in the development of periodontitis, such as interleukin-1 and interleukin-6, have their effects controlled by epigenetic mechanisms [156–158]. Because of their part in causing epigenetic changes that are important in the etiology of disease, environmental variables can affect a person's susceptibility to developing CVD risk factors [38].

13.11 CONCLUSION

The significance of oral bacteria in cardiac disease highlights the complex relationship of oral health and overall health. Certain oral infectious agents, such as *Porphyromonas gingivalis*, have been linked to an elevated likelihood of cardiac disease, according to scientific evidence. Inflammatory processes, bacterial migration, and immune response activation are among the mechanisms by which oral microbes may contribute to cardiac problems. While the link between oral health and cardiac diseases remains under review, maintaining good oral hygiene is an important aspect of preventive health care. There is no uncertainty that a number of oral bacteria and pathogens associated with periodontal disease have been associated with atherosclerotic lesions and bacteremia, in addition to a number of processes that have been suggested to be linked to the inflammatory reactions that cause atherosclerosis at all stages. There is evidence to support the theory that periodontal diseases are associated with low-grade generalized inflammation, as indicated by raised plasma CRP levels. However, more evidence is needed to link oral bacterial access to this raised state of widespread inflammation. Systemic inflammation and oral bacterial invasion of the bloodstream are two phenomena that are difficult to study in conjunction due to the complexity of the oral microbiota and the challenges in establishing an accurate exposure threshold to the cumulative microbial burden. Regular dental checkups, effective plaque control, and timely treatment of oral infections may not only promote oral health but also potentially mitigate the risk of cardiovascular complications. The holistic approach to health care should recognize the oral cavity as an integral part of the body, acknowledging the significance of oral hygiene practices in preventing and managing systemic conditions, including cardiac diseases. Further research is needed to elucidate the precise mechanisms and establish concrete causal relationships, but current evidence highlights the importance of considering oral health as a contributing factor in the broader context of cardiovascular well-being.

REFERENCES

1. Kilian, M.; Chapple, I. L. C.; Hannig, M.; Marsh, P. D.; Meuric, V.; Pedersen, A. M. L.; Tonetti; M.S.; Wade; W.G.; Zaura; E. The oral microbiome–an update for oral healthcare professionals. *British Dental Journal* **2016**, *221*(10), 657–666.
2. Ford, P.; Gemmell, E.; Walker, P.; West, M.; Cullinan, M.; Seymour, G. Characterization of heat shock protein-specific T cells in atherosclerosis. *Clinical and Vaccine Immunology* **2005**, *12*(2), 259–267.
3. Yamashita, Y.; Takeshita, T. The oral microbiome and human health. *Journal of Oral Science* **2017**, *59*(2), 201–206.
4. Bruno, A.; Fumagalli, S.; Ghisleni, G.; Labra, M. The microbiome of the built environment: The nexus for urban regeneration for the cities of tomorrow. *Microorganisms* **2022**, *10*(12), 2311.
5. Larson; B. T. Perspectives on principles of cellular behavior from the biophysics of protists. *Integrative and Comparative Biology* **2023**, *23*(1), 1405–1421.
6. Yamamoto, A.; Kambara, Y.; Fujiwara, H. Impact of oral microbiota on pathophysiology of GVHD. *Frontiers in Immunology* **2023** *14*, 1132983.
7. Deo, P. N.; Deshmukh, R. Oral microbiome: Unveiling the fundamentals. *Journal of Oral and Maxillofacial Pathology: JOMFP* **2019**, *23*(1), 122.

8. Zhang, L.; Liu, Y.; Zheng, H. J.; Zhang, C. P. The oral microbiota may have influence on oral cancer. *Frontiers in Cellular and Infection Microbiology* **2020**, *9*, 476.
9. Chattopadhyay, I.; Verma, M.; Panda, M. Role of oral microbiome signatures in diagnosis and prognosis of oral cancer. *Technology in Cancer Research & Treatment* **2019**, *18*, 1533033819867354.
10. Omidi, F.; Zangiabadian, M.; Shahidi Bonjar, A. H.; Nasiri, M. J.; Sarmastzadeh, T. Influenza vaccination and major cardiovascular risk: a systematic review and meta-analysis of clinical trials studies. *Scientific Reports* **2023**, *13*(1), 20235.
11. Ley, K. Role of the adaptive immune system in atherosclerosis. *Biochemical Society Transactions*, **2020**, *48*(5), 2273–2281.
12. Beck, J. D.; Offenbacher, S. Systemic effects of periodontitis: Epidemiology of periodontal disease and cardiovascular disease. *Journal of Periodontology* **2005**, *76*, 2089–2100.
13. Tonetti, M. S.; Van Dyke, T. E.; working group 1 of the joint EFP/AAP workshop. Periodontitis and atherosclerotic cardiovascular disease: consensus report of the Joint EFP/AAP Workshop on Periodontitis and Systemic Diseases. *Journal of Periodontology* **2013**, *84*, S24–S29.
14. Persson, G. R.; Persson, R. E. Cardiovascular disease and periodontitis: an update on the associations and risk. *Journal of Clinical Periodontology* **2008**, *35*, 362–379.
15. Forner, L.; Nielsen, C. H.; Bendtzen, K.; Larsen, T.; Holmstrup, P. Increased plasma levels of IL-6 in bacteremic periodontis patients after scaling. *Journal of Clinical Periodontology* **2006**, *33*(10), 724–729.
16. Zheng, X. Y.; Mao, C. Y.; Qiao, H.; Zhang, X.; Yu, L.; Wang, T. Y.; Lu, E. Y. Plumbagin suppresses chronic periodontitis in rats via down-regulation of TNF-α, IL-1β and IL-6 expression. *Acta Pharmacologica Sinica* **2017**, *38*(8), 1150–1160.
17. Marietta, E.; Horwath, I.; Balakrishnan, B.; Taneja, V. Role of the intestinal microbiome in autoimmune diseases and its use in treatments. *Cellular Immunology* **2019** *339*, 50–58.
18. Zardawi, F.; Gul, S.; Abdulkareem, A.; Sha, A.; Yates, J. Association between periodontal disease and atherosclerotic cardiovascular diseases: revisited. *Frontiers in Cardiovascular Medicine* **2021** *7*, 625579.
19. Herrera, D.; Molina, A.; Buhlin, K.; Klinge, B. Periodontal diseases and association with atherosclerotic disease. *Periodontology* **2020**, *83*(1), 66–89.
20. Kouijzer, J. J.; Noordermeer, D. J.; van Leeuwen, W. J.; Verkaik, N. J.; Lattwein, K. R. Native valve, prosthetic valve and cardiac device-related infective endocarditis: A review and update on current innovative diagnostic and therapeutic strategies. *Frontiers in Cell and Developmental Biology* **2022**, *10*, 995508.
21. Seo, H.; Hyun, J.; Kim, H.; Park, S.; Chung, H.; Bae, S.; Chong, Y. P. Risk and outcome of infective endocarditis in streptococcal bloodstream infections according to *Streptococcal* species. *Microbiology Spectrum* **2023**, e0104923.
22. Schöffer, C.; Oliveira, L. M.; Santi, S. S.; Antoniazzi, R. P.; Zanatta, F. B. C-reactive protein levels are associated with periodontitis and periodontal inflamed surface area in adults with end-stage renal disease. *Journal of Periodontology* **2021**, *92*(6), 793–802.
23. Han, P.; Moran, C. S.; Sulugodu Ramachandra, S.; Walsh, L. J.; Ivanovski, S. Antibody response to BNT162b2 mRNA vaccine in gingival crevicular fluid. *Journal of Periodontology* **2023**, *94*(1), 77–87.
24. Abdulkareem, A. A.; Al-Taweel, F. B.; Al-Sharqi, A. J.; Gul, S. S.; Sha, A.; Chapple, I. L. Current concepts in the pathogenesis of periodontitis: from symbiosis to dysbiosis. *Journal of Oral Microbiology* **2023**, *15*(1), 2197779.
25. Amato, M.; Di Spirito, F.; D'Ambrosio, F.; Boccia, G.; Moccia, G.; De Caro, F. Probiotics in periodontal and peri-implant health management: biofilm control, dysbiosis reversal and host modulation. *Microorganisms* **2022**, *10*(11), 2289.
26. Davis, E. M.; Weese, J. S. Oral microbiome in dogs and cats: dysbiosis and the utility of antimicrobial therapy in the treatment of periodontal disease. *Veterinary Clinics: Small Animal Practice* **2022**, *52*(1), 107–119.
27. Girija, A. S.; Ganesh, P. S. Functional biomes beyond the bacteriome in the oral ecosystem. *Japanese Dental Science Review* **2022** *58*, 217–226.
28. Santhosh Kumar, K. *Evaluation of the microbiome in Periodontal Health and Plaque Covering Supragingival calculus using Next Generation Sequencing Technology* **2019** (Doctoral dissertation, Ragas Dental College and Hospital, Chennai).
29. Chang, L. U.; Yi, C. H. U.; Jian Ru, L. I. U.; Wen Yi, L. I. U. Subgingival microbial profiles of young Chinese adults with stage I/II periodontitis, gingivitis and periodontal health status. *The Chinese Journal of Dental Research* **2021** *24*(3), 167–175.

30. Duran-Pinedo, A. E.; Solbiati, J.; Teles, F.; Frias-Lopez, J. Subgingival host–microbiome metatranscriptomic changes following scaling and root planing in grade II/III periodontitis. *Journal of Clinical Periodontology* **2023**, *50*(3), 316–330.
31. Ramberg, P.; Sekino, S.; Uzel, N. G.; Socransky, S.; Lindhe, J. Bacterial colonization during de novo plaque formation. *Journal of Clinical Periodontology* **2003** *30*(11), 990–995.
32. Herekar, M.; Sethi, M.; Prithviraj, D. R.; Bhat, K.; Fernandes, A.; Patil, V. A clinical study evaluating changes in the microbial flora around dental implants during various stages of implant restoration. *Implant Dentistry* **2015**, *24*(5), 527–532.
33. Wallin-Bengtsson, V.; Scherdin-Almhöjd, U.; Roos-Jansåker, A. M. Supra- and sub-gingival instrumentation of periodontitis with the adjunctive treatment of a chloramine–a one-year randomized clinical trial study. *Acta Odontologica Scandinavica* **2023**, *83*, 1–6.
34. Vaithilingam, R. D.; Cheah, C. W. Oral Microbiome and Periodontitis. *Immunology for Dentistry* **2023**, 100.
35. Ortíz de Urbina Comerón, P.; Zubizarreta-Macho, Á.; Lobo Galindo, A. B.; Montiel-Company, J. M.; Lorenzo-Gómez, M. F.; Flores Fraile, J. Relationship between prostate inflammation and periodontal disease—A systematic review and meta-analysis. *Journal of Clinical Medicine* **2023**, *12*(18), 6070.
36. Hbibi, A.; Bouziane, A.; Lyoussi, B.; Zouhdi, M.; Benazza, D. *Aggregatibacter actinomycetemcomitans*: From Basic to Advanced Research. In Santi-Rocca, J. (eds) *Periodontitis: Advances in Experimental Research* (pp. 45–67). Cham: Springer International Publishing, **2022**.
37. Haraguchi, A.; Yoshida, S.; Takeshita, M.; Sumi, Y.; Mitarai, H.; Yuda, A.; Wada, N. Effects of ultraviolet irradiation equipment on endodontic disease–related bacteria. *Lasers in Dental Science* **2022**, *6*(1), 31–40.
38. Mahajan, G.; Chauhan, V.; Sharma, V.; Gupta, R. Combining in vitro and in silico studies to unravel the antifungal potential of chitinase from a marine bacterial isolate *Cellulosimicrobium cellulans* RS7. *Process Biochemistry* **2023**, *134*, 88–100.
39. Michalowicz, B. S.; Ronderos, M.; Camara-Silva, R.; Contreras, A.; Slots, J. R. Human herpesviruses and *Porphyromonas gingivalis* are associated with juvenile periodontitis. *Journal of Periodontology* **2000**, *71*(6), 981–988.
40. Slots, J. *Herpesviruses*, the missing link between gingivitis and periodontitis? *Journal of International Academy of Periodontology* **2004**, *6*, 113–119.
41. Benslama, L. Oral and maxillofacial manifestations of human immunodefficiency virus infection. *Journal of Stomatology, Oral and Maxillofacial Surgery* **2022** *123*(6), 622–633.
42. Marsh, P. D.; Lewis, M. A.; Williams, D. W.; Martin, M. V. Oral microbiology E-book. Elsevier Health Sciences, **2009**.
43. Chauhan, V.; Dhiman, V.; Kanwar, S. S. Combination of classical and statistical approaches to enhance the fermentation conditions and increase the yield of Lipopeptide(s) by Pseudomonas sp. OXDC12: its partial purification and determining antifungal property. *Turkish Journal of Biology* **2021** *45*(6), 695–710.
44. Nguyen, T. H.; Turek, I.; Meehan-Andrews, T.; Zacharias, A.; Irving, H. Analysis of interleukin-1 receptor associated kinase-3 (IRAK3) function in modulating expression of inflammatory markers in cell culture models: A systematic review and meta-analysis. *PLOS ONE*, **2020**, *15*(12), e0244570.
45. Martin, C.; Celis, B.; Ambrosio, N.; Bollain, J.; Antonoglou, G. N.; Figuero, E. Effect of orthodontic therapy in periodontitis and non-periodontitis patients: a systematic review with meta-analysis. *Journal of Clinical Periodontology* **2021** *49*, 72–101.
46. Chauhan, V.; Dhiman, V. K.; Kanwar, S. S. Purification and characterization of a novel bacterial Lipopeptide(s) biosurfactant and determining its antimicrobial and cytotoxic properties. *Process Biochemistry* **2022**, *120*, 114–125.
47. Chauhan, V.; Gautam, P.; Kanwar, S. S. Nanoparticle-Associated Lipopeptides: A New Class of Antimicrobials. In *Biomedical Translational Research: Drug Design and Discovery* (pp. 483–494). Singapore: Springer Nature Singapore, **2022**.
48. Dixon, D. R.; Bainbridge, B. W.; Darveau, R. P. Modulation of the innate immune response within the periodontium. *Periodontology* **2004**, *35*(1), 53–74.
49. Paralkar, S.; Benjamin, A.; Naik, P. Matrix metalloproteinase-8 levels in peri-implant sulcular fluid in peri-implantitis patients as a point-of-care biomarker: A systematic review. *Journal of Dental Research & Review* **2022**, *9*(2), 95–103.
50. Deepika, B. A.; Ramamurthy, J. Role of Triggered Receptor Expressed in Myeloid Cells (TREM) in Periodontal Disease-A Systematic Review. *Journal of Population Therapeutics and Clinical Pharmacology* **2023**, *30*(10), 125–146.

51. Mohammed, M. A.; Abbas, R. F. IL-17 and IL-10 as potential biomarkers for diagnosing periodontitis patients: A review. *Diyala Journal of Medicine* **2023**, *25*(1), 108–120.
52. Neculae, E.; Gosav, E. M.; Valasciuc, E.; Dima, N.; Floria, M.; Tanase, D. M. The oral microbiota in valvular heart disease: Current knowledge and future directions. *Life* **2023**, *13*(1), 182.
53. Emelyanov, D. V.; Petrushanko, T. O.; Emelyanova, N. Y. Porphyromonas gingivalis and non-alcoholic fatty liver disease as combined factors of periodontitis. *International Journal of Bone Fragility* **2022**, *2*(2), 72–76.
54. Chauhan, V.; Mazumdar, S.; Pandey, A.; Kanwar, S. S. Pseudomonas lipopeptide: An excellent biomedical agent. *MedComm–Biomaterials and Applications* **2023**, *2*(1), e27.
55. Gualtero, D. F.; Lafaurie, G. I.; Buitrago, D. M.; Castillo, Y.; Vargas-Sanchez, P. K.; Castillo, D. M. Oral microbiome mediated inflammation, a potential inductor of vascular diseases: a comprehensive review. *Frontiers in Cardiovascular Medicine* **2023**, *10*, 1250263.
56. Mahajan, G.; Kaur, M.; Gupta, R. Green Functionalized Nanomaterials: Fundamentals and Future Opportunities. In *Green Functionalized Nanomaterials for Environmental Applications* (pp. 21–41). Elsevier, **2022**.
57. Wei, E. S.; Kavitha, R.; Sa'ad, M. A.; Lalitha, P.; Fuloria, N. K.; Ravichandran, M.; Fuloria, S. Development of a simple protocol for zymogram-based isolation and characterization of gingipains from *Porphyromonas gingivalis*: The causative agent of periodontitis. *Applied Sciences* **2023**, *13*(7), 4314.
58. Qi, M.; Chi, M.; Sun, X.; Xie, X.; Weir, M. D., Oates, T. W.; Xu, H. H. Novel nanomaterial-based antibacterial photodynamic therapies to combat oral bacterial biofilms and infectious diseases. *International Journal of Nanomedicine* **2019**, *14*, 6937–6956.
59. Arnav, W. Role of protease in virulence of periodontal bacteria. *Research & Reviews: A Journal of Dentistry* **2023**, *13*(2), 9–15.
60. Zou, H.; Zhou, N.; Cheng, X.; Qiu, Y., Hou, W.; Sun, J. Gingipains are the important virulence factors of *Porphyromonas gingivalis* downregulating B10 cells. *Molecular Oral Microbiology* **2023**, *38*(4), 275–288.
61. Kadowaki T; Nakayama K; Yoshimura F; Okamoto K; Abe N; Yamamoto K. Arg-gingipain acts as a major processing enzyme for various cell surface proteins in *Porphyromonas gingivalis. The Journal of Biological Chemistry* **1998**, *273*, 29072–29076.
62. Oleksy, A.; Banbula, A.; Bugno, M.; Travis, J.; Potempa, J. Proteolysis of interleukin-6 receptor (IL-6R) by *Porphyromonas gingivalis* cysteine proteinases (gingipains) inhibits interleukin-6-mediated cell activation. *Microbial Pathogenesis* **2002**, *32*(4), 173–181.
63. Sansores-España, L. D.; Morales, F.; Arriola-Pacheco, F.; Astorga, J.; Paula-Lima, A.; Carrillo-Ávila, A.; Díaz-Zúñiga, J. Gingival crevicular fluid as Biomarker's source for Alzheimer's disease. *Odovtos-International Journal of Dental Sciences* **2022**, *24*(1), 156–176.
64. Imamura, T; Potempa, J; Tanase, S; Travis, J. Activation of blood coagulation factor X by arginine-specific cysteine proteinases (gingipain-Rs) from *Porphyromonas gingivalis. The Journal of Biological Chemistry* **1997**, *272*, 16062–16067.
65. Hammers, D.; Carothers, K.; Lee, S. The role of bacterial proteases in microbe and host-microbe interactions. *Current Drug Targets* **2022**, *23*(3), 222–239.
66. Rajkumar, K.; Vijayalakshmi, D.; Kumar, N. V.; Sundarrajan, A. Comparison of the efficiency of pulsed electromagnetic field and transcutaneous electric nerve stimulation in reducing pain during initial orthodontic teeth alignment: A single-blind, randomized control trial. *Annals of the Romanian Society for Cell Biology* **2023**, *27*(01), 1–17.
67. Rahman, S.; Bunning, R. A. D.; Dobson, P. R. M.; Evans, D. B.; Chapman, K.; Jones, T. H.; Russell, R. G. G. Bradykinin stimulates the production of prostaglandin E2 and interleukin-6 in human osteoblast-like cells. *Biochimica et Biophysica Acta (BBA)-Molecular Cell Research* **1992**, *1135*(1), 97–102.
68. Sugawara, S.; Nemoto, E.; Tada, H.; Miyake, K.; Imamura, T.; Takada, H. Proteolysis of human monocyte CD14 by cysteine proteinases (gingipains) from *Porphyromonas gingivalis* leading to lipopolysaccharide hyporesponsiveness. *The Journal of Immunology* **2000**, *165*(1), 411–418.
69. Mahajan, G.; Sharma, V.; Gupta, R. Chitinase: A potent biocatalyst and its diverse applications. *Biocatalysis and Biotransformation* **2023**, 1–25.
70. Mahajan, G.; Sharma, V.; Ray, R. C.; Gupta, R. Lactic acid bacteria as potential sources of enzymes: From genes to industrial applications. In *Lactic Acid Bacteria as Cell Factories* (pp. 183–197). Woodhead Publishing, **2023**.
71. Smart, C. D.; Fehrenbach, D. J.; Wassenaar, J. W.; Agrawal, V.; Fortune, N. L.; Dixon, D. D.; Madhur, M. S. Immune profiling of murine cardiac leukocytes identifies triggering receptor expressed on myeloid cells 2 as a novel mediator of hypertensive heart failure. *Cardiovascular Research* **2023**, *119*(13), 2312–2328.

72. Tokuda, M.; Karunakaran, T.; Duncan, M.; Hamada, N.; Kuramitsu, H. Role of Arg-gingipain A in virulence of *Porphyromonas gingivalis. Infection and Immunity* **1998**, *66*(3), 1159–1166.
73. Kuramitsu, H.; Tokuda, M.; Yoneda, M.; Duncan, M.; Cho, M. I. Multiple colonization defects in a cysteine protease mutant of *Porphyromonas gingivalis. Journal of Periodontal Research.* **1997**, *32*, 140–142.
74. Yoneda, M.; Hirofuji, T.; Motooka, N.; Anan, H.; Hamachi, T.; Miura, M.; Maeda, K. Antibody responses to *Porphyromonas gingivalis* infection in a murine abscess model–involvement of gingipains and responses to re-infection. *Journal of Periodontal Research* **2003**, *38*(6), 551–556.
75. Watanabe, K.; Gomez, A. M.; Kuramitsu, S.; Siurala, M.; Da, T.; Agarwal, S.; June, C. H. Identifying highly active anti-CCR4 CAR T cells for the treatment of T-cell lymphoma. *Blood Advances* **2023**, *7*(14), 3416–3430.
76. Zou, H.; Zhou, N.; Cheng, X.; Qiu, Y.; Hou, W.; Sun, J. Gingipains are the important virulence factors of *Porphyromonas gingivalis* downregulating B10 cells. *Molecular Oral Microbiology* **2023**, *38*(4), 275–288.
77. Brown, J. L.; Johnston, W.; Delaney, C.; Short, B.; Butcher, M. C.; Young, T.; Ramage, G. Polymicrobial oral biofilm models: simplifying the complex. *Journal of Medical Microbiology* **2019**, *68*(11), 1573–1584.
78. Postolache, T. T.; Upadhyaya, S. K.; Spector, A. M.; Mohyuddin, I.; Constantine, N.; Ernst, R. K.; ... Reynolds, M. A. Decreasing tryptophan and increasing neopterin plasma levels during pregnancy are associated with high first trimester *Porphyromonas gingivalis* K-serotype IgG serointensity in a cohort of Hispanic women. *Current Topics in Medicinal Chemistry* **2022**, *22*(25), 2126–2144.
79. O'Brien-Simpson, N. M.; Paolini, R. A.; Reynolds, E. C. RgpA-Kgp peptide-based immunogens provide protection against *Porphyromonas gingivalis* challenge in a murine lesion model. *Infection and Immunity* **2000**, *68*(7), 4055–4063.
80. Mi, Y.; He, Y.; Mi, J.; Huang, Y.; Fan, H.; Song, L.; Tong, Y. Genetic and phenotypic analysis of phage-resistant mutant fitness triggered by phage–host interactions. *International Journal of Molecular Sciences* **2023**, *24*(21), 15594.
81. Groeger, S.; Meyle, J. Oral mucosal epithelial cells. *Frontiers in Immunology* **2019**, *10*, 208.
82. Cai, R.; Wang, L.; Zhang, W.; Liu, B.; Wu, Y.; Pang, J.; Ma, C. The role of extracellular vesicles in periodontitis: Pathogenesis, diagnosis and therapy. *Frontiers in Immunology* **2023**, *14*, 1151322.
83. Liu, X. B.; Gao, Z. Y.; Sun, C. T.; Wen, H.; Gao, B.; Li, S. B.; Tong, Q. The potential role of *P. gingivalis* in gastrointestinal cancer: A mini review. *Infectious Agents and Cancer* **2019**, *14*(1), 1–7.
84. Bregaint, S.; Boyer, E.; Fong, S. B.; Meuric, V.; Bonnaure-Mallet, M.; Jolivet-Gougeon, A. *Porphyromonas gingivalis* outside the oral cavity. *Odontology* **2022**, *110*(1), 1–19.
85. Amano, A.; Nakagawa, I.; Okahashi, N.; Hamada, N. Variations of *Porphyromonas gingivalis* fimbriae in relation to microbial pathogenesis. *Journal of Periodontal Research* **2004**, *39*(2), 136–142.
86. Manjunatha, V. A. Full-mouth disinfection-A literature update. *International Dental Journal of Students' Research* **2022**, *10*(1), 1–4.
87. Lupo, J.; Truffot, A.; Reani, J.; Habib, M.; Epaulard, O.; Morand, P.; Germi, R. Virological markers in *Epstein-Barr* virus-associated diseases. *Viruses* **2023**, *15*(3), 656.
88. Nakagawa, I.; Amano, A.; Kimura, R. K.; Nakamura, T.; Kawabata, S.; Hamada, S. Distribution and molecular characterization of *Porphyromonas gingivalis* carrying a new type of fimA gene. *Journal of Clinical Microbiology* **2000**, *38*(5), 1909–1914.
89. Nakagawa, S.; Schielzeth, H. Repeatability for Gaussian and non-Gaussian data: A practical guide for biologists. *Biological Reviews* **2010**, *85*(4), 935–956.
90. Bodet, C.; Chandad, F.; Grenier, D. *Porphyromonas gingivalis*-induced inflammatory mediator profile in an ex vivo human whole blood model. *Clinical & Experimental Immunology* **2006**, *143*(1), 50–57.
91. Pérez-Chaparro, P. J.; Lafaurie, G. I.; Gracieux, P.; Meuric, V.; Tamanai-Shacoori, Z.; Castellanos, J. E.; Bonnaure-Mallet, M. Distribution of *Porphyromonas gingivalis* fimA genotypes in isolates from subgingival plaque and blood sample during bacteremia. *Biomedica* **2009**, *29*(2), 298–306.
92. Pérez-Chaparro, P. J.; Gonçalves, C.; Figueiredo, L. C.; Faveri, M.; Lobão, E.; Tamashiro, N.; Feres, M. Newly identified pathogens associated with periodontitis: A systematic review. *Journal of Dental Research* **2014**, *93*(9), 846–858.
93. Boumpas, P.; Merabet, S.; Carnesecchi, J. Integrating transcription and splicing into cell fate: Transcription factors on the block. *Wiley Interdisciplinary Reviews: RNA* **2023**, *14*(2), e1752.
94. Nomura, R.; Nagasawa, Y.; Misaki, T.; Ito, S.; Naka, S.; Okunaka, M.; Nakano, K. Distribution of periodontopathic bacterial species between saliva and tonsils. *Odontology* **2023**, *111*(3), 719–727.
95. Bista, P. K.; Pillai, D.; Narayanan, S. K. Outer-membrane vesicles of *Fusobacterium necrophorum*: A proteomic, lipidomic and functional characterization. *Microorganisms* **2023**, *11*(8), 2082.

96. Shahoumi, L. A.; Saleh, M. H.; Meghil, M. M. Virulence factors of the periodontal pathogens: Tools to evade the host immune response and promote carcinogenesis. *Microorganisms* **2023**, *11*(1), 115.
97. Mangan, D. F.; Taichman, N. S.; Lally, E. T.; Wahl, S. M. Lethal effects of *Actinobacillus actinomycetemcomitans* leukotoxin on human T lymphocytes. *Infection and Immunity* **1991**, *59*(9), 3267–3272.
98. Pucar, A.; Milasin, J.; Lekovic, V.; Vukadinovic, M.; Ristic, M.; Putnik, S.; Kenney, E. B. Correlation between atherosclerosis and periodontal putative pathogenic bacterial infections in coronary and internal mammary arteries. *Journal of Periodontology* **2007**, *78*(4), 677–682.
99. Jacobs, H. M.; O'Neal, L.; Lopatto, E.; Wozniak, D. J.; Bjarnsholt, T.; Parsek, M. R. Mucoid *Pseudomonas aeruginosa* can produce calcium-gelled biofilms independent of the matrix components Psl and CdrA. *Journal of Bacteriology* **2022**, *204*(5), e0056821.
100. Moussa, D. G.; Ahmad, P.; Mansour, T. A.; Siqueira, W. L. Current state and challenges of the global outcomes of dental caries research in the meta-omics era. *Frontiers in Cellular and Infection Microbiology* **2022**, *12*, 887907.
101. Leme, A. P.; Koo, H.; Bellato, C. M.; Bedi, G.; Cury, J. A. The role of sucrose in cariogenic dental biofilm formation—new insight. *Journal of Dental Research* **2006**, *85*(10), 878–887.
102. Alwithanani, N. Periodontal diseases and heart diseases: A systemic review. *Journal of Pharmacy and Bioallied Sciences* **2023**, *15*(Suppl 1), S72–S78.
103. Bhatiaa, S.; Balotiab, T.; Gautamb, D. Inflammation and atherosclerosis in cardiovascular diseases: From mechanisms to therapeutic opportunities. *Journal of Cardiology* **2022**, 1–4.
104. Kim, S.; Lee, Y. R.; Lee, E. O.; Jin, H.; Choi, Y. H.; Joo, H. K.; Jeon, B. H. Capsanthin inhibits atherosclerotic plaque formation and vascular inflammation in ApoE−/− mice. *Biomedicines* **2022**, *10*(8), 1780.
105. Chang, Y. Q.; Fang, F. F.; Qin, S. S.; Dong, Y. C.; Chen, B. Effect of scaling and root planing on serum C-reactive protein levels in patients with moderate to severe chronic periodontitis: A systematic review and meta-analysis. Hua xi kou Qiang yi xue za zhi= Huaxi Kouqiang Yixue Zazhi= West *China Journal of Stomatology* **2020**, *38*(4), 385–392.
106. Buhlin, K.; Hultin, M.; Norderyd, O.; Persson, L.; Pockley, A. G.; Rabe, P.; Gustafsson, A. Risk factors for atherosclerosis in cases with severe periodontitis. *Journal of Clinical Periodontology* **2009**, *36*(7), 541–549.
107. Gani, D. K.; Lakshmi, D.; Krishnan, R.; Emmadi, P. Evaluation of C-reactive protein and interleukin-6 in the peripheral blood of patients with chronic periodontitis. *Journal of Indian Society of Periodontology* **2009**, *13*(2), 69.
108. Rashid, R. Evaluation of C-reactive proteins in patients with peri-implantitis. *Journal of Advanced Medical and Dental Sciences Research* **2022**, *10*(10), 98–100.
109. Yoshii, S.; Tsuboi, S.; Morita, I.; Takami, Y.; Adachi, K.; Inukai, J.; Nakagaki, H. Temporal association of elevated C-reactive protein and periodontal disease in men. *Journal of Periodontology* **2009**, *80*(5), 734–739.
110. Andriankaja, O. M.; Pérez, C. M.; Modi, A.; Suaréz, E. L.; Gower, B. A.; Rodríguez, E.; Joshipura, K. Systemic inflammation, endothelial function and risk of periodontitis in overweight/obese adults. *Biomedicines* **2023**, *11*(6), 1507.
111. Helvaci, M. R.; Pappel, V.; Arslanoglu, Z.; Caylar, A.; Sencan, H.; Davran, R.; Pocock, L. Atherosclerotic background of disseminated teeth losses in sickle cell diseases. *Journal of Clinical Case Studies Reviews & Reports* **2023**, *1*(2), 1–17.
112. Iannuzzi, J.; Conte, M. Peripheral Arterial Disease. In *Geriatric Medicine: A Person Centered Evidence Based Approach* (pp. 1–23). Cham: Springer International Publishing, **2023**.
113. Sun, L.; Zheng, H.; Qiu, M.; Hao, S.; Liu, X.; Zhu, X.; Huang, Y. *Helicobacter pylori* infection and risk of cardiovascular disease. *Helicobacter* **2023**, *28*(3), e12967.
114. Ibraheem, W. I.; Bhati, A. K.; Alhomood, M. A.; Allayl, O. A.; Qsadi, A. A.; Mohammed, T. I. M.; Aljabri, Y. S. Relationship between carotid artery calcification and periodontitis in the Saudi population: A retrospective study. *Applied Sciences* **2023**, *13*(22), 12370.
115. Fiehn, N. E.; Larsen, T.; Christiansen, N.; Holmstrup, P.; Schroeder, T. V. Identification of periodontal pathogens in atherosclerotic vessels. *Journal of Periodontology* **2005**, *76*(5), 731–736.
116. Aimetti, M.; Romano, F.; Nessi, F. Microbiologic analysis of periodontal pockets and carotid atheromatous plaques in advanced chronic periodontitis patients. *Journal of Periodontology* **2007**, *78*(9), 1718–1723.
117. Mahendra, J.; Mahendra, L.; Kurian, V. M.; Jaishankar, K.; Mythilli, R. 16S rRNA-based detection of oral pathogens in coronary atherosclerotic plaque. *Indian Journal of Dental Research* **2010**, *21*(2), 248–252.

118. Kozarov, E.; Sweier, D.; Shelburne, C.; Progulske-Fox, A.; Lopatin, D. Detection of bacterial DNA in atheromatous plaques by quantitative PCR. *Microbes and Infection* **2006**, *8*(3), 687–693.
119. Gaetti-Jardim Jr, E.; Marcelino, S. L.; Feitosa, A. C.; Romito, G. A.; Avila-Campos, M. J. Quantitative detection of periodontopathic bacteria in atherosclerotic plaques from coronary arteries. *Journal of Medical Microbiology* **2009**, *58*(12), 1568–1575.
120. Kozarov, E. V.; Dorn, B. R.; Shelburne, C. E.; Dunn Jr, W. A.; Progulske-Fox, A. Human atherosclerotic plaque contains viable invasive *Actinobacillus actinomycetemcomitans* and *Porphyromonas gingivalis*. *Arteriosclerosis, Thrombosis and Vascular Biology* **2005**, *25*(3), e17–e18.
121. Elkaim, R.; Dahan, M.; Kocgozlu, L.; Werner, S.; Kanter, D.; Kretz, J. G.; Tenenbaum, H. Prevalence of periodontal pathogens in subgingival lesions, atherosclerotic plaques and healthy blood vessels: A preliminary study. *Journal of Periodontal Research* **2008**, *43*(2), 224–231.
122. Cavrini, F.; Sambri, V.; Moter, A.; Servidio, D.; Marangoni, A.; Montebugnoli, L.; Cevenini, R. Molecular detection of *Treponema denticola* and *Porphyromonas gingivalis* in carotid and aortic atheromatous plaques by FISH: Report of two cases. *Journal of Medical Microbiology* **2005**, *54*(1), 93–96.
123. Baddour, L. M.; Wilson, W. R.; Bayer, A. S.; Fowler Jr, V. G.; Bolger, A. F.; Levison, M. E.; Taubert, K. A. Infective endocarditis: diagnosis, antimicrobial therapy and management of complications: A statement for healthcare professionals from the Committee on Rheumatic Fever, Endocarditis and Kawasaki Disease, Council on Cardiovascular Disease in the Young and the Councils on Clinical Cardiology, Stroke and Cardiovascular Surgery and Anesthesia, American Heart Association: endorsed by the Infectious Diseases Society of America. *Circulation* **2005**, *111*(23), e394–e434.
124. Nakano, K.; Inaba, H.; Nomura, R.; Nemoto, H.; Takeda, M.; Yoshioka, H.; Ooshima, T. Detection of cariogenic *Streptococcus mutans* in extirpated heart valve and atheromatous plaque specimens. *Journal of Clinical Microbiology* **2006**, *44*(9), 3313–3317.
125. Forner, L.; Larsen, T.; Kilian, M.; Holmstrup, P. Incidence of bacteremia after chewing, tooth brushing and scaling in individuals with periodontal inflammation. *Journal of Clinical Periodontology* **2003**, *33*(6), 401–407.
126. Alhasan, M. A.; Salih, M. B. Investigating the role of heat shock protein HSP60 in coronary artery disease patients infected with *Helicobacter pylori*. *Advancements in Life Sciences* **2023**, *10*, 73–78.
127. Olsen, I.; Yilmaz, Ö. Possible role of *Porphyromonas gingivalis* in orodigestive cancers. *Journal of Oral Microbiology* **2019**, *11*(1), 1563410.
128. Shetty, B.; Fazal, I.; Khan, S. F.; Nambiar, M.; Prasad, R.; Raj, A. Association between cardiovascular diseases and periodontal disease: more than what meets the eye. *Drug Target Insights* **2023**, *17*, 31.
129. Naji, A.; Al Hatem, A.; van der Hoeven, R. Studying oxidative stress caused by the mitis group *Streptococci* in *Caenorhabditis elegans*. *Journal of Visualized Experiments* **2019**, (145), e59301.
130. Abranches, J.; Zeng, L.; Bélanger, M.; Rodrigues, P. H.; Simpson-Haidaris, P. J.; Akin, D.; Burne, R. A. Invasion of human coronary artery endothelial cells by *Streptococcus mutans* OMZ175. *Oral Microbiology and Immunology* **2009**, *24*(2), 141–145.
131. Coats, S. R.; Kantrong, N.; To, T. T.; Jain, S.; Genco, C. A.; McLean, J. S.; Darveau, R. P. The distinct immune-stimulatory capacities of *Porphyromonas gingivalis* strains 381 and ATCC 33277 are determined by the fimB allele and gingipain activity. *Infection and Immunity* **2019**, *87*(12), e00319–19.
132. Roth, G. A.; Moser, B.; Huang, S. J.; Brandt, J. S.; Huang, Y.; Papapanou, P. N.; Lalla, E. Infection with a periodontal pathogen induces procoagulant effects in human aortic endothelial cells. *Journal of Thrombosis and Haemostasis* **2006**, *4*(10), 2256–2261.
133. Saito, Y.; Fujii, R.; Nakagawa, K. I.; Kuramitsu, H. K.; Okuda, K.; Ishihara, K. Stimulation of *Fusobacterium nucleatum* biofilm formation by *Porphyromonas gingivalis*. *Oral Microbiology and Immunology* **2008**, *23*(1), 1–6.
134. Li, L.; Michel, R.; Cohen, J.; DeCarlo, A.; Kozarov, E. Intracellular survival and vascular cell-to-cell transmission of *Porphyromonas gingivalis*. *BMC Microbiology* **2008**, *8*(1), 1–11.
135. Amar, S.; Wu, S. C.; Madan, M. Is *Porphyromonas gingivalis* cell invasion required for atherogenesis? Pharmacotherapeutic implications. *The Journal of Immunology* **2009**, *182*(3), 1584–1592.
136. Takayanagi, Y.; Kikuchi, T.; Hasegawa, Y.; Naiki, Y.; Goto, H.; Okada, K.; Mitani, A. *Porphyromonas gingivalis* Mfa1 induces chemokine and cell adhesion molecules in mouse gingival fibroblasts via toll-like receptors. *Journal of Clinical Medicine* **2020**, *9*(12), 4004.
137. Takahashi, Y.; Davey, M.; Yumoto, H.; Gibson Iii, F. C.; Genco, C. A. Fimbria-dependent activation of pro-inflammatory molecules in *Porphyromonas gingivalis* infected human aortic endothelial cells. *Cellular Microbiology* **2006**, *8*(5), 738–757.

138. Kim, S. Y.; Kim, S. H.; Choi, E. K.; Paek, Y. W.; Kang, I. C. Clathrin and lipid raft-dependent internalization of *Porphyromonas gingivalis* in endothelial cells. *International Journal of Oral Biology*, *39*(3) **2014**, 131–136.
139. Miyashita, H.; Honda, T.; Maekawa, T.; Takahashi, N.; Aoki, Y.; Nakajima, T.; Yamazaki, K. Relationship between serum antibody titres to *Porphyromonas gingivalis* and hs-CRP levels as inflammatory markers of periodontitis. *Archives of Oral Biology* **2012**, *57*(6), 820–829.
140. Hočevar, K.; Potempa, J.; Turk, B. Host cell-surface proteins as substrates of gingipains, the main proteases of *Porphyromonas gingivalis*. *Biological chemistry* **2018**, *399*(12), 1353–1361.
141. Groeger, S.; Hudel, M.; Zechel, S.; Chakraborty, T.; Lochnit, G.; Meyle, J.; Domann, E. Generation and functional characterization of recombinant *Porphyromonas gingivalis* W83 FimA. *Journal of Biotechnology* **2021**, *340*, 22–29.
142. Madrigal, A. G.; Barth, K.; Papadopoulos, G.; Genco, C. A. Pathogen-mediated proteolysis of the cell death regulator RIPK1 and the host defense modulator RIPK2 in human aortic endothelial cells. *PLOS Pathogens* **2012**, *8*(6), e1002723.
143. Zhang, T.; Kurita-Ochiai, T.; Hashizume, T.; Du, Y.; Oguchi, S.; Yamamoto, M.; *Aggregatibacter actinomycetemcomitans* accelerates atherosclerosis with an increase in atherogenic factors in spontaneously hyperlipidemic mice. *FEMS Immunology & Medical Microbiology* **2010**, *59*(2), 143–151.
144. Cho, Y. D.; Kim, K. H.; Lee, Y. M.; Ku, Y.; Seol, Y. J. Oral microbiome and host health: review on current advances in genome-wide analysis. *Applied Sciences* **2021**, *11*(9), 4050.
145. Nakamura, N.; Yoshida, M.; Umeda, M.; Huang, Y.; Kitajima, S.; Inoue, Y.; Iwai, T. Extended exposure of lipopolysaccharide fraction from *Porphyromonas gingivalis* facilitates mononuclear cell adhesion to vascular endothelium via Toll-like receptor-2 dependent mechanism. *Atherosclerosis* **2008**, *196*(1), 59–67.
146. Epstein, S. E.; Zhu, J.; Najafi, A. H.; Burnett, M. S. Insights into the role of infection in atherogenesis and in plaque rupture. *Circulation* **2009**, *119*(24), 3133–3141.
147. Ford, P. J.; Gemmell, E.; Hamlet, S. M.; Hasan, A.; Walker, P. J.; West, M. J.; Seymour, G. J. Cross-reactivity of GroEL antibodies with human heat shock protein 60 and quantification of pathogens in atherosclerosis. *Oral Microbiology and Immunology* **2005**, *20*(5), 296–302.
148. Ford, P. J.; Gemmell, E.; Chan, A.; Carter, C. L.; Walker, P. J.; Bird, P. S.; Seymour, G. J. Inflammation, heat shock proteins and periodontal pathogens in atherosclerosis: an immunohistologic study. *Oral Microbiology and Immunology* **2006**, *21*(4), 206–211.
149. Huang, X.; Xie, M.; Lu, X.; Mei, F.; Song, W.; Liu, Y.; Chen, L. The roles of periodontal bacteria in atherosclerosis. *International Journal of Molecular Sciences* **2023**, *24*(16), 12861.
150. Mahajan, G.; Sharma, V.; Gupta, R. Chitinase: A potent biocatalyst and its diverse applications. *Biocatalysis and Biotransformation* **2024**, *42*(2), 85–109.
151. Cheng, W.; Shao, F.; Mehta, J. L.; Wang, X. Renin–angiotensin–aldosterone system and LOX-1 interaction in hypertension with a focus on modulation of the immune system. *The Renin Angiotensin System in Cardiovascular Disease* **2023**, 3–21.
152. Dagar, K.; Samtiya, M.; Krishnan, V.; Srivastava, S. K.; Rana, A.; Kumari, A.; Dhewa, T. Nutritional biomarkers in metabolic disorders. *Nutritional Science and Technology: Concept to Application* **2023**, 165–198.
153. Rao, A.; Kumar, B. K. Role of periodontal pathogens in atherosclerotic plaque development and progression: An overview. *Acta Microbiologica et Immunologica Hungarica* **2023**, *70*(4), 272–277.
154. Tariq, M.; Iqbal, Z.; Ali, J.; Baboota, S.; Talegaonkar, S.; Ahmad, Z.; Sahni, J. K. Treatment modalities and evaluation models for periodontitis. *International Journal of Pharmaceutical Investigation* **2012**, *2*(3), 106.
155. Kiyomoto, K.; Matsuo, I.; Suita, K.; Ohnuki, Y.; Ishikawa, M.; Ito, A.; Okumura, S. Oral angiotensin-converting enzyme inhibitor captopril protects the heart from *Porphyromonas gingivalis* LPS-induced cardiac dysfunction in mice. *PLOS ONE* **2023**, *18*(11), e0292624.
156. Kuramitsu, H. K.; Qi, M.; Kang, I. C.; Chen, W. Role for periodontal bacteria in cardiovascular diseases. *Annals of Periodontology* **2001**, *6*(1), 41–47.
157. Nabel, E. G. Cardiovascular disease. *New England Journal of Medicine* **2003**, *349*(1), 60–72.
158. Wolf, D.; Ley, K. Immunity and inflammation in atherosclerosis. *Circulation Research* **2019**, *124*(2), 315–327.

14 Effect of the Oral Microbiome on Health

Wamik Azmi, Priyanka, and Bhanu Krishan
Department of Biotechnology, Himachal Pradesh University, Shimla, Himachal Pradesh, India

Aishwarya Jain
Department of Life Sciences, Shiv Nadar Institution of Eminence, Greater Noida, Uttar Pradesh, India

14.1 INTRODUCTION

Microorganisms have a very crucial role in the living system from generating oxygen and carbon dioxide as well as contributing to the nitrogen fixation cycle. Also, apart from their contribution to the environment, microbes play an important role in human health. The community of microbes present inside the human body is termed the microbiome; a term coined by Nobel Prize laureate Joshua Lederberg [1]. This microbial ecosystem comprises different communities, namely, archaea, bacteria, fungi, protists, and other microorganisms [2]. The microbiota exhibits a mutual relationship with humans, wherein the humans provide a suitable environment and nutrients for these microbes to survive. These microbes also perform crucial functions that are necessary for absorption, digestion, immune system regulation, protecting gut muscles (intestinal walls), and producing neurotransmitters [3]. The microorganisms present inside the human body vary within the human body and among individuals as well. The variation is associated with the organs, tissues, individuals age, diet, and personal hygiene. Due to different environments, lifestyle, and nutrition availability leads to the existence of different communities [4]. The external factors such as exposure to antibiotics and vaccines have also been shown to effect significant changes in the microbiome of the human body.

The mouth serves as a biome for different unique microbial habitats, each having its microbial community. Being diverse and enriched with various microbes, the human oral microbiome comprises bacteria, fungi, viruses, and archaea. Around 700 species have been isolated and characterized from the oral cavity, mainly from saliva, the tongue, buccal mucosa, enamel of teeth, gums, palate, and gingival plaque [2, 5]. The pathfinding microbes of the oral cavity include *Streptococcus mitis*, *Streptococcus sanguinis*, *Streptococcus gordonii*, *Streptococcus salivarius*, *Rothia mucilaginosa*, and certain uncharacterized cultivable species of *Eubacterium* strain FTB41 [2, 6]. The significant tropism in the oral cavity is due to the presence of two different surfaces: shedding surfaces (mucosal) and solid surfaces (teeth). Saliva acts as a fingerprint for the oral microbiome since microbes adhere to the solid surface of the mouth and discharge into the saliva. Around 1.4×10^8 CFU bacteria are present in a milliliter of saliva, wherein the major groups of microbes identified constitute 96% of the total oral bacteria population: Firmicutes, 36.7%; Bacteriodetes, 17.1%; Proteobacteria, 17.1%; Actinobacteria, 11.6%; Spirochaetes, 7.9%; and *Fusobacterium*, 5.2% [6, 7].

Certain specific and non-specific factors present in the saliva provide nutrition to oral microflora and also assist in defending the biofilms. However, numerous other microorganisms have been identified through the introduction of the Human Microbiome Project, which solved the limitation of polymerase chain reaction (PCR) amplification and 16S rRNA gene sequencing.

DOI: 10.1201/9781003478638-14

14.2 ORAL MICROFLORA, ENVIRONMENT, AND FUNCTION

The various sections of the oral cavity such as saliva, teeth, tongue, cheek, hard and soft palates, lips, and gingival sulcus are inhabited by numerous bacterial and non-bacterial microorganisms. The distribution of microorganisms in these sites varies due to different physical and chemical environments influenced by the changing external environment. The gingival crevices, teeth, tongue, and tonsils possess a greater diversity of microflora, namely, spirochetes and vibrio's as compared to other surfaces such as the lips, buccal mucosa, and hard and soft palates due to an anaerobic environment and ideal redox potential for the optimum growth of these microorganisms [4, 8]. The tongue, having a highly papillated surface, shows the greatest diversity of microbiota since the surface of the tongue constitutes desquamated epithelium, and the non-shedding surface (teeth) also assists in the accumulation of microorganisms (Figure 14.1).

The predominant microorganisms reported on the tooth surface are *S. sanguinis*, *S. gordonii*, *S. cristatus*, *Rothia dentocariosa*, *Gemella haemolysans*, *G. adiacens*, and *Actinomyces* spp. (Table 14.1), whereas microflora belonging to Saccharomycetales are also identified along with certain protozoa like *Entamoeba gingivalis* and *Trichomonas tenax*, however, these fungi and protozoans are present in a very low percentage of the total microbiome, localized to the hard palate, gingival plaque, and oral rinse specimens rather than tooth surface [4, 9, 10]. Further, the healthy oral microbiome niche comprises the core and variable oral microbiome. Microorganisms like *Actinomyces*, *Atopobium*, *Corynebacterium*, *Bergeyella*, *Capnocytophaga*, *Granulicatella*, *Streptococcus*, *Veillonella*, *Campylobacter*, *Cardiobacterium*, *Haemophilus*, *Neisseria*, and *Fusobacterium* are part of the core microbiome [11, 12]. The variable microbiome of the oral cavity is represented by *Anaeroglobus*, *Filifactor*, *Lactobacillus*, *Scardovia*, *Olsenella*, *Johnsonella*, *Shuttleworthia*, *Brachymonas*, *Propionibacterium*, and *Cryptobacterium* [10–12].

The bacterial component is distinctive compared to the mycobiome or fungal oral microbiome, since the mycobiome is understudied due to several challenges as it constitutes <0.1% colony-forming units (CFU) of the total microbiota. Further, the isolation of the genetic material from fungi is a sophisticated process, and due to unavailability of cultivation methods for such fungi, the isolation is a challenge. However, with the advent of shotgun metagenomic sequencing and the sequencing of internal transcribed spacer (ITS) of rRNA of the fungi, studies have shown that these rare fungal species are known to have an important correlation

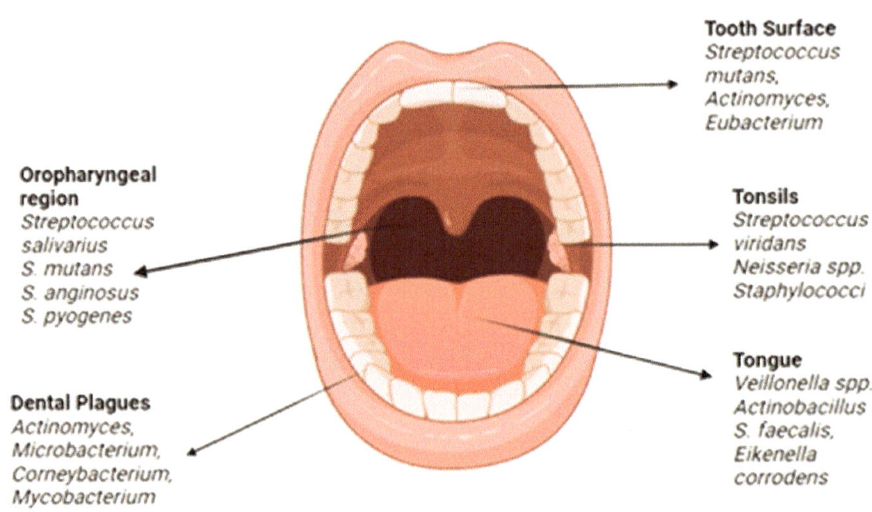

FIGURE 14.1 Diagram representing the human oral cavity, highlighting various regions where specific bacteria are commonly found.

TABLE 14.1
Oral Microbiome Bacteria and Oral Bacteria Related to Oral Diseases

	Microorganisms	References
	Oral Cavity Region	
Saliva	Actinobacteria, *Bacteriodes*, Firmicutes, *Fusobacterium*, Proteobacteria, and Spirochaetes	[6, 13, 19]
Dental plaques	Firmicutes and Actinobacteria	[6, 13, 19, 20]
Oral mucosa	*S. salivarius*, *R. mucilaginosa*, and *Eubacterium* strain FTB41	[6, 10, 11, 13, 19]
Tooth surface	*S. sanguinis*, *S. gordonii*, *S. cristatus*, *R. dentocariosa*, *Gemella haemolysans*, *G. adiacens*, *Actinomyces* spp., and *Abiotrophia defectiva*	[6, 11–13, 19]
	Oral Disease	
Dental caries	*Streptococcus*, *Veillonella*, *Actinomyces*, *Granulicatella*, *Leptotrichia*, *Lactobacillus*, and *Propionibacterium*	[6, 13, 19]
Periapical infection	Actinobacteria, *Bacteroides*, *Fusobacterium*, Firmicutes, *Olsenella*, Proteobacteria, and Spirochaetes	[6, 13, 19]

with certain diseases, which include inflammatory bowel syndrome, Crohn's disease, chronic respiratory disease, and hepatitis B [13, 14]. It was demonstrated that the fungal component of the oral microbiome is not limited to certain species but also certain species like *S. cerevisiae*, *Penicillium*, *Geotrichum*, *Aspergillus*, *Scopulariopsis*, *Hemispora*, and *Hormodendrum* have also been reported. In 20% of individuals, four common pathogenic fungi were found to be *Candida*, *Aspergillus*, *Fusarium*, and *Cryptococcus*, where *Candida* was the most abundant in healthy individuals [15].

Viruses are the most prominent and abundant biological entities on our planet. The human body surface does comprise bacteriophages, but in the case of the oral microbiome, the viral component still remains undiscovered similar to the mycobiome, since viruses cannot exist in the absence of the host [13, 16]. The viral microbiome (virome) of the human gut and oral cavity is highly personalized, and several studies reported only a single virus with different genotypes predominating the body surface [16, 17]. The viruses in the human oral cavity are associated with the sex of the human host; the viral population in the oral cavity consists largely of bacteriophage and eukaryotic viruses such as herpesviruses. The correlation is mainly due to relative hormone levels in the different human sexes and can be used for examining unrelated individuals and comparing them with the control [16].

These microorganisms have been designated to specific and important roles in the oral cavity, with certain microbes being niche-specific. Since the emanating teeth acquire polymicrobial communities that form a layer of biofilm on teeth termed dental plaques, these communities play a crucial role in maintaining oral health while also being implicated in the development of dental diseases such as caries and periodontal disease. *Streptococcus gordonii* is studied for its role in reducing dental plaques due to its ability to produce hydrogen peroxide, thus inhibiting the growth of microbial biofilms responsible for oral disease. Commensal bacteria like *Actinomyces* and *Veillonella* produce hydrogen sulfide in healthy individuals [4, 8]. Although it is understandable that diet influences the oral microbiome, it has been found that the microbiome influences the dietary preference of the host. The correlation of *Clostridium* and *Prevotella* spp. with taste thresholds shows that they influence dietary preferences to secure their membership in the oral microbiome and remain persistent in the oral cavity [2, 18]. The microbial biofilms possess quorum-sensing activity, which is involved in the virulence and pathogenicity of native microorganisms in the oral cavity and plays an important role in controlling bacterial infections. In one of the studies, the function of microbiota is reported to catalyze the conversion of dietary nitrates to nitrite.

14.3 ORAL HEALTH AND THE MICROBIOME

The oral cavity serves as the initial segment of the digestive tract, followed by the esophagus, stomach, intestine, and colon. Distinct microbial compositions characterize each segment, shaped by the unique environment of the corresponding digestive organ. The oral mucosa encompasses covering mucosa (oral floor, buccal region, labial region, and soft palate), masticatory mucosa (gingival region and hard palate), and specialized mucosa (tongue dorsum). Antonie van Leeuwenhoek, a Dutchman, made the pioneering observation of the oral microbiome in the late 1670s using a microscope of his own design. Reporting to the British Royal Society, he highlighted diverse microbes in plaque on tooth surfaces. This revelation occurred during a period when bacteriology was not established, and Van Leeuwenhoek was captivated by the dynamic nature of the microbiome. While his report documented individual variations in the oral microbiome, he implicitly acknowledged its impact on oral health. Unfortunately, it took two centuries before a comprehensive theory emerged, elucidating the relationship between the microbiome and oral diseases. The oral microbiome, a collective assembly of over 700 bacterial species, plays a pivotal role in understanding bacterial virulence. Variation in bacterial plaque composition on tooth surfaces between supragingival and subgingival regions is positively linked to periodontal pocket depth and the progression of periodontitis [7]. Saliva carries a bacterial composition resembling that of the tongue coating.

The oral microbial ecosystem, constantly exposed to external substances, creates unique microbe-host relationships under selective pressures. Diet not only supplies nutritional resources but also serves as a selective pressure, favoring organisms best adapted to host-derived dietary resources. Significant changes in the oral microbiota accompany major historical dietary shifts throughout evolution. Historical dietary shifts, especially during the Neolithic period and industrial revolutions, led to the "Westernized diet", causing notable changes in the oral microbiota [21]. Modern dietary staples, like meats, dairy, and processed grains, have altered the oral microbiota, contributing to oral health challenges. Today, such dietary constituents are staples of the American diet and that of many developed and developing countries. These shifts parallel pathological alterations, including an increase in acid-producing organisms and periodontal pathogens. Recent data reveal a reciprocal relationship: The oral microbiome influences the host's dietary preferences [22]. Certain bacteria, like some *Clostridia* and *Prevotella* species, are associated with taste thresholds, influencing preferences for sweet, sour, salty, and bitter flavors. This reciprocal relationship suggests a mechanism by which the oral microbiota influences dietary choices to sustain its presence in the oral cavity. Moreover, oral cavity–associated microbes extend beyond the mouth to distant organs, including the small intestines, lungs, heart, placenta, and brain. Pioneer colonizers such as *S. mitis*, *S. sanguinis*, *S. gordonii*, and *S. salivarius* exhibit core characteristics suited for this niche, selectively binding to tongue and cheek cells before teeth emerge, outcompeting other microbes. The complex interplay between microbes and diet highlights the oral microbiome's influence on both oral and overall health.

Numerous connections have been identified between oral microbes, particularly those linked to periodontal disease and prevalent chronic conditions such as cardiovascular disease and high blood pressure [20]. Despite limited data on the mechanistic links to diseases occurring outside the oral cavity, early research indicates that oral cavity–related microbes could impact immune responses and disease progression beyond their initial site, contingent upon the health status of that specific site. Throughout one's life, factors like dietary habits, age, hygiene practices, and behaviors (such as tobacco and alcohol use) also contribute to alterations in the oral microbiome. A study analyzing oral microbial swabs from a diverse population revealed an age-related decline in bacterial diversity, a trend observed similarly in lower gastrointestinal microbiomes concerning diet and health status. As children mature, their oral microbiomes tend to stabilize, attributed to the establishment of independent oral hygiene practices, the emergence of permanent dentition, and the adoption of adult dietary patterns [23]. While the human oral microbiota displays marked diversity among individuals, functional signatures within the oral microbiota are often conserved [24]. The Human

Microbiome Project identified *Streptococcus*, specifically *S. oralis*, *S. mitis*, and *S. peroris*, as the predominant genus among intraoral sites, at greater than 10% abundance and in more than 75% of samples at intraoral sites. Other core members with greater than 1% abundance across at least 80% of samples from one or more sites included members of the Pasteurellaceae family and genera such as *Gemella*, *Veillonella*, *Prevotella*, *Fusobacterium*, *Porphyromonas*, *Neisseria*, *Capnocytophaga*, *Corynebacterium*, and *Actinomyces*, as well as those from the orders Lactobacillales and Lachnospiraceae. However, explicitly identifying the exact composition of the oral microbiota among individuals remains challenging due to the nature of the oral cavity, a dynamic, open system regularly exposed to food, drink, exogenous microbes, air, and human contact.

The oral microbiome is personalized and dependent upon an individual's health, hygiene, and food. Such factors play a crucial role in maintaining a constant equilibrium. However, any disturbance in the equilibrium of microbiota can lead to diseases associated with oral health [12, 25, 26]. Several factors such as food habits, smoking, tobacco, alcohol consumption, psychological factors such as stress and anxiety, hormonal imbalance, poor oral hygiene, and diabetes are reported to disturb the native microbiota. Tobacco exhibits certain inhabitant microorganisms that generate toxins responsible for altering the normal conserved microbiota of the oral cavity [12, 27]. The processed tobacco leaves are reported to contain 15 different classes of bacteria, including *Burkholderia*, *Bacillus*, *Acinetobacter*, *Klebsiella*, *Clostridium*, *Serratia*, and *Pseudomonas aeruginosa*. Upon the intake of tobacco, these microorganisms enter the oral cavity along with several narcotic and carcinogenic factors. The next-generation sequencing (NGS) analysis of tobacco-based cigarettes has shown a high superiority in *Pseudomonas*, *Pantoea*, and *Bacillus*. However, *Pseudomonas* and *Pantoea* were not reported to be a part of a healthy oral microbiome. *Capnocytophaga gingivalis*, *Prevotella melaninogenica*, and *S. mitis* act as diagnostic markers for detecting oral cancer through saliva [12, 27, 28]. Further, a study using PCR restriction fragment length polymorphism (RFLP) of the 16S rDNA gene from the oral wash of smokers and non-smokers also reported a significant increase in gram-negative bacteria in the oral cavity [29]. A study was conducted to compare the proportions of bacterial genera in smokers and non-smokers. The saliva samples from smokers were reported to be enriched in genera *Treponema* and *Selenomonas* and a decrease was observed in the genera *Capnocytophaga* and *Cardiobacterium* compared to those of non-smokers. In the tongue samples, the smoker's samples were found to be enriched with *Dialister* and *Atopobium* and to have a decrease in the population of *Haemophilus*, *Gemella*, *Peptostreptococcus*, *Granulicatella*, *Catonella*, and Peptostreptococcaceae [30].

Al-Marzooq et al. (2022) also studied the consequences of smoking different tobacco products and their relation to alterations in the supragingival microbiome. The study was based on determining the variation in microbiota due to dental caries and smoking through pyrosequencing, where it was found that Bacteroidetes and Firmicutes were depleted in smokers with moderate to high dental caries [31]. The colonization pattern of a healthy mucosa was investigated to observe how smoking, tooth loss, and oral hygiene affect the normal microbiota of the oral cavity. The colonization difference was analyzed by sequencing the V1–V3 region of the 16S rRNA gene where a significant abundance in *Neisseria*, *Lautropia*, and *Haemophilus* spp. was observed in the case of the smoker's palate and tongue. *Prevotella* spp. Were observed high in population in the buccal mucosa, whereas the tongue and gingival palate had a significant increase in *Streptococcus* spp. The study also reported the impact of frequent tooth brushing on mucosal community and tooth loss resulting in the reduction of mucosal biodiversity [32]. A study conducted by Ganesan et al. (2020) found that 6% of the total population of Americans uses e-cigarettes, where toxic substances, volatile compounds, and metals in e-cigarettes were responsible for inducing stress in microorganisms, namely, *Streptococcus oralis*, *S. mitis*, *Veillonella parvula*, and *Streptococcus sanguis* causing these microorganisms to increase in biofilm mass, thereby responsible for implications in both dental caries and periodontitis [33].

The oral cavity consists of the native microbial community essential for maintaining oral health, and any disruption in the microbiota can lead to dental caries, periodontitis, and many other diseases

TABLE 14.2
Microorganisms Associated with Oral and Systemic Disease

Disease	Microorganisms Associated	Microorganisms Inhibited	References
	Oral Diseases		
Periodontitis	*Porphyromonas gingivalis*, *Treponema denticola*, *Tannerella forsythia*, *Filifactor alocis*, *Methanobrevibacter oralis*, *Methanobacterium curvum/congolense*, and *Methanosarcina mazei*	*Streptococcus*, *Actinomyces*, and *Granulicatella*	[5, 34, 35]
Dental caries	*Neisseria*, *Propionibacterium*, *Streptococcus mutans*, and *Candida albicans*	*Streptococcus*, *Corynebacterium matruchotii*, *C. gingivalis*, and *Lachnospiraceae* spp.	[5, 36, 37]
Oral cancer	*C. gingivalis*, *P. melaninogenica*, *S. mitis*, *Peptostreptococcus stomatis*, *S. salivarius*, *Gemella morbillorum*, and *Streptococcus parasanguinis*	*Granulicatella adiacens*	[5, 38–40]
	Systemic Diseases		
Alzheimer's disease	Spirochaetes, *P. gingivalis*		[41–43]
Cystic fibrosis	*S. oralis*, *S. mitis*, *S. gordonii*, and *S. sanguinis*	*Streptococcus oralis*	[5, 44]
Cardiovascular disease	*Campylobacter rectus*, *P. gingivalis*, *Porphyromonas endodontalis*, and *Prevotella nigrescens*		[5, 45, 46]
Diabetes	*Aggregatibacter*, *Neisseria*, *Gemella*, *Eikenella*, *Selenomonas*, *Actinomyces*, *Fusobacterium*, *Streptococcus*, and *Veillonella*	*Porphyromonas*, *Filifactor*, *Eubacterium*, *Treponema*	[5, 47]
Rheumatoid arthritis	*Veillonella*, *Atopobium*, *Prevotella*, and *Leptotrichia*		[5, 48, 49]

(Table 14.2). The subsequent reason for the alteration and disruption in the microflora of oral health is the presence of toxic compounds, nicotine, and volatile compounds that either inhibit certain conserved microorganisms or consist of microorganisms that are not part of a healthy oral microbiota.

14.3.1 The Salivary Microbiome and Its Relationship with Periodontal Diseases

When exploring the direct links between the microbiome and oral lesions like dental caries and periodontitis, it is essential to focus on the bacterial composition of dental plaque. These lesions are directly exposed to dental plaque, and the pathogens associated with dental caries and periodontitis are believed to reside in supragingival and subgingival plaque, respectively. Removing dental plaque is fundamental for preventing oral diseases. However, it's crucial to acknowledge that the oral microbiome may influence the virulence of specific bacteria within dental plaque. Moreover, the accumulation of virulent bacteria should be considered, even for bacteria with relatively low virulence but substantial relative abundance. Understanding how the differences in the salivary microbiome impact periodontal health remains limited. While effective measures to control the microbiome in saliva are lacking, focusing on controlling tongue coating shows promise as a potential periodontal therapy, given the similarity between the salivary microbiome and tongue coating [5].

Emerging teeth envelop themselves in a shield of protective glycoproteins, initiating a cascade of microbial colonization that orchestrates the formation of intricate polymicrobial biofilm communities—famously known as dental plaque. Within these sophisticated dental plaque matrices, distinctive microenvironments flourish, fostering acidity and anaerobic conditions that selectively nurture organisms distinct from those thriving directly on the tooth surface. Periodontal disease emerges from the dysbiosis of subgingival microbial communities, wielding a profound impact on the host immune system. This dysregulation instigates persistent inflammation in gingival and periodontal tissues, thus preventing immune subversion and tissue recovery. Critical players in oral biofilm, notably the red complex bacteria (*T. forsythia*, *T. denticola*, and *P. gingivalis*), possess the capacity to drive periodontal disease pathogenesis by orchestrating microbiota restructuring and fostering inflammation. Inflammatory and immune-mediated processes, ignited by microbes within the local plaque environment, serve as the architects of gingivitis and periodontal disease development. While the inflammation associated with gingivitis is largely reversible, periodontitis inflicts enduring and irreversible damage upon periodontal tissues and alveolar bone. This damage manifests as tissue detachment from the tooth, the formation of pockets facilitating further proliferation of pathogenic bacteria, and eventual tooth loss in advanced stages of the disease.

Beyond its impact on local (periodontal) disease, oral infection holds profound implications for systemic health. Three pivotal mechanisms intertwining oral infection with systemic pathology have come to light: the transmission of infection from the oral cavity through transient bacteremia, the dissemination of microbial toxins, and the provocation of systemic inflammation due to adverse immunologic responses to oral microbes. Saliva, a reservoir containing 10^9 bacteria per milliliter, is derived from the tongue, cheeks, and dental and gingival plaque [50]. The oral cavity generates between 0.75 and 1.5 liters of saliva daily, predominantly swallowed independently or with food and drink, thereby frequently transferring oral bacteria to the gut. Although most oral bacteria are ill-suited for survival in a healthy lower gastrointestinal tract, heightened levels of oral-associated microbes are evident in the gut of individuals with conditions such as inflammatory bowel disease, HIV, liver cirrhosis, and colon cancer [17].

14.3.2 Changes in Oral Microbiota Due to Tooth Eruption

The eruption of teeth brings about shifts in the relative abundance of *Streptococcus*, *Gemella*, Granulicatella and *Veillonella* species [40]. This transition is marked by an expansion in microbial functions, reflecting changes in the oral ecosystem during tooth eruption. These alterations include increased gene expression related to adhesion, biofilm formation, membrane transport, cell mobility, secretion systems, chemotaxis, flagella assembly, and oxidative phosphorylation [51]. The dynamics of the oral microbiome are further influenced by events like the exfoliation of primary dentition, the presence of mixed dentition, and the emergence of permanent teeth during early life and childhood. Findings from the NIH Human Microbiome Project, which examined microbial compositions across nine intra-oral sites in 200 participants, reveal that the oral microbiome possesses the most extensive core set of microbes shared among unrelated individuals compared to other colonized body sites, such as the gut and skin.

14.3.3 Effect of the Oral Microbiome on Cardiovascular Health

The oral microbiome plays a crucial role in systemic nutrient cycling, particularly regarding nitrate metabolism [2]. Approximately 25% of ingested nitrate is transported through the enterosalivary circuit to the oral cavity. Here, oral microbes convert nitrate to nitrite, which enters the bloodstream during digestion and is subsequently transformed into nitric oxide [52]. Nitric oxide is vital for cardiovascular health, exerting vasodilation and antihypertensive effects. Dysregulation of nitric oxide signaling is linked to conditions like pulmonary hypertension, obesity, and cardiovascular disease [53]. Notably, studies have demonstrated that the use of antibacterial mouthwash can elevate blood pressure due to its inhibitory impact on the oral microbiome. The "Stephan response" outlines the restoration of pH homeostasis after carbohydrate intake in the oral cavity. This process stimulates pathways that generate alkali, including those producing arginine, ornithine, citrulline, glutamate, serine, threonine, and urea. Key organisms involved in reestablishing pH homeostasis include *S. gordonii*, *S. sanguinis*, and *Lactobacillus casei*. Subgingival biofilms harbor a substantial bacterial load and owing to their proximity to deeper periodontal tissues, serve as reservoirs for lipopolysaccharide and gram-negative pathogens with the potential to enter systemic circulation. In systemic circulation, lipopolysaccharide triggers pathological responses in the vasculature, including the generation of inflammatory infiltrates within vessel walls and intravascular coagulation. Furthermore, lipopolysaccharide prompts the secretion of inflammatory factors that foster platelet aggregation and adhesion, the development of lipid foam cells, and the deposition of cholesterol. These proinflammatory molecules can directly impact the circulatory system, leading to systemic effects such as coagulation, thrombosis, platelet aggregation and adhesion, and the accumulation of cholesterol deposits [51]. Consequently, there is a distinct correlation between cardiovascular disease and oral infection. Periodontal disease increases the susceptibility to cardiovascular pathologies like atherosclerosis and myocardial infarction by elevating systemic levels of proinflammatory cytokines, inflammatory cells and infiltrates, and white blood cell counts.

14.3.4 Connection between the Oral Microbiome and Rheumatoid Arthritis

Rheumatoid arthritis, a chronic autoimmune disease, is characterized by synovial inflammation that can lead to damage in articular cartilage and bone. Studies have unveiled a connection between rheumatoid arthritis and dysbiosis in the gut and oral cavity microbiomes, establishing a long-term correlation between periodontal disease and rheumatoid arthritis. Notably, when treating rheumatoid arthritis, there is a partial restoration of the oral microbiome and vice versa. Elevated antibodies against *P. gingivalis* virulence factors have been observed in patients with rheumatoid arthritis, showing a correlation with disease severity [54]. Recent advancements have significantly contributed to uncovering the mechanistic underpinnings of this association. The genesis of rheumatoid arthritis predominantly stems from a breakdown in tolerance towards proteins undergoing citrullination, marked by the posttranslational modification of a positively charged arginine into a neutral citrulline group [54]. This citrullination process, facilitated by peptidyl arginine deiminase—an enzyme present in both humans and *P. gingivalis*—plays a pivotal role [49]. Infection with *P. gingivalis* and the ensuing development of microbial dysbiosis trigger a proinflammatory immune response in the gingiva, responding to lipopolysaccharides and endotoxins produced by proliferating bacterial pathogens. This immune response involves heightened activation of pathologic T-cell populations by antigen-presenting complexes, leading to the production of proinflammatory cytokines. Notably, *P. gingivalis* also expresses peptidyl arginine deaminase, resulting in the citrullination of gingival proteins and bacterial peptides, including auto-citrullination. This local protein citrullination in the gingiva prompts the production of antibodies against both citrullinated peptidyl arginine deiminase and *P. gingivalis*. Furthermore, human peptidyl arginine deaminase initiates similar actions in joints. The citrullination of α-enolase by *P. gingivalis* peptidyl arginine deaminase can generate antibodies that cross-react with host-citrullinated α-enolase, invoking autoantibody activity through molecular mimicry. Ultimately, prolonged inflammation evolves into chronic rheumatoid arthritis [55].

14.3.5 Role of the Oral Microbiome in Alzheimer's Disease

Porphyromonas gingivalis along with its association with periodontal disease emerges as a significant risk factor in the development of amyloid beta plaques, dementia, and Alzheimer's disease [43]. Gingipains, another virulence factor attributed to *P. gingivalis*, plays a pivotal role in Alzheimer's disease pathology. The diagnosis of Alzheimer's disease correlates not only with the gingipain load in the brain but also with other specific virulence factors of *P. gingivalis*, including its lipopolysaccharide. Gingipains exhibit co-localization with tau protein in the brain, a protein that undergoes misfolding and becomes insoluble in patients with Alzheimer's disease. Studies have demonstrated that gingipains cleave tau protein, contributing to the formation of tau tangles and subsequent disease progression. Studies in mice further reveal that oral *P. gingivalis* consistently invades the brain in a gingipain-assisted manner. Injection of gingipains into the brain intensifies neuronal degeneration beyond levels observed in controls [56]. This intricate connection underscores the potential impact of oral health on neurodegenerative diseases.

14.3.6 Oral Carcinoma

Periodontitis is linked to an increased risk [57] of oral squamous cell carcinoma (OSCC). In a recent study, Sedghi and colleagues delved into the role of *T. denticola*, *P. gingivalis*, and *Fusobacterium nucleatum* in promoting oral carcinogenesis. Their findings revealed that these oral pathogens significantly enhance cell migration, invasiveness, stemness, and oral tumorigenesis in OSCC, with minimal impact on cell proliferation and apoptosis [2]. Significantly, the detailed exploration of mechanisms revealed an intricate communication between oral pathogens and the signaling pathways of integrin/focal adhesion kinase and toll-like receptor/Myd88. Interestingly, these connections were reversed when oral pathogens were treated with the bacteriocin, nisin. This highlights the intricate interplay between oral health and the development of OSCC.

14.3.7 Effect of Type 2 Diabetes Mellitus on the Oral Microbiome

The interaction between the oral microbiome and systemic health operates in a two-way manner, indicating that systemic diseases can also impact the oral microbiome. Type 2 diabetes mellitus, along with its associated hyperglycemia, is linked to an elevated risk of periodontitis due to diabetes-related vascular complications and heightened levels of proinflammatory molecules [58]. Notably, variations in the subgingival plaque composition have been observed in patients with periodontitis, distinguishing those with or without type 2 diabetes. Specifically, individuals with diabetes exhibit higher proportions of *Capnocytophaga* species, *F. nucleatum*, *Eikenella corrodens*, and *Aggregatibacter actinomycetemcomitans* than their counterparts without diabetes [47]. *Capnocytophaga* species, recognized as saccharolytic organisms, demonstrate increased proteolytic potential in the presence of elevated glucose levels. Intriguingly, a surge in saccharolytic bacteria, including *Neisseria*, *Streptococcus*, and *Veillonella,* is theorized to occur among diabetes patients due to heightened glucose content in their serum and gingival crevicular fluid. This, in turn, creates an environment conducive to the proliferation of these organisms.

14.3.8 Effect of Psychological Health on the Composition of the Oral Microbiota

Emerging evidence indicates that persistent stress and associated conditions like depression and anxiety could play significant roles in fostering oral dysbiosis, advancing periodontal/peri-implant diseases and impeding consistent wound healing after periodontal therapy [59]. Additionally, cortisol, the stress hormone, has been demonstrated to directly influence shifts in the composition of the oral microbial community and its gene expression profiles *in vitro.* These findings parallel results observed in the gene expression profiles of the oral microbial community during periodontal disease and its progression [2].

14.3.9 Oral Microflora and Atherosclerotic Disease

Atherosclerotic disease, which includes myocardial infarction and stroke, is the leading global cause of mortality. While widely acknowledged risk factors such as smoking, obesity, high blood pressure, and diabetes contribute to atherosclerosis, studies suggest that up to half of individuals with atherosclerosis lack these conventional risk factors [60], hinting at the presence of additional causative factors. The widely accepted notion of oral bacteria playing a role in the initiation and progression of atherosclerosis is supported by findings [61]. The initial report of the detection of oral bacterial DNA in human atheromatous plaques dates back to 2000 [62]. More research confirmed the presence of oral bacterial DNA in coronary artery biopsies from individuals diagnosed with coronary artery disease and in endarterectomy specimens from those undergoing surgical treatment for atherosclerosis. The identified bacterial species include *A. actinomycetemcomitans*, *F. nucleatum*, *P. gingivalis*, *Prevotella intermedia*, and *T. forsythia*, with some samples exhibiting the coexistence of multiple bacterial types [61].

14.3.10 Adverse Pregnancy Outcomes Due to the Oral Microbiome

Adverse pregnancy outcomes (APO) is a comprehensive term encompassing various conditions, including preterm labor, preterm premature rupture of membranes, pre-eclampsia, miscarriage, intra-uterine growth retardation, low birth weight, stillbirth, and neonatal sepsis. Among these, *F. nucleatum* stands out as the most prevalent oral species and has been identified in diverse placental and fetal tissues, such as amniotic fluid, fetal membranes, cord blood, neonatal gastric aspirates, and fetal lung and stomach. Its presence is associated with preterm birth, stillbirth, and early-onset neonatal sepsis. An illustrative case report of term stillbirth linked to oral *F. nucleatum* provided the first human evidence, highlighting its origin in the mother's subgingival plaque and subsequent translocation to the placenta and fetus, resulting in acute inflammation and fetal demise [63]. In murine studies, most translocated oral species were commensals, suggesting the potential transition of oral commensals into pathogens elsewhere. Identifying the specific species/subspecies involved in APO holds promise for early and precise diagnosis, enabling the identification of individuals at risk [64].

14.3.11 Inflammatory Bowel Disease and Colorectal Cancer

Inflammatory bowel disease (IBD) encompasses a group of chronic gastrointestinal disorders, including ulcerative colitis (UC) and Crohn's disease (CD). Research indicates a multifaceted interplay of genetic and environmental factors in IBD development. Dysbiosis within the intestinal microbiota emerges as a pivotal factor in IBD pathogenesis. Notably, oral bacteria, such as *F. nucleatum* and *Campylobacter concisus*, have recently been linked to IBD [65]. *Fusobacterium nucleatum* strains sourced from inflamed biopsy tissues of IBD patients exhibit significantly higher invasiveness than those isolated from healthy tissues, irrespective of IBD or control status [61]. Furthermore, enteric invasive *C. concisus* oral strains were identified in 50% of IBD patients, in contrast to healthy controls. Considering IBD as a recognized risk factor for colorectal cancer (CRC), it is noteworthy that the same microorganisms are implicated in both conditions. A substantial elevation in *F. nucleatum* levels in colorectal carcinomas compared to normal colon tissues from the same patients has also been reported [66]. The causal relationship between *F. nucleatum* and CRC is currently under investigation, opening avenues for potential therapies such as antibiotics and/or vaccines if a conclusive link is established.

14.3.12 Role of Oral Microbiota in Lung Cancer

Lung cancer poses a significant global threat as a leading cause of cancer-related mortality. A recent comprehensive study has revealed a link between diminished oral microbiota diversity in the oral cavity and the subsequent risk of lung cancer among non-smokers. The intricate interplay between the oral microbiome and the host's immune-inflammatory response is crucial for maintaining equilibrium. Smoking is implicated in disrupting this balance, leading to oral microbiome dysbiosis, respiratory tract irritation, increased susceptibility to oral microbe aspiration into the lungs, and subsequent inflammation, contributing to various lung diseases, including lung cancer. Notably, dysbiosis in the oral microbiome has been identified in non-smoking female lung cancer patients, and this dysbiosis aligns with immunohistochemistry markers indicative of lung cancer. Examination of the oral microbiota in lung cancer patients has revealed a higher inoculum of genera *Blastomonas* and *Sphingomonas*, while healthy controls exhibit a higher inoculum of *Acinetobacter* and *Streptococcus* [67].

14.4 PREVENTION OF DISEASE CAUSED BY THE ORAL MICROBIOME

Emerging strategies for manipulating oral microbiomes are being explored. For instance, certain oral probiotics aim to enhance the alkalinity of the oral cavity and plaque, while others target specific pathogenic species like *S. mutans* [68]. The administration of supplements, such as arginine, can significantly impact the composition and metabolic output of oral microbial communities, providing an additional modulatory avenue. Consistent oral hygiene practices exert a notable influence on the oral microflora. Activities like toothbrushing and flossing serve as potent means to disrupt plaque, whose microbial residents can lead to long-term tooth demineralization and gingival inflammation. However, toothbrushes can also function as reservoirs for pathogenic bacteria, underscoring the importance of proper sanitization and storage of personal dental hygiene equipment [69]. Innovative toothpaste formulations have entered the market with the explicit goal of shaping the oral microbiota through proteins designed to promote species associated with healthy oral bacterial communities. Simultaneously, other products possess more general antimicrobial properties. Mouthwashes are engineered to emulate the effects of toothpaste, reducing microbial load through antimicrobial and bactericidal mechanisms [70].

14.4.1 The Commensal Microbiota and Oral Health

The significance of commensal oral microbiota in preserving oral health is paramount. One pivotal mechanism through which commensals contribute to oral well-being is by resisting colonization from pathogens. In this process, commensals outcompete pathogenic species for colonization substrates, limiting opportunities for integration by external pathogens. Certain oral commensals, including *S. sanguinis*, actively engage in direct antagonism against oral pathogens. *Streptococcus sanguinis*, *Streptococcus cristatus*, *S. salivarius*, *S. mitis*, *Actinomyces naeslundii*, and *Haemophilus parainfluenzae* have demonstrated the capability to diminish the adhesive capacity of *P. gingivalis* to substrates. Notably, *Lactococcus lactis*, a member of the oral microbiota, produces nisin—a bacteriocin with proven efficacy in reducing oral tumorigenesis [2]. This underscores the intricate and beneficial interactions within the commensal oral microbiota that contribute to the maintenance of oral health. Oral probiotics play a direct role in introducing specific species into the oral microbiome, aiming to either manipulate the existing microbiota or trigger a biochemical and/or physiological response [71]. Diverse strategies have been employed in crafting probiotic formulations with anticaries properties. One primary objective involves leveraging the alkalinogenicity of particular species to counteract acidification and mitigate tooth demineralization caused by pathogenic organisms. A significant contributor to oral cavity alkalinity is the enzymatic breakdown of arginine into ornithine, carbon dioxide, and two molecules of ammonia. This highlights the multifaceted approaches undertaken to develop probiotics targeting oral health in the context of caries prevention.

14.5 CONCLUSION

Oral bacteria have developed robust mechanisms for invading and persisting within host cells, evading immune surveillance, adapting to extra-oral niches, and triggering inflammatory responses with potential systemic repercussions. Given the current knowledge, certain recommendations stand justified. Firstly, recognizing periodontal infection as a catalyst for dental bacteremia suggests the likelihood of systemic dissemination of oral bacteria. Thus, maintaining optimal oral hygiene, particularly for immunocompromised individuals, proves crucial in controlling overall bacterial load and preventing dissemination. Secondly, the development of precise bacterial detection techniques becomes paramount for targeted patient management. Traditional microbiological cultures, focusing on cultivated bacterial species, present an incomplete view, given the significant portion of uncultivated bacteria. For instance, 16S rRNA gene profiling has unveiled a broader microbial diversity in amniotic fluid and cord blood than culture-dependent methods indicate. It becomes evident that not all oral bacteria are equal, with certain subtypes exhibiting a higher prevalence in extra-oral infections. Therefore, accurate bacterial detection methods are essential and employing multiple techniques for cross-validation becomes necessary due to inherent limitations in each method. Thirdly, while numerous studies link oral bacteria to extra-oral infections, many remain at the association stage. In-depth mechanistic studies are indispensable to elucidate the roles of these bacteria in systemic health. As novel virulence components and mechanisms continually emerge, a comprehensive understanding is essential for designing effective therapies against extra-oral infections and inflammation associated with oral bacteria. In summary, the oral microbiome's impact extends far beyond oral health, influencing systemic conditions and revealing a complex interplay that underscores the importance of comprehensive health management. Ongoing research and evolving strategies continue to deepen our understanding of this intricate relationship.

REFERENCES

1. Deo, P. N.; Deshmukh, R. Oral Microbiome: Unveiling the Fundamentals. *J Oral Maxillofac Pathol* **2019**, *23* (1), 122–128. https://doi.org/10.4103/jomfp.JOMFP_304_18.
2. Sedghi, L.; DiMassa, V.; Harrington, A.; Lynch, S. V.; Kapila, Y. L. The Oral Microbiome: Role of Key Organisms and Complex Networks in Oral Health and Disease. *Periodontol.* **2021**, *87* (1), 107–131. https://doi.org/10.1111/prd.12393.
3. Groussin, M.; Mazel, F.; Alm, E. J. Co-Evolution and Co-Speciation of Host-Gut Bacteria Systems. *Cell Host Microbe* **2020**, *28* (1), 12–22. https://doi.org/10.1016/j.chom.2020.06.013.
4. Jain, S.; Gannepalli, A.; Baghirath, P.; Vinay, Bh.; Krishna, Ab. Oral Microflora: Varied Habitats, Niche and Their Disparity in Systemic Health. *J Dr NTR Univ Health Sci* **2021**, *10* (4), 209. https://doi.org/10.4103/jdrntruhs.jdrntruhs_188_20.
5. Willis, J. R.; Gabaldón, T. The Human Oral Microbiome in Health and Disease: From Sequences to Ecosystems. *Microorganisms* **2020**, *8* (2), 1–28. https://doi.org/10.3390/microorganisms8020308.
6. He, J.; Li, Y.; Cao, Y.; Xue, J.; Zhou, X. The Oral Microbiome Diversity and Its Relation to Human Diseases. *Folia Microbiol (Praha)* **2015**, *60* (1), 69–80. https://doi.org/10.1007/s12223-014-0342-2.
7. Huttenhower, C.; Gevers, D.; Knight, R.; Abubucker, S.; Badger, J. H.; Chinwalla, A. T.; Creasy, H. H.; Earl, A. M.; Fitzgerald, M. G.; Fulton, R. S.; et al. Structure, Function and Diversity of the Healthy Human Microbiome. *Nature* **2012**, *486* (7402), 207–214. https://doi.org/10.1038/nature11234.
8. Sharma, N.; Bhatia, S.; Singh Sodhi, A.; Batra, N. Oral Microbiome and Health. *AIMS Microbiol* **2018**, *4* (1), 42–66. https://doi.org/10.3934/microbiol.2018.1.42.
9. Xu, X.; He, J.; Xue, J.; Wang, Y.; Li, K.; Zhang, K.; Guo, Q.; Liu, X.; Zhou, Y.; Cheng, L.; et al. Oral Cavity Contains Distinct Niches with Dynamic Microbial Communities. *Environ Microbiol* **2015**, *17* (3), 699–710. https://doi.org/10.1111/1462-2920.12502.
10. Samaranayake, L.; Matsubara, V. H. Normal Oral Flora and the Oral Ecosystem. *Dent Clin North Am* **2017**, *61* (2), 199–215. https://doi.org/10.1016/j.cden.2016.11.002.
11. Bik, E. M.; Long, C. D.; Armitage, G. C.; Loomer, P.; Emerson, J.; Mongodin, E. F.; Nelson, K. E.; Gill, S. R.; Fraser-Liggett, C. M.; Relman, D. A. Bacterial Diversity in the Oral Cavity of 10 Healthy Individuals. *ISME J* **2010**, *4* (8), 962–974. https://doi.org/10.1038/ismej.2010.30.
12. Verma, D.; Garg, P. K.; Dubey, A. K. Insights into the Human Oral Microbiome. *Arch Microbiol* **2018**, *200* (4), 525–540. https://doi.org/10.1007/s00203-018-1505-3.
13. Baker, J. L.; Bor, B.; Agnello, M.; Shi, W.; He, X. Ecology of the Oral Microbiome: Beyond Bacteria. *Trends Microbiol* **2017**, *25* (5), 362–374. https://doi.org/10.1016/j.tim.2016.12.012.

14. Delhaes, L.; Monchy, S.; Fréalle, E.; Hubans, C.; Salleron, J.; Leroy, S.; Prevotat, A.; Wallet, F.; Wallaert, B.; Dei-Cas, E.; et al. The Airway Microbiota in Cystic Fibrosis: A Complex Fungal and Bacterial Community–Implications for Therapeutic Management. *PLoS One* **2012**, *7* (4), e36313. https://doi.org/10.1371/journal.pone.0036313.
15. Ghannoum, M. A.; Jurevic, R. J.; Mukherjee, P. K.; Cui, F.; Sikaroodi, M.; Naqvi, A.; Gillevet, P. M. Characterization of the Oral Fungal Microbiome (Mycobiome) in Healthy Individuals. *PLoS Pathog* **2010**, *6* (1), e1000713. https://doi.org/10.1371/journal.ppat.1000713.
16. Abeles, S. R.; Pride, D. T. Molecular Bases and Role of Viruses in the Human Microbiome. *J Mol Biol* **2014**, *426* (23), 3892–3906. https://doi.org/10.1016/j.jmb.2014.07.002.
17. Minot, S.; Sinha, R.; Chen, J.; Li, H.; Keilbaugh, S. A.; Wu, G. D.; Lewis, J. D.; Bushman, F. D. The Human Gut Virome: Inter-Individual Variation and Dynamic Response to Diet. *Genome Res* **2011**, *21* (10), 1616–1625. https://doi.org/10.1101/gr.122705.111.
18. Cattaneo, C.; Gargari, G.; Koirala, R.; Laureati, M.; Riso, P.; Guglielmetti, S.; Pagliarini, E. New Insights into the Relationship between Taste Perception and Oral Microbiota Composition. *Sci Rep* **2019**, *9* (1), 3549. https://doi.org/10.1038/s41598-019-40374-3.
19. Mystkowska, J.; Niemirowicz-Laskowska, K.; Łysik, D.; Tokajuk, G.; Dąbrowski, J. R.; Bucki, R. The Role of Oral Cavity Biofilm on Metallic Biomaterial Surface Destruction–Corrosion and Friction Aspects. *Int J Mol Sci* **2018**, *19* (3). https://doi.org/10.3390/ijms19030743.
20. Bikel, S.; Valdez-Lara, A.; Cornejo-Granados, F.; Rico, K.; Canizales-Quinteros, S.; Soberón, X.; Del Pozo-Yauner, L.; Ochoa-Leyva, A. Combining Metagenomics, Metatranscriptomics and Viromics to Explore Novel Microbial Interactions: Towards a Systems-Level Understanding of Human Microbiome. *Comput Struct Biotechnol J* **2015**, *13*, 390–401. https://doi.org/10.1016/j.csbj.2015.06.001.
21. Adler, C. J.; Malik, R.; Browne, G. V.; Norris, J. M. Diet May Influence the Oral Microbiome Composition in Cats. *Microbiome* **2016**, *4* (1), 23. https://doi.org/10.1186/s40168-016-0169-y.
22. Moye, Z. D.; Zeng, L.; Burne, R. A. Fueling the Caries Process: Carbohydrate Metabolism and Gene Regulation by *Streptococcus mutans*. *J Oral Microbiol* **2014**, *6* (1), 24878. https://doi.org/10.3402/jom.v6.24878.
23. Hall, M. W.; Singh, N.; Ng, K. F.; Lam, D. K.; Goldberg, M. B.; Tenenbaum, H. C.; Neufeld, J. D.; Beiko, R. G.; Senadheera, D. B. Inter-Personal Diversity and Temporal Dynamics of Dental, Tongue, and Salivary Microbiota in the Healthy Oral Cavity. *NPJ Biofilms Microbiomes* **2017**, *3* (1), 2. https://doi.org/10.1038/s41522-016-0011-0.
24. Burcham, Z. M.; Garneau, N. L.; Comstock, S. S.; Tucker, R. M.; Knight, R.; Metcalf, J. L.; Miranda, A.; Reinhart, B.; Meyers, D.; Woltkamp, D.; et al. Patterns of Oral Microbiota Diversity in Adults and Children: A Crowdsourced Population Study. *Sci Rep* **2020**, *10* (1), 2133. https://doi.org/10.1038/s41598-020-59016-0.
25. Kistler, J. O.; Arirachakaran, P.; Poovorawan, Y.; Dahlén, G.; Wade, W. G. The Oral Microbiome in Human Immunodeficiency Virus (HIV)-Positive Individuals. *J Med Microbiol* **2015**, *64* (9), 1094–1101. https://doi.org/10.1099/jmm.0.000128.
26. Krishnan, K.; Chen, T.; Paster, B. A Practical Guide to the Oral Microbiome and Its Relation to Health and Disease. *Oral Dis* **2017**, *23* (3), 276–286. https://doi.org/10.1111/odi.12509.
27. Chopyk, J.; Chattopadhyay, S.; Kulkarni, P.; Smyth, E. M.; Hittle, L. E.; Paulson, J. N.; Pop, M.; Buehler, S. S.; Clark, P. I.; Mongodin, E. F.; et al. Temporal Variations in Cigarette Tobacco Bacterial Community Composition and Tobacco-Specific Nitrosamine Content Are Influenced by Brand and Storage Conditions. *Front Microbiol* **2017**, *8*. https://doi.org/10.3389/fmicb.2017.00358.
28. Mager, D. L.; Ximenez-Fyvie, L. A.; Haffajee, A. D.; Socransky, S. S. Distribution of Selected Bacterial Species on Intraoral Surfaces. *J Clin Periodontol* **2003**, *30* (7), 644–654. https://doi.org/10.1034/j.1600-051X.2003.00376.x.
29. Chakraborty, S.; Persaud, V.; Vanegas, S.; Gautier, G.; Esiobu, N. Analysis of the Human Oral Microbiome of Smokers and Non-Smokers Using PCR-RFLP and Ribotyping. *Adv Microbiol* **2014**, *04* (10), 681–691. https://doi.org/10.4236/aim.2014.410073.
30. Suzuki, N.; Nakano, Y.; Yoneda, M.; Hirofuji, T.; Hanioka, T. The Effects of Cigarette Smoking on the Salivary and Tongue Microbiome. *Clin Exp Dent Res* **2022**, *8* (1), 449–456. https://doi.org/10.1002/cre2.489.
31. Al-Marzooq, F.; Al Kawas, S.; Rahman, B.; Shearston, J. A.; Saad, H.; Benzina, D.; Weitzman, M. Supragingival Microbiome Alternations as a Consequence of Smoking Different Tobacco Types and Its Relation to Dental Caries. *Sci Rep* **2022**, *12* (1), 2861. https://doi.org/10.1038/s41598-022-06907-z.

32. Galvin, S.; Anishchuk, S.; Healy, C. M.; Moran, G. P. Smoking, Tooth Loss and Oral Hygiene Practices Have Significant and Site-Specific Impacts on the Microbiome of Oral Mucosal Surfaces: A Cross-Sectional Study. *J Oral Microbiol* **2023**, *15* (1). https://doi.org/10.1080/20002297.2023.2263971.
33. Ganesan, S. M.; Ganesan, S. M.; Dabdoub, S. M.; Nagaraja, H. N.; Scott, M. L.; Pamulapati, S.; Pamulapati, S.; Berman, M. L.; Shields, P. G.; Wewers, M. E.; et al. Adverse Effects of Electronic Cigarettes on the Disease-Naive Oral Microbiome. *Sci Adv* **2020**, *6* (22). https://doi.org/10.1126/sciadv.aaz0108.
34. Matarazzo, F.; Ribeiro, A. C.; Feres, M.; Faveri, M.; Mayer, M. P. A. Diversity and Quantitative Analysis of Archaea in Aggressive Periodontitis and Periodontally Healthy Subjects. *J Clin Periodontol* **2011**, *38* (7), 621–627. https://doi.org/10.1111/j.1600-051X.2011.01734.x.
35. Vartoukian, S. R.; Palmer, R. M.; Wade, W. G. Diversity and Morphology of Members of the Phylum "Synergistetes" in Periodontal Health and Disease. *Appl Environ Microbiol* **2009**, *75* (11), 3777–3786. https://doi.org/10.1128/AEM.02763-08.
36. Gross, E. L.; Leys, E. J.; Gasparovich, S. R.; Firestone, N. D.; Schwartzbaum, J. A.; Janies, D. A.; Asnani, K.; Griffen, A. L. Bacterial 16S Sequence Analysis of Severe Caries in Young Permanent Teeth. *J Clin Microbiol* **2010**, *48* (11), 4121–4128. https://doi.org/10.1128/JCM.01232-10.
37. Koo, H.; Bowen, W. H. *Candida albicans* and *Streptococcus mutans*: A Potential Synergistic Alliance to Cause Virulent Tooth Decay in Children. *Future Microbiol* **2014**, 1295–1297. https://doi.org/10.2217/fmb.14.92.
38. Pushalkar, S.; Ji, X.; Li, Y.; Estilo, C.; Yegnanarayana, R.; Singh, B.; Li, X.; Saxena, D. Comparison of Oral Microbiota in Tumor and Non-Tumor Tissues of Patients with Oral Squamous Cell Carcinoma. *BMC Microbiol* **2012**, *12* (1), 144. https://doi.org/10.1186/1471-2180-12-144.
39. Mager, D. L.; Haffajee, A. D.; Devlin, P. M.; Norris, C. M.; Posner, M. R.; Goodson, J. M. The Salivary Microbiota as a Diagnostic Indicator of Oral Cancer: A Descriptive, Non-Randomized Study of Cancer-Free and Oral Squamous Cell Carcinoma Subjects. *J Transl Med* **2005**, *3* (1), 27. https://doi.org/10.1186/1479-5876-3-27.
40. Wang, L.; Ganly, I. The Oral Microbiome and Oral Cancer. *Clin Lab Med* **2014**, *34* (4), 711–719. https://doi.org/10.1016/j.cll.2014.08.004.
41. Aguayo, S.; Schuh, C. M. A. P.; Vicente, B.; Aguayo, L. G. Association between Alzheimer's Disease and Oral and Gut Microbiota: Are Pore Forming Proteins the Missing Link? *J Alzheimers Dis* **2018**, *65* (1), 29–46. https://doi.org/10.3233/JAD-180319.
42. Miklossy, J. Bacterial Amyloid and DNA Are Important Constituents of Senile Plaques: Further Evidence of the Spirochetal and Biofilm Nature of Senile Plaques. *J Alzheimers Dis* **2016**, *53* (4), 1459–1473. https://doi.org/10.3233/JAD-160451.
43. Dominy, S. S.; Lynch, C.; Ermini, F.; Benedyk, M.; Marczyk, A.; Konradi, A.; Nguyen, M.; Haditsch, U.; Raha, D.; Griffin, C.; et al. *Porphyromonas gingivalis* in Alzheimer's Disease Brains: Evidence for Disease Causation and Treatment with Small-Molecule Inhibitors. *Sci Adv* **2019**, *5* (1). https://doi.org/10.1126/sciadv.aau3333.
44. Whiley, R. A.; Fleming, E. V; Makhija, R.; Waite, R. D. Environment and Colonisation Sequence Are Key Parameters Driving Cooperation and Competition between *Pseudomonas aeruginosa* Cystic Fibrosis Strains and Oral Commensal Streptococci. *PLoS One* **2015**, *10* (2), e0115513. https://doi.org/10.1371/journal.pone.0115513.
45. Teles, R.; Wang, C.-Y. Mechanisms Involved in the Association between Periodontal Diseases and Cardiovascular Disease. *Oral Dis* **2011**, *17* (5), 450–461. https://doi.org/10.1111/j.1601-0825.2010.01784.x.
46. Chhibber-Goel, J.; Singhal, V.; Bhowmik, D.; Vivek, R.; Parakh, N.; Bhargava, B.; Sharma, A. Linkages between Oral Commensal Bacteria and Atherosclerotic Plaques in Coronary Artery Disease Patients. *NPJ Biofilms Microbiomes* **2016**, *2* (1), 7. https://doi.org/10.1038/s41522-016-0009-7.
47. Casarin, R. C. V; Barbagallo, A.; Meulman, T.; Santos, V. R.; Sallum, E. A.; Nociti, F. H.; Duarte, P. M.; Casati, M. Z.; Gonçalves, R. B. Subgingival Biodiversity in Subjects with Uncontrolled Type-2 Diabetes and Chronic Periodontitis. *J Periodontal Res* **2013**, *48* (1), 30–36. https://doi.org/10.1111/j.1600-0765.2012.01498.x.
48. Zhang, X.; Zhang, D.; Jia, H.; Feng, Q.; Wang, D.; Liang, D.; Wu, X.; Li, J.; Tang, L.; Li, Y.; et al. The Oral and Gut Microbiomes Are Perturbed in Rheumatoid Arthritis and Partly Normalized after Treatment. *Nat Med* **2015**, *21* (8), 895–905. https://doi.org/10.1038/nm.3914.
49. Roszyk, E.; Puszczewicz, M. Role of Human Microbiome and Selected Bacterial Infections in the Pathogenesis of Rheumatoid Arthritis. *Rheumatology* **2017**, *55* (5), 242–250. https://doi.org/10.5114/reum.2017.71641.

50. Li, X.; Kolltveit, K. M.; Tronstad, L.; Olsen, I. Systemic Diseases Caused by Oral Infection. *Clin Microbiol Rev* **2000**, *13* (4), 547–558. https://doi.org/10.1128/CMR.13.4.547.
51. Koo, H.; Xiao, J.; Klein, M. I.; Jeon, J. G. Exopolysaccharides Produced by *Streptococcus mutans* Glucosyltransferases Modulate the Establishment of Microcolonies within Multispecies Biofilms. *J Bacteriol* **2010**, *192* (12), 3024–3032. https://doi.org/10.1128/JB.01649-09.
52. Hyde, E. R.; Andrade, F.; Vaksman, Z.; Parthasarathy, K.; Jiang, H.; Parthasarathy, D. K.; Torregrossa, A. C.; Tribble, G.; Kaplan, H. B.; Petrosino, J. F.; et al. Metagenomic Analysis of Nitrate-Reducing Bacteria in the Oral Cavity: Implications for Nitric Oxide Homeostasis. *PLoS One* **2014**, *9* (3), e88645. https://doi.org/10.1371/journal.pone.0088645.
53. Koch, C. D.; Gladwin, M. T.; Freeman, B. A.; Lundberg, J. O.; Weitzberg, E.; Morris, A. Enterosalivary Nitrate Metabolism and the Microbiome: Intersection of Microbial Metabolism, Nitric Oxide and Diet in Cardiac and Pulmonary Vascular Health. *Free Radic Biol Med* **2017**, *105*, 48–67. https://doi.org/10.1016/j.freeradbiomed.2016.12.015.
54. Paksoy, T. Assessment of Epstein–Barr Virus, *Candida albicans*, and Some Periodontal Pathogens in Rheumatoid Arthritis Patients with Periodontitis. *North Clin Istanb* **2023**. https://doi.org/10.14744/nci.2023.58998.
55. Bingham, C. O.; Moni, M. Periodontal Disease and Rheumatoid Arthritis. *Curr Opin Rheumatol* **2013**, *25* (3), 345–353. https://doi.org/10.1097/BOR.0b013e32835fb8ec.
56. Hayashi, C.; Gudino, C. V.; Gibson III, F. C.; Genco, C. A. REVIEW: Pathogen-Induced Inflammation at Sites Distant from Oral Infection: Bacterial Persistence and Induction of Cell-Specific Innate Immune Inflammatory Pathways. *Mol Oral Microbiol* **2010**, *25* (5), 305–316. https://doi.org/10.1111/j.2041-1014.2010.00582.x.
57. Chattopadhyay, I.; Verma, M.; Panda, M. Role of Oral Microbiome Signatures in Diagnosis and Prognosis of Oral Cancer. *Technol Cancer Res Treat* **2019**, *18*, 153303381986735. https://doi.org/10.1177/1533033819867354.
58. Xiao, E.; Mattos, M.; Vieira, G. H. A.; Chen, S.; Corrêa, J. D.; Wu, Y.; Albiero, M. L.; Bittinger, K.; Graves, D. T. Diabetes Enhances IL-17 Expression and Alters the Oral Microbiome to Increase Its Pathogenicity. *Cell Host Microbe* **2017**, *22* (1), 120–128.e4. https://doi.org/10.1016/j.chom.2017.06.014.
59. Castro, M. M. L.; Ferreira, R. de O.; Fagundes, N. C. F.; Almeida, A. P. C. P. S. C.; Maia, L. C.; Lima, R. R. Association between Psychological Stress and Periodontitis: A Systematic Review. *Eur J Dent* **2020**, *14* (01), 171–179. https://doi.org/10.1055/s-0039-1693507.
60. Xue, H.; Chen, X.; Yu, C.; Deng, Y.; Zhang, Y.; Chen, S.; Chen, X.; Chen, K.; Yang, Y.; Ling, W. Gut Microbially Produced Indole-3-Propionic Acid Inhibits Atherosclerosis by Promoting Reverse Cholesterol Transport and Its Deficiency Is Causally Related to Atherosclerotic Cardiovascular Disease. *Circ Res* **2022**, *131* (5), 404–420. https://doi.org/10.1161/CIRCRESAHA.122.321253.
61. Fan, Z.; Tang, P.; Li, C.; Yang, Q.; Xu, Y.; Su, C.; Li, L. Fusobacterium Nucleatum and Its Associated Systemic Diseases: Epidemiologic Studies and Possible Mechanisms. *J Oral Microbiol* **2023**, *15* (1). https://doi.org/10.1080/20002297.2022.2145729.
62. Jie, Z.; Xia, H.; Zhong, S.-L.; Feng, Q.; Li, S.; Liang, S.; Zhong, H.; Liu, Z.; Gao, Y.; Zhao, H.; et al. The Gut Microbiome in Atherosclerotic Cardiovascular Disease. *Nat Commun* **2017**, *8* (1), 845. https://doi.org/10.1038/s41467-017-00900-1.
63. Han, Y. W. Oral Health and Adverse Pregnancy Outcomes – What's Next? *J Dent Res* **2011**, *90* (3), 289–293. https://doi.org/10.1177/0022034510381905.
64. Cobb, C.; Kelly, P.; Williams, K.; Babbar, S.; Angolkar, M.; Derman, R. The Oral Microbiome and Adverse Pregnancy Outcomes. *Int J Womens Health* **2017**, *9*, 551–559. https://doi.org/10.2147/IJWH.S142730.
65. Sampaio-Maia, B.; Monteiro-Silva, F. Acquisition and Maturation of Oral Microbiome throughout Childhood: An Update. *Dent Res J (Isfahan)* **2014**, *11* (3), 291–301.
66. Docktor, M. J.; Paster, B. J.; Abramowicz, S.; Ingram, J.; Wang, Y. E.; Correll, M.; Jiang, H.; Cotton, S. L.; Kokaras, A. S.; Bousvaros, A. Alterations in Diversity of the Oral Microbiome in Pediatric Inflammatory Bowel Disease. *Inflamm Bowel Dis* **2012**, *18* (5), 935–942. https://doi.org/10.1002/ibd.21874.
67. Pathak, J. L.; Yan, Y.; Zhang, Q.; Wang, L.; Ge, L. The Role of Oral Microbiome in Respiratory Health and Diseases. *Respir Med* **2021**, *185*, 106475. https://doi.org/10.1016/j.rmed.2021.106475.
68. Beighton, D. The Complex Oral Microflora of High-risk Individuals and Groups and Its Role in the Caries Process. *Community Dent Oral Epidemiol* **2005**, *33* (4), 248–255. https://doi.org/10.1111/j.1600-0528.2005.00232.x.

69. Bunetel, L.; Tricot-Doleux, S.; Agnani, G.; Bonnaure-Mallet, M. In Vitro Evaluation of the Retention of Three Species of Pathogenic Microorganisms by Three Different Types of Toothbrush. *Oral Microbiol Immunol* **2000**, *15* (5), 313–316. https://doi.org/10.1034/j.1399-302x.2000.150508.x.
70. Bondonno, C. P.; Liu, A. H.; Croft, K. D.; Considine, M. J.; Puddey, I. B.; Woodman, R. J.; Hodgson, J. M. Antibacterial Mouthwash Blunts Oral Nitrate Reduction and Increases Blood Pressure in Treated Hypertensive Men and Women. *Am J Hypertens* **2015**, *28* (5), 572–575. https://doi.org/10.1093/ajh/hpu192.
71. Bui, F. Q.; Almeida-da-Silva, C. L. C.; Huynh, B.; Trinh, A.; Liu, J.; Woodward, J.; Asadi, H.; Ojcius, D. M. Association between Periodontal Pathogens and Systemic Disease. *Biomed J* **2019**, *42* (1), 27–35. https://doi.org/10.1016/j.bj.2018.12.001.

15 Role of the Oral Microbiome as Probiotics

Neha Gautam
Department of Food Science and Technology, Dr. Y. S. Parmar University of Horticulture and Forestry, Solan, Himachal Pradesh, India

Stuti Sharma
Department of Chemistry, Sardar Patel University, Mandi, Himachal Pradesh, India

Nivedita Sharma
Former Professor and Head, Department of Basic Sciences, Dr. Y. S. Parmar University of Horticulture and Forestry, Solan, Himachal Pradesh, India

15.1 INTRODUCTION

Probiotics are live microorganisms, which when consumed in adequate amounts, confer a health effect on the host [1]. Traditionally, probiotics have been linked with gut health and mainly utilized for the prevention or treatment of gastrointestinal diseases. However, in recent years several investigators have suggested the use of probiotics for oral health purposes. The interaction of probiotics with the oral microbiome sustains oral vigor and maintains a healthy microbial equilibrium. The most common genera of the oral cavity are *Lactobacillus*, *Bifidobacteria*, and *Streptococcus* [2]. These genera have been reported to control chronic diseases such as dental caries and periodontitis and recurring problems like candidal infections and halitosis. Among the three genera, *Lactobacillus* plays a noteworthy role in the oral biosphere and contributes in maintaining oral health. Probiotic lactobacilli could be considered as a means of inhibiting expansion of the oral biofilm and decreasing the surge of the detrimental immunoinflammatory response.

Unlike probiotic-containing food or supplement product, probiotic-containing toothpaste is a fairly new concept in the avoidance of oral infectious diseases. Probiotic milk drinks, chewing gums, toothpastes, and tablets have been shown to reduce severe gingival inflammation in some clinical trials. However, scientific verifications are scarce in this area, and their commendation for oral health purposes is not yet defensible. There is a need to look for potential microbes with probiotic attributes that can resist oral ecological conditions and can stay there for long enough to have a positive result. The exploration of the use of probiotics in treatment and management of some oral cavity illness and the potential of developing designer probiotics for the next generation of oral complementary health care is needed.

15.1.1 Oral Microbiome

There is a diversity of bacteria residing in the oral cavity. In 1674, Antonie van Leeuwenhoek observed his own dental plaque and described it as, "Many very little living animalcules prettily a-moving." According to Van Leeuwenhoek, "There are more animals living in the scum on the teeth in a man's mouth than there are men in a whole kingdom." The oral cavity's microbial ecology is intricate, rich in biological context, and home to a variety of niches that offer a special setting

DOI: 10.1201/9781003478638-15

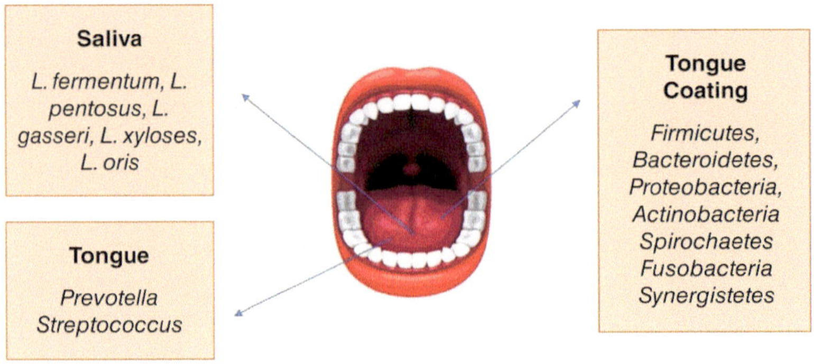

FIGURE 15.1 Microbial diversity of the oral cavity.

for microbial colonization. According to [3], these niches include the cheek, tongue, gingival sulcus, hard and soft palates, base of the mouth, throat, teeth, and saliva. Because some adhesins on bacterial surfaces bind to corresponding receptors on oral surfaces, certain surfaces in the mouth are colonized by oral bacteria preferentially [4]. A healthy microbiome is composed of bacteria, fungi, archaea, viruses, and protozoa. There are, however, relatively few data on the mycobiome (fungal microbiome), and the observations on a normal microbiome are limited to the bacteriome. One of the microbiomes that has been investigated the most to date is the oral cavity [5]. Bacteria inhabiting saliva are *Lactobacillus fermentum*, *L. pentosus*, *L. gasseri*, *L. brevis*, *L. casei*, *L. xylosus, and L. oris.* In the tongue coating (a feature on the tongue that provides it a distinct texture), Firmicutes, Actinobacteria, Bacteroidetes, Fusobacteria, Proteobacteria, Spirochaetes, and Synergistetes microbial phyla dominate, while *Prevotella* and *Streptococcus* inhabit the tongue as shown in Figure 15.1. Along with beneficial bacteria, harmful bacteria also inhabit the oral cavity. When the balance between these bacteria shifts because of poor hygiene, poor diet, and other health issues, it can cause the harmful bacteria to take a lead. This imbalance is the reason for bad breath, cavities, gum disease, and tooth loss.

The oral cavity contains over 700 species of prokaryotes that have been identified. These species are divided into 185 genera and 12 phyla, of which around 54% have official names, 14% are unidentified but have been cultivated, and 32% are only recognized as phylotypes that have not been farmed [6]. According to [7], the 12 phyla are Bacteroidetes, Fusobacteria, Proteobacteria, Actinobacteria, Firmicutes, Chlamydiae, Chloroflexi, Spirochaetes, Absconditabacteria, Synergistetes, Saccharibacteria, and Gracilibacteria. A few lesser-known microbes in the oral cavities are Chloroflexi, Chlorobi, and Synergistetes [8]. In healthy mouths, the oral microbial inhabitants are recognized at the genus level. In spite of the commonalities, microbiome diversity varies by individual and location. The tongue is home to a wide-ranging microbiota, including anaerobes, due to its papillae and sparse anaerobic locations. The buccal and palatal mucosae have the smallest microbial diversity [9]. There is a mutual relationship between the microorganisms in our oral cavity based on benefits. The commensal populations maintain a check on the pathogenic species by preventing them from adhering to the mucosa. The bacteria become pathogenic only after they break the barrier of the commensals and cause infection and disease [10]. The prominent bacterial genera present in the healthy oral cavity are listed in Table 15.1 [11].

The Archaea domain encompasses microorganisms with unique structural, physiologic, and genetic features. There are indications that archaea are part of the indigenous microbiota in various oral sites, such as periodontal and endodontic niches, carious dentine, saliva, and the tongue. Information about archaea and their participation in oral diseases is still inadequate. Oral archaea are far less numerous and diverse than oral bacteria. Methanogens are exclusive archaea thought to be present in the oral cavity [39]. Among methanogens, *Methanobrevibacter oralis* marks its dominance in oral cavity. Recently, advanced molecular studies revealed the presence of

TABLE 15.1
Prominent Bacterial Genera of the Oral Cavity

S. No.	Bacterial Genera	References
1	*Abiotrophia*	[12]
2	*Peptostreptococcus*	[13]
3	*Streptococcus*	[14]
4	*Stomatococcus*	[15]
5	*Actinomyces*	[16]
6	*Bifidobacterium*	[17]
7	*Corynebacterium*	[18]
8	*Eubacterium*	[19]
9	*Lactobacillus*	[20]
10	*Propionibacterium*	[21]
11	*Rothia*	[22]
12	*Pseudoramibacter*	[23]
13	*Moraxella*	[24]
14	*Neisseria*	[25]
15	*Veillonella*	[26]
16	*Campylobacter*	[27]
17	*Capnocytophaga*	[28]
18	*Desulfobacter*	[29]
19	*Desulfovibrio*	[29]
20	*Eikenella*	[30]
21	*Fusobacterium*	[31]
22	*Haemophilus*	[32]
23	*Leptotrichia*	[33]
24	*Prevotella*	[34]
25	*Selenomonas*	[35]
26	*Simonsiella*	[36]
27	*Treponema*	[37]
28	*Wolinella*	[38]

non-methanogenic archaea in the oral cavity [40, 41]. Unexpectedly, archaea associated with the oral cavity are recognized as non-pathogenic.

Other than bacteria and archaeal bacteria, the oral cavity contains varied forms of other microbes such as protozoa, fungi, and viruses. *Entamoeba gingivalis* and *Trichomonas tenax* are the most commonly found saprophytic protozoa. *Candida* species are the most prevalent fungi seen associated with the oral cavity. Fungi account for 0.004% of the overall oral microorganisms and have been detected in specimens from the hard palate, supragingival plaque, and oral rinses [42]. The major fungal species observed were those belonging to *Aspergillus, Aureobasidium*, *Candida*, *Cryptococcus, Cladosporium*, *Fusarium*, and *Saccharomycetales* [43]. Recently *Malassezia* and *Candida* genera were detected in saliva samples [44]. Among viruses, Anelloviridae, Herperviridae, and Papillomaviridae are reported in the oral cavity [45, 46].

15.2 WHY SPECIFIC BACTERIA IN THE ORAL MICROBIOME ACT AS PROBIOTICS

The oral microbiome comprises several microorganisms. But there are specific bacteria called probiotics that confer health benefits to the oral cavity. There are several species of probiotic bacteria that can be used to fight against several pathogens and diseases, playing a vital role in protecting the human oral microbiome. The two most common are *Lactobacillus* and *Bifidobacteria*. According to [47], high-quality research may be necessary to establish the effectiveness of "precision probiotics," which are probiotics that are particularly chosen or even developed for a disease, to address the variability inherent to probiotic strains, the hosts, and their microbiomes. The effectiveness of probiotics for oral health may be diminished by choosing an inappropriate probiotic strain, improper probiotic product manufacturing, and a lack of time for activity and adherence to oral surfaces. The biological characteristics of the probiotic determine just a portion of these interrelated aspects. The preparation of probiotic goods has the most impact on them. The development of a shelf-stable product is the source of the first problem. It can be difficult to produce a product that contains live bacteria since living things need to be continuously supplied with nutrients to exist. One solution is to make the probiotic bacteria temporarily dormant. This can be achieved by completely eliminating intracellular water, which leads to an intentional reversible stop of biological activity. Freeze-drying, or lyophilization, is the most popular method for doing this [48, 49].

Regarding commensal oral microorganisms, there are a number of factors that lend credence to the possibility of identifying bacteria that might be helpful in the treatment or prevention of oral illnesses. Indeed, several probiotic effects have been reported to be shared by multiple species (Table 15.2) rather than being exclusive to a small number of well-researched strains [50, 51]. According to [52], the ecological plaque hypothesis posits that the equilibrium between dental health and illness may be altered by selection pressure in environmental settings. Bacteria may contribute to the environmental pressure mentioned in the ecological plaque theory as they have the ability to affect their surroundings and are known to interact both synergistically and antagonistically in dental plaque.

15.3 BENEFICIAL PROBIOTIC SPECIES

In the last decade, several investigators have reported beneficial probiotic species for oral health. The most notable bacteria with probiotic characteristics and their effect on oral health have been discussed in this section.

1. *Streptococcus salivarius*: Suppresses oral pathogens such as *Streptococcus pyogenes* and *Streptococcus pneumonia* which cause infection and autoimmune disorders; provides strong immune support to help reduce symptoms of oral and respiratory viral infections; maintains ear, nose, and throat health; and can crowd out other potentially harmful oral pathogens. It produces bacteriocins and enzymes to remove plaque, metabolizes lactic acid and reduces pro-inflammatory immune response (i.e., cytokine storms).

TABLE 15.2
Probiotic Strains and Their Reported Effect

S. No.	Strains	Reported Effects	References
1	*L. rhamnosus* and *Saccharomyces cerevisiae*	Reduction in the count of *S. mutans* in saliva.	[53]
2	*L. reuteri* ATCC 55730	Notable decrease observed in salivary *S. mutans*.	[54]
3	*L. rhamnosus* SP1	Decrease in the severity and development of caries.	[55]
4	*L. rhamnosus* GG ATCC 53103	Decrease observed in the number of *S. mutans*, but the number of *Lactobacillus* remained unchanged.	[56]
5	*L. brevis* CD2	Noteworthy decline observed in the count of salivary *S. mutans*, plaque acidogenicity, and bleeding on probing.	[57]
6	*L. casei*	Reduction of periodontal pathogens.	[58]
7	*L. brevis*	Amelioration of periodontitis-associated signs and symptoms.	[59]

2. *Lactobacillus brevis*: This is reported to inhibit chemotherapy induced oral mucositis.
3. *Limosilactobacillus reuteri*: Reduces proinflammatory immune responses (i.e., cytokine storms), *Candida,* and periodontal pocket depth.
4. *Ligilactobacillus salivarius*: Reduces pathogenic bacteria population, supports tooth and gum health by improving resistance to cavities, produces sufficient hydrogen peroxide to kill certain bacterial species, and improves bad breath.
5. *Lacticaseibacillus paracasei*: Binds to oral pathogens that cause tooth decay and protect against *Candida.*

15.4 THERAPEUTIC EFFECT OF ORAL PROBIOTICS ON ORAL HEALTH

As infants, our oral microbiome is assembled in the first year and maintained by the right nutrient sources. However, over time, factors such as a poor acidic diet (refined carbohydrates and soft drinks), stress, smoking, and antibacterial agents (medication) can negatively impact the diversity and balance of the oral microbiome. We can take steps to reintroduce beneficial bacteria into our oral cavity by taking oral probiotic supplements in the morning after brushing our teeth. These supplements are available as lozenges, mouth rinses, and chewable tablets, which deliver beneficial bacterial strains directly into the mouth. From there, they can colonize in the oral cavity into biofilms that start competing and eliminating oral pathogens. Fortunately, while pathogenic oral bacteria are damaging to health, an abundance of probiotic oral bacteria in the oral cavity has the opposite effect. These friendly germs can protect our oral health from decay and disease. Probiotics adhere to dental tissues as a part of the biofilm (or plaque) and compete with the growth of cariogenic bacteria or periodontal pathogens, and thus when consumed regularly, they play a vital role in the avoidance and treatment of oral diseases such as dental caries, periodontitis, halitosis, candida infections, and gingivitis. Mucosal and systemic immunity can be developed and enhanced; the lifespan of a voice prosthesis device can be increased by regulating the composition of oral microflora. Probiotics such as *Lactobacillus, Streptococcus,* and *Bifidobacterium* reduce the harmful properties of pathogenic microorganisms through coaggregation, production of toxic by-products, and competing for substrates.

15.5 THERAPEUTIC EFFECT OF ORAL PROBIOTICS ON OVERALL HEALTH

The oral microbiome may have a positive effect on the health of the gastrointestinal (GI) tract. Actually, a huge variety of oral species can enter the GI tract via swallowing, but only few of these bacteria seems to colonize the gut. It is well narrated in the literature that these good bacteria and their metabolites can maintain health and well-being and potentially reduce long-term risk of diseases like GI disease, cancer, cardiovascular disease, lactose intolerance, autoimmune disease, and neurodegenerative disease.

15.5.1 Gastrointestinal Disease

Different types of diarrheas result due to GI disorders. One type of diarrhea is antibiotic-associated diarrhea, which is basically a common side effect of antibiotic treatment due to the disruption of the normal gut microbiota these medications cause. The other type is traveler's diarrhea, which occurs when a patient visits a place where the climate or sanitary practices are different from home, thus inviting an increased risk of developing traveler's diarrhea by consuming contaminated food and water. The use of probiotics to treat these GI disorders is an apparent measure and perhaps the most usual application of probiotics because most health effects attributed to them are related directly or indirectly (i.e., mediated by the immune system) to the GI tract [60]. During gastrointestinal disturbances, the gut bacterial composition becomes imbalanced, and the normal population of good bacteria is disrupted. This problem of unbalanced gut microflora can be overcome by supplementing food with probiotics that may help to repopulate and maintain beneficial gut bacteria and correct the imbalance. Multiple probiotics have been shown to shorten the duration of watery diarrhea by about 24 hours and have demonstrated the ability to decrease the number of diarrheal stools even earlier in the course. The most commonly studied beneficial probiotic strains include *Lactobacillus reuteri* and *Saccharomyces boulardii*. For families of children with acute infectious watery diarrhea, probiotics should be routinely discussed as a potential means of symptom improvement.

15.5.2 Cancer

Probiotic bacteria exhibit antitumor properties and play a significant role in immunomodulation. Bacterial strains could be responsible for the detection and degradation of potential carcinogens that affect cell death [61]. The anticancerous activity (ACA) and antimutagenic activity (AMA) of probiotics is due to the following:

1. Mutagen binding and degradation
2. Preventing conversion of nontoxic procarcinogens to toxic and highly reactive carcinogens.
3. Lowering intestinal pH by short-chain fatty acids (SCFAs) produced during non-digestible carbohydrate degradation.

Modulation and development of the host's innate immunity through the secretion of anti-inflammatory molecules [62] demonstrate the association between a diet enriched with *Lactobacillus* and a reduced occurrence of colon cancer.

15.5.3 Cardiovascular Diseases

Cardiovascular disease is a major cause of death around the globe. Several studies indicate the effects of probiotics on the composition of the gut microbiome and its relationship to cardiovascular diseases and risks. Probiotics may offer an alternative treatment for diabetes, obesity, hypercholesterolemia, hypertension, chronic kidney disease (CKD), cardiomyopathy, and atherosclerosis. Several clinical reports in humans have also confirmed the ability of probiotics to reduce unusually high blood pressure (BP) levels. Blood pressure is at the root of many different cardiovascular diseases. For example, an extract of *L. casei*, shown to reduce BP was able to bring on a drop in systolic/diastolic BP and heart rate in hypertensive patients. The bacterial strain *L. plantarum* as well as a lactic acid bacteria (LAB) blend containing four strains of bacteria have been found to decrease the level of cholesterol entering the bloodstream, potentially reducing the risk of heart disease [63].

15.5.4 Lactose Intolerance

Lactose intolerance is genetically determined β-galactosidase deficiency, which results in an inability to hydrolyze lactose into glucose and galactose. Unabsorbed lactose is then metabolized by probiotic bacteria. Probiotic administration after ingestion of milk or dairy foods is useful to reduce a variety of symptoms of lactose intolerance.

15.5.5 Autoimmune Disease

Autoimmune disease is a condition where the body's natural immune system cannot differentiate between its own cells and foreign cells, causing the body to attack its own normal cells. Several studies suggest that probiotics influence systemic immune responses, ensure the homeostasis of the

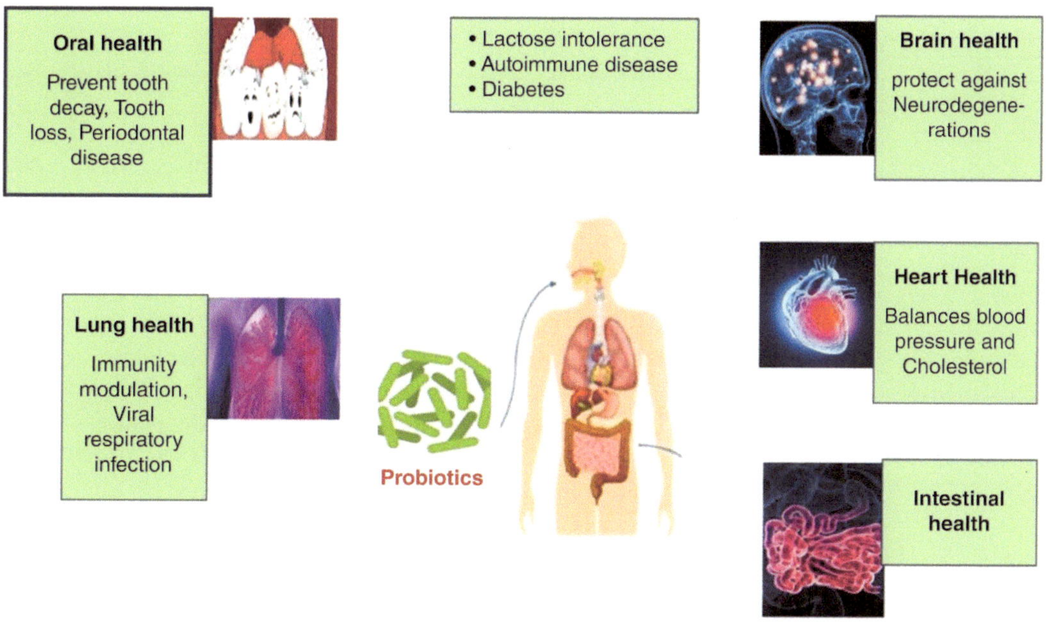

FIGURE 15.2 Therapeutic effect of probiotics on overall health.

healthy microbiota in the intestinal mucosa, and could, therefore, be used as adjuvant therapy to treat immune-mediated diseases.

15.5.6 Neurodegenerative Disease

Neurodegenerative disorders affect neurons in the spinal cord and brain leading to degeneration and loss of function. The most common neurodegenerative disorders are Alzheimer's and Parkinson's diseases. Probiotics are recommended in neurodegenerative disease to prevent cognitive dysfunction, which is a major symptom of these diseases. Many *in vivo* and clinical trials have revealed that probiotics (*Lactobacillus acidophilus*, *Bifidobacterium bifidum*, *Lactobacillus casei*, etc.) are effective candidates against the progression of neurodegeneration. It has been proven that the inflammatory process and oxidative stress can be modulated by modifying the gut microbiota with the help of probiotics [64] (Figure 15.2).

15.6 REQUIREMENTS FOR PROBIOTIC STATUS

A bacterial probiotic strain may be considered suitable if it tends to meet the following criteria:

1. Can be identified by genus, species, and strain level.
2. Has been screened for efficacy, for example, for resistance to digestive enzymes and adherence and colonization with oral surfaces by *in vitro* testing.
3. Demonstrates antagonism against potentially pathogenic bacteria.
4. Can delay colonization by pathogenic strains.
5. Is able to obstruct the organization of the extracellular matrix responsible for biofilm formation and limit the production of cytotoxic products by pathogenic bacteria.
6. Does not ferment sugars so as to avoid a pH decrease and demineralization of tooth enamel.
7. Beneficially alters the biochemical parameters that influence the dental plaque (e.g., salivary components and buffer capacity).

8. Meets requirements for proof that a probiotic strain is safe and without contamination in its delivery form.
9. Has the potential to secrete bacteriocins (This is one of the most important traits in the selection of a probiotic.).

The bacteriocins are multifunctional, ribosomally produced, proteinaceous substances with pronounced antimicrobial activity at certain concentrations. They are produced by bacteria and certain members of archaea to inhibit the growth of similar or closely related bacterial strains [65]. Bacteriocins may have a broad range or narrowed range inhibitory spectrum. If the bacteriocins secreted by a bacterium suppresses bacteria belonging to the same species, they are considered to be narrow-spectrum bacteriocins. In contrary, if they inhibit bacteria belonging to another genus, they are categorized to be broad-spectrum bacteriocins. The genes encoding bacteriocin production are plasmid encoded and chromosomal encoded, sometimes localized on transposons [66]. Bacteriocins have various applications in the food, pharmaceutical, and agriculture industries. Use of bacteriocins in toothpastes to prevent dental caries have been well documented [67].

15.6.1 Role of Probiotics in the Oral Cavity: Their Mechanism of Action

Probiotics can prevent or interfere with the proliferation of oral pathogens in a number of ways, including:

1. Binding to bad bacteria that limit their ability to attach to teeth and gums.
2. Suppressing the growth of bad bacteria—and in some cases killing certain pathogenic species—using antimicrobial actions via the release of hydrogen peroxide, acids, and bacteriocins (antimicrobial proteins).
3. Supporting the immune system by eliminating viruses and reducing inflammation.
4. Raising pH so that saliva is more pH neutral, which decreases the risk of tooth decay and erosion.

The mechanism of action of probiotic bacteria is shown in Figure 15.3.

15.6.2 Pathogens of the Oral Cavity and Their Therapeutic Probiotics

Our mouth is the gateway to our body If there is an overgrowth of pathogens in our mouth, it causes inflammation and damages the tiny blood vessels. Pathogenic bacteria can transfer to distant parts of our body through the bloodstream and can contribute to a number of other diseases like diabetes, heart diseases, and Alzheimer's disease. On the other side, patients with these diseases are more likely to have periodontal disease than people who do not have these diseases. Common pathogens of the oral cavity are discussed here with their therapeutic probiotics.

15.6.2.1 *Streptococcus mutans*

Streptococcus mutans is a gram-positive anaerobic coccus. It is the main component of the oral microbiota, and it is one of the main components of dental plaque [68]. It is also the main pathogen of caries, which is a bacterial infectious disease that occurs in hard tissues of the teeth and has the highest incidence among oral diseases [69]. In the year 1924, J. Clarke isolated an organism from carious lesions and called it *Streptococcus mutans*.

By the mid-1960s, clinical and animal-based laboratory studies depicted *S. mutans* as an important etiologic agent in dental caries [70]. The natural habitat of *S. mutans* is the human oral cavity, more specifically the dental plaque, a multispecies biofilm formed on the hard surfaces of the tooth. It has been largely accepted that the cariogenic potential of *S. mutans* resides in three core

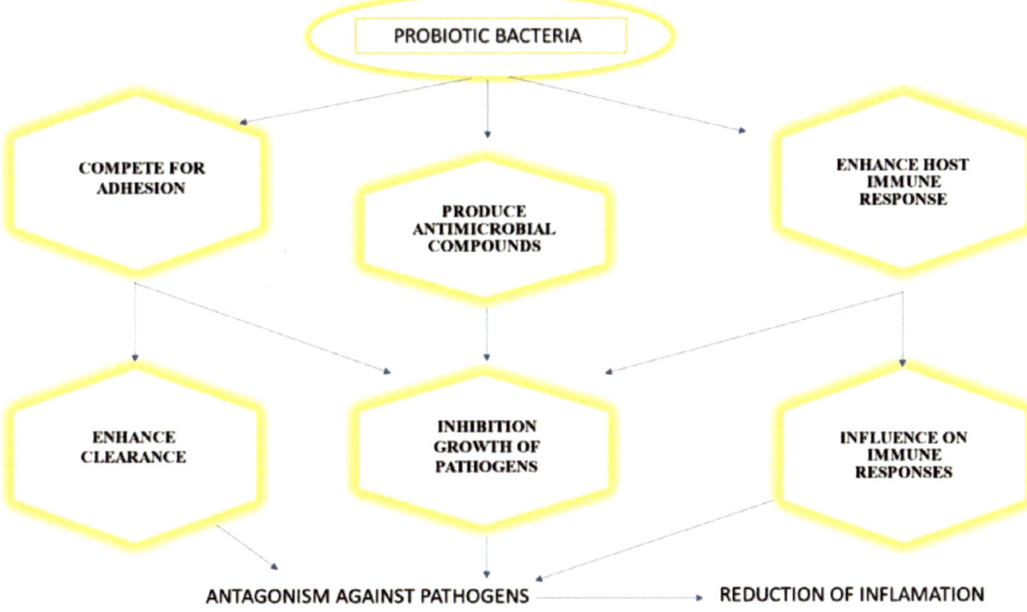

FIGURE 15.3 Mechanism of action of probiotic bacteria.

attributes: (1) the ability to synthesize large quantities of extracellular polymers of glucan from sucrose that aid in the permanent colonization of hard surfaces and in the development of the extracellular polymeric matrix in situ, (2) the ability to transport and metabolize a wide range of carbohydrates into organic acids (acidogenicity), and (3) the ability to thrive under environmental stress conditions, particularly low pH, i.e., aciduricity [71]. Strains of *S. mutans* produce up to three glucosyltransferases, GtfB, GtfC, and GtfD, that utilize the glucose moiety of sucrose as the substrate to synthesize glucose polymers of glucans (also known as mutans). These polymers, especially the α-1,3-1-linked water-insoluble glucans, are major constituents of plaque biofilm matrices [72]. *Lactobacillus acidophilus*, *L. reuteri*, *L. paracasei*. *L. rhamnosus*, and *Bifidobacterium* have been reported to have antagonistic activity against *S. mutans* which causes dental caries [73]. Reported a probiotic inhibitory effect of *L. rhamnosus* and *L. acidophilus* on clinical isolates of *Streptococcus mutans* and suggested that probiotics, especially *L. rhamnosus* and *L. acidophilus*, are effective in the prevention of dental caries.

15.6.2.2 *Porphyromonas gingivalis*

Porphyromonas gingivalis is a non-glycolytic gram-negative anaerobic bacterium that is a periodontal pathogen. According to the World Health Organization, periodontal disease affects 10%–15% of adult populations worldwide [74]. Periodontitis is a disease in which oral microbial species capable of thriving in inflamed microenvironments predominate [75]. *Porphyromonas gingivalis* is a keystone pathogen of periodontal disease as it has profound effects on the oral microbial community structure [76]. It inhabits the subgingival sulcus of the human oral cavity and survives in periodontal pockets where sugar levels are low, and it relies on the fermentation of amino acids for energy production [77]. *Porphyromonas gingivalis* serves as the secondary colonizer of dental plaque as it is an obligate anaerobe and often adheres to primary colonizers such as *Streptococcus gordonii* and *P. intermedia* [78]. Studied the effect of live probiotic *L. rhamnosus* Lr-32, probiotic lysate, and spent culture medium on *P. gingivalis*-infected OBA-9 cells. *Lactobacillus rhamnosus* Lr-32 cell components improved the expression of inflammatory mediators in *P. gingivalis*–infected gingival epithelial cells.

15.6.2.3 *Fusobacterium nucleatum*

Fusobacterium nucleatum is one of the most abundant species in the oral cavity [79, 80]. It is implicated in various forms of periodontal diseases including chronic periodontitis, localized aggressive periodontitis, and generalized aggressive periodontitis [79, 81]. It is also frequently associated with endodontic infections such as pulp necrosis and periapical periodontitis [82, 83]. The prevalence of *F. nucleatum* increases with the severity of disease, progression of inflammation, and pocket depth [84, 85].

15.7 DESIGNER PROBIOTICS

Designer probiotics are microorganisms produced by modifying natural probiotics through gene editing. Over the past few years, genome sequencing has become more inexpensive and tools for editing and modifying microbial genomes have successfully evolved that allow us to modify probiotics according to our requirement so that we can develop adapted probiotics. Designer probiotics have immunogenic properties, with sensors and genetic circuits, enabling them to identify a range of diseases with significant accuracy. Gene editing technology has evolved from homologous recombination to the zinc-finger nuclease (ZFN) technology, to the transcription activator-like effector nuclease (TALEN) technology, and then to the more popular clustered regularly interspaced short palindromic repeats (CRISPR) associated (CRISPR-Cas) technology in recent years [86]. The advantages of modified probiotics in terms of stability, specificity, safety, and low cost make them an innovative option for the treatment of numerous oral diseases. However, there are many challenges in the application of engineered probiotics that have to be overcome. Although many side effects like immune system hyperactivation, mutagenesis, and undesirable metabolic activities have also been reported, few studies also reported the stimulation of the immune system in susceptible hosts. Although, the intrinsic antibiotic resistance is a desired attribute in probiotic bacteria for maintaining the microbial balance within the oral cavity, the transmission of these resistant genes may also cause serious threats in terms of the development of multidrug-resistant pathogenic strains. The elimination of antibiotic-resistant genes is important to improve the safety of the probiotic. The approach was applied for a genetically modified probiotic strain, specifically *Lactobacillus*, which is of great importance and is widely used in industry to produce many useful metabolites [87, 88] used engineered *Bacillus subtilis* as oral probiotics to express lactate oxidase that could increase circulating lactic acid catabolism after delivery to the gut. The engineered probiotics may be helpful to reinstate oral health with ease, efficiency, and site specificity, and further research should explore the microbiota for the development of potential engineered probiotic strains for good oral health.

15.8 MARKET OF PROBIOTICS IN ORAL HEALTH: GLOBAL REPORT 2023–2032

Probiotics in the oral health market were valued at US $2.5 billion in 2022 and are anticipated to register a compound annual growth rate (CAGR) of over 6.5% between 2023 and 2032. Nowadays, people are susceptible to oral health issues. With the process of getting old, people are suffering with gum disease, tooth decay, dry mouth, etc. Probiotic oral products can be a better way to address all these concerns. It is evident from scientific research that *Lactobacillus* and *Bifidobacterium* species have the potential to reduce the risk of dental caries, gum disease, and bad breath. Based on distribution channel, the retail pharmacies segment accounted for 34.5% of the probiotics for oral health market share in 2022. Retail pharmacies are evolving into comprehensive health care centers. They offer a wide range of services beyond just dispensing medications, including immunizations, health screenings, medication therapy management, and chronic disease management. This expansion of services makes retail pharmacies more attractive and convenient for consumers. Major companies operating in the probiotics for oral health industry are listed in Figure 15.4. There is a growing awareness among adults about the importance of maintaining good oral health. This

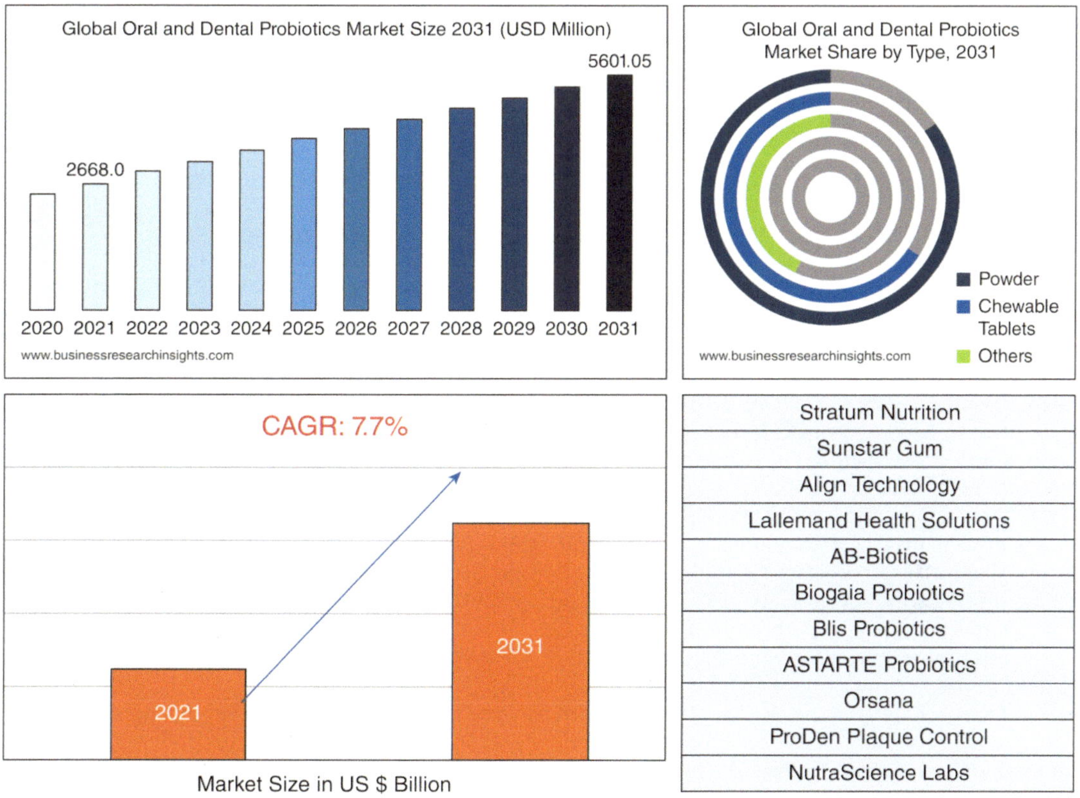

FIGURE 15.4 Market of probiotics in oral health. *Abbreviation*: CAGR, compound annual growth rate.

awareness has led to an increased interest in probiotics that can promote oral health. In future, to maintain the quality of probiotic products, there is a dire need for market regulations [89].

15.9 CONCLUSION

The increased worldwide curiosity in oral probiotics has paved the way for expanded marketing of these products. Although, numerous health-promoting effects of probiotic bacteria are well documented, there is a further need to understand their authentic role in oral health. Thus, research to untangle the mechanism of probiotic action and long-term clinical trials are needed to scientifically prove the role of probiotics in averting and treating oral diseases.

REFERENCES

1. Guarner, F.; Schaafsma, G.J. Probiotics. *Int. J. Food Microbiol.* **1998**, *39*(3), 237–8.
2. Rastogi, P.; Saini, H.; Dixit, J.; Singhal, R. Probiotics and oral health. *Nat. J. Maxillofac Surg.* **2011**, *2*(1), 6–9.
3. Dewhirst, F.E.; Chen, T.; Izard, J.; Paster, B.J.; Tanner, A.C.; Yu, W.H.; Lakshmanan, A.; Wade, W.G. The human oral microbiome. *J. Bacteriol.* **2010**, *192*(19), 5002–17.
4. Aas, J.A.; Paster, B.J.; Stokes, L.N.; Olsen, I.; Dewhirst, F.E. Defining the normal bacterial flora of the oral cavity. *J. Clin. Microbiol.* **2005**, *43*(11), 5721–5732.
5. Zaura, E.; Nicu, E.A.; Krom, B.P.; Keijser, B.J. Acquiring and maintaining a normal oral microbiome: current perspective. *Front. Cell Infect. Microbiol.* **2014**, 4, 85.
6. Zhao, H.; Chu, M.; Huang, Z. Variations in oral microbiota associated with oral Cancer. *Sci. Rep.* **2017**, *7*, 11773.

7. Perera, M.; Al-Hebshi, N.N.; Speicher, D.J.; Perera, I.; Johnson, N.W. Emerging role of bacteria in oral carcinogenesis: a review with special reference to perio-pathogenic bacteria. *J Oral Microbiol.* **2016**, *8*(1), 32762.
8. Camanocha, A.; Dewhirst, F.E. Host-associated bacterial taxa from Chlorobi.; Chloroflexi.; GN02.; Synergistetes.; SR1.; TM7.; and WPS-2 Phyla/candidate divisions. *J. Oral Microbiol.* **2014**, *6*(1), 25468
9. Sultan, A.S.; Kong, E.F.; Rizk, A.M.; Jabra-Rizk, M.A. The oral microbiome:A Lesson in Coexistence. *PLoS Pathog.* **2018**, *14*(1), 1006719.
10. Avila, M.; Ojcius, D.M.; Yilmaz, Ö. The oral microbiota: living with a permanent Guest. *NA Cell Biol.* **2009**, *28*(8), 405–411.
11. Marsh, P.D. Role of the Oral Microflora in Health. *Microb. Ecol. Health Dis.* **2009**, *12*(3), 130–137.
12. Kanamoto, T.; Terakubo, S.; Nakashima, H. Antimicrobial Susceptibilities of Oral Isolates of Abiotrophia and Granulicatella according to the Consensus Guidelines for Fastidious Bacteria. *Medicines (Basel)* **2018**, *5*(4), 129.
13. Downes, J.; Wade, W.G. *Peptostreptococcus stomatis* sp, nov, isolated from the human oral cavity. *Int. J. Syst. Evol. Microbiol.* **2006**, *56*(Pt 4), 751–754.
14. Kreth, J.; Merritt, J.; Qi, F. Bacterial and host interactions of oral streptococci. *DNA Cell Biol.* **2009**, *28*(8), 397–403.
15. Henwick, S.; Koehler, M.; Patrick, C.C. Complications of bacteremia due to *Stomatococcus mucilaginosus* in neutropenic children. *Clin. Infect. Dis.* **1993**, *17*(4), 667–71.
16. Thukral, R.; Shrivastav, K.; Mathur, V.; Barodiya, A.; Shrivastav, S. Actinomyces: a deceptive infection of oral cavity. *J Korean Assoc Oral Maxillofac Surg.* **2017**, *43*(4), 282–285.
17. Jayachandra, M.Y.; Gayathiri, R.; Aruna, C.N.; Bhat, P.K.; Arumugam, P.M. Clinical effects of Bifidobacterium as a probiotic on oral health: A systematic review. *Dent. Res. J. (Isfahan)* **2023**, *28*(20), 32.
18. Treerat, P.; Redanz, U.; Redanz, S. Synergism between Corynebacterium and Streptococcus sanguinis reveals new interactions between oral commensals. *PMC* **2020**, 14, 1154–1169.
19. Spratt, D.A.; Weightman, A.J.; Wade, W.G. Diversity of oral *Asaccharolytic Eubacterium* species in periodontitis–identification of novel phylotypes representing uncultivated taxa. *Oral Microbiol. Immunol.* **1999**, *14*(1), 56–9.
20. Badet, C.; Thebaud, N.B. Ecology of lactobacilli in the oral cavity: a review of Literature. *Open Microbiol. J.* **2008**, 2, 38–48.
21. Niazi, S.A.; Al Kharusi, H.S.; Patel, S.; Bruce, K.; Beighton, D. Foschi F and Mannocci F. Isolation of Propionibacterium acnes among the microbiota of primary endodontic infections with and without intraoral communication. *Clin. Oral Investig.* **2016**, *20*(8), 2149–2160.
22. Tsuzukibashi, O.; Uchibori, S.; Kobayashi, T.; Umezawa, K.; Mashimo, C.; Nambu, T.; Saito, M.; Hashizume-Takizawa, T.; Ochiai, T. Isolation and identification methods of Rothia species in oral cavities. *J. Microbiol. Methods* **2017**, *134*, 21–26.
23. Kawamoto, D.; Borges, R.; Ribeiro, R.A.; de Souza, R.F.; Amado, P.P.P.; Saraiva, L.; Horliana, A.C.R.T.; Faveri, M. Mayer MPA. Oral dysbiosis in severe forms of periodontitis is associated with gut dysbiosis and correlated with salivary inflammatory mediators: A preliminary study. *Front Oral Health* **2021**, *2*, 722495.
24. Narinesingh, S.P.; Whitby, D.J.; Davenport, P.J. Moraxella catarrhalis: an unrecognized pathogen of the oral cavity? *CPCJ* **2011**, *48*(4), 462–464.
25. Bradford, P.A.; Miller, A.A.; O'Donnell, J.; Mueller, J.P. Zoliflodacin: an oral spiropyrimidinetrione antibiotic for the treatment of *Neisseria gonorrheae*.; including multi- drug-resistant isolates. *ACS Infectious Diseases* **2020**, *6*(6), 1332–1345.
26. Luo, Y.X.; Sun, M.L.; Shi, P.L.; Liu, P.; Chen, Y.Y.; Peng, X. Research progress in the relationship between Veillonella and oral diseases. *Hua Xi Kou Qiang Yi Xue Za Zhi* **2020**, *38*(5), 576–582.
27. Etoh, Y.; Dewhirst, F.E.; Paster, B.J.; Yamamoto, A.; Goto, N. Campylobacter ssp isolated from the human oral cavity. *Int. J. Syst. Evol. Microbiol.* **1993**, *43*(4), 631–639.
28. Chesdachai, S.; Tai, D.B.G.; Yetmar, Z.A.; Misra, A.; Ough, N.; Abu Saleh, O. The characteristics of Capnocytophaga infection: 10 years of experience. *Open Forum Infect Dis.* **2021**, *8*(7), 175.
29. Langendijk, P.S.; Kulik, E.M.; Sandmeier, H.; Meyer, J.; Van Der Hoeven, J.S. Isolation of Desulfomicrobium oral sp, nov, and Desulfovibrio strain NY682, oral sulfate-reducing bacteria involved in human periodontal disease. *Int. J. Syst. Evol. Microbiol.* **2001**, *51*(3), 1035–1044.
30. Chen, C.C.; Wilson, M.E. Eikenellacorrodens in human oral and non-oral infections: a review. *J. Periodontol.* **1992**, *63*(12), 941–953.

31. McIlvanna, E.; Linden, G.J.; Craig, S.G.; Lundy, F.T; James, J.A. *Fusobacterium nucleatum* and oral cancer: a critical review. *BMC Cancer* **2021**, 21, 1–11.
32. Kurono, Y.; Shimamura, K.; Mogi, G. Inhibition of nasopharyngeal colonization of Hemophilus influenzae by oral immunization. *Ann. Otol. Rhinol. Laryngol.* **1992**, *101*(10), 11–15.
33. Eribe, E.R.K.; Olsen, I. Leptotrichia species in human infections. *Anaerobe* **2008**, *14*(3), 131–137.
34. Könönen, E.; Fteita, D.; Gursoy, U.K.; Gursoy, M. Prevotella species as oral residents and infectious agents with potential impact on systemic conditions. *J. Oral Microbiol.* **2022**, *14*(1), 2079814.
35. Kolenbrander, P.E.; Andersen, R.N.; Moore, L.V. Coaggregation of *Fusobacterium nucleatum.; Selenomonasflueggei.; Selenomonasinfelix.; Selenomonasnoxia.; and Selenomonassputigena* with strains from 11 genera of oral bacteria. *Infect. Immune.* **1998**, *57*(10), 3194–3203.
36. Kuhn, D.A.; Gregory, D.A.; Buchanan, G.E.; Nyby, M.D.; Daly, K.R. Isolation.; characterization.; and numerical taxonomy of *Simonsiella* strains from the oral cavities of cats.; dogs.; sheep.; and humans. *Arch. Microbiol.* **1978**, *118*, 235–241.
37. Belkacemi, S.; Bou Khalil, J.; Ominami, Y.; Hisada, A.; Fontanini, A.; Caputo, A.; Levasseur, A.; Lagier, J.C.; Khelaifia, S.; Raoult, D. Passive filtration.; rapid scanning electronmicroscopy.; and matrix-assisted laser desorption ionization–time of flight mass spectrometry for Treponema culture and identification from the oral cavity. *J. Clin. Microbiol.* **2019**, *57*(10), 10–1128.
38. Craven, M.; Recordati, C.; Gualdi, V.; Pengo, G.; Luini, M.; Scanziani, E.; Simpson, K.W. Evaluation of the Helicobacteraceae in the oral cavity of dogs. *Am. J. Vet. Res.* **2011**, *72*(11), 1476–1481.
39. Wade, W.G. The oral microbiome in health and disease. *Pharmacol. Res.* **2013**, *69*(1), 137–143.
40. Li, X.; Liu, Y.; Yang, X.; Li, C.; Song, Z. The oral microbiota: community composition.; influencing factors.; pathogenesis.; and interventions. *Front. Microbiol.* **2022**, *13*, 895537.
41. Dame-Teixeira, N.; de Cena, JA.; Côrtes, D.A.; Belmok, A.; dos Anjos Borges, L.G.; Marconatto, L.; Giongo, A.; Kyaw, C.M. Presence of Archaea in dental caries biofilms. *Arch Oral Biology* **2020**, *110*, 104606.
42. Caselli, E.; Fabbri, C.; D'Accolti, M.; Soffritti, I.; Bassi, C.; Mazzacane, S.; Franchi, M. Defining the oral microbiome by whole-genome sequencing and resistome analysis:the complexity of the healthy picture. *BMC Microbiol.* **2020**, *20*(1), 1–19.
43. Sharma, N.; Bhatia, S.; Sodhi, A.S. and Batra, N. Oral microbiome and health. *AIMS Microbiol.* **2018**, *4*(1), 42.
44. Hong, B.Y.; Hoare, A.; Cardenas, A.; Dupuy, A.K.; Choquette, L.; Salner, A.L.; Schauer, P.K.; Hegde, U.; Peterson, D.E.; Dongari-Bagtzoglou, A.; Strausbaugh, L.D. The salivary mycobiome contains 2 ecologically distinct mycotypes. *J. Dent. Res.* **2020**, *99*(6), 730–738.
45. Bottalico, D.; Chen, Z.; Dunne, A.; Ostoloza, J.; McKinney, S.; Sun, C.; Schlecht, N.F.; Fatahzadeh, M.; Herrero, R.; Schiffman, M.; Burk, R.D. The oral cavity contains abundant known and novel human papillomaviruses from the Beta papilloma virus and Gamma papilloma virus genera. *J. Infect. Di.* **2011**, *204*(5), 787–792.
46. Wylie, K.M.; Mihindukulasuriya, K.A.; Zhou, Y.; Sodergren, E.; Storch, G.A.; Weinstock, G.M. Metagenomic analysis of double-stranded DNA viruses in healthy adults. *BMCM Biology* **2014**, *12*(1), 1–10.
47. Veiga, P.; Suez, J.; Derrien, M. and Elinav, E. Moving from probiotics to precision probiotics. *Nat. Microbiol.* **2020**, 5, 878–880.
48. Wolkers, W.F.; Oldenhof, H. Cryopreservation and freeze-drying protocols. *Meth. Mol. Biol.* **2015**, 1257, 505.
49. Kieps, J.; Dembczynski, R. Current trends in the production of probiotic formulations. *Foods* **2022**, *11*, 2330.
50. Haukioja, A.; Yli-Knuuttila, H.; Loimaranta, V.; Kari, K.; Ouwehand, A.C.; Meurman, J.H.; et al. Oral adhesion and survival of probiotic and other lactobacilli and bifidobacteria in vitro. *Oral Microbiol. Immunol.* **2006**, *21*, 326–332.
51. Haukioja, A.; Loimaranta, V.; Tenovuo, J. Probiotic bacteria affect the composition of salivary pellicle and streptococcal adhesion in vitro. *Oral Microbiol. Immunol.* **2008**, 23, 336–343
52. Marsh P.D. Are dental diseases examples of ecological catastrophes? *Microbiology* **2003**, *149*(2), 279–294.
53. Jindal, G.; Pandey, R.; Agarwal, J.; and Singh, M. A comparative evaluation of probiotics on salivary mutans streptococci counts in Indian children. *Eur. Arch. Paediatric Dentistry* **2011**, *12*(4), 211–215.
54. Çaglar, E.; KavalogluCildir, S.; Ergeneli, S.; Sandalli, N. and Twetman, S. Salivary mutans streptococci and lactobacilli levels after ingestion of the probiotic bacterium *Lactobacillus reuteri* ATCC 55730 by straws or tablets. *Acta Odontol Scand.* **2006**, *64*(5), 314–318.

55. Kaye E.K. Daily intake of probiotic Lactobacilli may reduce caries risk in young children.; J.; Evidence.; Based. *Dental Pract.* **2017**, *17*(3), 284–286.
56. Glavina, D.; Gorseta, K.; Skrinjarić, I.; NegovetićVranić.; D.; Mehulić, K. and Kožul, K. Effect of LGG yoghurt on streptococcus mutans and lactobacillus spp.; salivary counts in children. *Collegium Antropol.* **2012**, *36*(1), 129–132.
57. Campus, G.; Cocco, F.; Carta, G.; Cagetti, M.G.; Simark-Mattson, C.; Strohmenger, L.; et al. Effect of a daily dose of lactobacillus brevis CD2 lozenges in high caries risk schoolchildren. *Clin. Oral Invest.* **2014**, *18*(2), 555–561.
58. Volozhin, Ai.; Il'in V.K.; MaksimovskiiIu, M.; Sidorwnko, A.B.; Istranov, L.P.; Tsarev VP et al. Development and use of periodontal dressing of collagen and Lactobacillus casei 37 cell suspension in combined treatment of periodontal disease of inflammatory origin. *Stomatologiia (Mosk)* **2004**, *83*, 6–8.
59. Riccia, D.N.; Bizzini, F.; Perilli, M.G.; Polimeni, A.; Trinchieri, V.; Amicosante, G.; et al. Anti- inflammmatory effects of L.; brevis (DC2) on periodontal disease. *Oral Dis.* **2007**, 13, 376–85.
60. Marteau P.; Shanahan F. Basic aspects and pharmacology of probiotics: an overview of pharmacokinetics.; mechanisms of action and side-effects. *Best Pract. Res. Clin. Gastroenterol.* **2003**, *17*(5), 725–740.
61. Xiao, S.D.; Zhang, D.Z.; Lu, H.; Jiang, S.H.; Liu, H.Y.; Wang, G.S.; Xu, G.M.; Zhang, Z.B.; Lin, G.J.; Wang, G.L. Multicenter randomized controlled trial of heat-killed *Lactobacillus acidophilus* lb in patients with chronic diarrhea. *Chin. J. Dig. Dis.* **2003**, *3*, 167–171.
62. Gao Jie. et al. A Novel Postbiotic from *Lactobacillus rhamnosus* GG With a Beneficial Effect on Intestinal Barrier Function. *Front. Microbiol.* **2019**, *10*, 477.
63. Ejtahed. et al. Effect of probiotic foods and supplements on blood pressure:a systematic review of meta-analysis studies of controlled trials. *J. Diabetes Metab. Disord.* **2020**, *19*, 617–623.
64. Ojha, S.; Patil, N.; Jain, M.; Kole, C.; Kaushik, P. Probiotics for Neurodegenerative Diseases: A Systemic Review. *Microorganisms* **2023**, *11*(4), 1083.
65. Chikindas, M.L.; Weeks, R.; Drider, D.; Chistyakov, V.A. & Dicks, L.M. Functions and emerging applications of bacteriocins. *Curr. Opin. Biotechnol.* **2018**, *49*, 23–28.
66. Nagesh, A.W.; Tsehei, B.A. Current Applications of Bacteriocin. *Int. J. Microbiol.* **2020**, *3*, 4374891.
67. Thapar, P.; Malik, R. K.; Salooja, M.K. Application of Bacteriocins of *Lactobacillus brevis* as Biomedicine. *Res. J. Biotechnol.* **2021**, *16*(5), 109–117.
68. Gomez-Armayones, S.; Chimenos-Küstner, E.; Marí, A.; Tous, S.; Penin, R.; Clavero, O.; Quirós, B.; Pavon, M.A.; Taberna, M.; Alemany, L. and Servitje, O. Human papillomavirus in premalignant oral lesions: No evidence of association in a Spanish cohort. *PLoS One* **2019**, *14*(1), 0210070.
69. Hujoel, P.P.; Hujoel, M.L.A. and Kotsakis, G.A. Personal oral hygiene and dental caries:A systematic review of randomised controlled trials. *Gerodontol.* **2018**, *35*(4), 282–289.
70. Loesche, W.J. The identification of bacteria associated with periodontal disease and dental caries by enzymatic methods. *Oral Microbiol. Immunol.* **1986**, *1*(1), 65–70.
71. Lemos-Júnior, C.A.; Villoria, G.E.M. Reviewed evidence about the safety of the daily use of alcohol-based mouthrinses. *Braz. Oral Res.* **2008**, *22*, 24–31.
72. Lemos, J.A.; Palmer, S.R.; Zeng, L.; Wen, Z.T.; Kajfasz, J.K.; Freires, I.A.; Abranches, J. and Brady, L.J. The biology of Streptococcus mutans. *Microbiol. Spectr.* **2019**, *7*(1), 10–1128.
73. Venugopal, M.; Khosla, E.; Abraham, K.; Vinitha, Alex.; Nishna, T.; Kumar, H. Evaluation of Probiotic Effects of Lactobacilli on Mutans Streptococci: An In Vitro Study. *J. Contemp. Dent. Pract.* **2022**, *23*(10), 984–990.
74. Petersen, P.E. and Ogawa, H. Prevention of dental caries through the use of fluoride–the WHO approach. *Community Dent. Health.* **2016**, *33*(2), 66–68.
75. Abusleme, L.; Dupuy, A.K.; Dutzan, N.; Silva, N.; Burleson, J.A.; Strausbaugh, L.D.; Gamonal, J. and Diaz, P.I. The subgingival microbiome in health and periodontitis and its relationship with community biomass and inflammation. *ISME J.* **2013**, *7*(5), 1016–1025.
76. Olsen, I.; Lambris, J.D. and Hajishengallis, G. Porphyromonasgingivalis disturbs host–commensal homeostasis by changing complement function. *J. Oral Microbiol.* **2017**, *9*(1), 1340085.
77. Bostanci N and Belibasakis GN. Porphyromonasgingivalis:an invasive and evasive opportunistic oral pathogen. *FEMS Microbiol Lett.* **2012**, *333*(1), 1–9.
78. Vale, G.C.; Mayer, M.P.A. Effect of probiotic *Lactobacillus rhamnosus* by-products on gingival epithelial cells challenged with *Porphyromonas gingivalis. Arch. Oral. Biol.* **2021**, 128, 105174.
79. Moore, W.E.; Moore, L.V. The bacteria of periodontal diseases. *Periodontol.* **1994**, *5*, 66–77.
80. Field, C.A.; Gidley, M.D.; Preshaw, P.M.; Jakubovics, N. Investigation and quantification of key periodontal pathogens in patients with type 2 diabetes. *J. Periodontal Res.* **2012**, *47*, 470–478.

81. Kistler, J.O.; Booth, V.; Bradshaw, D.J.; Wade, W.G. Bacterial community development in experimental gingivitis. *PLoS One.* **2013**, *8*, 71227.
82. Siqueira, J.F.; Rocas, I.N. The microbiota of acute apical abscesses. *J. Dent. Res.* **2009**, *88*, 61–65.
83. Didilescu, A.C.; Rusu, D.; Anghel, A.; Nica, L.; Iliescu, A.; Greabu, M.; Bancescu, G.; Stratul, S.I. Investigation of six selected bacterial species in endo-periodontal lesions. *Int. Endod. J.* **2012**, *45*, 282–293.
84. Riep, B.; Edesi-Neuss, L.; Claessen, F.; Skarabis, H.; Ehmke, B.; Flemmig, T.F.; Bernimoulin, J.P.; Gobel, U.B.; Moter, A. Are putative periodontal pathogens reliable diagnostic markers? *J. Clin. Microbiol.* **2009**, *47*, 1705–1711.
85. Yang N.Y.; Zhang, Q.; Li, J.L.; Yang, S.H.; Shi, Q. Progression of periodontal inflammation in adolescents is associated with increased number of Porphyromonas gingivalis.; Prevotella intermedia.; Tannerella for sythensis.; and *Fusobacterium nucleatum. Int. J. Paediatr. Dent.* **2014**, *24*, 226–233.
86. Ma, J.; Lyu, Y.; Liu, X.; Jia, X.; Cui, F.; Wu, X.; Deng, S.; Yue, C. Engineered probiotics. *Microb Cell Fac.t* **2022**, *21*(1), 72.
87. Mazhar S.F. et al. The prospects for the therapeutic implications of genetically engineered probiotics. *J Food Qual.* **2020**, 11.
88. Yang, M.; Hutchinson, N.; Ye, N.; Yin, J.; Guan, M.; Wang, Z.; Chen, P.; Yang, S.; Crane, J.D. Zhang K.; He X.; Li J. Engineered *Bacillus subtilis* as oral probiotics to enhance clearance of blood lactate. *BioRxiv [Preprint]* **2024**, 11.30:569300.
89. Global Health Market, **2023**, Report ID: GMI6984. https://www.gminsights.com/industry-analysis/probiotics-for-oral-health-market

16 Role of the Oral Microbiome in Prognosis and Diagnosis

Rahul Shrivastava, Sourav Kumar, and Diksha Suman
Department of Biotechnology and Bioinformatics, Jaypee University of Information Technology, Waknaghat, Himachal Pradesh, India

16.1 INTRODUCTION

The Dutchman Antonie van Leeuwenhoek first identified the oral microbiome by using a microscope he made himself. The late 1670s saw him report to the British Royal Society that the plaque that was found on tooth surfaces contained a variety of microbes. He made this discovery at a time when the field of bacteriology was still developing, and he was deeply intrigued by the kineticism of the microbiome [1]. Although he did not specifically mention it in his report, he understood that variations in the oral microbiome among individuals affected the condition of the oral cavity [1]. The next two centuries saw significant advancements in microscopy along with other technologies, leading to an ever-deeper understanding of microorganisms [2]. The human oral cavity is home to a variety of habitats, like the tongue, soft and hard palates, tonsils, and teeth. It also serves as a tube connecting the body's respiratory and digestive systems to the outside world, creating an optimal environment for the colonization of microorganisms. As food enters the mouth, it is chewed and combined with saliva before traveling through the digestive tract and into the stomach. On its journey to the trachea and lungs, air travels via the mouth and nose. There is a considerable chance that microorganisms that colonize one area of the oral cavity will spread to nearby locations via adjacent epithelial surfaces [3].

The terms oral microflora, oral microbiota, and oral microbiome have all been used to describe the microorganisms that are present in the human oral cavity. There are more than 700 microbiological species in the mouth cavity including fungi, bacteria, archaea, and opportunistic viruses as well as protozoa. Every animal engages in its unique function and closely interacts with Actinobacteria, the host, and the other species. Proteobacteria, Firmicutes, and Bacteroidetes are presumably the most important for dental health. Among the major genera most prevalent in oral cavities are the following: *Haemophilus, Rothia, Prevotella, Streptococcus, Fusobacterium, Neisseria*, Veillonellaceae, *and Porphyrin* [4–7]. Table 16.1 shows the major prevailing microorganisms in the oral cavity [4].

TABLE 16.1
Distribution of Microorganisms in the Oral Cavity

Oral Cavity	Microorganisms
Saliva and hard palate	*Prevotella, Streptococcus, Veillonella*, Pasteurellaceae, Lactobacillales
Subgingival plaque	*Fusobacterium, Streptococcus, Prevotella, Corynebacterium, Capnocytophaga*
Palatine tonsil	*Streptococcus, Veillonella, Prevotella*, Pasteurellaceae, *Fusobacterium*
Throat	*Fusobacterium, Veillonella, Streptococcus*, Pasteurellaceae, *Prevotella*, Lactobacillales, *Actinomyces*
Buccal mucosa	*Streptococcus, Gemella*, Pasteurellaceae
Keratinized gingiva	*Streptococcus*, Pasteurellaceae
Supragingival plaque	Neisseriaceae, *Streptococcus, Capnocytophaga*, Pasteurellaceae, *Corynebacterium*
Tongue dorsum	*Streptococcus, Veillonella, Prevotella*, Pasteurellaceae, *Actinomyces*
Lips	*Candida albicans, Streptococcus*
Dentures	*Streptococcus, Staphylococcus epidermidis*

DOI: 10.1201/9781003478638-16

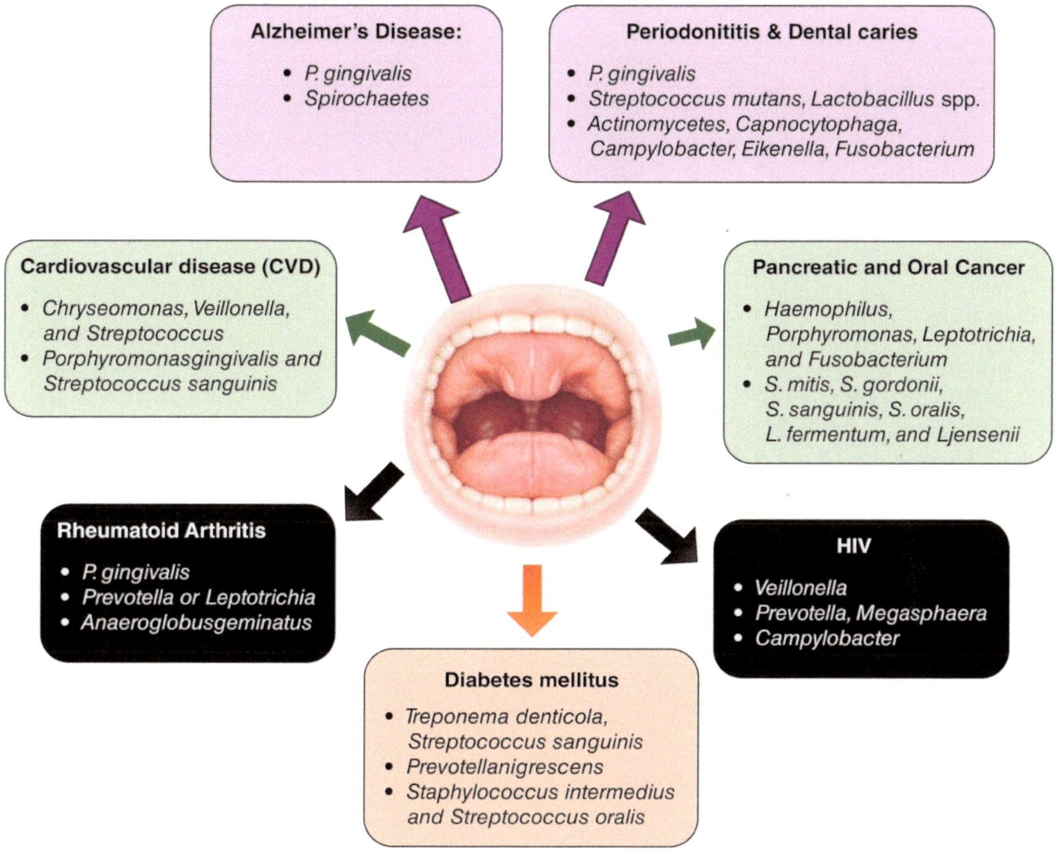

FIGURE 16.1 Relation of oral microbes with various diseases.

The interaction of different oral microorganisms protects the human body from outside stimuli that are not desired [3]. However, an imbalance in the microbial flora is linked to systemic illnesses, like nervous and gastrointestinal systemic diseases, and oral diseases, like periodontitis, dental caries, and diseases of the oral mucosa, and is associated with various other systemic diseases, namely, pneumonia, rheumatoid arthritis, cardiovascular disease, pancreatic cancer, esophageal cancer, stroke, and colorectal cancer, Figure 16.1. Human health and the human microbial community are significantly influenced by oral microbiomes [8–12]; there has been consideration of oral microbiota as a possible biomarker for human illnesses. Relations between systemic diseases and oral microbiota are crucial and must be clarified to offer a solid diagnostic foundation for illness prevention and management [4].

16.2 ORAL DISEASES

16.2.1 Dental Caries

Among chronic infectious diseases, caries is the most common disease. It is primarily caused by bacteria and can result in the progressive and long-term degeneration of dental hard tissue for a variety of reasons [13]. Caries occur frequently and can happen to people of any age, even the elderly. And among preschoolers worldwide, early childhood caries is the most damaging and has emerged as a major public health issue. A variety of factors, including the oral microbiome, can affect the incidence of this condition [14]. Dental caries results in pulp and periapical infection in addition to tooth disintegration [15]. To examine the microbial diversity of dental caries, specimens from saliva, infected dentin, and supragingival plaque are typically taken from adults with caries-active conditions, elderly patients with root caries, and children with severe early childhood caries (S-ECC) [16]. The bacteria *Veillonella*, *Streptococcus*, *Granulicatella*, *Actinomyces*, *Thiomonas*, *Leptotrichia*, *Prevotella*, and *Bifidobacterium* were also found at higher frequencies in the plaques of children with S-ECC than those of children without caries, and it is thought that these bacteria are related to S-ECC [11, 17–19]. In the plaque of adults with active dental caries, high relative abundances of the genera *Lactobacillus*, *Streptococcus*, *Propionibacterium*, *Actinomycetes*, *and Veillonella* were also found [20].

16.2.2 Periodontal Diseases

Periodontal diseases, which fall into two categories—periodontitis and gingival diseases—occur frequently in the human mouth. The periodontium, which includes tissues supporting teeth, like gingiva and alveolar bone, is destroyed by periodontal diseases, which can also be a risk factor for other systemic conditions. Since periodontal tissue in the oral cavity has a complex anatomical and organizational structure as well as physical and chemical properties, it offers favorable conditions for the growth of microbes. The oral cavity serves as a natural microbial culture medium [21, 22]. As gingivitis progresses, *Actinomycetes*, *Capnocytophaga*, *Campylobacter*, *Eikenella*, *Fusobacterium*, and *Prevotella* gradually replace *Streptococcus* as the predominant bacteria in subgingival plaque. Additionally, there were notable variations in the saliva microbiota between those who had gingivitis and those who did not. In addition, periodontitis is the primary cause of tooth loss globally and a chronic inflammatory disease of the tissues supporting teeth. In contrast to gingivitis, it results in the progressive loss of alveolar bone and periodontal membrane in addition to the gums [5, 23].

16.2.3 Halitosis

Oral malodor, or halitosis, is the term for an offensive stench that is exhaled during breathing. It is divided into two categories: extra-oral and intra-oral halitosis [24]. Intra-oral halitosis is thought to be directly caused by volatile sulfur compounds and malodorous fatty acids that are created when oral bacteria break down sulfur-containing amino acids, proteins, and peptides [24]

Microorganisms, mainly bacteria, on the tongue dorsum, particularly the posterior dorsum, have been gradually determined to be the main reason for intra-oral halitosis in individuals having healthy periodontal tissues and full dentition [25]. There are differences in the microbial diversity of the tongue dorsum between patients with intra-oral halitosis and healthy controls. The most common bacteria in the tongue dorsum of patients with intra-oral halitosis are *Atopobium parvulum*, *Solobacterium moorei*, and *Eubacterium sulci*, whereas *Streptococcus salivarius*, *Rothia mucilaionosa*, and an unnamed cultivable *Eubacterium* species are common in healthy controls [26].

16.3 ORAL MICROBIOMES: DIAGNOSIS AND PROGNOSIS IN DISEASES

16.3.1 Cardiovascular Diseases

A group of illnesses known as cardiovascular diseases include coronary artery disease (which includes atherosclerosis and myocardial infarction), heart failure, arrhythmias, heart valvular disease, along with stroke [27–29]. While there are currently insufficiently strong correlations between cardiovascular system diseases, researchers have demonstrated some potential associations between oral microbiomes and atherosclerosis. The buildup of cholesterol along with the attraction of macrophages to the walls of arteries are the hallmarks of atherosclerosis. As a result, it can be classified as an inflammatory and metabolic illness [30]. Some research used 16S rRNA sequencing to identify the predominant microbiomes of atherosclerotic patients as having *Chryseomonas*, *Veillonella*, *and Streptococcus*. Furthermore, the oral or gut samples from the same person and the atherosclerotic plaque shared several additional bacterial phylotypes [31]. It's interesting to note that the levels of plasma cholesterol were correlated with several bacterial taxa found in the gut and mouth. Oral cavities and possibly even gastrointestinal bacteria may be correlated with atherosclerosis disease markers [32]. Periodontitis is associated with atherosclerosis and thromboembolic events based on both epidemiological and mechanistic evidence [33]. A long-term study of patients with periodontitis found them to be 25% more likely to develop atherosclerotic plaque [34]. There is a positive and consistent correlation between the development of incident atherosclerotic cardiovascular disease (ASCVD) and different measures of periodontal disease. The Atherosclerosis Risk in Communities (ARIC) study, which included 10,362 healthy participants followed for 15 years, revealed a significant correlation between periodontal disease and the incidence of stroke (hazard ratio - HR 2.6 for cardioembolic stroke and HR 2.2 for thrombotic stroke) [35]. Maintaining dental health was protective, as it was linked to a lower risk of stroke (HR 0.77) [36]. Gum disease–related tooth loss was linked to a 30% higher possibility of venous thromboembolic incidents in another investigation of the ARIC study participants [37]. A meta-analysis of prospective cohort studies indicates that the risk of cardiovascular disease is increased by 24% and 34%, respectively, by periodontal disease and tooth loss [38]. Since the late 1980s, periodontal disease has been associated with a higher risk of acute myocardial infarction [39]. Comorbid periodontitis increases the risk of peripheral artery disease (PAD) by 5 times [40]. Individuals with elevated serum levels of TNF and IL-6 have the highest risk of PAD [41]. In a small Mexican investigation ($n = 30$) [42], severe periodontal infection was linked to a six-fold increase in the risk of PAD, and patients with PAD had significantly worse tooth loss than control subjects. Moreover, periodontitis has been linked to hyperlipidemia [43], a significant risk factor for ASCVD, and elevated levels of circulating fibrinogen, which may indicate a higher risk of thrombosis [44, 45].

According to studies, oral pathogens that modify platelet aggregation, such as *Porphyromonas gingivalis* and *Streptococcus sanguinis*, are linked to atherosclerotic thromboembolism. In rabbits, these bacteria cause fatty streaks and hasten the development of atheroma plaques [46]. In atherosclerotic plaques from patients receiving interventional procedures, researchers found 23 different types of oral commensal bacteria. Thirty-six secretory proteins that may be involved in the advancement of ASCVD were found from these [47]. Oral dysbiosis contributes to the development of ASCVD through various mechanisms. Bacteria from pathogenic oral microbiota invade arterial walls, promoting plaque formation [48]. Oral inflammatory lesions release proinflammatory cytokines, while autoimmunity targets self-proteins similar to oral pathogenic antigens. Oral pathogenic bacteria release toxins, and chronic or aggressive periodontitis increases the risk of ASCVD [49].

Researchers discovered that both atherosclerosis and periodontal diseases were associated with significantly higher levels of salivary inflammatory cytokines, such as IL-1β, IL-6, TNF-α, and prostaglandin E2 [50]. These cytokines may have diagnostic value as biomarkers for atherosclerosis and periodontal disease. It was found that the majority accurate biomarker of acute myocardial infarction was C-reactive protein (CRP). An 80.0% sensitivity and 100% specificity combination of CRP levels and an electrocardiogram (ECG) predicted an acute myocardial infarction. These findings suggested that salivary biomarkers and an ECG could be used to diagnose acute myocardial infarction. Furthermore, patients with CVD had lower salivary levels of α-2-HS-glycoprotein, suggesting that the peptidome may offer a potential means of early patient diagnosis [51].

16.3.2 Diabetes Mellitus

The clinical syndrome known as diabetes mellitus (DM) is typified by hyperglycemia brought on by insufficient insulin secretion or decreased insulin action. Type 1 diabetes, which is caused by the body's inability to produce insulin, and type 2 diabetes, which is caused by insulin resistance, are the two main forms [52]. The sixth complication of DM is considered to be periodontitis, which is three- to four-fold higher in patient with type 1 and type 2 DM. There has been some research on the differences between diabetics' and non-diabetics' subgingival dental plaque composition, but no consensus has been reached on how diabetes affects the subgingival microbiota in the periodontal system. Research has revealed that type 2 diabetics with supragingival plaque had higher frequencies of *Treponema denticola, Prevotella nigrescens, Streptococcus sanguinis, Staphylococcus intermedius, and Streptococcus oralis* when compared to non-diabetics. However, there were no noteworthy variations observed in subgingival plaque samples [53]. Following periodontal severity control, comparable subgingival infection patterns were also noted in type 1 diabetics and non-diabetics. In contrast, a study found that *P. gingivalis, Aggregatibacter actinomycetemcomitans,* and *Campylobacter* spp. were more frequently found in the periodontitis sites of type 1 DM patients [54]. In type 2 diabetics, there was also evidence of a lower frequency of *T. forsythia* and a higher prevalence of *P. gingivalis* and *Candida* spp. (primarily *Candida dubliniensis* and *albicans*) [55]. Significant differences were found in the subgingival microbiota between subjects with type 2 diabetes and those without the disease in a much more recent study that used 16S rRNA gene sequencing. The diabetic subjects showed lower percentages of *Tannerella*, Synergistetes, *Filifactor, Treponemas, Eubacterium, and Porphyromonas* and higher abundances of total clones of TM7, *Aggregatibacter, Neisseria, Actinomyces, Gemella, Eikenella, Selenomonas, Fusobacterium, Veillonella*, and *Streptococcus*, while the diabetic subjects showed higher abundances of these species. There may be differences in the subgingival microbiome between diabetic patients and non-diabetics due to the increased glucose content in the subgingival microenvironment along with changed or impeded immune responses of hosts [20].

A positive correlation was observed between HbA1c and α-2-macroglobulin, indicating that salivary α-2-macroglobulin levels may represent glycemic control in individuals with type 2 diabetes mellitus. On the other hand, individuals with periodontitis and type 2 diabetes had lower salivary melatonin concentrations [56]. This suggests that salivary melatonin plays a significant role in the pathophysiology of periodontal and diabetes disorders and may develop into a valuable biomarker for diagnosis and management of these conditions. Patients with diabetes and/or periodontitis had saliva that contained 475 distinct metabolites, according to research [57]. Patients without diabetes who had gingivitis and periodontitis had significantly higher levels of cellular energetic stress, glutathione metabolism, purine degradation, cysteine glutathione disulfide, oxidized glutathione, markers of oxidative stress, amino acids, and ω-3 (docosapentaenoate) and ω-6 (linoleate and arachidonate) fatty acid signs. On the other hand, individuals diagnosed with diabetes exhibited markedly elevated glucose and α-hydroxybutyrate levels, along with a noteworthy alteration in their levels of carbohydrate, lipid, and oxidative stress. As a result, the metabolites may be helpful for diabetes and periodontal disease diagnosis, treatment, and prognostic evaluation. Patients with diabetes and the concentrations of salivary glucose and HbA1c showed a significant correlation. This suggested that individuals with diabetes mellitus could use their saliva to track their blood glucose levels.

16.3.3 Rheumatoid Arthritis

Rheumatoid arthritis (RA) is an autoimmune disorder, linked with increased mortality owing to cardiovascular along with other systemic problems. The cause of arthritis is still unknown. Although research on genetic susceptibility to RA reveal genes such as *TNFAIP3, HLA-DRB1, PTPN22*, and *PADI4*, environmental variables have also been shown to contribute to the pathophysiology of the illness. There is evidence linking microbial triggers to RA [58]. Similarities between the oral and gut microbiomes were found, indicating that species' functions and abundances at various body regions overlap. Patients with RA were found to have dysbiosis in their oral and gut microbiomes, although this condition was only partially treated with medication. In RA, microbial triggers have been linked [59]. Saliva, tooth, or gut microbiome changes distinguished arthritic patients from healthy controls, related to clinical indicators, and might be used to stratify patients based on therapy response [60].

Autoantibodies generated in rheumatoid arthritis are known as rheumatoid factor and anti-citrullinated protein antibodies (ACPAs), which are the most powerful and early diagnostic indicators. Although the exact cause of RA is still unknown, smoking and periodontal disease (PD) seem to be linked to the generation of ACPA [61]. The high prevalence of periodontal inflammatory disorders in RA patients and their correlation with anti-cyclic citrullinated peptide (anti-CCP) antibody levels lend credence to the theory that the oral microbiota plays a role in initiating or accelerating the progression of RA. Chronic inflammation brought on by oral bacteria and leukocyte infiltration that gradually destroys the alveolar bone is the hallmark of periodontitis [62]. As the most thoroughly researched pathogen in RA and the keystone pathogen in periodontitis, *P. gingivalis* has been shown to induce dysbiosis and deregulate local immune responses. Numerous studies have investigated the clinical relationship between *P. gingivalis* and RA. These investigations usually employ bacterial culture or molecular identification of bacterial DNA from gingival crevicular fluid or plaque. Serum antibody reactivity has also been studied in animal models to examine the role of *P. gingivalis* in the development of experimental arthritis [63]. *Porphyromonas gingivalis,* which expresses the bacterial protein arginine deiminase (PPAD), which can citrullinate free *L*-arginine and C-terminal arginine residues in cleaved peptides, was initially suggested as a key player in the onset of RA. It has been proposed that the C-terminal citrullination of both host and bacterial proteins breaks immunological tolerance and triggers the ACPA response in RA. The study also demonstrated that *Porphyromonas gingivalis* worsens arthritis by altering the gut microbiota and raising the percentage of T-helper 17 (Th17) cells in the lymph nodes of the mesenteric region. The evidence showing that treating periodontal disease may lessen RA symptoms is another indication of the connection between RA and oral inflammation [64].

A study comparing the oral microbiota of patients with early versus advanced phases of RA revealed that patients with early arthritis had a higher prevalence of Parkinson's disease at the onset of the disease, and the microbiota of their subgingival biofilm was comparable to that of patients with chronic RA. According to a study, patients with newly diagnosed RA had a different subgingival microbiota than healthy controls [65].

The subgingival microbiota of patients with recent onset of RA did not contain *Prevotella* or *Leptotrichia,* and this finding was not related to periodontal disease [66]. *Anaeroglobus geminatus* correlated with both the rheumatoid factor and ACPAs, whereas *P. gingivalis* was directly linked to the severity of periodontitis but not with ACPAs. A Chinese study showed that RA patients could be distinguished from healthy controls by alterations in their gut microbiome, saliva, or teeth. In particular, treatment-naive RA patients showed an increased abundance of *Lactobacillus salivarius* in the microbiota of all three sites, especially those with high disease activity, while *Haemophilus* species were depleted in RA patients and negatively correlated with serum levels of autoantibodies. According to a study, treatment with anti-rheumatic medications partially resolved dysbiosis, highlighting the reciprocal interactions between the hosts and their microbiota [67]. Fascinatingly, a study found a substantial correlation between the amounts of *Porphyromonas gingivalis* in the entire tongue biofilm along with the activity of the RA disease (DAS28).

Additionally, research has demonstrated that patients with RA have altered mouth and gut microbiomes, which may be used to predict treatment response and partially normalize after taking disease-modifying anti-rheumatic drugs (DMARDs) [68]. According to Zhang, X., et al. [10], patients who showed a good response to treatment had higher levels of virulence factors prior to treatment, and their levels of *Bacteroides* spp. *and Holdemania filiformis* decreased following treatment. Additionally, they stated that the impact on the oral microbiome was greater than that on the gut microbiome [69].

16.3.4 Human Immunodeficiency Virus Infection

Individuals who are positive for human immunodeficiency virus (HIV) have previously been found to have higher concentrations of disease-associated microbial species. HIV infection is linked to a variety of oral conditions. Significantly higher proportions of potentially pathogenic *Prevotella*, *Veillonella*, *Campylobacter*, and *Megasphaera* species are linked to elevated viremia in untreated patients compared to healthy controls [70]. According to a different study, oral cavity microbial diversity in HIV-positive persons was less than in healthy controls, and it decreased even more after starting anti-retroviral therapy (ART) [71]. There were no discernible differences between HIV-negative controls and well-managed HIV-positive patients, indicating that oral flora in HIV-positive patients and non-patients are essentially similar. The evidence presented here indicates a possible shift in the oral microbiome, which may be related to HIV infection, treatment, or other oral manifestations of diseases [72].

16.4 CANCER

16.4.1 Pancreatic Cancer

Pancreatic cancer (PC) has an estimated overall 5-year survival rate of 11%, making it one of the deadliest cancers even with significant advances in cancer therapies. In North America, it is anticipated that 47,050 people will pass away from PC in just the upcoming year. It is projected that rates will double by 2030, moving it up to the second position as the leading cause of death from cancer. PC is the sixth leading cause of cancer-related deaths worldwide, accounting for an estimated 441,083 deaths [73].

Due to advanced disease, most patients (80%–90%) with PC are incurable at the time of presentation [74]. Furthermore, early recurrence and death are common outcomes for the small number of patients who qualify for curative surgical resection [75]. More than 90% of pancreatic cancers are caused by pancreatic ductal adenocarcinoma (PDAC), which is also the third most common cause of cancer-related mortality [76].

Both PDAC patients and healthy controls have distinct oral microbiota. Lu and his associates looked at the tongue-covering microbiota in 25 healthy controls and 30 PDAC patients [77]. It was discovered that there is a notable difference in the bacterial composition of the tongue covering between PDAC patients and controls, and that the diversity of microbiota of the tongue coat was significantly higher in PDAC patients. *Haemophilus*, *Porphyromonas*, *Leptotrichia*, *and Fusobacterium* are a few bacterial genera that could be used to differentiate PDAC patients from healthy people [78]. Furthermore, several studies have demonstrated that the saliva microbiota profiles of PDAC patients differ from those of healthy controls, albeit the different studies' findings have identified distinct species of bacteria as the distinguishing factor [79–81].

16.4.2 Oral Microbiome in Diagnosis and Prognosis of PDAC

As of right now, there is no suggested pancreatic cancer screening program. Prompt diagnosis enables potentially curative intervention, which could improve patient outcomes [81]. Serum tests are not the best approach for screening since carbohydrate antigen 19-9 (CA 19-9) has a low positive predictive value in people who are asymptomatic [82].

PDAC patients have a distinct microbiome, which presents novel possibilities for the development of screening and diagnostic biomarkers [83]. Additionally, because fecal and salivary samples are simpler to acquire and less intrusive than tissue samples from metastatic tumors, the tests may be conducted in a clinic setting. One study investigated the potential salivary microbiome profile as a biomarker for diagnosis comparing the salivary microbiota of ten PDAC patients and ten matched controls using the Human Oral Microbe Identification Microarray

(HOMIM) to find bacterial candidates [84]. To confirm this, a cohort consisting of 27 samples with chronic pancreatitis, 28 matched controls, and 28 PDAC samples was utilized. The salivary microbiota of PDAC patients differed significantly from that of controls, according to the authors' observations. *Streptococcus mitis* and *Neisseria elongata* levels were specifically considerably lower in pancreatic cancer patients, and the combination of these two microbial biomarkers demonstrated 96.4% sensitivity and 82.1% specificity in separating healthy subjects from PDAC patients. Furthermore, no notable changes were seen in any of the bacterial possibilities from the PDAC investigation when these biomarkers were contrasted with results from a lung cancer HOMIM profiling analysis [85]. Moreover, Torres et al. [79] examined the salivary microbiota of individuals with PDAC, patients with various medical conditions, and healthy individuals (8, 78, and 22 patients, respectively). By employing high-throughput sequencing, the scientists discovered a considerably larger ratio of *Leptotrichia* to *Porphyromonas* (LP ratio) in the saliva of PDAC patients compared to the other two groups. As a result, they suggested that the LP ratio might be a useful biomarker for PDAC diagnosis [86].

It's interesting to note that there was no significant difference in the relative abundances of *Neisseria* and *Streptococcus* between the groups of patients in this study. In contrast to the study conducted by Farrell et al. [78], the authors ascribe this to a different study approach. Mendez et al. assessed the gut microbiota and its metabolic products in KPC mice and PDAC patients in an effort to diagnose PDAC early [87]. They discovered that during the early stages of PDAC development, the gut microbiota was enriched in the bacterial phyla Firmicutes and Proteobacteria. Increased level of serum polyamine as a result of active metabolic pathways were linked to this [88]. As a result, monitoring microbial dysbiosis metabolites may offer yet another non-invasive diagnostic for PDAC early identification. Many obstacles still need to be overcome, even though these studies have revealed viable approaches for creating microbiota-based diagnostic biomarkers. The variability in the bacterial populations found in these studies is one of the main obstacles. While several studies have found that patients with PDAC have higher concentrations of certain bacterial phyla, such as Firmicutes, Bacteroidetes, and Proteobacteria, common genera or species of bacteria have been regularly changed in PDAC patients, according to relatively recent investigations [87]. Divergent baseline microbiome composition can result from variations in host dietary intake, geographic location, lifestyle, and ethnicity, which probably accounts for some of the variability observed across these studies. This complicates the process of choosing a broadly applicable biomarker. Therefore, more research will be required to verify these analytical approaches and find universal biomarkers for clinical use [89].

PDAC patients' biological characteristics must be considered when stratifying treatment plans, and this requires accurate prognostic biomarkers. Numerous studies have shown how the microbiota profile can be used to predict PDAC patients' outcomes [90].

Tumor samples were examined in a PDAC study to check for *Fusobacterium* spp. MicroRNA expression, epigenetic changes, and particular mutations were also examined in the samples. *Fusobacterium* spp. were discovered in 8.8% of the tumor samples. Ignorant of other clinical and molecular features, *Fusobacterium* species positive was associated with a substantially reduced cancer-specific survival, suggesting that intra-tumoral *Fusobacterium* status may be a viable predictive biomarker [91]. The microbial profiles of short-term and long-term survivors were compared, and the bacterial composition of duodenal fluid collections from 74 PDAC patients was examined by Kohi et al., followed by addition of *Fusobacterium, Rothia, and Neisseria* to samples from short-term survivors to enhance their growth [90]. A retrospective investigation of intraoperative bile cultures from 211 patients with PDAC who were either locally progressed or had a borderline resectable polygenic risk score (PFS) revealed a correlation between a shorter PFS and larger numbers of microbial species [91].

Not all internal tumoral bacterial colonization, though, is linked to unfavorable consequences. The internal tumoral microbiome composition of 68 resected PDAC tumors was studied by Riquelme et al. Of these patients, 36 were deemed long-term survivors (greater than 5 years) and

32 were deemed short-term survivors (less than 5 years). When comparing the intra-tumoral microbiome diversity of long-term survivors to that of short-term survivors, they discovered a significant difference. Long-term survival was significantly predicted by a particular microbiome signature (*Streptomyces*, *Bacillus clausii*, *Saccharopolyspora*, and *Pseudoxanthomonas*) [92].

All such findings point to the possibility that PDAC patients' microbiota composition and diversity could provide predictive data about their prognosis and their likelihood of survival. To better define the biomarkers and explain inconsistent results, more research is required [93].

16.4.3 Oral Cancer

Lip, mouth, or oropharynx cancer is referred to as oral cancer. The oral microbiota has gained attention and been proposed as a potential mediator in three distinct pathways for the pathophysiology of oral cancers [94]. This may have occurred as a result of the chronic inflammatory reactions that bacteria induce, since these reactions promote oncogene activation, mutagenesis, and cell proliferation [95]. It is recognized that anaerobic oral bacteria induce long-term inflammatory processes in periodontal tissue by elevating interleukin-1β (IL-1β), IL-6, tumor necrosis factor-α, and matrix metalloproteinases MMP-8 and MMP-9. Type 3 or type 4 secretion systems used by the bacterial effector proteins may affect cytoskeletal reorganizations, NF-κB activation, inhibition of cellular apoptosis, and cell proliferation [96]. *Porphyromonas gingivalis* in particular may have anti-apoptotic properties. Furthermore, *P. gingivalis* has an adenosine triphosphate (ATP)–consuming enzyme called nucleoside-diphosphate-kinase homolog, which influences the purinergic P2X7 receptors that are activated by ATP and linked to apoptosis and cell death [96].

Carcinogenic substances and processes may be aided by substances like volatile sulfur compounds (VSCs) produced by bacteria, reactive oxygen species (ROS), reactive nitrogen species (RNS), and acetaldehyde produced from ethanol. Alcohol dehydrogenase is an enzyme that *S. sanguinis*, *S. mitis*, *S. oralis*, *S. salivarius*, and *S. gordonii* as well as *Candida* can use to convert alcohol to acetaldehyde [97]. Numerous oral cancers have been linked to reactive oxygen species and to reactive nitrogen species produced by peroxigenic oral bacteria, such as *Streptococcus gordonii*, *Streptococcus mitis*, *Streptococcus oralis*, *Lactobacillus fermentum*, and *L. jensenii*. Genomic mutations can be caused by VSCs, and they have genotoxic effects. *Aggregatibacter actinomycetemcomitans*, *F. nucleatum*, *P. intermedia*, and *P. gingivalis* are primarily in charge of producing methyl mercaptan (CH_3SH), dimethyl sulfide ($(CH_3)2S$), and hydrogen sulfide (H_2S). There is a correlation between human cytomegalovirus and carcinogenesis, according to recent data. Human cytomegalovirus is commonly found in periodontal tissues and is a member of the Herpesviridae beta subfamily [96, 98].

Human cytomegalovirus has the ability to cause cancer due to its proangiogenic and anti-apoptotic mechanisms as well as its immunosuppressive effects [99]. The virus can also cause non-lytic infection and encourage the cancerous transformation of epithelial cells. Although specific proof of the virus's involvement in the carcinogenic process of oral cancers has not yet been shown, these data imply that it may [100]. There is increasing evidence that oral cancer surfaces contain much greater concentrations of oral aerobes and anaerobes than the surfaces of healthy mucosa, even if additional study on the pathophysiology of oral malignancies is required to demonstrate the relationship between oral cancer and oral microbiota [96]. Consequently, the prevention of oral cancer depends on the early identification and treatment of the oral microenvironment [101].

16.5 CONCLUSION

The oral microbiome is a dynamic and varied microbial population found in the human mouth. This chapter has examined the growing importance of the oral microbiome in systemic disorders ranging from diabetes to cancer, as well as oral diseases including periodontitis and dental caries. We are learning important things about how diseases start and spread by examining the makeup and role of the oral microbiome. This is a very promising new area in diagnostics. Oral samples are simple to obtain and require little intrusion, which makes them perfect for non-traditional diagnostic instruments. Early diagnosis and focused therapies may be made possible by identifying certain microbial signatures linked to disease states. To completely comprehend the intricate interactions between human health and the oral microbiome, however, a great deal of study is still required. Subsequent research endeavors have to concentrate on verifying these first results through more extensive clinical trials and creating uniform testing protocols. In the end, using the oral microbiome's capability to improve illness detection, prognosis, and maybe even therapy might usher in a new era of customized medicine.

REFERENCES

1. Y. Yamashita and T. Takeshita, "The oral microbiome and human health," *J Oral Sci*, vol. 59, no. 2, pp. 201–206, 2017, doi: 10.2334/josnusd.16-0856.
2. L. Gao, T. Xu, G. Huang, S. Jiang, Y. Gu, and F. Chen, "Oral microbiomes: more and more importance in oral cavity and whole body," *Protein Cell*, vol. 9, no. 5, pp. 488–500, 2018, doi: 10.1007/s13238-018-0548-1.
3. F. E. Dewhirst *et al.*, "The human oral microbiome," *J Bacteriol*, vol. 192, no. 19, pp. 5002–5017, 2010, doi: 10.1128/JB.00542-10.
4. G. Jia *et al.*, "The oral microbiota - A mechanistic role for systemic diseases," *Br Dent J*, vol. 224, no. 6, pp. 447–455, 2018, doi: 10.1038/sj.bdj.2018.217.
5. E. Zaura, B. J. Keijser, S. M. Huse, and W. Crielaard, "Defining the healthy 'core microbiome' of oral microbial communities," *BMC Microbiol*, vol. 9, 2009, doi: 10.1186/1471-2180-9-259.
6. P. Gholizadeh, H. Eslami, M. Yousefi, M. Asgharzadeh, M. Aghazadeh, and H. S. Kafil, "Role of oral microbiome on oral cancers, a review," *Biomed Pharmacother*, vol. 84, pp. 447–455, 2016, doi: 10.1016/j.biopha.2016.09.082.
7. B. Sampaio-Maia, I. M. Caldas, M. L. Pereira, D. Pérez-Mongiovi, and R. Araujo, "The Oral Microbiome in Health and Its Implication in Oral and Systemic Diseases," *Adv Appl Microbiol*, vol. 97, pp. 171–210, 2016, doi: 10.1016/bs.aambs.2016.08.002.
8. B. Liu *et al.*, "Deep sequencing of the oral microbiome reveals signatures of periodontal disease," *PLoS One*, vol. 7, 2012, doi: 10.1371/journal.pone.0037919.
9. N. Philip, B. Suneja, and L. J. Walsh, "Ecological Approaches to Dental Caries Prevention: Paradigm Shift or Shibboleth?," *Caries Res*, vol. 52, pp. 153–165. 2018, doi: 10.1159/000484985.
10. X. Zhang *et al.*, "The oral and gut microbiomes are perturbed in rheumatoid arthritis and partly normalized after treatment," *Nat Med*, vol. 21, no. 8, pp. 895–905, 2015, doi: 10.1038/nm.3914.
11. C. Blod *et al.*, "The oral microbiome—the relevant reservoir for acute pediatric appendicitis?," *Int J Colorectal Dis*, vol. 33, no. 2, pp. 209–218, 2018, doi: 10.1007/s00384-017-2948-8.
12. M. F. Zarco, T. J. Vess, and G. S. Ginsburg, "The oral microbiome in health and disease and the potential impact on personalized dental medicine," *Oral Dis*, vol. 18, no. 2, pp. 109–120, 2012. doi: 10.1111/j.1601-0825.2011.01851.x.
13. H. F. Jenkinson and R. J. Lamont, "Oral microbial communities in sickness and in health," *Trends Microbiol*, vol. 13, no.12, 589–595, 2005. doi: 10.1016/j.tim.2005.09.006.
14. C. Ma *et al.*, "Comparison of oral microbial profiles between children with severe early childhood caries and caries-free children using the human oral microbe identification microarray," *PLoS One*, vol. 10, no. 3, 2015, doi: 10.1371/journal.pone.0122075.
15. R. H. Selwitz, A. I. Ismail, and N. B. Pitts, "Dental caries," *The Lancet*, vol. 369, no. 9555, pp. 51–59, 2007.
16. M. Balakrishnan, R. S. Simmonds, and J. R. Tagg, "Dental caries is a preventable infectious disease," *Aust Dent J*, vol. 45, no. 4, 235–245, 2000, doi: 10.1111/j.1834-7819.2000.tb00257.x.
17. A. C. R. Tanner *et al.*, "Microbiota of severe early childhood caries before and after therapy," *J Dent Res*, vol. 90, no. 11, pp. 1298–1305, 2011, doi: 10.1177/0022034511421201.
18. E. Kanasi *et al.*, "Clonal analysis of the microbiota of severe early childhood caries," *Caries Res*, vol. 44, no. 5, pp. 485–497, 2010, doi: 10.1159/000320158.
19. M. R. Becker *et al.*, "Molecular analysis of bacterial species associated with childhood caries," *J Clin Microbiol*, vol. 40, no. 3, pp. 1001–1009, 2002, doi: 10.1128/JCM.40.3.1001-1009.2002.

20. J. He, Y. Li, Y. Cao, J. Xue, and X. Zhou, "The oral microbiome diversity and its relation to human diseases," *Folia Microbiol (Praha)*, vol. 60, no. 1, 69–80, 2015, doi: 10.1007/s12223-014-0342-2.
21. B. L. Pihlstrom, B. S. Michalowicz, and N. W. Johnson, "Periodontal diseases," *Lancet*, vol. 366, no. 9499, pp. 1809–1820, 2005, doi: 10.1016/S0140-6736(05)67728-8.
22. M. Agnello *et al.*, "Microbiome Associated with Severe Caries in Canadian First Nations Children," *J Dent Res*, vol. 96, no. 12, pp. 1378–1385, 2017, doi: 10.1177/0022034517718819.
23. S. Huang *et al.*, "Preliminary characterization of the oral microbiota of Chinese adults with and without gingivitis," *BMC Oral Health*, vol. 11, 2011, doi: 10.1186/1472-6831-11-33.
24. T. Murata, T. Yamaga, T. Iida, H. Miyazaki, and K. Yaegaki, "Classification and examination of halitosis," *Int Dent J*, vol. 52, no. S5P1, pp. 181–186, 2002.
25. S. R. Porter and C. Scully, "Oral malodour (halitosis)," *BMJ*, vol. 333, no. 7569, pp. 632–635, 2006.
26. H. Mouridsen, "Aromatase inhibitors in the early adjuvant setting - the latest evidence," *Eur J Cancer*, vol. 4, no. 9, pp. 3–9, 2006, doi: 10.1016/j.ejcsup.2006.06.001.
27. R. Ross, "Atherosclerosis--an inflammatory disease," *N Engl J Med*, vol. 340, no. 2, pp. 115–126, 1999, doi: 10.1056/NEJM199901143400207.
28. G. A. Mensah, G. A. Roth, and V. Fuster, "The global burden of cardiovascular diseases and risk factors: 2020 and beyond," *J Am Coll Cardiol*, vol. 74, no. 20, pp. 2529–2532, 2019, doi: 10.1016/j.jacc.2019.10.009.
29. A. D. K. Bowry, J. Lewey, S. B. Dugani, and N. K. Choudhry, "The burden of cardiovascular disease in low- and middle-income countries: epidemiology and management," *Can J Cardiol*, vol. 31, no. 9, pp. 1151–1159, 2015, doi: 10.1016/j.cjca.2015.06.028.
30. P. Libby, P. M. Ridker, and A. Maseri, "Inflammation and atherosclerosis," *Circulation*, vol. 105, no. 9, pp. 1135–1143, 2002, doi: 10.1161/hc0902.104353.
31. O. Koren *et al.*, "Human oral, gut, and plaque microbiota in patients with atherosclerosis," *Proc Natl Acad Sci USA*, vol. 108, no. SUPPL. 1, pp. 4592–4598, 2011, doi: 10.1073/pnas.1011383107.
32. G. K. Hansson, "Inflammation, atherosclerosis, and coronary artery disease," *N Engl J Med*, vol. 352, no. 16, 1685–1695, 2005.
33. T. Dietrich, P. Sharma, C. Walter, P. Weston, and J. Beck, "The epidemiological evidence behind the association between periodontitis and incident atherosclerotic cardiovascular disease," *J Periodontol*, vol. 84, no. 4S, 2013, doi: 10.1902/jop.2013.134008.
34. S. Sen *et al.*, "Periodontal Disease, Regular Dental Care Use, and Incident Ischemic Stroke," *Stroke*, vol. 49, no. 2, pp. 355–362, 2018, doi: 10.1161/strokeaha.117.018990.
35. M. Sanz *et al.*, "Periodontitis and cardiovascular diseases: Consensus report," *J Clin Periodontol*, vol. 47, no. 3, pp. 268–288, 2020, doi: 10.1111/jcpe.13189.
36. F. DeStefano, R. F. Anda, H. S. Kahn, D. F. Williamson, and C. M. Russell, "Dental disease and risk of coronary heart disease and mortality," *British Medical Journal*, 1993, vol. 306, no. 6879, pp. 688–691.
37. Y. W. Chen *et al.*, "Periodontitis May Increase the Risk of Peripheral Arterial Disease," *Eur J Vasc Endovasc Surg*, vol. 35, no. 2, pp. 153–158, 2008, doi: 10.1016/j.ejvs.2007.08.016.
38. L. L. Humphrey, R. Fu, D. I. Buckley, M. Freeman, and M. Helfand, "Periodontal disease and coronary heart disease incidence: A systematic review and meta-analysis," *J Gen Intern Med*, vol. 23, no. 12, pp. 2079–2086, 2008, doi: 10.1007/s11606-008-0787-6.
39. L. T. Cowan *et al.*, "Periodontal disease and incident venous thromboembolism: The Atherosclerosis Risk in Communities study," *J Clin Periodontol*, vol. 46, no. 1, pp. 12–19, 2019, doi: 10.1111/jcpe.13029.
40. K. J. Mattila *et al.*, "Association between dental health and acute myocardial infarction," *BMJ*, 1989, vol. 298, no. 6676, pp. 779–781.
41. J. A. Cauley, A. M. Kassem, N. E. Lane, and S. Thorson, "Prevalent peripheral arterial disease and inflammatory burden," *BMC Geriatr*, vol. 16, no. 1, 2016, doi: 10.1186/s12877-016-0389-9.
42. H. Rashid, "The effect of surface roughness on ceramics used in dentistry: A review of literature," *Eur J Dent*, vol. 8, no. 04, pp. 571–579, 2014.
43. U. Soto-Barreras *et al.*, "Peripheral Arterial Disease Associated With Caries and Periodontal Disease," *J Periodontol*, vol. 84, no. 4, pp. 486–494, 2013, doi: 10.1902/jop.2012.120051.
44. W. Ma et al., "Exploring the bi-directional relationship between periodontitis and dyslipidemia: A comprehensive systematic review and meta-analysis," *BMC Oral Health, vol.* 24, no. 1, 508, 2023, doi: 10.21203/rs.3.rs-2241257/v1.
45. S. Chandy *et al.*, "Evaluation of C-reactive protein and fibrinogen in patients with chronic and aggressive periodontitis: A clinico-biochemical study," *J Clin Diagn Res*, vol. 11, no. 3, pp. ZC41–ZC45, 2017, doi: 10.7860/JCDR/2017/23100.9552.

46. H. A. Schenkein, P. N. Papapanou, R. Genco, and M. Sanz, "Mechanisms underlying the association between periodontitis and atherosclerotic disease," *Periodontol 2000*, vol. 83, no. 1, pp. 90–106, 2020, doi: 10.1111/prd.12304.
47. A. Radaic and Y. L. Kapila, "The oralome and its dysbiosis: New insights into oral microbiome-host interactions," *Comput Struct Biotechnol J*, vol. 19, pp. 1335–1360, 2021, doi: 10.1016/j.csbj.2021.02.010.
48. R. Saini, S. Saini, and S. R. Saini, "Periodontitis and psychological stress: a dental view," *Ind Psychiatry J*, 2010, vol. 19, no. 1, pp. 66–67.
49. A. Tonelli, E. N. Lumngwena, and N. A. B. Ntusi, "The oral microbiome in the pathophysiology of cardiovascular disease," *Nat Res*, vol. 20, no. 6, pp. 386–403, 2023, doi: 10.1038/s41569-022-00825-3.
50. C. S. Miller *et al.*, "Utility of salivary biomarkers for demonstrating acute myocardial infarction," *J Dent Res*, vol. 93, pp. 72S–79S, 2014, doi: 10.1177/0022034514537522.
51. C. Z. Zhang *et al.*, "Saliva in the diagnosis of diseases," *Int J Oral Sci*. vol. 8, no. 3, 133–137, 2016, doi: 10.1038/ijos.2016.38.
52. J. Hintao, R. Teanpaisan, V. Chongsuvivatwong, C. Ratarasan, and G. Dahlen, "The microbiological profiles of saliva, supragingival and subgingival plaque and dental caries in adults with and without type 2 diabetes mellitus," *Oral Microbiology and Immunology*, 2007, vol. 22, no. 3, pp. 175–181.
53. H. Löe, "Periodontal disease: the sixth complication of diabetes mellitus," *Diabetes Care*, 1993, vol. 16, no. 1, pp. 329–334.
54. E. Lalla *et al.*, "Periodontal infection profiles in type 1 diabetes," *J Clin Periodontol*, vol. 33, no. 12, pp. 855–862, 2006, doi: 10.1111/j.1600-051X.2006.00996.x.
55. J. L. Ebersole, S. C. Holt, R. Hansard, and M. J. Novak, "Microbiologic and Immunologic Characteristics of Periodontal Disease in Hispanic Americans With Type 2 Diabetes," *J Periodontol*, vol. 79, no. 4, pp. 637–646, 2008, doi: 10.1902/jop.2008.070455.
56. V. M. Barnes *et al.*, "Global metabolomic analysis of human saliva and plasma from healthy and diabetic subjects, with and without periodontal disease," *PLoS One*, vol. 9, 2014, doi: 10.1371/journal.pone.0105181.
57. H. Abdolsamadi *et al.*, "Reduction of melatonin level in patients with type II diabetes and periodontal diseases," *J Dent Res Dent Clin Dent Prospect*, vol. 8, no. 3, pp. 160–165, 2014, doi: 10.5681/joddd.2014.029.
58. R. Lira-Junior and E. A. Boström, "Oral-gut connection: One step closer to an integrated view of the gastrointestinal tract?," *Mucosal Immunol*, vol. 11, no. 2, pp. 316–318, 2018, doi: 10.1038/mi.2017.116.
59. S. Raychaudhuri *et al.*, "Five amino acids in three HLA proteins explain most of the association between MHC and seropositive rheumatoid arthritis," *Nat Genet*, vol. 44, no. 3, pp. 291–296, 2012, doi: 10.1038/ng.1076.
60. I. B. McInnes and G. Schett, "The pathogenesis of rheumatoid arthritis," *N Eng J Med*, vol. 365, no. 23, pp. 2205–2219, 2011.
61. M. F. Konig, "The microbiome in autoimmune rheumatic disease," *Best Pract Res Clin Rheumatol*, 2020, vol. 34, no. 1, p. 101473.
62. A. Goris and A. Liston, "The immunogenetic architecture of autoimmune disease," *Cold Spring Harb Perspect Biol*, vol. 4, 2012, doi: 10.1101/cshperspect.a007260.
63. L. A. Yurkovetskiy, J. M. Pickard, and A. V. Chervonsky, "Microbiota and autoimmunity: Exploring new avenues," Cell Host Microbe, vol. 17, no. 5, pp. 548–552, 2015, doi: 10.1016/j.chom.2015.04.010.
64. A. Amedei *et al.*, "Molecular mimicry between helicobacter pylori antigens and H +, K+-adenosine triphosphatase in human gastric autoimmunity," *J Exp Med*, vol. 198, no. 8, pp. 1147–1156, 2003, doi: 10.1084/jem.20030530.
65. J. Potempa, P. Mydel, and J. Koziel, "The case for periodontitis in the pathogenesis of rheumatoid arthritis," *Nat Rev Rheumatol*, vol. 13, no. 10, pp. 606–620, 2017, doi: 10.1038/nrrheum.2017.132.
66. M. F. Konig *et al.*, "Aggregatibacteractinomycetemcomitans-induced hypercitrullination links periodontal infection to autoimmunity in rheumatoid arthritis," *Sci Transl Med*, vol. 8, no. 369, 2016, doi: 10.1126/scitranslmed.aaj1921.
67. N. Wegner *et al.*, "Peptidylarginine deiminase from Porphyromonasgingivalis citrullinates human fibrinogen and α-enolase: Implications for autoimmunity in rheumatoid Arthritis," *Arthritis Rheum*, vol. 62, no. 9, pp. 2662–2672, 2010, doi: 10.1002/art.27552.
68. K. Białowąs, M. Radwan-Oczko, I. Duś-Ilnicka, L. Korman, and J. Świerkot, "Periodontal disease and influence of periodontal treatment on disease activity in patients with rheumatoid arthritis and spondyloarthritis," *Rheumatol Int*, vol. 40, no. 3, pp. 455–463, 2020, doi: 10.1007/s00296-019-04460-z.

69. S. Bellando-Randone *et al.*, "Exploring the oral microbiome in rheumatic diseases, state of art and future prospective in personalized medicine with an AI approach," *J Pers Med*, vol. 11, no. 7, 625, 2021, doi: 10.3390/jpm11070625.
70. A. T. Dang *et al.*, "Evidence of an increased pathogenic footprint in the lingual microbiome of untreated HIV infected patients," *BMC Microbiol*, vol. 12, 2012, doi: 10.1186/1471-2180-12-153.
71. Y. Li *et al.*, "HIV infection and microbial diversity in saliva," *J Clin Microbiol*, vol. 52, no. 5, pp. 1400–1411, 2014, doi: 10.1128/JCM.02954-13.
72. S. E. Heron and S. Elahi, "HIV infection and compromised mucosal immunity: Oral manifestations and systemic inflammation," *Front Immunol*, vol. 8, 2017, doi: 10.3389/fimmu.2017.00241.
73. G. Benassai *et al.*, "Long-term survival after curative resection for pancreatic ductal adenocarcinoma - Surgical treatment," *Int J Surg*, vol. 21 no. Suppl 1, pp. S1–S3, 2015, doi: 10.1016/j.ijsu.2015.06.050.
74. G. Lippi and C. Mattiuzzi, "The global burden of pancreatic cancer," *Archives of Medical Science*, vol. 16, no. 4, pp. 820–824, 2020, doi: 10.5114/aoms.2020.94845.
75. P. Rawla, T. Sunkara, and V. Gaduputi, "Epidemiology of pancreatic cancer: Global trends, etiology and risk factors," *World J Oncol*, vol. 10, no. 1, pp. 10–27, 2019, doi: 10.14740/wjon1166.
76. R. L. Siegel, K. D. Miller, H. E. Fuchs, and A. Jemal, "Cancer statistics, 2022," *CA Cancer J Clin*, vol. 72, no. 1, pp. 7–33, 2022, doi: 10.3322/caac.21708.
77. E. Vogtmann *et al.*, "Oral microbial community composition is associated with pancreatic cancer: A case-control study in Iran," *Cancer Med*, vol. 9, no. 2, pp. 797–806, 2020, doi: 10.1002/cam4.2660.
78. J. J. Farrell *et al.*, "Variations of oral microbiota are associated with pancreatic diseases including pancreatic cancer," *Gut*, vol. 61, no. 4, pp. 582–588, 2012, doi: 10.1136/gutjnl-2011-300784.
79. P. J. Torres, E. M. Fletcher, S. M. Gibbons, M. Bouvet, K. S. Doran, and S. T. Kelley, "Characterization of the salivary microbiome in patients with pancreatic cancer," *PeerJ*, 2015, doi: 10.7717/peerj.1373.
80. H. Sun *et al.*, "Characterization of Oral Microbiome and Exploration of Potential Biomarkers in Patients with Pancreatic Cancer," *Biomed Res Int*, 2020, doi: 10.1155/2020/4712498.
81. A. L. Wei *et al.*, "Oral microbiome and pancreatic cancer," *World J Gastroenterol*, vol. 26, no. 48, pp. 7679–7692, 2020, doi: 10.3748/WJG.V26.I48.7679.
82. J. E. Kim, K. T. Lee, J. K. Lee, S. W. Paik, J. C. Rhee, and K. W. Choi, "Clinical usefulness of carbohydrate antigen 19-9 as a screening test for pancreatic cancer in an asymptomatic population," *J Gastroenterol Hepatol*, 2004, vol. 19, no. 2, pp. 182–186.
83. L. Grzeskowiak *et al.*, "Distinct Gut microbiota in southeastern African and northern European infants," *J Pediatr Gastroenterol Nutr*, vol. 54, no. 6, pp. 812–816, 2012, doi: 10.1097/MPG.0b013e318249039c.
84. T. Yatsunenko *et al.*, "Human gut microbiome viewed across age and geography," *Nature*, vol. 486, no. 7402, pp. 222–227, 2012. doi: 10.1038/nature11053.
85. A. J. Obregon-Tito *et al.*, "Subsistence strategies in traditional societies distinguish gut microbiomes," *Nat Commun*, vol. 6, 2015, doi: 10.1038/ncomms7505.
86. C. De Filippo *et al.*, "Impact of diet in shaping gut microbiota revealed by a comparative study in children from Europe and rural Africa," *Proc Natl Acad Sci U S A*, vol. 107, no. 33, pp. 14691–14696, 2010, doi: 10.1073/pnas.1005963107.
87. R. Mendez *et al.*, "Microbial dysbiosis and polyamine metabolism as predictive markers for early detection of pancreatic cancer," *Carcinogenesis*, vol. 41, no. 5, pp. 561–570, 2020, doi: 10.1093/carcin/bgz116.
88. J. Li *et al.*, "Comparative analysis of the human saliva microbiome from different climate zones: Alaska, Germany, and Africa," *BMC Microbiol*, vol. 14, no. 1, 2014, doi: 10.1186/s12866-014-0316-1.
89. J. J. Li et al., "The role of microbiome in pancreatic cancer," *Cancer Metastasis Rev*, vol. 40, pp. 777–789, 2021.
90. M. Weniger *et al.*, "Influence of Klebsiella pneumoniae and quinolone treatment on prognosis in patients with pancreatic cancer," *Br J Surg*, vol. 108, no. 6, pp. 709–716, 2021, doi: 10.1002/bjs.12003.
91. K. Mitsuhashi *et al.*, "Association of Fusobacterium species in pancreatic cancer tissues with molecular features and prognosis," *Oncotarget*, vol. 6, no. 9, pp. 7209–7220, 2015.
92. E. Riquelme *et al.*, "Tumor Microbiome Diversity and Composition Influence Pancreatic Cancer Outcomes," *Cell*, vol. 178, no. 4, pp. 795–806.e12, 2019, doi: 10.1016/j.cell.2019.07.008.
93. S. Kohi *et al.*, "Alterations in the Duodenal Fluid Microbiome of Patients With Pancreatic Cancer," *Clin Gastroenterol Hepatol*, vol. 20, no. 2, pp. e196–e227, 2022, doi: 10.1016/j.cgh.2020.11.006.
94. J. H. Meurman and J. Uittamo, "Oral micro-organisms in the etiology of cancer," *Acta Odontol Scand*, vol. 66, no. 6, pp. 321–326, 2008. doi: 10.1080/00016350802446527.
95. E. R. Herrero *et al.*, "Dysbiotic Biofilms Deregulate the Periodontal Inflammatory Response," *J Dent Res*, vol. 97, no. 5, pp. 547–555, 2018, doi: 10.1177/0022034517752675.

96. Y. Zhang, X. Wang, H. Li, C. Ni, Z. Du, and F. Yan, "Human oral microbiota and its modulation for oral health," *Biomed Pharmacother*, vol. 99, pp. 883–893, 2018, doi: 10.1016/j.biopha.2018.01.146.
97. E. Marttila *et al.*, "Fermentative 2-carbon metabolism produces carcinogenic levels of acetaldehyde in Candida albicans," *Mol Oral Microbiol*, vol. 28, no. 4, pp. 281–291, 2013, doi: 10.1111/omi.12024.
98. N. Suzuki, M. Yoneda, T. Takeshita, T. Hirofuji, and T. Hanioka, "Induction and inhibition of oral malodor," *Mol Oral Microbiol*, vol. 34, no. 3, pp. 85–96, 2019, doi: 10.1111/omi.12259.
99. J. Lee, J. A. S. Roberts, K. R. Atanasova, N. Chowdhury, and Ö. Yilmaz, "A novel kinase function of a nucleoside-diphosphate-kinase homologue in Porphyromonasgingivalis is critical in subversion of host cell apoptosis by targeting heat-shock protein 27," *Cell Microbiol*, vol. 20, 2018, doi: 10.1111/cmi.12825.
100. M. S. Attene-Ramos, E. D. Wagner, M. J. Plewa, and H. R. Gaskins, "Evidence that hydrogen sulfide is a genotoxic agent," *Mol Cancer Res*, vol. 4, no. 1, pp. 9–14, 2006, doi: 10.1158/1541-7786.MCR-05-0126.
101. C. Thomas et al., "Oral microbiota: A major player in the diagnosis of systemic diseases," *Diagnostics*, vol. 11, no. 8, p. 1376, 2021.

17 Oral Microbiome and Longevity

Duni Chand, Arti, Alka Kumari, Lalita Kaushal, and Ekta Arya
Department of Biotechnology, Himachal Pradesh University, Shimla, Himachal Pradesh, India

Pradeep Kumar
Department of Forensic Science, Himachal Pradesh University, Shimla, Himachal Pradesh, India

17.1 INTRODUCTION

The term "microbiome" denoting the collective microbial inhabitants within our body, was coined by Nobel Prize laureate Joshua Lederberg. This term encompasses the intricate ecological community comprising symbiotic, commensal, and pathogenic microorganisms. These microorganisms actively cohabitate and share the spatial environment within our bodies. Notably, the quantity of microbes present in our bodies is nearly equivalent to or even surpasses that of our own cells.

The exploration of the oral microbiome stands as a dynamic and burgeoning realm of scientific inquiry. This intricate microbial community plays a pivotal role in maintaining overall health, exerting influence not only on oral well-being but also on systemic health, with the potential to contribute to various diseases. Distributed within resilient biofilms throughout the oral cavity, the oral microbiome forms a complex ecosystem that sustains a delicate equilibrium crucial to health. The oral microbiome, also known as the oral microbiota or oral microflora, encompasses the diverse array of microorganisms residing in the human oral cavity (Gao et al., 2018). The discovery of the oral microbiome is credited to the Dutch scientist Antonie van Leeuwenhoek, widely acknowledged as the father of microbiology. Utilizing a microscope of his own construction, Van Leeuwenhoek identified the oral microbiome, marking a pioneering moment in microbiological exploration. In 1674, he observed his own dental plaque, reporting the presence of "little living animalcules prettily moving," thus laying the foundation for our understanding of oral microorganisms (Gest, 2004).

The oral microbiome is defined as the aggregate genome of microorganisms residing within the oral cavity. Following the gut, it represents the second-largest microbial community in humans. In comparison to other bodily sites, the oral microbiome showcases a remarkable diversity in predicted protein functions. The human microbiome comprises both a core and a variable component. The core microbiome is shared among all individuals, while the variable microbiome is distinctive to each person, influenced by lifestyle and physiological differences. The oral cavity provides two distinct surfaces for bacterial colonization: the hard and soft tissues of the teeth and the oral mucosa (Zaura et al., 2014). Various structures, including the teeth, tongue, cheeks, gingival sulcus, tonsils, hard palate, and soft palate, create a conducive environment for the flourishing of microorganisms (Dewhirst et al., 2010). These oral cavity surfaces are adorned with a myriad of bacteria, forming what is colloquially referred to as the bacterial biofilm (Zhao et al., 2017).

In exploring the connection between oral health and overall well-being, a journey unfolds that intertwines our oral health with systemic implications for our body. This exploration deeply explores scientific discoveries, illuminating the relationship between oral health, diseases, and far-reaching systemic complications (Fiorillo, 2019). Beyond the absence of disease, oral health encapsulates a person's ability to eat, speak, smile, and actively contribute to society. This concept of health-related

DOI: 10.1201/9781003478638-17

quality of life, woven into the biopsychosocial model, highlights the need to consider individual perceptions of oral health in clinical practice and research (Baiju, 2017). The oral cavity, a complex realm of anatomical structures, harbors microorganisms and faces challenges from consumed substances. The mucosal immune system safeguards this complicated environment, where T lymphocytes play pivotal roles. Understanding these immune responses in health and disease becomes paramount, shedding light on T-cell subsets and their functions (Wu et al., 2012). Oral inflammation, triggered by microbial factors, triggers host responses that can either resolve or become chronic, a fate determined by the phenotype of macrophages. Unraveling the molecular intricacies within this complex system becomes crucial, revealing the roles of innate immune system cells beyond mere defense mechanisms (Cekici et al., 2014). The concept of oral frailty emerges, highlighting the decline in oral function paralleled by mental and physical decline. Research links oral frailty to heightened risks of physical decline, severe conditions, and mortality, reshaping perspectives on aging and its impact on overall health (Watanabe et al., 2020). As global demographics evolve, the challenges of an aging population necessitate a redefined vision of oral health in the broader context of global health policy. Despite being preventable or treatable, oral conditions often receive inadequate attention, marking a failure in delivering fundamental human rights to older individuals (Patel et al., 2021). The interaction between oral inflammation and the immune system is crucial in determining oral longevity. Inflammation triggered by microbial factors is influenced by immune responses, particularly macrophages, impacting the resolution or chronicity of oral issues. This delicate balance between inflammation and the immune system significantly influences the long-term health and longevity of the oral cavity.

17.2 UNDERSTANDING THE ORAL MICROBIOME

The oral microbiome refers to a complex ecosystem of microorganisms, primarily bacteria, thriving within the oral cavity of both humans and animals. The unique conditions within the mouth create an optimal habitat for specific microorganisms, setting the oral microbiome apart from the microbiomes found in the gut and skin. Key factors contributing to the composition of the oral microbiome encompass abundant moisture, ample access to nutrients, and a moderate temperature Radaic and Kapila, 2021) (Figure 17.1).

17.3 ROLE OF THE ORAL MICROBIOME IN MAINTAINING ORAL HEALTH

The significance of the oral microbiome in maintaining health extends beyond the oral cavity, influencing various body sites. The shift from a harmonious, commensal oral microbiome to a pathogenic state disrupts oral homeostasis, leading to dysbiosis. This concept is pivotal in comprehending the development of numerous oral diseases. The human mouth harbors an intricate community of microorganisms collectively known as the oral microbiome. While essential for optimal oral health, their influence reaches beyond teeth and gums. The oral microbiome can significantly impact overall well-being, potentially contributing to seemingly unrelated conditions such as gut diseases and certain cancers. Health experts recommend surprisingly simple steps that can be implemented immediately to enhance dental hygiene and maintain the health of the oral microbiome (Kilian et al., 2016).

Recent research indicates that the condition of the oral microbiota and the existence of periodontal disease contribute to non-oral diseases and conditions, such as type 2 diabetes and stroke. A 2018 study published in *AIMS Microbiology* emphasizes the pivotal role of the oral microbiome in shaping the host's health profile. The study notes, "The oral microbiome plays a key role in shaping up the host's health profile, and there is an intricate relationship between the oral microbiome and the occurrence of other diseases such as heart and liver-related disorders." Therefore, the identification and characterization of causative organisms may serve as a crucial gateway for maintaining overall health (Alsafi et al., 2024). Indications of an imbalanced oral microbiome encompass manifestations like persistent bad breath, non-healing mouth

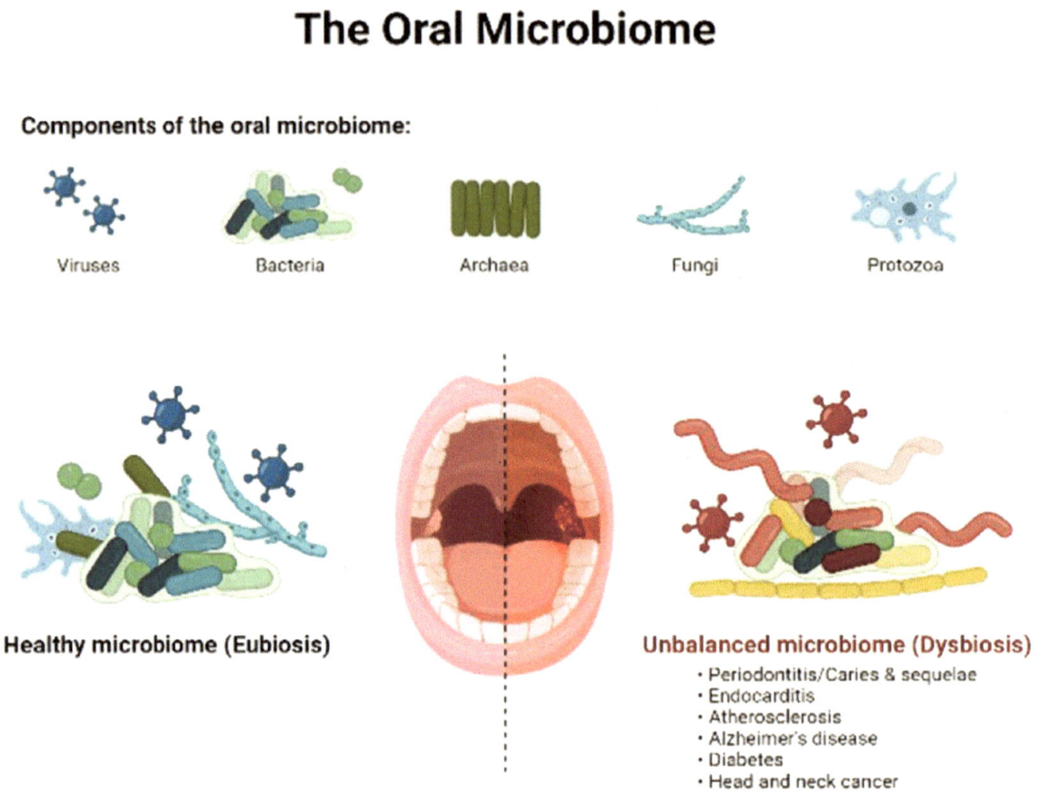

FIGURE 17.1 **Role of the oral microbiome in maintaining oral health**. (From Radaic and Kapila, 2021.)

ulcers extending beyond two weeks, the presence of oral thrush or fungal growth in the mouth, inflamed and bleeding gums, and occurrences of tonsillitis, noted Kathleen Navarro, RDH, DMD, (a licensed dental hygienist Licensed Dentist, Dental Hygienist, Marketing Strategist, San Juan del Monte, National Capital Region, Philippines). Further, an additional sign is the heightened formation of plaque and tartar on teeth.

17.4 FIVE WAYS TO KEEP THE ORAL MICROBIOME HEALTHY

17.4.1 Floss and Brush Your Teeth Every Day, Twice a Day

To prioritize oral health, one must first establish good oral hygiene practices, which are paramount for maintaining a healthy oral microbiome. According to Navarro, effective brushing doesn't entail applying excessive force; rather, it involves thoroughly and efficiently cleaning all tooth surfaces, preferably with a soft-bristled toothbrush. Equally important is daily flossing between teeth to prevent food from decaying beneath the gums. Additionally, scheduling regular dental checkups every 6 months is essential to promptly identify and address potential issues, including diseases, as well as the accumulation of plaque and tartar.

17.4.2 Eat a Balanced Diet, Limiting Processed Foods and Added Sugars

Maintaining a well-balanced diet rich in whole foods is crucial. Advises Navarro, "Steer clear of processed foods and those laden with added sugars." Hydration is equally vital, as signs of dehydration, such as dry mouth and bad breath, should not be overlooked. Given the interconnectedness of our mouths and stomachs, Navarro underscores that many beneficial bacteria crucial for our stomachs originate from our mouths. Some of these oral microbes actively contribute to reducing inflammation, balancing acidity in our mouths, and eliminating harmful bacteria. Moreover, they play a pivotal role in regulating blood pressure by converting nitrate into nitric oxide, a function our bodies cannot perform independently but rely on these vital bacteria. Proper fueling is essential for optimal bodily function, and an excess of added and artificial sugars can foster the growth of harmful organisms throughout our bodies.

17.4.3 Double-check the Ingredient Labels on Oral Care Products

Exercise caution regarding the ingredients in oral care products. With hundreds of bacterial types in our mouths, not all should be eradicated. One must refrain from using mouthwashes containing alcohol, as it may compromise a significant portion of the beneficial bacteria necessary alongside the harmful ones it eliminates. Be mindful of the ingredients in toothpaste and oral hygiene products, as some may have adverse effects. Steer clear of bisphenol A (BPA), exercise caution with oral chlorhexidine, and be vigilant against harsh detergents like sodium lauryl sulfate. Additionally, avoid high concentrations of fluoride, which have the potential to alter the microbiome as well as can cause dental fluorosis, which is a discoloration of the teeth.

17.4.4 Avoid Smoking and Vaping

Navarro emphasizes that smoking must be strongly discouraged, not only due to its detrimental effects on the lungs and respiratory system but also because it depletes oxygen levels and enhances bacterial adherence. Smoking exacerbates gum disease. Early research suggests that, concerning oral health, vaping is equally as harmful as traditional cigarettes.

17.4.5 Don't Skip Your Regular Dental Checkups

Even if you believe your teeth are staying bright and fresh, it's essential to recognize that certain common issues may go unnoticed without a trained eye. Don't wait for pain or bleeding as signals to see a dentist; instead, proactively schedule two appointments each year to uphold optimal oral care. Dentists, equipped to identify potential issues, can recommend probiotics and dietary adjustments to enhance your dental health for the long term.

17.5 MAINTAINING A HEALTHY ORAL MICROBIOME

After its establishment, the oral microbiome is upheld through a combination of host- and microbe-derived factors, encompassing processes that remain incompletely understood. The resident bacteria exhibit both pro- and anti-inflammatory activities, playing a vital role in preserving homeostasis, especially in densely colonized areas like the oral cavity (Ames et al., 2017). The interaction between the host's immune system and microbial symbionts mitigates the occurrence of acute infections in the oral mucosa, despite the dense microbial colonization. The significance of these host-microbe interactions becomes evident in observations of immunosuppressed patients, where life-threatening viral and fungal infections of the mucous membranes, as well as oral infections by non-oral species, can occur (Deo et al., 2019). The role of saliva in promoting oral health is well-established (Gevers et al., 2012). Apart from its roles in facilitating mastication, swallowing, speech, and aiding digestion, saliva serves as a crucial reservoir of enzymes and proteins essential for maintaining a balanced microbiota. Up to 108 microorganisms per milliliter of saliva have been identified, predominantly originating from oral mucosal surfaces, such as the tongue (Moon and Lee, 2016). The primary nutritional source for microorganisms is derived from salivary components, crucial for the establishment of a balanced microbiome. Numerous salivary components, such as secretory immunoglobulin A, lactoferrin, lactoperoxidase, lysozyme, statherin, and histatins, play both direct and indirect roles in regulating the microbiome, ensuring its equilibrium.

As an illustration, lactoperoxidase plays a role in catalyzing the production of hypothiocyanite from hydrogen peroxide as a by-product of bacterial metabolism and thiocyanate, which is secreted in saliva. Hypothiocyanite exerts direct antimicrobial effects by inhibiting bacterial glycolysis. Yet another salivary component with antimicrobial potential is nitrite, which is converted from dietary nitrates by oral bacteria. Nitrite undergoes further reduction to nitric oxide, which is capable of inhibiting the growth of cariogenic bacteria, potentially offering protection against caries (Goodrich et al., 2014).

Proteins, encompassing enzymes, lipids, and other components such as carbohydrates and nucleic acids, primarily sourced from saliva but also derived from gingival crevicular fluid (GCF), the oral mucosa, and bacteria, collectively form the acquired pellicle. This pellicle plays a crucial role in modulating the attachment of bacteria to dental and epithelial surfaces, offering protection against acid attacks on tooth surfaces. Enzymes vital for regulating the microbiome are immobilized within the acquired pellicle in an active conformation. The individually composed acquired pellicle initiates and facilitates bacterial adherence to non-shedding tooth surfaces through various interactions (Bik, 2016).

Saliva serves a dual role by not only fostering an environment conducive to the flourishing of biofilms but also regulating the layers of plaque through a multitude of proteins, including enzymes and glycoproteins, along with minerals. These components effectively control the buildup and activity of biofilms. Additionally, the movement of the oral muscles of the cheeks and tongue during speech and mastication, as well as the flow of saliva, contribute to the dislodgment of plaque biofilm.

17.6 HOW TO PROMOTE A HEALTHY ORAL MICROBIOME

Mouths stand out as extraordinary, intricate ecosystems flourishing with life: A lone tooth may serve as a habitat for more than 500 million oral bacteria. After several hours without tooth brushing, the microbial community in your mouth could surpass the entire human population on Earth. Though specific oral bacteria may play a role in conditions such as cavities, gingivitis, and bad breath, it's important to note that not all of them are detrimental. A substantial portion of the bacteria inhabiting your mouth are actually beneficial, serving as allies in safeguarding your teeth and gums, refreshing your breath, and aiding in the digestive process.

Researchers have uncovered over 700 distinct species of oral bacteria residing in the mouth, collectively constituting the oral microbiome. Astonishingly, the unique amalgamation of these microbes could be as distinctive as an individual's fingerprint. Within the oral microbiome, these bacteria coexist, collaborate, and jointly influence overall health. In the realm of oral well-being, fostering the proliferation of beneficial bacteria might be of greater significance than simply eradicating harmful ones. Scientists are in the early stages of unraveling the intricate interplay of oral bacteria and its implications for individual health.

17.7 BACTERIA IN THE ORAL MICROBIOME

Certain bacteria play a crucial role in food digestion and offer protection against harmful microbes present in the food you consume. While the intricacies of these processes are still under investigation, it is established that the bacteria in the human mouth operate synergistically. Collaboratively, these beneficial microbes generate substances that facilitate the growth and flourishing of their neighboring counterparts, creating a cooperative environment that supports overall oral health.

It's important to acknowledge that numerous bacteria in your mouth may not be working in one's favor. Take *Streptococcus mutans*, for instance: *S. mutans* thrives on sugar and releases acid compounds that can erode tooth enamel, contributing to tooth decay. If one is grappling with bad breath, *Porphyromonas gingivalis*, *Treponema denticola*, or *Prevotella intermedia* might be the causative agents. These bacteria take up residence in the spaces between the teeth and just beneath the gumline, thus creating an environment conducive to the production of unpleasant odors. These "bad" bacteria team up and work together just like their friendlier counterparts. Some bacterial strains predigest substances found in the food one eats so it can be more readily absorbed by their odor-emitting neighbors.

17.8 THE BALANCING ACT: HOW TO GET A HEALTHY MICROBIOME

The collective composition of microbes in your oral microbiome holds greater significance than any individual species. While achieving an instantaneous germ makeover for your mouth may be impractical, there are a few straightforward actions you can take to empower the bacteria to work in your favor.

17.8.1 Avoid Sugar

When people consume food, they are not just nourishing themselves; they are also providing sustenance to the bacteria residing in their bodies. The bacteria responsible for tooth decay, particularly in the mouth, thrive on sugar. The greater the sugar intake, the more they produce enamel-dissolving acid, contributing to the risk of dental decay.

17.8.2 Oral Hygiene

An excellent oral care regimen involves brushing and using mouthwash twice a day, coupled with daily cleaning between the teeth. The bacteria in the mouth are continually generating plaque that can lead to gingivitis, making consistent removal by thorough brushing a crucial aspect of maintaining oral health.

17.8.3 Visit Your Dentist

An oral examination enables the dentist to assess whether one's microbial balance is disrupted. Professional cleaning is essential for removing buildup that may be challenging to eliminate independently. If needed, the dentist can provide recommendations to enhance and refine your oral hygiene routine.

17.9 HOW TO BALANCE THE ORAL MICROBIOME

Promoting homeostasis in the oral microbiome is key to preventing certain bacteria and microbes from causing issues and contributing to disease. If one is wondering, "How can I enhance my oral microbiome?", consider the following steps to foster the well-being of the second-largest microbiome.

17.9.1 Practice Smart Oral Hygiene

Excessive cleaning inside the mouth can potentially lead to issues but engaging in proper "interdental cleansing"—incorporating flossing, brushing with natural toothpastes, and occasionally using picks—is believed to be beneficial in maintaining balance within the oral environment. Some aggressive products used in oral care, like certain toothpaste and mouthwash formulations, have the potential to eliminate beneficial bacteria in the mouth, disrupting the microbiome. Experts advise against using products that contain harsh ingredients, including detergent-based toothpaste and alcohol-containing/antiseptic mouthwashes. It's recommended to steer clear of formulations that include the following questionable ingredients: sodium laurel sulphate (SLS), sodium fluoride, triclosan, artificial sweeteners (such as sodium saccharin, aspartame, xylitol, and erythritol), artificial color dyes (often made from coal tar), propylene glycol, diethanolamine (DEA), and microbeads (tiny solid plastic particles).

17.9.2 Consume Probiotics

Probiotic supplements and foods play a role in populating the mouth with friendly bacteria, similar to their support for gut health. They operate by establishing a protective biofilm that enhances the well-being of teeth and gums, reducing inflammation, and preventing the intrusion of harmful bacteria onto enamel or vulnerable tissues. Additionally, they may assist in adjusting the pH of the mouth to hinder cavity growth and enhance saliva quality, promoting the remineralization of enamel.

What is the best oral probiotic to take? Dentists believe that oral microbiome probiotics are most effective when delivered directly to the mouth in lozenge, milk, or chewable form. Probiotic strains deemed highly beneficial for oral health and recommended for a duration of 2 to 3 months include: *Lactobacillus salivarius, L. reuteri, Streptococcus salivarius* K12 and M18, *L. sakei, L. brevis, L. acidophilus, L. rhamnosus, L. reuteri, L. casei, Bifidobacillus*, and *S. thermophilus.* The primary distinctions between gut probiotics and oral probiotics lie in the specific types of organisms they contain, with some degree of overlap. Typically, gut probiotics are consumed in tablet or capsule form, while oral probiotics are available in various forms such as drinks, mouthwashes, lozenges, and chewable tablets. For optimal results, consider selecting an oral probiotic supplement with a substantial strain count, such as a minimum of 3 billion colony-forming units (CFUs). Taking it in the morning before eating is recommended.

17.9.3 Embrace a Nutrient-Dense Diet

Our dietary choices significantly impact the microorganisms we introduce from our surroundings. For example, the consumption of plant-based foods like vegetables and fruits not only supplies essential nutrients but also introduces elements such as dirt, water, and bacteria into our system. Certain foods play important roles in fostering the thriving of beneficial bacteria, while others contribute to the growth of pathogens and unfriendly bacteria. One must prioritize foods in the diet that are alkalizing, anti-inflammatory, and rich in antioxidants. To promote oral health, make sure to regularly include the following foods in your routine diet: green leafy vegetables; berries, citrus fruits, and other fruits; cruciferous vegetables such as broccoli, kale, and cauliflower, along with a variety of fresh vegetables; organically raised meat, poultry, and eggs; wild-caught fish; healthy fats like olive oil, avocado, nuts, and seeds; **and** algae and seaweed.

Also, one must incorporate the following into the diet to support oral health. Probiotic, fermented foods like yogurt, kimchi, kefir, kombucha, sauerkraut, and dill pickles, Prebiotic foods including onions, leeks, dandelion greens, garlic, artichokes, chicory root, asparagus, jicama, apples, and flaxseeds, ample water, herbal teas, green/black tea, and moderate amounts of coffee. It's crucial to limit the intake of sugar and refined carbohydrates, particularly from sugary beverages and highly processed foods. Certain bacteria flourish in a high-sugar environment, contributing to oral health issues such as enamel erosion and tooth decay. A diet rich in refined carbohydrates and sugar can lead to a shift in the oral microbiome from a slightly alkaline state to a more acidic environment. This alteration makes the oral microbiome more susceptible to infections and various issues.

17.9.4 Approach Oral Health Holistically

While the connection between lack of exercise, chronic stress, and poor oral health may not be immediately apparent, it's understood that an overall unhealthy lifestyle and high stress levels can contribute to a reduction in saliva production. This decrease in saliva can influence the types of microbes that inhabit the mouth. Alterations in saliva production affect the ability of microbes to transport minerals like calcium and phosphorus from saliva to the enamel, influencing the health of the teeth. Additionally, changes in saliva composition can create a more acidic environment in the mouth, altering the overall oral ecosystem. Stress may induce conditions such as dry mouth, loss of appetite, temporomandibular joint (TMJ) dysfunction, teeth grinding, and other issues that impact the health of the mouth and jaw. Effectively managing stress and maintaining regular exercise are crucial for controlling inflammation. Engaging in activities like exercise, yoga, and meditation can aid in reducing the stress-induced, inflammatory fight-or-flight response, ultimately preventing a compromised immune system. Exercise offers additional benefits by enhancing circulation and promoting detoxification through the lymphatic system. This contributes to maintaining the health of the teeth, gums, and oral tissues.

17.10 ORAL INFLAMMATION, IMMUNE RESPONSE, AND LONGEVITY

17.10.1 Overview of Oral Inflammation and Disease

Inflammation, a response to injury, can become chronic if left unresolved. Acute inflammation involves rapid responses, while chronic inflammation entails a complex interplay involving cells like lymphocytes and macrophages. This process relies on eliminating invaders and mediators. Chronic inflammatory conditions like diabetes and heart disease have a profound impact on oral health. They interact reciprocally, influencing periodontal health. Macrophages are crucial in steering inflammation toward chronic changes or resolution without scar tissue (Hasturk et al., 2012). Periodontitis, a consequence of interactions between viruses, bacteria, and immune responses, results in the loss of periodontal ligament and alveolar bone. Diagnosis involves clinical assessment, with treatments ranging from scaling to surgical intervention, although the emphasis has shifted away from surgery in recent times (Slots, 2017). Periodontitis affects over 40% of U.S. adults and about 11% globally, causing tooth loss and reduced quality of life. It's triggered by subgingival dental biofilm, leading to irreversible damage to the periodontium. Effective management involves understanding its causes, reducing risks, and using treatments like scaling and root planing. Professional plaque control and interdental brushes are effective for managing inflammation, while chemical agents benefit those with gingivitis. Regular, thorough brushing with fluoridated toothpaste is crucial, especially for periodontitis patients who may need extra cleaning time (Kwon et al., 2021).

Maintaining periodontal health necessitates a mix of self-care and professional interventions. From careful plaque control to adjunctive rinses, the focus is on preserving oral health while avoiding overuse, particularly concerning dental implants. However, the effectiveness of laser therapy remains uncertain (Chapple et al., 2015). Gingivitis, an inflammatory condition, demands early management through oral hygiene and professional cleaning. The use of advanced tools like power toothbrushes aids in plaque reduction, contributing to early intervention and care (Rathee and Jain, 2023). Multiple processes drive the shift from a healthy microbial community to a dysbiotic state. Changes in the host's immune system or diet impact the community's composition, increasing the production of harmful elements. As the community evolves, the interplay between microbial activities and the host's immune response alters the local environment, favoring the growth of dysbiotic microorganisms. A healthy microbiota tends to be diverse, while disease-associated ones are influenced by specialized microorganisms with increased virulence. Once a community reaches a dysbiotic state, specialized components maintain the condition, leading to chronic oral diseases like periodontitis and dental caries, although acute episodes can occur under specific conditions compromising the host's health (Figure 17.2) (Lamont et al., 2018).

The interplay between oral health and systemic well-being is significant. Periodontitis, a prevalent chronic inflammatory disease, is interconnected with various non-communicable diseases. Collaborative efforts between physicians and dentists are essential for early detection and holistic care (Dörfer et al., 2017). Despite their substantial impact on overall health and socioeconomic

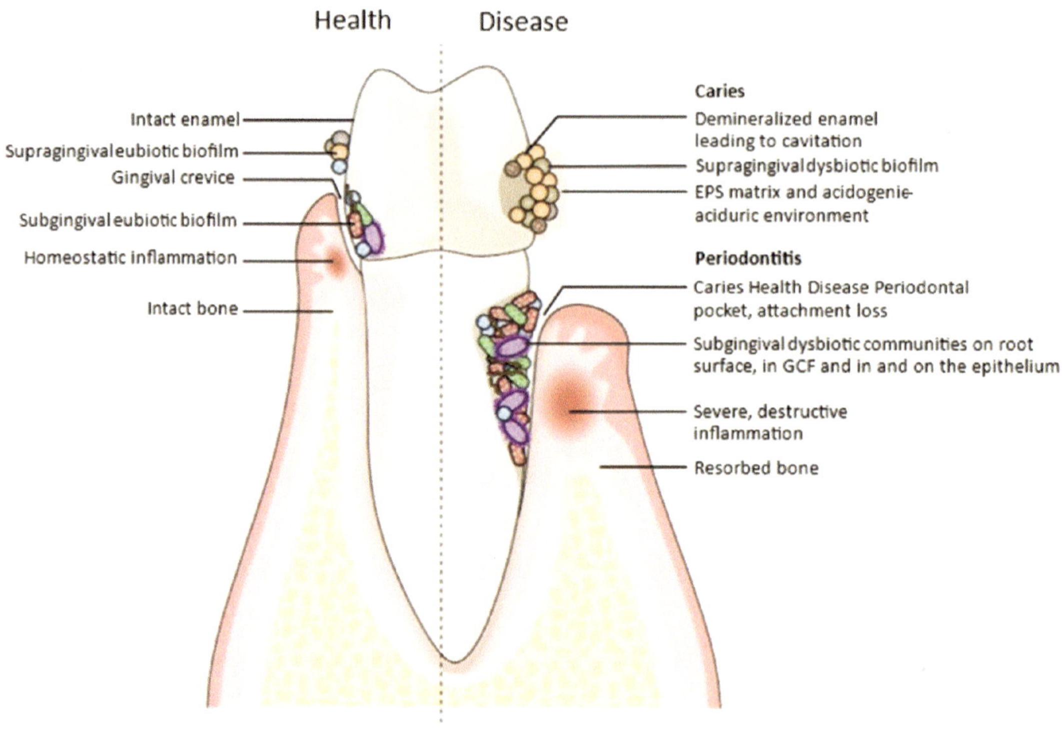

FIGURE 17.2 The oral microbiota's biogeography involves colonization in various areas within the oral cavity. *Abbreviations*: EPS, extracellular polymeric substance; GCF, gingival crevicular fluid. (From Lamont et al., 2018.)

consequences, oral diseases often receive inadequate attention in public health policies. However, understanding their shared risk factors and interconnected inflammatory pathways with major diseases highlights their importance in public health (Jin et al., 2016). Understanding the link between oral inflammation and systemic health is crucial. It highlights the interplay between chronic inflammatory conditions, like periodontal diseases, and overall well-being. Effective management involves comprehensive diagnostics, treatments including scaling and professional plaque control, and recognizing the role of the oral microbiome in diseases. Collaboration between dental and medical professionals is essential for early detection and holistic care. Recognizing shared risk factors between oral and non-communicable diseases is pivotal for comprehensive preventive strategies, addressing the global burden of oral diseases and promoting overall health.

17.10.2 Immune Response in Oral Health

Inside the human mouth and throat, there is a busy place filled with lots of tiny living things that we eat. But in this busy place, there is a strong protector called the mucosal immune system. It is super important because it keeps our mouth safe, helped by special cells like T lymphocytes that protect and help us tolerate things. Understanding how these cells work is important for keeping our mouths healthy (Wu et al., 2014). When microbial triggers set off inflammatory processes within the mouth, the body responds in a cascading manner. The fate of these responses hinges upon the behavior of key cells like macrophages, determining whether issues resolve or become chronic. Beyond their role in defense, innate immune cells such as neutrophils and monocytes/macrophages play a broader role in ensuring tissue health, repair, and regeneration. They hold the keys to maintaining tissue equilibrium and fostering healing, an aspect critical to understanding oral health, both in sickness and in health (Hasturk et al., 2012). The oral cavity functions like a battleground, coordinating a range of defense mechanisms to fend off infections and diseases. Local and systemic immunities join forces here, weaving a protective shield against harm. Within the oral arsenal, saliva holds elements that guard not only the mouth but also extend their protective prowess to other mucosal areas when triggered showing a shared immunity between the gut, mouth, and even the vagina. The unique landscape of teeth, with their specialized epithelia, presents a distinct challenge within the oral realm. Issues such as cavities and gum diseases dominate in this area. However, the crevicular fluid surrounding teeth acts as a gateway to understanding systemic immunity, aiding in understanding the complexities of oral health. Reflecting a microcosm of various diseases affecting different mucosal areas and autoimmune conditions, the oral cavity responds uniquely to infections caused by viruses, bacteria, or fungi. These responses, whether unique or similar, form the crux of immunological reactions within the oral environment (Challacombe et al., 2015). Biofilms, housing friendly bacteria, silently contribute to dental problems within the oral cavity. The studies acknowledge that diseases like cavities and gum infections find their origins in the thriving biofilms. Exploring the complexities of these microbial communities within biofilms reveals a different aspect of our immune responses, distinct from those observed in laboratory cultures (Lang et al., 2010). Unlocking the potential of oral fluids, such as gingival crevicular fluid, emerges as a promising diagnostic tool. Laden with biomarkers reflecting local and systemic conditions, these fluids aid in pinpointing periodontal diseases. Advanced analytical methods, including metabolomics, promise insights into the transition from health to inflammation and disease progression. These innovative approaches underscore the interplay between microbial communities and the body's responses, potentially revolutionizing the landscape of diagnostics and treatments for oral health challenges (Barros et al., 2015). In the story of oral health, how germs interact with our body's defenses is an important conversation. It helps us learn a lot about how our mouths stay healthy. This ongoing story shows how we stay strong and change to fight oral problems, opening up new ways to make oral health care better.

17.10.3 Impact on Longevity

The mouth isn't just performing the functions of eating and talking; it is essential for our overall health. How we eat, breathe, and swallow can affect how long one lives. Studies show that when our ability to chew declines, our health suffers, often linked to not getting enough nutrients. While having more teeth is vital, it's impossible to go back to having lost teeth. But improving other aspects of oral health, like how well we chew, eat, and keep our mouth healthy, can still make a big difference (Watanabe et al., 2020). With global life expectancy rising and an increasing elderly population, promoting healthy aging is of prime concern. Comprehensive policies aim to support this across various levels. Recognizing oral health as integral to overall well-being is a crucial part of combating non-communicable diseases outlined in the World Health Organization's (WHO's) sustainable development goals. As people age, the risk of oral issues and other diseases rises significantly. Oral disorders alone accounted for 8.9 million years of life with disability in those over 60 by 2019. Encouragement for research into aging-related physical and cognitive decline and its impact on oral health is vital for understanding these processes (Colombo and Wu, 2023). The risk of dental issues like cavities and gum diseases increases as we get older, especially as more people retain their teeth for longer. It's essential to take action to prevent these problems. Proper care can help us keep the natural teeth longer, leading to better overall health, a happier life, and the ability to eat well, which is crucial as we age. Taking care of oral health in older age needs everyone's efforts (Tonetti et al., 2017). Human mouths are more than just for eating and speaking, they're essential for our overall health and longevity. As we age, oral health becomes increasingly crucial, impacting our ability to eat well and stay healthy. Recognizing this, global efforts are focusing on promoting healthy aging, with oral health playing a key role. Preventing dental issues, especially as people keep their teeth longer, is vital for better overall health and a happier life as we grow older. It's not just about our mouths; it's about our well-being as a whole, requiring everyone's attention and care.

Oral health, including the composition of the oral microbiome, may have implications for overall health and longevity. There is increasing recognition of the connection between oral health and systemic health. Poor oral health has been associated with various systemic conditions, including cardiovascular diseases, diabetes, and inflammatory diseases. Chronic inflammation, often linked to oral health issues, is considered a potential contributor to the aging process. The oral microbiome, when imbalanced, can contribute to inflammation in the oral cavity. Chronic inflammation is implicated in various age-related diseases and may impact overall health and longevity. The gut-oral axis, which represents the bidirectional relationship between the gut and oral microbiomes, is gaining attention. The balance of these microbial communities may influence systemic health and, indirectly, longevity. The relationship between lifestyle factors and oral health is well-established, as various aspects of one's lifestyle can significantly impact the health of the oral cavity.

17.11 ROLE OF DIET, HYGIENE PRACTICE, AND OTHER LIFESTYLE FACTORS IN SHAPING THE ORAL MICROBIOME

The oral microbiome, a complex ecosystem of microorganisms residing in the oral cavity, is shaped by various lifestyle factors, including diet, oral hygiene practices, and other behaviors. Understanding the interplay between these factors and the oral microbiome is crucial for promoting oral health. Here, we explore the role of diet, hygiene practices, and other lifestyle factors in shaping the oral microbiome.

17.11.1 Diet and the Oral Microbiome

The term "diet" typically encompasses all the food consumed by an individual, implying that it provides essential nutrients crucial for the body. Nutrition plays a pivotal role in fulfilling the fundamental requirements for overall well-being, exerting a substantial impact on both general health and oral health. Essential nutrients are chemical substances in food that our body cannot synthesize adequately but are vital for life, growth, and tissue repair. Remarkably, maintaining optimal levels of essential nutrients, especially before undergoing surgery, has been observed to significantly enhance treatment effectiveness. Consequently, diet and nutrition emerge as indispensable components in patient care. Various dietary nutrients actively contribute to the preservation of oral health. Poor dietary habits, particularly high sugar intake and a lack of essential nutrients, contribute to dental caries and gum diseases. Diet influences the composition and metabolic activities of the oral microbiome. High sugar intake, for example, can promote the growth of acid-producing bacteria, contributing to dental caries (Shetty et al., 2023).

17.11.2 Oral Hygiene Practices and the Oral Microbiome

Inadequate oral hygiene practices, such as infrequent tooth brushing and poor flossing habits, contribute to the accumulation of plaque and increase the risk of dental diseases. Irregular toothbrushing heightened the likelihood of severe periodontal disease. Various mouth rinse formulations (such as chlorhexidine, essential oils, cetylpyridinium and delpominol, triclosan) as supplements to toothbrushing in managing gingivitis are being assessed. Proper oral hygiene practices, including regular brushing and flossing, help maintain a balance in the oral microbiome by preventing the accumulation of plaque and pathogenic bacteria (Welti et al., 2023).

17.11.3 Smoking and Tobacco Use

Smoking and tobacco use are linked to various oral health issues, including gum disease, tooth decay, dental caries, and an increased risk of oral cancers. Tobacco use alters the oral microbiome, increasing the prevalence of pathogenic bacteria and contributing to conditions such as periodontal diseases and oral cancers. Tobacco smokers exhibited elevated plaque index, decreased bleeding on probing, and deeper probing depths in comparison to non-smokers. Similar impacts on periodontal health are observed in individuals using e-cigarettes. Current tobacco smokers faced a twofold risk of heightened tooth loss compared to never-smokers, along with increased odds of oral squamous cell carcinoma. Additionally, there seems to be an association between smoking and an elevated risk of dental caries (Welti et al., 2023).

17.11.4 Stress

Chronic stress may contribute to oral health issues, including bruxism (teeth grinding) and an increased susceptibility to periodontal diseases. Chronic stress can impact the oral microbiome, potentially leading to conditions like bruxism (teeth grinding) and an altered microbial balance.

17.11.5 Other Lifestyle Factors

Excessive alcohol consumption is associated with an increased risk of oral cancers and can contribute to periodontal diseases. Alcohol consumption and systemic diseases, can influence the oral microbiome, impacting oral health. Alcohol consumption was linked to an increased likelihood of oral squamous cell carcinoma and was associated with periodontal disease and tooth loss. The study found that individuals aged over 60 who consumed alcohol were particularly associated with an increased risk of periodontal disease and tooth loss (Welti et al., 2023).

17.12 PRACTICAL TIPS FOR MAINTAINING A HEALTHY ORAL MICROBIOME

Certainly, nurturing a robust oral microbiome is crucial for maintaining optimal oral health. Primary prevention initiatives should proactively identify and forestall the onset of oral problems even before the initial manifestation of any signs of the disease (Krol and Whelan, 2023). Actionable suggestions to foster a harmonious and flourishing oral microbiome are given here.

17.12.1 Adopt a Balanced Diet

Meals act as nutritional sources for oral bacteria, and the nutrients ingested play a role in shaping the types of bacteria that can thrive in the oral cavity. Oral bacteria have the potential to impact the dietary preferences of their human hosts by adjusting thresholds for sweet, sour, salty, and bitter tastes. A balanced diet is as important to oral health as it is to overall health. One must consume a diet rich in fruits, vegetables, whole grains, and lean proteins while limiting intake of sugary snacks and beverages (Moh. Khafid et al., 2023; Nagakubo and Kaibori, 2023).

17.12.2 Practice Good Oral Hygiene

The significance of maintaining excellent oral hygiene resides in managing the levels and activity of bacteria that can cause diseases in the oral cavity, as well as in facilitating the delivery of fluoride to the tooth's surface. Supplementing toothbrushing with flossing has shown the potential to decrease gingivitis at 1, 3, or 6 months. Furthermore, the use of interdental brushes has demonstrated a more significant reduction in gingivitis compared to floss at 1 and 3 months. Uphold optimal oral hygiene practices, such as consistent brushing, flossing, and establishing a connection with a personal dental care provider as suggested by dentist (Krol and Whelan, 2023; Welti et al., 2023).

17.12.3 Limit Sugar Intake

Limiting the consumption of free sugars to less than 5% of total energy intake has been proven to lower the incidence of dental caries. Additionally, there is an indication that the consumption of food and drinks close to bedtime in children might be associated with an increased risk of dental caries. The consumption of sugars significantly contributes to the risk of dental caries. Pediatricians can seamlessly integrate anticipatory guidance related to dental caries prevention into discussions about dietary habits and nutritional intake during consultations. Reducing consumption of sugary foods and beverages will help prevent the growth of acid-producing bacteria that contribute to tooth decay (Krol and Whelan, 2023; Welti et al., 2023).

17.12.4 Stay Hydrated

Drink plenty of water throughout the day, as it helps to rinse away food particles and maintains saliva production.

17.12.5 Avoid Tobacco Products

There is a connection between tobacco smoking and a heightened risk of dental caries, as well as a correlation between tobacco smoking and an increased risk of oral squamous cell carcinoma. Quit smoking and avoid the use of tobacco products to reduce the risk of periodontal diseases and oral cancers (Welti et al., 2023).

17.12.6 Manage Stress

Practice stress-management techniques such as deep breathing, meditation, or exercise to reduce stress-related impacts on the oral microbiome. Physical exercise has the potential to enhance microbiome diversity, elevate the Bacteroidetes/Firmicutes ratio, potentially aiding in weight reduction, mitigating obesity-related diseases, and addressing gut disorders. Additionally, exercise can promote the growth of bacteria that contribute to the regulation of mucosal immunity and enhancement of barrier functions (Elzayat et al., 2023).

17.12.7 Use of Xylitol and Sorbitol Gums

Chewing gum containing xylitol and sorbitol has been shown to offer protection against dental caries in both children and adults. However, there is insufficient evidence to endorse the use of xylitol gum specifically for preventing gingival bleeding (Welti et al., 2023). High doses of xylitol can cause digestive issues like bloating or diarrhea because it is a sugar alcohol that ferments in the gut.

17.12.8 Use of Mouthguard

The utilization of mouthguards during contact sports has been demonstrated to effectively prevent traumatic dental injuries (Welti et al., 2023). By adopting these practical tips, we can promote a healthy oral microbiome and contribute to overall oral health. One must remember to visit a dentist regularly for checkups and professional cleanings to ensure comprehensive oral care.

17.13 ORAL MICROBIOME AND SYSTEMIC HEALTH

The oral cavity serves as the entry point to the gastrointestinal tract and the microorganisms residing within it play pivotal roles in shaping the balance between health and disease, not only locally but also systemically. Within the oral cavity, an array of microorganisms, including bacteria, archaea, fungi, protozoa, and viruses, coexist, with each having distinct roles. These microorganisms engage in intricate interactions with one another and with the host, influencing health outcomes. Fungi interact with bacteria, exogenous or indigenous viruses interact with bacteria and the host, and protozoa prey upon microbes (Nobbs et al., 2011). These diverse taxonomic groups have coevolved, contributing to the formation of the modern human oral ecosystem.

Genetic analysis of microbial communities within oral and other samples, without relying on cultivation, can be conducted through next-generation sequencing (NGS) or community metagenomics. Originally, the term "metagenomics" referred to the comprehensive analysis of functional microbial genomes within samples. However, it has since been adopted in studies focusing on the amplification of specific genes of interest (Oulas et al., 2015). NGS enables the characterization of microbial communities by addressing questions about the composition of the sample, identifying organism lineages (such as taxonomic family and genera), conducting functional composition analysis, and exploring the interactions among microbes within the oral cavity (Dewhirst et al., 2010; Duran-Pinedo and Frias-Lopez, 2015; Oulas et al., 2015). Currently, NGS technology has inherent limitations in accurately describing lower taxonomic levels, typically providing results at higher taxonomic ranks such as the phylum, class, and family levels. In some instances, it may achieve genus-level resolution, and very rarely, species-level identification. However, there are scenarios, both in the context of human and environmental samples, where information at low taxonomic levels proves to be crucial (Sampaio-Maia et al., 2016; Zaura et al., 2014).

However, the volume of information generated by NGS technology is vast, and frequently, many of the reported sequences do not align with existing information in databases. To address this, the operational taxonomic unit (OTU) was established for the definition and classification of closely related sequences obtained through the amplification of specific genes, such as 16S for profiling bacterial communities or an internal transcribed spacer for profiling fungal communities. OTUs are often derived by clustering sequences based on their similarity. Metagenomics has proven valuable in studying and elucidating specific metabolic pathways during oral diseases, as well as assessing population changes associated with infections in the oral cavity (Duran-Pinedo and Frias-Lopez, 2015).

The examination of supragingival and subgingival calculus samples from the teeth of prehistoric European human skeletons, spanning from before the Mesolithic period (prior to farming) to the Medieval period, reveals that the oral cavity has long functioned as a reservoir for microbes. The shift from a hunter-gatherer to farming lifestyle altered the oral microbial community, resulting in a less diverse microbiota. This transformation may contribute to chronic oral and systemic diseases observed in post-industrial lifestyles (Adler et al., 2017). Presently, we recognize that oral microbiomes offer valuable insights into personal health status and are associated with nutritional deficiencies as well as various chronic and acute diseases. In this context, we offer an updated overview of the "normal" oral microbiome, delineating the microbial communities that characterize oral health. We explore the alterations that occur in oral diseases, specifically in dental caries, endodontic infections, and periodontal disease. Additionally, we provide insights into how the oral microbiome may contribute to oral cancer. Furthermore, we delve into the role of the oral microbiome in systemic diseases, encompassing infections, diabetes, cardiovascular disease, and adverse pregnancy outcomes (APOs).

17.13.1 Cardiovascular Diseases

Oral pathogens may also play important roles in the development of non-infectious systemic diseases. Numerous studies have indicated an increased risk of cardiovascular disease in individuals with periodontal disease (Badewy et al., 2022; Kholy et al., 2015). Currently, the proposed pathophysiological mechanisms linking the oral microbiome and cardiovascular disease revolve around the "colonization" of atherosclerotic plaques by oral microorganisms and the induction of systemic inflammation due to oral infections (Kholy et al., 2015).

Various studies have identified oral microorganisms in atherosclerotic plaques, including *Streptococcus* spp., *Veillonella* spp., *P. gingivalis, Aggregatibacter actinomycetemcomitans, T. denticola, Fusobacterium nucleatum, Tannerella forsythia*, and *Neisseria* spp. (Chiu, 1999; Libby et al., 2002; Ford et al., 2010; Pucar et al., 1991; Koren et al., 2011; Figuero et al., 2013; Ohki and Kikuchi, 2017). A recent study utilizing 16S rDNA sequencing identified 84 different bacterial taxa in atheromas and vascular endothelia of patients with atherosclerotic vascular disease and periodontal disease, compared to only 18 different taxa in patients with little or no periodontal disease (Armingohar et al., 2014). This underscores the proposition of the "colonization" of atherosclerotic plaques with oral microorganisms.

Hence, the "colonization" of atherosclerotic plaques appears to play a role in plaque inflammation and instability, contributing to endothelial dysfunction and, consequently, cardiovascular disease (Luo et al., 2015). Moreover, there is evidence supporting two modes of invasion of cardiovascular tissues by periodontal pathogens, namely, bacteremia and phagocyte-mediated bacterial translocation from the periodontal lesion to the vascular tissues (Carrión et al., 2012; Kinane et al., 2005).

Systemic inflammation resulting from oral infections is another proposed pathophysiological mechanism contributing to cardiovascular disease. Oral infections, particularly periodontal diseases, are believed to indirectly stimulate increased production of inflammatory mediators in the systemic circulation, thereby accelerating atherosclerosis (Slocum et al., 2016). Supporting this correlation, a recent study demonstrated that both cardiovascular and periodontal inflammations were reduced after inflammation was lowered in patients receiving statin therapy (Subramanian et al., 2013).

17.13.2 Diabetes

It is becoming increasingly evident that oral microbiome dysbiosis is associated with other systemic inflammatory diseases, including diabetes (Ohlrich et al., 2010). Although the biological mechanisms linking oral infections and diabetes are not fully understood, the epidemiological association between periodontitis and diabetes is becoming relatively clear. Notably, this relationship appears to be bidirectional: On one side, diabetes impacts the subgingival environment and, consequently, the subgingival microbiome, and on the other side, the host response to periodontal disease may play a role in the progression of diabetes (Ohlrich et al., 2010).

Compared to healthy controls, differences have been identified in the subgingival microbiome of patients with type 1 and 2 diabetes (Ebersole et al., 2008; Casarin et al., 2017). The elevated glucose content in the subgingival microenvironment and the compromised immune system of the host may account for these distinctions (Ficara et al., 1975; Ohlrich et al., 2010). Additionally, diabetes induces alterations in connective tissue metabolism, reducing the tissue's capacity to remodel and thus influencing the progression of periodontal disease (Willershausen-Zönnchen et al., 1991). Moreover, it has been suggested that the inflamed periodontium may function as an endocrine-like source of inflammatory mediators, potentially contributing to increased insulin resistance (Preshaw et al., 2007).

17.13.3 Non-Oral Infectious Diseases

The oral microbiome can serve as a reservoir for opportunistic infections, not only within the oral cavity but also in distant body sites, particularly in immune-compromised hosts (Li et al., 2000). Given that oral microorganisms can traverse oral mucous membranes and periodontal pockets to enter the bloodstream, they can rapidly reach the heart, lungs, and peripheral blood capillary system within minutes. This dissemination of oral microorganisms to various body locations constitutes the initial step toward the potential development of systemic or focal infections.

Naturally, this bacteremia implies the overcoming of several protective barriers within the oral system to facilitate microbial penetration. These barriers include the surface epithelium physical barrier, host-derived peptide antibiotics known as defensins, an electrical barrier reflecting the oxidation-reduction potential (Eh) difference between the host cell and the microbial layer, an immunological barrier involving antibody-forming cells, and the reticuloendothelial system (phagocyte barrier) (Li et al., 2000). Under normal conditions, only a small number of microorganisms reach the bloodstream. However, in instances of oral mucosal trauma, such as during invasive dental procedures or in immunologically compromised hosts, oral microorganisms can easily disseminate to distant locations within the human body. It is important to note that even toothbrushing alone has the potential to disrupt a substantial surface area of gingival crevicular tissue and promote bacteremia. Lockhart et al. (2008) assessed this process and demonstrated the persistence of pathogenic species for at least 60 minutes after toothbrushing. Additionally, poor oral hygiene significantly elevates the number of oral bacteria, consequently increasing the prevalence and magnitude of bacteremia (Li et al., 2000). The notion that bacteremia originating from the oral cavity could lead to infective endocarditis was proposed over 100 years ago (Lockhart et al., 2012). While invasive dental procedures, such as dental extractions, were initially considered significant contributors to the development of infective endocarditis related to oral bacteria, recent data suggest that poor oral hygiene and periodontal diseases are more likely to be responsible for the majority of cases of oral bacteria-related infective endocarditis (Lockhart et al., 2011).

The oral cavity has been recognized as a potential reservoir for respiratory pathogens (Mojon, 2002). Aspiration of oral bacteria into the respiratory tract is a daily occurrence; however, it may become particularly critical in the elderly, immunocompromised individuals, or those on artificial ventilation. For instance, toothbrushing has been shown to reduce the incidence of pneumonia and mortality rates in the elderly residing in nursing homes (Yoneyama et al., 2002). Additionally, maintaining oral hygiene has been suggested as an important preventive measure for ventilator-associated pneumonia (Ames, 2011). Supporting these observations are studies that have detected oral bacteria using molecular methods in intraoperative bronchial fluids of patients with pulmonary carcinoma (Hasegawa et al., 2014), in the lungs of cystic fibrosis patients (Rogers et al., 2006), and even in healthy individuals (Bassis and Ma, 2015).

Numerous reports in scientific literature suggest an odontogenic etiology for non-oral infections (Corson et al., 2001; Li et al., 1999; Mills et al., 2008). However, most of these reports do not include supportive genotypic analyses between the infection and oral isolates. Da Silva et al. (2004) utilized both phenotypic and three genetic fingerprinting techniques to strongly indicate that the oral cavity was the origin of the *S. constellatus* responsible for a brain abscess. This study provided a genetic link between the oral microbiome and the etiological agents responsible for non-oral infections.

17.14 CURRENT INTERVENTIONS FOR PROMOTING ORAL HEALTH AND LONGEVITY

Maintaining a healthy mouth and strong teeth is essential for a lifetime of good oral health. Here are some tips to help you achieve this:

1. *Opt for fluoridated water and fluoride toothpaste*: Ensure that you drink fluoridated water and use fluoride toothpaste when brushing your teeth.
2. *Embrace good oral hygiene habits*: Practice thorough oral hygiene by brushing your teeth twice a day and flossing daily to remove dental plaque.
3. *Regular dental checkups*: Visit your dentist at least once a year, even if you have dentures or no natural teeth. Regular checkups are crucial for maintaining oral health.
4. *Say "no" to tobacco*: Avoid using tobacco products, and if you smoke, consider quitting for the sake of your oral and overall health.

5. *Moderate alcohol consumption*: Limit your intake of alcoholic drinks to promote better oral health.
6. *Manage diabetes effectively*: If you have diabetes, work on controlling the disease. This not only reduces the risk of complications but also helps prevent gum disease. Treating gum disease can contribute to better blood sugar control.
7. *Address dry mouth issues*: If your medication causes dry mouth, consult your doctor about alternative medications. If dry mouth persists, stay hydrated, chew sugarless gum, and avoid tobacco products and alcohol.
8. *Seek professional help for taste and smell changes*: If you experience sudden changes in taste and smell, consult your doctor or dentist for appropriate guidance.
9. *Assist older individuals with oral care*: When acting as a caregiver for older individuals, assist them with brushing and flossing if they are unable to do so independently.

By incorporating these habits into your routine, you can promote oral health and keep your teeth for a lifetime. The emergence of microbiome research has unequivocally established the microbiome as a pivotal contributor to human health. The microbiome encompasses the intricate assembly of microorganisms residing both within and on the human body, incorporating their genomes and collective functionalities.

The field has experienced profound advancements through the genomic revolution, enabling the DNA-based identification of non-culturable bacteria inhabiting diverse body sites. This breakthrough has provided valuable insights into microbial communities and their potential implications for human health. Alterations in these communities, often characterized as dysbiosis, have demonstrated associations with a spectrum of diseases. From infectious conditions like *Clostridium difficile* infection to inflammatory disorders such as inflammatory bowel disease and rheumatoid arthritis, and extending to metabolic diseases like diabetes and obesity, microbial dysbiosis has been identified as a significant player in the pathogenesis of multifactorial health conditions (Jameson JL and Longo DL, 2015).

This paradigm shift in understanding emphasizes the critical role of the microbiome and its potential as a target for precision therapeutic interventions. By unraveling the intricate relationships between the microbiome and human health, precision medicine can harness this knowledge to develop tailored approaches that consider the individualized microbial landscape for improved health outcomes. A crucial characteristic of the microbiome lies in its remarkable resilience and plasticity, rendering it more adaptable than human cells. Despite the apparent contradiction between these two concepts, the microbiome's resilience becomes evident in instances of health. Even when subjected to transient disturbances such as travel, changes in diet, or antibiotic use, the microbiome exhibits the ability to sustain a relatively stable and consistent state (Marchesi JR and Ravel J, 2015).

This resilience underscores the microbiome's capacity to withstand external challenges and return to a balanced state conducive to overall health. Simultaneously, the microbiome functions as a malleable organ, susceptible to modification through targeted interventions, particularly dietary adjustments and other directed therapies (Figure 17.3). This duality emphasizes the dynamic nature of the microbiome adapting to disturbances while remaining open to intentional alterations based on external influences (Marchesi JR et al., 2016).

Understanding the interplay between the resilience and plasticity of the microbiome is essential for developing strategies that leverage its inherent flexibility for therapeutic purposes. This knowledge opens avenues for interventions aimed at promoting a resilient and balanced microbiome, thereby contributing to the overall well-being of the host. Moreover, the inter-individual variability in both composition and metabolic capacity of the microbiome plays a pivotal role in shaping interactions with the environment, influencing the development of diseases, as well as dictating responses to treatment and the occurrence of adverse events. While the microbiome is influenced to some extent by the host genome, the magnitude of this contribution appears relatively

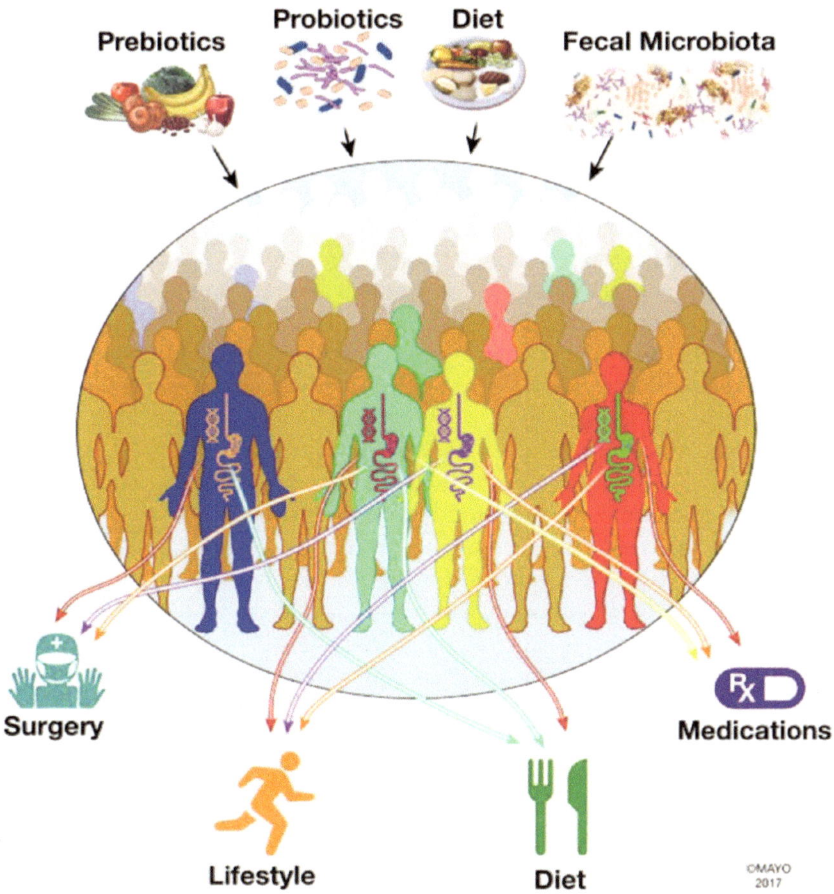

FIGURE 17.3 Gut microbiome and human health. (Kashyap et al., 2017.)

minor in comparison to the substantial impact of environmental factors on microbiome modulation. Consequently, the critical role of the microbiome in human health, coupled with its inter-individual variability and its significant contribution to host function in health, underscores its adaptability as a targetable factor.

Notably, the microbiome's plasticity, rendering it amenable to deliberate modifications, further emphasizes its potential as a focal point in precision medicine initiatives (Figure 17.3). Incorporating the microbiome into the framework of precision medicine becomes imperative, given its dynamic nature and its ability to influence health outcomes. By recognizing and harnessing the diverse and individualized aspects of the microbiome, precision medicine can strive to optimize therapeutic approaches tailored to the unique microbial landscape of each individual, thereby advancing the precision and efficacy of medical interventions.

Distinguished from host genes, the gut microbiome stands out as a malleable factor that can be intentionally altered and addressed through interventions like probiotics, prebiotics, dietary adjustments, and innovative strategies such as fecal microbiota transplantation. Recognizing the dynamic nature of the gut microbiome opens avenues for therapeutic approaches that extend beyond conventional methods, offering a spectrum of interventions that can be tailored to an individual's unique microbial profile. This holistic consideration underscores the importance of embracing a personalized and microbiome-informed approach to optimize health outcomes.

17.15 MICROBIOME SEQUENCING

The transformative breakthrough in our capacity to comprehend the significance of the microbiome coincided with the advent of next-generation sequencing. This technology has enabled an in-depth characterization of the gut microbiota through multi-omics approaches, eliminating the necessity for culturing individual microbes, a task that can be notably challenging in certain instances. A widely embraced method for characterizing microbial communities involves the use of the marker gene approach, specifically targeting the 16S rRNA gene. This gene is highly conserved among bacteria, with minimal evidence of horizontal gene transfer.

While the marker gene approach is popular, it comes with limitations, particularly in terms of species and strain-level resolution. Addressing this limitation necessitates the utilization of metagenomic sequencing and the de novo assembly of genomes. This advanced approach not only offers improved compositional insights into the microbiome but also enhances functional resolution. By employing metagenomic sequencing, researchers can delve deeper into the intricate details of microbial communities, discerning not only their composition but also their genetic makeup and functional capabilities. This comprehensive methodology provides a more nuanced understanding of the microbiome at both the taxonomic and functional levels (Marchesi JR and Ravel J, 2015). Metatranscriptomics serves as a valuable complement to metagenomics, enabling the identification of microbial genes expressed under varying conditions. Meanwhile, metabolomics and metaproteomics contribute to the identification of metabolites and proteins resulting from microbe-host co-metabolism. These end products serve as reliable biomarkers, representing the culmination of metabolic interactions between microbes and the host. The integration of these multi-omic technologies not only enhances confidence in the identification of diagnostic and therapeutic biomarkers but also generates testable hypotheses.

To experimentally validate and explore emergent hypotheses arising from these technologies, as well as to elucidate the mechanisms through which microbes influence the host, germ-free animal models offer a highly controllable experimental system. These models reduce complexity and provide a controlled environment to study interactions between the host and its resident microbiota. Leveraging germ-free animal models allows researchers to isolate and manipulate variables, facilitating a more precise understanding of the intricate dynamics between the host and its microbiota. This experimental approach enhances the robustness of findings and contributes significantly to advancing our comprehension of the impact of the microbiome on host physiology and health.

17.16 MICROBIOME AS A DETERMINANT OF HUMAN THERAPEUTICS

The ecology of a microbial population, akin to any ecosystem, is characterized by intricate interactions among various species. Microbial survival and growth are profoundly influenced by their chemical environment, and it comes as no surprise that these microorganisms have evolved gene cassettes for engaging in chemical warfare (Shank and Kolter, 2009; Cornforth and Foster, 2015). Certainly, the initial discovery of antibiotics occurred within microbial cultures, showcasing a distinctive trait of microbial colonies (Jones et al., 1944). Since that milestone, extensive surveys of the soil microbiota have unveiled an even more extensive array of antibiotic compounds (Aigle et al., 2014; Bachmann et al., 2014). Moreover, the exploration has been extended to the human microbiome, spanning multiple sites across the human body. This implies that the source of compounds essential for gaining control over our microbiome might already exist within us (Donia et al., 2014).

In addition to antibiotics and signaling agents, the identification of the bacterial immune system, known as the CRISPR-Cas system, provides bacteria with the capability to resist and exclude bacteriophages from the population. This is achieved by specifically targeting and cleaving particular sequences (Barrangou et al., 2007). In providing adaptive immunity against viruses, the industrial applications of this biological system have garnered widespread recognition, leading to the creation of synthetic CRISPR-Cas systems (Cong et al., 2013). These advancements have paved the way for the implementation of species-specific antimicrobial agents. This innovative approach holds the promise of preserving the majority of the microbiome while enabling crucial alterations (Bikard et al., 2014).

In addition to being a source of therapeutics with implications for human disease, the microbiome serves both as a modulator of traditional therapies and as a target for therapies. The variability in individual responses to therapy and the development of adverse events are often attributed to individual-specific disease phenotypes and host genetics, with the gut microbiota frequently overlooked. However, the gut microbiota plays a crucial role in drug transformation, impacting their efficacy.

For instance, acetaminophen, a commonly used analgesic, may engage in competition with bacteria-generated *p*-cresol for *O*-sulfonation, resulting in acetaminophen glucuronidation. This interaction can, in part, explain the inter-individual variability in analgesic response. Moreover, it sheds light on differences in adverse events due to the accumulation of its toxic metabolite *N*-acetyl-*p*-benzoquinone imine (NAPQI). This underscores the significance of considering the gut microbiota as a determinant in therapeutic outcomes and adverse reactions (Clayton et al., 2009). Microbiome markers of drug efficacy, spanning from chemotherapeutic agents to statins, have been extensively documented. For instance, *Bifidobacterium* has been identified as a contributor to enhanced tumor control in mouse models of melanoma treated with anti-programmed death-ligand 1 (PD-L1). Similarly, in humans, *Bacteroides* has been implicated in the antitumor effects observed with cytotoxic T-lymphocyte associated protein 4 (CTLA-4) blockade, a common approach in cancer immunotherapy.

In the case of irinotecan (CPT-11), a frequently used chemotherapeutic agent for colorectal cancer, gut bacteria can catalyze beta-glucuronidation, leading to the formation of an active metabolite associated with severe diarrhea. It is crucial to acknowledge that host genetic variation also exerts a substantial influence on shaping the composition of the gut microbiome. This interplay between host genetics and the microbiome underscores the multifaceted factors contributing to drug responses and potential side effects (Blekhman et al., 2015). Levels of bacteria-derived coprostanol have been linked to the clinical response to statins, commonly employed as low-density lipoprotein (LDL) cholesterol-lowering agents. Notably, digoxin, a cardiac glycoside with a narrow therapeutic window, can undergo inactivation by the gut bacterium *Eggerthella lenta*. Lastly, a recent study underscores the significance of the gut microbiota in mediating the antidiabetic effects of metformin (Wu et al., 2017). These examples unequivocally underscore the significance of taking into account the gut microbiota when assessing drug responses, akin to the principles of pharmacogenomics (Figure 17.3). Integrating both approaches holds the potential to enhance the precision and effectiveness of therapeutics while concurrently reducing overall adverse events.

17.17 CONCLUSION

Growing evidence highlights the critical role of the oral microbiome in maintaining the balance between health and disease, influencing both oral and systemic conditions. As research in this field rapidly advances, more factors contributing to the onset, progression, and mechanisms of oral diseases are expected to emerge. This review focuses on the current understanding of the oral microbiome in health and its broader implications for both oral and systemic diseases. The oral cavity hosts a diverse array of microorganisms, including bacteria, archaea, fungi, protozoa, and viruses. These microorganisms engage in complex interactions with one another and the host, playing pivotal roles in both health and disease. Interactions among microorganisms in the oral cavity can begin as early as the intrauterine environment. The composition of the oral microbiome is dynamic, evolving throughout an individual's life. It undergoes changes during childhood development, in response to environmental factors, and in association with oral and systemic diseases. Maintaining a healthy oral microbiome is essential for overall oral health. Practical steps such as regular dental check-ups and professional cleanings are key to ensuring comprehensive oral care and promoting a balanced oral microbiome.

Lifestyle factors play a critical role in determining oral health outcomes. Adopting a healthy lifestyle, including a balanced diet, tobacco cessation, moderate alcohol consumption, proper oral hygiene practices, and stress management can significantly contribute to maintaining optimal oral health. The oral microbiome is a dynamic and complex community influenced by various lifestyle factors.

REFERENCES

Adler, C. J.; Browne, G. V.; Sukumar, S.; Hughes, T. Evolution of the oral microbiome and dental caries. *Current Oral Health Reports*, **2017**, *4*, 264–269.

Aigle, B.; Lautru, S.; Spiteller, D.; Dickschat, J. S.; Challis, G.L.; Leblond, P.; Pernodet, J.L. Genome mining of Streptomyces ambofaciens. *Journal of Industrial Microbiology & Biotechnology*, **2014**, 41(2), 251–263. doi: 10.1007/s10295-013-1379-y.

Alsafi, W.M.; Ashwaq Al Eed; Ahmed A. H.; Abdullah Al-N.; Jaber A. Ishag A. Prevalence of and factors associated with pre-diabetes among adolescents in Eastern Sudan: A community-based cross-sectional study. *BMJ Open*, **2024**, *14*(10), e086197.

Ames, N. J. Evidence to support tooth brushing in critically ill patients. *American Journal of Critical Care*, **2011**, 20(3), 242–250.

Ames, N.J.; Ranucci, A.; Moriyama, B.; Wallen, G.R. The human microbiome and understanding the 16S rRNA gene in translational nursing science. *Nursing Research*, **2017**, 66, 184–197.

Armingohar, Z.; Jørgensen, J. J.; Kristoffersen, A. K.; Abesha-Belay, E.; Olsen, I. Bacteria and bacterial DNA in atherosclerotic plaque and aneurysmal wall biopsies from patients with and without periodontitis. *Journal of Oral Microbiology*, **2014**, *6*(1), 23408.

Bachmann, B.O.; Van Lanen, S.G.; Baltz, R.H. Microbial genome mining for accelerated natural products discovery: Is a renaissance in the making? *Journal of Industrial Microbiology & Biotechnology*, **2014**, 41(2), 175–184.

Badewy, R.; Azarpazhooh, A.; Tenenbaum, H.; Connor, K. L.; Lai, J. Y.; Sgro, M.; Glogauer, M. The association between maternal oral inflammation and neutrophil phenotypes and poly-unsaturated fatty acids composition in human milk: A prospective cohort study. *Cells*, **2022**, *11*(24), 4110.

Baiju, R. Oral Health and Quality of Life: Current Concepts. *Journal of Clinical and Diagnostic Research*, **2017**, *11*(6), ZE21–ZE26.

Barrangou, R.; Fremaux, C.; Deveau, H.; Richards, M.; Boyaval, P.; Moineau, S.; Romero, D.A.; Horvath, P. CRISPR provides acquired resistance against viruses in prokaryotes. *Science*, **2007**, *315*(5819), 1709–1712. doi: 10.1126/science.1138140.

Barros, S. P.; Williams, R.; Offenbacher, S.; Morelli, T. Gingival crevicular fluid as a source of biomarkers for periodontitis. *Periodontology 2000*, **2015**, *70*(1), 53–64.

Bassis, J. N.; Ma, Y. Evolution of basal crevasses links ice shelf stability to ocean forcing. *Earth and Planetary Science Letters*, **2015**, 409, 203–211.

Bik, E.M. The hoops, hopes and hypes of human microbiome research. *The Yale Journal of Biology and Medicine*, **2016**, 89, 363–373.

Bikard, D.; Euler, C.W.; Jiang, W.; Nussenzweig, P.M.; Goldberg, G.W.; Duportet, X.; Fischetti, V.A.; Marraffini, L.A. Exploiting CRISPR-Cas nucleases to produce sequence-specific antimicrobials. *Nature Biotechnology*, **2014**, 32(11), 1146–1150. doi: 10.1038/nbt.3043.

Blekhman, R.; Goodrich, J. K.; Huang, K.; Sun, Q.; Bukowski, R.; Bell, J. T.; Spector, T. D.; Keinan, A.; Ley, R. E.; Gevers, D.; Clark, A. G. Host genetic variation impacts microbiome composition across human body sites. *Genome Biology*, **2015**, *16*(1), 191. doi: 10.1186/s13059-015-0759-1.

Carrión, S.; Cabré, M.; Monteis, R.; Roca, M.; Palomera, E.; Serra-Prat, M.; Clavé, P. Oropharyngeal dysphagia is a prevalent risk factor for malnutrition in a cohort of older patients admitted with an acute disease to a general hospital. *Clinical Nutrition*, **2015**, *34*(3), 436–442.

Casarin, M.; Antoniazzi, R. P.; Vaucher, R. A.; Feldens, C. A.; Zanatta, F. B. RT-PCR quantification of periodontal pathogens in crack users and non-users. *Oral Diseases*, **2017**, *23*(3), 324–330.

Cekici, A.; Kantarci, A.; Hasturk, H.; Van Dyke, T. E. Inflammatory and immune pathways in the pathogenesis of periodontal disease. *Periodontology 2000*, **2014**, *64*(1), 57–80.

Challacombe, S. J.; Shirlaw, P. J.; Thornhill, M. H. Immunology of diseases of the oral cavity. *Mucosal Immunology*, **2015**, 1943–1983.

Chapple, I. L. C.; Van der Weijden, F.; Doerfer, C.; Herrera, D.; Shapira, L.; Polak, D.; Madianos, P.; Louropoulou, A.; Machtei, E.; Donos, N.; Greenwell, H.; Van Winkelhoff, A. J.; Eren Kuru, B.; Arweiler, N.; Teughels, W.; Aimetti, M.; Molina, A.; Montero, E.; Graziani, F. Primary prevention of periodontitis: managing gingivitis. *Journal of Clinical Periodontology*, **2015**, 42(S16), S71–S76.

Chiu, B. Multiple infections in carotid atherosclerotic plaques. *American Heart Journal*, **1999**, *138*(5), S534–S536.

Clayton, T.A.; Baker, D.; Lindon, J.C.; Everett, J.R.; Nicholson, J.K.; Pharmacometabonomic identification of a significant host-microbiome metabolic interaction affecting human drug metabolism. *Proceedings of the National Academy of Sciences of the United State of America*, **2009**, *106*(34), 14728–14733.

Colombo, A. P.; Wu, B.; Aging and oral health: Biological and sociobehavioral perspectives. *Journal of Dental Research*, **2023**, 102(8), 841–843.

Cong L, Ran FA, Cox D, et al. Multiplex genome engineering using CRISPR/Cas systems. *Science*, **2013**, *339*(6121), 819–823.

Cornforth, D. M.; Foster, K. R. Antibiotics and the art of bacterial war. *Proceedings of the National Academy of Sciences of the United State of America*, **2015**, 112(35), 10827–10828.

Corson, M. A.; Postlethwaite, K. P.; Seymour, R. A. Are dental infections a cause of brain abscess? Case report and review of the literature. *Oral diseases*, **2001**, 7(1), 61–65.

da Silva, R. M.; Caugant, D. A.; Josefsen, R.; Tronstad, L.; Olsen, I. Characterization of Streptococcus constellatus strains recovered from a brain abscess and periodontal pockets in an immunocompromised patient. *Journal of Periodontology*, **2004**, 75(12), 1720–1723.

Deo, P. N.; Deshmukh, R. Oral microbiome: Unveiling the fundamentals. *Journal of Oral and Maxillofacial Pathology*, **2019**, *23*(1), 122–128. doi: 10.4103/jomfp.JOMFP_304_18.

Dewhirst, F. E.; Chen, T.; Izard, J.; Paster, B. J.; Tanner, A. C.; Yu, W. H.; Lakshmanan, A.; Wade, W. G. The human oral microbiome. *Journal of Bacteriology*, **2010**, *192*(19), 5002–5017.

Donia, M. S.; Cimermancic, P.; Schulze, C. J.; Brown, L. C. W.; Martin, J.; Mitreva, M.; Fischbach, M. A. A systematic analysis of biosynthetic gene clusters in the human microbiome reveals a common family of antibiotics. *Cell*, **2014**, *158*(6), 1402–1414.

Dörfer, C.; Benz, C.; Aida, J.; Campard, G. The relationship of oral health with general health and NCDs: A brief review. *International Dental Journal*, **2017**, 67, 14–18.

Duran-Pinedo, A. E.; Frias-Lopez, J. Beyond microbial community composition: Functional activities of the oral microbiome in health and disease. *Microbes and Infection*, **2015**, 17(7), 505–516. doi: 10.1016/j.micinf.2015.03.014.

Ebersole, J. L.; Holt, S. C.; Hansard, R.; Novak, M. J. Microbiologic and immunologic characteristics of periodontal disease in Hispanic Americans with type 2 diabetes. *Journal of Periodontology*, **2008**, *79*(4), 637–646.

El Kholy, K.; Genco, R. J.; Van Dyke, T. E. Oral infections and cardiovascular disease. *Trends in Endocrinology & Metabolism*, **2015**, *26*(6), 315–321.

Elzayat, H.; Mesto, G.; Al-Marzooq, F. Unraveling the impact of gut and oral microbiome on gut health in inflammatory bowel diseases. *Nutrients*, **2023**, *15*(15), 3377.

Ficara, A. J.; Levin, M. P.; Grower, M.; Kramer, G. D. A comparison of the glucose and protein content of gingival fluid from diabetics and nondiabetics. *Journal of Periodontal Research*, **1975**, *10*(3), 171–175.

Figuero, E.; Carrillo-de-Albornoz, A.; Martín, C.; Tobías, A.; Herrera, D. Effect of pregnancy on gingival inflammation in systemically healthy women: A systematic review. *Journal of Clinical Periodontology*, **2013**, *40*(5), 457–473.

Fiorillo, L. Oral health: The first step to well-being. *Medicina*, **2019**, *55*(10), 676.

Ford, P. J.; Raphael, S. L.; Cullinan, M. P.; Jenkins, A. J.; West, M. J.; Seymour, G. J. Why should a doctor be interested in oral disease? *Expert Review of Cardiovascular Therapy*, **2010**, *8*(10), 1483–1493.

Gao, L.; Xu, T.; Huang, G.; Jiang, S.; Gu, Y.; Chen, F. Oral microbiomes: more and more importance in oral cavity and whole body. *Protein & Cell*, **2018**, *9*(5), 488–500.

Gest H. The discovery of microorganisms by Robert Hooke and Antoni van Leeuwenhoek, Fellows of The Royal Society. *Notes and Records of the Royal Society of London*, **2004**, *58*, 187–201. doi: 10.1098/rsnr.2004.0055.

Gevers, D.; Knight, R.; Petrosino, J. F.; Huang, K.; McGuire, A.L.; Birren, B. W.; Huttenhower, C. The human microbiome project: A community resource for the healthy human microbiome. *PLoS Biology*, **2012**, *10*, e1001377.

Goodrich, J. K.; Di Rienzi, S. C.; Poole, A. C.; Koren, O.; Walters, W. A.; Caporaso, J. G.; Ley, R. E. Conducting a microbiome study. *Cell*, **2014**, *158*, 250–262.

Hasegawa, E.; Yanagisawa, M.; Sakurai, T.; Mieda, M. Orexin neurons suppress narcolepsy via 2 distinct efferent pathways. *The Journal of Clinical Investigation*, **2014**, *124*(2), 604–616.

Hassan, A. A.; Al-Nafeesah, A.; AlEed, A.; Adam, I. Prevalence and factors associated with hypertension among adolescents in Sudan: a cross-sectional school-based study. *BMJ Open*, **2023**, *13*(12), e078234.

Hasturk, H.; Kantarci, A.; Van Dyke, T. E. Oral Inflammatory Diseases and Systemic Inflammation: Role of the Macrophage. *Frontiers in Immunology*, **2012**, *3*, 118.

Jameson, J. L.; Longo, D. L. Precision medicine–personalized, problematic, and promising. *New England Journal of Medicine*, **2015**, *372*(23), 2229–2234.

Jin, L.; Lamster, I.; Greenspan, J.; Pitts, N.; Scully, C.; Warnakulasuriya, S. Global burden of oral diseases: Emerging concepts, management and interplay with systemic health. *Oral Diseases*, **2016**, *22*(7), 609–619.

Jones, D.; Metzger, H.J.; Schatz, A.; Waksman, S.A. Control of gram-negative bacteria in experimental animals by streptomycin. *Science*, **1944**, *100*(2588), 103–105.

Kashyap, P. C.; Chia, N.; Nelson, H.; Segal, E.; Elinav, E. Microbiome at the frontier of personalized medicine. *Mayo Clinic Proceedings*, **2017**, *92*(12), 1855–1864. doi: 10.1016/j.mayocp.2017.10.004.

Kilian, M.; Chapple, I. L. C.; Hannig, M.; Marsh, P. D.; Meuric, V.; Pedersen, A. M. L.; Zaura, E. The oral microbiome–an update for oral healthcare professionals. *British Dental Journal*, **2016**, *221*(10), 657–666.

Kinane, D. F.; Riggio, M. P.; Walker, K. F.; MacKenzie, D.; Shearer, B. Bacteraemia following periodontal procedures. *Journal of Clinical Periodontology*, **2005**, *32*(7), 708–713.

Koren, O.; Spor, A.; Felin, J.; Fåk, F.; Stombaugh, J.; Tremaroli, V.; Bäckhed, F. Human oral, gut, and plaque microbiota in patients with atherosclerosis. *Proceedings of the National Academy of Sciences*, **2011**, *108*(supplement_1), 4592–4598.

Krol, D. M.; Whelan, K. Maintaining and improving the oral health of young children. *Pediatrics*, **2023**, *151*, e2022060417.

Kwon, T.; Lamster, I. B.; Levin, L. Current Concepts in the Management of Periodontitis. *International Dental Journal*, **2021**, *71*(6), 462–476.

Lamont, R. J.; Koo, H.; Hajishengallis, G. The oral microbiota: dynamic communities and host interactions. *Nature Reviews Microbiology*, **2018**, *16*(12), 745–759.

Lang, M. L.; Zhu, L.; Kreth, J. Keeping the bad bacteria in check: Interactions of the host immune system with oral cavity biofilms. *Endodontic Topics*, **2010**, *22*(1), 17–32.

Li, X.; Kolltveit, K. M.; Tronstad, L.; Olsen, I. Systemic diseases caused by oral infection. *Clinical Microbiology Reviews*, **2000**, *13*(4), 547–558.

Li, X.; Tronstad, L.; Olsen, I. Brain abscesses caused by oral infection. *Dental Traumatology*, **1999**, *15*(3), 95–101.

Libby, P.; Ridker, P. M.; Maseri, A. Inflammation and atherosclerosis. *Circulation*, **2002**, *105*(9), 1135–1143.

Lockhart, P. B.; Bolger, A. F.; Papapanou, P. N.; Osinbowale, O.; Trevisan, M.; Levison, M. E.; Baddour, L. M. Periodontal disease and atherosclerotic vascular disease: Does the evidence support an independent association? A scientific statement from the American Heart Association. *Circulation*, **2012**, *125*(20), 2520–2544.

Lockhart, P. B.; Brennan, M. T.; Sasser, H. C.; Fox, P. C.; Paster, B. J.; Bahrani-Mougeot, F. K. Bacteremia associated with toothbrushing and dental extraction. *Circulation*, **2008**, *117*(24), 3118–3125.

Lockhart, S. R.; Iqbal, N.; Cleveland, A. A.; Farley, M. M.; Harrison, L. H.; Bolden, C. B.; Chiller, T. Species identification and antifungal susceptibility testing of Candida bloodstream isolates from population-based surveillance studies in two US cities from 2008 to 2011. *Journal of Clinical Microbiology*, **2011**, *50*(11), 3435–3442.

Luo, A. L.; Zhao, Y. H.; Zhao, G.; Deng, L. C.; Liu, X. W.; Jing, Y. P.; Zuo, F. The first data release (DR1) of the LAMOST regular survey. *Research in Astronomy and Astrophysics*, **2015**, *15*(8), 1095.

Marchesi, J. R.; Adams, D. H.; Fava, F.; Hermes, G. D.; Hirschfield, G. M.; Hold, G.; Hart, A. The gut microbiota and host health: A new clinical frontier. *Gut*, **2016**, *65*(2), 330–339.

Marchesi, J. R.; Ravel, J. The vocabulary of microbiome research: A proposal. *Microbiome*, **2015**, *3*, 31.

Mills, J. M.; Lofthouse, E.; Roberts, P.; Karas, J. A. A patient with bacteraemia and possible endocarditis caused by a recently-discovered genomospecies of Capnocytophaga: Capnocytophaga genomospecies AHN8471: A case report. *Journal of Medical Case Reports*, **2008**, 2, 1–3.

Moh. Khafid.; Alvi, A. A.; Putri T.; Danik I. P. Prevention dental caries by improving oral hygiene through a high-fiber diet in children with mixed dentition: Experimental study in Kediri, East Java, Indonesia. *World Journal of Advanced Research and Reviews*, **2023**, *18*(3), 204–208.

Mojon, P. Oral health and respiratory infection. *Journal-Canadian Dental Association*, **2002**, *68*(6), 340–345.

Moon, J.H.; Lee, J.H. Probing the diversity of healthy oral microbiome with bioinformatics approaches. *BMB Reports*, **2016**, *49*, 662–670.

Nagakubo, D.; Kaibori, Y. Oral microbiota: The influences and interactions of saliva, IgA, and dietary factors in health and disease. *Microorganisms*, **2023**, *11*(9), 2307.

Nobbs, A. H.; Jenkinson, H. F.; Jakubovics, N. S. Stick to your gums: Mechanisms of oral microbial adherence. *Journal of Dental Research*, **2011**, *90*(11), 1271–1278.

Ohki, M.; Kikuchi, S. Nasal, oral, and pharyngolaryngeal manifestations of pemphigus vulgaris: Endoscopic ororhinolaryngologic examination. *Ear, Nose & Throat Journal*, **2017**, *96*(3), 120–127.

Ohlrich, E. J.; Cullinan, M. P.; Leichter, J. W. Diabetes, periodontitis, and the subgingival microbiota. *Journal of Oral Microbiology*, **2010**, *2*(1), 5818.

Oulas, A.; Pavloudi, C.; Polymenakou, P.; Pavlopoulos, G. A.; Papanikolaou, N.; Kotoulas, G.; Iliopoulos, L. Metagenomics: tools and insights for analyzing next-generation sequencing data derived from biodiversity studies. *Bioinformatics and Biology Insights*, **2015**, *9*, BBI-S12462.

Patel, J.; Wallace, J.; Doshi, M.; Gadanya, M.; Ben Yahya, I.; Roseman, J.; Srisilapanan, P. Oral health for healthy ageing. *The Lancet Healthy Longevity*, **2021**, *2*(8), e521–e527.

Patil, S.; Rao, R. S.; Amrutha, N.; Sanketh, D. S. Oral microbial flora in health. *World Journal of Dentistry*. **2013**, *4*, 262–266.

Preshaw, P. M.; Foster, N.; Taylor, J. J. Cross-susceptibility between periodontal disease and type 2 diabetes mellitus: an immunobiological perspective. *Periodontology*, **2007**, *45*(1), 138–157.

Pucar, A.; Milasin, J.; Lekovic, V.; Vukadinovic, M.; Ristic, M.; Putnik, S.; Kenney; E. B. Correlation between atherosclerosis and periodontal putative pathogenic bacterial infections in coronary and internal mammary arteries. *Journal of Periodontology*, **1991**, *78*(4), 677–682.

Radaic, A.; Kapila, Y. L. The oralome and its dysbiosis: New insights into oral microbiome-host interactions. *Computational and Structural Biotechnology Journal*, **2021**, *19*, 1335–1360.

Rathee M.; Jain P. Gingivitis. In: *StatPearls*. Treasure Island (FL): StatPearls Publishing; **2023**.

Rogers, M. K.; Seigfried, K.; Tidke, K.; Self-reported computer criminal behavior: A psychological analysis. *Digital Investigation*, **2006**, *3*, 116–120.

Sampaio-Maia, B.;Simões-Silva, L.; Pestana, M.; Araujo, R.; Soares-Silva; I. J. The role of the gut microbiome on chronic kidney disease. *Advances in Applied Microbiology*, **2016**, *96*, 65–94.

Sankaranarayanan, R.; Ramadas, K.; Amarasinghe, H.; Subramanian, S.; Johnson, N. Oral cancer: Prevention, early detection, and treatment. In H. Gelband et al. (Eds.), *Cancer: Disease Control Priorities* (3rd edition, volume 3), **2015**. Washington, DC: World Bank. doi: 10.1596/978-1-4648-0349-9_ch5.

Shank, E.A; Kolter, R. New developments in microbial interspecies signaling. *Current Opinion in Microbiology*, **2009**, *12*(2), 205–214.

Shetty, A. S.; Shenoy, R.; Bajaj, P. D.; Rao, A.; Aparna, K. S.; Pai, M.; Avinash, B. R.; Jodalli, P. Role of nutritional supplements on oral health in adults–A systematic review. *F1000Research*, **2023**, *12*, 492.

Slocum, C.; Kramer, C.; Genco, C. A. Immune dysregulation mediated by the oral microbiome: potential link to chronic inflammation and atherosclerosis. *Journal of Internal Medicine*, **2016**, *280*(1), 114–128.

Slots, J. Periodontitis: Facts, fallacies and the future. *Periodontology 2000*, **2017**, *75*(1), 7–23.

Subramaniam, P.; Kumar, K. Oral mucosal status and salivary IgA levels of HIV-infected children. *Journal of Oral Pathology & Medicine*, **2013**, *42*(9), 705–710.

Tonetti, M. S.; Bottenberg, P.; Conrads, G.; Eickholz, P.; Heasman, P.; Huysmans, M.; López, R.; Madianos, P.; Müller, F.; Needleman, I.; Nyvad, B.' Preshaw, P. M.; Pretty, I.; Renvert, S.; Schwendicke, F.; Trombelli, L.; van der Putten, G.' Vanobbergen, J.; West, N.; Paris, S. Dental caries and periodontal diseases in the ageing population: call to action to protect and enhance oral health and well-being as an essential component of healthy ageing – Consensus report of group 4 of the joint EFP-ORCA workshop on the boundaries between caries and periodontal diseases. *Journal of Clinical Periodontology*, **2017**, *44*(S18), S135–S144.

Watanabe, Y.; Okada, K.; Kondo, M.; Matsushita, T.; Nakazawa, S.; Yamazaki, Y.; Oral health for achieving longevity. *Geriatrics & Gerontology International*, **2020**, *20*(6), 526–538. Portico.

Welti, R.; Jones B.; Paula Moynihan, P.; Silva, M. Evidence pertaining to modifiable risk factors for oral diseases: An umbrella review to inform oral health messages for Australia. *Australian Dental Journal*, **2023**, *68*(4), 222–237. https://doi.org/10.1111/adj.12972.

Willershausen-Zönnchen, B.; Lemmen, C.; Hamm, G. Influence of saliva components on periodontal disease in insulin-dependent diabetics. *Deutsche Zahnarztliche Zeitschrift*, **1991**, *46*(4), 281–284.

Wu, H.; Esteve, E.; Tremaroli, V.; et al. Metformin alters the gut microbiome of individuals with treatment-naive type 2 diabetes, contributing to the therapeutic effects of the drug. *Nature Medicine*. **2017**, *23*(7), 850–858.

Wu, R-Q.; Zhang, D. F.; Tu, E., Chen, Q. M.; Chen, W. The mucosal immune system in the oral cavity an orchestra of T cell diversity. *International Journal of Oral Science*, **2014**, *6*(3), 125–132.

Wu, S.; Huang, Z.; Yang, X.; Zhou, Y.; Wang, A.; Chen, L.; ... Cai, J. Prevalence of ideal cardiovascular health and its relationship with the 4-year cardiovascular events in a northern Chinese industrial city. *Circulation: Cardiovascular Quality and Outcomes*, **2012**, *5*(4), 487–493.

Yoneyama, T.; Yoshida, M.; Ohrui. T.; Mukaiyama, H.; Okamoto, H.; Hoshiba, K.; Ihara, S.; Yanagisawa, S.; Ariumi, S.; Morita, T.; Mizuno, Y.; Ohsawa, T.; Akagawa, Y.; Hashimoto, K.; Sasaki, H.; Oral Care Working Group. Oral care reduces pneumonia in older patients in nursing homes. *Journal of the American Geriatrics Society*, **2002**, *50*(3), 430–433.

Yost, S.; Duran-Pinedo.; A. E.; Teles, R. Krishnan.; Frias-Lopez, J. Functional signatures of oral dysbiosis during periodontitis progression revealed by microbial metatranscriptome analysis. *Genome Medicine*, **2015**, *7*, 1–19.

Zaura, E.; Nicu, E. A.; Krom, B. P.; Keijser, B. J.; Acquiring and maintaining a normal oral microbiome: Current perspective. *Frontiers in Cellular and Infection Microbiology*, **2014**, *4*, 85.

Zhao, H.; Chu, M.; Huang, Z.; Yang, X.; Ran, S.; Hu, B.; Liang, J. Variations in oral microbiota associated with oral cancer. *Scientific Reports*, **2017**, *7*(1), 11773.

18 Oral Microbiome and Pediatric Health

Duni Chand, Ekta Arya, Lalita Kaushal, Alka Kumari, and Arti
Department of Biotechnology, Himachal Pradesh University, Shimla, Himachal Pradesh, India

Pradeep Kumar
Department of Forensic Science, Himachal Pradesh University, Shimla, Himachal Pradesh, India

18.1 INTRODUCTION

In the convoluted world of our mouths resides an entire universe: the oral microbiome. The oral microbiome refers to the variety of microorganisms that inhabit the oral cavity, including bacteria, viruses, fungi, and other microbes. This microscopic ecosystem acts as a shield, defending against harmful bacteria (Arweiler & Netuschil, 2016). The overall composition and balance of the oral microbiome play an essential role in maintaining oral health, and recent research suggests that it can also influence general health, especially pediatric health. Yet, it's a delicate balance. Within this community lies the root of prevalent oral diseases like caries (tooth decay) and periodontitis (gum disease). Interestingly, the defense against these diseases often starts with something as simple as regular brushing. The mechanical removal of microbial buildup through oral hygiene routines stands as our primary defense (Zhang et al., 2018). But the story of our oral microbiome begins long before we even open our mouths for the first time. Even before birth, maternal influences start shaping our immune system through early exposure to microbes. From birth and through the stages of childhood, various factors including feeding, hygiene, diet, antibiotics, and genetics, all have a hand in sculpting the oral microbiota (Kaan et al., 2021).

The "microbiota," is not a recent discovery but traces back to the early 1900s, recognizing the rich tapestry of microorganisms inhabiting various human body sites. Astonishingly, the oral cavity alone houses the second-largest microbial community within the human body. It's not a monolithic entity but rather a collection of diverse habitats, each hosting unique microbial populations. These populations constantly evolve, influenced by pH changes, gene mutations, and interactions between different bacteria. Despite similarities across seven oral sites, subtle differences persist in the oral microbiota. Firmicutes, Proteobacteria, Bacteroidetes, Actinobacteria, and Fusobacteria dominate this bustling community (Hou et al., 2022). Recognizing that each person's oral microbiome is unique, as it is shaped by genetics, immune response, and environment, highlights the need for personalized preventive measures. Early detection and customized interventions empower dental professionals to manage these diseases effectively. It's through this understanding and personalized approach that we can navigate the complex world thriving within our mouths and safeguard our oral health (Mosaddad et al., 2019).

The development of children's oral microbiomes is a dynamic and complex process that depends on a variety of factors. Oral cavity colonization begins shortly after birth and changes significantly throughout infancy and early childhood. Exploring the prenatal to the postnatal continuum of the oral microbiome reveals a dynamic narrative that emphasizes the necessity of understanding and fostering this microbial ecology for lifelong oral health (Figure 18.1).

DOI: 10.1201/9781003478638-18

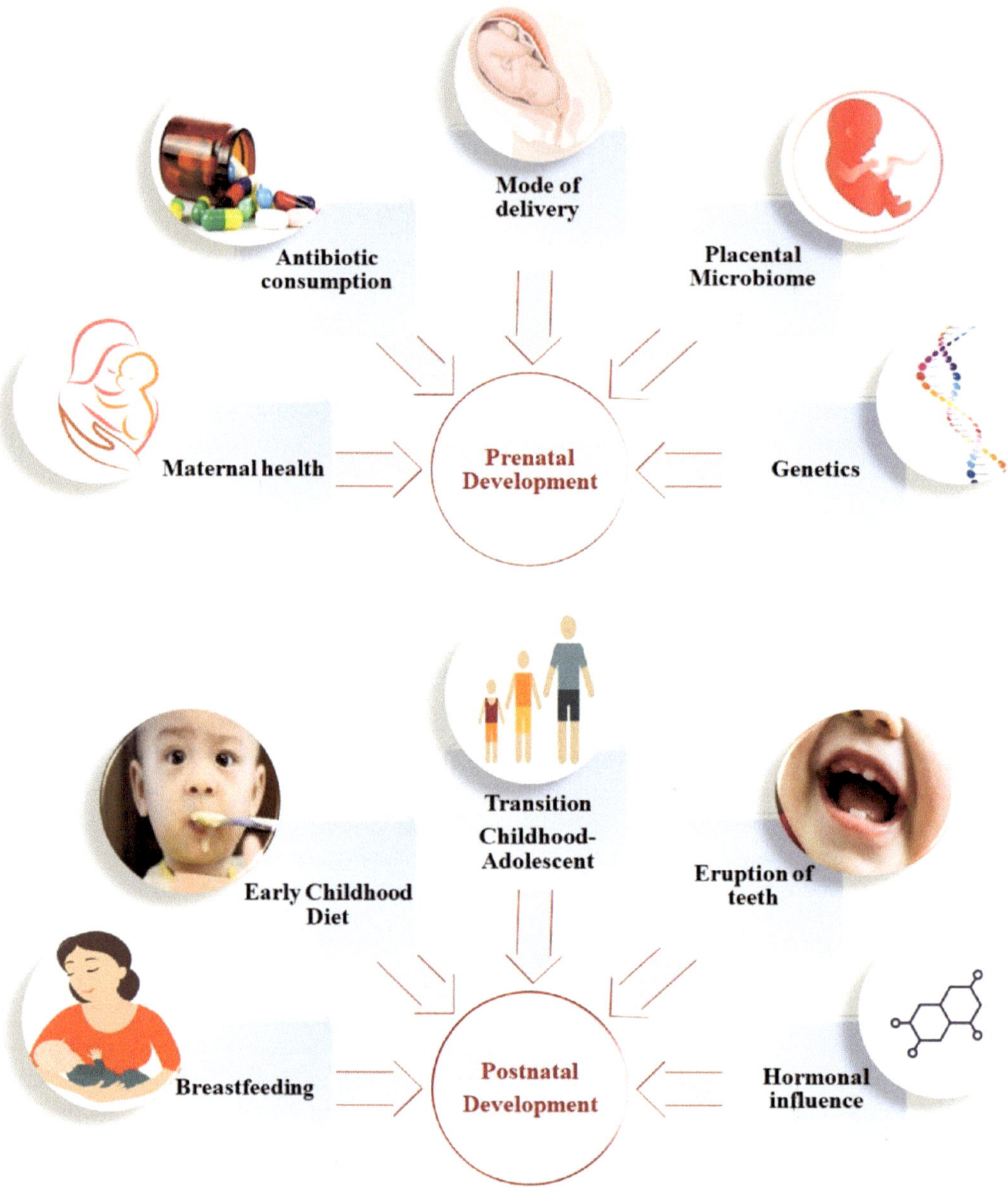

FIGURE 18.1 Prenatal and postnatal development of the oral microbiome.

18.2 PRENATAL ORAL MICROBIOTA DEVELOPMENT

18.2.1 Maternal Wellness

There has been growing evidence that maternal health and nutrition during pregnancy can influence the developing infant microbiota (Lundgren et al., 2018). While much of this data is derived from studies on animals or focuses on the baby gut microbiome, there are likely implications for the oral microbiome. Maternal health during pregnancy has demonstrated its impact on the infant's gut microbiome. After examining oral swabs collected at the time of birth, differences in oral microbial communities were noted among infants born to mothers with gestational diabetes compared to those born to mothers without gestational diabetes (He et al., 2019). The exact mechanisms responsible for the prenatal influence on the infant oral microbiome remain unclear, but they probably involve the transfer of various factors across the placenta. The placenta plays a crucial role in supplying the growing fetus with oxygen, nutrients, and an array of maternal factors, such as immunoglobulins. It collaborates with the amniotic fluid, which is in direct contact with the developing fetus (Gude et al., 2004).

18.2.2 Placental Microbiome

The placental microbiota is believed to initiate the colonization of the fetus with microbes (Aagaard et al., 2014). Upon a more detailed analysis of this research, it becomes evident that the presence of bacteria in meconium correlates with the time elapsed since delivery. This suggests that microbial introduction occurs at birth rather than during pregnancy. Divergences in infant microbiota resulting from different delivery methods contradict the intrauterine seeding hypothesis. If the microbiome were indeed seeded *in utero*, delivering germ-free mammals, including humans, through cesarean section would be implausible (Perez-Munoz et al., 2017). Therefore, one may dismiss the hypothesis suggesting that the biological function of the placental microbiome is intrauterine microbial seeding.

What potential role would bacteria or their fragments play if they were transported to the placenta but did not initiate colonization of the fetus? During pregnancy, the placenta serves as a site for gathering antigens for the fetal immune system and undergoes subsequent "training" in developing antigen tolerance (Zaura et al., 2014). Pregnant women frequently encounter increased gingival inflammation, commonly known as pregnancy gingivitis. This occurrence is triggered by pregnancy-related hormones, which result in the dilation of the vascular bed. As a result, oral microbes have the potential to reach the placenta through hematogenic pathways, either by direct transmission or through uptake and transportation by the mother's immune cells (Jeurink et al., 2013). Microbial cells or fragments become trapped in placental tissue and are presented to the developing embryonic immune system. During the prenatal stage, immune cells tasked with presenting antigens to the fetus may interact with microbial antigens from the mother, subsequently migrating back to the fetal peripheral lymphoid organs. Studies have shown that when exposed to maternal alloantigenes, regulatory T lymphocytes in human fetuses experience functional suppression and continue to persist, at least until early adulthood (Mold et al., 2008). Consequently, the fetus would establish prenatal tolerance to the mother's microbiome, perceiving it as "safe" during subsequent postnatal encounters with these bacteria. Hence, the establishment of fetal tolerance towards the mother's microbiome during pregnancy might play a crucial role in the successful acquisition of a normal microbiome. In summary, during the prenatal phase of microbiome acquisition, the maternal microbiome doesn't directly transfer to the fetus. Instead, it readies the fetus for postnatal microbial interactions by training and fine-tuning the developing immune system. The promotion of a child's health is significantly influenced by maintaining good maternal dental health and ensuring a balanced oral microbiota that aligns with the mother's body.

18.2.3 Birth Delivery Mode

The mode of birth, whether vaginal or cesarean section (C-section), may influence the oral microbiome development in infants. Several studies investigated the colonization patterns of microbial species in neonates delivered vaginally with those delivered by cesarean section. Some bacteria, such as *Streptococcus* spp., appear to be present in the oral cavity of all children, regardless of delivery mode, whereas other species appear to colonize the oral cavity of infants in proportions that vary between the two different forms of birth delivery (Lif Holgerson et al., 2011). In one of the studies, distinct colonization patterns were found between the two delivery modes, with significantly more Actinobacteria and Firmicutes and less Bacteroidetes and Proteobacteria in the vaginal delivery group compared with C-section infants (Li et al., 2019). Another study on infant oral microbiota development found that the influence of birth mode was transient. At 1 week of age, newborns delivered via C-section had significantly more diverse oral microbiota than the infants delivered vaginally, but this difference did not persist 4 weeks later (Hurley et al., 2019). While the biological significance of fewer oral microbial species in infancy is unknown, it is speculated that C-section delivery, as opposed to vaginal birth, reduces exposure to commensal, protective bacteria from the mother. As a result, the natural barrier to colonization by other, potentially more harmful microbes is reduced. Given the inconsistency of the data, more research is needed to determine the effects of different modes of delivery on the oral microbiota and whether these effects persist over time.

18.2.4 Maternal Intrapartum Antibiotic Consumption

Antibiotics administered to mothers during the intrapartum (during labor) or preoperative period (before a cesarean section) can impact the oral microbial colonization pattern of neonates within a few days of birth. Although antibiotics are commonly utilized for treating infections, their impact is often broad-spectrum, affecting not only pathogenic bacteria but also non-pathogenic and beneficial bacteria. The families of Streptococcaceae, Gemellaceae, and Lactobacillaceae were significantly enriched among neonates born to mothers who did not receive intrapartum antibiotics. There was a reduction in the abundance of the Streptococcaceae family in neonates whose mothers received antibiotics, specifically cephazolin and benzylpenicillin. These are the primary antibiotics administered during the intrapartum period, targeting a variety of *Streptococci* strains. Because *Streptococcus* spp. play an important role as early colonizers of the infant oral cavity and contribute to the adhesion of surfaces within the oral cavity, they are believed to aid the development of future colonizers. This shows that *Streptococcus* spp. may have a role in encouraging stable community growth and the creation of biofilms during the early stages of human life. The oral microbiomes of neonates whose mothers received antibiotics showed greater diversity than those of neonates whose mothers received no antibiotics, and this may be partly attributable to the decreased abundance of *Streptococcus* spp. (Gomez and Nelson, 2017). Maternal intrapartum antibiotic consumption is likely to impact neonatal oral microbial community development; however, the long-term impact on the oral microbiome is not known.

18.2.5 Genetics

There are currently no twin studies that investigate the interaction of genetics and environment in forming the oral microbiome during infancy. While there is a lot of research on older children and adults, these studies typically look at the oral microbiome in terms of disease rather than wellness. Although host genetic factors undoubtedly influence the human oral microbiome, results from a limited number of twin and genetic studies show that variation in oral microbiome composition is significantly driven by environmental exposures (Gomez et al., 2017). While twins may initially share a similar oral microbiome, this similarity tends to diminish as they age and no longer live together. The overall composition of the human oral microbiome is likely influenced by a combination of individual genetic variation and environmental factors. Nonetheless, the prevailing consensus suggests that the environment plays a more substantial role in determining the variation observed in the oral microbiome.

18.3 POSTNATAL ORAL MICROBIOME DEVELOPMENT

The postnatal oral microbiome development in infants is a dynamic and complex process influenced by various factors. Understanding the postnatal development of the oral microbiome is essential as alterations in microbial composition during this critical period may have implications for the infant's health, including the risk of oral diseases, allergies, and other immune-related conditions.

From birth, a neonate's oral cavity is exposed to a variety of bacteria originating from the neonate's family, particularly the mother's saliva, as well as their diet, other people, and the surrounding environment. Even though the oral cavity of humans is primarily an open environment, most of these bacteria do not colonize the infant's mouth. For bacteria to effectively establish colonization in the oral cavity, they must adhere to epithelial surfaces, acquire adequate nutrition, and proliferate as part of polymicrobial biofilm communities (Kolenbrander et al., 2006). Once established, these resident bacterial communities in the oral cavity resist colonization by external organisms by occupying binding sites, efficiently utilizing nutrients, producing antimicrobial agents and end products, and modifying the environment. The oral microbiome can be seen as a functional element of the host's immune response. Indeed, differences in the oral microbial composition observed in children with allergic symptoms compared to those without support the notion that the evolving oral microbial community during childhood influences the host immune response maturation (Dzidic et al., 2018).

18.3.1 Trafficking of Microbes from the Oral Cavity to Other Sites in the Body

In the early stages after birth, the microbiomes of neonates exhibit uniformity across gut, oral, nasal, and skin communities. However, microbial composition changes over time, with distinct body sites emerging as crucial factors influencing both the composition and functional capacity of the microbial community. At 6 months of age, the oral microbiota of infants closely resembles the microbiota found in their mother's oral cavity, breast milk, and mammary areola (Chu et al., 2017). Similar patterns were observed between the skin and gingival microbiota, with the microbiota of the nasal passages acting as a connecting link between the two. These findings underscore the impact of continuous interaction on the composition of microbial communities. Saliva, owing to its liquid nature, is believed to act as a carrier of microbiota, facilitating the transportation of microorganisms from one body site to another. In infants, the elevated gastric pH is influenced by the substantially high pH of human milk and buffering capacity, potentially contributing to the inflow and colonization of bacteria in the gastrointestinal tract.

18.3.2 Breastfeeding

Breastfeeding plays a significant role in the postnatal development of the oral microbiome in infants. Human milk, particularly breast milk, is not only a source of essential nutrients but also provides various bioactive compounds that support the growth of beneficial microorganisms and influence the development of a healthy oral microbiome. During the first few days after birth, the mother produces colostrum, a thin yellowish fluid rich in protein and antibodies that provides the newborn with passive immunity. This gradually turns into mature milk. Human milk contains 1.2% protein, 3.6% fat, and 7.4% carbohydrates (most of which are lactose). Casein, α-lactalbumin, lactoferrin, and secretory immunoglobulin IgA (SIgA) are the main proteins. Breast milk is not sterile, but it does have its microbiota, proving at the molecular level that breast milk is a key source of bacteria for the newborn (Martın et al., 2007). The protective qualities of maternal SIgA in breast milk may modulate the formation of the oral microbiome, restricting colonization by potentially pathogenic species. The efficacy of SIgA inhibition is known to decrease as lactation progresses, implying that the protective characteristics of human milk may be most influential to the newborn in the period immediately following birth. The secondary effects of this protein, along with the other protective factors in breast milk that promote the colonization of commensal bacteria, are likely to play a critical role in laying the necessary foundations for microbial community assembly that is directed toward a health-associated oral profile.

The first research on bacteria as a natural component of breast milk surfaced more than 2 decades ago. The sources of bacteria in breast milk are considered to be both extrinsic, originating from the skin surfaces of the mother and her infant, as well as the infant's oral cavity, and intrinsic, stemming from the mother via the entero-mammary pathway (Rodriguez, 2014). It has been estimated that breast milk contains about 10^6 bacterial cells per mL (Boix-Amoros et al., 2016), and the most predominant genera being *Streptococcus*, *Staphylococcus*, *Bifidobacterium*, *Propionibacterium*, and *Lactobacillus* (Fitzstevens et al., 2017). Several factors, including the mother's body mass index (BMI), birth style, and breastfeeding duration, have been observed to influence the species of bacteria present in breast milk. Several studies have shown that certain bacterial strains are transferred from a mother to her newborn via breast milk, implying that the natural job of breast microbiota is to seed the infant's gastrointestinal system. Human milk oligosaccharides, the major carbohydrates in breast milk that are indigestible to the newborn, play a role in providing nutrients to these newly acquired bacteria.

It has been established that newborn saliva reacts with breast milk during breastfeeding, producing reactive oxygen species. Purine metabolites (xanthine, hypoxanthine, adenosine, inosine, and guanosine) are abundant in neonatal saliva and gradually decline with maturity, reaching adult levels when the infant is around 6 months old. Xanthine and hypoxanthine are substrates for xanthine oxidase, an enzyme found in high concentrations in breast milk that, together with the enzyme lactoperoxidase, is involved in the generation of hydrogen peroxide. This proposes a novel biochemical synergy between the newborn saliva and breast milk components, which could influence the oral and gut microbes (Al-Shehri et al., 2015).

18.3.3 Teeth Eruption

During the pre-dentate stage when newborns lack teeth, their oral microbiota closely mirrors that of their mother. This microbial interchange occurs through activities like kissing, sharing utensils, and other close contact between the mother and the infant. Nevertheless, as a newborn's teeth start to emerge and their diet expands to include solid foods alongside milk, the oral microbiome undergoes significant changes. The introduction of new foods alters the oral environment, providing diverse substrates for the development of bacteria. The shift in nutrition, coupled with the mechanical action of chewing and the emergence of teeth, creates a more diverse and dynamic environment within the oral cavity (Mason et al., 2018). Teeth start to appear in the infant oral cavity around 6 months of age, and a complete set of deciduous dentitions is typically present by around 30 months of age. The emergence of teeth brings about ecological shifts, facilitated by the non-shedding tooth surface that allows for the formation of intricate microbial biofilms if not disrupted. Additionally, gingival sulci and the spaces between teeth add to the varied environments within the oral cavity, promoting microbial colonization and growth (Welch et al., 2020). The eruption of teeth and the introduction of solid foods, often concurrent events, pose challenges in distinguishing the individual impacts of tooth emergence and the availability of new nutrients on the development of oral microbial communities. The nature of the introduced foods can shape the types of microorganisms thriving in the oral cavity. For instance, the breakdown of carbohydrates from solid diets may foster the growth of specific bacterial species specialized in metabolizing these compounds.

Analyzing the functions of additional species observed post-tooth eruption suggests that the tooth environment may exert a greater influence on shaping the composition of the oral microbiome during this life stage. This is evident in the enrichment of genes associated with biofilm development in bacterial samples collected at this stage. It's important to note that tooth eruption creates fresh surfaces for microbial colonization, introducing new niches and habitats for microbes to establish and form biofilms. This complexity adds to the intricate nature of the oral microbiome during this developmental phase. The core microbiome changes with tooth eruption, accompanied by the entrance of new habitats and the presence of taxa such as *Rothia*, *Fusobacterium*, and *Neisseria*. The shift from deciduous (baby) teeth to permanent dentition may also be linked to additional maturation of the oral microbiome later in childhood (Shi et al., 2018). Understanding the alterations in the oral microbiome during the transition from an exclusively milk-based diet to a more varied one is crucial for optimizing newborn oral health. With the oral environment becoming more intricate, the implementation of proper oral hygiene practices, including cleaning the erupting teeth and gums, becomes progressively pronounced. This phase of microbial transformation is also considered pivotal in influencing the child's long-term oral health.

18.3.4 Microbial Changes from Childhood to Adolescence

The composition of the oral microbiota undergoes substantial changes throughout a person's life, with noteworthy transformations taking place during the transition from childhood to adolescence. Around the age of 6, primary teeth begin to shed, paving the way for the eruption of permanent teeth. This transitional period, referred to as the mixed-dentition phase, involves the simultaneous presence of both primary and permanent teeth in dental development. The commencement of permanent dentition begins with the appearance of the mandibular central incisors and the first mandibular and maxillary molars. This process reaches its conclusion with the eruption of the second molars, usually around the age of 12. During the mixed dentition period, characterized by the simultaneous presence of primary and permanent teeth, the oral environment experiences dynamic changes influenced by the eruption of permanent teeth and the onset of puberty. Analyzing microbial profiles in plaque from children at different dental stages—primary, mixed, and permanent dentition—reveals distinct patterns. In permanent dentition, supragingival plaque exhibits elevated levels of both gram-positive and gram-negative anaerobic species compared to primary dentition. Moreover, in the mixed dentition stage, there is a noticeable change in the microbial composition of subgingival plaque, showing a significantly higher prevalence of gram-negative anaerobes compared to subgingival plaque from both primary and permanent dentition. Additional studies employing targeted microbiological methods further validate an increase in anaerobic taxa during the transition to the mixed dentition phase. When occlusion is incomplete, dental plaque and leftover food particles tend to accumulate on the occlusal surfaces of partially erupted permanent molars (Brailsford et al., 2005). The enlargement of the gingiva around erupting teeth can impede effective oral hygiene practices, causing the accumulation of more mature dental plaque. Concurrently, diminished tissue resistance near erupting teeth may lead to increased gingival bleeding. While preschool children with primary dentition typically exhibit low susceptibility to gingivitis, this condition gradually intensifies around the ages of 6–8 years with the eruption of the first permanent teeth. This susceptibility may further increase around 10–13 years of age, aligning with heightened hormonal activity during puberty.

18.3.5 Early Childhood Diet

The World Health Organization suggests starting complementary foods in the infant's diet at 6 months of age. The introduction of food to a newborn is a gradual process, with the mouth cavity only being exposed to small amounts of different foods for short periods. It emphasizes the significance of eating a balanced and varied diet early in infancy for the formation of a healthy mouth microbiota. The meals that a child consumes have a direct impact on the availability of nutrients in the mouth cavity. Different bacteria thrive on different nutrients, and a varied diet provides a greater spectrum of nutrients that promote a more balanced microbial community. A fiber- and prebiotic-rich diet can have a favorable impact on the oral microbiome. Fiber encourages chewing, which stimulates saliva flow and aids in the maintenance of a healthy microbial balance. Prebiotics serve as food for beneficial microorganisms, encouraging their proliferation. Diet affects the pH levels in the mouth cavity. Acidic foods and beverages may encourage the growth of acid-producing bacteria, eventually leading to enamel erosion and other oral health issues. Further, high sugar consumption, particularly from sugary snacks and beverages, can promote the growth of cariogenic (cavity-causing) bacteria such as *Streptococcus mutans*. These bacteria use sugar to generate acids that might cause tooth decay. Promoting a healthy oral microbiome in early childhood entails providing a well-balanced and nutritious diet, minimizing sugary snacks and beverages, and supporting useful dental hygiene practices. These elements work together to produce an extensive and beneficial microbial population in the oral cavity, promoting general oral wellness throughout a child's life.

18.3.6 Hormonal Influence

Hormones are important in the growth and development of several tissues and organs, including those in the mouth cavity. Hormonal changes occur during childhood as part of the natural growth and development process. However, the direct influence of hormones on the oral microbiota hasn't been studied as thoroughly as it has been in other parts of the body. One aspect of hormonal impact on children's oral microbiome is related to oral tissue development. Hormones, including growth hormone and sex hormones, can influence the development of salivary glands and other oral structures, which in turn can influence the oral microbiota. Saliva, which is produced by the salivary glands, is essential for maintaining dental health because it flushes away food particles, neutralizes acids, and has antibacterial qualities. Furthermore, hormonal changes throughout puberty might alter the composition of saliva. The rise in sex hormones, such as estrogen and testosterone, might affect saliva secretion and composition. These alterations may have an indirect effect on the oral microbiome by altering the oral environment.

Recognizing the function of the oral microbiome in pediatric health highlights the importance of early interventions and preventive measures to create a balanced and healthy microbial community in the mouth cavity. Good oral hygiene, a balanced diet, and frequent dental checkups are essential for children to have a healthy oral microbiota. Limiting sugar intake and encouraging practices that promote varied microbiota contribute to overall oral and systemic health.

18.4 NURTURING CHILDHOOD ORAL HEALTH

There's a subtle problem in children's happy world: dental decay. It is a major problem that affects about half of all children worldwide. Those small holes in the teeth may appear to be harmless, but if ignored, they can cause more than simply chewing problems (Kazeminia et al., 2020). Imagine, tiny teeth succumbing to decay, often triggered by night feeds or prolonged bottle use. Early childhood caries (ECC) strikes swiftly, causing agony—abscesses, pain, and even psychological disturbances. The majority of the clinical findings in decayed and healthy teeth in children are shown in Figure 18.2 The toll isn't just physical; it extends to the emotional fabric of families. Treatment, complex and expensive, yields uncertain outcomes (Mathur & Dhillon, 2018). Preventing this invasive enemy is simple, yet crucial. Educating parents, instilling good oral hygiene, and tweaking diets away from sugars and stickiness form the frontline defense. Physicians, with their keen eyes, can identify early signs, counsel parents, and steer toward specialized care. The gravity of this issue demands a united front. Health care providers, armed with knowledge, must guide parents on dental care routines, especially after sugar-based medications. Identifying dental caries should prompt swift referrals to pediatric dentists (Elkhodary et al., 2023). Oral hygiene practices vary across countries and homes. Factors like age, sex, dental visits, maternal education, and residential settings shape these practices. Understanding these influences lays the foundation for customized oral health promotion programs. The culprit isn't just sugar; it's the societal and behavioral patterns ingrained within us. Dental health professionals advocate for comprehensive strategies, including taxes, warning labels, and policy changes to curb added sugar intake and, in turn, mitigate tooth decay's impact (Chi & Scott, 2019). The tale doesn't end with teeth. Oral bacteria can trigger respiratory complications, especially in the elderly and critically ill. Adequate oral care becomes a linchpin in preventing such dire consequences (Winning et al., 2021). Childhood oral health isn't a solitary battle; it's a collective responsibility. Educate, prevent, and intervene early; these are the threads weaving a tapestry of oral well-being for future generations.

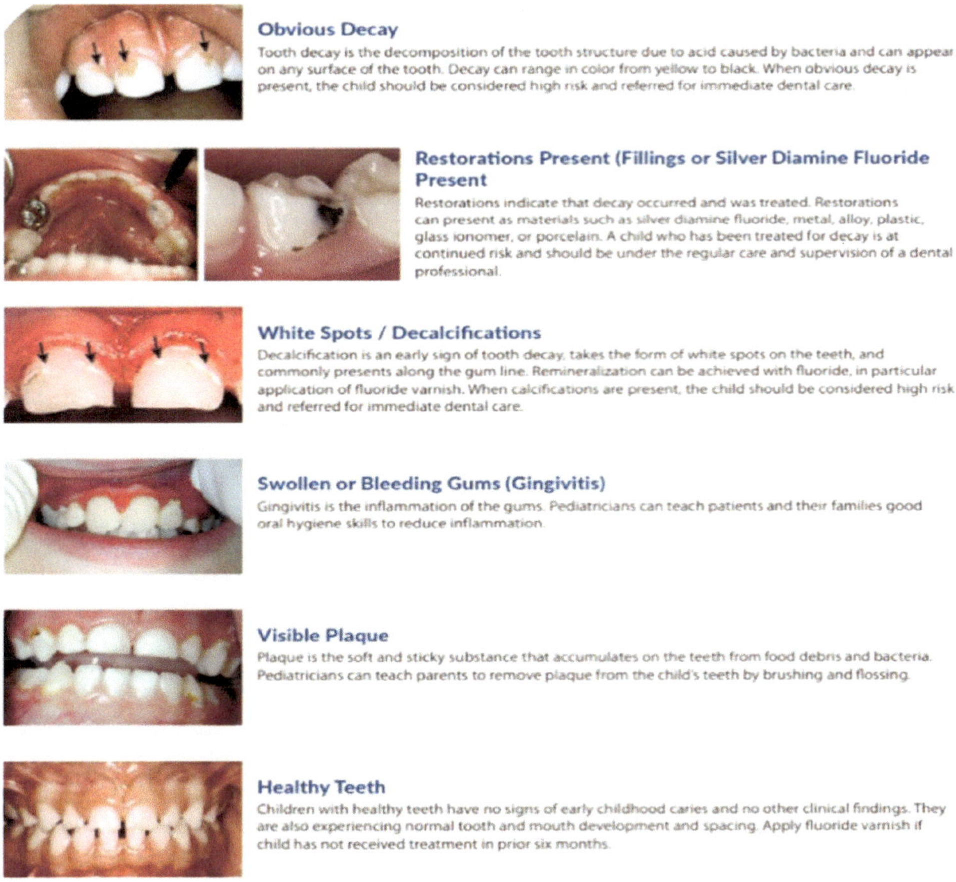

FIGURE 18.2 Clinical findings in decayed and healthy teeth in children.

18.5 FACTORS IMPACTING CHILDHOOD ORAL MICROBES

Within the complex interplay of oral health, a fragile equilibrium is maintained between our bodies and the tiny inhabitants within them. This equilibrium, crucial for oral well-being, faces disruption from various fronts. Think of our habits: the way we eat, consume sugars, or even partake in tobacco. These actions wield the power to upset the balance between us and the microbial world within. Alterations in this internal community often pave the path for oral diseases, including the notorious periodontal disease. Diet emerges as a crucial factor influencing our oral microbiome. Studies link certain foods like farmed animal meat, refined oils, and processed cereals to a shift in microbial composition, favoring acid-producing organisms and periodontal pathogens. Proteins and simple carbohydrates, seemingly harmless, create acidic environments that breed inflammation. Amidst this complex interplay, strategies emerge for promoting oral well-being. Beyond routine oral hygiene, dietary control, and vitamin supplements, interventions involving probiotics, prebiotics, or symbiotic agents show promise in fostering a balanced microbial environment (Santonocito et al., 2022). The early years, a critical period for imprinting health behaviors, pave the way for future oral health. Studies illuminate the link between proper oral hygiene and reduced gingivitis prevalence among children. Fair or poor oral hygiene amplifies the risk of bleeding, signaling the need for widespread oral hygiene promotion among schoolchildren (Pawlaczyk-Kamieńska et al., 2018).

ECC, a scourge affecting marginalized communities, stems from factors like frequent sugar intake, inadequate toothbrushing, and environmental stressors altering caregiver behaviors. Despite advancements in preventive dentistry, ECC persists, rooted in complex societal and environmental issues (Seow, 2018). Enter saliva, the often-overlooked force regulating oral microbial composition. Saliva's molecules shape surfaces for bacterial attachment while delivering vital defenses. Changes in saliva flow can swiftly disrupt the balance, leading to microbial imbalance (Marsh et al., 2015). Dental caries, a chronic ailment, originates from a combination of factors: microbial activity, dietary carbohydrates, and host characteristics.

Societal and behavioral influences further worsen its impact. The biofilm formed by microorganisms causes significant harm by demineralizing enamel and leading to irreversible damage if unchecked (Strużycka, 2014). Risk factors affecting a child's susceptibility to caries include maternal oral health, maternal access to dental care, continual use of bottle/sippy cups with sugary liquids, frequent snacking, and special health care needs. Children with caregivers experiencing dental issues or lacking regular dental care are at higher risk. Continuous consumption of sugary drinks without neutralization by saliva increases the risk. Frequent snacking on sugary and refined foods also heightens susceptibility. Additionally, children with special health needs, including specific medical conditions and premature birth, face increased caries risk due to various factors such as dry mouth, difficulties in oral hygiene, and enamel issues.

18.6 ROLE OF PARENTS AND CAREGIVERS IN PROMOTING PEDIATRIC HEALTH

If parents brush their children's teeth with fluoride toothpaste every morning and before bedtime, the youngster is more likely to continue the practice as an adult. Similarly, if parents feed their children healthy, tooth-friendly meals, the child is more likely to eat healthful foods as an adult. Furthermore, if parents take their child to the dentist regularly, the child is more likely to seek oral health treatment as an adult. Educating parents and caregivers about their children's oral health is critical for promoting healthy dental practices and preventing future dental problems. Some key points parents and caregivers should do for the good oral health of their children are given in this section.

18.6.1 Early Care

Stressing the need for oral care before the first tooth appears is an important part of early dental hygiene that lays the groundwork for a child's oral health. Even before the first tooth emerges, bacteria from milk or formula can be found in the baby's mouth. Cleaning the gums after each meal helps remove these leftovers, reducing bacteria buildup that could lead to dental health problems. Gum cleaning aids in the prevention of gum problems such as inflammation and irritation. It keeps the baby's oral tissues healthy, allowing the teeth to emerge in a healthy environment. Providing oral care for babies sets a good example for parents and caregivers. Children are more likely to embrace these activities as they grow older if they see their parents doing so.

18.6.2 First Dental Visit

Scheduling a child's first dental visit by his or her first birthday, or within 6 months of the first tooth eruption, is an important step toward promoting lifelong oral health. The primary purpose of the first dental visit is to identify and manage any potential oral health issues as soon as possible. Even if a kid has only a few teeth at this age, a dental professional can check tooth and gum growth, finding any irregularities or concerns. Parents can receive advice on preventing ECC through proper oral hygiene habits, diet suggestions, and fluoride use by scheduling a dental visit early. Teething is a difficult stage for both infants and parents. The first dental visit provides an opportunity for parents to learn about their child's oral health. Dental specialists can advise on proper brushing and flossing procedures, the use of fluoride, and the necessity of eating a balanced diet for good oral health. Regular checkups can prevent difficulties by recognizing and resolving dental diseases in their early stages. Treating a minor cavity early, for example, can avoid the need for more invasive operations such as root canals or extractions.

18.6.3 Educating Parents

It is critical to educate parents on the impact of nutrition on oral health to instill good oral habits in their children. There is a need to emphasize the importance of sugar consumption frequency. Frequent snacking or drinking sugary drinks throughout the day exposes teeth to sustained acid assaults, increasing the risk of decay. Parents should be encouraged to consult with doctors or dentists for individualized advice on their child's nutritional needs and any specific oral health concerns. They should be aware of the link between diet and oral hygiene. Even if one eats a healthy diet, frequent brushing and flossing are necessary for good oral health.

18.7 ROLE OF THE DIET IN SHAPING THE ORAL MICROBIOME

The significance of food and nutrients in oral health is paramount. Diet wields a dual impact, with the potential for both positive and negative effects on the development and maintenance of human dentition. The balance lies in the appropriate consumption of specific micronutrients (such as vitamins and minerals) and macronutrients (including carbohydrates, proteins, and lipids). Understanding the role of these elements is crucial for fostering overall oral health. The oral microbiome, a diverse community of microorganisms residing in the oral cavity, is strongly influenced by diet. The foods and nutrients that individuals consume play a crucial role in shaping the composition and function of the oral microbiome. This discussion explores the dynamic relationship between diet and the oral microbiome, highlighting the impact on oral health (Chandan & Rao 2023).

18.7.1 Dietary Influences on Microbial Diversity

18.7.1.1 Fiber and Plant-Based Diets

Diets rich in fiber and plant-based foods promote microbial diversity in the oral cavity. Fiber acts as a prebiotic, supporting the growth of beneficial bacteria. A diet abundant in fiber, fruits, and vegetables can serve as a protective shield against oral diseases through various mechanical actions. This includes cleansing tooth surfaces by removing plaque and promoting increased chewing. The latter stimulates the parotid salivary gland, enhancing salivary flow and eliminating harmful bacteria. Ultimately, this dietary approach acts as a safeguard, preventing the demineralization of tooth substance (Khafid et al., 2023).

18.7.1.2 Fermented Foods, Probiotics, and Postbiotics

Fermented foods, such as kombucha and yogurt, contribute beneficial bacteria that positively influence gut microbiota diversity, reducing inflammatory biomarkers. Additionally, fermented foods show promise in addressing various health conditions, from insomnia to colitis, through modulation of the gut microbiota. Certain fermented dairy products, like probiotic fermented milk and yogurt, have been found to regulate the oral microbiota, aiding in the prevention of periodontal disease and dental caries, while also exhibiting anti-inflammatory effects (Nagakubo & Kaibori, 2023). Probiotics, commonly found in fermented foods, play crucial roles in maintaining a balanced oral microbiome and have therapeutic potential in preventing and treating various diseases (Figure 18.3). The emerging interest in postbiotics, particularly their durability and ease of incorporation into food formulations, offers a promising alternative to live probiotics. Research indicates that both probiotics and postbiotics can significantly reduce oral health issues, including tooth decay and fungal

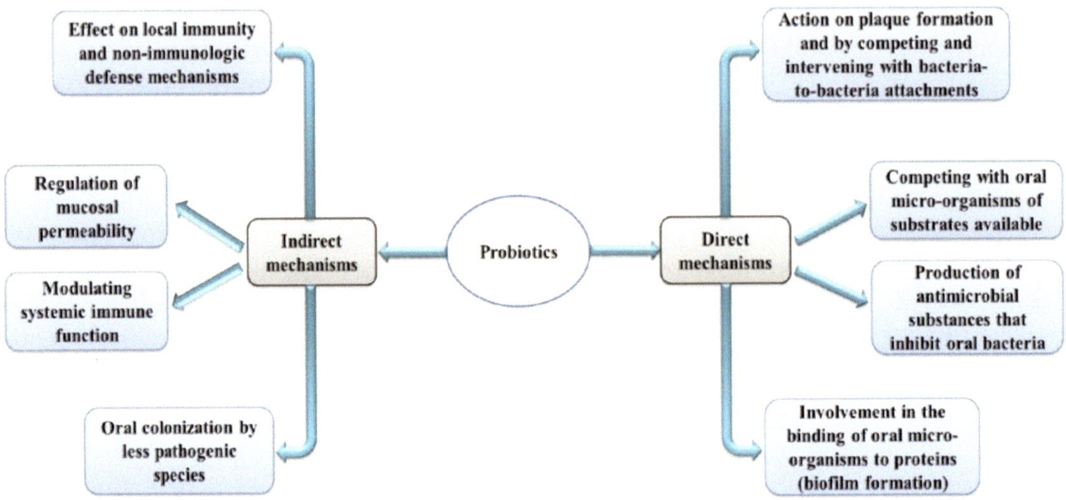

FIGURE 18.3 Direct and indirect effects of probiotics on human health. (From Homayouni Rad et al., 2023.)

infections. However, challenges remain in establishing the durability of postbiotics compared to probiotics, necessitating further studies (Homayouni Rad et al., 2023).

18.7.2 Impact of Sugars on the Oral Microbiome

18.7.2.1 Sucrose and Dental Caries

High sucrose intake promotes the growth of acid-producing bacteria linked to dental caries, as sucrose acts as a substrate for the production of enamel-damaging acids (Azevedo et al., 2023). Dental caries, a prevalent oral health concern, is linked to dietary carbohydrates. The degradation of these carbohydrates by oral bacteria forms organic acids, leading to tooth surface solubilization. Sugar intake, especially free sugars, plays a significant role, with the World Health Organization (WHO) recommending their consumption to be less than 10% of total energy intake. The cariogenic potential depends on factors such as frequency, stickiness, and form of sugar consumed. Sucrose, a disaccharide, is particularly cariogenic due to its rapid glycolytic pathway entry. In contrast, lactose, found in milk, is less cariogenic due to calcium's protective effect. Sugar alcohols like xylitol and sorbitol have shown cariostatic effects by reducing acidogenic potential and inhibiting *Streptococcus mutans* growth. Adopting a low-sugar or sugar-free diet and focusing on low-glycemic, unprocessed, and complex carbohydrates contributes to both oral and overall health. A diet low in free sugars is beneficial for both oral and overall health. Recommendations include limiting high-glycemic, processed, fermentable carbohydrates (such as sugar-sweetened foods and beverages) and opting for low-glycemic, unprocessed, complex carbohydrates (e.g., apples, cranberries, tea, and high-fiber wholegrain foods). Including modest amounts of legumes, lean meat, and ample fluids, including milk and water, is encouraged for a well-balanced approach (Chandan and Rao, 2023).

18.7.3 Macronutrients and Micronutrients

18.7.3.1 Fat and Protein

Dietary fats and proteins play crucial roles in shaping the oral microbial environment. Fats serve as an energy source and contribute to essential components like cell membranes, while proteins, as fundamental building blocks, support cellular growth and repair. Triglycerides, phospholipids, and sterols constitute dietary lipids, with an emphasis on reducing saturated fats by lowering levels of low-density lipoproteins (LDL). Research suggests that saturated fats and omega-6 fatty acids may induce inflammation, while omega-3 fatty acids exhibit anti-inflammatory effects. Disruptions in fat and protein intake can impact the oral microbiome. Proteins, composed of amino acid chains, are essential for growth and repair. Essential amino acids, replenished through dietary proteins, must be considered for a balanced diet. Protein deficiency, often due to malnutrition, can affect immune function and wound healing and increase susceptibility to oral infections. Plant products may lack certain amino acids, and animal products, while offering complete proteins, may be associated with increased insulin-like growth factor 1, linked to carcinogenesis. Protein-rich foods are generally non-cariogenic (Santonocito et al., 2021).

18.7.3.2 Vitamins and Minerals

Adequate vitamin and mineral intake are crucial for a healthy oral microbiome and overall well-being. The bidirectional relationship between vitamins and oral health emphasizes that vitamin deficiency can lead to tooth or tissue degradation, while oral issues can impact dietary choices, causing vitamin imbalances. Fat-soluble vitamins (A, D, E, and K) are stored in the liver and fatty tissues and can become toxic with excess intake. Water-soluble vitamins (B-complex and C) require regular replenishment in the diet to prevent deficiency, as they are not stored in the body. Understanding functions, deficiency manifestations, recommended daily allowances, and achieving a balanced vitamin intake in the diet is imperative (Chandan & Rao, 2023).

Animal studies show that prenatal and maternal vitamin A deficiency impacts cranial and dental development, with additional roles in oral epithelium and periodontium health. Vitamin D is essential for tooth and bone development; vitamin D and its metabolites exhibit antiproliferative, prodifferentiative, and immunomodulatory effects. It reduces enamel defects, influences hydroxyapatite crystal formation, and demonstrates anti-caries effects. Vitamin D deficiency correlates with increased dental caries in children. Vitamin E deficiency is rare; it plays a role in tissue regeneration and positively affects periodontal healing when used as a supplement. The impact of vitamin K on oral health is less clear; its deficiency can affect clotting mechanisms, leading to an increased risk of hemorrhage, especially after oral surgery. B-complex vitamins are essential for cell metabolic activity; they include thiamine (B1), riboflavin (B2), niacin (B3), pantothenic acid (B5), pyridoxin (B6), biotin (B8), folic acid (B9), and cobalamin (B12). Deficiencies manifest in various oral issues, such as beriberi, neurodegenerative disorders, cheilitis, angular stomatitis, glossitis, and severe gingivitis. Vitamin C is crucial for collagen synthesis in gingival and periodontal connective tissue; vitamin C deficiency causes scurvy, characterized by bleeding gums, joint effusions, impaired wound healing, and compromised periodontal health (Carvalho et al., 2020).

Micronutrients, essential for supporting the immune system and oral tissues, are divided into major (macro) minerals and trace elements (microminerals). A balanced diet should provide these requirements. These minerals play vital roles in enzymatic actions and biological reactions. Calcium, magnesium, and phosphorus form the major components of bones and teeth, crucial for tooth mineralization and formation. Calcium regulates tooth development, muscle contraction, nerve conduction, and immune health. Vitamin D aids in their regulation, preventing bone fragility, hypo-mineralization, osteoporosis, and tooth decay. Fluoride strengthens bones and enamel, making them resistant to demineralization. However, an excess leads to fluorosis, causing discolored teeth and bone issues. Iron, vital for hemoglobin and immune function, in deficiency anemia manifests as oral burning, redness, angular stomatitis, dysphagia, and pallor. Zinc acts as a cofactor for cell growth, DNA/RNA stabilization, and collagen synthesis. Deficiency results in a loss of taste, loss of appetite, and xerostomia, contributing to malnutrition and infections. Other minerals (sodium, potassium, iodine, copper, selenium, and manganese) are vital for fluid balance, growth, and enzymatic action. Oral manifestations of their deficiency are rare (Dodhia et al., 2021).

18.8 IMPACT OF NUTRITION ON PREVENTING ORAL HEALTH ISSUES IN CHILDREN

Nutrition is foundational to pediatric oral health, shaping the development of teeth, gums, and overall oral structures. Ensuring adequate nutrient intake is crucial for preventing dental problems and promoting optimal oral health in children. This exploration delves into the nutritional factors influencing pediatric oral health, underscoring the significance of a balanced diet in averting dental issues. Adequate intake of essential nutrients plays a pivotal role in the development and maintenance of oral structures, immune function, and the prevention of conditions like dental caries and periodontal diseases. A well-balanced diet, low in sugars and rich in fruits, vegetables, and whole grains, serves as a preventive measure against dental caries by minimizing the availability of substrates for acid-producing bacteria. High fiber intake is a marker of overall dietary quality and a healthy lifestyle. The childhood period is an important time to develop an oral healthy lifestyle, and providing a high fiber intake is an easy way to help children maintain good oral hygiene. Figure 18.4 shows the intra-oral conditions before and after chewing high-fiber food (Khafid et al., 2023).Probiotics found in fermented foods contribute to a balanced oral microbiome, reducing the risk of oral health issues. Antioxidants, obtained through a diet rich in fruits and vegetables, help protect oral tissues from oxidative stress and inflammation.

18.9 PROMOTING A HEALTHY ORAL MICROBIOME IN CHILDREN: STRATEGIES AND INTERVENTIONS

Schools are key in promoting children's health as per the Ottawa Charter for Health Promotion. While measures exist to boost oral health at different levels, their comprehensive adoption in schools worldwide is limited. Sharing local insights, navigating budget constraints, and integrating health promotion into school life is vital. This involves crafting health policies, implementing interventions, and exchanging experiences. Collaboration, especially in low- and middle-income countries, is essential for global oral health promotion (Jürgensen & Petersen, 2013). The WHO defines probiotics as beneficial microorganisms that confer health benefits when administered in adequate amounts. While traditionally used for gastrointestinal health, interest has surged in exploring their potential impact on oral health. Studies have shown promising results in inhibiting harmful oral bacteria like *Streptococcus mutans* and *Candida albicans*. However, the long-term effects and specific strains' efficacy in preventing dental caries warrant further extensive clinical studies before widespread recommendations can be made (Hasslöf & Stecksén-Blicks, 2019). Despite various preventive measures available, dental caries remains a prevalent childhood chronic disease. Disparities in oral health and access to care persist, particularly affecting low-income and minority children. Initiatives such as water fluoridation, topical fluorides, dental sealants, and fluoride toothpaste aid in preventing caries but encounter challenges, including the lack of widespread fluoridation in water supplies. Addressing pediatric oral care gaps requires a collaborative approach involving pediatricians, prioritized Medicaid services, integrated medical and dental training, and national leadership to emphasize preventive oral care in primary pediatric care (Mouradian, 2000). The impact

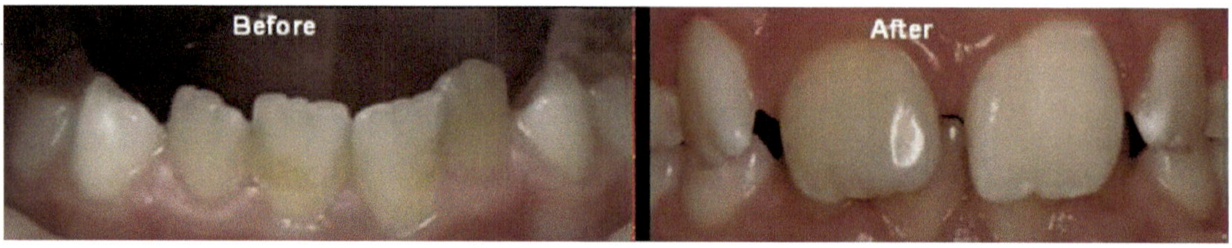

FIGURE 18.4 Intra-oral conditions before and after chewing high-fiber food. (From Khafid et al., 2023.)

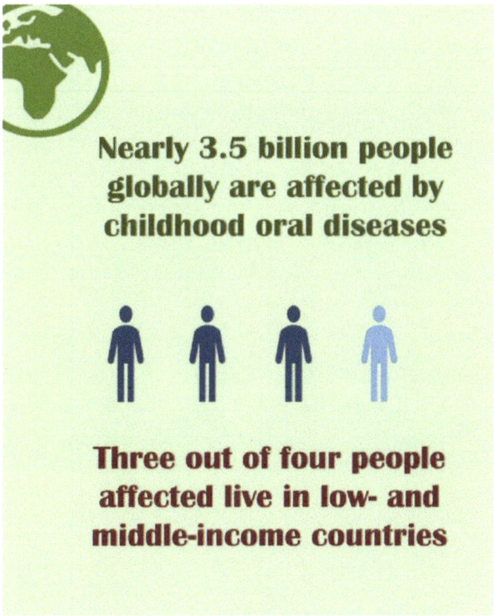

FIGURE 18.5 Prevalence of oral diseases. (Oral Health WHO, 2024)

of dental problems in early childhood extends beyond oral health, affecting overall quality of life and potentially contributing to long-term health issues. Novel approaches in dental care, such as partial removal or sealing of carious lesions, deserve further exploration and assessment to improve outcomes and ensure healthier permanent dentitions for children. Oral diseases persist as a global public health concern, with dental caries, periodontal disease, tooth loss, and oral cancers among the most prevalent and impactful. These diseases significantly burden individuals and health systems, particularly in low- and middle-income countries, due to social and economic disparities and inadequate funding for prevention and treatment (Drummond et al., 2013) (Figure 18.5).

Despite being largely preventable, untreated oral diseases lead to severe consequences, including pain, reduced quality of life, disruptions in daily routines, and economic strains. Efforts to address oral health disparities, especially among marginalized groups and in resource-limited settings, remain a critical challenge in the global health care landscape (Peres et al., 2019). In summary, fostering oral health in children requires concerted efforts at various levels, from schools to health care systems, focusing on preventive measures, research advancements, and equitable access to care, aiming to alleviate the substantial burdens of oral diseases on individuals and societies worldwide.

18.10 ADVANCING ORAL HEALTH INITIATIVES

The current Global Strategy on Oral Health and the forthcoming Global Oral Health Action Plan mark significant milestones, delineating a path toward comprehensive oral health for all by 2030. The WHO Global Oral Health Status Report offers a comprehensive assessment of oral disease burdens, available resources, and upcoming challenges (Global oral health status report WHO, 2024).

Regionally, the WHO Southeast Asia Region took a pioneering step by developing an action plan aligning with the *Global Strategy and Action Plan on Oral Health 2023–2030*. This plan, endorsed in 2022, aims for universal oral health coverage by 2030, focusing on reducing premature mortality from oral cancer and the prevalence of untreated dental caries. In the United States, the American Academy of Pediatrics (AAP) initiated an interdisciplinary project, engaging pediatric dentists, dental hygienists, pediatricians, ob-gyns, and practice managers. This project, spanning seven months, focused on pediatric oral health improvement across eight states. Through a learning collaborative format, virtual meetings facilitated education and collaboration, emphasizing skills like caries risk assessments, fluoride varnish application, and oral health counseling within medical practices. The project empowered participating physicians to earn certification points while distributing oral health starter kits comprising toothbrushes, fluoridated toothpaste, and educational materials to patients and caregivers. An infographic related to Protect Tiny Teeth Implementation Project (Centers for Disease Control and Prevention) is presented in Figure 18.6. The ongoing initiatives and collaborative efforts, both globally and regionally, underscore the commitment to advancing oral health through comprehensive strategies, education, and interdisciplinary collaborations. The

FIGURE 18.6 Infant oral health infographics. (From American Academy of Pediatrics, 2023.)

focus on prevention and intervention signals a promising direction in mitigating oral disease burdens worldwide.

18.11 CONCLUSION

The oral microbiome, consisting of diverse microorganisms in the oral cavity, is integral to overall health, particularly in pediatric populations. Understanding its development is pivotal for pediatric health. A balanced oral microbiome in early childhood is linked to preventing oral diseases and influencing systemic health. Disruptions can lead to issues like early childhood caries and gingivitis, impacting a child's well-being. The research underscores the intricate link between the oral microbiome and systemic health in children, including respiratory infections, diabetes, and cardiovascular diseases. This emphasizes the importance of adopting a holistic approach to pediatric health care. Good oral hygiene, a balanced diet, and frequent dental checkups are essential for children to have a healthy oral microbiota. Limiting sugar intake and encouraging practices that promote varied microbiota contribute to overall oral and systemic health. The mode of delivery, breastfeeding, and the introduction to solid foods are critical factors shaping the infancy oral microbiome. These early-life exposures have a lasting impact on the microbiome's diversity and stability, affecting susceptibility to both oral and systemic diseases in later stages of life. Ongoing research in this field will likely uncover novel insights, paving the way for innovative approaches to preventive and therapeutic interventions in pediatric oral and systemic health.

REFERENCES

Aagaard, K.; Ma, J.; Antony, K.M.; Ganu, R.; Petrosino, J.; Versalovic, J. The placenta harbors a unique microbiome. *Sci Transl Med.* **2014**, 6(237):237ra65.

Al-Shehri, S.S.; Knox, C.L.; Liley, H.G.; et al. Breastmilk-saliva interactions boost innate immunity by regulating the oral microbiome in early infancy. *PLoS One.* **2015**, 10(9):e0135047

American Academy of Pediatrics. **2023**. Oral Health Campaign Toolkit. https://www.aap.org/en/news-room/campaigns-and-toolkits/oral-health

Arweiler, N. B.; Netuschil, L. The oral microbiota. *Adv Exp Med Biol.* **2016**, 45–60. https://doi.org/10.1007/978-3-319-31248-4_4

Azevedo, M. J.; Garcia, A.; Costa, C. F.; Ferreira, A. F.; Falcão-Pires, I.; Brandt, B. W.; Sampaio-Maia, B. The contribution of maternal factors to the oral microbiota of the child: Influence from early life and clinical relevance. *Jpn Dent Sci Rev.* **2023**, 59:191–202.

Boix-Amoros, A.; Collado, M.C.; Mira, A. Relationship between milk microbiota, bacterial load, macronutrients, and human cells during lactation. *Front Microbiol.* **2016**, 7:492.

Brailsford, S.R.; Sheehy, E.C.; Gilbert, S.C.; et al. The microflora of the erupting first permanent molar. *Caries Res.* **2005**, 39:78–84.

Carvalho S.C.; Mendes, R.; Manso, M.D.C.; Gavinha, S.; Melo, P. Prenatal or childhood serum levels of vitamin D and dental caries in pediatric patients: a systematic review. *Oral Health Prev Dent.* **2020**, 18:653–667. https://doi.org/10.3290/j.ohpd.a45089

Centers for Disease Control and Prevention. Protect Tiny Teeth Implementation Project: A prenatal oral health medical-dental integration quality improvement project, 2021–2022. Available from https://www.cdc.gov Accessed on 06/12/2023

Chandan, S. N.; Rao, S. Dietary interventions and nutritional impact on oral health and development: A review. *J Food Sci Technol.* **2023**. https://doi.org/10.1007/s13197-022-05423-2

Chi, D.L.; Scott, J. M. Added Sugar and Dental Caries in Children. *Dent Clin North Am.* **2019**, 63(1):17–33. https://doi.org/10.1016/j.cden.2018.08.003

Chu, D.M.; Ma, J.; Prince, A.L.; Antony, K.M.; Seferovic, M.D.; Aagaard K.M. Maturation of the infant microbiome community structure and function across multiple body sites and about mode of delivery. *Nat Med.* **2017**, 23:314–326.

Dodhia, S.A.; West, N.X.; Thomas, S.J.; Timpson, N.J.; Johansson, I.; Lif Holgerson, P.; Dudding, T.; Haworth, S. Examining the causal association between 25-hydroxyvitamin D and caries in children and adults: a two-sample Mendelian randomization approach. *Wellcome Open Res.* **2021**, 5:281. https://doi.org/10.12688/wellcomeopenres.16369.2.

Drummond, B. K.; Meldrum, A. M.; Boyd, D. Influence of dental care on children's oral health and wellbeing. *Br Dent J.* **2013**, 214(11):E27.

Dzidic, M.; Collado, M. C.; Abrahamsson, T.; Artacho, A.; Stensson, M.; Jenmalm, M. C.; Mira, A. Oral microbiome development during childhood: An ecological succession influenced by postnatal factors and associated with tooth decay. *ISME J.* **2018**, 12(9):2292–2306.

Elkhodary, H. M.; Abdelnabi, M. H.; Swelem, A. A.; Sabbagh, H. J.; El Meligy, O. A. E. S.; Talaat, I. M.; Abdellatif, E. B.; Khader, Y.; Al-Batayneh, O. B.; Al-Khanati, N. M.; Nurelhuda, N. M.; Alhabli, S.; Mostafa, M. H.; Qureshi, S.; Qureshi, N.; Yousaf, M. A.; Taha, D.; Marafi, Y. F.; Al Harris, S. N.; El Tantawi, M. Individual, familial and country-level factors associated with oral hygiene practices in children: An international survey. *BMC Oral Health.* **2023**, 23(1). https://doi.org/10.1186/s12903-023-02746-0.

Fitzstevens, J.L.; Smith, K.C.; Hagadorn, J.I.; Caimano, M.J.; Matson, A.P.; Brownell, E.A. Systematic review of the human milk microbiota. *Nutr Clin Pract.* **2017**, 32(3):354–364.

Gomez, A.; Espinoza, J. L.; Harkins, D. M.; Leong, P.; Saffery, R.; Bockmann, M.; Nelson, K. E. Host genetic control of the oral microbiome in health and disease. *Cell Host Microbe.* **2017**, 22(3):269–278.

Gomez, A.; Nelson, K.E. The oral microbiome of children: Development, disease, and implications beyond oral health. *Microb Ecol.* **2017**, 73(2):492–503

Gude, N.M.; Roberts, C.T.; Kalionis, B.; King, R.G. Growth and function of the normal human placenta. *Thromb Res.* **2004**, 114(5-6):397–407.

Hasslöf, P.; Stecksén-Blicks, C. Chapter 10: Probiotic Bacteria and Dental Caries. The Impact of Nutrition and Diet on Oral Health **2019**, 99–107. https://doi.org/10.1159/000455377

He, Z.; Wu, J.; Xiao, B.; Xiao, S.; Li, H.; Wu, K. The initial oral microbiota of neonates among subjects with gestational diabetes mellitus. *Front Pediatr.* **2019**, 7, 513.

Homayouni Rad, A.; Pourjafar, H.; Mirzakhani, E. A comprehensive review of the application of probiotics and postbiotics in oral health. *Front Cell Infect Microbiol.* **2023**, 13:1120995.

Hou, K.; Wu, Z.-X.; Chen, X.-Y.; Wang, J.-Q.; Zhang, D.; Xiao, C.; Zhu, D.; Koya, J. B.; Wei, L.; Li, J.; Chen, Z.-S. Microbiota in health and diseases *Signal Transd Target Ther.* **2022**, 7(1):135. https://doi.org/10.1038/s41392-022-00974-4

Hurley, E.; Mullins, D.; Barrett, M.P.; O'Shea, C.A.; Kinirons, M.; Ryan, C.A.; Stanton, C.; Whelton, H.; Harris, H.M.; O'Toole, P.W. The microbiota of the mother at birth and its influence on the emerging infant oral microbiota from birth to 1 year of age: A cohort study. *Oral Microbiol.* **2019**, 11(1):1599652

Jeurink, P.V.; Bergenhenegouwen, J.; Jimenez, E.; et al. Human milk: A source of more life than we imagine. *Benef Microbes.* **2013**, 4(1):17–30.

Jürgensen, N.; Petersen, P.E. Promoting oral health of children through schools–results from a WHO global survey 2012. *Community Dent Health.* **2013**, 30(4):204–218. https://doi.org/10.1922/CDH_3283Petersen15

Kaan, A. M. Marije; Kahharova, D.; Zaura, E. Acquisition and establishment of the oral microbiota. *Periodontology 2000.* **2021**, 86(1):123–141. https://doi.org/10.1111/prd.12366

Kazeminia, M.; Abdi, A.; Shohaimi, S.; Jalali, R.; Vaisi-Raygani, A.; Salari, N.; Mohammadi, M. Dental caries in primary and permanent teeth in children worldwide, 1995 to 2019: A systematic review and meta-analysis. *Head Face Med.* **2020**, 16(1). https://doi.org/10.1186/s13005-020-00237-z

Khafid, M.; Triharsono, A. A. A. P.; Prasiska, D. I. Prevention dental caries by improving oral hygiene through a high-fiber diet in children with mixed dentition: Experimental study in Kediri, East Java, Indonesia. *World J Adv Res Rev.* **2023**, *18*(3):204–208.

Kolenbrander, P.E.; Palmer, R.J. Jr.; Rickard, A.H.; Jakubovics, N.S.; Chalmers, N.I.; Diaz, P.I. Bacterial interactions and successions during plaque development. *Periodontol 2000* **2006**, 42:47–79.

Li, H.; Xiao, B.; Zhang, Y.; Xiao, S.; Luo, J.; Huang, W. Impact of maternal intrapartum antibiotics on the initial oral microbiome of neonates. *Pediatr Neonatol.* **2019**, 60(6):654–661.

Lif Holgerson, P.; Harnevik, L.; Hernell, O.; Tanner, A.; Johansson, I. Mode of birth delivery affects oral microbiota in infants. *J Dent Res.* **2011**, 90(10):1183–1188.

Lundgren, S.N.; Madan, J.C.; Emond, J.A.; Morrison, H.G.; Christensen, B.C.; Karagas, M.R.; Hoen, A.G. Maternal diet during pregnancy is related to the infant stool microbiome in a delivery mode-dependent manner. *Microbiome.* **2018**, 6(1):109

Marsh, P. D.; Do, T.; Beighton, D.; Devine, D. A. Influence of saliva on the oral microbiota. *Periodontol 2000.* **2015**, 70(1):80–92. https://doi.org/10.1111/prd.12098

Martın, R.; Heilig, G.; Zoetendal, E.; Smidt, H.; Rodrıguez, J. Diversity of the Lactobacillus group in breast milk and vagina of healthy women and potential role in the colonization of the infant gut. *J Appl Microbiol.* **2007**, 103(6):2638–2644

Mason, M.R.; Chambers, S.; Dabdoub, S.M.; Thikkurissy, S.; Kumar, P.S. Characterizing oral microbial communities across dentition states and colonization niches. *Microbiome.* **2018**, 6(1):67

Mathur, V. P.; Dhillon, J. K. Dental caries: A disease which needs attention. *Indian J Pediatr.* **2018**, 85:202–206.

Mold, J.E.; Michaelsson, J.; Burt, T.D.; et al. Maternal alloantigens promote the development of tolerogenic fetal regulatory T cells in utero. *Science.* **2008**, 322(5907):1562–1565.

Mosaddad, S. A.; Tahmasebi, E.; Yazdanian, A.; Rezvani, M. B.; Seifalian, A.; Yazdanian, M.; Tebyanian, H. Oral microbial biofilms: An update. *Eur J Clin Microbiol Infect Dis.* **2019**, 38(11):2005–2019. https://doi.org/10.1007/s10096-019-03641-9

Mouradian, W. E. Disparities in Children's Oral Health and Access to Dental Care. *JAMA.* **2000**, 284(20):2625. https://doi.org/10.1001/jama.284.20.2625

Nagakubo, D.; Kaibori, Y. Oral microbiota: The influences and interactions of saliva, IgA, and dietary factors in health and disease. *Microorganisms.* **2023**, 11(9):2307.

Pawlaczyk-Kamieńska, T.; Torlińska-Walkowiak, N.; Borysewicz-Lewicka, M. The relationship between oral hygiene level and gingivitis in children. *Adv Clin Exp Med.* **2018**, 27(10):1397–1401. https://doi.org/10.17219/acem/70417

Peres, M. A.; Macpherson, L. M. D.; Weyant, R. J.; Daly, B.; Venturelli, R.; Mathur, M. R.; Listl, S.; Celeste, R. K.; Guarnizo-Herreño, C. C.; Kearns, C.; Benzian, H.; Allison, P.; Watt, R. G. Oral diseases: A global public health challenge. *Lancet.* **2019**, 394(10194):249–260. https://doi.org/10.1016/s0140-6736(19)31146-8

Perez-Munoz, M.E.; Arrieta, M.C.; Ramer-Tait, A.E.; Walter, J. A critical assessment of the "sterile womb" and "in utero colonization" hypothesis: Implications for research on the pioneer infant microbiome. *Microbiome.* **2017**, 5(1):48.

Rodriguez, J.M. The origin of human milk bacteria: Is there a bacterial entero-mammary pathway during late pregnancy and lactation. *Adv Nutr.* **2014**, 5(6):779–784

Santonocito, S.; Giudice, A.; Polizzi, A.; Troiano, G.; Merlo, E. M.; Sclafani, R.; Grosso, G.; Isola, G. A Cross-Talk between Diet and the Oral Microbiome: Balance of Nutrition on Inflammation and Immune System's Response during Periodontitis. *Nutrients.* **2022**, 14(12):2426. https://doi.org/10.3390/nu14122426

Santonocito, S.; Polizzi, A.; Palazzo, G.; Indelicato, F.; Isola, G. Dietary factors affecting the prevalence and impact of periodontal disease. *Clin Cosmet Investig Dent.* **2021**, 13:283–292. https://doi.org/10.2147/CCIDE.S288137

Seow, W. K. Early Childhood Caries. *Pediatr Clin North Am.* **2018**, 65(5):941–954. https://doi.org/10.1016/j.pcl.2018.05.004

Shi, W.; Tian, J.; Xu, H.; Zhou, Q.; Qin, M. Distinctions and associations between the microbiota of saliva and supragingival plaque of permanent and deciduous teeth. *PLoS One.* **2018**, 13(7):e0200337

Strużycka, I. The Oral Microbiome in Dental Caries. *Po J Microbiol.* **2014**, 63(2):127–135. https://doi.org/10.33073/pjm-2014-018

Welch, J.L.M.; Ramırez-Puebla, S.T.; Borisy, G.G. Oral microbiome geography: Micron-scale habitat and niche. *Cell Host Microbe.* **2020**, 28(2):160–168.

Winning, L.; Lundy, F. T.; Blackwood, B.; McAuley, D. F.; El Karim, I. Oral health care for the critically ill: A narrative review. *Crit Care.* **2021**, 25(1). https://doi.org/10.1186/s13054-021-03765-5

World Health Organization (WHO). Global oral health status report: Towards universal health coverage for oral health by 2030: Regional summary of the Region of the Americas Management of Noncommunicable Diseases (MND), Noncommunicable diseases. Available from: https://www.who.int/health-topics/oral-health#tab=tab_1 Accessed on 27/12/2024.

Zaura, E.; Nicu, E.A.; Krom, B.P.; Keijser, B.J. Acquiring and maintaining a normal oral microbiome: current perspective. *Front Cell Infect Microbiol.* **2014**, 4(85). https://doi.org/10.3389/fcimb.2014.00085.

Zhang, Y.; Wang, X.; Li, H.; Ni, C.; Du, Z.; Yan, F. Human oral microbiota and its modulation for oral health. *Biomed Pharmacother.* **2018**, 99:883–893. https://doi.org/10.1016/j.biopha.2018.01.146

19 Diagnostic Techniques Related to the Oral Microbiome

Rajesh Kumar, Ankita Thakur, and Divya
Department of Biosciences, Himachal Pradesh University, Shimla, Himachal Pradesh, India

Younis Ahmad Hajam
Department of Life Sciences and Allied Health, Sant Baba Bhag Singh University, Jalandhar, Punjab, India

Indu Bala
Sukhjinder Group of Institutes, VPO, Gurdaspur, Punjab, India

Neelam, and Shweta Singh
Department of Life Sciences and Allied Health, Sant Baba Bhag Singh University, Jalandhar, Punjab, India

19.1 INTRODUCTION TO ORAL MICROBIOME ANALYSIS

The oral microbiome, a multifaceted and dynamic microbial community inhabiting surfaces like teeth, gums, the tongue, and mucosal linings, plays a pivotal role in maintaining oral health and influencing systemic well-being, while also contributing to disease pathogenesis [1]. Within the oral cavity, a diverse array of microorganisms, including bacteria, viruses, fungi, and archaea, create a symbiotic relationship with the host. This relationship contributes to crucial physiological processes such as digestion, immune system modulation, and the maintenance of tissue homeostasis, involving notable bacterial [2] phyla like Firmicutes, Bacteroidetes, Actinobacteria, Proteobacteria, and Fusobacteria (Figure 19.1).

Microbial colonization is not uniform, as distinct microenvironments—dental surfaces, periodontal pockets, and mucosal linings—offer unique ecological niches for microbial communities. Dental plaque, a biofilm on teeth, provides a substrate for microbial adherence, while gingival crevices and mucosal surfaces host additional communities [3]. The oral microbiome is subject to dynamic changes influenced by factors like diet, hygiene practices, and systemic health, with a delicate balance among microbial species in a state of health. Disruptions, such as poor oral hygiene or systemic conditions, can lead to dysbiosis, an imbalance associated with oral diseases like periodontitis, dental caries, and gingivitis [4].

The oral microbiome closely interacts with the host's immune system, influencing immune responses and defense mechanisms. Commensal microorganisms contribute to immune system development, while pathogenic species can trigger inflammatory responses. Beyond its localized impact, the oral microbiome has implications for systemic health, with associations noted between oral health and conditions like cardiovascular disease, diabetes, and respiratory infections [5]. This underscores the role of the oral microbiome as a potential modulator of overall health. Hence, the oral microbiome represents a complex and dynamic ecosystem with profound implications for oral and systemic health. A comprehensive understanding of its composition, spatial distribution,

DOI: 10.1201/9781003478638-19

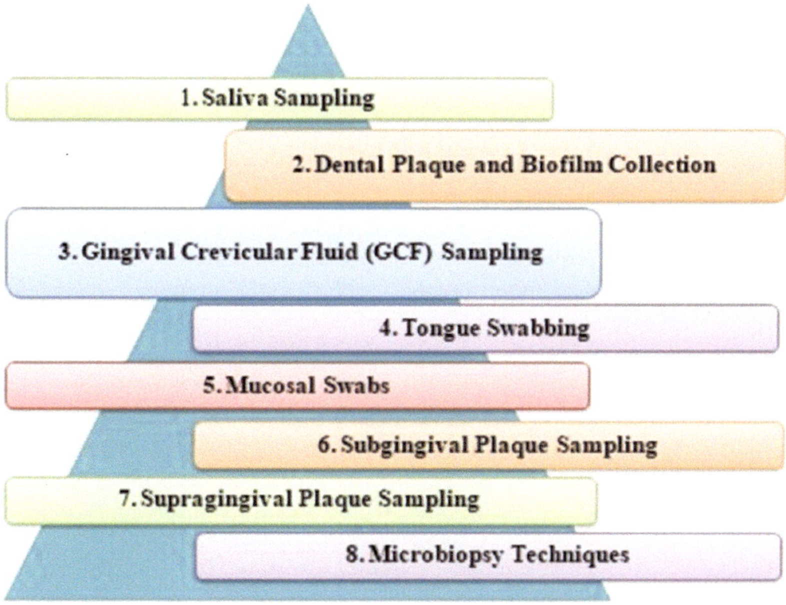

FIGURE 19.1 Different types of microbes present in oral cavity.

dynamics, and interactions with the host is imperative for the development of effective diagnostic and therapeutic strategies in the realms of oral health and beyond [6].

19.1.1 Significance of Studying the Oral Microbiome

The study of the oral microbiome is of paramount importance as it plays a central role in maintaining health, influencing systemic well-being, and contributing to the development of various diseases. A comprehensive understanding of the intricacies of the oral microbiome is crucial for advancing knowledge in both oral and systemic health [7]. Investigating the oral microbiome provides valuable insights into the delicate balance between microbial communities and the host, offering a basis for identifying biomarkers and patterns associated with prevalent oral conditions such as periodontal diseases, dental caries, and gingivitis. This knowledge forms the foundation for developing targeted preventive and therapeutic interventions [8].

The influence of the oral microbiome extends beyond the oral cavity to impact systemic health, with established connections to conditions like cardiovascular diseases, diabetes, and respiratory infections. The study of the oral microbiome provides a platform to explore the mechanisms underpinning these associations, potentially leading to innovative strategies for preventing and managing systemic diseases through interventions in oral health [9]. A nuanced understanding of the oral microbiome opens avenues for developing novel therapeutic approaches, including the exploration of probiotics, prebiotics, and targeted antimicrobial strategies to positively modulate the oral microbial community. Personalized medicine initiatives can leverage insights into individual variations in the oral microbiome to tailor treatment plans, optimizing therapeutic outcomes for diverse patient populations [10]. Furthermore, the oral microbiome serves as a potential diagnostic reservoir for assessing overall health, offering non-invasive avenues for early disease detection through specific microbial signatures and dysbiosis patterns. Diagnostic techniques targeting the oral microbiome present a promising frontier in precision medicine, enabling timely intervention and improved patient outcomes. Knowledge derived from studying the oral microbiome contributes to the development of public health initiatives and preventive strategies, encompassing educational programs, community-based interventions, and policies addressing factors influencing the oral microbiome [11]. These holistic approaches have the potential to enhance oral health on a population scale, with ripple effects on systemic health. Thus, the study of the oral microbiome is essential for unraveling the complexities of oral and systemic health, providing insights into disease mechanisms and ways to overcome pathogenic infections [12].

19.1.2 Importance of Diagnostic Techniques

The importance of diagnostic techniques in studying the oral microbiome cannot be overstated, as these methods play pivotal roles in unraveling the intricate dynamics of microbial communities within the oral cavity. Indispensable for identifying, characterizing, and understanding the diverse array of microorganisms constituting the oral microbiome, these techniques provide critical insights into both oral and systemic health [5]. Specifically, these diagnostic tools enable precise identification of microbial species, facilitating a comprehensive analysis of their abundance, diversity, and interactions. This precision is particularly crucial for distinguishing between health and disease states, aiding in the early detection and monitoring of oral conditions such as periodontal diseases, dental caries, and gingivitis, thereby forming the basis for targeted therapeutic interventions [13]. Moreover, advanced diagnostic techniques contribute significantly to the identification of microbial biomarkers linked to specific oral and systemic health conditions, serving as indicators of microbial dysbiosis or pathogenic presence. This information is valuable for prognostic assessments and the development of personalized treatment strategies, enhancing diagnostic precision and prognostic potential [14]. The dynamic nature of the oral microbiome necessitates continuous monitoring, and diagnostic techniques facilitate longitudinal studies to track alterations in microbial composition and abundance over time in response to various factors. This temporal dimension is crucial for understanding the progression of oral diseases and assessing the efficacy of treatment regimens. Diagnostic techniques also play profound roles in advancing therapeutic approaches by providing insights into how specific interventions impact microbial communities [15]. This knowledge is instrumental in tailoring therapeutic strategies to positively modulate the oral microbiome, promoting health and preventing disease progression. As the era of personalized medicine unfolds, diagnostic techniques contribute significantly to tailoring oral health interventions based on individual variations in the oral microbiome [16]. Understanding unique microbial profiles allows clinicians to design personalized treatment plans, optimizing therapeutic outcomes and minimizing adverse effects, representing a paradigm shift in oral healthcare towards more effective and patient-centric interventions. Additionally, these diagnostic techniques inform public health strategies by providing data on the prevalence and distribution of specific microbial taxa within populations [17]. This information guides the development of targeted public health interventions, educational programs, and policies aimed at improving oral health on a broader scale. Further, the multifaceted role of diagnostic techniques in studying the oral microbiome, from precise disease identification and uncovering microbial biomarkers to monitoring dynamic changes and advancing therapeutic approaches, is indispensable for our understanding of the oral microbiome role in health and disease. Moreover, these techniques pave the way for personalized medicine and informed public health strategies, marking them indispensable tools in oral microbiome research and healthcare [18].

19.2 COMPOSITION AND DIVERSITY OF THE ORAL MICROBIOME

The oral microbiome, a dynamic and intricate ecosystem, constitutes a complex amalgamation of microorganisms populating surfaces within the oral cavity, including the teeth, gums, tongue, and mucosal linings. This diverse microbial community encompasses bacteria, viruses, fungi, and archaea, collectively contributing to the overall balance and functioning of the oral environment. A comprehensive exploration of the oral microbiome's composition and diversity is imperative for a thorough comprehension of its role in sustaining oral health and its broader implications for systemic well-being [19]. Within the oral cavity, an estimated 700–1000 different bacterial species, alongside viruses, fungi, and archaea, predominate, with key bacterial phyla like Firmicutes, Bacteroidetes, Actinobacteria, Proteobacteria, and Fusobacteria playing significant roles. The taxonomic distribution and relative abundance of these microbial species are pivotal in unraveling the intricate dynamics of the oral microbiome. The diversity of the oral microbiome is intricately shaped by various factors, including host genetics, oral hygiene practices, diet, and environmental exposures, contributing to individual variations in microbial composition [20]. Understanding these influencing factors provides valuable insights into the mechanisms governing microbial colonization and interactions within the oral cavity. In a state of health, the oral microbiome maintains a delicate balance among diverse microbial species, establishing a symbiotic relationship with the host. However, disruptions to this equilibrium, triggered by factors such as poor oral hygiene, dietary changes, or systemic conditions, can lead to dysbiosis, an imbalance associated with oral diseases like periodontitis, dental caries, and gingivitis [21]. The exploration of dynamic shifts in microbial composition during health and disease is crucial for identifying biomarkers and therapeutic targets. Beyond taxonomic composition, understanding the functional contributions of microbial communities is essential, as the oral microbiome actively participates in physiological processes such as digestion, immune system modulation, and tissue homeostasis. Elucidating the functional profiles of specific microbial taxa provides a comprehensive view of their roles in health and disease, guiding the development of targeted therapeutic interventions [22]. Furthermore, microbial colonization in the oral cavity is non-uniform, with different surfaces offering distinct ecological niches for microbial communities. Dental surfaces, coated with the biofilm dental plaque, provide a substrate for microbial adherence and growth, while gingival crevices and mucosal surfaces host additional microbial communities. Understanding the spatial distribution of these populations is integral to characterizing their specific roles and interactions within the oral ecosystem [23]. Thus, a detailed exploration of the composition and diversity of the oral microbiome serves as a foundational cornerstone for understanding its multifaceted role in maintaining oral health and influencing systemic well-being. Ranging from the taxonomic distribution of microbial species to factors shaping diversity, oral microbiome dynamics in health and disease, functional contributions, and spatial distribution, this comprehensive understanding provides a crucial framework for advancing research and developing targeted interventions in oral healthcare [24].

19.2.1 Microbial Species in the Oral Cavity

The oral cavity serves as a rich and diverse habitat hosting an extensive array of microbial species, collectively constituting a complex ecosystem crucial for both oral and systemic health. Viruses, including bacteriophages and eukaryotic viruses, also contribute significantly to the oral microbial community, although their role is not as well understood as that of bacteria. Fungi, predominantly of the *Candida* genus, and archaea, though less studied, are present in the oral cavity, influencing oral health dynamics. Microbial colonization within the oral cavity is non-uniform, influenced by distinct ecological niches on surfaces like teeth, gums, the tongue, and mucosal linings [25, 26]. Dental surfaces, coated with dental plaque, act as primary substrates for bacterial adherence and growth, adding to the complexity of the oral microbiome. The oral microbiome is characterized by intricate interactions and symbiotic relationships among microbial species, where commensal bacteria contribute significantly to oral health by aiding in digestion, modulating the immune system, and preventing the overgrowth of potential pathogens [27]. Understanding these symbiotic interactions is vital for comprehending the delicate balance that exists within the oral environment. Various factors including host genetics, oral hygiene practices, diet, and environmental exposures, influence the composition of microbial species, shaping individual variations. Investigating these factors provides valuable insights into the mechanisms governing microbial colonization and their implications for oral health [28]. Interestingly, the diverse and dynamic microbial community in the oral cavity underscores its significance in maintaining overall health. The intricate balance among bacteria, viruses, fungi, and archaea highlights the complexity of the oral microbiome, necessitating a comprehensive understanding of their composition, interactions, and influencing factors for advancing oral microbiology research and developing targeted interventions for oral health [5].

19.2.2 Factors Influencing Oral Microbial Diversity

The diversity of the oral microbiome is intricately shaped by a multitude of factors, reflecting the dynamic interplay between the host and the surrounding environment. Understanding these influencing factors is essential for unraveling the complexities of microbial composition within the oral cavity and holds significant implications for oral health and systemic well-being [29]. Host genetics, with its unique variations, plays a fundamental role in shaping the personalized microbial landscape of the oral cavity, impacting susceptibility to specific oral conditions and contributing to the overall structure of the oral microbiome. Concurrently, oral hygiene practices directly influence microbial biofilms and the abundance of specific bacterial species, with poor oral hygiene potentially leading to dysbiosis and the development of oral diseases. Dietary habits and nutritional intake significantly shape microbial diversity by influencing the availability of nutrients and the growth of specific bacteria, while also impacting the acidity of the oral environment [30]. External environmental factors, such as pollutants, air quality, and lifestyle choices, can alter microbial diversity and community structure, further influencing oral health. Additionally, age-related changes, including hormonal fluctuations, developmental milestones, and alterations in dietary patterns, contribute to dynamic shifts in the oral microbiome from infancy through adulthood and into the elderly years [31]. Furthermore, systemic health conditions like diabetes, autoimmune disorders, and immunodeficiency conditions have notable impacts on the oral microbiome, showcasing the bidirectional relationship between oral and overall health. This comprehensive understanding of how these factors intricately influence microbial composition provides a foundational basis for developing targeted interventions to promote oral health and prevent the onset of oral diseases. Such nuanced knowledge underscores the intricate connections between the oral microbiome and systemic well-being, paving the way for holistic approaches to healthcare [32].

19.2.3 Oral Microbiome Dynamics in Health and Disease

19.2.3.1 Sampling Techniques for Oral Microbiome Analysis

Accurate and representative sampling is crucial for gaining insights into the complex and diverse oral microbiome. Various sampling techniques are employed to capture microbial communities from different oral surfaces (Figure 19.2), thereby ensuring a comprehensive understanding of the oral ecosystem [3].

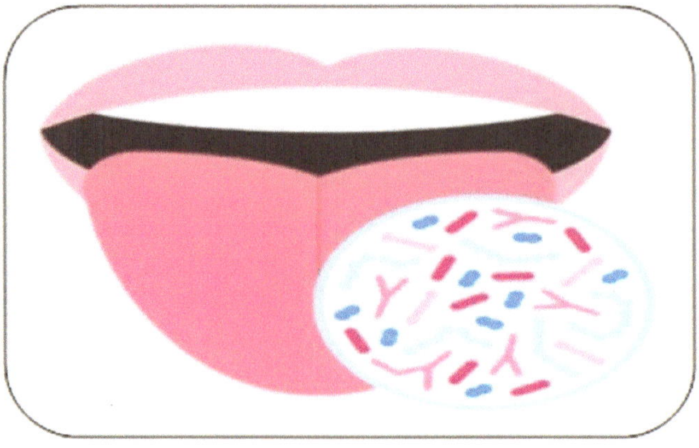

FIGURE 19.2 The microbial examination of saliva under microscope.

Saliva serves as a readily accessible and non-invasive source for oral microbiome analysis. Collection methods include passive drool, spit, or use of specialized collection devices. Saliva provides a snapshot of the overall oral microbiome, reflecting contributions from various oral surfaces, making it suitable for large-scale studies and population-level analyses [33].

19.2.3.2 Dental Plaque and Biofilm Collection

Dental plaque, a biofilm that forms on tooth surfaces, is a prime habitat for microbial communities. Sampling involves gentle scraping or swabbing of tooth surfaces to collect plaque. This method allows for the analysis of microorganisms directly associated with dental health or diseases such as caries and periodontitis [34].

19.2.3.3 Gingival Crevicular Fluid Sampling

Gingival crevicular fluid (GCF), the fluid surrounding the teeth at the gum line, is collected using absorbent paper strips or microcapillary pipettes. GCF sampling enables the study of microbial communities associated with the gingival crevice and is particularly relevant for understanding periodontal health and diseases [35].

19.2.3.4 Tongue Swabbing

The tongue harbors a diverse microbial community, and swabbing is a simple method to collect samples. Tongue swabs capture microbiota involved in processes like taste sensation and contribute to a more holistic understanding of the oral microbiome [36].

19.2.3.5 Mucosal Swabs

Swabbing mucosal surfaces, such as the buccal mucosa or the soft palate, provides insights into microbial communities associated with these specific regions. Mucosal swabs are valuable for studying interactions between the oral microbiome and mucosal tissues [37].

19.2.3.6 Subgingival Plaque Sampling

Subgingival plaque, located below the gum line, is crucial for studying periodontal diseases. Sampling involves inserting a sterile paper point or curette into the gingival crevice to collect microbial communities associated with periodontal health or disease [38].

19.2.3.7 Supragingival Plaque Sampling

Supragingival plaque, located above the gum line, can be collected using a curette or a swab. This method helps study microbial communities on tooth surfaces and is relevant for understanding oral health and diseases like dental caries [39].

19.2.3.8 Microbiopsy Techniques

In-depth analysis may involve microsampling techniques, where tiny biopsies of oral tissues are obtained for microbial profiling. This approach provides a more localized perspective on the oral microbiome and its interactions with host tissues [40].

19.3 MOLECULAR TECHNIQUES FOR MICROBIAL DNA EXTRACTION

19.3.1 Microbial DNA Extraction

Regardless of the sampling method, efficient microbial DNA extraction is a critical step for downstream analysis. Various commercial kits and laboratory techniques are employed to isolate high-quality DNA for next-generation sequencing (NGS) or other molecular analyses. Choosing appropriate sampling techniques is vital for accurately characterizing the oral microbiome. Each method offers unique insights into specific niches within the oral cavity, enabling researchers and clinicians to unravel the intricacies of microbial communities associated with oral health and diseases [41].

19.3.2 Polymerase Chain Reaction in Oral Microbiome Analysis

Polymerase chain reaction (PCR) is a powerful molecular biology technique widely employed in the analysis of the oral microbiome. This method enables the amplification of specific DNA regions, allowing for the study of microbial diversity, identification of species, and investigation of microbial functions within the oral cavity [42].

19.3.3 Target Gene Selection

PCR in oral microbiome analysis begins with the selection of target genes that are indicative of microbial diversity. Commonly targeted regions include the 16S ribosomal RNA gene for bacteria and the internal transcribed spacer (ITS) region for fungi. These genes provide the necessary variability for discriminating between different microbial taxa [43].

19.3.4 DNA Extraction

Before PCR amplification, the DNA must be extracted from collected oral samples, such as saliva, dental plaque, or swabs. Efficient DNA extraction methods are crucial for obtaining high-quality DNA suitable for PCR. Commercial kits or laboratory protocols are employed to break down cell membranes and release genomic DNA from microbial cells [44].

19.3.5 PCR Primer Design

Primer design is a critical step in PCR, as primers are short DNA sequences that flank the target region. The specificity of primers determines which microbial taxa will be amplified. In oral microbiome analysis, primers are designed to target conserved regions of microbial genes, ensuring broad coverage across different species [45].

19.3.6 PCR Amplification

The PCR reaction mix contains the DNA template, primers, DNA polymerase, nucleotides, and buffer. The reaction goes through a series of temperature cycles, including denaturation, annealing, and extension. Denaturation separates the DNA strands, annealing allows primers to bind to their complementary sequences, and extension involves the synthesis of new DNA strands. This process results in the exponential amplification of the target DNA region [45].

19.3.7 Quantitative PCR

Quantitative PCR (qPCR), or real-time PCR, allows for the quantification of the initial amount of target DNA in the sample. This is achieved by incorporating fluorescent dyes or probes into the PCR reaction mix. As the DNA amplifies, the fluorescence increases proportionally, allowing for real-time monitoring and quantification of the PCR product. qPCR is valuable for determining the abundance of specific microbial taxa in a given sample [45].

19.3.8 Next-Generation Sequencing

PCR products generated from oral microbiome samples can be further analyzed using next-generation sequencing technologies. NGS platforms provide high-throughput sequencing of millions of DNA fragments, enabling the identification of a wide range of microbial species present in the sample. This approach facilitates a comprehensive and detailed assessment of oral microbial diversity [46].

19.3.9 Phylogenetic Analysis

The amplified DNA sequences can be subjected to phylogenetic analysis to understand the evolutionary relationships among different microbial taxa. This analysis helps in categorizing and identifying the microbes present in the oral microbiome, contributing to a deeper understanding of microbial community structure and dynamics. Therefore, PCR plays a pivotal role in oral microbiome analysis by enabling the targeted amplification of specific DNA regions, facilitating the study of microbial diversity, and providing insights into the composition and functions of the oral microbiome. The combination of PCR with advanced techniques such as quantitative PCR and next-generation sequencing has revolutionized our ability to comprehensively explore the intricacies of the oral microbial ecosystem [47].

19.3.10 NGS Technologies

NGS technologies have revolutionized the field of genomics and have been instrumental in advancing our understanding of the oral microbiome. These high-throughput sequencing methods allow for the simultaneous analysis of millions of DNA fragments, providing unprecedented insights into microbial diversity, functional potential, and community dynamics within the oral cavity [48].

19.3.11 DNA Library Preparation

NGS begins with the preparation of a DNA library, which involves fragmenting the DNA extracted from oral samples into smaller, manageable pieces. Adapters are then added to the ends of these fragments, allowing for the attachment of the DNA to a solid support during the sequencing process [49].

19.3.12 Sequencing Platforms

Various NGS platforms are available, each with its unique sequencing chemistry and workflow. Common platforms include Illumina, Ion Torrent, Pacific Biosciences (PacBio), and Oxford Nanopore Technologies. Illumina platforms, for instance, employ reversible terminator chemistry, while Ion Torrent relies on semiconductor sequencing. PacBio and Oxford Nanopore Technologies use single-molecule real-time (SMRT) sequencing and nanopore sequencing, respectively [50].

19.3.13 Amplification and Clonal Generation

In some NGS workflows, prior to sequencing, the DNA fragments are amplified to generate clusters or colonies of identical DNA molecules. This process, known as clonal amplification, facilitates the detection of individual sequences during sequencing. Each cluster or colony represents a unique DNA fragment from the original sample [48].

19.3.14 Sequencing by Synthesis

Illumina platforms utilize a sequencing-by-synthesis (SBS) approach. During SBS, fluorescently labeled nucleotides are added to the growing DNA strands and the emitted fluorescence is captured by a camera. This process is repeated for multiple cycles, allowing for the determination of the DNA sequence [51].

19.3.15 Ion Torrent Sequencing

Ion Torrent sequencing is based on the detection of hydrogen ions released during DNA synthesis. As nucleotides are incorporated into the growing DNA strand, a hydrogen ion is released, leading to changes in pH. These pH changes are detected by a semiconductor chip, thereby enabling the determination of the DNA sequence [52].

19.3.16 Single-Molecule Real-Time Sequencing

PacBio platforms use SMRT sequencing, where individual DNA molecules are sequenced in real time. DNA polymerase is immobilized on a zero-mode waveguide, and the incorporation of fluorescently labeled nucleotides is monitored. This single-molecule approach eliminates the need for clonal amplification [53].

19.3.17 Nanopore Sequencing

Oxford Nanopore Technologies' platforms leverage nanopore sequencing. DNA strands pass through a nanopore, and as each nucleotide passes, it produces a characteristic electrical signal. The sequence is deduced based on the unique signal for each nucleotide, allowing for real-time sequencing without the need for amplification [54].

19.3.18 Data Analysis

NGS generates vast amounts of raw sequence data, requiring sophisticated bioinformatics tools for analysis. Bioinformatic pipelines are employed for tasks such as quality control, alignment, taxonomic classification, and functional annotation. These analyses provide comprehensive insights into the composition and functional potential of the oral microbiome [50].

19.4 APPLICATIONS IN ORAL MICROBIOME RESEARCH

NGS technologies have significantly advanced oral microbiome research by enabling in-depth profiling of microbial communities. They facilitate the identification of bacterial, viral, fungal, and archaeal taxa, as well as the exploration of functional genes and pathways. NGS has been instrumental in uncovering the associations between the oral microbiome and various oral and systemic diseases. NGS technologies have transformed our ability to study the oral microbiome, providing a comprehensive and high-resolution view of microbial communities. These platforms have accelerated discoveries in microbial diversity, community dynamics, and functional potential, contributing to a deeper understanding of the intricate relationship between the oral microbiome and human health [55].

19.4.1 Metagenomic Approaches

Metagenomics, a powerful and holistic approach, has become a cornerstone in the study of the oral microbiome. This methodology involves the direct analysis of genetic material recovered from environmental samples, providing a comprehensive view of the collective genomes of the microbial community within the oral cavity [56].

19.4.2 DNA Extraction

Metagenomic studies begin with the extraction of total DNA from oral samples, such as saliva, dental plaque, or mucosal swabs. Efficient DNA extraction methods are crucial to obtaining high-quality genomic material that accurately represents the diverse microbial populations present [44].

19.4.3 Shotgun Sequencing

One of the primary metagenomic approaches is shotgun sequencing, wherein the entire genomic content of the microbial community is sequenced without prior target-specific amplification. This unbiased method captures the collective DNA of bacteria, viruses, fungi, and archaea, allowing for a thorough examination of the oral microbiome's genetic diversity [57].

19.4.4 Functional Genomics

Metagenomics facilitates the exploration of functional genes and pathways within the oral microbiome. By analyzing the entire genetic content, researchers can gain insights into the metabolic potential of the microbial community. This approach is valuable for understanding the functional contributions of microbes to processes such as digestion, immune modulation, and biofilm formation [58].

19.4.5 Taxonomic Profiling

Metagenomic data enable taxonomic profiling to identify and characterize the diversity of microbial species present in the oral cavity. By comparing sequence data to reference databases, researchers can assign taxonomic classifications to the genetic material, providing a comprehensive overview of the oral microbiome's composition [59].

19.4.6 Comparative Metagenomics

Comparative metagenomics involves the analysis of multiple oral microbiome samples to identify commonalities and differences between individuals or groups. This approach is particularly useful for studying variations in microbial composition and functional potential associated with factors such as oral health, disease, or demographic variables [60].

19.4.7 Assembly and Notation

Shotgun metagenomic data often undergo assembly, where short DNA sequences are reconstructed into longer contigs, representing genomic fragments. Annotation involves identifying and labeling genes within these contigs, providing information about the genetic elements present in the oral microbiome [61].

19.4.8 Metatranscriptomics and Metaproteomics

Beyond DNA analysis, metagenomics can be complemented by metatranscriptomics and metaproteomics. Metatranscriptomics examines the expression of genes, providing insights into which genes are actively transcribed. Metaproteomics analyzes the proteins produced by the microbial community, offering a direct view of functional activities [62].

19.4.9 Bioinformatic Analysis

Metagenomic data necessitate advanced bioinformatic analyses to interpret the complex genetic information. Bioinformatics tools are employed for tasks such as taxonomic classification, functional annotation, pathway analysis, and the integration of multi-omic datasets. These analyses contribute to a comprehensive understanding of the oral microbiome [63].

19.4.10 Applications in Disease Research

Metagenomic approaches have significantly advanced our understanding of the role of the oral microbiome in health and disease. These studies have revealed associations between microbial dysbiosis and conditions such as periodontal diseases, dental caries, and systemic diseases, paving the way for targeted interventions and personalized healthcare strategies. Metagenomic approaches provide a powerful means to explore the genetic landscape of the oral microbiome. From taxonomic profiling and functional genomics to comparative analyses and multi-omic investigations, metagenomics has enriched our understanding of the intricate microbial communities within the oral cavity and their impact on human health [64].

19.5 BIOINFORMATICS ANALYSIS OF ORAL MICROBIOME DATA

The analysis of oral microbiome data heavily relies on sophisticated bioinformatics tools and computational methodologies to decipher the complex genetic information generated from high-throughput sequencing technologies. This pivotal role of bioinformatics extends across various critical stages of the analysis process. Initial steps involve quality control and preprocessing of raw sequencing data, ensuring the removal of low-quality reads, adapter sequences, and potential contaminants for reliable downstream analyses [65]. Taxonomic classification tools like QIIME and mothur employ algorithms to accurately assign DNA sequences to specific microbial taxa, revealing the identity and abundance of species within the oral microbiome. Functional annotation tools such as HUMAnN and PICRUSt predict the functional potential of genes, shedding light on the metabolic activities of the microbial community. Alpha and beta diversity analyses, facilitated by tools like UniFrac and Bray-Curtis indices, quantitatively evaluate diversity, aiding in the characterization of microbial community structures [66]. Differential abundance analysis tools like DESeq2 and edgeR detect significant differences in microbial abundance between sample groups, providing insights into dysbiosis associated with oral conditions. Network analysis tools like CoNet and SparCC construct co-occurrence networks, offering valuable insights into microbial community dynamics and ecological relationships. Prediction of metabolic pathways using tools like KEGG and MetaCyc enhances our understanding of microbial contributions to host metabolism and systemic health [67]. Integration of oral microbiome data with clinical information through bioinformatics platforms facilitates a contextual understanding of microbial signatures associated with health and disease. Visualization tools like Krona, MetaPhlAn (Metagenomic Phylogenetic Analysis), and GraPhlAn (Graphical Phylogenetic Analysis) enable graphical representation, aiding in the creation of informative figures and visual summaries, thereby enhancing the communication of research findings. Hence, bioinformatics analysis serves as an indispensable toolkit, unraveling the intricacies of the oral microbiome's functional and taxonomic landscape and its relevance to human health [68].

19.5.1 Taxonomic Profiling of the Oral Microbiome

Taxonomic profiling stands as a foundational and indispensable step in elucidating the intricate microbial diversity within the oral cavity, providing a nuanced understanding of the identity and abundance of microbial taxa. This intricate process involves the systematic classification of DNA sequences obtained through high-throughput sequencing technologies, allowing for a thorough exploration of the composition of the oral microbiome [69].

19.5.1.1 DNA Sequencing and Sample Collection

The initiation of taxonomic profiling involves the meticulous collection of oral samples, ranging from saliva to dental plaque and mucosal swabs. Subsequent DNA extraction from these samples is followed by the application of high-throughput sequencing technologies, such as NGS. These advanced technologies yield vast amounts of sequence data that encapsulate the genetic material of the diverse microbial community within the oral sample [70].

19.5.1.2 Bioinformatics Tools for Classification

Central to taxonomic profiling is the utilization of specialized bioinformatics tools dedicated to the accurate classification of DNA sequences into specific microbial taxa. Notable tools in this domain include QIIME, mothur, and the SILVA database. These tools employ sophisticated algorithms that compare the obtained DNA sequences with reference databases, ensuring precise and reliable taxonomic assignment [71].

19.5.1.3 Taxonomic Units and Hierarchy

The taxonomic profiling process unveils the organizational hierarchy of microbial taxa, spanning phylum, class, order, family, genus, and species. This hierarchical arrangement provides valuable information regarding the relative abundance of each taxonomic unit, offering insights into the richness and diversity characterizing the oral microbiome [72].

19.5.1.4 Visualization of Taxonomic Composition

Powerful visualization tools such as Krona, MetaPhlAn (Metagenomic Phylogenetic Analysis) and GraPhlAn play pivotal roles in representing the taxonomic composition of the oral microbiome. These tools generate interactive plots and visual summaries, facilitating effective communication of complex taxonomic data to diverse audiences, including researchers and clinicians [73].

19.5.1.5 Alpha and Beta Diversity Measures

Taxonomic profiling significantly contributes to the assessment of alpha diversity, gauging the diversity within a single oral microbiome sample. Simultaneously, beta diversity measures the differences in microbial composition between multiple samples, offering valuable insights into community structures. Tools like UniFrac and Bray-Curtis dissimilarity indices are instrumental in conducting these diversity analyses [74].

19.5.1.6 Identification of Key Taxa

Through taxonomic profiling, researchers can identify key microbial taxa associated with specific oral health or disease states. Differential abundance analysis tools, including DESeq2 and edgeR, enable the detection of significant differences in the abundance of specific taxa between distinct sample groups, thus highlighting potential biomarkers crucial for further investigations [75].

19.5.1.7 Temporal and Spatial Dynamics

Taxonomic profiling extends its impact to the exploration of temporal and spatial dynamics within the oral cavity. Tracking changes in taxonomic composition over time and understanding the spatial distribution of microbial communities provide invaluable information regarding the stability and adaptability of the oral microbiome in response to various environmental and physiological factors [76].

19.5.1.8 Integration with Functional Annotations

The integration of taxonomic profiling data with functional annotations enhances the holistic understanding of the oral microbiome. This integration establishes links between the identity of microbial taxa and their potential functional contributions, shedding light on the intricate metabolic activities within the oral ecosystem. Taxonomic profiling in oral microbiome analysis serves as a cornerstone, laying the groundwork for characterizing the diverse microbial communities residing within the oral cavity. From meticulous sample collection and DNA sequencing to sophisticated bioinformatics analysis and visualization, taxonomic profiling contributes crucial information that forms the basis for further exploration of the functional and ecological aspects of the oral microbiome in health and disease [77].

19.6 FUNCTIONAL ANALYSIS AND METABOLIC PATHWAYS

The functional analysis of the oral microbiome is a critical endeavor, offering essential insights into the metabolic activities and potential contributions of microbial communities to both oral and systemic health. This intricate process involves decoding the functional genes encoded in the microbial genomes within the oral cavity and exploring the associated metabolic pathways. Beginning with metagenomic approaches utilizing high-throughput sequencing technologies like NGS, researchers generate extensive genomic data that forms the foundation for functional insights [78]. Bioinformatics tools dedicated to functional annotation, such as HMP Unified Metabolic Analysis Network (HUMAnN) and Phylogenetic Investigation of Communities by Reconstruction of Unobserved States (PICRUSt), enable the prediction of the oral microbiome's functional potential based on taxonomic information. Further, tools like Kyoto Encyclopedia of Genes and Genomes (KEGG) and MetaCyc play a crucial role in predicting metabolic pathways, unveiling the metabolic capabilities of microbial communities and their potential impact on host health. Integrating functional analysis with taxonomic profiling enhances the contextual understanding of the oral microbiome, allowing for the identification of specific microbial taxa responsible for particular metabolic activities [79]. This correlation between microbial identity and functional contributions provides valuable insights into the intricate balance within the oral environment and its implications for health and disease. Functional analysis sheds light on diverse microbial metabolic activities, contributing to our understanding of biofilm formation, acid production, and immune modulation, crucial for comprehending the mechanisms underlying oral diseases. Identifying specific metabolic pathways and functions through functional analysis allows researchers to link insights to oral health outcomes, paving the way for potential therapeutic targets within the oral microbiome [80]. By pinpointing metabolic pathways associated with dysbiosis or disease, targeted interventions, including probiotics, prebiotics, or antimicrobial strategies, can be explored to modulate the functional landscape and promote oral health. The integration of functional analysis contributes to a holistic understanding of microbial functions within the oral cavity, providing a comprehensive exploration of the multifaceted roles of microbial communities, from contributing to host homeostasis to potential involvement in disease pathogenesis [81]. In conclusion, this multifaceted analysis of the oral microbiome, encompassing metagenomic approaches, functional annotation tools, pathway prediction, correlation with taxonomic data, insights into microbial metabolism, linkage to oral health, identification of therapeutic targets, and the holistic understanding of microbial functions, forms a crucial foundation for advancing our comprehension of microbial contributions to oral and systemic health, guiding targeted interventions for a balanced and healthy oral microbiome [80].

19.7 INTEGRATION OF MULTI-OMICS DATA

The integration of multi-omics data represents a sophisticated and comprehensive analytical approach that amalgamates information across various molecular levels, encompassing genomics, transcriptomics, proteomics, and metabolomics. This holistic perspective empowers researchers to unravel intricate interactions and gain profound insights into the dynamic nature of microbial communities within the oral microbiome, both in states of health and disease [82]. Genomic integration delves into the DNA content, enabling the identification of specific microbial taxa and assessment of genetic diversity, unraveling potential functional contributions encoded in microbial genomes. Transcriptomics integration explores active genes and expression patterns, providing a window into how microbial communities respond to environmental cues, diseases, or therapeutic interventions [83]. Proteomics integration reveals functional activities at the protein level, correlating protein abundance with genetic information and uncovering pivotal contributors to the oral microbiome's functional landscape. Metabolomics integration captures metabolites, offering a comprehensive view of metabolic pathways, fostering an understanding of the intricate interactions among microbial species and their impact on the host environment. Correlation and network analyses identify relationships between molecular layers, offering a systems-level understanding of the oral microbiome's complexity [84].

Functional annotation tools link genetic information to metabolic pathways, shedding light on how genetic variations translate into functional outcomes. Considering temporal and spatial dynamics allows researchers to track changes in microbial composition, gene expression, and metabolic activities over time and across different oral sites, enhancing insights into community stability and adaptability. Integrating multi-omics data with clinical information enhances translational relevance, identifying molecular signatures associated with oral health or disease conditions and paving the way for biomarkers and personalized therapeutic strategies. In conclusion, this integrative analysis not only advances our comprehension of microbial community dynamics within the oral microbiome but also opens avenues for targeted interventions and precision oral healthcare strategies, grounded in a holistic understanding of its functional and structural landscape [85].

19.8 OMICS APPROACHES IN ORAL MICROBIOME DIAGNOSIS

Omics approaches in oral microbiome diagnosis constitute a state-of-the-art paradigm, harnessing high-throughput technologies across diverse molecular levels to revolutionize our comprehension and evaluation of oral health. This comprehensive strategy encompasses genomics, transcriptomics, proteomics, and metabolomics, providing a detailed analysis of microbial communities within the oral cavity for diagnostic purposes [86]. Genomic approaches, facilitated by next-generation sequencing, enable the precise identification and characterization of microbial taxa based on their genetic makeup, contributing to an understanding of their potential pathogenicity or symbiotic roles. Transcriptomic analyses delve into the active genes and expression patterns within the oral microbiome, offering real-time insights into microbial community dynamics and responses to environmental changes or disease conditions. Proteomic investigations unravel functional signatures by studying the complete set of proteins expressed by the oral microbiome, aiding in biomarker discovery and enhancing our understanding of functional dynamics [87]. Metabolomic analyses focus on small-molecule metabolites, providing a detailed metabolic profile that identifies signatures associated with health and disease.

Integrating data from genomics, transcriptomics, proteomics, and metabolomics enhances diagnostic capabilities, offering a holistic understanding of the structural, functional, and metabolic aspects of the oral microbiome. This integrative approach enables the identification of robust diagnostic markers and signatures for various oral conditions [7]. Omics approaches significantly contribute to biomarker discovery, allowing for the identification of specific signatures associated with health or disease, serving as indicators for early detection, and monitoring. Aligned with precision medicine principles, the application of omics approaches in oral microbiome diagnosis considers the unique genomic and functional profiles of individual patients, paving the way for personalized diagnostic and therapeutic strategies [88]. Ongoing technological advances and the integration of multi-omics data with clinical parameters continue to refine diagnostic capabilities, enhancing the translational relevance of oral microbiome diagnosis and setting the stage for more effective and personalized oral healthcare interventions. In conclusion, omics approaches represent a transformative frontier in oral healthcare, unraveling molecular intricacies to provide unprecedented diagnostic insights and laying the foundation for precision medicine and personalized interventions in oral health [89].

19.9 METAGENOMICS FOR MICROBIAL GENOMIC ANALYSIS

Metagenomics, a revolutionary approach in the field of microbiology, has significantly transformed our understanding of microbial communities by facilitating the comprehensive study of their genomic content. This cutting-edge technique has particularly revolutionized microbial genomic analysis, moving away from the constraints of traditional studies that focus on individual isolates.

Metagenomics uniquely enables the exploration of the collective genetic material of entire microbial communities present in a given environment, proving invaluable in uncovering the vast diversity of microorganisms and elucidating their functional potential, thereby shedding light on their intricate roles in various ecosystems [56]. The key principles of metagenomics involve the direct extraction and sequencing of DNA from environmental samples, circumventing the need for cultivating and isolating individual microorganisms. This technique allows scientists to survey the genetic diversity within complex microbial communities, encompassing bacteria, archaea, viruses, and other microorganisms, providing a holistic view of the microbial landscape [90].

Beyond species identification, metagenomics enables functional profiling by analyzing the genes and pathways encoded in the metagenome, offering insights into the potential metabolic activities and functional capabilities of microbial consortia. Metagenomic studies often include environmental sampling from diverse habitats such as soil, water, and the human gut, providing context-specific information about microbial communities in their natural environments. Furthermore, the culture-independent nature of metagenomics, in contrast to traditional microbiology methods reliant on laboratory cultivation, is advantageous for studying elusive or unculturable microorganisms critical to various ecosystems [91]. Metagenomics has found widespread applications in microbial genomic analysis across multiple scientific domains. In environmental microbiology, it aids in understanding microbial composition and functional potential for purposes like environmental monitoring, bioremediation, and exploring microbial contributions to nutrient cycling. Human microbiome studies leverage metagenomics to gain unparalleled insights into complex interactions between microorganisms and their hosts, with implications for understanding and addressing various diseases [92]. The biotechnology and industrial applications of metagenomics have facilitated the discovery of novel enzymes, pathways, and bioactive compounds, contributing to the development of new biocatalysts and optimization of industrial processes. Moreover, metagenomics contributes significantly to ecological and evolutionary studies by examining the genomic content of microbial communities across diverse ecosystems, thereby enhancing our understanding of microbial evolution, adaptation, and ecological dynamics [93]. In conclusion, metagenomics emerges as a powerful approach for microbial genomic analysis, allowing researchers to delve into the genetic makeup of entire microbial communities and providing not only a comprehensive understanding of their diversity but also insights into their functional roles and interactions. As technology continues to advance, metagenomics is poised to play a pivotal role in shaping our understanding of microbial life and its myriad implications across various scientific disciplines [94].

19.10 METATRANSCRIPTOMICS FOR GENE EXPRESSION STUDIES

Metatranscriptomics, an innovative methodology in molecular biology, represents a cutting-edge approach to comprehending the dynamic world of gene expression within complex microbial communities. This sophisticated technique transcends the limitations of traditional gene expression studies, which typically focus on individual organisms, offering a holistic perspective on the active transcriptional profiles of entire microbial populations in real time. At its core, metatranscriptomics involves the extraction and sequencing of RNA directly from environmental samples, providing a comprehensive snapshot of actively expressed genes across diverse microorganisms [95]. Unlike genomic studies that capture potential capabilities, metatranscriptomics directly observes the actively transcribed genes, offering valuable insights into the immediate functional state of microbial communities. This approach facilitates the identification of functional pathways, metabolic activities, and responses to environmental stimuli within microbial communities, allowing researchers to uncover the adaptive strategies employed in dynamic environmental conditions [93]. The application of metatranscriptomics extends to environmental microbiology, where it serves as a powerful tool for monitoring and understanding microbial responses to environmental changes, as well as to human health studies, contributing significantly to unraveling the dynamics of the human microbiome and its implications for health and disease. Furthermore, metatranscriptomics holds promise in biotechnology, enabling the discovery of novel enzymes and metabolic pathways with industrial applications, thereby offering opportunities for innovative biotechnological solutions and optimization of industrial processes [96]. Looking ahead, as technology continues to advance, metatranscriptomics is poised to play a pivotal role in advancing our understanding of gene expression in complex microbial ecosystems, providing more nuanced insights into the ecological and functional relationships governing microbial communities across diverse environments. This ongoing refinement holds the potential to contribute to advancements in both basic science and applied research, solidifying metatranscriptomics as a powerful and transformative tool shaping the future of molecular microbial ecology [97].

19.11 METAPROTEOMICS FOR PROTEIN ANALYSIS

Metaproteomics, positioned at the forefront of molecular biology, stands as a powerful and innovative methodology for unraveling the intricate protein dynamics within diverse microbial communities. Overcoming the limitations of traditional protein analysis methods focused on individual organisms, metaproteomics provides a comprehensive lens to explore the collective protein composition of entire microbial populations, offering unprecedented insights into the functional roles and interactions of proteins within complex environmental and biological systems [98]. At its essence, metaproteomics involves the extraction, identification, and quantification of proteins directly from complex environmental or biological samples, enabling the simultaneous analysis of multiple species within a community and capturing the dynamic interplay of proteins across diverse microorganisms. Employing state-of-the-art mass spectrometry techniques, metaproteomics allows for the high-throughput identification and quantification of thousands of proteins simultaneously, offering a detailed snapshot of the protein diversity within microbial communities [99].

By elucidating the identity and abundance of proteins, researchers gain insights into the metabolic activities, cellular functions, and adaptive responses of microbial populations in their natural environments. This real-time perspective enhances our comprehension of how microbial populations dynamically adapt and interact with their surroundings, providing critical information for ecological and environmental studies. Metaproteomics finds extensive application in environmental microbiology, serving as a powerful tool for monitoring and understanding protein dynamics in response to environmental changes [100]. Additionally, in biotechnology, metaproteomics contributes to the discovery of novel enzymes, metabolic pathways, and biomolecules with industrial applications, offering valuable resources for optimizing bioprocesses and developing innovative biotechnological solutions. Beyond environmental applications, metaproteomics holds implications for the study of complex biological systems, including the human microbiome, contributing to understanding host-microbe interactions, disease mechanisms, and potential therapeutic targets. Looking ahead, as technology continues to advance, metaproteomics is poised to play a crucial role in advancing our understanding of protein dynamics within complex microbial ecosystems. Ongoing developments in mass spectrometry, bioinformatics, and sample preparation techniques are expected to enhance the depth and accuracy of metaproteomic analyses, unveiling new layers of information on the intricate interplay of proteins in diverse environments. Its applaudable that metaproteomics emerges as a transformative tool, contributing significantly to our understanding of microbial ecology, environmental adaptations, and biotechnological potential, while standing at the forefront of molecular biology with the promise of further breakthroughs in the study of proteins within complex biological and environmental systems [101].

19.12 METABOLOMICS FOR METABOLIC PROFILING

Metabolomics, situated at the dynamic intersection of systems biology, emerges as a powerful methodology for comprehensively unraveling the intricate metabolic landscapes inherent in biological systems. Commonly referred to as metabolic profiling, this advanced approach provides a holistic view of metabolites—small molecules—across cells, tissues, or organisms, serving as a crucial bridge between genomics, transcriptomics, and the phenotypic characteristics of living systems. At its core, metabolomics systematically identifies, quantifies, and analyzes the diverse array of metabolites within a biological sample, capturing the real-time physiological state and reflecting the end products of cellular processes [102]. This unique perspective allows researchers to discern patterns, pathways, and alterations in response to internal and external factors. Metabolomics employs sophisticated analytical techniques like mass spectrometry (MS) and nuclear magnetic resonance (NMR) spectroscopy, enabling high-throughput profiling and quantification of numerous metabolites. Bioinformatics analyses of resulting datasets unveil biomarkers, characterize metabolic pathways, and reveal associations between metabolite patterns and specific biological conditions [103, 104]. Offering profound insights into the biochemical and physiological status of living systems, metabolomics facilitates the deciphering of key metabolic pathways, understanding of energy dynamics, and identification of disease markers, thereby contributing to personalized medicine, drug discovery, and biomarker identification.

The integration of metabolomics with other omics technologies enhances our understanding of complex biological systems, providing a holistic perspective on the molecular interactions governing cellular functions across genes, proteins, and metabolites. In medical research, metabolomics contributes to the identification of diagnostic biomarkers, comprehension of disease mechanisms, and guidance of therapeutic interventions. In precision health, metabolomics holds promise for tailoring treatments based on individual metabolic profiles, fostering personalized and effective healthcare strategies [54]. Looking forward, with ongoing advancements in analytical techniques and computational tools, metabolomics is poised to play an increasingly pivotal role in elucidating metabolic pathways, biomarker discovery, and the intricate interplay of metabolites in health and disease. This evolution positions metabolomics at the forefront of modern biological research, continually shaping our understanding of metabolic dynamics and providing unprecedented insights into the molecular underpinnings of life [105].

19.13 DIAGNOSTIC BIOMARKERS IN ORAL MICROBIOME ANALYSIS

19.13.1 Identification of Biomarkers for Oral Diseases

The identification of biomarkers for oral diseases is a crucial undertaking in dental and medical research, aiming to enhance early detection, prognosis, and treatment monitoring for conditions affecting the oral cavity. Biomarkers, serving as measurable indicators of normal and pathogenic biological processes, play a pivotal role in advancing our understanding of oral diseases and improving clinical outcomes. This comprehensive approach involves exploring and validating specific molecules or molecular patterns as reliable indicators of oral health or disease states. The diverse array of biomarkers for oral diseases includes proteins, nucleic acids, metabolites, and microbial signatures, with proteomic analyses revealing associations with periodontal disease, oral cancers, and temporomandibular disorders [106]. Utilizing cutting-edge technologies such as MS, NGS, and advanced imaging modalities enables precise identification and quantification of biomarkers, facilitating their validation for clinical use. In the context of periodontal diseases, research focuses on indicators associated with inflammation, tissue destruction, and microbial dysbiosis, while genomic studies explore genetic variations influencing susceptibility to periodontitis [102]. For oral cancers, biomarker discovery involves identifying genetic mutations, altered protein expression, and changes in circulating nucleic acids, serving as valuable indicators for early detection and prognostication. Salivary biomarkers, encompassing proteins, nucleic acids, and metabolites, gained prominence due to their non-invasive accessibility, showing promise for oral cancer, periodontal diseases, and systemic health assessments. The identified and validated biomarkers have significant clinical implications, enabling early detection through diagnostic tools, timely intervention, and personalized treatment strategies tailored to individual patient profiles [107]. Despite substantial progress, challenges such as standardization, validation, and the establishment of large-scale biomarker databases persist. Future directions involve the integration of multi-omics approaches, including genomics, proteomics, and metabolomics, to comprehensively understand the molecular landscape associated with oral diseases. The ongoing integration of cutting-edge technologies and multi-omics approaches in biomarker research offers promising avenues for personalized and effective clinical interventions, shaping the future of oral disease diagnosis and management [108].

19.13.2 Host-Response Markers

Host-response markers, integral to understanding the dynamic interplay between the host organism and environmental stimuli, encompass biomolecules reflecting molecular and cellular responses crucial for indicating the body's reactions to challenges like pathogens, stressors, or diseases. This examination yields valuable insights into defense mechanisms, immune responses, and physiological adaptations, aiding in disease diagnosis, prognosis, therapeutic monitoring, and assessing environmental impacts on host health [109]. These markers span proteins, nucleic acids, metabolites, and cellular components, with proteins like cytokines, nucleic acids such as RNA, and metabolites providing multifaceted perspectives on host responses. Technological advancements in transcriptomics, proteomics, metabolomics, and flow cytometry facilitate comprehensive analyses of large biomolecule sets, enabling the identification of specific markers associated with infection stages, disease progression, or therapeutic responses. In infectious diseases, host-response markers are crucial for early detection, accurate diagnosis, and severity assessment and contribute to targeted therapies and vaccine development. In inflammatory and autoimmune disorders, markers like cytokines and autoantibodies aid in diagnosis, guide treatment strategies, and monitor disease activity [110]. In oncology, host-response markers, including gene expression profiles and immune cell infiltrates, play a vital role in cancer diagnosis, prognosis, personalized treatment, and therapeutic efficacy assessment. Challenges persist, requiring standardized methodologies, robust validation, and integration of multi-omics data. Future perspectives involve developing sophisticated analytical techniques, employing artificial intelligence for data interpretation, and identifying novel host-response markers to enhance diagnostic precision and therapeutic outcomes. Host-response markers stand central in unraveling intricate host reactions to diverse challenges, with ongoing technological and methodological advancements holding promise for deeper insights, improved diagnostics, and enhanced therapeutic strategies across diverse medical domains [111].

19.13.3 Microbial Markers of Health and Disease

Microbial markers, encompassing diverse microorganisms and their biomolecules, are pivotal indicators in the intricate interplay between the human microbiome and health or disease. This research field focuses on identifying and characterizing microbial species, community structures, and functional attributes that contribute to maintaining homeostasis or precipitate pathological conditions. The exploration of bacterial, archaeal, viral, and fungal markers, along with their genetic material, metabolites, and structural components, reveals the significance of a balanced microbial community in maintaining physiological functions [112]. Dysbiosis correlates with various diseases, including inflammatory conditions, metabolic disorders, and infectious diseases. Technological advancements in sequencing, metagenomics, metatranscriptomics, and metabolomics allow a comprehensive analysis of microbial communities and their functional potential. Culturomics further expands understanding by cultivating and characterizing previously unculturable microorganisms. In human health, microbial markers, especially in the gut microbiota, are implicated in metabolic health, immune function, and gut barrier integrity [113].

Specific markers like *Bifidobacterium* and *Lactobacillus* denote a healthy gut, while shifts indicate conditions like inflammatory bowel disease (IBD) and irritable bowel syndrome (IBS). In infectious diseases, microbial markers serve as indicators of pathogenic microorganisms, aiding accurate and rapid diagnostics. Monitoring microbial community changes offers insights into infection progression and guides treatment strategies. Metabolomic profiling of microbial communities, including short-chain fatty acids and lipids, provides unique perspectives on microbiota functional activities and their impact on host physiology [114]. Challenges persist in defining universal microbial markers due to the individualized nature of the microbiome. Further research is needed to refine our understanding of the microbiome's functional roles, exploring host-microbiome interactions and developing personalized approaches to microbiome-based diagnostics and therapeutics. Microbial markers stand as a burgeoning field, holding immense potential to advance our understanding of the intricate relationships between the human microbiome and health conditions. From the gut to other body sites, these markers offer valuable insights into maintaining homeostasis or signaling disruptions associated with diseases, and ongoing research and technological innovations are anticipated to unlock new dimensions in personalized medicine, diagnostics, and therapeutic interventions based on microbial signatures [112].

19.13.4 Challenges in Validating Biomarkers

The identification and validation of biomarkers, crucial in medicine, diagnostics, and biomedical research, involve complex processes. Biomarkers, immensely useful for disease detection, prognosis, and treatment monitoring, undergo multifaceted validation. Addressing challenges is paramount for translating promising biomarker candidates into clinically actionable tools. The inherent heterogeneity in biomarker expression across individuals and diverse patient populations poses a primary challenge, requiring extensive validation studies to ensure robustness and generalizability. Achieving standardized protocols and ensuring reproducibility across laboratories are significant hurdles, demanding rigorous efforts to enhance reliability [115]. The clinical heterogeneity of diseases necessitates distinct biomarkers for accurate characterization, adding complexity to identifying universal biomarkers applicable across diverse populations. Considering ethnic and geographical variability is crucial to ensure inclusivity in validation studies to avoid biases and validate biomarker performance across demographics. The dynamic nature of biomarker expression over time presents challenges, emphasizing the need for longitudinal studies and dynamic biomarkers for real-time insights. Navigating regulatory compliance is critical for successful translation, aligning with standards set by regulatory authorities like the U.S. Food and Drug Administration (FDA). Despite substantial resource requirements, balancing robust validation with cost considerations is essential for developing cost-effective and scalable biomarker assays for routine clinical practice. Collaborative efforts among researchers, clinicians, and regulatory bodies are vital for overcoming challenges and establishing standardized protocols. Successful biomarker validation holds the promise of transforming disease diagnosis, prognosis, and treatment strategies, ultimately advancing precision medicine and improving patient outcomes [116].

19.14 CLINICAL APPLICATIONS OF ORAL MICROBIOME DIAGNOSTICS

19.14.1 Periodontal Disease Diagnosis

Periodontal disease diagnosis is a critical facet of oral health assessment, involving the identification and evaluation of conditions impacting the supportive structures of teeth, encompassing the gums and bone. This diagnostic process is imperative for prompt intervention and the effective management of periodontal diseases, ranging from mild gingivitis to severe forms of periodontitis. A comprehensive clinical examination, typically conducted by a dental professional like a periodontist or general dentist, forms the foundation of periodontal disease diagnosis. This examination includes assessment of gingival health, probing the depth of periodontal pockets and examining signs of inflammation, bleeding, or other abnormalities. Visual inspection and palpation aid in identifying concerns such as gum recession, tooth mobility, and the presence of dental plaque and calculus

[117]. Periodontal probing, an essential diagnostic tool, measures pocket depth to determine attachment loss and potential infection or inflammation areas. Radiographic imaging, including dental X-rays and cone-beam computed tomography (CBCT), plays a pivotal role in assessing bone loss and revealing conditions like dental abscesses or cysts. Periodontal charting involves recording measurements related to each tooth's periodontal condition, serving as valuable baseline data for monitoring disease progression or treatment response. Microbiological analysis assesses microbial composition, aiding in tailoring treatment strategies [88]. Genetic and host response testing, part of advancements in personalized medicine, contribute to a more targeted approach to periodontal care. Periodontal disease diagnosis also includes risk assessment and consideration of the patient's medical history to determine if there are factors like smoking, diabetes, and systemic conditions that could impact periodontal health.

In complex cases, collaboration with dental specialists such as periodontists or oral surgeons provides additional insights for comprehensive treatment planning. This dynamic process integrates clinical, radiographic, and laboratory assessments to understand the patient's periodontal health comprehensively. Early and accurate diagnosis is essential for timely interventions, preventing disease progression, and preserving overall oral health, emphasizing the importance of regular dental checkups and patient education on oral hygiene practices for effective periodontal disease management and long-term oral well-being [118].

19.14.2 Dental Caries Assessment

Dental caries assessment is a crucial element in evaluating oral health, focusing on the identification and evaluation of the presence and severity of cavities or tooth decay. This prevalent oral health issue results from the demineralization of tooth enamel, primarily due to acid produced by bacteria in dental plaque. The assessment process involves a comprehensive examination of the patient's teeth using diagnostic tools and techniques. Dentists conduct a thorough clinical examination, employing dental mirrors, explorers, and probes to visually inspect each tooth's surfaces, looking for visible signs of decay such as discoloration, enamel softening, or the presence of cavities. Radiographic evaluation, through X-rays like bitewing and periapical, aids in assessing areas not easily visible during a clinical examination, enabling the detection of early decay stages, and assessment of damage extent [119]. Additional diagnostic aids, such as laser fluorescence devices and transillumination, enhance the identification of hidden caries or lesions at an early stage when intervention can be less invasive. Dentists also conduct risk assessments by evaluating factors like diet, oral hygiene practices, fluoride exposure, and overall health to develop personalized preventive strategies and treatment plans. The classification of caries, based on location, extent, and severity, guides the choice of treatment, whether it be restorative procedures or preventive measures [120]. Systematic recording and documentation of findings in the patient's dental records serve as a reference for future assessments and facilitate effective communication with patients and other healthcare providers. Moreover, the multifaceted process of dental caries assessment combines clinical expertise, diagnostic tools, and patient-specific information to create a comprehensive understanding of oral health. Early detection and intervention are pivotal in preventing the progression of dental caries, promoting overall oral well-being, and ensuring a healthy smile for years to come. Regular dental checkups and adherence to preventive measures are key components in maintaining optimal oral health and preventing the development of dental caries [121].

19.14.3 Oral Cancer Screening

Oral cancer screening is an integral part of comprehensive dental examinations, aiming to identify early signs of oral cancer, a potentially life-threatening condition affecting the mouth and throat. This screening process is paramount for detecting abnormalities, lesions, or precancerous conditions at an early stage, optimizing the effectiveness of intervention. Incorporating various examination methods, the screening thoroughly evaluates the oral cavity and surrounding tissues, facilitating timely diagnosis and treatment [122]. Dentists conduct a visual examination, inspecting the lips, tongue, gums, inner cheeks, roof of the mouth, and the back of the throat for changes in color, texture, or the presence of sores, lumps, or persistent ulcers. Palpation of oral tissues involves hands-on assessment to identify abnormalities, swelling, tenderness, or unusual masses in the oral tissues, neck, and lymph nodes. Some dentists utilize specialized screening devices like the VEL scope or oral cancer screening lights, highlighting abnormal tissues not visible to the naked eye and aiding in early lesion detection. Through risk assessment, dentists evaluate factors such as tobacco and alcohol use, family history, and human papillomavirus (HPV) exposure, determining the screening frequency and offering targeted preventive guidance [123]. Patient education is a crucial component, as dentists educate individuals on oral cancer signs and symptoms, encouraging self-examinations between dental visits, thereby empowering patients to seek prompt attention for any unusual changes in oral health. Establishing a regular screening schedule is emphasized, particularly for individuals with higher risk factors, as routine screenings contribute to early detection and improved oral cancer prognosis. Thus, oral cancer screening represents a proactive approach to safeguarding oral health and overall well-being. By combining visual examinations, palpation, specialized screening devices, risk assessment, and patient education, dental professionals can detect and address potential issues in their early stages, significantly enhancing the chances of successful treatment. This underscores the importance of regular dental checkups in maintaining optimal oral health while promoting overall health and wellness [124].

19.14.4 Systemic Health Connections

The concept of systemic health connections emphasizes the intricate interplay between oral health and overall well-being. Scientific research underscores the bidirectional relationship between oral and systemic health, highlighting the significant impact of oral health on various aspects of the body. This understanding is paramount for healthcare professionals, enabling a comprehensive approach to patient care that considers the holistic nature of health. A key link between oral and systemic health lies in the prevalence of inflammation, with periodontal diseases like gingivitis and periodontitis contributing to systemic inflammation, which is recognized as a potential factor in the development and exacerbation of conditions such as cardiovascular diseases, diabetes, and rheumatoid arthritis. Furthermore, the oral cavity, acting as a gateway to the body, can lead to systemic infections, as bacteria from gum disease may enter the bloodstream, contributing to infections in other organs [125]. Notably, oral health significantly impacts cardiovascular health, with research suggesting that inflammation and infection associated with periodontal disease may contribute to atherosclerosis, increasing the risk of heart disease and stroke. An observed association exists between periodontal health and diabetes, where individuals with diabetes are more susceptible to gum disease and untreated gum disease can complicate blood sugar level management.

Oral health also affects pregnant women, as poor oral health is linked to adverse pregnancy outcomes, including preterm birth and low birth weight. In the realm of respiratory health, oral issues leading to chronic infections are linked to conditions such as pneumonia and chronic obstructive pulmonary disease (COPD) [126]. Dental professionals play indispensable roles in recognizing these systemic health connections during patient assessments, addressing oral health issues, and collaborating with other healthcare providers to manage and prevent conditions with systemic implications. Encouraging patients to maintain good oral hygiene practices, adopt a healthy lifestyle, and seek regular dental checkups contributes to overall well-being by mitigating potential systemic health risks. Therefore, recognizing and understanding the systemic health connections associated with oral health is integral to providing comprehensive healthcare, enabling a holistic approach that emphasizes the importance of maintaining optimal oral health as a key component of overall well-being. Regular dental checkups, preventive measures, and effective collaboration between dental and medical professionals are essential for promoting a healthy balance between oral and systemic health [127].

19.15 MAJOR CHALLENGES AND FUTURE DIRECTIONS

Understanding the oral microbiome, a complex community of microorganisms in the mouth, is crucial for unraveling its role in both oral and systemic health. Despite technological advancements enabling in-depth exploration of its composition and function, several limitations in oral microbiome diagnostics hinder a comprehensive understanding. The absence of standardized methods for sample collection, DNA extraction, and sequencing protocols poses a significant challenge, leading to variability in processes and impeding data comparison across studies [9]. Achieving high taxonomic resolution in identifying oral microbial species remains challenging due to classification difficulties with certain taxa, limiting the capture of the full microbial diversity. While identifying microbial species is essential, assessing their functional roles encounters limitations, restricting insights into their impact on health and disease. The dynamic nature of the oral microbiome, influenced by factors such as diet and oral hygiene, contrasts with the static snapshots provided by current diagnostic tools, complicating the capture of temporal changes. The intricate interactions between microbial species in the oral cavity are not fully understood, as diagnostic methods often focus on individual species without considering broader community dynamics [5]. Integrating oral microbiome data with clinical information is crucial, yet the effective integration of microbiome data with clinical parameters remains an ongoing challenge. Additionally, the cost and limited accessibility of advanced diagnostic technologies, like high-throughput sequencing, hinder widespread implementation. Ethnic and geographic variability in the oral microbiome is not fully accounted for by current diagnostics, potentially limiting applicability across diverse demographic groups. Addressing these limitations necessitates collaborative efforts to develop standardized protocols, enhance taxonomic resolution, and integrate functional analyses, ultimately paving the way for more precise and informative oral microbiome diagnostics and thereby advancing our understanding of its health implications [13].

19.16 CONCLUSION

The future of oral microbiome diagnostics is characterized by promising prospects driven by advancements in technology, research methodologies, and interdisciplinary collaboration, collectively paving the way for a more profound comprehension of the intricate microbial communities within the oral cavity. These anticipated developments are strategically poised to address existing limitations, promising enhanced precision and heightened clinical relevance in the assessment of oral microbiome dynamics. Foremost among these advancements is the commitment to establishing standardized methods for sample collection, DNA extraction, and sequencing protocols, with the integration of cutting-edge technologies like metatranscriptomics poised to revolutionize the field, providing heightened resolution and accuracy in profiling the oral microbiome. Concurrent efforts to refine taxonomic resolution, leveraging enhanced sequencing techniques and bioinformatics tools, promise to unveil a more nuanced understanding of the diverse oral microbial landscape, capturing previously undetected taxa. The evolution towards comprehensive functional profiling through advanced methods such as metagenomics and metabolomics is a prominent shift in the future of oral microbiome diagnostics, offering profound insights into the functional roles of oral microbial communities and a deeper comprehension of their contributions to health and disease. Furthermore, emerging technologies facilitating dynamic monitoring over time and conducting longitudinal studies are expected to provide a holistic view, allowing researchers and clinicians to capture fluctuations influenced by various factors. The integration of systems biology approaches is anticipated to play a pivotal role in interpreting the oral microbiome more holistically, unraveling the intricate interactions between microbial species and their host environments.

The application of artificial intelligence has poised to decipher complex oral microbiome data, facilitating the integration of large datasets with clinical, genetic, and environmental factors to offer personalized health insights. As future developments also focus on making advanced diagnostic technologies more accessible and cost effective, innovations in portable sequencing devices and point-of-care diagnostics may democratize oral microbiome assessments, ensuring broader implementation in both research and clinical settings. Tailoring oral microbiome diagnostics to consider population-specific variations, including ethnic and geographic diversity, emerges as a key aspect of future developments, enhancing the applicability and relevance of diagnostic tools across different demographics. The major advancements hold great promise for the future of oral microbiome diagnostics, and we can anticipate an era where these innovations will revolutionize personalized oral healthcare. By providing clinicians with valuable tools for early detection, prevention, and tailored interventions, these advancements are poised to contribute significantly to the improvement of overall oral and systematic health.

REFERENCES

1. Deo, P. N.; Deshmukh, R. Oral microbiome: Unveiling the fundamentals. *Journal of Oral and Maxillofacial Pathology* **2019**, *23*(1), 122.
2. Wade, W. G. The oral microbiome in health and disease. *Pharmacological Research* **2013**, *69*(1), 137–143.
3. Sedghi, L.; DiMassa, V.; Harrington, A.; Lynch, S. V.; Kapila, Y. L. The oral microbiome: Role of key organisms and complex networks in oral health and disease. *Periodontology* **2021**, *87*(1), 107–131.
4. Irfan, M.; Delgado, R. Z. R.; Frias-Lopez, J. The oral microbiome and cancer. *Frontiers in Immunology* **2020**, *11*, 591088.
5. Willis, J. R.; Gabaldón, T. The human oral microbiome in health and disease: from sequences to ecosystems. *Microorganisms* **2020**, *8*(2), 308.
6. Belibasakis, G. N.; Bostanci, N.; Marsh, P. D.; Zaura, E. Applications of the oral microbiome in personalized dentistry. *Archives of Oral Biology* **2019**, *104*, 7–12.
7. Chattopadhyay, I.; Panda, M. Recent trends of saliva omics biomarkers for the diagnosis and treatment of oral cancer. *Journal of Oral Biosciences* **2019**, *61*(2), 84–94.
8. Caselli, E.; Fabbri, C.; D'Accolti, M.; Soffritti, I.; Bassi, C.; Mazzacane, S.; Franchi, M. Defining the oral microbiome by whole-genome sequencing and resistome analysis: the complexity of the healthy picture. *BMC Microbiology* **2020**, *20*(1), 1–19.
9. Lee, Y. H.; Chung, S. W.; Auh, Q. S.; Hong, S. J.; Lee, Y. A.; Jung, J.; Hong, J. Y. Progress in oral microbiome related to oral and systemic diseases: an update. *Diagnostics* **2021**, *11*(7), 1283.
10. Pathak, J. L.; Yan, Y.; Zhang, Q.; Wang, L.; Ge, L. The role of oral microbiome in respiratory health and diseases. *Respiratory Medicine* **2021**, *185*, 106475.
11. Morrison, A. G.; Sarkar, S.; Umar, S.; Lee, S. T.; Thomas, S. M. The Contribution of the Human Oral Microbiome to Oral Disease: A Review. *Microorganisms* **2023**, *11*(2), 318.
12. Thomas, C.; Minty, M.; Vinel, A.; Canceill, T.; Loubières, P.; Burcelin, R.; Laurencin-Dalicieux, S. Oral microbiota: a major player in the diagnosis of systemic diseases. *Diagnostics* **2021**, *11*(8), 1376.
13. Xiao, J.; Fiscella, K. A.; Gill, S. R. Oral microbiome: possible harbinger for children's health. *International Journal of Oral Science* **2020**, *12*(1), 12.
14. Hajjo, R.; Sabbah, D. A.; Al Bawab, A. Q. Unlocking the potential of the human microbiome for identifying disease diagnostic biomarkers. *Diagnostics* **2022**, *12*(7), 1742.
15. Lim, Y. K. Oral microbiome: *A novel biomarker for oral cavity and oropharyngeal cancer detection and a potential surveillance tool for post-treatment monitoring* **2019** (Doctoral dissertation, Queensland University of Technology).
16. Cullen, C. M.; Aneja, K. K.; Beyhan, S.; Cho, C. E.; Woloszynek, S.; Convertino, M.; Rosen, G. L. Emerging priorities for microbiome research. *Frontiers in microbiology* **2020**, *11*, 136.
17. Merali, N.; Chouari, T.; Kayani, K.; Rayner, C. J.; Jiménez, J. I.; Krell, J.; Frampton, A. E. A comprehensive review of the current and future role of the microbiome in pancreatic ductal adenocarcinoma. *Cancers* **2022**, *14*(4), 1020.
18. Tan, A. H.; Lim, S. Y.; Lang, A. E. The microbiome–gut–brain axis in Parkinson disease—from basic research to the clinic. *Nature Reviews Neurology* **2022**, *18*(8), 476–495.

19. Jain, T.; Sharma, P.; Are, A. C.; Vickers, S. M.; Dudeja, V. New insights into the cancer–microbiome–immune axis: decrypting a decade of discoveries. *Frontiers in Immunology* **2021**, *12*, 622064.
20. Baker, J. L.; Mark Welch, J. L.; Kauffman, K. M.; McLean, J. S.; He, X. The oral microbiome: diversity, biogeography and human health. *Nature Reviews Microbiology* **2023**, 1–16.
21. Verma, D.; Garg, P. K.; Dubey, A. K. Insights into the human oral microbiome. *Archives of Microbiology* **2018**, *200*, 525–540.
22. Yamashita, Y.; Takeshita, T. The oral microbiome and human health. *Journal of Oral Science* **2017**, *59*(2), 201–206.
23. Kitamoto, S.; Nagao-Kitamoto, H.; Hein, R.; Schmidt, T. M.; Kamada, N. The bacterial connection between the oral cavity and the gut diseases. *Journal of Dental Research* **2020**, *99*(9), 1021–1029.
24. Torralba, M. G.; Aleti, G.; Li, W.; Moncera, K. J.; Lin, Y. H.; Yu, Y., ... Nelson, K. E. Oral microbial species and virulence factors associated with oral squamous cell carcinoma. *Microbial Ecology* **2021**, *82*, 1030–1046.
25. Su, S. C.; Chang, L. C.; Huang, H. D.; Peng, C. Y.; Chuang, C. Y.; Chen, Y. T., ... Yang, S. F. Oral microbial dysbiosis and its performance in predicting oral cancer. *Carcinogenesis* **2021**, *42*(1), 127–135.
26. Kadota, T.; Hamada, M.; Nomura, R.; Ogaya, Y.; Okawa, R.; Uzawa, N.; Nakano, K. Distribution of Helicobacter pylori and periodontopathic bacterial species in the oral cavity. *Biomedicines* **2020**, *8*(6), 161.
27. Sogodogo, E.; Doumbo, O.; Aboudharam, G.; Kouriba, B.; Diawara, O.; Koita, H.; Drancourt, M. First characterization of methanogens in oral cavity in Malian patients with oral cavity pathologies. *BMC Oral Health* **2019**, *19*, 1–6.
28. Černáková, L.; Rodrigues, C. F. Microbial interactions and immunity response in oral Candida species. *Future Microbiology* **2020**, *15*(17), 1653–1677.
29. Khan, R. A.; Sarmah, R. Factors influencing microbial growth in the human oral cavity. *Reviews and Research in Medical Microbiology* **2023**, *34*(3), 123–129
30. Li, X.; Liu, Y.; Yang, X.; Li, C.; Song, Z. The oral microbiota: community composition, influencing factors, pathogenesis, and interventions. *Frontiers in Microbiology* **2022**, *13*, 895537
31. He, J.; Li, Y.; Cao, Y.; Xue, J.; Zhou, X. The oral microbiome diversity and its relation to human diseases. *Folia Microbiologica* **2015**, *60*, 69–80.
32. Rozas, N. S.; Tribble, G. D.; Jeter, C. B. Oral factors that impact the oral microbiota in Parkinson's disease. *Microorganisms* **2021**, *9*(8), 1616.
33. Sampaio-Maia, B.; Caldas, I. M.; Pereira, M. L.; Pérez-Mongiovi, D.; Araujo, R. The oral microbiome in health and its implication in oral and systemic diseases. *Advances in Applied Microbiology* **2016**, *97*, 171–210.
34. Duran-Pinedo, A.; Solbiati, J.; Teles, F.; Teles, R.; Zang, Y.; Frias-Lopez, J. Long-term dynamics of the human oral microbiome during clinical disease progression. *BMC Biology* **2021**, *19*, 1–17.
35. Bibi, T.; Khurshid, Z.; Rehman, A.; Imran, E.; Srivastava, K. C.; Shrivastava, D. Gingival crevicular fluid (GCF): a diagnostic tool for the detection of periodontal health and diseases. *Molecules* **2021**, *26*(5), 1208.
36. Ealand, C. S.; Sewcharran, A.; Peters, J. S.; Gordhan, B. G.; Kamariza, M.; Bertozzi, C. R.; Kana, B. D. The performance of tongue swabs for detection of pulmonary tuberculosis. *Frontiers in Cellular and Infection Microbiology* **2023**, *13*.
37. Sun, S.; Zhu, X.; Huang, X.; Murff, H. J.; Ness, R. M.; Seidner, D. L.; Fodor, A. A. On the robustness of inference of association with the gut microbiota in stool, rectal swab and mucosal tissue samples. *Scientific Reports* **2021**, *11*(1), 14828.
38. Jiang, Y.; Song, B.; Brandt, B. W.; Cheng, L.; Zhou, X.; Exterkate, R. A.; Deng, D. M. Comparison of red-complex bacteria between saliva and subgingival plaque of periodontitis patients: A systematic review and meta-analysis. *Frontiers in cellular and infection microbiology* **2021**, *11*, 727732.
39. Oba, P. M.; Carroll, M. Q.; Alexander, C.; Valentine, H.; Somrak, A. J.; Keating, S. C.; Swanson, K. S. Microbiota populations in supragingival plaque, subgingival plaque, and saliva habitats of adult dogs. *Animal Microbiome* **2021**, *3*(1), 1–18.
40. Deschrevel, J.; Maes, K.; Andries, A.; Beukelaer, N. D.; Corvelyn, M.; Costamagna, D.; Gayan-Ramirez, G. Fine-needle percutaneous muscle microbiopsy technique as a feasible tool to address histological analysis in young children with cerebral palsy and age-matched typically developing children. *PLoS One* **2023**, *18*(11), e0294395.
41. Hagan, R. W.; Hofman, C. A.; Hübner, A.; Reinhard, K.; Schnorr, S.; Lewis Jr, C. M.; Warinner, C. G. Comparison of extraction methods for recovering ancient microbial DNA from paleofeces. *American Journal of Physical Anthropology* **2020**, *171*(2), 275–284.

42. Sano, H.; Wakui, A.; Kawachi, M.; Washio, J.; Abiko, Y.; Mayanagi, G.; Sato, T. Profiling system of oral microbiota utilizing polymerase chain reaction-restriction fragment length polymorphism analysis. *Journal of oral biosciences* **2021**, *63*(3), 292–297.
43. Covert, I., Gala, R., Wang, T., Svoboda, K., Sümbül, U., &Lee, S. I. Predictive and robust gene selection for spatial transcriptomics. *Nature Communications* **2023**, 14(1), 2091
44. Li, S.; Liu, X.; Li, Z.; Liu, H.; Hu, D. Combination of direct boiling and glass beads increases the purity and accuracy of bacterial DNA extraction. *Biotechnology Journal* **2023**, *18*(11), 2300135.
45. Li, D.; Zhang, J.; Li, J. Primer design for quantitative real-time PCR for the emerging Coronavirus SARS-CoV-2. *Theranostics* **2020**, *10*(16), 7150.
46. Mosele, F.; Remon, J.; Mateo, J.; Westphalen, C. B.; Barlesi, F.; Lolkema, M. P.; André, F. Recommendations for the use of next-generation sequencing (NGS) for patients with metastatic cancers: a report from the ESMO Precision Medicine Working Group. *Annals of Oncology* **2020**, *31*(11), 1491–1505.
47. Morel, B.; Barbera, P.; Czech, L.; Bettisworth, B.; Hübner, L.; Lutteropp, S.; Stamatakis, A. Phylogenetic analysis of SARS-CoV-2 data is difficult. *Molecular biology and evolution* **2021**, *38*(5), 1777–1791
48. Hu, T.; Chitnis, N.; Monos, D.; Dinh, A. Next-generation sequencing technologies: An overview. *Human Immunology* **2021**, *82*(11), 801–811.
49. Kapp, J. D.; Green, R. E.; Shapiro, B. A fast and efficient single-stranded genomic library preparation method optimized for ancient DNA. *Journal of Heredity* **2021**, *112*(3), 241–249.
50. Pereira, R.; Oliveira, J.; Sousa, M. Bioinformatics and computational tools for next-generation sequencing analysis in clinical genetics. *Journal of Clinical Medicine* **2020**, *9*(1), 132.
51. Uhlen, M.; Quake, S. R. Sequential sequencing by synthesis and the next-generation sequencing revolution. *Trends in Biotechnology* **2023**.
52. Carbo, E. C.; Mourik, K.; Boers, S. A.; Munnink, B. O.; Nieuwenhuijse, D.; Jonges, M., … de Vries, J. J. A comparison of five Illumina, Ion Torrent, and nanopore sequencing technology-based approaches for whole genome sequencing of SARS-CoV-2. *European Journal of Clinical Microbiology & Infectious Diseases* **2023**, *42*(6), 701–713
53. Lou, F.; Song, N.; Han, Z.; Gao, T. Single-molecule real-time (SMRT) sequencing facilitates *Tachypleus tridentatus* genome annotation. *International Journal of Biological Macromolecules* **2020**, 147, 89–97.
54. Wang, H.; de Carvalho, L. P. S. Metabolomic profiling reveals bacterial metabolic adaptation strategies and new metabolites. *Current Opinion in Chemical Biology* **2023**, *74*, 102287.
55. Do, T.; Dame-Teixeira, N.; Deng, D. Applications of Next Generation Sequencing (NGS) Technologies to Decipher the Oral Microbiome in Systemic Health and Disease. *Frontiers in Cellular and Infection Microbiology* **2021**, *11*, 801122.
56. Pérez-Cobas, A. E.; Gomez-Valero, L.; Buchrieser, C. Metagenomic approaches in microbial ecology: an update on whole-genome and marker gene sequencing analyses. *Microbial Genomics* **2020**, *6*(8).
57. Ranjan, R.; Rani, A.; Metwally, A.; McGee, H. S.; Perkins, D. L. Analysis of the microbiome: Advantages of whole genome shotgun versus 16S amplicon sequencing. *Biochemical and Biophysical Research Communications* **2016**, *469*(4), 967–977.
58. Li, Y.; Oosting, M.; Smeekens, S. P.; Jaeger, M.; Aguirre-Gamboa, R.; Le, K. T.; Netea, M. G. A functional genomics approach to understand variation in cytokine production in humans. *Cell* **2016**, *167*(4), 1099–1110.
59. Kumar Awasthi, M.; Ravindran, B.; Sarsaiya, S.; Chen, H.; Wainaina, S.; Singh, E.; Zhang, Z. RETRACTED ARTICLE: Metagenomics for taxonomy profiling: tools and approaches. *Bioengineered* **2020**, *11*(1), 356–374.
60. Kang, K.; Hu, Y.; Wu, S.; Shi, S. Comparative metagenomic analysis of chicken gut microbial community, function, and resistome to evaluate noninvasive and cecal sampling resources. *Animals* **2021**, *11*(6), 1718.
61. Jung, H.; Ventura, T.; Chung, J. S.; Kim, W. J.; Nam, B. H.; Kong, H. J.; Eyun, S. I. Twelve quick steps for genome assembly and annotation in the classroom. *PLoS Computational Biology* **2020**, *16*(11), e1008325.
62. Dubey, R. K.; Tripathi, V.; Prabha, R.; Chaurasia, R.; Singh, D. P.; Rao, C. S.; Abhilash, P. C. Metatranscriptomics and metaproteomics for microbial communities profiling. *Unravelling the Soil Microbiome: Perspectives for Environmental Sustainability* **2020**, 51–60.
63. Kanzi, A. M.; San, J. E.; Chimukangara, B.; Wilkinson, E.; Fish, M.; Ramsuran, V.; De Oliveira, T. Next generation sequencing and bioinformatics analysis of family genetic inheritance. *Frontiers in Genetics* **2020**, *11*, 544162.

64. Si, Z.; Wang, X. Stem cell therapies in Alzheimer's disease: applications for disease modeling. *Journal of Pharmacology and Experimental Therapeutics* **2021**, *377*(2), 207–217.
65. Sierra, M. A.; Li, Q.; Pushalkar, S.; Paul, B.; Sandoval, T. A.; Kamer, A. R.; Saxena, D. The influences of bioinformatics tools and reference databases in analyzing the human oral microbial community. *Genes* **2020**, *11*(8), 878.
66. Moon, J. H.; Lee, J. H. Probing the diversity of healthy oral microbiome with bioinformatics approaches. *BMB Reports* **2016**, *49*(12), 662.
67. Kumar, R.; Eipers, P.; Little, R. B.; Crowley, M.; Crossman, D. K.; Lefkowitz, E. J.; Morrow, C. D. Getting started with microbiome analysis: sample acquisition to bioinformatics. *Current Protocols in Human Genetics* **2014**, *82*(1), 18–8
68. Ai, D.; Huang, R.; Wen, J.; Li, C.; Zhu, J.; Xia, L. C. Integrated metagenomic data analysis demonstrates that a loss of diversity in oral microbiota is associated with periodontitis. *BMC Genomics* **2017**, *18*(1), 1–15.
69. Szafranski, S. P.; Wos-Oxley, M. L.; Vilchez-Vargas, R.; Jáuregui, R.; Plumeier, I.; Klawonn, F.; Wagner-Döbler, I. High-resolution taxonomic profiling of the subgingival microbiome for biomarker discovery and periodontitis diagnosis. *Applied and Environmental Microbiology* **2015**, *81*(3), 1047–1058
70. Heather, J. M.; Chain, B. The sequence of sequencers: The history of sequencing DNA. *Genomics* **2016**, *107*(1), 1–8.
71. Showalter, A. M.; Keppler, B.; Lichtenberg, J.; Gu, D.; Welch, L. R. A bioinformatics approach to the identification, classification, and analysis of hydroxyproline-rich glycoproteins. *Plant Physiology* **2010**, *153*(2), 485–513.
72. Kotov, R.; Krueger, R. F.; Watson, D.; Achenbach, T. M.; Althoff, R. R.; Bagby, R. M., ... Zimmerman, M. The Hierarchical Taxonomy of Psychopathology (HiTOP): A dimensional alternative to traditional nosologies. *Journal of Abnormal Psychology* **2017**, *126*(4), 454.
73. Foster, Z. S.; Sharpton, T. J.; Grünwald, N. J. Metacoder: An R package for visualization and manipulation of community taxonomic diversity data. *PLoS Computational Biology* **2017**, *13*(2), e1005404.
74. Tuomisto, H. A diversity of beta diversities: straightening up a concept gone awry. Part 1. Defining beta diversity as a function of alpha and gamma diversity. *Ecography* **2010**, *33*(1), 2–22.
75. Rewald, B.; Meinen, C.; Trockenbrodt, M.; Ephrath, J. E.; Rachmilevitch, S. Root taxa identification in plant mixtures–current techniques and future challenges. *Plant and Soil* **2012**, *359*, 165–182.
76. Bagstad, K. J.; Johnson, G. W.; Voigt, B.; Villa, F. Spatial dynamics of ecosystem service flows: A comprehensive approach to quantifying actual services. *Ecosystem Services* **2013**, *4*, 117–125
77. Huerta-Cepas, J.; Szklarczyk, D.; Forslund, K.; Cook, H.; Heller, D.; Walter, M. C.; Bork, P. eggNOG 4.5: a hierarchical orthology framework with improved functional annotations for eukaryotic, prokaryotic and viral sequences. *Nucleic Acids Research* **2016**, *44*(D1), D286–D293.
78. Guo, Y.; Darshi, M.; Ma, Y.; Perkins, G. A.; Shen, Z.; Haushalter, K. J.; Taylor, S. S. Quantitative proteomic and functional analysis of liver mitochondria from high fat diet (HFD) diabetic mice. *Molecular & Cellular Proteomics* **2013**, *12*(12), 3744–3758.
79. Caspi, R.; Altman, T.; Billington, R.; Dreher, K.; Foerster, H.; Fulcher, C. A., ... Karp, P. D. The MetaCyc database of metabolic pathways and enzymes and the BioCyc collection of Pathway/Genome Databases. *Nucleic Acids Research* **2014**, *42*(D1), D459–D471.
80. Pang, Z.; Chong, J.; Zhou, G.; de Lima Morais, D. A.; Chang, L.; Barrette, M.; Xia, J. MetaboAnalyst 5.0: narrowing the gap between raw spectra and functional insights. *Nucleic Acids Research* **2021**, *49*(W1), W388–W396.
81. Li, D.; Gaquerel, E. Next-generation mass spectrometry metabolomics revives the functional analysis of plant metabolic diversity. *Annual Review of Plant Biology* **2021**, *72*, 867–891.
82. Argelaguet, R.; Velten, B.; Arnol, D.; Dietrich, S.; Zenz, T.; Marioni, J. C., ... Stegle, O. Multi-Omics Factor Analysis—a framework for unsupervised integration of multi-omics data sets. *Molecular Systems Biology* **2018**, *14*(6), e8124.
83. Subramanian, I.; Verma, S.; Kumar, S. Jere, A.; Anamika, K. Multi-omics data integration, interpretation, and its application. *Bioinformatics and Biology Insights* **2020**, *14*, 1177932219899051.
84. Bersanelli, M.; Mosca, E.; Remondini, D.; Giampieri, E.; Sala, C.; Castellani, G.; Milanesi, L. Methods for the integration of multi-omics data: mathematical aspects. *BMC Bioinformatics* **2016**, 17, 167–177.
85. Canzler, S.; Schor, J.; Busch, W.; Schubert, K.; Rolle-Kampczyk, U. E.; Seitz, H., ... Hackermüller, J. Prospects and challenges of multi-omics data integration in toxicology. *Archives of Toxicology* **2020**, *94*, 371–388.
86. Mira, A. Oral microbiome studies: potential diagnostic and therapeutic implications. *Advances in Dental Research* **2018**, *29*(1), 71–77.

87. Lin, Y.; Liang, X.; Li, Z.; Gong, T.; Ren, B.; Li, Y.; Peng, X. Omics for deciphering oral microecology. *International Journal of Oral Science* **2024**, *16*(1), 2.
88. Bostanci, N.; Belibasakis, G. N. Precision periodontal care: from omics discoveries to chairside diagnostics. *Clinical Oral Investigations* **2023**, *27*(3), 971–978.
89. Granato, D. C.; Neves, L. X.; Trino, L. D.; Carnielli, C. M.; Lopes, A. F.; Yokoo, S., … Leme, A. F. P. Meta-omics analysis indicates the saliva microbiome and its proteins associated with the prognosis of oral cancer patients. *Biochimica et Biophysica Acta (BBA)-Proteins and Proteomics* **2021**, *1869*(8), 140659.
90. Taş, N.; de Jong, A. E.; Li, Y.; Trubl, G.; Xue, Y.; Dove, N. C. Metagenomic tools in microbial ecology research. *Current Opinion in Biotechnology* **2021**, *67*, 184–191.
91. Ko, K. K.; Chng, K. R.; Nagarajan, N. Metagenomics-enabled microbial surveillance. *Nature Microbiology* **2022**, *7*(4), 486–496.
92. Calle, M. L. Statistical analysis of metagenomics data. *Genomics & Informatics* **2019**, *17*(1).
93. Bharagava, R. N.; Purchase, D.; Saxena, G.; Mulla, S. I. Applications of metagenomics in microbial bioremediation of pollutants: from genomics to environmental cleanup. *In Microbial diversity in the genomic era* **2019**, (pp. 459–477). Academic Press.
94. Zhang, L.; Loh, K. C.; Lim, J. W.; Zhang, J. Bioinformatics analysis of metagenomics data of biogas-producing microbial communities in anaerobic digesters: A review. *Renewable and Sustainable Energy Reviews* **2019**, *100*, 110–126
95. Zhang, Y.; Thompson, K. N.; Huttenhower, C.; Franzosa, E. A. Statistical approaches for differential expression analysis in metatranscriptomics. *Bioinformatics* **2021**, *37*(Supplement_1), i34–i41.
96. Shakya, M.; Lo, C. C.; Chain, P. S. Advances and challenges in metatranscriptomic analysis. *Frontiers in Genetics* **2019**, *10*, 904.
97. Carvalhais, L. C.; Dennis, P. G.; Tyson, G. W.; Schenk, P. M. Application of metatranscriptomics to soil environments. *Journal of Microbiological Methods* **2012**, *91*(2), 246–251.
98. Leary, D. H.; Hervey IV, W. J.; Deschamps, J. R.; Kusterbeck, A. W.; Vora, G. J. Which metaproteome? The impact of protein extraction bias on metaproteomic analyses. *Molecular and Cellular Probes* **2013**, *27*(5-6), 193–199.
99. Heyer, R.; Schallert, K.; Zoun, R.; Becher, B.; Saake, G.; Benndorf, D. Challenges and perspectives of metaproteomic data analysis. *Journal of Biotechnology* **2017**, *261*, 24–36.
100. Schiebenhoefer, H.; Schallert, K.; Renard, B. Y.; Trappe, K.; Schmid, E.; Benndorf, D., … Fuchs, S. A complete and flexible workflow for metaproteomics data analysis based on MetaProteome Analyzer and Prophane. *Nature Protocols* **2020**, *15*(10), 3212–3239.
101. Abiraami, T. V.; Singh, S.; Nain, L. Soil metaproteomics as a tool for monitoring functional microbial communities: promises and challenges. *Reviews in Environmental Science and Bio/Technology* **2020**, *19*, 73–102.
102. Zhang, S.; Kong, C.; Yang, Y.; Cai, S.; Li, X.; Cai, G.; Ma, Y. Human oral microbiome dysbiosis as a novel non-invasive biomarker in detection of colorectal *cancer. Theranostics* **2020**, *10*(25), 11595.
103. Huang, S.; Guo, Y.; Li, Z.; Zhang, Y.; Zhou, T.; You, W., … Li, W. A systematic review of metabolomic profiling of gastric cancer and esophageal cancer. *Cancer Biology &Medicine* **2020**, *17*(1), 181.
104. Klåvus, A.; Kokla, M.; Noerman, S.; Koistinen, V. M.; Tuomainen, M.; Zarei, I.; Hanhineva, K. "Notame": workflow for non-targeted LC–MS metabolic profiling. *Metabolites* **2020**, *10*(4), 135.
105. Hou, X. W.; Wang, Y.; Ke, C.; Pan, C. W. Metabolomics facilitates the discovery of metabolic profiles and pathways for myopia: *A systematic Review* **2023**, *37*(4), 670–677.
106. Na, H. S.; Kim, S. Y.; Han, H.; Kim, H. J.; Lee, J. Y.; Lee, J. H.; Chung, J. Identification of potential oral microbial biomarkers for the diagnosis of periodontitis. *Medicine* **2020**, *9*(5), 1549.
107. Negrut, R. L.; Cote, A.; Maghiar, A. M. Exploring the Potential of Oral Microbiome Biomarkers for Colorectal Cancer Diagnosis and Prognosis: A Systematic Review. *Microorganisms* **2023**, *11*(6), 1586.
108. Lim, M. Y.; Kim, J. H.; Nam, Y. D. Oral microbiome correlates with selected clinical biomarkers in individuals with no significant systemic disease. *Frontiers in Cellular and Infection Microbiology* **2023**, *13*, 366.
109. Shi, M.; Zong, X.; Hur, J.; Birmann, B. M.; Martinez-Maza, O.; Epeldegui, M.; Cao, Y. Circulating markers of microbial translocation and host response to bacteria with risk of colorectal cancer: a prospective, nested case-control study in men. *EBioMedicine* **2023**, *91*.
110. Jennaro, T. S.; Puskarich, M. A.; Evans, C. R.; Karnovsky, A.; Flott, T. L.; McLellan, L. A.; Stringer, K. A. Sustained Perturbation of Metabolism and Metabolic Subphenotypes Are Associated with Mortality and Protein Markers of the Host Response. *Critical Care Explorations* **2023**, *5*(4).

111. Jeffrey, M.; Denny, K. J.; Lipman, J.; Conway Morris, A. Differentiating infection, colonisation, and sterile inflammation in critical illness: the emerging role of host-response profiling. *Intensive Care Medicine* **2023**, 1–12.
112. Watson, A. R.; Füssel, J.; Veseli, I.; DeLongchamp, J. Z.; Silva, M.; Trigodet, F.; Eren, A. M. Metabolic independence drives gut microbial colonization and resilience in health and disease. *Genome Biology* **2023**, *24*(1), 78.
113. Kleinstein, S. E.; Nelson, K. E.; Freire, M. Inflammatory networks linking oral microbiome with systemic health and disease. *Journal of Dental Research* **2020**, *99*(10), 1131–1139.
114. Manor, O.; Dai, C. L.; Kornilov, S. A.; Smith, B.; Price, N. D.; Lovejoy, J. C.; Magis, A. T. Health and disease markers correlate with gut microbiome composition across thousands of people. *Nature Communications* **2020**, *11*(1), 5206.
115. Davis, K. D.; Aghaeepour, N.; Ahn, A. H.; Angst, M. S.; Borsook, D.; Brenton, A.; Pelleymounter, M. A. Discovery and validation of biomarkers to aid the development of safe and effective pain therapeutics: challenges and opportunities. *Nature Reviews Neurology* **2020**, *16*(7), 381–400.
116. Moqri, M.; Herzog, C.; Poganik, J. R.; Justice, J.; Belsky, D. W.; Higgins-Chen, A.; Gladyshev, V. N. Biomarkers of aging for the identification and evaluation of longevity interventions. *Cell* **2023**, *186*(18), 3758–3775.
117. Salvi, G. E.; Roccuzzo, A.; Imber, J. C.; Stähli, A.; Klinge, B.; Lang, N. P. Clinical periodontal diagnosis. *Periodontology* **2023**.
118. Ray, R. R. Periodontitis: An oral disease with severe consequences. *Applied Biochemistry and Biotechnology* **2023**, *195*(1), 17–32.
119. Frencken, J. E.; Giacaman, R. A.; Leal, S. C. An assessment of three contemporary dental caries epidemiological instruments: a critical review. *British Dental Journal* **2020**, *228*(1), 25–31.
120. Cheng, L.; Zhang, L.; Yue, L.; Ling, J.; Fan, M.; Yang, D.; Zhou, X. Expert consensus on dental caries management. *International Journal of Oral Science* **2022**, *14*(1), 17.
121. Manohar, N.; Hayen, A.; Fahey, P.; Arora, A. Obesity and dental caries in early childhood: A systematic review and meta-analyses. *Obesity Reviews* **2020**, *21*(3), e12960.
122. Warnakulasuriya, S.; Kerr, A. R. Oral cancer screening: past, present, and future. *Journal of Dental Research* **2021**, *100*(12), 1313–1320.
123. Nagao, T.; Warnakulasuriya, S. Screening for oral cancer: Future prospects, research and policy development for Asia. *Oral Oncology* **2020**, *105*, 104632.
124. Ilhan, B.; Lin, K.; Guneri, P.; Wilder-Smith, P. Improving oral cancer outcomes with imaging and artificial intelligence. *Journal of Dental Research* **2020**, *99*(3), 241–248.
125. Karim, F.; Oyewande, A. A.; Abdalla, L. F.; Ehsanullah, R. C.; Khan, S. Social media use and its connection to mental health: a systematic review. *Cureus* **2020**, *12*(6).
126. Phillips-Beck, W.; Eni, R.; Lavoie, J. G.; Avery Kinew, K.; Kyoon Achan, G.; Katz, A. Confronting racism within the Canadian healthcare system: systemic exclusion of First Nations from quality and consistent care. *International Journal of Environmental Research and Public Health* **2020**, *17*(22), 8343.
127. Braveman, P. A.; Arkin, E.; Proctor, D.; Kauh, T.; Holm, N. Systemic and Structural Racism: Definitions, Examples, Health Damages, and Approaches to Dismantling: Study examines definitions, examples, health damages, and dismantling systemic and structural racism. *Health Affairs* **2022**, *41*(2), 171–178.

Index

U

V

W

Y

Z